medical
biotechnology

medical biotechnology

Bernard R. Glick
Department of Biology
University of Waterloo
Waterloo, Ontario, Canada

Terry L. Delovitch
Department of Microbiology and Immunology
Western University
London, Ontario, Canada

Cheryl L. Patten
Biology Department
University of New Brunswick
Fredericton, New Brunswick, Canada

ASM
PRESS

WASHINGTON, DC

Library of Congress Cataloging-in-Publication Data

Glick, Bernard R., author.
Medical biotechnology / Bernard R. Glick, Department of Biology, University of Waterloo, Waterloo, Ontario, Canada; Terry L. Delovitch, Department of Microbiology and Immunology, Western University, London, Ontario, Canada; Cheryl L. Patten, Biology Department, University of New Brunswick, Fredericton, New Brunswick, Canada.
pages cm
Includes bibliographical references and index.
ISBN 978-1-55581-705-3 (hardcover)—ISBN 978-1-55581-889-0 (e-book) 1. Biotechnology. 2. Medical technology. I. Delovitch, T. L., author. II. Patten, Cheryl L., author. III. Title.
TP248.2.G57 2014
660.6—dc23
2013027259

10 9 8 7 6 5 4 3 2 1
Printed in the United States of America

Address editorial correspondence to ASM Press, 1752 N St. NW, Washington, DC 20036-2904, USA
E-mail: books@asmusa.org
Send orders to ASM Press, P.O. Box 605, Herndon, VA 20172, USA
Phone: (800) 546-2416 or (703) 661-1593; Fax: (703) 661-1501
Online: http://www.asmscience.org

doi:10.1128/9781555818890

Cover and interior design: Susan Brown Schmidler
Illustrations: Patrick Lane, ScEYEnce Studios

Image credits for cover and section openers
Cell image on cover: A dendritic cell infected with human immunodeficiency virus (HIV), showing projections called filopodia (stained red) with HIV particles (white) at their ends. Reproduced from the cover of *PLoS Pathogens*, June 2012. Courtesy of Anupriya Aggarwal and Stuart Turville (Kirby Institute, University of New South Wales).
DNA image on cover: majcot/Shutterstock
DNA image on chapter and section opener pages: Mopic/Shutterstock
Section opener images: mouse, Sergey Galushko/Shutterstock; lab equipment, Vasiliy Koval/Shutterstock; vaccine, Nixx Photography/Shutterstock

Some figures and tables in this book are reprinted or modified from Glick et al., *Molecular Biotechnology: Principles and Applications of Recombinant DNA*, 4th ed. (ASM Press, Washington, DC, 2010).

*To our spouses, Marcia Glick, Regina Delovitch, and Patrick Patten,
for their omnipresent love and tolerance, support, wisdom, and humor*

Contents

SECTION  II

Production of Therapeutic Agents 327

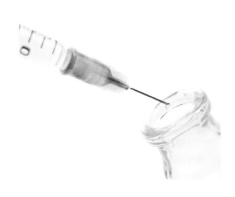

Preface

From the very beginning of the biotechnology revolution in the early 1970s, many scientists understood that this new technology would radically change the way that we think about health care. They understood early on, well before any products were commercialized, that medical science was about to undergo a major paradigm shift in which all of our previous assumptions and approaches would change dramatically. Forty years later, biotechnology has delivered on much of its early promise. Hundreds of new therapeutic agents, diagnostic tests, and vaccines have been developed and are currently available in the marketplace. Moreover, it is clear that we are presently just at the tip of a very large iceberg, with many more products in the pipeline. It is likely that, in the next 10 to 15 years, biotechnology will deliver not only new products to diagnose, prevent, and treat human disease but also entirely new approaches to treating a wide range of hitherto difficult-to-treat or untreatable diseases.

We have written *Medical Biotechnology* with the premise that it might serve as a textbook for a wide range of courses intended for premedical and medical students, dental students, pharmacists, optometrists, nurses, nutritionists, genetic counselors, hospital administrators, and other individuals who are stakeholders in the understanding and advancement of biotechnology and its impact on the practice of modern medicine. The book is intended to be as jargon-free and as easy to read as possible. In some respects, our goal is to demystify the discipline of medical biotechnology. This is not a medical textbook per se. However, a discussion of some salient features of selected diseases is presented to illustrate the applications of many biotechniques and biochemical mechanisms. Thus, this book may be considered a biomedical road map that provides a fundamental understanding of many approaches being pursued by scientists to diagnose, prevent, and treat a wide range of ailments. Indeed, this presents a large challenge, and the future is difficult to predict. Nevertheless, we hope that this volume will provide a useful introduction to medical biotechnology for a wide range of individuals.

About the Authors

Bernard R. Glick is a professor of biology at the University of Waterloo in Waterloo, Ontario, Canada, where he received his PhD in biochemistry in 1974. His current research is focused on the biochemical and genetic mechanisms used by plant growth-promoting bacteria to facilitate plant growth. In addition to his nearly 300 research publications, Dr. Glick is a coauthor of the textbook *Molecular Biotechnology: Principles and Applications of Recombinant DNA*, published by ASM Press. According to Google Scholar, his work has been cited more than 15,000 times. In addition to having served two terms as chair of the Department of Biology at Waterloo, Dr. Glick has taught 10 courses in five countries on various aspects of biotechnology.

Terry L. Delovitch obtained his BSc in chemistry (1966) and PhD in chemistry/immunology (1971) at McGill University. He received postdoctoral training at the Massachusetts Institute of Technology and Stanford University and then joined the faculty of the University of Toronto. In 1994, he was appointed senior scientist and director of the Autoimmune Disease Group on Type 1 Diabetes at the Robarts Research Institute, Western University, London, Ontario, Canada. After a 45-year research career, he retired from Western in 2011. He has received several academic awards and published about 200 research papers, review articles, and book chapters. He is the former chief scientific advisor to the Juvenile Diabetes Research Foundation Canada and past president of the International Immunology of Diabetes Society, and he is a consultant or advisor for several biotechnology and pharmaceutical companies, granting agencies, and journal editorial boards and a national allergy, asthma, and immunology research network. He and his wife, Regina, live in Toronto.

Cheryl L. Patten is an associate professor of microbiology and associate chair of the Department of Biology at the University of New Brunswick (UNB) in Fredericton, New Brunswick, Canada. Dr. Patten received her PhD from the University of Waterloo in 2001 and did postdoctoral work at McMaster University before joining the UNB faculty in 2004. Her research aims to understand how bacteria respond to the host environment at the biochemical and genetic levels. In particular, she is interested in secreted bacterial metabolites that may impact host health. As well as teaching introductory and advanced courses in microbiology, she enjoys introducing first-year science students to the wonders of biochemistry and molecular biology. She is a coauthor of another ASM Press textbook, *Molecular Biotechnology: Principles and Applications of Recombinant DNA*.

SECTION I

The Biology behind the Technology

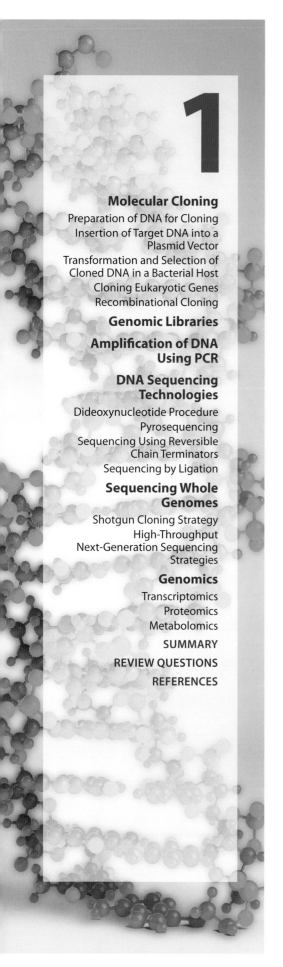

1

Fundamental Technologies

Molecular Cloning

Molecular biotechnology uses a variety of techniques for isolating genes and transferring them from one organism to another. At the root of these technologies is the ability to join a sequence of deoxyribonucleic acid (DNA) of interest to a vector that can then be introduced into a suitable host. This process is known as **recombinant DNA technology** or molecular cloning. A vast number of variations on this basic process has been devised. Development of the core technologies depended on an understanding of fundamental processes in molecular biology, bacterial genetics, and nucleic acid enzymology (Box 1.1). The beginning of the application of these technologies for the purpose of manipulating DNA has been credited to Stanley Cohen of Stanford University, Stanford, California, who was developing methods to transfer plasmids, small circular DNA molecules, into bacterial cells, and Herbert Boyer at the University of California at San Francisco, who was working with enzymes that cut DNA at specific nucleotide sequences. They hypothesized that Boyer's enzymes could be used to insert a specific segment of DNA into a plasmid and then the recombinant plasmid could be introduced into a host bacterium using Cohen's method. Within a few years, the method was used successfully to produce human insulin, which is used in the treatment of diabetes, in *Escherichia coli*. In the 25 years since the first commercial production of recombinant human insulin, more than 200 new drugs produced by recombinant DNA technology have been used to treat over 300 million people for diseases such as cancer, multiple sclerosis, cystic fibrosis, and cardiovascular disease and to provide protection against infectious diseases. Moreover, over 400 new drugs are in the process of being tested in human trials to treat a variety of serious human diseases.

Preparation of DNA for Cloning

In theory, DNA from any organism can be cloned. The target DNA may be obtained directly from genomic DNA, derived from messenger ribonucleic acid (mRNA), subcloned from previously cloned DNA, or synthesized in vitro. The target DNA may contain the complete coding sequence

doi:10.1128/9781555818890.ch1

3

box 1.1
The Development of Recombinant DNA Technology

Most important technologies are developed in small steps, and recombinant DNA technology is no exception. The ability to join DNA molecules from different sources to produce life-changing therapeutic agents like human insulin depends on the contributions of many researchers. The early 1970s were ripe for the development of recombinant DNA technology following the milestone discoveries of the structure of DNA by Watson and Crick (Watson and Crick, 1953) and the cracking of the genetic code by Nirenberg, Matthaei, and Jones (Nirenberg and Matthaei, 1961; Nirenberg et al., 1962). Building on this, rapid progress was made in understanding the structure of genes and the manner in which they are expressed. Isolating and preparing genes for cloning would not be possible without type II restriction endonucleases that cut DNA in a sequence-specific and highly reproducible manner (Kelly and Smith, 1970). Advancing the discovery by Herbert Boyer and colleagues (Hedgpeth et al., 1972), who showed that the RI restriction endonuclease from *E. coli* (now known as EcoRI) made a staggered cut at a specific nucleotide sequence in each strand of double-stranded DNA, Mertz and Davis (Mertz and Davis, 1972) reported that the complementary ends produced by EcoRI could be rejoined by DNA ligase in vitro. Of course, joining of the restriction endonuclease-digested molecules required the discovery of DNA ligase (Gellert et al., 1968). In the meantime, Cohen and Chang (Cohen and Chang, 1973) had been experimenting with constructing plasmids by shearing large plasmids into smaller random pieces and introducing the mixture of pieces into the bacterium *E. coli*. One of the pieces was propagated. However, the randomness of plasmid fragmentation reduced the usefulness of the process. During a now-legendary lunchtime conversation at a scientific meeting in 1973, Cohen and Boyer reasoned that EcoRI could be used to splice a specific segment of DNA into a plasmid, and then the recombinant plasmid could be introduced into and maintained in *E. coli* (Cohen et al., 1973). Recombinant DNA technology was born. The potential of the technology was immediately evident to Cohen and others: "It may be possible to introduce in *E. coli*, genes specifying metabolic or synthetic functions such as photosynthesis, or antibiotic production indigenous to other biological classes." The first commercial product produced using this technology was human insulin.

DNA
deoxyribonucleic acid

mRNA
messenger ribonucleic acid

for a protein, a part of the protein coding sequence, a random fragment of genomic DNA, or a segment of DNA that contains regulatory elements that control expression of a gene. Prior to cloning, both the source DNA that contains the target sequence and the cloning vector must be cut into discrete fragments, predictably and reproducibly, so that they can be joined (ligated) together to form a stable molecule. Bacterial enzymes known as type II **restriction endonucleases**, or (more commonly) restriction enzymes, are used for this purpose. These enzymes recognize and cut DNA molecules at specific base pair sequences and are produced naturally by bacteria to cleave foreign DNA, such as that of infecting bacterial viruses (**bacteriophage**). A bacterium that produces a specific restriction endonuclease also has a corresponding system to modify the sequence recognized by the restriction endonuclease in its own DNA to protect it from being degraded.

A large number of restriction endonucleases from different bacteria is available to facilitate cloning. The sequence and length of the recognition site vary among the different enzymes and can be four or more nucleotide pairs. One example is the restriction endonuclease HindIII from the bacterium *Haemophilus influenzae*. HindIII is a homodimeric protein (made up of two identical polypeptides) that specifically recognizes and binds to the DNA sequence $^{AAGCTT}_{TTCGAA}$ (Fig. 1.1A). Note that the recognition sequence is a **palindrome**, that is, the sequence of nucleotides in each of the two strands of the binding site is identical when either is read in the same polarity, i.e., 5′ to 3′. HindIII cuts within the DNA-binding site between

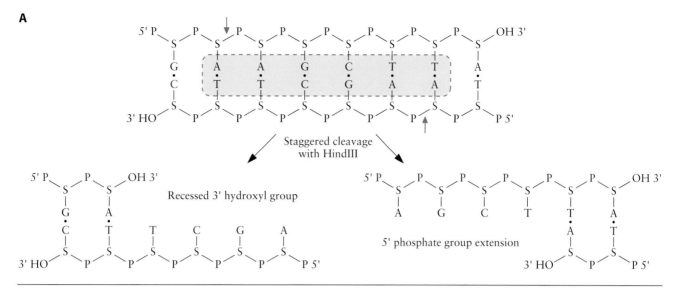

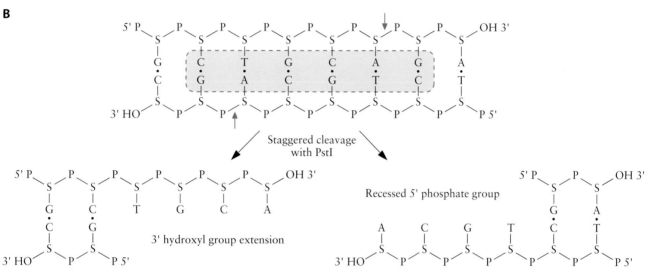

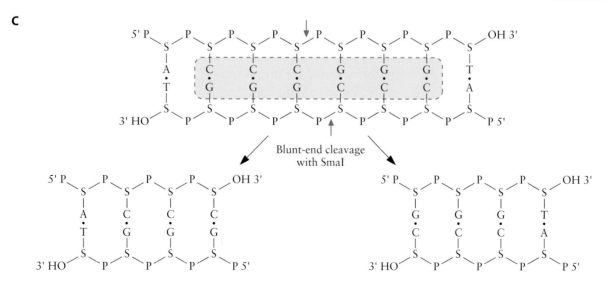

Figure 1.1 Type II restriction endonucleases bind to and cut within a specific DNA sequence. (**A**) HindIII makes a staggered cut in the DNA strands producing single-stranded, complementary ends (sticky ends) with a 5′ phosphate group extension. (**B**) PstI also makes a staggered cut in both strands but produces sticky ends with a 3′ hydroxyl group extension. (**C**) Cleavage of DNA with SmaI produces blunt ends. Arrows show the sites of cleavage in the DNA backbone. S, deoxyribose sugar; P, phosphate group; OH, hydroxyl group; A, adenine; C, cytosine; G, guanine; T, thymine. The restriction endonuclease recognition site is shaded. doi:10.1128/9781555818890.ch1.f1.1

the adjacent adenine nucleotides on each strand (Fig. 1.1A). Specifically, it cleaves the bond between the oxygen attached to the 3′ carbon of the sugar of one nucleotide and the phosphate group attached to the 5′ carbon of the sugar of the adjacent nucleotide. The symmetrical staggered cleavage of DNA by HindIII produces two single-stranded, complementary ends, each with extensions of four nucleotides, known as sticky ends. Each single-stranded extension terminates with a 5′ phosphate group, and the 3′ hydroxyl group of the opposite strand is recessed (Fig. 1.1A). Some other restriction enzymes, such as PstI, leave 3′ hydroxyl extensions with recessed 5′ phosphate ends (Fig. 1.1B), while others, such as SmaI, cut the backbone of both strands within a recognition site to produce blunt-ended DNA molecules (Fig. 1.1C).

Restriction enzymes isolated from different bacteria may recognize and cut DNA at the same site (Fig. 1.2A). These enzymes are known as **isoschizomers**. Some recognize and bind to the same sequence of DNA but cleave at different positions (**neoschizomers**), producing different single-stranded extensions (Fig. 1.2B). Other restriction endonucleases (**isocaudomers**) produce the same nucleotide extensions but have different recognition sites (Fig. 1.2C). In some cases, a restriction endonuclease will cleave a sequence only if one of the nucleotides in the recognition site is methylated. These characteristics of restriction endonucleases are considered when designing a cloning experiment.

Many other enzymes are used to prepare DNA for cloning. In addition to restriction endonucleases, nucleases that degrade single-stranded extensions, such as S1 nuclease and mung bean nuclease, are used to generate blunt ends for cloning (Fig. 1.3A). This is useful when the recognition sequences for restriction enzymes that produce complementary sticky ends are not available on both the vector and target DNA molecules. Blunt ends can also be produced by extending 3′ recessed ends using a DNA polymerase such as **Klenow polymerase** derived from *E. coli* DNA polymerase I (Fig. 1.3B). **Phosphatases** such as calf intestinal alkaline phosphatase cleave the 5′ phosphate groups from restriction enzyme-digested

Figure 1.2 Restriction endonucleases have been isolated from many different bacteria. (**A**) Isoschizomers such as BspEI from a *Bacillus* species and AccIII from *Acinetobacter calcoaceticus* bind the same DNA sequence and cut at the same sites. (**B**) Neoschizomers such as NarI from *Nocardia argentinensis* and SfoI from *Serratia fonticola* bind the same DNA sequence but cut at different sites. (**C**) Isocaudomers such as NcoI from *Nocardia corallina* and PagI from *Pseudomonas alcaligenes* bind different DNA sequences but produce the same sticky ends. Bases in the restriction enzyme recognition sequence are shown. Arrows show the sites of cleavage in the DNA backbone. doi:10.1128/9781555818890.ch1.f1.2

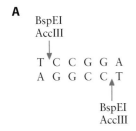

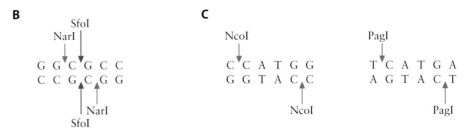

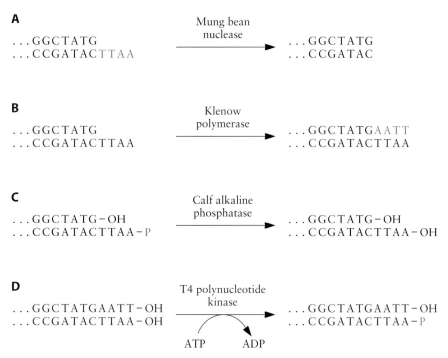

Figure 1.3 Some other enzymes used to prepare DNA for cloning. **(A)** Mung bean nuclease degrades single-stranded 5′ and 3′ extensions to generate blunt ends. **(B)** Klenow polymerase extends 3′ recessed ends to generate blunt ends. **(C)** Calf alkaline phosphatase removes the 5′ phosphate group from the ends of linear DNA molecules. **(D)** T4 polynucleotide kinase catalyzes the addition of a 5′ phosphate group to the ends of linear DNA fragments. Dotted lines indicate that only one end of the linear DNA molecule is shown. doi:10.1128/9781555818890.ch1.f1.3

DNA (Fig. 1.3C). A 5′ phosphate group is required for formation of a phosphodiester bond between nucleotides, and therefore, its removal prevents recircularization (self-ligation) of vector DNA. On the other hand, **kinases** add phosphate groups to the ends of DNA molecules. Among other activities, T4 polynucleotide kinase catalyzes the transfer of the terminal (γ) phosphate from a nucleoside triphosphate to the 5′ hydroxyl group of a polynucleotide (Fig. 1.3D). This enzyme is employed to prepare chemically synthesized DNA for cloning, as such DNAs are often missing a 5′ phosphate group required for ligation to vector DNA.

Insertion of Target DNA into a Plasmid Vector

When two different DNA molecules are digested with the same restriction endonuclease, that produces the same sticky ends in both molecules, and then mixed together, new DNA combinations can be formed as a result of base-pairing between the extended regions (Fig. 1.4). The enzyme **DNA ligase**, usually from the *E. coli* bacteriophage T4, is used to reform the phosphodiester bond between the 3′ hydroxyl group and the 5′ phosphate group at the ends of DNA strands that are already held together by the hydrogen bonds between the complementary bases of the extensions

(Fig. 1.4). DNA ligase also joins blunt ends, although this is generally much less efficient.

Ligation of restriction enzyme-digested DNA provides a means to stably insert target DNA into a vector for introduction and propagation in a suitable host cell. Many different vectors have been developed to act as carriers for target DNA. Most are derived from natural gene carriers, such as genomes of viruses that infect eukaryotic or prokaryotic cells and

Figure 1.4 Ligation of two different DNA fragments after digestion of both with restriction endonuclease BamHI. Complementary nucleotides in the single-stranded extensions form hydrogen bonds. T4 DNA ligase catalyzes the formation of phosphodiester bonds by joining 5′ phosphate and 3′ hydroxyl groups at nicks in the backbone of the double-stranded DNA. doi:10.1128/9781555818890.ch1.f1.4

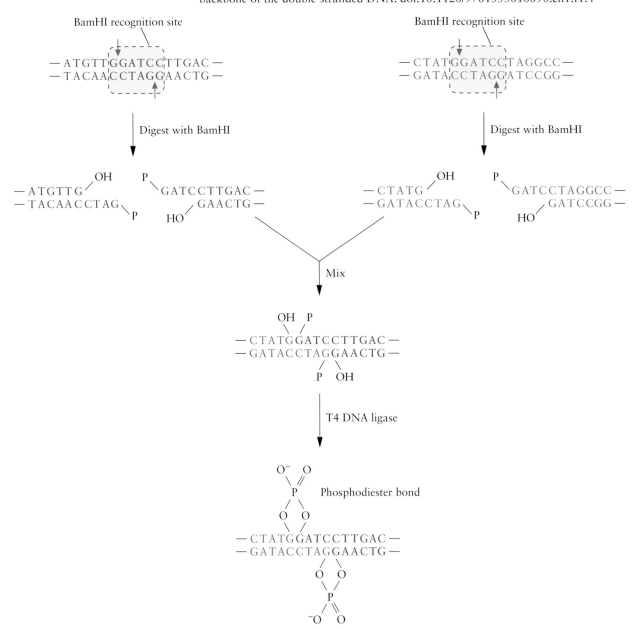

integrate into the host genome or plasmids that are found in bacterial or fungal cells. Others are synthetically constructed **artificial chromosomes** designed for delivery of large pieces of target DNA (>100 kilobase pairs [kb]) into bacterial, yeast, or mammalian host cells. Many different vectors that carry sequences required for specific functions, for example, for expression of foreign DNA in a host cell, are described throughout this book. Here, vectors based on bacterial **plasmids** are used to illustrate the basic features of a cloning vector.

Plasmids are small, usually circular, double-stranded DNA molecules that are found naturally in many bacteria. They can range in size from less than 1 kb to more than 500 kb and are maintained as extrachromosomal entities that replicate independently of the bacterial chromosome. While they are not usually essential for bacterial cell survival under laboratory conditions, plasmids often carry genes that are advantageous under particular conditions. For example, they may carry genes that encode resistance to antibiotics or heavy metals, genes for the degradation of unusual organic compounds, or genes required for toxin production. Each plasmid has a sequence that functions as an origin of DNA replication that is required for it to replicate in a host cell. Some plasmids carry information for their own transfer from one cell to another.

The number of copies of a plasmid that are present in a host cell is controlled by factors that regulate plasmid replication and are characteristic of that plasmid. High-copy-number plasmids are present in 10 to more than 100 copies per cell. Other, low-copy-number plasmids are maintained in 1 to 4 copies per cell. When two or more different plasmids cannot coexist in the same host cell, they are said to belong to the same **plasmid incompatibility group**. But plasmids from different incompatibility groups can be maintained together in the same cell. This coexistence is independent of the copy numbers of the individual plasmids. Some microorganisms have been found to contain as many as 8 to 10 different plasmids. In these instances, each plasmid can carry out different functions and have its own unique copy number, and each belongs to a different incompatibility group. Some plasmids can replicate in only one host species because they require very specific proteins for their replication as determined by their origin of replication. These are generally referred to as **narrow-host-range plasmids**. On the other hand, **broad-host-range plasmids** have less specific origins of replication and can replicate in a number of bacterial species.

As autonomous, self-replicating genetic elements, plasmids are useful vectors for carrying cloned DNA. However, naturally occurring plasmids often lack several important features that are required for a good cloning vector. These include a choice of unique (single) restriction endonuclease recognition sites into which the target DNA can be cloned and one or more selectable genetic markers for identifying recipient cells that carry the cloning vector–insert DNA construct. Most of the plasmids that are currently used as cloning vectors have been genetically modified to include these features.

An example of a commonly used plasmid cloning vector is pUC19, which is derived from a natural *E. coli* plasmid. The plasmid pUC19 is

2,686 bp long, contains an origin of replication that enables it to replicate in *E. coli*, and has a high copy number, which is useful when a large number of copies of the target DNA or its encoded protein is required (Fig. 1.5A). It has been genetically engineered by addition of a short (54-bp) DNA sequence with many unique restriction enzyme sites which is called a **multiple-cloning site** (also known as a polylinker) (Fig. 1.5B). A segment of the lactose operon of *E. coli* has also been added that includes the β-galactosidase gene (*lacZ′*) under the control of the *lac* promoter and a *lacI* gene that produces a repressor protein that regulates the expression of the *lacZ′* gene from the *lac* promoter (Fig. 1.5A). The multiple-cloning site has been inserted within the β-galactosidase gene in a manner that does not disrupt the function of the β-galactosidase enzyme when it is expressed (Fig. 1.5B). In addition, pUC19 carries the *bla* gene (Amp^r gene) encoding β-lactamase that renders the cell resistant to ampicillin and can therefore be used as a selectable marker to identify cells that carry the vector.

To clone a gene of interest into pUC19, the vector is cut with a restriction endonuclease that has a unique recognition site within the multiple-cloning site (Fig. 1.6). The source DNA carrying the target gene is digested with the same restriction enzyme, which cuts at sites flanking the sequence of interest. The resulting linear molecules, which have the same sticky ends, are mixed together and then treated with T4 DNA ligase. A number of different ligated combinations are produced by this reaction, including the original circular plasmid DNA (Fig. 1.6). To reduce the amount of this unwanted ligation product, prior to ligation, the

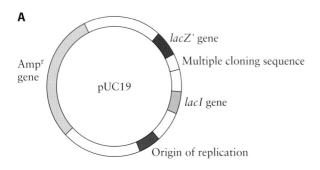

A

Amp^r gene

pUC19

lacZ′ gene

Multiple cloning sequence

lacI gene

Origin of replication

Figure 1.5 Plasmid cloning vector pUC19. (**A**) The plasmid contains an origin of replication for propagation in *E. coli*, an ampicillin resistance gene (Amp^r) for selection of cells carrying the plasmid, and a multiple-cloning site for insertion of cloned DNA. (**B**) The multiple-cloning site (nucleotides in uppercase letters) containing several unique restriction endonuclease recognition sites (indicated by horizontal lines) was inserted into the *lacZ′* gene (nucleotides in lowercase letters) in a manner that does not disrupt the production of a functional β-galactosidase (LacZα fragment). The first 26 amino acids of the protein are shown. Expression of the *lacZ′* gene is controlled by the LacI repressor encoded by the *lacI* gene on the plasmid. The size of the plasmid is 2,686 bp. doi:10.1128/9781555818890.ch1.f1.5

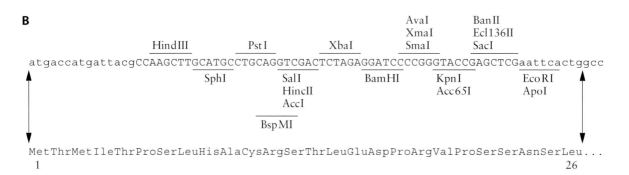

B

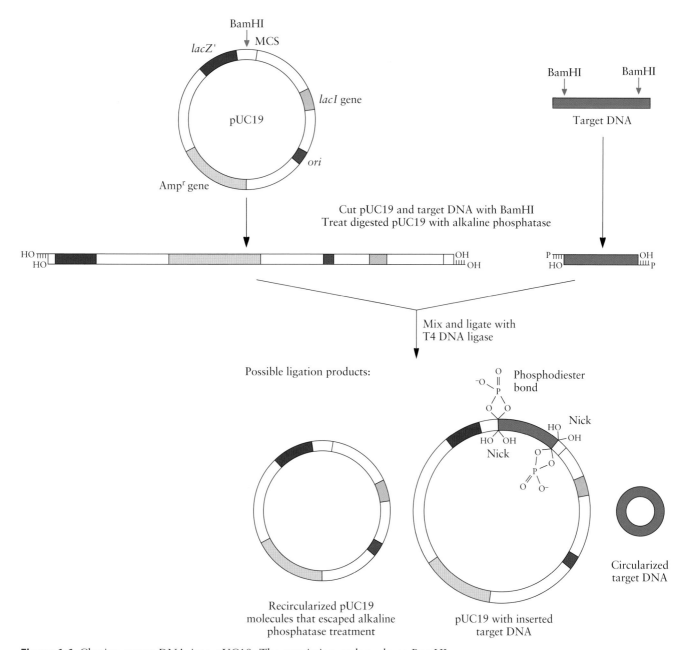

Figure 1.6 Cloning target DNA into pUC19. The restriction endonuclease BamHI cleaves pUC19 at a unique site in the multiple-cloning site (MCS) and at sequences flanking the target DNA. The cleaved vector is treated with alkaline phosphatase to remove 5′ phosphate groups to prevent vector recircularization. Digested target DNA and pUC19 are mixed to join the two molecules via complementary single-stranded extensions and treated with T4 DNA ligase to form a phosphodiester bond between the joined molecules. Several ligation products are possible. In addition to pUC19 inserted with target DNA, undesirable circularized target DNA molecules and recircularized pUC19 that escaped treatment with alkaline phosphatase are produced. doi:10.1128/9781555818890.ch1.f1.6

cleaved plasmid DNA is treated with the enzyme alkaline phosphatase to remove the 5′ phosphate groups from the linearized plasmid DNA (Fig. 1.3C). T4 DNA ligase cannot join the ends of the dephosphorylated linear plasmid DNA. However, the target DNA is not treated with alkaline phosphatase and therefore provides phosphate groups to form two phosphodiester bonds with the alkaline phosphatase-treated vector DNA (Fig. 1.6). Following treatment with T4 DNA ligase, the two phosphodiester bonds are sufficient to hold the circularized molecules together, despite the presence of two nicks (Fig. 1.6). After introduction into a host bacterium, these nicks are sealed by the host cell DNA ligase system.

Transformation and Selection of Cloned DNA in a Bacterial Host

After ligation, the next step in a cloning experiment is to introduce the vector–target DNA construct into a suitable host cell. A wide range of prokaryotic and eukaryotic cells can be used as cloning hosts; however, routine cloning procedures are often carried out using a well-studied bacterial host, usually *E. coli*. The process of taking up DNA into a bacterial cell is called **transformation**, and a cell that is capable of taking up DNA is said to be competent. **Competence** occurs naturally in many bacteria, usually when cells are stressed in high-density populations or in nutrient-poor environments, and enables bacteria to acquire new sequences that may enhance survival. Although competence and transformation are not intrinsic properties of *E. coli*, competence can be induced by various treatments such as cold calcium chloride. A brief heat shock facilitates uptake of exogenous DNA molecules. Alternatively, uptake of free DNA can be induced by subjecting bacteria to a high-voltage electric field in a procedure known as **electroporation**. Generally, transformation is an inefficient process, and therefore, most of the cells will not have acquired a plasmid; at best, about 1 cell in 1,000 *E. coli* host cells is transformed. The integrity of the introduced DNA constructs is also more likely to be maintained in host cells that are unable to carry out exchanges between DNA molecules because the recombination enzyme RecA has been deleted from the host chromosome.

Cells transformed with vectors that carry a gene encoding resistance to an antibiotic can be selected by plating on medium containing the antibiotic. For example, cells carrying the plasmid vector pUC19, which contains the *bla* gene encoding β-lactamase, can be selected on ampicillin (Fig. 1.7A). Nontransformed cells or cells transformed with circularized target DNA cannot grow in the presence of ampicillin. However, cells transformed with the pUC19–target DNA construct and cells transformed with recircularized pUC19 that escaped dephosphorylation by alkaline phosphatase are both resistant to ampicillin. To differentiate cells carrying the desired vector–target DNA construct from those carrying the recircularized plasmid, loss of β-galactosidase activity that results from insertion of target DNA into the *lacZ′* gene is determined. Recall that the multiple-cloning site in pUC19 lies within the *lacZ′* gene (Fig. 1.5). An *E. coli* host is used that can synthesize the part of β-galactosidase (LacZω fragment) that combines with the product of the *lacZ′* gene (LacZα

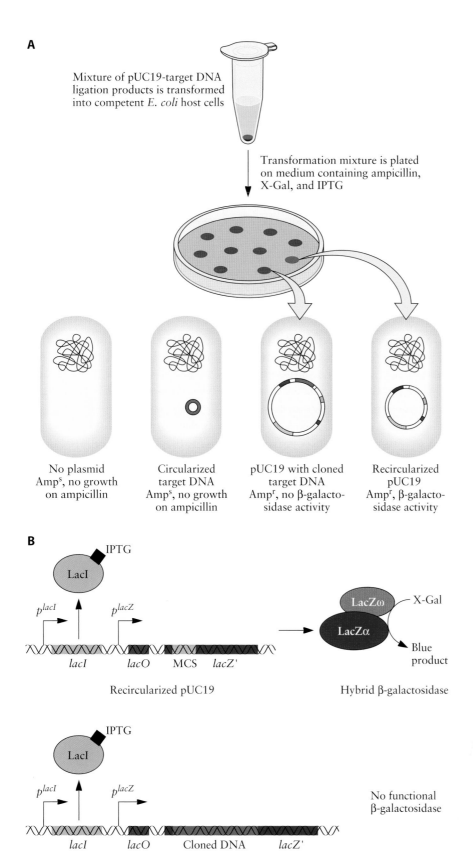

Figure 1.7 (**A**) Strategy for selecting host cells that have been transformed with pUC19 carrying cloned target DNA. *E. coli* host cells transformed with the products of pUC19–target DNA ligation are selected on medium containing ampicillin, X-Gal, and IPTG. Nontransformed cells and cells transformed with circularized target DNA do not have a gene conferring resistance to ampicillin and therefore do not grow on medium containing ampicillin (recircularized target DNA also does not carry an origin of replication, and therefore, the plasmids are not propagated in host cells even in the absence of ampicillin). However, *E. coli* cells transformed with recircularized pUC19 or pUC19 carrying cloned target DNA are resistant to ampicillin and therefore form colonies on the selection medium. The two types of ampicillin-resistant transformants are differentiated by the production of functional β-galactosidase. (**B**) IPTG in the medium induces expression of the *lacZ'* gene by binding to the LacI repressor and preventing LacI from binding to the *lacO* operator sequence. This results in production of the β-galactosidase LacZα fragment in cells transformed with recircularized pUC19. The LacZα fragment combines with the LacZω fragment of β-galactosidase encoded in the host *E. coli* chromosome to form a functional hybrid β-galactosidase. β-Galactosidase cleaves X-Gal, producing a blue product, and the colonies on the plate appear blue. Insertion of target DNA into the multiple-cloning site of pUC19 alters the reading frame of the *lacZ'* gene, thereby preventing production of a functional β-galactosidase in cells transformed with this construct. Colonies that appear white on medium containing X-Gal and IPTG carry the cloned target gene (**A**). doi:10.1128/9781555818890.ch1.f1.7

fragment) encoded on pUC19 to form a functional enzyme. When cells carrying recircularized pUC19 are grown in the presence of isopropyl-β-D-thiogalactopyranoside (IPTG), which is an inducer of the *lac* operon, the protein product of the *lacI* gene is prevented from binding to the promoter–operator region of the *lacZ'* gene, so the *lacZ'* gene in the plasmid is transcribed and translated (Fig. 1.7B). The LacZα fragment combines with a host LacZω fragment to form an active hybrid β-galactosidase. If the substrate 5-bromo-4-chloro-3-indolyl-β-D-galactopyranoside (X-Gal) is present in the medium, it is hydrolyzed by the hybrid β-galactosidase to form a blue product. Under these conditions, colonies containing unmodified pUC19 appear blue (Fig. 1.7A). In contrast, host cells that carry a plasmid-cloned DNA construct produce white colonies on the same medium. The reason for this is that target DNA inserted into a restriction endonuclease site within the multiple-cloning site usually disrupts the correct sequence of DNA codons (reading frame) of the *lacZ'* gene and prevents the production of a functional LacZα fragment, so no active hybrid β-galactosidase is produced (Fig. 1.7B). In the absence of β-galactosidase activity, the X-Gal in the medium is not converted to the blue compound, so these colonies remain white (Fig. 1.7A). The white (positive) colonies subsequently must be confirmed to carry a specific target DNA sequence.

A number of selection systems have been devised to identify cells carrying vectors that have been successfully inserted with target DNA. In addition to ampicillin, other antibiotics such as tetracycline, kanamycin, and streptomycin are used as selective agents for various cloning vectors. Some vectors carry a gene that encodes a toxin that kills the cell (Table 1.1). The toxin gene is under the control of a regulatable promoter, such as the promoter for the *lacZ'* gene that is activated only when the inducer IPTG is supplied in the culture medium. Insertion of a target DNA fragment into the multiple-cloning site prevents the production of a functional toxin protein in the presence of the inducer. Only cells that carry a vector with the target DNA survive under these conditions.

In addition to *E. coli*, other bacteria, such as *Bacillus subtilis*, often are the final host cells. For many applications, cloning vectors that function in

Table 1.1 Some toxin genes used to select for successful insertion of target DNA into a vector

Toxin gene	Source	Mode of action
sacB	*Bacillus subtilis*	Encodes the enzyme levansucrase, which converts sucrose to levans that are toxic to gram-negative bacteria
ccdB	*E. coli* F plasmid	Encodes a toxin that inhibits DNA gyrase activity and therefore DNA replication
barnase	*Bacillus amyloliquefaciens*	Encodes a ribonuclease that cleaves RNA
eco47IR	*E. coli*	Encodes an endonuclease that cleaves DNA at the sequence GGWCC (W = A or T) when it is not methylated
hsv-tk	Herpes simplex virus	Encodes thymidine kinase that converts the synthetic nucleotide analogue ganciclovir into a product that is toxic to insect cells

E. coli may be provided with a second origin of replication that enables the plasmid to replicate in the alternative host cell. With these **shuttle cloning vectors**, the initial cloning steps are conducted using *E. coli* before the final construct is introduced into a different host cell. In addition, a number of plasmid vectors have been constructed with a single broad-host-range **origin of DNA replication** instead of a narrow-host-range origin of replication. These vectors can be used with a variety of microorganisms.

Broad-host-range vectors can be transferred among different bacterial hosts by exploiting a natural system for transmitting plasmids known as **conjugation**. There are two basic genetic requirements for transfer of a plasmid by conjugation: (i) a specific **origin-of-transfer** (*oriT*) sequence on the plasmid that is recognized by proteins that initiate plasmid transfer and (ii) several genes encoding the proteins that mediate plasmid transfer that may be present on the transferred plasmid (Fig. 1.8A) or in the genome of the plasmid donor cell or supplied on a helper plasmid (Fig. 1.8B). Some of these proteins form a pilus that extends from the donor cell and, following contact with a recipient cell, retracts to bring the two cells into close contact. A specific endonuclease cleaves one of the two strands of the plasmid DNA at the *oriT*, and as the DNA is unwound, the displaced single-stranded DNA is transferred into the recipient cell through a conjugation pore made up of proteins encoded by the transfer genes. A complementary strand is synthesized in both the donor and recipient cells, resulting in a copy of the plasmid in both cells.

Cloning Eukaryotic Genes

Bacteria lack the molecular machinery to excise the **introns** from RNA that is transcribed from eukaryotic genes. Therefore, before a eukaryotic sequence is cloned for the purpose of producing the encoded protein in a bacterial host, the intron sequences must be removed. Functional eukaryotic mRNA does not contain introns because they have been removed by the eukaryotic cell splicing machinery. Purified mRNA molecules are used as a starting point for cloning eukaryotic genes but must be converted to double-stranded DNA before they are inserted into a vector that provides bacterial sequences for transcription and translation.

Purified mRNA can be obtained from eukaryotic cells by exploiting the tract of up to 200 adenine residues {polyadenylic acid [poly(A)] tail} that are added to the 3′ ends of mRNA before they are exported from the nucleus (Fig. 1.9). The poly(A) tail provides the means for separating the mRNA fraction of a tissue from the more abundant ribosomal RNA (rRNA) and transfer RNA (tRNA). Short chains of 15 thymidine residues (oligodeoxythymidylic acid [oligo(dT)]) are attached to cellulose beads, and the oligo(dT)–cellulose beads are packed into a column. Total RNA extracted from eukaryotic cells or tissues is passed through the oligo(dT)–cellulose column, and the poly(A) tails of the mRNA molecules bind by base-pairing to the oligo(dT) chains. The tRNA and rRNA molecules, which lack poly(A) tails, pass through the column. The mRNA is removed (eluted) from the column by treatment with a buffer that breaks the A:T hydrogen bonds.

poly(A)
polyadenylic acid

rRNA
ribosomal RNA

tRNA
transfer RNA

oligo(dT)
oligodeoxythymidylic acid

A Biparental conjugation

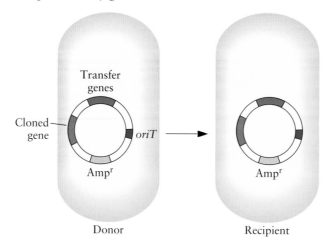

B Triparental conjugation

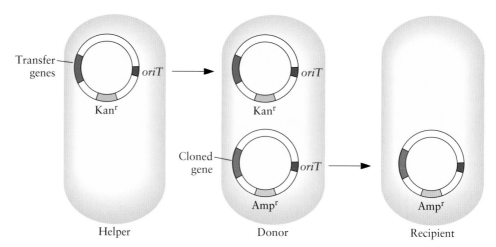

Figure 1.8 Plasmid transfer by biparental (**A**) or triparental (**B**) conjugation. A plasmid carrying a cloned gene, an origin of transfer (*oriT*), and transfer genes is transferred by biparental mating from a donor cell to a recipient cell (**A**). Proteins encoded by the transfer genes mediate contact between donor and recipient cells, initiate plasmid transfer by nicking one of the DNA strands at the *oriT*, and form a pore through which the nicked strand is transferred from the donor cell to recipient cell. In a cloning experiment, the donor is often a strain of *E. coli* that does not grow on minimal medium, allowing selection of recipient cells that grow on minimal medium. Acquisition of the plasmid by recipient cells is determined by resistance to an antibiotic, such as ampicillin in this example. If the plasmid carrying the cloned gene does not possess genes for plasmid transfer, these can be supplied by a helper cell (**B**). In triparental mating, the helper plasmid is first transferred to the donor cell, where the proteins that mediate transfer of the plasmid carrying the cloned gene are expressed. Although the plasmid carrying the cloned gene does not possess transfer genes, it must have an *oriT* in order to be transferred. Neither the helper nor donor cells grow on minimal medium, and therefore, the recipient cells can be selected on minimal medium containing an antibiotic such as ampicillin. To ensure that the helper plasmid was not transferred to the recipient cell, sensitivity to kanamycin is determined. doi:10.1128/9781555818890.ch1.f1.8

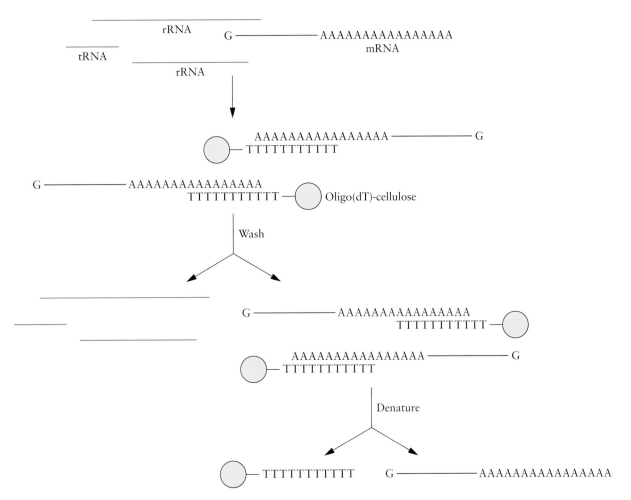

Figure 1.9 Purification of mRNA from total cellular RNA. Total RNA extracted from cells is added to a column of cellulose beads that carry a short chain of thymine nucleotides [oligo(dT)]. The poly(A) tails that are found on most eukaryotic mRNAs hybridize to the complementary oligo(dT), while other RNA molecules, such as tRNA and rRNA, that do not have poly(A) tails pass through the column. After a washing step, the mRNA molecules are eluted from the column by treatment with a solution that breaks the A:T hydrogen bonds. doi:10.1128/9781555818890.ch1.f1.9

To convert mRNA to double-stranded DNA for cloning, the enzyme **reverse transcriptase**, encoded by certain RNA viruses (**retroviruses**), is used to catalyze the synthesis of **complementary DNA (cDNA)** from an RNA template. If the sequence of the target mRNA is known, a short (~20 nucleotides), single-stranded DNA molecule known as an oligonucleotide primer that is complementary to a sequence at the 3′ end of the target mRNA is synthesized (Fig. 1.10A). The primer is added to a sample of purified mRNA that is extracted from eukaryotic cells known to produce the mRNA of interest. This sample of course contains all of the different mRNAs that are produced by the cell; however, the primer will specifically base-pair with the complementary sequence on the target mRNA. Not only is the primer important for targeting a specific mRNA, but also it provides an available 3′ hydroxyl group to prime the synthesis

cDNA
complementary DNA

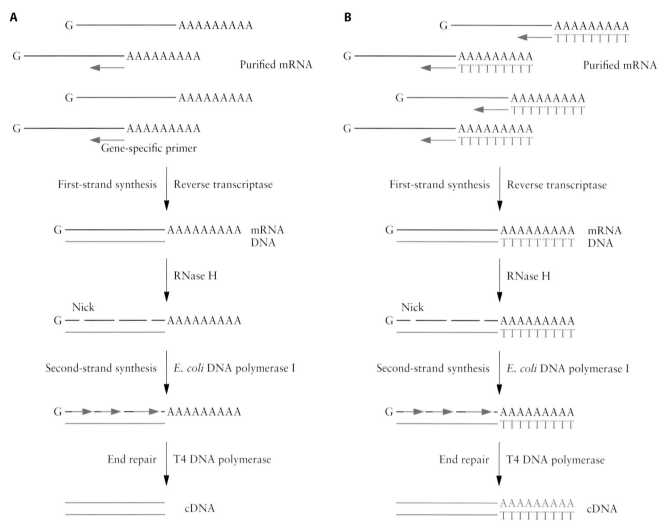

Figure 1.10 Synthesis of double-stranded cDNA using gene-specific primers (**A**) or oligo(dT) primers (**B**). A short oligonucleotide primer is added to a mixture of purified mRNA and anneals to a complementary sequence on the mRNA. Reverse transcriptase catalyzes the synthesis of a DNA strand from the primer using the mRNA as a template. To synthesize the second strand of DNA, the mRNA is nicked by RNase H, which creates initiation sites for *E. coli* DNA polymerase I. The 5′ exonuclease activity of DNA polymerase I removes both RNA sequences that are encountered as DNA synthesis proceeds. The ends of the cDNA are blunted using T4 DNA polymerase prior to cloning. doi:10.1128/9781555818890.ch1.f1.10

RNase
ribonuclease

of the first cDNA strand. In the presence of the four deoxyribonucleotides, reverse transcriptase incorporates a complementary nucleotide into the growing DNA strand as directed by the template mRNA strand. To generate a double-stranded DNA molecule, the RNA:DNA (heteroduplex) molecules are treated with **ribonuclease (RNase)** H, which nicks the mRNA strands, thereby providing free 3′ hydroxyl groups for initiation of DNA synthesis by DNA polymerase I. As the synthesis of the second DNA strand progresses from the nicks, the 5′ exonuclease activity of DNA polymerase I removes the ribonucleotides of the mRNA. After synthesis of the second DNA strand is completed, the ends of the cDNA molecules are blunted (end repaired, or polished) with T4 DNA polymerase, which removes 3′ extensions and fills in from 3′ recessed ends. The double-stranded cDNA carrying only the exon sequences encoding the eukaryotic protein can be cloned directly into a suitable vector by blunt-end ligation. Alternatively, chemically synthesized **adaptors** with extensions containing a restriction endonuclease recognition sequence can be ligated to the ends of the cDNA molecules for insertion into a vector via sticky-end ligation.

When the sequence of the target mRNA intended for cloning is not known or when several target mRNAs in a single sample are of interest, cDNA can be generated from all of the mRNAs using an oligo(dT) primer rather than a gene-specific primer (Fig. 1.10B). The mixture of cDNAs, ideally representing all possible mRNA produced by the cell, is cloned into a vector to create a cDNA library that can be screened for the target sequence(s) (described below).

Recombinational Cloning

Recombinational cloning is a rapid and versatile system for cloning sequences without restriction endonuclease and ligation reactions. This system exploits the mechanism used by bacteriophage λ to integrate viral DNA into the host bacterial genome during infection. Bacteriophage λ integrates into the *E. coli* chromosome at a specific sequence (25 bp) in the bacterial genome known as the attachment bacteria (*attB*) site. The bacteriophage genome has a corresponding attachment phage (*attP*) sequence (243 bp) that can recombine with the bacterial *attB* sequence with the help of the bacteriophage λ recombination protein **integrase** and an *E. coli*-encoded protein called integration host factor (Fig. 1.11A). Recombination between the *attP* and *attB* sequences results in insertion of the phage genome into the bacterial genome to create a **prophage** with attachment sites *attL* (100 bp) and *attR* (168 bp) at the left and right ends of the integrated bacteriophage λ DNA, respectively. For subsequent excision of the bacteriophage λ DNA from the bacterial chromosome, recombination between the *attL* and *attR* sites is mediated by **integration host factor**, integrase, and bacteriophage λ **excisionase** (Fig 1.11B). The recombination events occur at precise locations without either the loss or gain of nucleotides.

For recombinational cloning, a modified *attB* sequence is added to each end of the target DNA. The *attB* sequences are modified so that they will only recombine with specific *attP* sequences. For example, *attB1* recombines only with *attP1*, and *attB2* recombines with *attP2*. The target DNA with flanking *attB1* and *attB2* sequences is mixed with a vector (donor vector) that has *attP1* and *attP2* sites flanking a toxin gene that will be used for negative selection following transformation into a host cell (Fig. 1.12A). Integrase and integration host factor are added to the mixture of DNA molecules to catalyze in vitro recombination between the *attB1* and *attP1* sites and between the *attB2* and *attP2* sites. As a consequence of the two recombination events, the toxin gene sequence

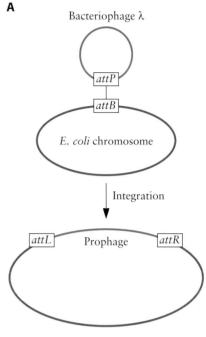

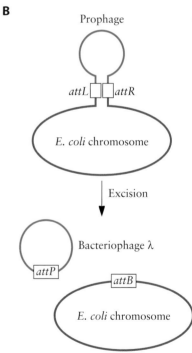

Figure 1.11 Integration (**A**) and excision (**B**) of bacteriophage λ into and from the *E. coli* genome via recombination between attachment (*att*) sequences in the bacterial and bacteriophage DNAs. doi:10.1128/9781555818890.ch1.f1.11

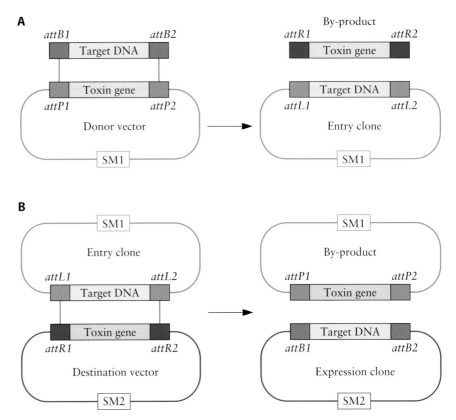

Figure 1.12 Recombinational cloning. (**A**) Recombination (thin vertical lines) between a target gene with flanking attachment sites (*attB1* and *attB2*) and a donor vector with *attP1* and *attP2* sites on either side of a toxin gene results in an entry clone where the target gene is flanked by *attL1* and *attL2* sites. The selectable marker (SM1) enables selection of cells transformed with an entry clone. The protein encoded by the toxin gene kills cells transformed with nonrecombined donor vectors. The origin of replication of the donor vector is not shown. (**B**) Recombination between the entry clone with flanking *attL1* and *attL2* sites and a destination vector with *attR1* and *attR2* sites results in an expression clone with *attB1* and *attB2* sites flanking the target gene. The selectable marker (SM2) enables selection of transformed cells with an expression clone. The second plasmid, designated as a by-product, has the toxin gene flanked by *attP1* and *attP2* sites. Cells with an intact destination vector that did not undergo recombination or that retain the by-product plasmid are killed by the toxin. Transformed cells with an entry clone, which lacks the SM2 selectable marker, are selected against. The origins of replication and the sequences for expression of the target gene are not shown. doi:10.1128/9781555818890.ch1.f1.12

between the *attP1* and *attP2* sites on the donor vector is replaced by the target gene. The recombination events create new attachment sites flanking the target gene sequence (designated *attL1* and *attL2*), and the plasmid with the *attL1*-target gene-*attL2* sequence is referred to as an entry clone. The mixture of original and recombinant DNA molecules is transformed into *E. coli*, and cells that are transformed with donor vectors that have not undergone recombination retain the toxin gene and therefore do not survive. Host cells carrying the entry clone are positively selected by the presence of a selectable marker.

The advantage of this procedure is the ability to easily transfer the target gene to a variety of vectors that have been developed for different purposes. For example, to produce high levels of the protein encoded on the cloned gene, the target DNA can be transferred to a destination vector that carries a promoter and other expression signals. An entry clone is mixed with a destination vector that has *attR1* and *attR2* sites flanking a toxin gene (Fig. 1.12B). In the presence of integration host factor, integrase, and bacteriophage λ excisionase, the *attL1* and *attL2* sites on the entry clone recombine with the *attR1* and *attR2* sites, respectively, on the destination vector. This results in the replacement of the toxin gene on the destination vector with the target gene from the entry clone, and the resultant plasmid is designated an expression clone. The reaction mixture is transformed into *E. coli*, and a selectable marker is used to isolate transformed cells that carry an expression clone. Cells that carry an intact destination vector or the exchanged entry plasmid (known as a by-product plasmid) will not survive, because these carry the toxin gene. Destination vectors are available for maintenance and expression of the target gene in various host cells such as *E. coli* and yeast, insect, and mammalian cells.

Genomic Libraries

A **genomic library** is a collection of DNA fragments, each cloned into a vector, that represents the entire genomic DNA, or cDNA derived from the total mRNA, in a sample. For example, the genomic library may contain fragments of the entire genome extracted from cells in a pure culture of bacteria or from tissue from a plant or animal. A genomic library can also contain the genomes of all of the organisms present in a complex sample such as from the microbial community on human tissue. Such a library is known as a **metagenomic library**. Whole-genome libraries may be used to obtain the genome sequence of an organism or to identify genes that contain specific sequences, encode particular functions, or interact with other molecules. A cDNA library constructed from mRNA may be used to identify all of the genes that are actively expressed in bacterial or eukaryotic cells under a particular set of conditions or to identify eukaryotic genes that have a particular sequence or function.

To create a genomic library, the DNA extracted from the cells (cell cultures or tissues) of a source organism (or a community of organisms for a metagenomic library) is first digested with a restriction endonuclease. Often a restriction endonuclease that recognizes a sequence of four nucleotides, such as Sau3AI, is used. Although four-cutters will theoretically cleave the DNA approximately once in every 256 bp, the reaction conditions are set to give a partial, not a complete, digestion to generate fragments of all possible sizes (Fig. 1.13). This is achieved by using either a low concentration of restriction endonuclease or shortened incubation times, and usually some optimization of these parameters is required to determine the conditions that yield fragments of suitable size. The range of fragment sizes depends on the goal of the experiment. For example, for genome sequencing, large (100- to 200-kb) fragments are often desirable. To identify genes that encode a particular enzymatic function, that is, genes that are expressed to produce proteins in the size range of an average protein, smaller (~5- to 40-kb) fragments are cloned.

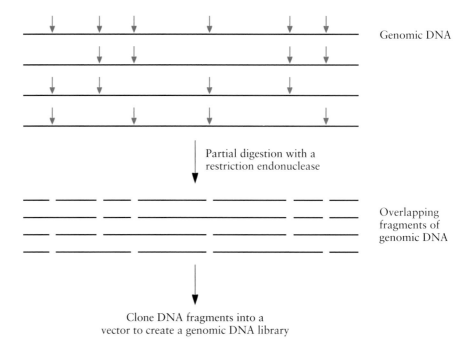

Genomic DNA

Partial digestion with a
restriction endonuclease

Overlapping
fragments of
genomic DNA

Clone DNA fragments into a
vector to create a genomic DNA library

Figure 1.13 Construction of a genomic DNA library. Genomic DNA extracted from cells or tissues is partially digested with a restriction endonuclease. Conditions are set so that the enzyme does not cleave at all possible sites. This generates overlapping DNA fragments of various lengths that are cloned into a vector. doi:10.1128/9781555818890.ch1.f1.13

The number of clones in a genomic library depends on the size of the genome of the organism, the average size of the insert in the vector, and the average number of times each sequence is represented in the library (**sequence coverage**). To ensure that the entire genome, or most of it, is contained within the clones of a library, the sum of the inserted DNA in the clones of the library should be at least three times the amount of DNA in the genome. For example, the size of the *E. coli* genome is approximately 4×10^6 bp; if inserts of an average size of 1,000 bp are desired, then 12,000 clones are required for threefold coverage, i.e., $3[(4 \times 10^6)/10^3]$. For the human genome, which contains 3.3×10^9 bp, about 80,000 clones with an average insert size of 150,000 bp are required for fourfold coverage, i.e., $4[(3.3 \times 10^9)/(15 \times 10^4)]$. Statistically, the number of clones required for a comprehensive genomic library can be estimated from the relationship $N = \ln(1 - P)/\ln(1 - f)$, where N is the number of clones, P is the probability of finding a specific gene, and f is the ratio of the length of the average insert to the size of the entire genome. On this basis, about 700,000 clones are required for a 99% chance of discovering a particular sequence in a human genomic library with an average insert size of 20 kb.

Several strategies can be used to identify target DNA in a genomic library. Genomic or metagenomic libraries can be screened to identify members of the library that carry a gene encoding a particular protein function. Many genes encoding enzymes that catalyze specific reactions have been isolated from a variety of organisms by plating the cells of a genomic library on medium supplemented with a specific substrate. For efficient screening of thousands of clones, colonies that carry a cloned gene encoding a functional catabolic enzyme must be readily identifiable, often by production of a colored product or a zone of substrate clearing around the colony. In one example, genes encoding enzymes that break down dietary fiber were isolated from a metagenomic library of DNA from microbes present in the human intestinal tract (Fig. 1.14). Dietary fiber is a complex mixture of carbohydrates derived from ingested plant material that is not digested in the small intestine but, rather, is broken down to metabolizable mono- and oligosaccharides in the colon by microorganisms. About 80% of these microbes have not been grown in the laboratory, and therefore screening a metagenomic library prepared in a host bacterium such as *E. coli* enables the isolation of novel carbohydrate catabolic enzymes that contribute to human nutrition without the need to first culture the natural host cell. A library of 156,000 clones with an average insert size of 30 to 40 kb representing 5.4×10^9 bp of metagenomic DNA was constructed in *E. coli* from DNA extracted from the feces of a healthy human who had consumed a diet rich in plant fiber. The library was screened by plating separately on media containing different polysaccharides commonly found in dietary fiber such as pectin, amylose, and β-glucan. The polysaccharides were tagged with a blue or red dye that is released when the polysaccharide is hydrolyzed. Extracellular diffusion of the dye enabled visualization of cells that expressed a carbohydrate catabolic enzyme from a cloned DNA fragment. More than 300 clones were identified that could degrade at least one of the polysaccharides tested.

The presence of particular proteins produced by a genomic library can also be detected using an immunological assay. Rather than screening for the function of a protein, the library is screened using an antibody that specifically binds to the protein encoded by a target gene. The colonies are arrayed on a solid medium, transferred to a matrix, and then lysed to release the cellular proteins (Fig. 1.15). Interaction of the primary antibody with the target protein (antigen) on the matrix is detected by applying a secondary antibody that is specific for the primary antibody. The secondary antibody is attached to an enzyme, such as alkaline phosphatase, that converts a colorless substrate to a colored or light-emitting (chemiluminescent) product that can readily identify positive interactions.

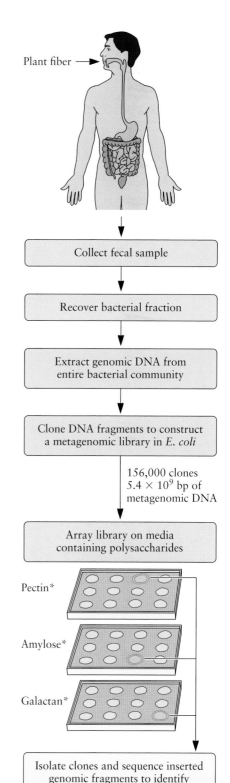

Figure 1.14 Isolation of carbohydrate catabolic genes from a metagenomic DNA library of human intestinal microorganisms. Genomic DNA extracted from fecal bacteria was fragmented and cloned to generate a metagenomic library of the human gut microbiome. *E. coli* host cells carrying the cloned DNA were arrayed on solid medium containing polysaccharides tagged with a blue dye (denoted with an asterisk). Colonies that hydrolyzed the polysaccharide were identified by diffusion of the blue dye. Positive clones were further characterized for specific enzyme activity, and the cloned DNA fragment was sequenced to identify the polysaccharide-degrading enzyme. Adapted from Tasse et al., *Genome Res.* **20**:1605–1612, 2010. doi:10.1128/9781555818890.ch1.f1.14

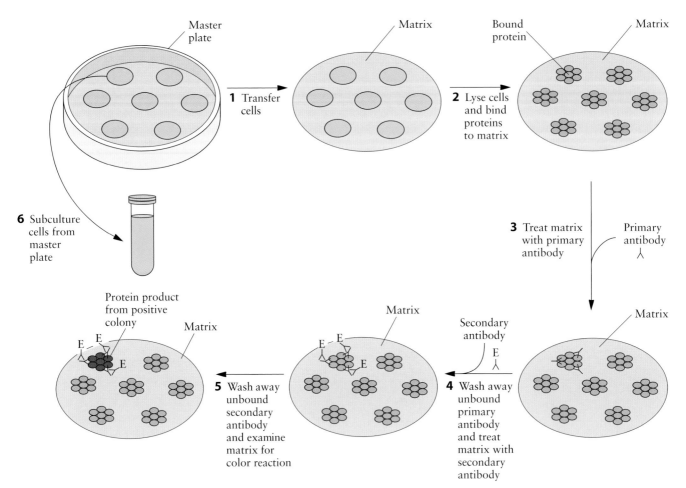

Figure 1.15 Screening of a genomic DNA library using an immunological assay. Transformed cells are plated onto solid agar medium under conditions that permit transformed but not nontransformed cells to grow. (1) From the discrete colonies formed on this master plate, a sample from each colony is transferred to a solid matrix such as a nylon membrane. (2) The cells on the matrix are lysed, and their proteins are bound to the matrix. (3) The matrix is treated with a primary antibody that binds only to the target protein. (4) Unbound primary antibody is washed away, and the matrix is treated with a secondary antibody that binds only to the primary antibody. (5) Any unbound secondary antibody is washed away, and a colorimetric (or chemiluminescent) reaction is carried out. The reaction can occur only if the secondary antibody, which is attached to an enzyme (E) that performs the reaction, is present. (6) A colony on the master plate that corresponds to a positive response on the matrix is identified. Cells from the positive colony on the master plate are subcultured because they may carry the plasmid–insert DNA construct that encodes the protein that binds the primary antibody. doi:10.1128/9781555818890.ch1.f1.15

Amplification of DNA Using PCR

The **polymerase chain reaction (PCR)** is a simple, efficient procedure for synthesizing large quantities of a specific DNA sequence in vitro (Fig. 1.16). The reaction exploits the mechanism used by living cells to accurately replicate a DNA template (Box 1.2). PCR can be used to produce millions of copies from a single template molecule in a few hours and to detect a specific sequence in a complex mixture of DNA even when other, similar

PCR
polymerase chain reaction

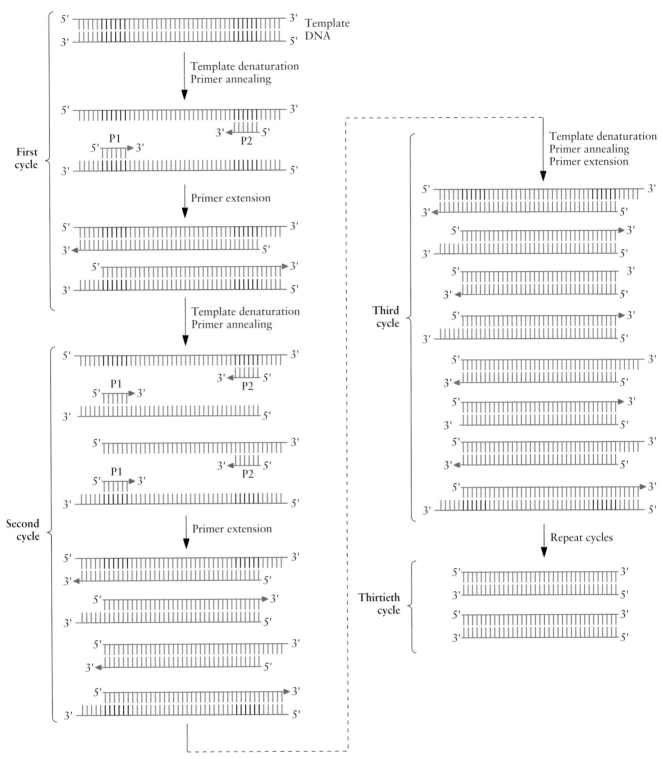

Figure 1.16 PCR. During a PCR cycle, the template DNA is denatured by heating and then slowly cooled to enable two primers (P1 and P2) to anneal to complementary (black) bases flanking the target DNA. The temperature is raised to about 70°C, and in the presence of the four deoxyribonucleotides, *Taq* DNA polymerase catalyzes the synthesis of a DNA strand extending from the 3′ hydroxyl end of each primer. In the first PCR cycle, DNA synthesis continues past the region of the template DNA strand that is complementary to the other primer sequence. The products of this reaction are two long strands of DNA that serve as templates for DNA synthesis during the second PCR cycle. In the second cycle, the primers hybridize to complementary regions in both the original strands and the long template strands, and DNA synthesis produces more long DNA strands from the original strands and short strands from the long template strands. A short template strand has a primer sequence at one end and the sequence complementary to the other primer at its other end. During the third PCR cycle, the primers hybridize to complementary regions of original, long template, and short template strands, and DNA synthesis produces long strands from the original strands and short strands from both long and short templates. By the end of the 30th PCR cycle, the products (amplicons) consist almost entirely of short double-stranded DNA molecules that carry the target DNA sequence delineated by the primer sequences. Note that in the figure, newly synthesized strands are differentiated from template strands by a terminal arrow.

doi:10.1128/9781555818890.ch1.f1.16

box 1.2
Polymerase Chain Reaction, a Powerful Method To Detect and Amplify Specific DNA Sequences

In the spring of 2008, strains of highly drug-resistant *E. coli* and *Klebsiella pneumoniae* were isolated from a Swedish patient. These strains were resistant to all β-lactam antibiotics, an important group of antibiotics that includes penicillin derivatives and cephalosporins, and carbapenem, a synthetic "last-resort" antibiotic that had previously been effective against β-lactam-resistant strains. Similar resistant strains began to appear in patients in the United States in June 2010, and soon physicians around the world reported infections with *E. coli*, *K. pneumoniae*, and other *Enterobacteriaceae* that resisted treatment with carbapenem. What was different about these strains? Although the *K. pneumoniae* isolate from the Swedish patient tested positive for the activity of metallo-β-lactamase, an enzyme that cleaves the β-lactam ring and confers resistance to these antibiotics, PCR failed to detect known metallo-β-lactamase genes. A screen for β-lactam resistance genes in a genomic library revealed a novel metallo-β-lactamase gene known as bla_{NDM-1}. Development of PCR primers that specifically target the bla_{NDM-1} gene have enabled rapid discrimination of strains carrying the gene in patients within a few hours. This is important

in the management of infected patients to administer an effective treatment and to prevent dissemination of the pathogen. Specific detection by PCR has enabled epidemiologists to track the spread of this superbug across the globe.

This epidemiological scenario is commonplace. But 25 years ago, tracking strains carrying a specific variant of a gene would have been a more difficult and time-consuming process. However, in the early 1980s, Kary Mullis was trying to solve the problem of using synthetic oligonucleotides to detect single nucleotide mutations in sequences that were present in low concentrations. He needed a method to increase the concentration of the target sequence. He reasoned that if he mixed heat-denatured DNA with two oligonucleotides that bound to opposite strands of the DNA at an arbitrary distance from each other and added some DNA polymerase and deoxynucleoside triphosphates, the polymerase would add the deoxynucleoside triphosphates to the hybridized oligonucleotides. In the first attempt, the reaction did not yield the expected products. Mullis then heated the reaction products to remove the extended oligonucleotides and then repeated the process with fresh polymerase, hypothesizing that

after each cycle of heat denaturation and DNA synthesis, the number of molecules carrying the specific sequence between the primers would double. Despite the skepticism of his colleagues, Mullis proved that his reasoning was correct, albeit the hard way. By manually cycling the reaction through temperatures required to denature the DNA and anneal and extend the oligonucleotides, each time adding a fresh aliquot of a DNA polymerase isolated from *E. coli*, he was able to synthesize unprecedented amounts of target DNA. Mullis received the Nobel Prize in chemistry in 1993 for his invention of the PCR method. Thermostable DNA polymerases that eliminate the need to add fresh polymerase after each denaturation step and automated cycling have since made PCR a routine and indispensable laboratory procedure.

Following PCR amplification of large amounts of bla_{NDM-1} from *K. pneumoniae*, the gene was sequenced, and by comparing the sequence to those of known *bla* genes, researchers identified a mutation that rendered the protein able to degrade carbapenem, in addition to other β-lactam antibiotics. Worryingly, the gene is found in a region of the *K. pneumoniae* genome and on a plasmid in *E. coli* that are readily transferred among bacteria. The spread of the bla_{NDM-1} gene poses a significant threat to human health because there are currently few treatments available that are effective against pathogens carrying the gene.

sequences are present. The essential components for PCR amplification are (i) two synthetic oligonucleotide primers (~20 nucleotides each) that are complementary to regions on opposite strands that flank the target DNA sequence and that, after annealing to the source DNA, have their 3′ hydroxyl ends oriented toward each other; (ii) a template sequence in a DNA sample that lies between the primer-binding sites and can be from 100 to 3,000 bp in length (larger regions can be amplified with reduced efficiency); (iii) a thermostable DNA polymerase that is active after repeated heating to 95°C or higher and copies the DNA template with high fidelity; and (iv) the four deoxyribonucleotides.

Replication of a specific DNA sequence by PCR requires three successive steps as outlined below. Amplification is achieved by repeating the three-step cycle 25 to 40 times. All steps in a PCR cycle are carried out in an automated block heater that is programmed to change temperatures after a specified period of time.

1. Denaturation. The first step in a PCR is the thermal **denaturation** of the double-stranded DNA template to separate the strands. This is achieved by raising the temperature of a reaction mixture to 95°C. The reaction mixture is comprised of the source DNA that contains the target DNA to be amplified, a vast molar excess of the two oligonucleotide primers, a thermostable DNA polymerase (e.g., *Taq* DNA polymerase, isolated from the bacterium *Thermus aquaticus*), and four deoxyribonucleotides.

2. Annealing. For the second step, the temperature of the mixture is slowly cooled. During this step, the primers base-pair, or anneal, with their complementary sequences in the DNA template. The temperature at which this step of the reaction is performed is determined by the nucleotide sequence of the primer that forms hydrogen bonds with complementary nucleotides in the target DNA. Typical annealing temperatures are in the range of 45 to 68°C, although optimization is often required to achieve the desired outcome, that is, a product consisting of fragments of target DNA sequence only.

3. Extension. In the third step, the temperature is raised to ~70°C, which is optimum for the catalytic activity of *Taq* DNA polymerase. DNA synthesis is initiated at the 3′ hydroxyl end of each annealed primer, and nucleotides are added to extend the complementary strand using the source DNA as a template.

To understand how the PCR protocol succeeds in amplifying a discrete segment of DNA, it is important to keep in mind the location of each primer annealing site and its complementary sequence within the strands that are synthesized during each cycle. During the synthesis phase of the first cycle, the newly synthesized DNA from each primer is extended beyond the endpoint of the sequence that is complementary to the second primer. These new strands form "long templates" that are used in the second cycle (Fig. 1.16).

During the second cycle, the original DNA strands and the new strands synthesized in the first cycle (long templates) are denatured and then hybridized with the primers. The large molar excess of primers in the reaction mixture ensures that they will hybridize to the template DNA before complementary template strands have the chance to reanneal to each other. A second round of synthesis produces long templates from the original strands as well as some DNA strands that have a primer sequence at one end and a sequence complementary to the other primer at the other end ("short templates") from the long templates (Fig. 1.16).

During the third cycle, short templates, long templates, and original strands all hybridize with the primers and are replicated (Fig. 1.16). In subsequent cycles, the short templates preferentially accumulate, and by the 30th cycle, these strands are about a million times more abundant than either the original or long template strands.

The specificity, sensitivity, and simplicity of PCR have rendered it a powerful technique that is central to many applications in medical biotechnology, as illustrated throughout this book. For example, it is used to detect specific mutations that cause genetic disease, to confirm biological relatives, to identify individuals suspected of committing a crime, and to diagnose infectious diseases (see chapter 8). Specific viral, bacterial, or fungal pathogens can be detected in samples from infected patients containing complex microbial communities by utilizing PCR primers that anneal to a sequence that is uniquely present in the genome of the pathogen. This technique is often powerful enough to discriminate among very similar strains of the same species of pathogenic microorganisms, which can assist in epidemiological investigations. Moreover, PCR protocols have been developed to quantify the number of target DNA molecules present in a sample. **Quantitative PCR** is based on the principle that under optimal conditions, the number of DNA molecules doubles after each cycle.

Commonly, PCR is used to amplify target DNA for cloning into a vector. To facilitate the cloning process, restriction enzyme recognition sites are added to the 5′ end of each of the primers that are complementary to sequences that flank the target sequence in a genome (Fig. 1.17). This is especially useful when suitable restriction sites are not available in the regions flanking the target DNA. Although the end of the primer containing the restriction enzyme recognition site lacks complementarity and therefore does not anneal to the target sequence, it does not interfere with DNA synthesis. Base-pairing between the 20 or so complementary nucleotides at the 3′ end of the primer and the template molecule is sufficiently stable for primer extension by DNA polymerase. At the end of the first cycle of PCR, the noncomplementary regions of the primer remain single stranded in the otherwise double-stranded DNA product. However, after the synthesis step of the second cycle, the newly synthesized complementary strand extends to the 5′ end of the primer sequence on the template strand and therefore contains a double-stranded restriction enzyme recognition site at one end (Fig. 1.17). Subsequent cycles yield DNA products with double-stranded restriction enzyme sites at both ends. Alternatively, PCR products can be cloned using the single adenosine triphosphate (ATP) that is added to the 3′ ends by *Taq* DNA polymerase, which lacks the **proofreading** activity of many DNA polymerases to correct mispaired bases. A variety of linearized vectors have been constructed that possess a single complementary 3′ thymidine triphosphate overhang to facilitate cloning without using restriction enzymes (Fig. 1.18).

ATP
adenosine triphosphate

DNA Sequencing Technologies

Determination of the nucleotide composition and order in a gene or genome is a foundational technology in modern medical biotechnology. Cloned or PCR-amplified genes, and, indeed, entire genomes, are routinely sequenced. DNA sequences can often reveal something about the function of the protein or RNA molecule encoded in a gene, for example,

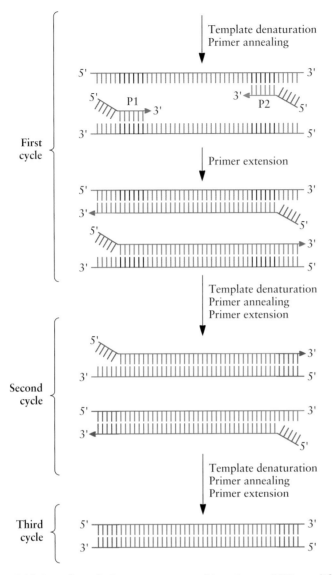

Figure 1.17 Addition of restriction enzyme recognition sites to PCR-amplified target DNA to facilitate cloning. Each of the two oligonucleotide primers (P1 and P2) has a sequence of approximately 20 nucleotides in the 3′ end that is complementary to a region flanking the target DNA (shown in black). The sequence at the 5′ end of each primer consists of a restriction endonuclease recognition site (shown in green) that does not base-pair with the template DNA during the annealing steps of the first and second PCR cycles. However, during the second cycle, the long DNA strands produced in the first cycle serve as templates for synthesis of short DNA strands (indicated by a terminal arrow) that include the restriction endonuclease recognition sequences at both ends. DNA synthesis during the third and subsequent PCR cycles produces double-stranded DNA molecules that carry the target DNA sequence flanked by restriction endonuclease recognition sequences. These linear PCR products can be cleaved with the restriction endonucleases to produce sticky ends for ligation with a vector. Note that not all of the DNA produced during each PCR cycle is shown. doi:10.1128/9781555818890.ch1.f1.17

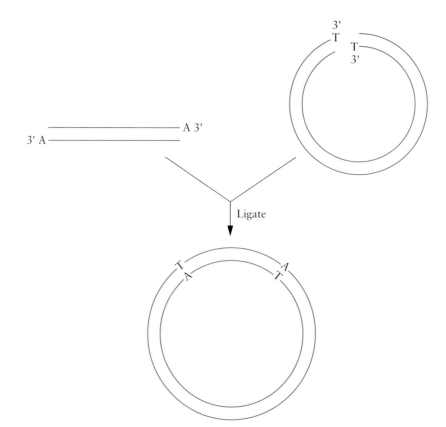

Figure 1.18 Cloning of PCR products without using restriction endonucleases. *Taq* DNA polymerase adds a single dATP (A) to the ends of PCR-amplified DNA molecules. These extensions can base-pair with complementary single thymine overhangs on a specially constructed linearized cloning vector. Ligation with T4 DNA ligase results in insertion of the PCR product into the vector.
doi:10.1128/9781555818890.ch1.f1.18

from predicted cofactor binding sites, transmembrane domains, receptor recognition sites, or DNA-binding regions. The nucleotide sequences in noncoding regions that do not encode a protein or RNA molecule may provide information about the regulation of a gene. Comparison of gene sequences among individuals can reveal mutations that contribute to genetic diseases (see chapter 3) or the relatedness among the individuals. Comparison of gene sequences among different organisms can lead to the development of hypotheses about the evolutionary relationships among organisms.

For more than three decades, the **dideoxynucleotide** procedure developed by the English biochemist Frederick Sanger has been used for DNA sequencing. This includes sequencing of DNA fragments containing one to a few genes and also the entire genomes from many different organisms, including the human genome. However, the interest in sequencing large numbers of DNA molecules in less time and at a lower cost has driven the recent development of new sequencing technologies that can process thousands to millions of sequences concurrently. Many different sequencing technologies have been developed, including **pyrosequencing**, sequencing using **reversible chain terminators**, and sequencing by ligation. In general, all of these methods involve (i) enzymatic addition of nucleotides to a primer based on complementarity to a template DNA fragment and (ii) detection and identification of the nucleotide(s) added. The techniques differ in the method by which the nucleotides are extended, employing either DNA polymerase to catalyze the addition of single nucleotides (sequencing by synthesis) or ligase to add a short, complementary oligonucleotide (sequencing by ligation), and in the method by which the addition is detected.

Dideoxynucleotide Procedure

The dideoxynucleotide procedure for DNA sequencing is based on the principle that during DNA synthesis, addition of a nucleotide triphosphate requires a free hydroxyl group on the 3′ carbon of the sugar of the last nucleotide of the growing DNA strand (Fig. 1.19A). However, if a synthetic dideoxynucleotide that lacks a hydroxyl group at the 3′ carbon of the sugar moiety is incorporated at the end of the growing chain, DNA

Figure 1.19 Incorporation of a dideoxynucleotide terminates DNA synthesis. (**A**) Addition of an incoming deoxyribonucleoside triphosphate (dNTP) requires a hydroxyl group on the 3′ carbon of the last nucleotide of a growing DNA strand. (**B**) DNA synthesis stops if a synthetic dideoxyribonucleotide that lacks a 3′ hydroxyl group is incorporated at the end of the growing chain because a phosphodiester bond cannot be formed with the next incoming nucleotide. doi:10.1128/9781555818890.ch1.f1.19

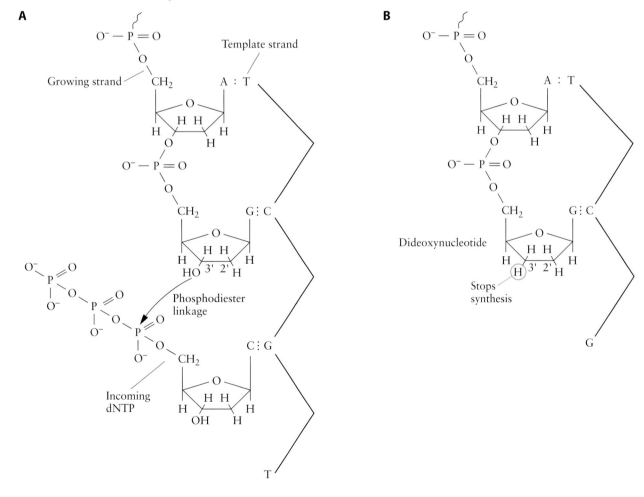

synthesis stops because a phosphodiester bond cannot be formed with the next incoming nucleotide (Fig. 1.19B). The termination of DNA synthesis is the defining feature of the dideoxynucleotide DNA sequencing method.

In a dideoxynucleotide DNA sequencing procedure, a synthetic oligonucleotide primer (~17 to 24 nucleotides) anneals to a predetermined site on the strand of the DNA to be sequenced (Fig. 1.20A). The oligonucleotide primer defines the beginning of the region to be sequenced and provides a 3′ hydroxyl group for the initiation of DNA synthesis. The reaction tube contains a mixture of the four deoxyribonucleotides (deoxyadenosine triphosphate [dATP], deoxycytidine triphosphate [dCTP], deoxyguanosine triphosphate [dGTP], and deoxythymidine triphosphate [dTTP]) and four dideoxynucleotides (dideoxyadenosine

dATP, dCTP, dGTP, and dTTP
deoxyadenosine triphosphate, deoxycytidine triphosphate, deoxyguanosine triphosphate, and deoxythymidine triphosphate

Figure 1.20 Dideoxynucleotide method for DNA sequencing. An oligonucleotide primer binds to a complementary sequence adjacent to the region to be sequenced in a single-stranded DNA template (**A**). As DNA synthesis proceeds from the primer, dideoxynucleotides are randomly added to the growing DNA strands, thereby terminating strand extension. This results in DNA molecules of all possible lengths that have a fluorescently labeled dideoxynucleotide at the 3′ end (**B**). DNA molecules of different sizes are separated by capillary electrophoresis, and as each molecule passes by a laser, a fluorescent signal that corresponds with one of the four dideoxynucleotides is recorded. The successive fluorescent signals are represented as a sequencing chromatogram (colored peaks) (**C**). doi:10.1128/9781555818890.ch1.f1.20

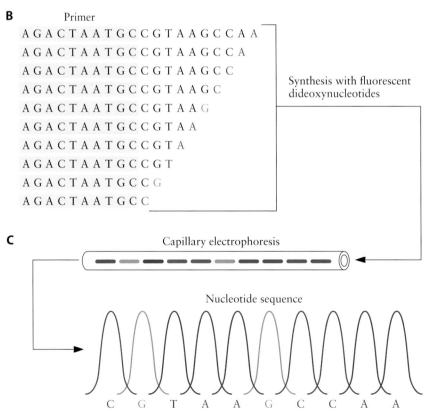

triphosphate [ddATP], ddCTP, ddGTP, and ddTTP). Each dideoxynucleotide is labeled with a different fluorescent dye. The concentration of the
dideoxynucleotides is optimized to ensure that during DNA synthesis a
modified DNA polymerase incorporates a dideoxynucleotide into the
mixture of growing DNA strands at every possible position. Thus, the
products of the reaction are DNA molecules of all possible lengths, each
of which includes the primer sequence at its 5′ end and a fluorescently
labeled dideoxynucleotide at the 3′ terminus (Fig. 1.20B).

PCR-based cycle sequencing is performed to minimize the amount
of template DNA required for sequencing. Multiple cycles of denaturation, primer annealing, and primer extension produce large amounts of
dideoxynucleotide-terminated fragments. These are applied to a polymer
in a long capillary tube that enables separation of DNA fragments that
differ in size by a single nucleotide. As each successive fluorescently labeled fragment moves through the polymeric matrix in an electric field
and passes by a laser, the fluorescent dye is excited. Each of the four different fluorescent dyes emits a characteristic wavelength of light that represents a particular nucleotide, and the order of the fluorescent signals
corresponds to the sequence of nucleotides (Fig. 1.20C).

The entire dideoxynucleotide sequencing process has been automated to increase the rate of acquisition of DNA sequence data. This
is essential for large-scale sequencing projects such as those involving
whole prokaryotic or eukaryotic genomes. Generally, automated DNA
cycle sequencing systems can read with high accuracy about 500 to 600
bases per run.

Pyrosequencing

Pyrosequencing was the first of the next-generation sequencing technologies to be made commercially available and has contributed to the rapid
output of large amounts of sequence data by the scientific community.
The basis of the technique is the detection of **pyrophosphate** that is released during DNA synthesis. When a DNA strand is extended by DNA
polymerase, the α-phosphate attached to the 5′ carbon of the sugar of
an incoming deoxynucleoside triphosphate forms a phosphodiester bond
with the 3′ hydroxyl group of the last nucleotide of the growing strand.
The terminal β- and γ-phosphates of the added nucleotide are cleaved
off as a unit known as pyrophosphate (Fig. 1.21A). The release of pyrophosphate correlates with the incorporation of a specific nucleotide in the
growing DNA strand.

To determine the sequence of a DNA fragment by pyrosequencing, a
short DNA adaptor that serves as a binding site for a sequencing primer
is first added to the end of the DNA template (Fig. 1.21B). Following annealing of the sequencing primer to the complementary adaptor sequence,
one deoxynucleotide is introduced at a time in the presence of DNA polymerase. Pyrophosphate is released only when the complementary nucleotide is incorporated at the end of the growing strand. Nucleotides that are
not complementary to the template strand are not incorporated, and no
pyrophosphate is formed.

ddATP
dideoxyadenosine triphosphate

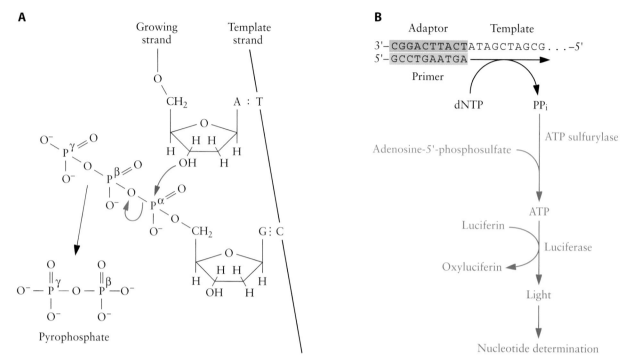

Figure 1.21 Pyrosequencing is based on the detection of pyrophosphate that is released during DNA synthesis. (**A**) A phosphodiester bond forms between the 3′ hydroxyl group of the deoxyribose sugar of the last incorporated nucleotide and the α-phosphate of the incoming nucleotide (blue arrow). The bond between the α- and β-phosphates is cleaved (green arrow), and pyrophosphate is released (black arrow). (**B**) An adaptor sequence is added to the 3′ end of the DNA sequencing template that provides a binding site for a sequencing primer. One nucleotide (deoxyribonucleoside triphosphate [dNTP]) is added at a time. If the dNTP is added by DNA polymerase to the end of the growing DNA strand, pyrophosphate (PPi) is released and detected indirectly by the synthesis of ATP. ATP is required for light generation by luciferase. The DNA sequence is determined by correlating light emission with incorporation of a particular dNTP. doi:10.1128/9781555818890.ch1.f1.21

The pyrophosphate released following incorporation of a nucleotide is detected indirectly after enzymatic synthesis of ATP (Fig. 1.21B). Pyrophosphate combines with adenosine-5′-phosphosulfate in the presence of the enzyme ATP sulfurylase to form ATP. In turn, ATP drives the conversion of luciferin to oxyluciferin by the enzyme luciferase, a reaction that generates light. Detection of light after each cycle of nucleotide addition and enzymatic reactions indicates the incorporation of a complementary nucleotide. The amount of light generated after the addition of a particular nucleotide is proportional to the number of nucleotides that are incorporated in the growing strand, and therefore sequences containing tracts of up to eight identical nucleotides in a row can be determined. Because the natural nucleotide dATP can participate in the luciferase reaction, dATP is replaced with deoxyadenosine α-thiotriphosphate, which can be incorporated into the growing DNA strand by DNA polymerase but is not a substrate for luciferase. Repeated cycles of nucleotide addition, pyrophosphate release, and light detection enable determination of sequences of 300 to 500 nucleotides per run (the **read length**).

Sequencing Using Reversible Chain Terminators

For pyrosequencing, each of the four nucleotides must be added sequentially in separate cycles. The sequence of a DNA fragment could be determined more rapidly if all the nucleotides were added together in each cycle. However, the reaction must be controlled to ensure that only a single nucleotide is incorporated during each cycle, and it must be possible to distinguish each of the four nucleotides. Synthetic nucleotides known as reversible chain terminators have been designed to meet these criteria and form the basis of some of the next-generation sequencing-by-synthesis technologies.

Reversible chain terminators are deoxynucleoside triphosphates with two important modifications: (i) a chemical blocking group is added to the 3′ carbon of the sugar moiety to prevent addition of more than one nucleotide during each round of sequencing, and (ii) a different fluorescent dye is added to each of the four nucleotides to enable identification of the incorporated nucleotide (Fig. 1.22A). The **fluorophore** is added at a position that does not interfere with either base-pairing or phosphodiester bond formation. Similar to the case with other sequencing-by-synthesis methods, DNA polymerase is employed to catalyze the addition of the modified nucleotides to an oligonucleotide primer as specified by the DNA template sequence (Fig. 1.22B). After recording fluorescent emissions, the

Figure 1.22 Sequencing using reversible chain terminators. (**A**) Reversible chain terminators are modified nucleotides that have a removable blocking group on the oxygen of the 3′ position of the deoxyribose sugar to prevent addition of more than one nucleotide per sequencing cycle. To enable identification, a different fluorescent dye is attached to each of the four nucleotides via a cleavable linker. Shown is the fluorescent dye attached to adenine. (**B**) An adaptor sequence is added to the 3′ end of the DNA sequencing template that provides a binding site for a sequencing primer. All four modified nucleotides are added in a single cycle, and a modified DNA polymerase extends the growing DNA chain by one nucleotide per cycle. Fluorescence is detected, and then the dye and the 3′ blocking group are cleaved before the next cycle. Removal of the blocking group restores the 3′ hydroxyl group for addition of the next nucleotide. doi:10.1128/9781555818890.ch1.f1.22

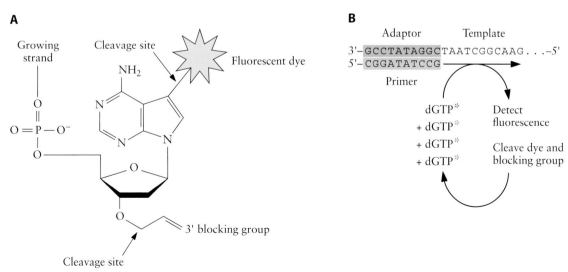

fluorescent dye and the 3′ blocking group are removed. The blocking group is removed in a manner that restores the 3′ hydroxyl group of the sugar to enable subsequent addition of another nucleotide in the next cycle. Cycles of nucleotide addition to the growing DNA strand by DNA polymerase, acquisition of **fluorescence** data, and chemical cleavage of the blocking and dye groups are repeated to generate short read lengths, i.e., 50 to 100 nucleotides per run.

Sequencing by Ligation

Pyrosequencing and sequencing using reversible terminators extend the growing DNA strand by a single base during each cycle. In contrast, sequencing by ligation extends the DNA strand by ligation of short oligonucleotides, in a template-dependent fashion, and utilizes the enzyme ligase rather than DNA polymerase. In one version of this technology, the oligonucleotides are eight nucleotides in length (octamers) with two known nucleotides at the 3′ (query) end, any nucleotide in the next three (degenerate) positions, and a sequence that is common (universal) to all of the oligonucleotides at the 5′ end (Fig. 1.23A). A set of 16 different oligonucleotides representing all possible combinations of two nucleotides in the query position is used. Each oligonucleotide is tagged at the 5′ end with a different fluorescent dye that corresponds to the query nucleotide composition.

A short nucleotide adaptor is joined to the 5′ ends of the DNA templates that are to be sequenced (Fig. 1.23B). After denaturation of the template DNA, a primer binds to the adaptor sequence and provides a 5′ phosphate end for ligation to the 3′ hydroxyl end of an adjacent hybridized octamer. Sequential cycles of octamer ligation extend the complementary strand, and each cycle enables determination of the query dinucleotide sequence (Fig. 1.23B). One ligation cycle consists of the following steps. (i) Pools of different octamers, each tagged with a different fluorescent dye, are added along with T4 DNA ligase. Reaction conditions are set so that ligation occurs only if the bases at the 3′ end of an octamer (i.e., the query and degenerate nucleotides) are complementary to the template sequence. (ii) After washing away of the nonligated octamers and other components, the fluorescence signal is recorded. The identity of the dinucleotides in the query position is determined by the distinctive fluorescence emitted. (iii) The fluorescent dye is removed after each cycle by cleaving the terminal universal nucleotides. Cleavage provides a free end for ligation of another octamer in the next ligation cycle.

Successive cycles of ligation extend the complementary strand and enable identification of discontiguous nucleotide pairs that are separated by three nucleotides (Fig. 1.23B). The length of the sequence read is determined by the number of ligation cycles (~50 bp). To determine the nucleotide sequence of the intervening regions (i.e., between the identified dinucleotides), the primer and octamers are removed from the template strand, and the process is repeated using a new primer that is set back on the template DNA by one nucleotide from the first primer (Fig. 1.23C). In total, the entire process of primer extension by serial octamer ligation is

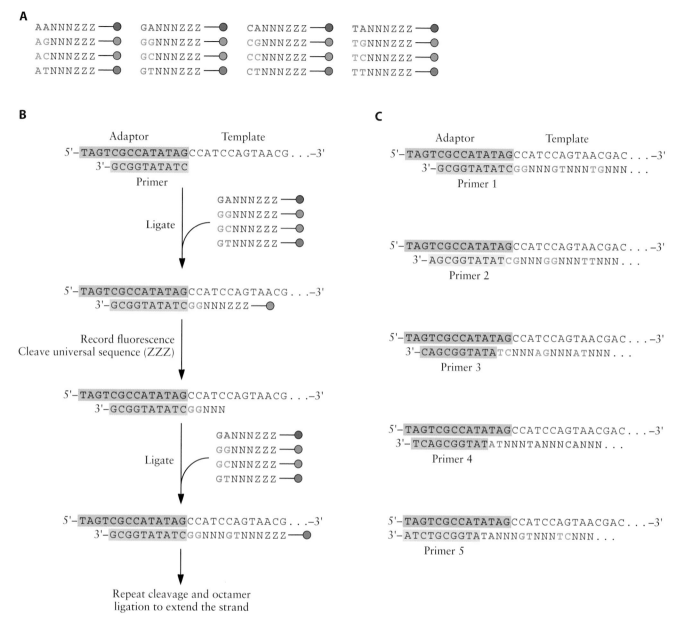

Figure 1.23 Sequencing by ligation. (**A**) Pools of octamers. Each octamer has two known nucleotides at the 3′ query end (colored letters), any nucleotide in the next three positions (NNN), and a universal sequence that is common to all of the octamers at the 5′ end (ZZZ). A set of 16 different octamers representing all possible combinations of two nucleotides in the query positions is shown. Each of the four octamers in a pool (each pool is shown in column) is tagged at the 5′ end with a different fluorescent dye that corresponds to the query nucleotide composition. (**B**) To determine a sequence by ligation, an adaptor is first added to the 5′ end of the template DNA molecule. The adaptor provides a binding site for a sequencing primer. Pools of octamers are added, and an octamer that contains a nucleotide sequence that is complementary to the first five bases immediately adjacent to the primer will hybridize. T4 DNA ligase catalyzes the formation a phosphodiester bond between the 5′ phosphate group of the primer and the 3′ hydroxyl group of the hybridized octamer. After unbound octamers are washed away, the fluorescent signal from the ligated octamer is recorded to identify the dinucleotides in the query position. The fluorescent dye is removed by cleavage of the universal sequence (ZZZ) to expose a 5′ phosphate for ligation of another octamer. The cycle of octamer ligation and dye removal is repeated to extend the DNA strand and identify discontiguous dinucleotides that are separated by three nucleotides (NNN). (**C**) To determine the identity of the intervening nucleotides, the primer and ligated octamers are removed and the entire process is repeated using a different primer that binds to a position on the adaptor one nucleotide from the previous primer-binding site. The process is repeated with five different primers. In this manner, each nucleotide in the template sequence is identified twice, once in two separate reactions. doi:10.1128/9781555818890.ch1.f1.23

repeated five times, each time using a different primer that is offset from the previous primer by one nucleotide (Fig. 1.23C). In this way, the nucleotide in each position in the template DNA is identified twice in separate reactions, which increases the accuracy of the sequence determination.

Sequencing Whole Genomes

Just as the sequence of a gene can provide information about the function of the encoded protein, the sequence of an entire genome can contribute to our understanding of the nature of an organism. Thousands of whole genomes have now been sequenced, from organisms from all domains of life. Initially, the sequenced genomes were relatively small, limited by the early sequencing technologies. The first DNA genome to be sequenced was from the *E. coli* bacteriophage φX174 (5,375 bp) in 1977, while the first sequenced genome from a cellular organism was that of the bacterium *Haemophilus influenzae* (1.8 Mbp) in 1995. Within 2 years, the sequence of the larger *E. coli* genome (4.6 Mbp) was reported, and the sequence of the human genome (3,000 Mbp), the first vertebrate genome, was completed in 2003.

Most of these first genome sequences were generated using a **shotgun cloning** approach. In this strategy, a clone library of randomly generated, overlapping genomic DNA fragments is constructed in a bacterial host. The plasmids are isolated, and then the cloned inserts are sequenced using the dideoxynucleotide method. Using this approach, the first human genome was sequenced in 13 years at a cost of $2.7 billion. The aspiration to acquire genome sequences faster and at a lower cost has driven the development of new genome sequencing strategies. Today, many large-scale sequencing projects have been completed and many more are underway, motivated by compelling biological questions. Some will contribute to our understanding of the microorganisms that cause infectious diseases and to the development of new techniques for their detection and treatment. For example, what makes a bacterium pathogenic? Why are some strains of influenza virus more virulent than others? How do microorganisms in the human gut influence susceptibility to infectious disease or response to drugs? Others are aimed at helping us to understand what it means to be human and how we evolved. For example, how did multicellularity arise? What were early humans like (e.g., by comparing the sequences of the modern human genome and the Neanderthal genome)? How are we different from one another? A project known as the 1000 Genomes Project aims to address the last question by sequencing the genomes from a large number of people from many different populations. Understanding the nucleotide differences (**polymorphisms**) among individuals, especially between those with and without a specific disease, will help us to determine the genetic basis of disease (see chapter 3). One goal, which is likely to be realized within the next few years, is to provide each of us with our genome sequence as part of our personal health care regimen (Box 1.3).

A large-scale sequencing project currently entails (i) preparing a library of source DNA fragments, (ii) amplifying the DNA fragments to increase

box 1.3
Personalized Genomic Medicine

One of the goals of the next-generation sequencing technologies is to make genome sequencing available to everyone. The idea is that each of us will have our genomes sequenced to assist our physicians in recommending disease prevention strategies, diagnosing our ailments, and determining a course of treatment tailored to our individual genotypes. This is personalized medicine, a concept not unlike the current practice of considering family medical history and personal social, economic, and behavioral factors to determine disease susceptibility, causation, and management.

An illustration of the power of a genome sequence in medical care is the case of a young child with a rare, life-threatening inflammatory bowel disease that defied physicians trying to determine the cause. A congenital immunodeficiency was suspected; however, immunological tests yielded inconclusive results, and single gene tests for known immune defects were not revealing. After obtaining the sequence of the child's genome, a novel mutation was apparent in a key position in the gene encoding XIAP (X-linked inhibitor of apoptosis protein), a protein that inhibits programmed cell death. The mutation enhanced the susceptibility of the child's cells to induced cell death. One result of this genetic defect is misregulation of the inflammatory response that can lead to inflammatory bowel disease. A bone marrow transplant prevented uncontrolled hyperinflammation and recurrence of the child's intestinal disease.

Can we afford this technology for the masses? Perhaps not at the moment, although there are currently several companies offering fee-for-service genome sequencing to consumers. However, the development of faster, cheaper sequencing technologies is on the threshold of making the technology more accessible. And perhaps we cannot afford not to offer this technology widely, considering that the cost of misdiagnoses and the trial-and-failure drug treatment approaches contribute to high medical costs and loss of human productivity. A great benefit of personal genome sequences is the potential to predict and prevent diseases, which, of course, is only as good as our knowledge of the genetic basis of disease. The health benefits must also be weighed against the possibility that the information encoded in our genomes may be subject to misuse by insurance companies, employers, educators, and law enforcement agencies.

the detection signal from the sequencing templates, (iii) sequencing the template DNA using one of the sequencing techniques describe above, and (iv) assembling the sequences generated from the fragments in the order in which they are found in the original genome. Sequencing massive amounts of DNA required not only the development of new technologies for nucleotide sequence determination but also new methods to reduce the time for preparation and processing of large libraries of sequencing templates. The high-throughput **next-generation sequencing** approaches have circumvented the cloning steps of the shotgun sequencing strategy by attaching, amplifying, and sequencing the genomic DNA fragments directly on a solid support. All of the templates are sequenced at the same time. The term used to describe this is **massive parallelization**. In the not-so-distant future, sensitive sequencing systems are anticipated that directly sequence single DNA molecules and therefore do not require a PCR amplification step (Box 1.4).

Shotgun Cloning Strategy

A shotgun library is constructed by fragmenting genomic DNA and inserting the fragments into a vector to generate sequencing templates. To obtain random, overlapping fragments, the genomic DNA is usually sheared physically by applying sound waves (sonication) or forcing the

box 1.4
Future Sequencing Technologies

New sequencing strategies are just beginning to emerge, most within last decade, propelled by the need for faster, cheaper genome sequencing. New developments aim to reduce the time and cost of template preparation and/or the time to acquire sequence data.

One general approach is to sequence single DNA molecules, circumventing the amplification step currently required. This eliminates the need for substantial template preparation and avoids introduction of mutations inherent in PCR that are interpreted as nucleotide variations. Random fragments of genomic DNA are ligated to adaptors, denatured, and then attached to a solid support directly or by hybridization of the adaptors to immobilized sequencing primers. During the sequencing stage, DNA polymerase extends the primer in a template-dependent fashion. Alternatively, DNA polymerase is immobilized on a solid support and captures the primed DNA fragments. Often, single-molecule sequencing strategies employ cycles of addition and detection of reversible chain terminators.

A promising technology for rapid acquisition of sequence data is real-time sequencing. For real-time sequencing, the nucleotides do not carry a blocking group on the 3′ hydroxyl group, and therefore, DNA synthesis is continuous. A fluorescent tag is attached to the terminal phosphate of each nucleoside triphosphate. With each nucleotide addition to the growing DNA chain, pyrophosphate is cleaved and with it the fluorescent tag. Tag cleavage therefore corresponds to nucleotide addition. In some approaches, the bases are covalently linked to a quencher group that suppresses fluorescence of the dye. However, release of the dye following pyrophosphate cleavage produces a fluorescent signal.

Nanopore technology aims to identify individual nucleotides in a DNA molecule as it passes through a pore. The pore is either a membrane protein or a synthetic structure that accommodates only one molecule of negatively charged, single-stranded DNA that is drawn through the pore as it moves toward a positive charge. The principle behind the technology is that each nucleotide in the DNA polymer momentarily blocks the pore as it passes through. The obstructions create fluctuations in electrical conductance that can be measured. Because they have different structures, each of the four nucleotides obstructs conductance to a different extent, and therefore, the amount of conductance corresponds to nucleotide sequence. With this technology, it may be possible to determine the sequence of a human genome in 20 h without the expense of DNA library preparation or sequencing chemicals.

DNA through a narrow tube using compressed gas (nebulization) or in a solution (hydrodynamic shearing). The shearing conditions are optimized as much as possible to obtain DNA fragments of a uniform size. Physical fragmentation tends to leave extended single-stranded ends that must be blunted (end repaired, or polished) by filling in 3′ recessed ends with DNA polymerase in the presence of the four deoxyribonucleotides and removing protruding 3′ ends with an exonuclease (Fig. 1.3A and B). The 5′ ends of the polished genomic fragments are phosphorylated with T4 polynucleotide kinase (Fig. 1.3D) and ligated with a vector. The library is introduced into *E. coli*, and plasmids are subsequently extracted and used as sequencing templates. Depending on the sequencer, either 96 or 384 sequencing templates are analyzed concurrently. Primers that anneal to complementary vector sequences flanking the insert are used to obtain the sequence of both ends of the cloned DNA fragment using the dideoxynucleotide method. In this manner, each template yields two "end reads" which are known as **paired end reads** (mate pair) (Fig. 1.24).

The large number of nucleotide reads that is generated from a genomic library is assembled using a computer program to align overlapping reads. The process of generating successive overlapping sequences produces long contiguous stretches of nucleotides called **contigs** (Fig. 1.24). The presence of repetitive sequences in a genome can result in erroneous matching

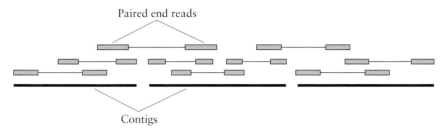

Figure 1.24 Genome sequence assembly. Sequence data generated from both ends of a DNA fragment are known as paired end reads. A large number of reads are generated and assembled into longer contiguous sequences (contigs) using a computer program that matches overlapping sequences. Paired end reads help to determine the order and orientation of contigs as they are assembled into scaffolds. Shown is a scaffold consisting of three contigs. doi:10.1128/9781555818890.ch1.f1.24

of overlapping sequences. This problem can be overcome by using the sequences from paired ends, which are a known distance apart, to order and orient the reads and to assemble the contigs into larger **scaffolds** (Fig. 1.24). Many overlapping reads are required to ensure that the nucleotide sequence is accurate and assembled correctly. For shotgun sequencing, each nucleotide site in a genome should be sequenced at least 6 to 10 different times from different fragments. The extent of sequencing redundancy is called sequence coverage or depth of coverage. The assembly process generates a draft sequence; however, small gaps may remain between contigs. Although a draft sequence is sufficient for many purposes, for example, in resequencing projects that map a sequence onto a **reference genome**, in some cases it is preferable to close the gaps to complete the genome sequence. For de novo sequencing of genomes from organisms that lack a reference genome, gap closure is desirable. The gaps can be closed by PCR amplification of high-molecular-weight genomic DNA across each gap, followed by sequencing of the amplification product, or by obtaining short sequences from primers designed to anneal to sequences adjacent to a gap. Sequencing of additional clone libraries containing fragments of different sizes may be required to complete the overall sequence.

High-Throughput Next-Generation Sequencing Strategies

Although shotgun sequencing has been used successfully to obtain the sequences of many whole genomes, preparation of clone libraries in bacterial cells is costly and time-consuming for routine sequencing of large amounts of genomic DNA that are required for many research and clinical applications. To reduce the time and cost of large-scale sequencing, high-throughput next-generation sequencing strategies have been developed that use cell-free methods to generate a library of genomic DNA fragments in a dense array on the surface of a glass slide or in picoliter-volume wells of a multiwell plate. This minimizes the volume of reagents for the sequencing reactions and enables hundreds of millions of sequences to be acquired simultaneously.

Most of the current commercially available, high-throughput next-generation sequencing strategies use PCR to generate clusters containing millions of copies of each DNA sequencing template. The clusters are spatially separated and are immobilized on a surface. In some strategies, the sequencing templates are captured and amplified on the surface of a small bead. Amplification of each DNA template occurs in a droplet of an aqueous solution contained within an oil coat; this is referred to as **emulsion PCR** (Fig. 1.25). For emulsion PCR, fragments of the source genomic DNA are first ligated at each end to two different adaptors that have specific sequences for binding primers for PCR amplification of the fragments and for binding sequencing primers. PCR primers that are complementary to a sequence on one of the adaptors are bound to DNA capture beads. After ligation of the adaptors, the genomic DNA fragments are melted and the single-stranded molecules anneal to the beads through complementary base-pairing. Each DNA capture bead carries more than 10^7 primer molecules; however, initially, the DNA fragments are mixed with the beads under conditions that result in binding of one DNA molecule per bead. For amplification by emulsion PCR, the DNA capture beads carrying the hybridized DNA templates are mixed with the PCR components and oil to create a water-in-oil emulsion. The conditions are set so that a single bead along with all of the components required for amplification of the attached DNA template are contained within an oil globule. Each oil globule is a separate reaction chamber, and the amplification products that remain bound to the bead by hybridization with the primers are contained within the globule. After PCR, the emulsion is broken, the DNA is denatured, and the beads are deposited on a glass slide or in wells of a plate that can have more than a million wells (one bead per well). The DNA immobilized on the beads is used as a template for pyrosequencing or sequencing by ligation. Repeated cycles of flooding the wells or slide with sequencing reagents and detecting the light or fluorescent signals that correspond to nucleotide additions generate sequence reads from all of the templates simultaneously.

A genome sequence can be assembled by aligning the nucleotide reads to the sequence of a highly related genome. For example, reads from resequenced human genomes, that is, genomes from different individuals, are mapped to a reference human genome. Alternatively, when a reference sequence is not available, the reads can be assembled de novo in a manner similar to that for shotgun cloning, by aligning the matching ends of different reads to construct contigs. For sequencing methods that generate short read lengths such as sequencing by ligation or using reversible chain terminators, the coverage must be greater. Generally, 30-fold coverage is required for correct assembly.

Genomics

Genome sequence determination is only a first step in understanding an organism. The next steps require identification of the features encoded in a sequence and investigations of the biological functions of the encoded RNA, proteins, and regulatory elements that determine the physiology and ecology of the organism. The area of research that generates,

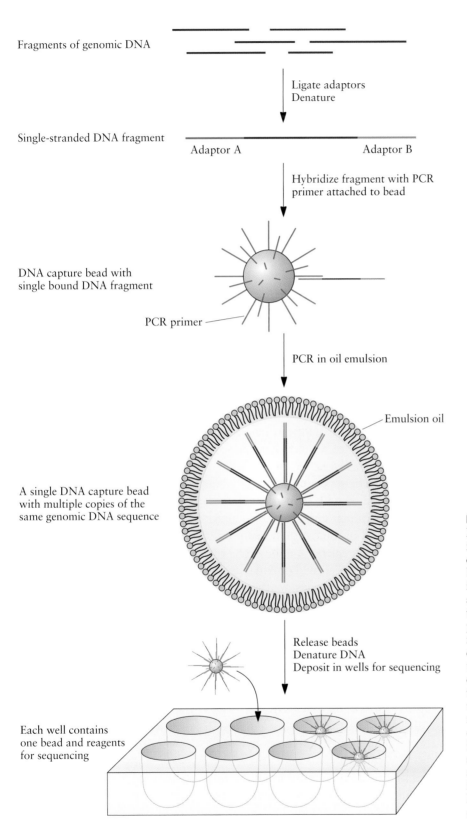

Fragments of genomic DNA

Ligate adaptors
Denature

Single-stranded DNA fragment

Adaptor A   Adaptor B

Hybridize fragment with PCR
primer attached to bead

DNA capture bead with
single bound DNA fragment

PCR primer

PCR in oil emulsion

Emulsion oil

A single DNA capture bead
with multiple copies of the
same genomic DNA sequence

Release beads
Denature DNA
Deposit in wells for sequencing

Each well contains
one bead and reagents
for sequencing

Figure 1.25 Emulsion PCR. Genomic DNA fragments are ligated to two different adaptors that have sequences for capturing the fragments on the surface of beads and for binding PCR and sequencing primers. The fragments are denatured and mixed with beads under conditions that favor binding of one DNA molecule per bead. The DNA-bound beads are mixed with PCR components and oil to create a PCR microreactor in an oil globule. Emulsion PCR results in millions of copies of the genomic DNA fragment attached to the bead within the globule. At the end of the PCR cycles, the beads are released from the emulsion, the DNA is denatured, and the beads are deposited into the wells of a multiwell plate. The single-stranded genomic DNA fragments attached to the beads serve as templates for pyrosequencing or sequencing by ligation.
doi:10.1128/9781555818890.ch1.f1.25

analyzes, and manages the massive amounts of information about genome sequences is known as **genomics**.

Sequence data are deposited and stored in databases that can be searched using computer algorithms to retrieve sequence information (data mining, or **bioinformatics**). Public databases such as GenBank (National Center for Biotechnology Information, Bethesda, MD), the European Molecular Biology Lab Nucleotide Sequence Database, and the DNA Data Bank of Japan receive sequence data from individual researchers and from large sequencing facilities and share the data as part of the International Nucleotide Sequence Database Collaboration. Sequences can be retrieved from these databases via the Internet. Many specialized databases also exist, for example, for storing genome sequences from individual organisms, protein coding sequences, regulatory sequences, sequences associated with human genetic diseases, gene expression data, protein structures, protein–protein interactions, and many other types of data.

One of the first analyses to be conducted on a new genome sequence is the identification of descriptive features, a process known as **annotation**. Some annotations are protein-coding sequences (**open reading frames**), sequences that encode functional RNA molecules (e.g., rRNA and tRNA), regulatory elements, and repetitive sequences. Annotation relies on algorithms that identify features based on conserved sequence elements such as translation start and stop codons, intron–exon boundaries, promoters, transcription factor-binding sites, and known genes (Fig. 1.26). It is important to note that annotations are often predictions of sequence function based on similarity (homology) to sequences of known functions. In many cases, the function of the sequence remains to be verified through experimentation.

Comparison of a genome sequence to other genome sequences can reveal interesting and important sequence features. Comparisons among closely related genomes may reveal polymorphisms and mutations based on sequence differences. Association of specific polymorphisms with diseases can be used to predict, diagnose, and treat human diseases. Traditionally, cancer genetic research has investigated specific genes that were hypothesized to play a role in tumorigenesis based on their known cellular functions, for example, genes encoding transcription factors that control expression of cell division genes. Although important, this gives an incomplete view of the genetic basis for cancers. Sequencing of tumor genomes and comparing the sequences to those of normal cells have revealed point mutations, copy number mutations, and structural rearrangements associated with specific cancers (see chapter 3). For instance, comparison of the genome sequences from acute myeloid leukemia tumor cells and normal skin cells from the same patient revealed eight previously unidentified mutations in protein coding sequences that are associated with the disease. In addition, comparison of the genomes of bacterial pathogens with those from closely related nonpathogens has led to the identification of **virulence** genes (see chapter 5). Unique sequences can be used for pathogen detection, and genes encoding proteins that are unique to a pathogen are potential targets for antimicrobial drugs and vaccine development.

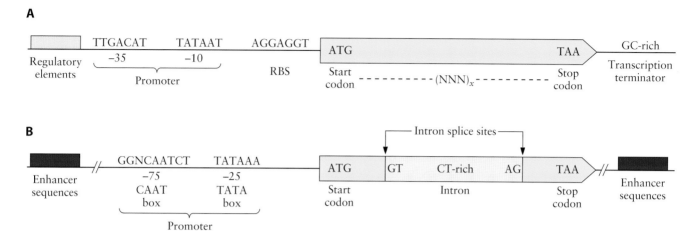

Figure 1.26 Genome annotation utilizes conserved sequence features. Predicting protein coding sequences (open reading frames) in prokaryotes (**A**) and eukaryotes (**B**) requires identification of sequences that correspond to potential translation start (ATG or, more rarely, GTG or TTG) and stop (TAA, shown; also TAG or TGA) codons in mRNA. The number of nucleotides between the start and stop codons must be a multiple of three (i.e., triplet codons) and must be a reasonable size to encode a protein. In prokaryotes, a conserved ribosome-binding site (RBS) is often present 4 to 8 nucleotides upstream of the start codon (**A**). Prokaryotic transcription regulatory sequences such as an RNA polymerase recognition (promoter) sequence and binding sites for regulatory proteins can often be predicted based on similarity to known consensus sequences. Transcription termination sequences are not as readily identifiable but are often GC-rich regions downstream of a predicted translation stop codon. In eukaryotes, protein coding genes typically have several intron sequences in primary RNA that are delineated by GU and AG and contain a pyrimidine-rich tract. Introns are spliced from the primary transcript to produce mRNA (**B**). Transcription regulatory elements such as the TATA and CAAT boxes that are present in the promoters of many eukaryotic protein coding genes can sometimes be predicted. Sequences that are important for regulation of transcription are often difficult to predict in eukaryotic genome sequences; for example, enhancer elements can be thousands of nucleotides upstream and/or downstream from the coding sequence that they regulate. doi:10.1128/9781555818890.ch1.f1.26

Genome comparisons among distantly related organisms enable scientists to make predictions about evolutionary relationships. For example, the Genome 10K Project aims to sequence and analyze the genomes of 10,000 vertebrate species, roughly 1 per genus. Comparison of these sequences will contribute to our understanding of the genetic changes that led to the diversity in morphology, physiology, and behavior in this group of animals.

Another goal of genomic analysis is to understand the function of sequence features. Gene function can sometimes be inferred by the pattern of transcription. **Transcriptomics** is the study of gene transcription profiles either qualitatively, to determine which genes are expressed, or quantitatively, to measure changes in the levels of transcription of genes. **Proteomics** is the study of the entire protein populations of various cell types and tissues and the numerous interactions among proteins. Some proteins, particularly enzymes, are involved in biochemical pathways that produce metabolites for various cellular processes. **Metabolomics** aims to characterize metabolic pathways by studying the metabolite profiles of cells. All of these "-omic" subdisciplines of genomics use a genome-wide approach to study the function of biological molecules in cells, tissues, or organisms, at different developmental stages, or under different physiological or environmental conditions.

Transcriptomics

Transcriptomics (gene expression profiling) aims to measure the levels of transcription of genes on a whole-genome basis under a given set of conditions. Transcription may be assessed as a function of medical conditions, as a consequence of mutations, in response to natural or toxic agents, in different cells or tissues, or at different times during biological processes such as cell division or development of an organism. Often, the goal of gene expression studies is to identify the genes that are up- or downregulated in response to a change in a particular condition. Two major experimental approaches for measuring RNA transcript levels on a whole-genome basis are DNA **microarray** analysis and a newer approach called high-throughput next-generation **RNA sequencing**.

DNA Microarrays

A DNA microarray (DNA chip or gene chip) experiment consists of hybridizing a nucleic acid sample (target) derived from the mRNAs of a cell or tissue to single-stranded DNA sequences (probes) that are arrayed on a solid platform. Depending on the purpose of the experiment, the probes on a microarray may represent an entire genome, a single chromosome, selected genomic regions, or selected coding regions from one or several different organisms. Some DNA microarrays contain sets of oligonucleotides as probes, usually representing thousands of genes that are synthesized directly on a solid surface. Thousands of copies of an oligonucleotide with the same specific nucleotide sequence are synthesized and then placed in a predefined position on the array surface (probe cell). The probes are typically 20 to 70 nucleotides, although longer probes can also be used, and several probes with different sequences for each gene are usually present on the microarray to minimize errors. Probes are designed to be specific for their target sequences, to avoid hybridization with non-target sequences, and to have similar melting (annealing) temperatures so that all target sequences can bind to their complementary probe sequence under the same conditions. A complete whole-genome oligonucleotide array may contain more than 500,000 probes representing as many as 30,000 genes.

For most gene expression profiling experiments that utilize microarrays, mRNA is extracted from cells or tissues and used as a template to synthesize cDNA using reverse transcriptase. Usually, mRNA is extracted from two or more sources for which expression profiles are compared, for example, from diseased versus normal tissue, or from cells grown under different conditions (Fig. 1.27). The cDNA from each source is labeled with a different fluorophore by incorporating fluorescently labeled nucleotides during cDNA synthesis. For example, a green-emitting fluorescent dye (Cy3) may be used for the normal (reference) sample and a red-emitting fluorescent dye (Cy5) for the test sample. After labeling, the cDNA samples are mixed and hybridized to the same microarray (Fig. 1.27). Replicate samples are independently prepared under the same conditions and hybridized to different microarrays. A laser scanner determines the intensities of Cy5 and Cy3 for each probe cell on a microarray. The ratio of red (Cy5) to green (Cy3) fluorescence intensity of a probe cell indicates the relative expression levels of the represented gene in the two samples. To avoid variation due to inherent and sequence-specific differences in labeling efficiencies between Cy3 and Cy5, reference and test samples are often reverse labeled and hybridized to another microarray. Alternatively, for some microarray platforms, the target sequences from reference and test samples are labeled with the same fluorescent dye and are hybridized to different microarrays. Methods to calibrate the data among microarrays in an experiment include using the fluorescence intensity of a gene that is not differentially expressed among different conditions as a reference point (i.e., a **housekeeping gene**), including spiked control sequences that are sufficiently different from the target sequences and therefore bind only to a corresponding control probe cell, and adjusting the total fluorescence intensities of all genes on each microarray to similar values under the assumption that a relatively small number of genes are expected to change among samples.

Genes whose expression changes in response to a particular biological condition are identified by comparing the fluorescence intensities for each gene, averaged among replicates, under two different conditions. The raw data of the fluoresence emissions of each gene are converted to a ratio, commonly expressed as fold change. Generally, positive ratios represent greater expression of the gene in the test sample than in the reference sample. Negative values indicate a lower level of expression in the test sample relative to the reference sample. The data are often organized into clusters of genes whose expression patterns are similar under different conditions or over a period of time (Fig. 1.28). This facilitates predictions of gene products that may function together in a pathway.

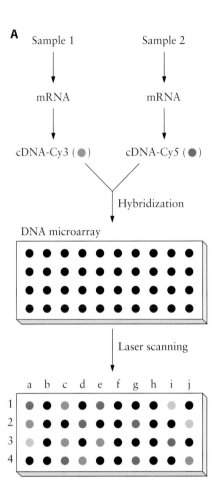

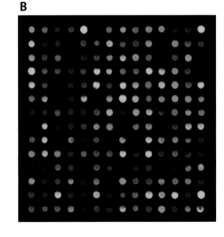

Figure 1.27 Gene expression profiling with a DNA microarray. (**A**) mRNA is extracted from two samples (sample 1 and sample 2), and during reverse transcription, the first cDNA strands are labeled with the fluorescent dyes Cy3 and Cy5, respectively. The cDNA samples are mixed and hybridized to an ordered array of either gene sequences or gene-specific oligonucleotides. After the hybridization reaction, each probe cell is scanned for both fluorescent dyes and the separate emissions are recorded. Probe cells that produce only a green or red emission represent genes that are transcribed only in sample 1 or 2, respectively; yellow emissions indicate genes that are active in both samples; and the absence of emissions (black) represents genes that are not transcribed in either sample. (**B**) Fluorescence image of a DNA microarray hybridized with Cy3- and Cy5-labeled cDNA. Reproduced with permission from the University of Arizona Biology Project. Courtesy of N. Anderson, University of Arizona. doi:10.1128/9781555818890.ch1.f1.27

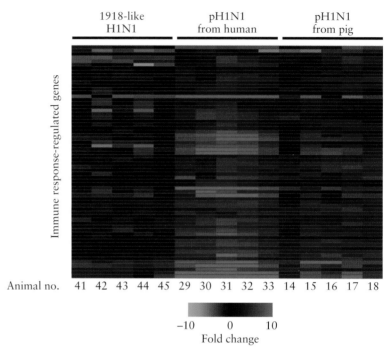

Figure 1.28 Microarray analysis of gene expression in pigs infected with three different strains of influenza virus. Two pandemic H1N1 (pH1N1) influenza viruses were isolated from a human patient and from a pig during the 2009 pandemic. The 2009 pH1N1 virus killed more than 18,000 people worldwide and was transmitted to several other species, including pigs. A third H1N1 virus isolated from a pig is a derivative of the virus that caused the 1918 pandemic that killed more than 50 million people. The clinical symptoms induced by the pH1N1 virus in pigs were more severe (coughing, sneezing, labored breathing, and nasal shedding of the virus) than those induced by the 1918-like H1N1 virus. Gene expression profiling showed that 3 days after infection, pigs mounted a stronger immune and inflammatory response to the pH1N1 viruses than was mounted to the 1918-like H1N1 virus. Each row represents the level of expression of an immune response-regulated gene in an infected pig (columns) relative to the level in uninfected control pigs, with red indicating a higher level of expression and green indicating a lower level of expression. Adapted with permission from Ma et al., *J. Virol.* **85:**11626–11637, 2011. doi:10.1128/9781555818890.ch1.f1.28

RNA Sequencing

Similar to microarrays, RNA sequencing is used to detect and quantify the complete set of gene transcripts produced by cells under a given set of conditions. In addition, RNA sequencing can delineate the beginning and end of genes, reveal posttranscriptional modifications such as variations in intron splicing that lead to variant proteins, and identify differences in the nucleotide sequence of a gene among samples. In contrast to microarray analysis, this approach does not require prior knowledge of the genome sequence, avoids high background due to nonspecific hybridization, and can accurately quantify highly expressed genes (i.e., probe saturation is not a concern as it is for DNA microarrays). Traditionally, RNA sequencing approaches required generating cDNA libraries from isolated RNA and sequencing the cloned inserts, or the end(s) of the cloned inserts (**expressed sequence tags**), using the dideoxynucleotide method. New developments in sequencing technologies circumvent the requirement for preparation of a clone library and enable high-throughput sequencing of cDNA.

For high-throughput RNA sequencing, total RNA is isolated and converted to cDNA using reverse transcriptase and a mixture of oligonucleotide primers composed of six random bases (random hexamers) that bind to multiple sites on all of the template RNA molecules (Fig. 1.29A). Because rRNA makes up a large fraction (>80%) of the total cellular RNA and levels are not expected to change significantly under different conditions, these molecules are often removed prior to cDNA synthesis by hybridization to complementary oligonucleotides that are covalently linked to a magnetic bead for removal. Long RNA molecules are fragmented to pieces of about 200 bp by physical (e.g., nebulization), chemical (e.g., metal ion hydrolysis), or enzymatic (e.g., controlled RNase digestion) methods either before (RNA fragmentation) or after (cDNA fragmentation) cDNA synthesis.

The cDNA fragments are ligated at one or both ends to an adaptor that serves as a binding site for a sequencing primer (Fig. 1.29A). High-throughput next-generation sequencing technologies are employed to sequence the cDNA fragments. The sequence reads are assembled in a manner similar to that for genomic DNA, that is, by aligning the reads to a reference genome or by aligning overlapping sequences to generate contigs for de novo assembly when a reference genome is not available. The reads are expected to align uniformly across the transcript (Fig. 1.29A). Gene expression levels are determined by counting the reads that correspond to each nucleotide position in a gene and averaging these across the length of the transcript (Fig. 1.29B). Expression levels are typically normalized between samples by scaling to the total number of reads per sample (e.g., reads per kilobase pair per million reads). Appropriate coverage (i.e., the number of cDNA fragments sequenced) is more difficult to determine for RNA sequencing than for genome sequencing because the total complexity of the transcriptome is not known before the experiment. In general, larger genomes and genomes that have more **RNA splicing variants** have greater transcriptome complexity and therefore require greater coverage. Also, accurate measurement of transcripts from genes with low expression levels requires sequencing of a greater number of transcripts. Quantification may be confounded by high GC content of some cDNA fragments which have a higher melting temperature and therefore are inefficiently sequenced, by overrepresentation of cDNA fragments from the 5' end of transcripts due to the use of random hexamers, and by reads that map to more than one site in a genome due to the presence of repeated sequences. However, because each transcript is represented by many different reads, these biases are expected to have minimal effects on quantification of a transcript.

RNA sequencing has been employed to profile gene expression in a variety of prokaryotes and eukaryotic cells and tissues. In addition to profiling human gene expression under several different conditions, for example, in brain tissue from individuals chronically exposed to alcohol or cocaine, RNA sequencing has identified hundreds of new intron splice

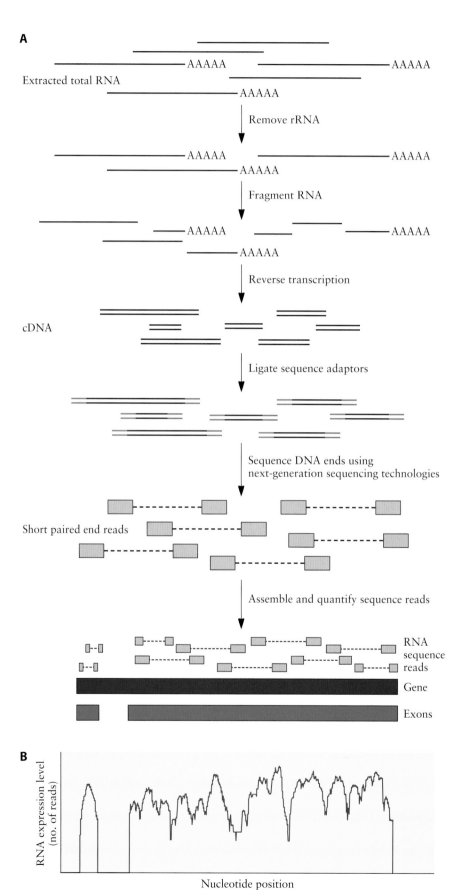

Figure 1.29 High-throughput RNA sequencing. (**A**) Total RNA is extracted from a sample and rRNA may be removed. The RNA is fragmented and then converted to cDNA using reverse transcriptase. Adaptors are added to the ends of the cDNA to provide binding sites for sequencing primers. High-throughput next-generation sequencing technologies are used to determine the sequences at the ends of the cDNA molecules (paired end reads). The sequence reads are aligned to a reference genome or assembled into contigs using the overlapping sequences. Shown is the alignment of paired end reads to a gene containing one intron. (**B**) RNA expression levels are determined by counting the reads that correspond to a gene. Adapted with permission from Macmillan Publishers Ltd. (Wang et al., *Nat. Rev. Genet.* **10:** 57–63, 2009), copyright 2009.
doi:10.1128/9781555818890.ch1.f1.29

variants and transcripts from regions of the human genome that were previously thought to be inactive (not transcribed). In addition, many small RNA molecules have been discovered that are not translated into protein but, rather, function as regulators of gene expression.

Proteomics

Proteins are the molecular machines of cells. They catalyze biochemical reactions, monitor the internal and external environments of the cell and mediate responses to perturbations, and make up the structural components of cells. Some proteins are present at more or less the same levels in all cells of an individual under most conditions, for example, proteins that make up ribosomes or the cytoskeleton. The levels of other proteins differ among cells according to the cells' functions or change in response to developmental or environmental cues. Thus, analysis of the proteins that are present under particular biological conditions can provide insight into the activities of a cell or tissue.

Proteomics is the comprehensive study of all the proteins of a cell, tissue, body fluid, or organism from a variety of perspectives, including structure, function, expression profiling, and protein–protein interactions. There are several advantages to studying the protein complement (proteome) of cells or tissues compared to other genomic approaches. Although analysis of genomic sequences can often identify protein coding sequences, in many cases the function of a protein, and the posttranslational modifications that influence protein activity and cellular localization, cannot be predicted from the sequence. On the other hand, it may be possible to infer a protein's function by determining the conditions under which it is expressed and active. While expression profiles of protein coding sequences can be determined using transcriptomics, mRNA levels do not always correlate with protein levels and do not indicate the presence of active proteins, and interactions between proteins cannot be assessed by these methods. Generally, mRNA is turned over rapidly, and therefore, transcriptomics measures actively transcribed genes, whereas proteomics monitors relatively more stable proteins. From a practical standpoint, proteomics can be used to identify proteins associated with a clinical disorder (protein **biomarkers**), especially in the early stages of disease development, that can aid in disease diagnosis or provide targets for treatment of disease.

Identification of Proteins

A cell produces a large number of different proteins that must first be separated in order to identify individual components of the proteome. To reduce the complexity, proteins are sometimes extracted from particular subcellular locations such as the cell membrane, nucleus, Golgi apparatus, endosomes, or mitochondria. Two-dimensional polyacrylamide gel **electrophoresis** (2D PAGE) is an effective method to separate proteins in a population (Fig. 1.30). Proteins in a sample are first separated on the basis of their net charge by electrophoresis through an immobilized pH gradient in one dimension (the first dimension) (Fig. 1.30A). Amino acids in a

2D PAGE
two-dimensional polyacrylamide gel electrophoresis

A

pH gradient

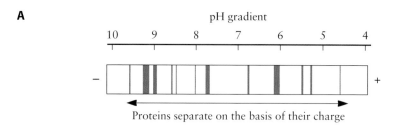

Proteins separate on the basis of their charge

B

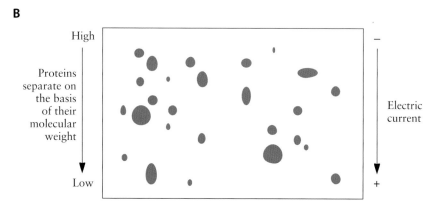

Figure 1.30 2D PAGE for separation of proteins. (**A**) First dimension. Isoelectric focusing is performed to first separate proteins in a mixture on the basis of their net charge. The protein mixture is applied to a pH gradient gel. When an electric current is applied, proteins migrate toward either the anode (+) or cathode (−) depending on their net charge. As proteins move through the pH gradient, they gain or lose protons until they reach a point in the gel where their net charge is zero. The pH in this position of the gel is known as the isoelectric point and is characteristic of a given protein. At that point, a protein no longer moves in the electric current. (**B**) Second dimension. Several proteins in a sample may have the same isoelectric point and therefore migrate to the same position in the gel as in the first dimension. Therefore, proteins are further separated on the basis of differences in their molecular weights by electrophoresis, at a right angle to the first dimension, through an SDS-polyacrylamide gel (SDS-PAGE). doi:10.1128/9781555818890.ch1.f1.30

polypeptide have ionizable groups that contribute to the net charge of a protein; the degree of ionization (protonation) is influenced by the pH of the solution. In a gel to which an electric current is applied, proteins migrate through a pH gradient until they reach a specific pH (the **isoelectric point**) where the overall charge of the protein is zero and they no longer move. A particular position in the pH gradient may be occupied by two or more proteins that have the same isoelectric point. However, the proteins often have different molecular weights and can be further separated according to their molecular mass by electrophoresis at right angles to the first dimension (the second dimension) through a sodium dodecyl sulfate (SDS)–polyacrylamide gel (Fig. 1.30B). The separated proteins form an array of spots in the gel that is visualized by staining with Coomassie blue or silver protein stain.

Depending on the size of the 2D polyacrylamide gel, approximately 2,000 different proteins can be resolved. The pattern of spots is captured by densitometric scanning of the gel. Databases have been established with images of 2D polyacrylamide gels from different cell types, and software is available for detecting spots, matching patterns between gels, and

SDS
sodium dodecyl sulfate

quantifying the protein content of the spots. Proteins with either low or high molecular weights, those with highly acidic or basic isoelectric points (such as ribosomal proteins and histones), those that are found in cellular membranes, and those that are present in small amounts are not readily resolved by 2D PAGE.

After separation, individual proteins are excised from the gel and the identity of the protein is determined, usually by **mass spectrometry (MS)**. A mass spectrometer detects the masses of the ionized form of a molecule. For identification, the protein is first fragmented into peptides by digestion with a protease, such as trypsin, that cleaves at lysine or arginine residues (Fig. 1.31). The peptides are ionized and separated according to their mass-to-charge (*m/z*) ratio, and then the abundance and *m/z* ratios of the ions are measured. Several mass spectrometers are available that differ in the type of sample analyzed, the mode of ionization of the sample, the method for generating the electromagnetic field that separates and sorts the ions, and the method of detecting the different masses. Peptide masses are usually determined by matrix-assisted laser desorption ionization–time of flight (MALDI-TOF) MS. To determine the *m/z* value of each peptide fragment generated from an excised protein by MALDI-TOF MS, the peptides are ionized by mixing them with a matrix consisting of an organic acid and then using a laser to promote ionization. The ions are accelerated through a tube using a high-voltage current, and the time required to reach the ion detector is determined by their molecular mass, with lower-mass ions reaching the detector first.

To facilitate protein identification, computer algorithms have been developed for processing large amounts of MS data. Databases have been established that contain the masses of peptides from trypsin digestion for all known proteins. The databases are searched to identify a protein whose peptide masses match the values of the peptide masses of an unknown protein that were determined by MALDI-TOF MS (Fig. 1.31). This type of analysis is called **peptide mass fingerprinting**.

Protein Expression Profiling

Protein expression profiling is important for cataloging differences between normal and diseased cells that can be used for diagnosis, tracking changes during disease processes, and monitoring the cellular responses to therapeutic drugs. Several methods have been developed to quantitatively compare the proteomes among samples. Two-dimensional differential in-gel electrophoresis is very similar to 2D PAGE; however, rather than separating proteins from different samples on individual gels and then comparing the maps of separated proteins, proteins from two different samples are differentially labeled and then separated on the same 2D polyacrylamide gel (Fig. 1.32). Typically, proteins from each sample are labeled with different fluorescent dyes (e.g., Cy3 and Cy5); the labeled samples are mixed and then run together in the same gel, which overcomes the variability between separate gel runs. The two dyes carry the same mass and charge, and therefore, a protein labeled with Cy3 migrates to the same position as the identical protein labeled with Cy5. The Cy3 and Cy5 protein patterns are visualized separately by fluorescent

MS
mass spectrometry

m/z
mass to charge

MALDI-TOF
matrix-assisted laser desorption ionization–time of flight

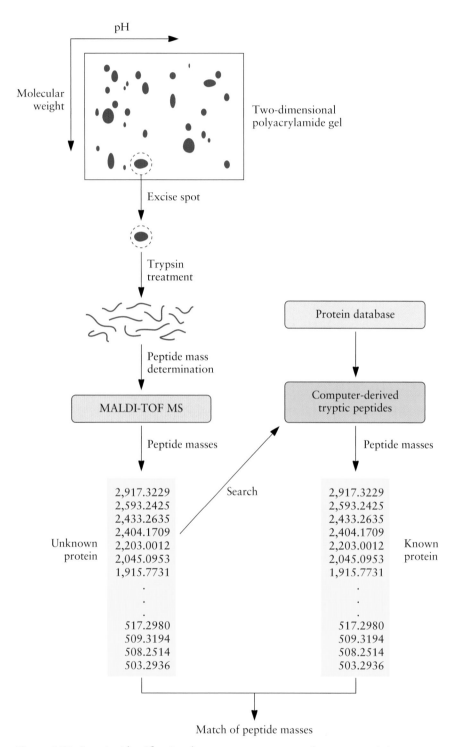

Figure 1.31 Protein identification by mass spectrometry. A spot containing an unknown protein that was separated by 2D PAGE is excised from the gel and digested with trypsin to generate peptides. The peptides are separated by MALDI-TOF MS. The set of peptide masses from the unknown protein is used to search a database that contains the masses of tryptic peptides for every known sequenced protein, and the best match is determined. Because trypsin cleaves proteins at specific amino acids, the trypsin cleavage sites of known proteins can be determined from the amino acid sequence, and consequently, the masses of the tryptic peptides are easy to calculate. Only some of the tryptic peptide masses for the unknown protein are listed in this example. doi:10.1128/9781555818890.ch1.f1.31

Figure 1.32 Protein expression profiling using 2D differential in-gel electrophoresis. The proteins of two proteomes are labeled with fluorescent dyes Cy3 and Cy5. The labeled proteins from the two samples are combined and separated by 2D PAGE. The gel is scanned for each fluorescent dye, and the relative levels of the two dyes in each protein spot are recorded. Each spot with an unknown protein is excised for identification by MS. Photo reproduced from Lee et al., *Appl. Environ. Microbiol.* **76:** 4655–4663, 2010, with permission. doi:10.1128/9781555818890.ch1.f1.32

excitation. The images are compared, and any differences are recorded. In addition, the ratio of Cy3 to Cy5 fluorescence for each spot is determined to detect proteins that are either up- or downregulated. Unknown proteins are identified by MS.

Another powerful technique for comparing protein populations among samples utilizes **protein microarrays**. Protein microarrays are similar to DNA microarrays; however, rather than arrays of oligonucleotides, protein microarrays consist of large numbers of proteins immobilized in a known position on a surface such as a glass slide in a manner that preserves the structure and function of the proteins. The proteins arrayed on the surface can be antibodies specific for a set of proteins in an organism, purified proteins that were expressed from a DNA or cDNA library, short synthetic peptides, or multiprotein samples from cell lysates or tissue specimens. The arrayed proteins are probed with samples that contain molecules that interact with the proteins. For example, the interacting molecules can be other proteins to detect protein-protein interactions, nucleic acid sequences to identify proteins that regulate gene expression by binding to DNA or RNA, substrates for specific enzymes, or small protein-binding compounds such as lipids or drugs.

Microarrays consisting of immobilized antibodies are used to detect and quantify proteins present in a complex sample. Antibodies directed against more than 1,800 human proteins have been isolated, characterized, and validated, and subsets of these that detect specific groups of proteins such as cell signaling proteins can be arrayed. To compare protein profiles in two different samples, for example, in normal and diseased tissues, proteins extracted from the two samples are labeled with two different fluorescent dyes (e.g., Cy3 and Cy5) and then applied to a single **antibody microarray** (Fig. 1.33). Proteins present in the samples bind to their cognate antibodies, and after a washing to remove unbound proteins, the antibody-bound proteins are detected with a fluorescence scanner. Interpretation of the fluorescent signals that represent the relative levels of specific proteins in the two samples on a protein microarray is very similar to analysis of a DNA microarray.

To increase the sensitivity of the assay and therefore the detection of low-abundance proteins, or to detect a specific subpopulation of proteins, a "sandwich"-style assay is often employed (Fig. 1.34). In this case, unlabeled proteins in a sample are bound to an antibody microarray, and then a second, labeled antibody is applied. This approach has been used to determine whether particular posttranslational protein modifications such as phosphorylation of tyrosine or glycosylation are associated with specific diseases. Serum proteins are first captured by immobilized antibodies on a microarray. Then, an antiphosphotyrosine antibody is applied that binds only to phosphorylated proteins (Fig. 1.34A). The antiphosphotyrosine antibody is tagged, for example, with a biotin molecule, and fluorescently labeled streptavidin, which binds specifically to biotin, is added to detect the phosphorylated protein. In a similar manner, glycosylated proteins can be detected with lectins (Fig. 1.34B). Lectins are plant glycoproteins that bind to specific carbohydrate moieties on the surface of

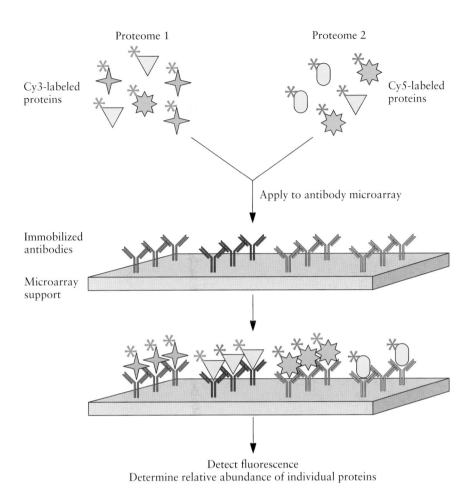

Figure 1.33 Protein expression profiling with an antibody microarray. Proteins extracted from two different samples are labeled with fluorescent dyes Cy3 and Cy5, respectively. The labeled proteins are mixed and incubated with an array of antibodies immobilized on a solid support. Proteins bound to their cognate antibodies are detected by measuring fluorescence, and the relative levels of specific proteins in each sample are determined. doi:10.1128/9781555818890.ch1.f1.33

proteins or cell membranes, and many different lectins with affinities for different glycosyl groups (glycans) are available.

In another type of microarray, purified proteins representing as many proteins of a proteome under study as possible are arrayed on a solid support and then probed with antibodies in serum samples collected from healthy (control) and diseased individuals. The purpose of these studies is to discover whether individuals produce antibodies that correlate with particular diseases or biological processes. For example, the differential expression of antibodies in serum samples from individuals with and without Alzheimer disease was tested using a microarray consisting of more than 9,000 unique human proteins (Fig. 1.35). After incubation of the serum samples with the protein microarray, bound antibodies were detected using a fluorescently labeled secondary antibody that interacts specifically with human antibodies. The screen resulted in the identification

A

1

Immobilized antibodies

Microarray support

2

Phosphotyrosine

Protein

3

Ⓑ Ⓑ Biotinylated
antiphosphotyrosine
antibody   Ⓑ

4

⭐ Fluorescent dye

Ⓑ Ⓑ Streptavidin Ⓑ

B

1

Immobilized antibodies

Microarray support

2

Glycan

Protein

3

Ⓑ Ⓑ Biotinylated
lectin   Ⓑ

4

⭐ Fluorescent dye

Ⓑ Streptavidin Ⓑ Ⓑ

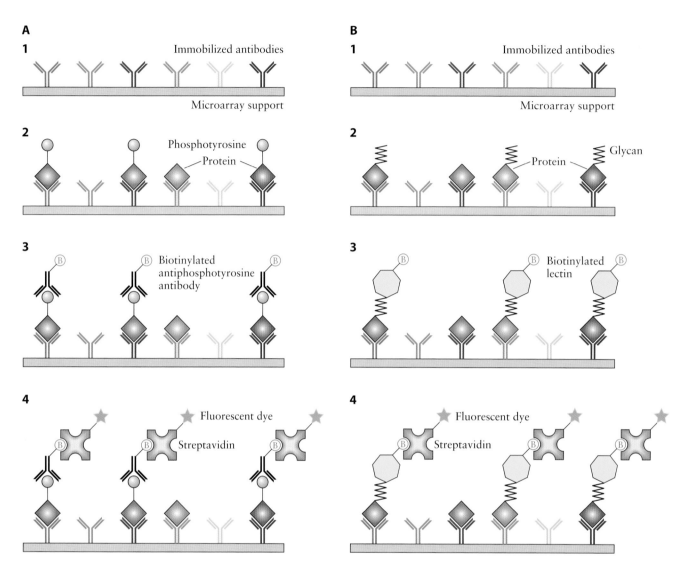

Figure 1.34 Detection of posttranslationally modified proteins with a sandwich-style antibody microarray. (**A**) Detection of phosphorylated proteins. An antibody microarray (1) is incubated with a protein sample (2). Biotinylated antiphosphotyrosine antibodies are added (3). Binding of the antibodies to phosphorylated tyrosine residues on some of the proteins is detected with fluorescently labeled streptavidin, which has a specific affinity for biotin (4). (**B**) Detection of glycosylated proteins. An antibody microarray (1) is incubated with a protein sample (2). Biotinylated lectin molecules that bind to a specific glycan group are added (3). Binding of lectin to the glycan groups on some of the proteins is detected with fluorescently labeled streptavidin (4). doi:10.1128/9781555818890.ch1.f1.34

of 10 autoantibodies (i.e., directed against an individual's own protein) that may be used as biomarkers to diagnose Alzheimer disease. Protein microarrays can also be used to identify proteins that interact with therapeutic drugs or other small molecules (Fig. 1.36). This can aid in determining the mechanism of action of a drug, for assessing responsiveness among various forms of a target protein (e.g., variants produced by different individuals), and for predicting undesirable side effects.

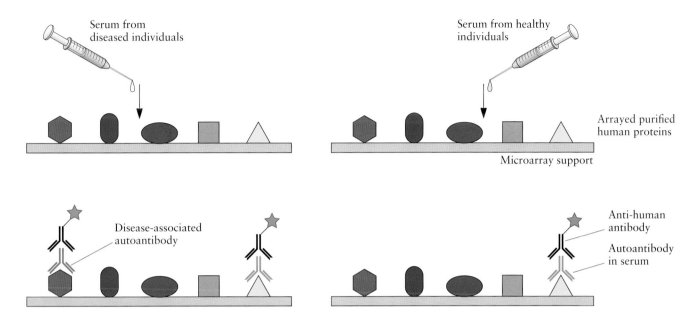

Figure 1.35 Identification of disease biomarkers with a human protein microarray. Serum samples are collected from diseased and healthy individuals and incubated with microarrays of purified human proteins. Serum autoantibodies bind to specific proteins on the microarray and are detected by applying a fluorescently labeled secondary antibody directed against human antibodies. Autoantibodies present in the serum from diseased individuals but not in serum from healthy individuals are potential biomarkers that can be used in diagnosis of the disease. doi:10.1128/9781555818890.ch1.f1.35

Protein–Protein Interactions

Proteins typically function as complexes comprised of different interacting protein subunits. Important cellular processes such as DNA replication, energy metabolism, and signal transduction are carried out by large multiprotein complexes. Thousands of protein–protein interactions occur in a cell. Some of these are short-lived, while others form stable multicomponent complexes that may interact with other complexes. Determining the functional interconnections among the members of a proteome is not an easy task. Several strategies have been developed to examine protein interactions, including protein microarrays, **two-hybrid systems**, and **tandem affinity purification** methods.

The two-hybrid method that was originally devised for studying the yeast proteome has been used extensively to determine pairwise protein–protein interactions in both eukaryotes and prokaryotes. The underlying principle of this assay is that the physical connection between two

Figure 1.36 Protein microarrays to detect protein-drug interactions. Therapeutic drugs or other small molecules tagged with a fluorescent dye are applied to purified proteins arrayed on a solid support. doi:10.1128/9781555818890.ch1.f1.36

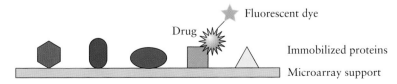

proteins reconstitutes an active transcription factor that initiates the expression of a reporter gene. The transcription factors employed for this purpose have two domains. One domain (DNA-binding domain) binds to a specific DNA site, and the other domain (activation domain) activates transcription (Fig. 1.37A). The two domains are not required to be part of the same protein to function as an effective transcription factor. However, the activation domain alone will not bind to RNA polymerase to activate transcription. Connection with the DNA-binding domain is necessary to place the activation domain in the correct orientation and location to initiate transcription by RNA polymerase.

For a two-hybrid assay, the coding sequences of the DNA-binding and activation domains of a specific transcription factor are cloned into separate vectors (Fig. 1.37). Often, the Gal4 transcriptional factor from *Saccharomyces cerevisiae* or the bacterial LexA transcription factor is used. A cDNA sequence that is cloned in frame with the DNA-binding domain sequence produces a fusion (hybrid) protein and is referred to as the "bait." This is the target protein for which interacting proteins are to be identified. Another cDNA sequence is cloned into another vector in frame with the activation domain coding sequence. A protein attached to the activation domain is called the "prey" and potentially interacts with the bait protein. Host yeast cells are transformed with both bait and prey DNA constructs. After expression of the fusion proteins, if the bait and prey do not interact, then there is no transcription of the reporter gene (Fig. 1.37B). However, if the bait and prey proteins interact, then the DNA-binding and activation domains are also brought together. This enables the activation domain to make contact with RNA polymerase and activate transcription of the reporter gene (Fig. 1.37C). The product of an active reporter gene may produce a colorimetric response or may allow a host cell to proliferate in a specific medium.

For a whole-proteome protein interaction study, two libraries are prepared, each containing thousands of cDNAs generated from total cellular mRNA (or genomic DNA fragments in a study of proteins from a prokaryote). To construct the bait library, cDNAs are cloned into the vector adjacent to the DNA sequence for the DNA-binding domain of the transcription factor Gal4 and then introduced into yeast cells. To construct the prey library, the cDNAs are cloned into the vector containing the sequence for the activation domain, and the constructs are transferred to yeast cells. The libraries are typically screened for bait-prey protein interactions in one of two ways. In one method, a prey library of yeast cells is arrayed on a grid. The prey library is then screened for the production of proteins that interact with a bait protein by introducing individual bait constructs to the arrayed clones by mating (Fig. 1.38A). Alternatively, each yeast clone in a bait library is mated en masse with a mixture of strains in the prey library, and then positive interactions are identified by screening for activation of the reporter gene (Fig. 1.38B). Challenges with using the two-hybrid system for large-scale determination of protein–protein interactions include the inability to clone all possible protein coding genes in frame with the activation and DNA-binding domains, which leads to missed interactions (false negatives), and the detection of interactions that

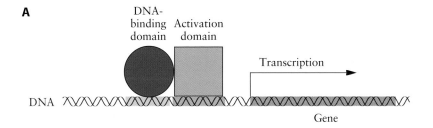

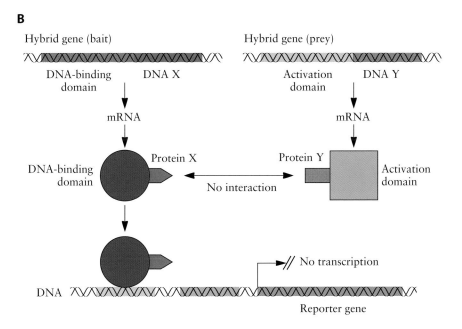

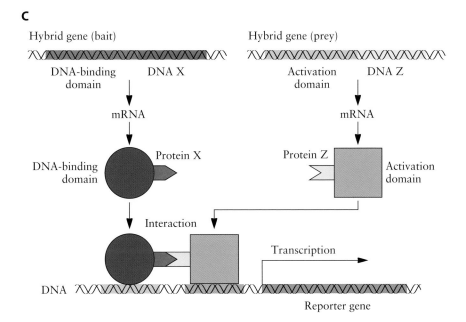

Figure 1.37 Two-hybrid analysis to identify protein–protein interactions. (**A**) The DNA-binding domain of a transcription factor binds to a specific sequence in the regulatory region of a gene, which orients the activation domain required for the initiation of transcription of the gene by RNA polymerase. (**B**) The coding sequences for the DNA-binding domain and the activation domain are fused to DNA X and DNA Y, respectively, in separate vectors, and both constructs (hybrid genes) are introduced into a cell. After translation, the DNA-binding domain–protein X fusion protein binds to the regulatory sequence of a reporter gene. However, protein Y (prey) does not interact with protein X (bait), and the reporter gene is not transcribed because the activation domain does not, on its own, associate with RNA polymerase. (**C**) The coding sequence for the activation domain is fused to the DNA for protein Z (DNA Z) and transformed into a cell containing the DNA-binding domain–DNA X fusion construct. The proteins encoded by the DNAs of the hybrid genes interact, and the activation domain is properly oriented to initiate transcription of the reporter gene, demonstrating a specific protein–protein interaction. doi:10.1128/9781555818890.ch1.f1.37

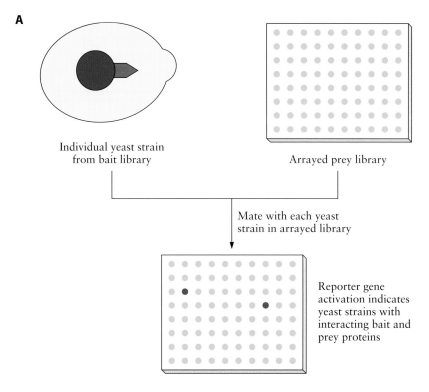

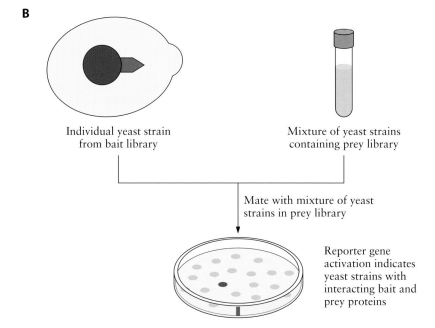

Figure 1.38 Whole-proteome screens for protein interactions using the yeast two-hybrid system. Two libraries are prepared, one containing genomic DNA fragments or cDNAs fused to the coding sequence for the DNA-binding domain of a transcription factor (bait library) and another containing genomic DNA fragments or cDNAs fused to the activation domain of the transcription factor (prey library). Two methods are commonly used to screen for pairwise protein interactions. (**A**) Individual yeast strains in the bait library are mated with each yeast strain in an arrayed prey library. Resulting strains in the array that produce bait and prey proteins that interact are detected by assaying for reporter gene activation (wells of a multiwell plate containing cells that express the reporter gene are indicated in green). (**B**) Yeast strains in the prey library are mated en masse with individual strains in the bait library. The mixture of strains is screened for reporter gene activity that identifies strains with interacting bait and prey proteins (green). doi:10.1128/9781555818890.ch1.f1.38

do not normally occur in their natural environments within the original cells and therefore are not biologically relevant (false positives). Nonetheless, this approach has been used to successfully identify interacting proteins in a wide range of organisms from bacteria to humans.

Instead of studying pairwise protein interactions, the tandem affinity purification tag procedure is designed to capture multiprotein clusters and then identify the components with MS (Fig. 1.39). In this method, a cDNA sequence that encodes the bait protein is fused to a DNA sequence that encodes two small peptides (tags) separated by a protease cleavage site. The peptide tags bind with a high affinity to specific molecules and facilitate purification of the target protein. A "two-tag" system allows two successive rounds of affinity binding to ensure that the target and its associated proteins are free of any nonspecific proteins. Alternatively, a "one-tag" system with a small protein tag that is immunoprecipitated with a specific antibody requires only a single purification step. In a number of trials, the tags did not alter the function of various test proteins.

A cDNA–two-tag construct is introduced into a host cell, where it is expressed and a tagged protein is synthesized (Fig. 1.39). The underlying assumption is that the cellular proteins that normally interact with the native protein in vivo will also combine with the tagged protein. After the cells are lysed, the tagged protein and any interacting proteins are purified using the affinity tags. The proteins of the cluster are separated according to their molecular weight by PAGE and identified with MS. Computer programs are available for generating maps of clusters with common proteins, assigning proteins with shared interrelationships to specific cellular activities, and establishing the links between multiprotein complexes.

Metabolomics

Metabolomics provides a snapshot of the small molecules present in a complex biological sample. The **metabolites** present in cells and cell secretions are influenced by genotype, which determines the metabolic capabilities of an organism, and by environmental conditions such as the availability of nutrients and the presence of toxins or other stressors. Metabolite composition varies depending on the developmental and health status of an organism, and therefore, a comprehensive metabolite profile can identify molecules that reflect a particular physiological state. For example, metabolites present in diseased cells but not in healthy cells are useful biomarkers for diagnosing and monitoring disease. Metabolic profiles can also aid in understanding drug metabolism, which may reduce the efficacy of a treatment, or in understanding drug toxicity, which can help to reduce adverse drug reactions. Metabolomic analysis can be used to determine the catalytic activity of proteins, for example, by quantifying changes in metabolite profiles in response to mutations in enzyme coding genes and to connect metabolic pathways that share common intermediates.

Biological samples for metabolite analysis may be cell or tissue lysates, body fluids such as urine or blood, or cell culture media that contain a great diversity of metabolites. These include building blocks for biosynthesis of cellular components such as amino acids, nucleotides, and

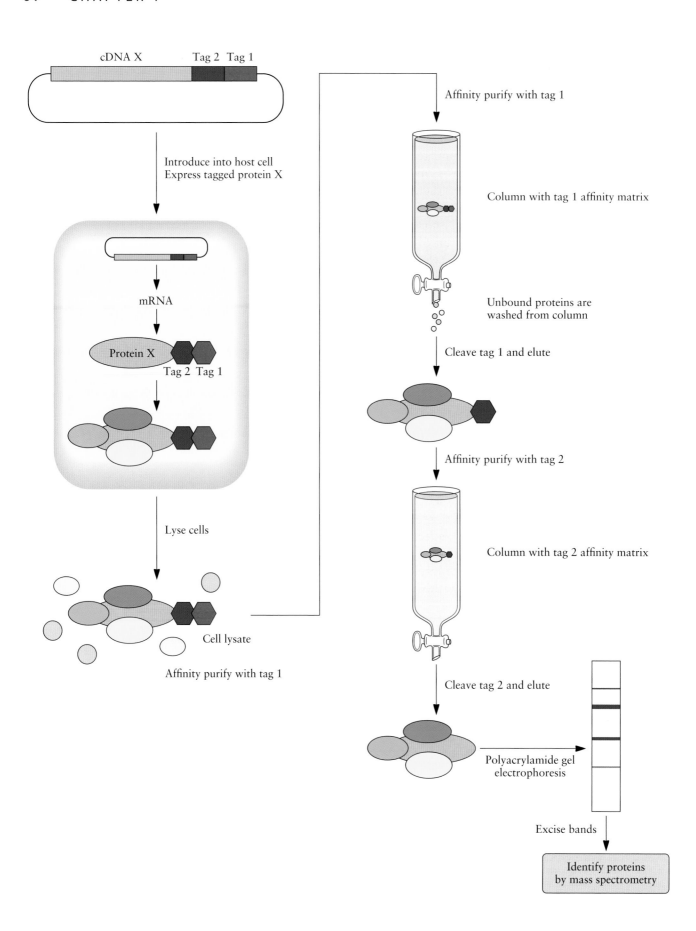

lipids. Also present are various substrates, cofactors, regulators, intermediates, and end products of metabolic pathways such as carbohydrates, vitamins, organic acids, amines and alcohols, and inorganic molecules. These molecules have very different properties, and therefore comprehensive detection and quantification using a single method based on chemical characteristics present a challenge.

Metabolomics employs spectroscopic techniques such as MS and **nuclear magnetic resonance (NMR)** spectroscopy to identify and quantify the metabolites in complex samples. Often, multiple methods are used in parallel to obtain a comprehensive view of a metabolome. In a manner similar to protein identification described above, MS measures the m/z ratio of charged metabolites. The molecules may be ionized by various methods before separation of different ions in an electromagnetic field. MS is typically coupled with chromatographic techniques that first separate metabolites based on their properties. For example, MS may be coupled with **gas chromatography** to separate volatile metabolites. Some nonvolatile metabolites, such as amino acids, are chemically modified (derivatized) to increase their volatility. **Liquid chromatography** separates metabolites dissolved in a liquid solvent based on their characteristic retention times as they move through an immobilized matrix.

NMR spectroscopy is based on the principle that in an applied magnetic field, molecules (more precisely, atomic nuclei with an odd mass number) absorb and emit electromagnetic energy at a characteristic resonance frequency that is determined by their structure. Thus, the resonance frequencies provide detailed information about the structure of a molecule and enable differentiation among molecules with different structures, even when the difference is very small, such as between structural isomers. In contrast to MS, an initial metabolite separation step is not required, and NMR measures different types of molecules. In addition, NMR is not destructive, and in fact, it has been adapted to visualize molecules in living human cells in the diagnostic procedure **magnetic resonance imaging (MRI)**. A drawback of NMR is low sensitivity, which means that it does not detect low-abundance molecules.

An illustration of the application of metabolome analysis is the identification of metabolites that are associated with the progression of prostate cancer to **metastatic disease**. Researchers compared more than 1,000

NMR
nuclear magnetic resonance

MRI
magnetic resonance imaging

Figure 1.39 Tandem affinity purfication to detect multiprotein complexes. The coding region of a cDNA (cDNA X) is cloned into a vector in frame with two DNA sequences (tag 1 and tag 2), each encoding a short peptide that has a high affinity for a specific matrix. The tagged cDNA construct is introduced into a host cell, where it is transcribed and the mRNA is translated. Other cellular proteins bind to the protein encoded by cDNA X (protein X). The cluster consisting of protein X and its interacting proteins (colored shapes) is separated from other cellular proteins by the binding of tag 1 to an affinity matrix which is usually fixed to a column that retains the cluster while allowing noninteracting proteins to flow through. The cluster is then eluted from the affinity matrix by cleaving off tag 1 with a protease, and a second purification step is carried out with tag 2 and its affinity matrix. The proteins of the cluster are separated by one-dimensional PAGE. Single bands are excised from the gel and identified by MS. doi:10.1128/9781555818890.ch1.f1.39

metabolites in benign prostate tissue, localized prostate tumors, and metastatic tumors from liver, rib, diaphragm, and soft tissues using MS combined with liquid and gas chromatography (different metastatic tumor tissues were analyzed to minimize identification of tissue-specific metabolites). Sixty metabolites were found in localized prostate and/or metastatic tumors but not in benign prostate tissue, and six of these were significantly higher in the metastatic tumors. The metabolite profile indicated that progression of prostate cancer to metastatic disease was associated with an increase in amino acid metabolism. In particular, levels of sarcosine, a derivative of the amino acid glycine, were much higher in the metastatic tumors than in localized prostate cancer tissue and were not detectable in noncancerous tissue (Fig. 1.40). Moreover, sarcosine levels were higher in the urine of men with prostate tissue biopsies that tested positive for cancer than in that of biopsy-negative controls, and higher in prostate cancer cell lines than in benign cell lines. Benign prostate epithelial cells became motile and more invasive upon exposure to sarcosine than did those treated with alanine as a control. From this analysis, sarcosine appears to play a key role in cancer cell invasion and shows promise as a biomarker for progression of prostate cancer and as a target for prevention.

Figure 1.40 Metabolite profiles of benign prostate, localized prostate cancer, and metastatic tumor tissues. The relative levels of a subset of 50 metabolites are shown in each row. Levels of a metabolite in each tissue (columns) were compared to the median metabolite level (black); shades of yellow represent increases, and shades of blue indicate decreases. Metastatic samples were taken from soft (**A**), rib or diaphragm (**B**), or liver (**C**) tissues. Modified with permission from Macmillan Publishers Ltd. from Sreekumar et al., *Nature* **457:** 910–914, 2009. doi:10.1128/9781555818890.ch1.f1.40.

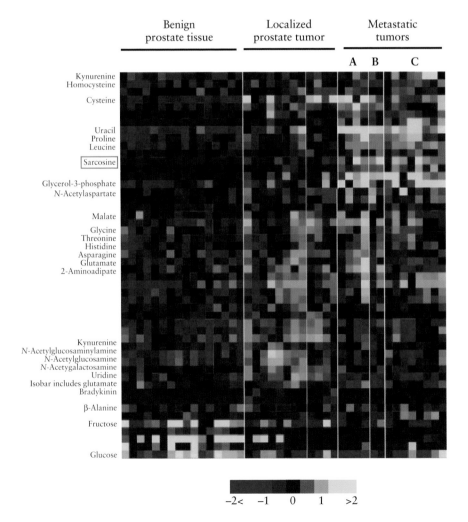

summary

Medical biotechnology is the application of molecular technologies to diagnose and treat human diseases. At the core of these technologies are strategies to identify and isolate specific genes and to propagate them in host organisms. The tools for these processes were developed from an understanding of the biochemistry, genetics, and molecular biology of cells, especially prokaryotic cells, and viruses. Molecular cloning is the process of inserting a gene or other DNA sequence from one organism into a vector and introducing it into a host cell. The discovery of restriction endonucleases was essential for this process, as it enabled predictable and reproducible cleavage of both target (insert) and vector DNAs in preparation for joining the two molecules. A restriction endonuclease is a protein that binds to DNA at a specific nucleotide sequence and cleaves a phosphodiester bond in each of the DNA strands within the recognition sequence. Digestion of target and vector DNA with the same restriction endonuclease generates compatible single-stranded extensions that can be joined by complementary base-pairing and the activity of the enzyme DNA ligase that catalyzes the formation of phosphodiester bonds. Another cloning method known as recombinational cloning does not utilize restriction endonucleases or DNA ligase for insertion of target DNA into a vector but, rather, exploits a system used by some viruses to integrate into the host genome via recombination at specific attachment sequences.

Cloned DNA is introduced into host bacterial cells that are competent to take up exogenous DNA, a process known as transformation. Vectors that carry the target DNA into the host cell are often derived from natural bacterial plasmids that have been genetically engineered with several endonuclease recognition sequences (multiple-cloning sites) to facilitate cloning. A vector can be propagated in a host cell if it possesses a DNA sequence (origin of replication) that enables it to replicate in the host. Transformation is generally inefficient; however, transformed cells may be distinguished from nontransformed cells by testing for the activity of genes that are present on the vector, including genes for resistance to antibiotics or synthesis of colored products.

To clone and express genes that encode eukaryotic proteins in a bacterial host, the introns must first be removed. Purified mRNA is used as a template for the synthesis of cDNA by the enzyme reverse transcriptase. Oligonucleotide primers can be designed to target a specific mRNA for cDNA synthesis or to anneal to the poly(A) tails present on most eukaryotic mRNAs to generate a cDNA library that contains all of the protein coding sequences from the genome of a source eukaryote. Construction of a genomic DNA library from a prokaryote is more straightforward and entails cleaving the DNA to obtain overlapping fragments for cloning. Libraries are screened by a variety of methods to identify clones with a particular sequence or that produce a target protein.

Amplification and sequencing of DNA are also fundamental tools of medical biotechnology. PCR is a powerful method for generating millions of copies of a specific sequence of DNA from very small amounts of starting material. Amplification is achieved in 30 or more successive cycles of template DNA denaturation, annealing of the two oligonucleotide primers to complementary sequences flanking a target gene in the single-stranded DNA, and DNA synthesis extending from the primer by a thermostable DNA polymerase. Among innumerable applications, PCR can be used to detect a specific nucleotide sequence in a complex biological sample or to obtain large amounts of a particular DNA sequence either for cloning or for sequencing.

The nucleotide sequence of a gene can reveal useful information about the function, regulation, and evolution of the gene. All of the sequencing technologies currently used involve (i) enzymatic addition of nucleotides to a primer based on complementarity to a template DNA fragment and (ii) detection and identification of the nucleotide(s) added. The techniques differ in the method by which the nucleotides are extended, employing either DNA polymerase to catalyze the addition of single nucleotides (sequencing by synthesis) or ligase to add a short, complementary oligonucleotide (sequencing by ligation), and in the method by which the addition is detected. The dideoxynucleotide method developed by Sanger and his colleagues has been used for several decades to sequence genes and whole genomes. This method relies on the incorporation of a synthetic dideoxynucleotide that lacks a 3′ hydroxyl group into a growing DNA strand, which terminates DNA synthesis. Conditions are optimized so that the dideoxynucleotides are incorporated randomly, producing DNA fragments of different lengths that terminate with one of the four dideoxynucleotides, each tagged with a different fluorescent dye. The fragments are separated according to their size by electrophoresis, and the sequence of fluorescent signals is determined and converted into a nucleotide sequence. Pyrosequencing entails correlating the release of pyrophosphate, which is recorded as the emission of light, with the incorporation of a particular nucleotide into a growing DNA strand. Sequencing using reversible chain terminators also reveals the sequence of a DNA fragment by detecting single-nucleotide extensions; however, in contrast to pyrosequencing, the four nucleotides are added to the reaction together in each cycle, and after the unincorporated nucleotides are washed away, the nucleotide incorporated by DNA polymerase is distinguished by its fluorescent signal. The fluorescent dye and a blocking group that prevents addition of more than one nucleotide during each cycle are chemically cleaved, and the

(continued)

summary *(continued)*

cycle is repeated. In another method, short sequences can be determined by ligating fluorescently tagged oligonucleotides that have a known nucleotide sequence in the query position to a primer in a DNA template-dependent fashion.

The dideoxynucleotide procedure in combination with shotgun cloning was used to sequence many whole genomes, including the first human genome sequence. In this approach, random, overlapping genomic DNA fragments are cloned into a vector and introduced in a bacterial host, and after isolation of the plasmids, the cloned DNAs serve as sequencing templates. However, pyrosequencing and sequencing using reversible chain terminators or ligated oligonucleotides form the basis of the high-throughput next-generation sequencing strategies. These sequencing technologies, together with cell-free methods to generate a library of genomic DNA sequencing templates in a dense array on a solid surface, have substantially reduced the time and cost to sequence whole genomes. Hundreds of millions of short nucleotide reads can now be acquired simultaneously (massive parallelization) and assembled into contigs. Using these approaches, the genome sequences of thousands of organisms from all domains of life have been completed or are in progress. The next steps are to annotate the sequence features and to determine the functions of the genes encoded in the genomes by investigating patterns of transcription (transcriptomics), protein synthesis (proteomics), and small-molecule production (metabolomics) using a variety of techniques such as DNA and protein microarray analysis, RNA sequencing, 2D PAGE, mass spectrometry, and NMR. Comparison of genome sequences can reveal the genetic basis of a disease, the mechanism of pathogenicity of a microbe, or the evolutionary relationships among organisms, while transcript, protein, and metabolite profiles can identify biomarkers for diagnosis and treatment of disease.

review questions

1. Describe a strategy using restriction endonucleases to clone a bacterial gene into a vector for propagation in *E. coli.* Assume that the target sequence is known. Describe the selection for *E. coli* cells that carry the cloned gene. Consider methods to minimize unwanted products.

2. Describe the features that make pUC19 a useful cloning vector.

3. Outline a strategy to clone a eukaryotic gene into a vector for expression in *E. coli.* Briefly describe the activity of the enzymes used in the process.

4. Describe how a library of open reading frames that represents a proteome is constructed by recombinational cloning.

5. A genomic DNA library of the bacterium *Pseudomonas putida* was constructed by partially digesting the genomic DNA with Sau3AI and inserting the fragments into pUC19 digested with BamHI. Why were two different restriction enzymes used in this experiment? How is the partial digestion performed, and what is the result? Why was a partial digestion used to construct the library?

6. Outline the steps in a PCR cycle. What component of a PCR determines the specificity of the amplified product?

7. Describe how PCR is used to clone a specific gene.

8. What is a dideoxynucleotide? How is it used to determine the sequence of a DNA molecule?

9. Outline the basic features of pyrosequencing.

10. How are incorporated nucleotides recognized after each cycle of sequencing using reversible chain terminators? How does this differ from pyrosequencing?

11. Why are several different primers used in sequencing by ligation?

12. Why are adaptors often ligated to DNA fragments prior to sequencing?

13. How does preparation of sequencing templates differ between the shotgun cloning and the high-throughput next-generation approaches to whole-genome sequencing?

14. Describe emulsion PCR.

15. What are some of the benefits of whole-genome sequencing to human medicine?

16. Outline a DNA microarray experiment. List some applications for this technology.

17. What are some of the advantages of using RNA sequencing rather than DNA microarrays to profile gene expression?

18. How are gene expression levels quantified using high-throughput RNA sequencing?

review questions *(continued)*

19. Explain why random hexamers used for RNA sequencing result in overrepresentation of sequences from the 5′ end of a gene.

20. How can 2D PAGE be used to identify proteins that are differentially expressed in two samples?

21. Describe some applications for protein microarrays.

22. What biological information may be provided by a protein microarray assay that is not provided by using a DNA microarray?

23. Describe three methods to determine protein–protein interactions.

24. What biological information may be provided by the tandem affinity purification tag system that is not provided by a two-hybrid assay?

25. Explain how metabolomics may be used to identify biomarkers of disease.

references

Benson, D. A., I. Karsch-Mizrachi, D. J. Lipman, J. Ostell, and D. L. Wheeler. 2007. GenBank. *Nucleic Acids Res.* **35:** D21–D25.

Berg, P., and J. E. Mertz. 2010. Personal reflections on the origins and emergence of recombinant DNA technology. *Genetics* **184:**9–17.

Church, G. M. 2006. Genomes for all. *Sci. Am.* **294**(January):47–54.

Cohen, S. N., and A. C. Chang. 1973. Recircularization and autonomous replication of a sheared R-factor DNA segment in *Escherichia coli* transformants. *Proc. Natl. Acad. Sci. USA* **70:** 1293–1297.

Cohen, S. N., A. C. Y. Chang, H. W. Boyer, and R. B. Helling. 1973. Construction of biologically functional bacterial plasmids in vitro. *Proc. Natl. Acad. Sci. USA* **70:**3240–3244.

Fodor, S. P., R. P. Rava, X. C. Huang, A. C. Pease, C. P. Holmes, and C. L. Adams. 1993. Multiplexed biochemical assays with biological chips. *Nature* **364:**555–556.

Gellert, M., J. W. Little, C. K. Oshinsky, and S. B. Zimmerman. 1968. Joining of DNA strands by DNA ligase of E. coli. *Cold Spring Harbor Symp. Quant. Biol.* **33:**21–26.

Genome 10K Community of Scientists. 2009. Genome 10K: a proposal to obtain whole-genome sequence for 10,000 vertebrate species. *J. Hered.* **100:**659–674.

Goeddel, D. V., D. G. Kleid, F. Bolivar, H. L. Heyneker, D. G. Yansura, R. Crea, T. Hirose, A. Kraszewski, K. Itakura, and A. D. Riggs. 1979. Expression in *Escherichia coli* of chemically synthesized genes for human insulin. *Proc. Natl. Acad. Sci. USA* **76:**106–110.

Hedgpeth, J., H. M. Goodman, and H. W. Boyer. 1972. DNA nucleotide sequence restricted by the RI endonuclease. *Proc. Natl. Acad. Sci. USA* **69:** 3448–3452.

Holmes, E., I. D. Wilson, and J. K. Nicholson. 2008. Metabolic phenotyping in health and disease. *Cell* **134:**714–717.

International Human Genome Sequencing Consortium. 2001. Initial sequencing and analysis of the human genome. *Nature* **409:**860–921.

Ito, T., T. Chiba, R. Ozawa, M. Yoshida, M. Hattori, and Y. Sakaki. 2001. A comprehensive two-hybrid analysis to explore the yeast protein interactome. *Proc. Natl. Acad. Sci. USA* **98:** 4569–4574.

Kelly, T. J., Jr., and H. O. Smith. 1970. A restriction enzyme from *Hemophilus influenzae*. II. *J. Mol. Biol.* **51:**393–409.

Ley, T. J., E. R. Mardis, L. Ding, et al. 2008. DNA sequencing of a cytogenetically normal acute myeloid leukaemia genome. *Nature* **456:**66–72.

Li, Y. 2011. The tandem affinity purification technology: an overview. *Biotechnol. Lett.* **33:**1487–1499.

Ma, W., S. E. Belisle, D. Mosier, X. Li, E. Stigger-Rosser, Q. Liu, C. Qiao, J. Elder, R. Webby, M. G. Katze, and J. A. Richt. 2011. 2009 Pandemic H1N1 influenza virus causes disease and up-regulation of genes related to inflammatory and immune responses, cell death, and lipid metabolism in pigs. *J. Virol.* **85:**11626–11637.

Mardis, E. R. 2011. A decade's perspective on DNA sequencing technology. *Nature* **470:**198–203.

Martin, J. A., and Z. Wang. 2011. Next-generation transcriptome assembly. *Nat. Rev. Genet.* **12:**671–682.

Mertz, J. E., and R. W. Davis. 1972. Cleavage of DNA by R_1 restriction endonuclease generates cohesive ends. *Proc. Natl. Acad. Sci. USA* **69:** 3370–3374.

Metzker, M. L. 2010. Sequencing technologies—the next generation. *Nat. Rev. Genet.* **11:**31–46.

Mullis, K. B., F. A. Faloona, S. J. Scharf, R. K. Saiki, G. T. Horn, and H. A. Erlich. 1986. Specific enzymatic amplification of DNA in vitro: the polymerase chain reaction. *Cold Spring Harbor Symp. Quant. Biol.* **51:**263–273.

Nagele, E., M. Han, C. DeMarshall, B. Belinka, and R. Nagele. 2011. Diagnosis of Alzheimer's disease based on disease-specific autoantibody profiles in human sera. *PLoS One* **6:**e23112.

Nirenberg, M. W., and J. H. Matthaei. 1961. The dependence of cell-free protein synthesis in *E. coli* upon naturally occurring or synthetic polyribonucleotides. *Proc. Natl. Acad. Sci. USA* **47:** 1588–1602.

Nirenberg, M. W., J. H. Matthaei, and O. W. Jones. 1962. An intermediate in the biosynthesis of polyphenylalanine directed by synthetic template RNA. *Proc. Natl. Acad. Sci. USA* **48:**104–109.

Noonan, J. P., G. Coop, S. Kudaravalli, D. Smith, J. Krause, J. Alessi, F. Chen, D. Platt, S. Pääbo, J. K. Pritchard, and E. M. Rubin. 2006. Sequencing and analysis of Neanderthal genomic DNA. *Science* **314:**1113–1118.

Parrish, J. R., K. D. Gulyas, and R. L. Finley, Jr. 2006. Yeast two-hybrid contributions to interactome mapping. *Curr. Opin. Biotechnol.* **17:**387–393.

Roberts, R. J. 2005. How restriction enzymes became the workhorses of molecular biology. *Proc. Natl. Acad. Sci. USA* **102:**5905–5908.

Saiki, R. K., D. H. Gelfand, S. Stoffel, S. Scharf, R. Higuchi, G. T. Horn, K. B. Mullis, and H. A. Erlich. 1988. Primer-directed enzymatic amplification of DNA with a thermostable DNA polymerase. *Science* **239:**487–491.

Sanger, F., S. Nicklen, and A. R. Coulson. 1977. DNA sequencing with chain-terminating inhibitors. *Proc. Natl. Acad. Sci. USA* **74:**5463–5467.

Shendure, J. A., G. J. Porreca, G. M. Church, A. F. Gardner, C. L. Hendrickson, J. Kieleczawa, and B. E. Slatko. 2011. Overview of DNA sequencing strategies. *Curr. Protoc. Mol. Biol.* **96:** 7.1.1–7.1.23.

Sreekumar, A., L. M. Poisson, T. M. Rajendiran, et al. 2009. Metabolic profiles delineate potential role for sarcosine in prostate cancer progression. *Nature* **457:**910–914.

Tasse, L., J. Bercovici, S. Pizzut-Serin, et al. 2010. Functional metagenomics to mine human gut microbiome for dietary fiber catabolic enzymes. *Genome Res.* 20:1605–1612.

The 1000 Genomes Project Consortium. 2010. A map of human genome variation from population scale sequencing. *Nature* 467:1061–1073.

Venter, J. C., M. D. Adams, E. W. Myer, et al. 2001. The sequence of the human genome. *Science* 291:1304–1351.

Vigil, P. D., C. H. Alteri, and H. L. T. Mobley. 2011. Identification of *in vivo*-induced antigens including an RTX family exoprotein required for uropathogenic *Escherichia coli* virulence. *Infect. Immun.* 79:2335–2344.

Walhout, A. J., G. F. Temple, M. A. Brasch, J. L. Hartley, M. A. Lorson, S. van den Heuvel, and M. Vidal. 2000. GATEWAY recombinational cloning: application to the cloning of large numbers of open reading frames or ORFeomes. *Methods Enzymol.* 328: 575–592.

Wang, Z., M. Gerstein, and M. Snyder. 2009. RNA-seq: a revolutionary tool for transcriptomics. *Nat. Rev. Genet.* 10:57–63.

Watson, J. D., and F. H. Crick. 1953. Molecular structure of nucleic acids; a structure for deoxyribose nucleic acid. *Nature* 171:737–738.

Wheeler, D. A., M. Srinivasan, M. Egholm, Y. Shen, L. Chen, A. McGuire, W. He, Y. J. Chen, V. Makhijani, G. T. Roth, X. Gomes, K. Tartaro, F. Niazi, C. L. Turcotte, G. P. Irzyk, J. R. Lupski, C. Chinault, X. Z. Song, Y. Liu, Y. Yuan, L. Nazareth, X. Qin, D. M. Muzny, M. Margulies, G. M. Weinstock, R. A. Gibbs, and J. M. Rothberg. 2008. The complete genome of an individual by massively parallel DNA sequencing. *Nature* 452:872–877.

Williams, R. J. 2003. Restriction endonucleases. *Mol. Biotechnol.* **23:** 225–243.

Wilson, D. S., and S. Nock. 2002. Functional protein microarrays. *Curr. Opin. Chem. Biol.* 6:81–85.

Worthey, E. A., A. N. Mayer, G. D. Syverson, D. Helbling, B. B. Bonacci, B. Decker, J. M. Serpe, T. Dasu, M. R. Tschannen, R. L. Veith, M. J. Basehore, U. Broeckel, A. Tomita-Mitchell, M. J. Arca, J. T. Casper, D. A. Margolis, D. P. Bick, M. J. Hessner, J. M. Routes, J. W. Verbsky, H. J. Jacob, and D. P. Dimmock. 2011. Making a definitive diagnosis: successful clinical application of whole exome sequencing in a child with intractable inflammatory bowel disease. *Genet. Med.* 13:255–262.

Yanisch-Perron, C., J. Vieira, and J. Messing. 1985. Improved M13 phage cloning vectors and host strains: nucleotide sequences of the M13mp18 and pUC19 vectors. *Gene* 33:103–119.

Yong, D., M. A. Toleman, C. G. Giske, H. S. Cho, K. Sundman, K. Lee, and T. R. Walsh. 2009. Characterization of a new metallo-β-lactamase gene, bla_{NDM-1}, and a novel erythromycin esterase gene carried on a unique genetic structure in *Klebsiella pneumoniae* sequence type 14 from India. *Antimicrob. Agents Chemother.* 53:5046–5054.

Zhou, Z., Q. Yuan, D. C. Mash, and D. Goldman. 2011. Substance-specific and shared transcription and epigenetic changes in the human hippocampus chronically exposed to cocaine and alcohol. *Proc. Natl. Acad. Sci. USA* 108: 6626–6631.

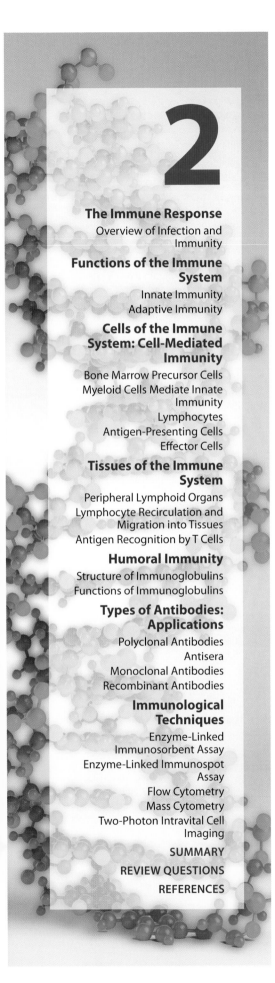

2

Fundamental Concepts in Immunology

The Immune Response

Overview of Infection and Immunity

A brief overview of some immunological terms, the important properties and principles of all immune responses, and the essential components (tissues, cells, and molecules) of an immune response is presented in this chapter. The objective is to provide the reader with a basic understanding of fundamental immunology concepts that may be applied for the development of novel biomedical techniques in the immunotherapy of infection and disease. Emphasis is placed on addressing the following questions:

- What are the important characteristics, functions, and mechanisms of immune responses?
- Which immune responses protect individuals from infections?
- How do the components of the immune system (tissues, cells, and molecules) recognize microbial pathogens and help to eliminate them?

Immunology is the study of the immune system and its responses to a host of invading pathogens and other harmful agents (e.g., insect toxins). The **immune system** is comprised of several types of cells, tissues, and molecules that mediate resistance to infection and accompanying infectious disease, and the interaction of these cells, tissues, and molecules to infectious microbes elicits an **immune response**. Thus, the raison d'être of the immune system is to initially prevent infection and then to arrest and eliminate established infection.

The immune system is essential to maintain the health of an individual, as demonstrated by the increased susceptibility of individuals with deficient immune responses to acute, invariably life-threatening infections (Table 2.1). In contrast, the stimulation of immune responses against

doi:10.1128/9781555818890.ch2

Table 2.1 Role of the immune system in health and disease

Defensive role	Consequence
Infection	Reduced immunity increases susceptibility to infection (e.g., HIV).
	Vaccination increases immunity and protects against infections.
Immune responses to tissue transplants and novel proteins	Barriers to transplantation and gene therapy
Tumors	Possible immunotherapy for cancer

Adapted with permission from Abbas and Lichtman, *Basic Immunology: Functions and Disorders of the Immune System*, 3rd ed. (Saunders Elsevier, Philadelphia, PA, 2011).

The defensive role of the immune system and the consequences in various immune responses are presented. Note that immune responses not only protect against disease but also elicit or exacerbate disease.

WHO
World Health Organization

AIDS
acquired immunodeficiency syndrome

microbes by vaccination is the most effective method of protection of individuals against infection (Table 2.2). Notably, by 1977, smallpox, long considered to be the most deadly and persistent human pathogenic disease, was eradicated by the World Health Organization (WHO). This was accomplished through a massive, worldwide outbreak search and vaccination program. The underlying benefit of such immunizations is that the vaccine triggers an immune response more rapidly than the natural infection itself. The emergence of acquired immunodeficiency syndrome (AIDS) since the 1980s has further emphasized the importance of the immune system for defending individuals against infection.

Importantly, the impact of immunology extends beyond infectious disease. For example, the immune response is the major barrier to successful organ transplantation, which is being used more frequently as a therapy

Table 2.2 Benefits of vaccination for some infectious diseases

Disease	Highest no. of cases (yr)	No. of cases in 2004	Decrease (%)
Diphtheria	206,939 (1921)	0	100.0
Haemophilus influenzae type b infection	20,000 (1984)	16	99.9
Hepatitis B	26,611 (1985)	6,632	75.1
Measles	894,134 (1941)	37	99.9
Mumps	152,209 (1968)	236	99.9
Pertussis	265,269 (1934)	18,957	96.8
Polio (paralytic)	21,269 (1952)	0	100.0
Rubella	57,686 (1969)	12	99.9
Tetanus	1,560 (1923)	26	98.3
TB	17,500 (1993)	14,511	96.7

Adapted with permission from Abbas and Lichtman, *Basic Immunology: Functions and Disorders of the Immune System*, 3rd ed. (Saunders Elsevier, Philadelphia, PA, 2011).

The incidence of certain infectious diseases is dramatically reduced in the general population after prior vaccination. A vaccine was recently developed for hepatitis B, and as a result, the incidence of hepatitis B is continuing to decrease worldwide. In 2011, about 9,350 new tuberculosis (TB) cases were reported in the United States, an incidence of 3.4 cases per 100,000 population, which is lower than the rate in 2010. This is the lowest rate recorded since reporting began in 1953. The percent decline is greater than the average 3.8% decline per year observed from 2000 to 2008 but is not as large as the record decline of 11.4% from 2008 to 2009. According to these rates, if current efforts are not improved or expanded, TB elimination in the United States is unlikely before 2100 (Centers for Disease Control and Prevention, 2012).

for organ failure. Attempts to treat cancers by stimulating immune responses against cancer cells are being tried for many human malignancies. Furthermore, abnormal immune responses are the causes of many inflammatory diseases with serious morbidity and mortality. Antibodies, protein products synthesized by B cells during an immune response, are highly specific reagents for detecting a wide variety of molecules in the circulation and in cells and tissues. Therefore, antibodies have emerged as valuable reagents for the development of laboratory diagnostic technologies for both clinical research and clinical trials.

Antibodies designed to block or eliminate potentially harmful molecules and cells are in extensive use for the treatment of immunological diseases, cancers, and other types of disorders. Accordingly, the field of immunology impacts highly on the interests of clinicians, scientists, and the lay public.

Functions of the Immune System

Four important functions are performed by the immune system (Table 2.3). First, **immunological recognition** is conducted by **leukocytes** (e.g., neutrophils, macrophages, and **natural killer cells**) of the **innate immune system** that provide an early rapid response and by the **lymphocytes** (T cells and B cells) of the later, more antigen-specific and more efficient **adaptive immune system**. Second, to contain and eliminate the infection, **immune effector functions** that involve the complement system of blood proteins, enzymes, and antibodies together with certain T cells and B cells are required. Third, while combating foreign pathogens, the immune system must learn not to elicit damage to self-tissues and other components of the body. It accomplishes this by **immune regulation**, a process in which regulatory lymphocytes control various immune responses against self-components. A failure in immune regulation could potentially result in allergy and autoimmune disease. Fourth, the immune system must be equipped to protect an individual against recurring disease upon exposure to the same or a closely related pathogen. This task is uniquely mediated by the adaptive immune system during a response known as **immunological memory**. After exposure to an infectious pathogen, a person will make a rapid and more vigorous response to the pathogen upon

Table 2.3 Functions of the immune system

Function	Cell type
Immune recognition	Leukocytes (neutrophils, macrophages, and natural killer cells) and lymphocytes (T cells and B cells)
Contain and eliminate infection	Effector cells (T cells and B cells) and molecules (complement proteins, enzymes, and antibodies)
Immune regulation	Regulatory lymphocytes
Immunological memory	Memory lymphocytes

The four main functions and cell types involved of the immune system are shown. The primary challenge of immunologists continues to be the development of novel immunotherapies that will optimally stimulate the relevant functions of the immune system in order to eliminate and/or prevent infection.

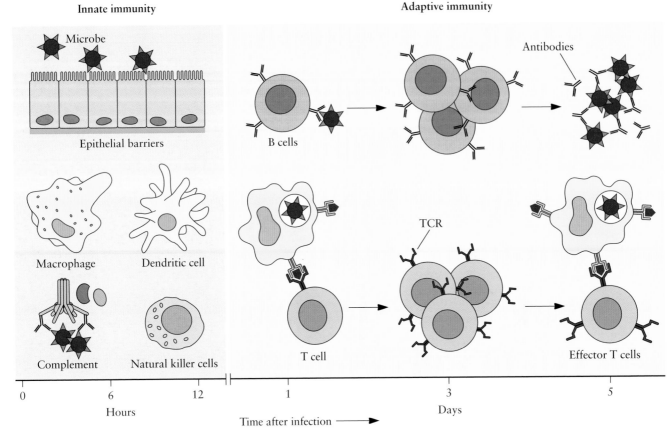

Figure 2.1 Mechanisms of innate and adaptive immunity. The initial defense against infections is mediated by mechanisms of innate immunity that either protect from infection (mucosal epithelial barriers) or eliminate microbes (macrophages, natural killer [NK] cells, and complement system). Adaptive immune responses develop later and are mediated by lymphocytes and their products. While antibodies block infections and eliminate microbes, T cells eradicate intracellular microbes. The kinetics of innate (early after infection) and adaptive (late after infection) immune responses are different and may vary with the type of infection incurred. Adapted with permission from Abbas and Lichtman, *Basic Immunology: Functions and Disorders of the Immune System*, 3rd ed. (Saunders Elsevier, Philadelphia, PA, 2010). doi:10.1128/9781555818890.ch2.f2.1

subsequent encounter with this pathogen and will acquire protective immunity against it. Thus, a very significant challenge facing immunologists today is the identification of relevant cellular pathways and molecules that generate long-lasting immunity (i.e., memory) to pathogens, with the aim of identifying novel **immunotherapies** using cells and/or antibodies that eradicate such pathogens.

Two major types of immunity are carried out by cells of the immune system in response to inflammation and infection (Fig. 2.1). First, **innate immunity** provides a mechanism of host defense that mediates the initial protection against microbial infections. Second, **adaptive immunity**, which develops more slowly and confers specificity against a foreign antigen, mediates the later and more vigorous defense against infections.

Innate Immunity

Invertebrates and vertebrates have developed systems to defend themselves against microbial infections. As these defense mechanisms are always poised to recognize and eliminate microbes, these mechanisms are believed to initiate innate immunity. Several important features of innate immunity are summarized below.

- It elicits responses specifically targeted to microbial pathogens but not nonmicrobial antigens.
- It can be triggered by host cells that are infected and damaged by microbes.
- It is a mechanism of an early defense that can control and eradicate infections before the emergence of adaptive immunity.
- It instructs the adaptive immune system to respond to and combat different microbes.
- It mediates bidirectional cross talk between innate and adaptive immunity.

Thus, increasing emphasis continues to be placed on elucidating the mechanisms of innate immunity and learning how to translate these mechanisms into preventing or eradicating infections in humans.

With this overall objective in mind, three important questions must be addressed.

1. How are microbes recognized by the innate immune system?
2. Which components of innate immunity differentially mediate responses to various microbes?
3. How are adaptive immune responses stimulated by innate immune responses?

Recognition of Microbes by the Innate Immune System

The components of innate immunity recognize structures that are shared by various classes of microbes and are not present on host cells (Fig. 2.2). Each component of innate immunity may recognize many bacteria, viruses, or fungi. For instance, phagocytes express receptors for bacterial lipopolysaccharide, also called endotoxin, which is present in the cell wall of many bacterial species but is not produced by mammalian cells. Other receptors of phagocytes recognize terminal mannose residues, which are typical of bacterial but not mammalian glycoproteins. Phagocytes recognize and respond to double-stranded RNA, which is found in many viruses but not in mammalian cells, and to unmethylated CpG oligonucleotides that are short C- and G-rich stretches of DNA common in microbial DNA but scarce in mammalian DNA. The microbial target molecules of innate immunity are called **pathogen-associated molecular patterns (PAMPs)**, since they have many structural features that are expressed by the same types of microbes. The receptors present on cells that mediate innate immunity and recognize these shared structures are called **pattern recognition receptors**. Some components of innate immunity can bind to

PAMP
pathogen-associated molecular pattern

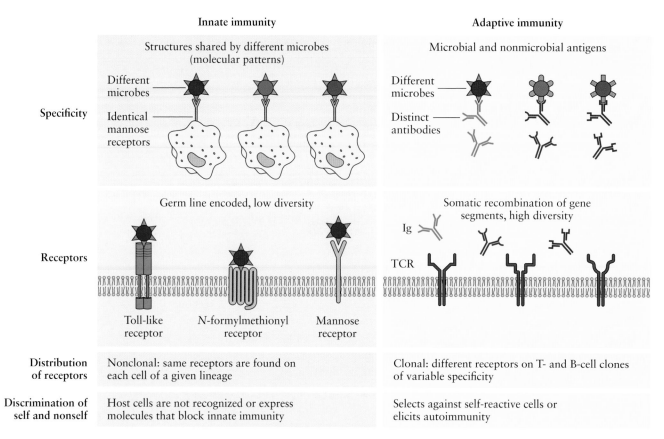

Figure 2.2 Specificity of innate and adaptive immunity. Examples of the specificity and receptors of innate and adaptive immunity are shown. Ig, immunoglobulin (antibody). Adapted with permission from Abbas and Lichtman, *Basic Immunology: Functions and Disorders of the Immune System*, 3rd ed. (Saunders Elsevier, Philadelphia, PA, 2010). doi:10.1128/9781555818890.ch2.f2.2

host cells but are prevented from being activated by these cells. The different types of PAMPs are shown in Fig. 2.3.

Interactions between PAMPs and pattern recognition receptors have evolved to be recognized by components of innate immunity and thereby control the survival and infectivity of these microbes. Thus, innate immunity comprises a highly effective defense mechanism because a microbe cannot evade innate immunity simply by mutating or not expressing the targets of innate immune recognition. Microbes that do not express functional forms of these structures are unable to infect and colonize the host. In contrast, microbes frequently evade adaptive immunity by mutating the antigens that are recognized by lymphocytes, because these antigens are usually not required for the survival of the microbes.

Stressed or necrotic cells release molecules that are recognized by the innate immune system, and these cells are eliminated by the subsequent innate immune response. Such molecules are classified as **damage-associated molecular patterns (DAMPs)**. The receptors of the innate immune system are encoded in the germ line and are not produced by somatic recombination of genes. Somatic recombination is a mechanism used to generate

DAMP
damage-associated molecular pattern

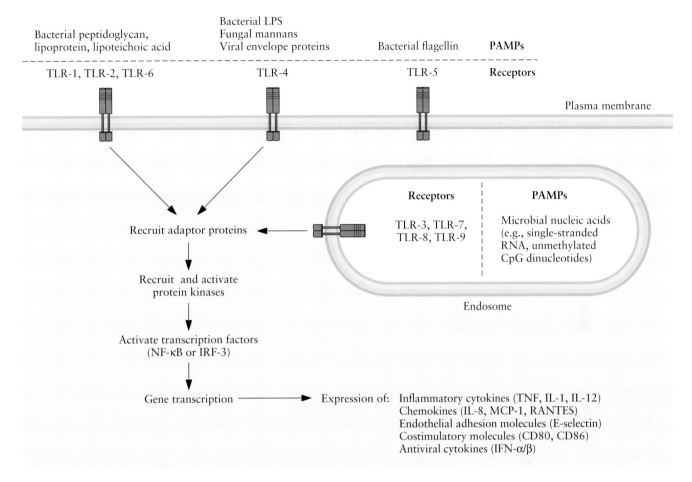

Figure 2.3 Ligands and activation pathways of TLRs. Different microbial antigens engage and stimulate different TLRs. These various TLRs activate similar signaling mechanisms, which elicit innate immune responses. IFN, interferon; IRF-3, interferon response factor 3; LPS, lipopolysaccharide; NF-κB, nuclear factor κB; TNF, tumor necrosis factor. MCP-1 (monocyte chemoattractant protein 1) and RANTES are two chemokines. Adapted with permission from Abbas and Lichtman, *Basic Immunology: Functions and Disorders of the Immune System*, 3rd ed. (Saunders Elsevier, Philadelphia, PA, 2010). doi:10.1128/9781555818890.ch2.f2.3

diversity in antibody production by the rearrangement of DNA segments in B cells during their differentiation, a process that involves the cutting and splicing of immunoglobulin genes (Fig. 2.4; Milestone 2.1). The germ line-encoded pattern recognition receptors for DAMPs have evolved as a protective mechanism against potentially harmful microbes. In contrast, the antigen receptors of lymphocytes, i.e., antibodies on B cells and T-cell antigen receptors on T cells, are produced by random recombination of receptor genes during the maturation of these cells. Gene recombination can generate many more structurally different receptors than can be expressed by inherited germ line genes, but these different germ line-encoded receptors cannot harbor a predetermined specificity for microbes. Therefore, the specificity of adaptive immunity is much more diverse than that of innate immunity, and the adaptive immune system is capable of recognizing many more chemically distinct structures (Fig. 2.1). Whereas the total

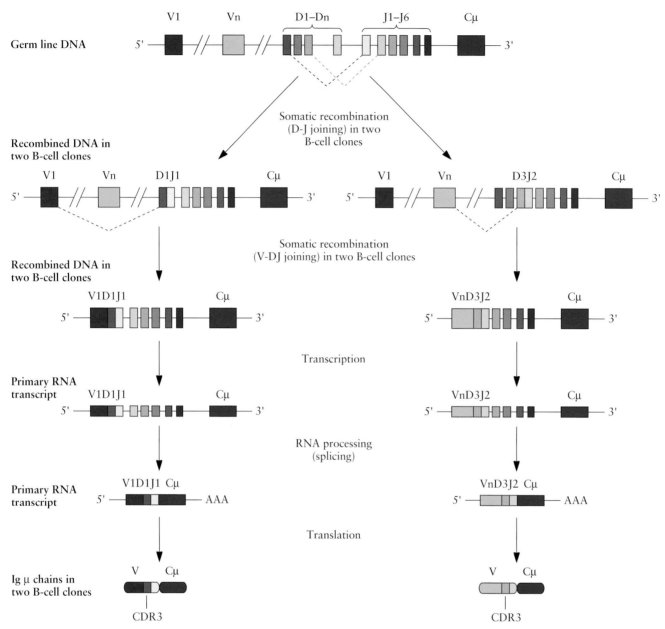

Figure 2.4 Recombination and expression of immunoglobulin (Ig) genes. The expression of an Ig H chain involves two gene recombination events (D-J joining, followed by joining of a V region to the DJ complex, with deletion and loss of intervening gene segments). The recombined gene is transcribed, and the VDJ segment is spliced onto the first H chain RNA (which is μ), giving rise to the μ mRNA. The mRNA is translated to produce the μ H-chain protein. The recombination of other antigen receptor genes, that is, the Ig L chain and the TCR α and β chains, follows essentially the same sequence, except that in loci lacking D segments (Ig L chains and TCR α), a V gene recombines directly with a J gene segment. Adapted with permission from Abbas and Lichtman, *Basic Immunology: Functions and Disorders of the Immune System*, 3rd ed. (Saunders Elsevier, Philadelphia, PA, 2010). doi:10.1128/9781555818890.ch2.f2.4

Evidence for Somatic Rearrangement of Immunoglobulin Genes Coding for Variable and Constant Regions

N. HOZUMI AND S. TONEGAWA
Proc. Natl. Acad. Sci. USA 73:3628–3632, 1976

How are antigen receptors with an infinite range of specificities encoded by a finite number of genes? This question was answered in 1976, when Nobumichi Hozumi and Susumu Tonegawa discovered that the genes for immunoglobulin V regions are inherited as sets of gene segments, each encoding a part of the variable region of one of the immunoglobulin polypeptide chains. During B-cell development in the bone marrow, these gene segments are irreversibly joined by DNA recombination to form a stretch of DNA encoding a complete V region. Since there are many different gene segments in each set, and different gene segments are joined together in different cells, each cell generates unique genes for the V regions of the H and L chains of the immunoglobulin molecule. If these recombination events successfully generate a functional receptor, no additional rearrangement can occur. In this way, each lymphocyte expresses only one immunoglobulin receptor specificity.

Three important consequences emerged from this elegant study. First, it enables a limited number of gene segments to generate a large number of different proteins. Second, as each cell assembles a different set of gene segments, each cell is endowed with the ability to express a receptor specificity that is unique. Third, because each gene rearrangement involves an irreversible change in a cell's DNA, all the progeny of that cell will inherit genes encoding the same receptor specificity. A similar scheme was subsequently shown to be operative for TCRs on T cells.

Thus, Susumu Tonegawa and his coworkers elucidated the mechanism of antibody gene randomization predicted by Burnet in 1957. He showed the somatic rearrangement of genes coding for antibodies and their stabilization in mature B cells. For this work, Susumu Tonegawa received the Nobel Prize in Physiology or Medicine in 1987.

population of lymphocytes is estimated to recognize $>10^9$ different antigens, all the receptors of innate immunity are thought to recognize $<10^3$ microbial PAMPs. Moreover, while the receptors of adaptive immunity (T-cell and B-cell antigen receptors; see below) are clonally distributed, the pattern recognition receptors on cells of innate immunity are nonclonally distributed. Indeed, identical receptors are expressed on all innate immune cells of a given type, e.g., macrophages. Accordingly, many innate immune cells may respond to the same microbe, which may explain why the expression of relatively few ($<10^3$) nonclonal receptors is sufficient to engage a microbe during an innate immune response.

The innate immune system is not permitted to react against self, as it is specific for microbial antigens and mammalian cells express regulatory molecules that prevent innate immune reactions (Fig. 2.2). In the adaptive immune system, lymphocytes evolved to be functionally specific for foreign antigens, while they die after encountering self-antigens. Equivalent responses by innate immune cells are obtained upon primary and subsequent encounters with microbial antigens, whereas the adaptive immune system responds more efficiently to each successive encounter with a microbe during repeated or persistent infections. Thus, immunological memory is a key property of the adaptive, but not innate, immune system.

Inflammation and antiviral defense are the dominant responses governed by the innate immune system. Inflammation depends on the recruitment and activation of leukocytes. Defense against intracellular viruses is mediated primarily by natural killer cells and selected cytokines (interferons).

Upon the first encounter with an infectious agent, the initial defenses of an individual are physical (skin and mucosal epithelia) and chemical (natural antibiotics in our body) barriers that prevent microbes from entering the body. Only after these barriers are overcome does the immune system become activated. The first cells that respond are **macrophages**, which form part of the innate immune system (Fig. 2.1). These cells, termed **phagocytes**, are phagocytic in their action and can ingest and kill microbes by producing an array of toxic chemicals and degradative enzymes. This antimicrobial attack is mounted by macrophages in collaboration with natural killer cells and plasma proteins, including complement proteins. The various components of innate immunity may differ according to the different classes of microbes under attack. A form of innate immunity is found in all plants and animals.

Cell Receptors for Microbes

Receptors found on cells in the innate immune system that react against microbes are expressed on phagocytes, **dendritic cells**, lymphocytes, and epithelial and endothelial cells, which all participate in defense against many microbes (Fig. 2.5). These receptors are expressed in different cellular compartments where microbes may be located. Some are present on the cell surface; others are present in the endoplasmic reticulum and are rapidly recruited to vesicles (endosomes) into which microbial products are engulfed and enzymatically digested, a process known as phagocytosis. This process provides an important defense against infection. Other

Figure 2.5 Cellular localization of innate immune receptors. Different TLRs may be expressed either at the cell surface or in endosomes. Other receptors, such as those for viral RNA or bacterial peptides, are found in the cytoplasm. LPS, lipopolysaccharide. Adapted with permission from Abbas and Lichtman, *Basic Immunology: Functions and Disorders of the Immune System*, 3rd ed. (Saunders Elsevier, Philadelphia, PA, 2010). doi:10.1128/9781555818890.ch2.f2.5

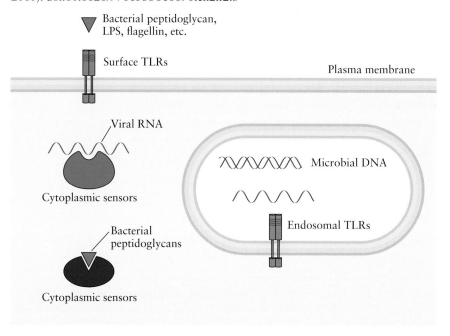

receptors are present in the cytoplasm, where they function as sensors of cytoplasmic microbes. Several classes of these receptors are specific for different types of microbial products or so-called molecular patterns.

Toll-like receptors (TLRs) are homologous to a *Drosophila* protein called Toll, which mediates protection of fruit flies against infections. TLRs are specific for different components of microbes, and they function as pattern recognition receptors for PAMPs (Fig. 2.3). TLR-1, -2, and -6 facilitate responses to several bacterial peptidoglycans and lipoproteins; TLR-3, -7, and -8 facilitate responses to viral nucleic acids (such as double-stranded RNA); TLR-4 facilitates responses to bacterial lipopolysaccharide (endotoxin); TLR-5 facilitates responses to flagellin (a component of bacterial flagella); and TLR-9 facilitates responses to unmethylated CpG oligonucleotides, which are more abundant in bacteria than in mammalian cells. Some of these TLRs are associated with the cell surface and recognize products of extracellular microbes. Other TLRs are in endosomes, into which microbes are ingested. Signals generated by engagement of TLRs activate transcription factors that stimulate expression of genes encoding cytokines, enzymes, and other proteins involved in the antimicrobial functions of activated phagocytes and dendritic cells. Two rather important transcription factors activated by TLR signals are nuclear factor κB, which promotes the expression of various cytokines and endothelial adhesion molecules, and interferon response factor 3, which stimulates the production of type I interferons, cytokines that block viral replication during an antiviral response. Cytokines are a family of proteins that are synthesized and secreted by phagocytic cells and dendritic cells and that function as cell signaling molecules of the immune system to activate and promote intercellular communication (thus the term interleukin) between various subsets of T cells and B cells. Several cytoplasmic receptors recognize viral nucleic acids or bacterial peptides (Fig. 2.5). Other cytoplasmic receptors that participate in innate immune reactions recognize microbes and components of dead cells, including uric acid and DNA. Some of these receptors associate with an **inflammasome**, a multiprotein complex which transmits signals that activate an enzyme that cleaves a precursor of the cytokine interleukin-1 (IL-1) to generate its biologically active form. IL-1 is a powerful inducer of the inflammatory reaction to microbes and damaged tissues. Mutations that modify components of the inflammasome may cause a group of rare human diseases known as auto-inflammatory syndromes. In these diseases, the clinical manifestations are the result of excessive IL-1 production, and IL-1 antagonists are highly effective therapies.

TLRs
Toll-like receptors

IL-1
interleukin-1

Adaptive Immunity

Although innate immunity effectively protects against infections, many microbes that are pathogenic and result in disease in humans have evolved to resist innate immunity. Since an adaptive immune response promotes defense against these pathogens, it is reasonable to think that a defect(s) in this type of response may enhance susceptibility to infections. While innate immune responses provide early defense against infections, they also enhance adaptive immune responses against infectious agents. Thus, the adaptive immune response represents a second line of defense in vertebrates.

Microbes or their antigens can pass through epithelial barriers, be transported to lymphoid tissues, and then be recognized by lymphocytes resident in these tissues. These events can trigger adaptive immune responses that are specialized to combat different types of infections. For instance, antibodies modulate the activity of microbes in extracellular fluids, and activated T cells eliminate microbes located intracellularly. Adaptive immune responses often use components of the innate immune system to eliminate microbes, and conversely, adaptive immunity can provide components that augment antimicrobial innate immune responses. Thus, antibodies can bind to microbes, and the **avidity**, or overall strength of binding, between the antibody-coated microbes and phagocytes is sufficiently high to activate the phagocytes to destroy the microbes. This avidity of an antibody for an antigen is a function of **affinity** (strength of interaction between a single antigen-binding site of an antibody and its specific antigenic determinant or **epitope**) and **valence** (number of antigen-binding sites available on an antibody for binding epitopes on an antigen).

The two types of adaptive immunity, **humoral immunity** and **cell-mediated immunity**, are mediated by different cells and molecules and provide defense against extracellular microbes and intracellular microbes, respectively (Table 2.4). Humoral immunity is mediated by **antibodies**, which are a family of blood-derived glycoproteins known as **immunoglobulins** that are synthesized by **B cells** in response to a specific antigen. Antibodies are secreted into the circulation and mucosal fluids, and they neutralize and eliminate host cell extracellular microbes and microbial toxins in the blood and the lumens of mucosal organs, such as the gastrointestinal and respiratory tracts. An important prophylactic function of antibodies is to block extracellular microbes present at mucosal surfaces and in the blood from localizing to and colonizing host cells and connective tissues, thereby preventing infections from being established. However, the relatively large molecular size (molecular mass $\geq$ 150,000 kilodaltons [kDa]) of antibodies generally prohibits them from being transported across cell plasma membranes to gain access to intracellular microbes in infected host cells. Thus, once a cell is infected with a microbe, antibodies cannot be used to inactivate such microbes localized in infected cells of the host.

Table 2.4 Humoral and cell-mediated adaptive immunity

Immune parameter	Humoral immunity	Cell-mediated immunity	
Microbe	Extracellular	Phagocytosed by macrophages	Intracellular (e.g., viruses)
Lymphocyte responses	B cells	Th cells	CTLs
Effector function	Secreted antibodies block infection and destroy extracellular microbes	Activated macrophages kill intracellular microbes	CTLs kill microbe infected cells and eliminate infection

Adapted with permission from Abbas and Lichtman, *Basic Immunology: Functions and Disorders of the Immune System*, 3rd ed. (Saunders Elsevier, Philadelphia, PA, 2011).

Extracellular microbes are inactivated by antibodies, while intracellular microbes are phagocytosed (engulfed) and destroyed by macrophages. B cells are effector cells in humoral immunity and secrete antibodies that destroy extracellular microbes. Th cells and CTLs are effector cells in cell-mediated immunity and indirectly activate macrophages to kill intracellular microbes or directly kill microbe infected cells and eliminate infection, respectively.

In this instance, defense against such intracellular microbes requires immunity that is mediated by T cells. Some T cells activate phagocytes to destroy microbes that have been ingested by the phagocytes into intracellular vesicles. Other T cells kill any type of host cells that harbor infectious microbes in the cytoplasm. Thus, the message is that the antibodies produced by B cells recognize extracellular microbial antigens, whereas T cells recognize antigens produced by intracellular microbes. Subsets of T cells and B cells can recognize protein, carbohydrate, and lipid antigens. It is this elegant division of labor and synergy between antibodies, B cells, macrophages, and T cells that constitutes a full-blown adaptive immune response.

Immunity may be induced in an individual by infection or vaccination **(active immunity)** or conferred on an individual by passive transfer of antibodies or lymphocytes from an actively immunized individual **(passive immunity)**. Individuals exposed to the antigens of a microbe mount an active response to eradicate infection and develop resistance to subsequent infection by that microbe. These people are actively immune to that microbe, in contrast to those healthy individuals who were not previously exposed to those microbial antigens. As it is desirable to actively immunize a subpopulation of individuals to infectious microbial (e.g., bacterial) antigens such as diphtheria, pneumonia, and tetanus antigens, it is important to identify the mechanisms of active immunity and apply them to treat and maintain the health of individuals.

In passive immunity, a host individual may receive cells capable of mounting an immune response (e.g., lymphocytes) or molecules (e.g., antibodies) from another genetically identical or related individual previously rendered immune to an infection. The recipient can combat the infection for the duration that the donor's transferred antibodies or cells remain functionally active. Accordingly, passive immunity may rapidly transfer immunity to an individual even before that individual is able to mount an active response, but it may not induce long-lived resistance to the infection. An excellent example in nature of the passive transfer of immunity is evident in newborn infants, whose immune systems are not sufficiently mature to respond to various pathogens but who are nonetheless protected against infections by the acquisition of antibodies from their mothers via the placenta and **colostrum**, the first lacteal secretion produced by the mammary gland of a mother prior to the production of milk.

Properties of Adaptive Immune Responses

Clonal selection of lymphocytes (Milestone 2.2) is the single most important principle in adaptive immunity. Its four main postulates are listed in Table 2.5. The last postulate, the mechanism of generation of diversity of lymphocyte antigen receptors, was solved by Susumu Tonegawa and his colleagues in the 1970s when advances in molecular biology made it possible to clone the genes encoding antibody molecules (Milestone 2.1).

Certain properties of adaptive immune responses are essential to resist infection effectively, including specificity, diversity, memory, clonal expansion, specialization, contraction and homeostasis, and nonreactivity to self (Table 2.6).

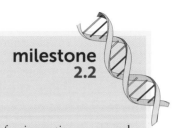

A Modification of Jerne's Theory of Antibody Production Using the Concept of Clonal Selection

milestone 2.2

F. M. BURNET
Aust. J. Sci. **20**:67–69, 1957

A turning point in immunology came in the 1950s with the introduction of a Darwinian view of the cellular basis of specificity in the immune response. The now universally accepted clonal selection theory was proposed and developed by Niels K. Jerne and Sir Macfarlane Burnet (both subsequently awarded the Nobel Prize) and by David Talmage. The postulates of this theory (Table 2.4) proposed a mechanism whereby randomization of the coding for an antibody in each lymphocyte (clone) makes it unique and provides enough diversity to create the millions of different specificities (clones) envisaged by Jerne. Further, this theory explained how elimination of self-reactive clones ("forbidden clones") accounts for self-tolerance. Later exposure to complementary molecular shapes leads to proliferation of specific clones and the secretion of the receptor as a soluble antibody. Once the pathogen is eliminated, the expanded clone contributes to memory cells that ensure that immunity occurs on reinfection.

The clonal selection theory had a truly revolutionary effect on the field of immunology. Although the original paper cited here was written in only 3 days and was published in an obscure scientific journal, it dramatically changed our approach to studying the immune system and affected all research carried out during the last half of the 20th century. Indeed, this work was very informative about the possible molecular mechanisms of activation and regulation of cellular immune responses. Remarkably, the main principles of the clonal selection theory still hold today. Sir Macfarlane Burnet and Peter Medawar collaborated on understanding immunological tolerance, a phenomenon explained by clonal selection. This is the organism's ability to tolerate the transplantation of cells without an immune response as long as this occurs early in the organism's development. Only those cells that are tolerant to one's own (self) tissues survive the embryonic stage (neonatal tolerance). For this work, Sir Macfarlane Burnet and Peter Medawar shared the Nobel Prize in Physiology or Medicine in 1960.

The specificity of immune responses is based on the ability of B and T cells to recognize foreign antigens, respond to them, and, when required, eliminate them. Clonal expansion of these cells is highly efficient, but there is always the rare chance that errors or mutations will occur. Such errors can result in the generation of B and T cells with receptors that bind to self-antigens and therefore display self-reactivity. Under normal conditions, nonfunctioning cells may survive or be aborted with no deleterious consequences to the individual. In contrast, the rare self-reactive cells are clonally deleted or suppressed by other regulatory cells of the immune system. If such a mechanism were absent, autoimmune responses leading to autoimmune diseases might occur routinely. It is noteworthy that during the early stages of development, lymphocytes with receptors that bind to self-antigens are produced, but fortunately, they are eliminated or functionally inactivated. This process gives rise to the initial repertoire of mature lymphocytes that are programmed to generate antigen-specific responses.

Specificity

The adaptive immune system is capable of distinguishing among millions of different antigens or portions of antigens. Specificity for many different antigens implies that the total collection of lymphocyte specificities, termed the **lymphocyte repertoire**, is extremely diverse. The basis of this precise

Table 2.5 Four main postulates of clonal selection theory

No.	Postulate
1	Each lymphocyte bears a single type of receptor with a unique specificity.
2	Interaction between a foreign molecule and a lymphocyte receptor capable of binding that molecule with high affinity leads to lymphocyte activation.
3	Differentiated effector cells derived from an activated lymphocyte express receptors of identical specificity to that of the parental cell from which that lymphocyte was derived.
4	Lymphocytes bearing receptors specific for ubiquitous self-molecules are deleted at an early stage in lymphoid cell development and are therefore absent from the repertoire of mature lymphocytes.

The four main ideas of the clonal selection theory, as originally postulated, are listed.

Table 2.6 Characteristics of adaptive immune responses

Characteristic	Relevance
Specificity	Different antigens stimulate specific responses
Diversity	The immune system responds to a wide array of antigens
Memory	Repeated exposure to the same antigens stimulates enhanced responses
Clonal expansion	Augments number of antigen-specific cells to regulate the microbial load
Specialization	Yields optimal responses against various types of microbes
Contraction and homeostasis	Triggers immune responses to newly encountered antigens
Nonreactivity to self	Minimizes host injury during responses to foreign antigens

Adapted with permission from Abbas and Lichtman, *Basic Immunology: Functions and Disorders of the Immune System*, 3rd ed. (Saunders Elsevier, Philadelphia, PA, 2011).

The different characteristics of adaptive immunity that elicit the destruction of microbes and protection from infection are listed.

specificity and diversity is that lymphocytes express clonally distributed receptors for antigens, meaning that the total population of lymphocytes consists of many different **clones** (each of which is made up of one cell and its progeny), and each clone expresses an antigen receptor that is different from the receptors of all other clones. The **clonal selection hypothesis** formulated by Sir Macfarlane Burnet in the 1950s (Milestone 2.2), correctly predicted that clones of lymphocytes specific for different antigens arise before encounter with these antigens, and each antigen elicits an immune response by selecting and activating the lymphocytes of a specific clone (Fig. 2.6). Adaptive immune responses depend on the specific recognition of pathogens by lymphocytes that use their highly specialized surface antigen receptors to bind and respond to individual antigens (soluble and membrane-bound self and nonself components). For example, T cells express surface **T-cell antigen receptors (TCRs)**, while B cells express **B-cell antigen receptors (BCRs)** in the form of surface immunoglobulins. These surface immunoglobulins as well as B-cell-secreted soluble immunoglobulins of the same antigen specificity function as antibodies during infectious responses. Both membrane-bound TCRs and BCRs present on the surface of $>10^9$ lymphocytes in the body comprise a very large repertoire of antigen receptors, which enables the immune system to recognize and respond to a vast array of antigen specificities borne by virtually any pathogen (bacterial, viral, or other microbial) a person may be exposed to during a lifetime. By the specific recognition and response to a particular antigen, the adaptive immune response enables the body to inactivate and/or eliminate pathogens that have escaped surveillance or removal by an innate immune response.

The diversity of lymphocytes means that very few cells, perhaps 1 in 100,000 lymphocytes, are specific for any one antigen. To defend against microbes effectively, these few cells must proliferate to generate a large number of cells capable of combating the microbes. The high effectiveness of immune responses is attributable to three features of adaptive immunity: (i) marked expansion of the pool of lymphocytes specific for any antigen after exposure to that antigen, (ii) positive-feedback loops that amplify immune responses, and (iii) selection mechanisms that preserve the most desirable lymphocytes.

TCR
T-cell antigen receptor

BCR
B-cell antigen receptor

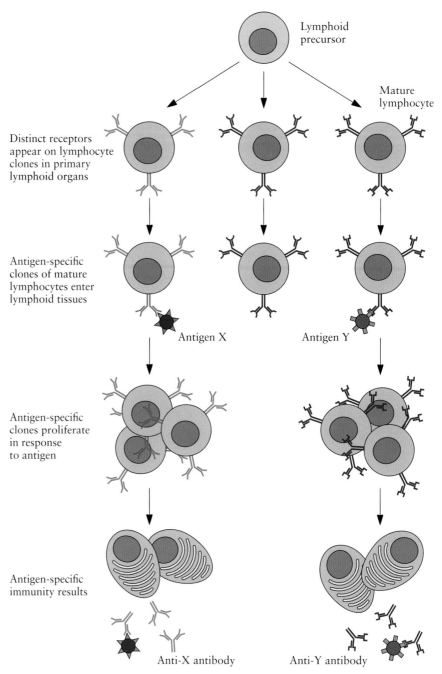

Figure 2.6 Clonal selection. Mature lymphocytes express receptors for many antigens and develop before these receptors encounter their specific antigens. Clones are populations of cells that are derived from a single precursor cell, and thus cells in this clone express identical receptors and specificities. Examples of B-cell clonal selection are shown, but the same principles apply for T-cell clonal selection. Clonal selection also applies for both soluble and surface-associated microbial antigens. Adapted with permission from Abbas and Lichtman, *Basic Immunology: Functions and Disorders of the Immune System*, 3rd ed. (Saunders Elsevier, Philadelphia, PA, 2010). doi:10.1128/9781555818890.ch2.f2.6

Memory

Antibodies and lymphocytes activated against pathogens during an adaptive immune response generally persist after the pathogenic infection subsides and then synergize to prevent reinfection. Previously activated antigen-specific T and B cells provide immunological memory to a pathogen such that a more rapid and vigorous secondary response is elicited upon reexposure to this pathogen even if it occurs several years later (Fig. 2.7). The immune system mounts larger and more effective responses to repeated exposures to the same antigen. A **primary immune response** elicited upon the first exposure to antigen is mediated by **naïve lymphocytes**, since these cells are not experienced immunologically and have not previously responded to antigens (Fig. 2.7). Subsequent stimulation by the same antigen leads to a **secondary immune response**, which is usually more rapid, larger, and better able to eliminate the antigen than a primary response. Secondary responses result from the activation of **memory lymphocytes**, which are long-lived cells that are induced during the primary immune response. Immunological memory conditions the immune system to combat persistent and recurrent infections, because each encounter with a microbe generates more memory cells and activates previously generated memory cells. Memory also explains why vaccines confer long-lasting protection against infections.

Expansion, Specialization, Homeostasis, and Nonself Reactivity

Following antigen-induced activation, lymphocytes undergo proliferation and give rise to many thousands of clonal progeny cells, all with the same

Figure 2.7 Immunological memory. Different antibodies with different specificities are produced in response to antigens X and Y. Secondary responses to antigen X are more rapid and of greater amplitude than the primary response, indicating that immunological memory was achieved. After each immunization, circulating antibody levels are reduced with time. Adapted with permission from Abbas and Lichtman, *Basic Immunology: Functions and Disorders of the Immune System*, 3rd ed. (Saunders Elsevier, Philadelphia, PA, 2010). doi:10.1128/9781555818890.ch2.f2.7

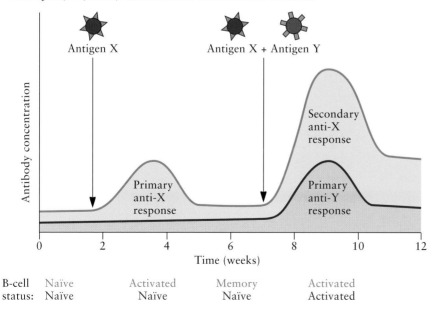

antigen specificity. This process, called **clonal expansion**, ensures that adaptive immunity can balance and keep the rate of proliferation of microbes in check; otherwise, infection will occur. Immune responses are specialized, and different responses are designed to best defend against different classes of microbes. All immune responses decline as the infection is eliminated, allowing the system to return to a resting state known as **homeostasis**, prepared to respond to another infection. Generally, the above-described properties of the immune system enable it to react against a vast number and variety of microbes and other foreign antigens without reacting against the host's own antigens, i.e., self-antigens. If **self-reactivity** does arise, built-in control mechanisms mediated by **regulatory cells** suppress the possible exacerbation of antiself responses and resultant disease.

Knowledge of how the components of the adaptive immune system are induced and function to protect the body from an infection is critical to the design, route of administration, and potential efficacy and success of vaccines. Fortunately, we now have a more extensive understanding of how an antibody-mediated adaptive immune response can lead to either a desirable (effective destruction of pathogen) or an undesirable (autoimmune disease) outcome. This knowledge provides a basis to produce vaccines against a wide array of diseases with high morbidity and mortality, including AIDS, gonorrhea, chlamydial disease, tuberculosis, cholera, and malaria.

Cells of the Immune System: Cell-Mediated Immunity

Prior to discussing the functions of the immune system, it is important to first identify the various cell types of the immune system, where and how they are generated, and how they differentiate and become functionally mature. The immune system is comprised of various cell types of white blood cells, including leukocytes and lymphocytes. Together, these cells mediate the capture and display of microbial antigens and the activation of effector cells that eliminate microbes. The patterns of differentiation and functional maturation of these leukocytes and lymphocytes in the immune system are summarized below.

Bone Marrow Precursor Cells

All immune responses are mediated by different types of white blood cells called leukocytes, which originate from precursor cells or **stem cells** in the **bone marrow**, where some of them develop and mature. Certain of these cell types migrate to other lymphoid and nonlymphoid peripheral tissues, where they may reside or otherwise circulate in the periphery via the bloodstream or **lymphatic system**, a specialized system of blood vessels. The latter system drains extracellular fluid and free cells from tissues, transports them through the body as **lymph**, and ultimately recirculates them back to the blood system.

The cells in blood, including red blood cells (which transport oxygen), platelets (which trigger blood clotting in damaged tissues), and leukocytes (which mediate immune responses), are derived from **hematopoietic stem cells** in the bone marrow. As these stem cells can develop into different types of blood cells, they are known as pluripotent hematopoietic stem cells. In turn, hematopoietic stem cells can give rise to stem cells of more limited developmental potential, such as progenitors of red blood cells, platelets, and the **myeloid** and lymphoid lineages of leukocytes. The different types of blood cells and their lineage relationships are presented in Fig. 2.8.

Figure 2.8 Differentiation of bone marrow-derived cells of the immune system. Pluripotent stem cells in the bone marrow develop into different cell types of the immune system. A common lymphoid progenitor differentiates into the lymphoid lineage (T, B, and natural killer cells [blue]). After stimulation by an antigen, T cells differentiate into various subsets of effector T cells, whereas B cells differentiate into antibody-secreting plasma cells. Unlike T and B cells, natural killer cells lack antigen specificity. A common myeloid progenitor develops into the myeloid lineage that includes the megakaryocytes (red) and leukocytes, i.e., monocytes, macrophages, dendritic cells, mast cells, neutrophils, eosinophils, and basophils (yellow). The last three circulate in the blood and are termed granulocytes, since they contain cytoplasmic granules. Immature dendritic cells are phagocytic migratory cells that enter tissues and mature functionally after encounter with a pathogen. Monocytes and mast cells also mature in tissues after antigen stimulation. Adapted from Wilson et al., *Bacterial Pathogenesis: A Molecular Approach*, 3rd ed. (ASM Press, Washington, DC, 2011). doi:10.1128/9781555818890.ch2.f2.8

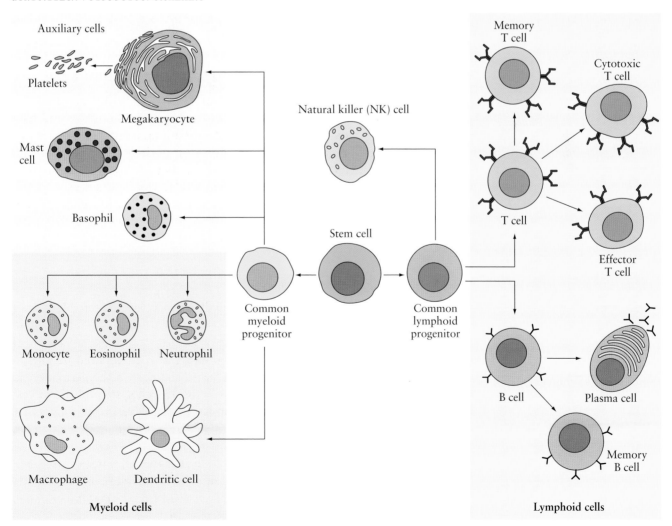

Myeloid Cells Mediate Innate Immunity

The common myeloid progenitor is the precursor of the macrophages, granulocytes, mast cells, and dendritic cells of the innate immune system, and also of megakaryocytes and red blood cells, which are not discussed here. The cells of the myeloid lineage are shown in Fig. 2.9.

Monocytes mature into macrophages, which circulate in the blood and continually migrate into almost all tissues, where they differentiate and reside. While monocytes and macrophages represent one type of phagocytes, granulocytes (neutrophils, eosinophils, and basophils) and dendritic cells comprise the other two types of phagocytes. Macrophages are relatively long-lived cells that can multitask and function during both innate and adaptive immunity; i.e., they ingest and kill bacteria as a first line of defense during innate immunity, and interestingly, during adaptive immunity, they eliminate pathogens and target cells in infected tissues. A crucial role of macrophages is to induce inflammation and trigger immune responses by the secretion of various cytokines and chemokines that activate other immune system cells and recruit them into an immune

Figure 2.9 Myeloid cells in innate and adaptive immunity. The cells of the myeloid lineage and their functions during innate and adaptive immune responses are shown. Adapted from Murphy et al., *Janeway's Immunobiology*, 7th ed. (Garland Science, New York, NY, 2008). doi:10.1128/9781555818890.ch2.f2.9

Cell		Activated function
Macrophage	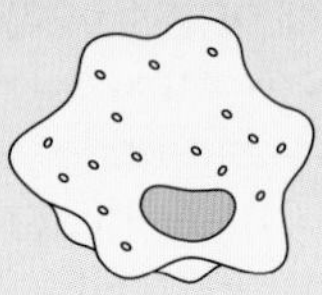	Phagocytosis and activation of bacteriocidal mechanisms Antigen presentation
Dendritic cell	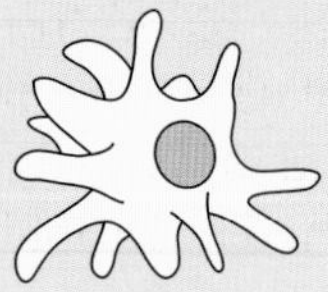	Antigen uptake in peripheral sites Antigen presentation in lymph nodes
Neutrophil	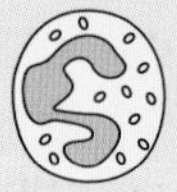	Phagocytosis and activation of bacteriocidal mechanisms
Eosinophil	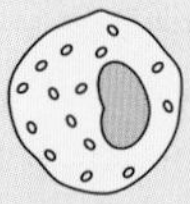	Killing of antibody-coated parasites
Basophil	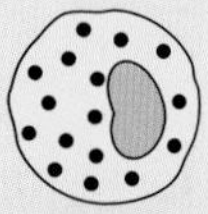	Function and granule content similar to those of mast cells
Mast cell	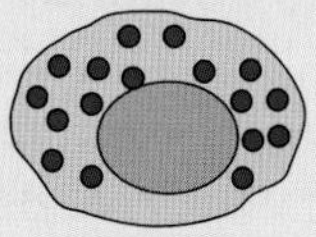	Release of granules containing histamine and other active agents

response, respectively. Finally, macrophages fulfill a more general role by functioning as scavenger cells that clear dead cells and cell debris from the body.

Granulocytes are characterized by their densely staining granules in their cytoplasm, and they are also called **polymorphonuclear leukocytes** because they have irregularly shaped nuclei. The three types of granulocytes—neutrophils, eosinophils, and basophils—are distinguished by their different patterns of granule staining. These granulocytes are shorter lived than macrophages and survive for only a few days. They are increased in number during immune responses, when they exit the blood and circulate to sites of infection and inflammation. Neutrophils are the most numerous and active type of granulocyte during an antimicrobial adaptive immune response. An inheritable deficiency in neutrophil function may result in an overwhelming bacterial function, which, if left untreated, can become fatal.

Upon activation, eosinophils and basophils secrete granules that contain a host of enzymes and toxic proteins, which elicit damage rather than protection of target tissues and cells, particularly during an allergic inflammatory response. The protective roles of eosinophils and basophils in innate immunity are currently less well understood than the roles of other innate immune cell types and require further study. Mast cells also have large granules in their cytoplasm that are released upon activation, and the net effect is an induced inflammatory response. Although they mediate allergic inflammatory responses, like eosinophils and basophils, mast cells also protect the internal epithelial surfaces of the body against pathogens and are involved in responses to parasitic worms.

A third type of phagocytic cell of the immune system possesses long protruding finger-like processes similar to the dendrites of nerve cells; hence, cells of this type are termed dendritic cells. Immature dendritic cells are migratory cells and travel via the blood from the bone marrow into tissues. They both take up particulate matter by phagocytosis and ingest large amounts of extracellular fluid and its contents by the process of macropinocytosis. Like macrophages and neutrophils, they degrade and clear their ingested pathogens. Nevertheless, the main functional role of dendritic cells is not to clear the body of microbes but, rather, to encounter and enzymatically degrade a pathogen, mature, and then present fragments of this pathogen to a T cell. This enables the subsequent activation of T cells that mediate adaptive immune responses. Importantly, then, the primary role of dendritic cells is to function as antigen-presenting cells in the activation and regulation of T-cell-mediated immune responses. Dendritic cells are able to display pathogen-derived antigens on their surface in a manner that facilitates recognition by specific TCRs and stimulation of an adaptive immune response. Thus, dendritic cells may be thought of as antigen-presenting cells that bridge an innate immune response with an adaptive immune response. Accordingly, dendritic cells have immediate application in the regulation of desirable and undesirable immune responses to be considered in the design of clinical trials for immune therapy.

The principal cells of the immune system are lymphocytes, antigen-presenting cells, and effector cells (Fig. 2.10).

Cell type	Major functions
Lymphocytes: B cells, T cells, natural killer cells 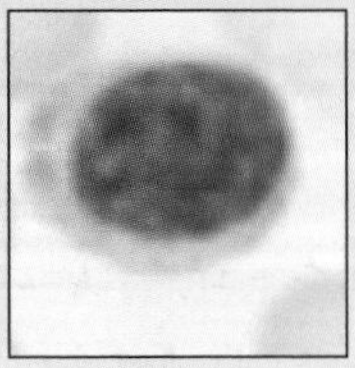	Specific antigen recognition B cells: mediate humoral immunity T cells: mediate cell-mediated immunity NK cells: mediate innate and adaptive immunity
Antigen-presenting cells: Dendritic cells 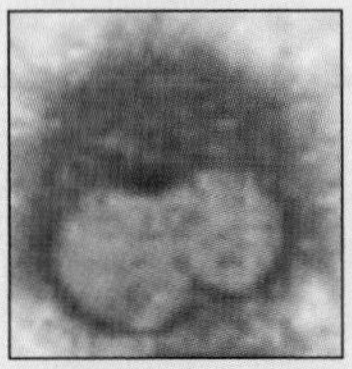 / Macrophages, follicular dendritic cells 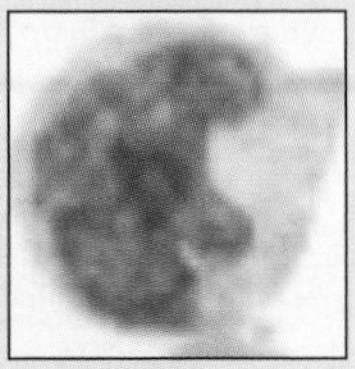	Capture antigens for presentation to T, B, and natural killer cells Dendritic cells: initiation of T-cell responses Macrophages: initiation and effector phases of cell-mediated immunity Follicular dendritic cells: present antigens to B cells in humoral immunity
Effector cells: T cells, macrophages, granulocytes 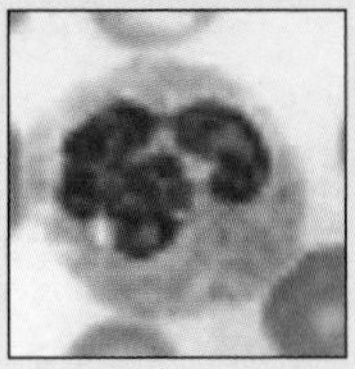	Antigen elimination Th cells and CTLs Macrophages and monocytes Neutrophils and eosinophils

Figure 2.10 Cells of the immune system. The major cell types of the immune system and their functions during innate and adaptive (humoral and cell-mediated) immunity are presented. The morphology of each cell type is shown in the micrographs (left panels). Photos reprinted with permission from Abbas and Lichtman, *Basic Immunology: Functions and Disorders of the Immune System*, 3rd ed. (Saunders Elsevier, Philadelphia, PA, 2010). doi:10.1128/9781555818890.ch2.f2.10

Lymphocytes

The antigen-specific TCRs and BCRs, expressed on T cells and B cells, respectively, confer specificity to adaptive immune responses. Most lymphocytes are heterogeneous in lineage, function, and phenotype and can also mediate many biological responses and activities (Fig. 2.11). Subsets of lymphocytes may be distinguished by their different cell surface antigen phenotypes, as defined by expression of characteristic patterns of reactivity with panels of **monoclonal antibodies (MAbs)** (see below). The nomenclature for these proteins is the CD (cluster of differentiation) numerical designation, which is used to delineate surface proteins that define a particular cell type or stage of cell differentiation and are recognized by a cluster or group of antibodies. A complete list of CD molecules mentioned in this book may be viewed online at http://www.hcdm.org (Human Cell Differentiation Molecules workshop).

MAb
monoclonal antibody

CD
cluster of differentiation

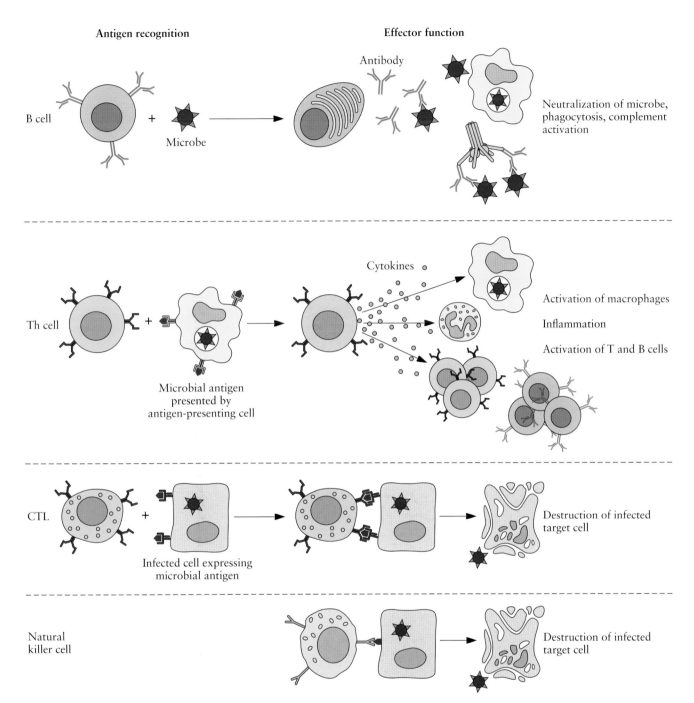

Antigen recognition

Effector function

Figure 2.11 Classes of lymphocytes. Different classes of lymphocytes recognize different types of microbial antigens and differentiate into effector cells that eliminate the antigens. B cells recognize soluble or cell surface antigens and differentiate into antibody-secreting plasma cells. Th cells recognize antigens presented by MHC molecules on the surface of antigen-presenting cells and secrete cytokines that stimulate immune and inflammatory responses. CTLs recognize antigens on infected cells and kill these cells. Natural killer cells recognize changes on the surface of infected cells and kill these cells. Adapted with permission from Abbas and Lichtman, *Basic Immunology: Functions and Disorders of the Immune System*, 3rd ed. (Saunders Elsevier, Philadelphia, PA, 2010). doi:10.1128/9781555818890.ch2.f2.11

B cells produce antibodies and mediate humoral immunity. In addition, B cells express specific BCRs, which are membrane-bound forms of antibodies that function as receptors that bind to soluble antigens and antigens on the surface of microbes and other cells. Upon engagement of antigens by relevant BCRs, B cells are triggered to become activated for antibody synthesis and secretion and in this manner can elicit humoral immune responses against the specific antigens under analysis. T cells help and collaborate with B cells during cell-mediated immune responses. Unlike BCRs on B cells that may bind to antigenic determinants (epitopes) of intact whole proteins, TCRs on T cells recognize only peptide fragments of such protein antigens that are bound to specialized peptide display molecules called **major histocompatibility complex (MHC)** molecules. Among T-cell subsets, CD4$^+$ T cells function as **T helper cells (Th cells)** because they help B cells to produce antibodies and help phagocytes to destroy ingested microbes. In addition, a subset of CD4$^+$ T cells that can prevent or limit immune responses are called **T regulatory cells**. CD8$^+$ T cells are called **cytotoxic (or cytolytic) T lymphocytes (CTLs)**, because they destroy (i.e., lyse) cells harboring intracellular microbes. A third class of cells called natural killer cells also kill microbe-infected host cells, but they do not express the kinds of clonally distributed antigen receptors that B cells and T cells do and are components of innate immunity, capable of rapidly attacking infected cells. Nonetheless, recent studies indicate that natural killer cells can persist for as long as 2 months after primary exposure to antigen. Secondary exposure to antigen results in higher levels of secretion of the proinflammatory cytokine γ-interferon, and memory-like natural killer cells have been described in response to murine cytomegalovirus, herpesvirus, **human immunodeficiency virus (HIV)** type 1, poxvirus, and influenza virus. Thus, the ability of natural killer cells to express suggests that they can mediate both innate immunity and adaptive immunity.

When naïve lymphocytes recognize microbial antigens and also receive additional signals induced by microbes, the antigen-specific lymphocytes proliferate and differentiate into **effector cells** and **memory cells** (Fig. 2.12). Naïve lymphocytes express antigen receptors but do not perform the functions required to eliminate antigens. These cells reside in and circulate between peripheral lymphoid organs and survive for several weeks or months, waiting to encounter and respond to an antigen. If they are not activated by an antigen, naïve lymphocytes die by a process known as apoptosis and are replaced by new cells that have developed in the generative lymphoid organs. This balanced cycle of cell loss and replacement maintains a stable number of lymphocytes, a phenomenon called **immune homeostasis**. The differentiation of naïve lymphocytes into effector cells and memory cells is initiated by antigen recognition, thus ensuring that the immune response that develops is specific for the antigen. Effector cells are the differentiated progeny of naïve cells that are able to be activated and produce molecules capable of eliminating antigens. Such effector cells in the B-cell lineage are antibody-secreting **plasma cells**. Effector CD4$^+$ Th cells produce proteins, called cytokines, which activate B cells and macrophages, thereby mediating the helper function of this lineage. Further, effector CD8$^+$ T CTLs can kill infected host cells. Most effector lymphocytes are short lived and die as the antigen becomes

MHC
major histocompatibility complex

CTL
cytotoxic (or cytolytic) T lymphocyte

HIV
human immunodeficiency virus

A

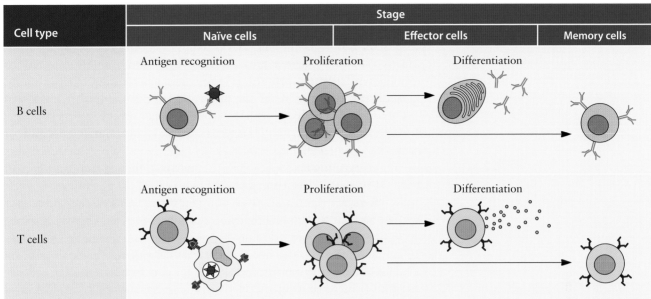

Cell type	Stage		
	Naïve cells	**Effector cells**	**Memory cells**
B cells	Antigen recognition Proliferation Differentiation		
T cells	Antigen recognition Proliferation Differentiation		

B

Property	Stage		
	Naïve cells	**Effector cells**	**Memory cells**
Antigen receptor	Yes	B cells: reduced T cells: yes	Yes
Lifespan	Weeks or months	Usually short (days)	Long (years)
Effector function	None	Yes B cells: antibody secretion Helper T cells: cytokine secretion CTLs: cell killing	None
Special characteristics B cells Affinity of Ig	Low	Variable	High (affinity maturation)
Isotype of Ig	Membrane-associated IgM, IgD	Membrane-associated and secreted IgM, IgD, IgG, IgA, IgE (class switching)	Various
T cells Migration	To lymph nodes	To peripheral tissue (sites of infection)	To lymph nodes and mucosal and other tissues

Figure 2.12 Functions of lymphocytes at different stages of their life cycle. (**A**) Naïve B cells may recognize and be stimulated to proliferate by a foreign antigen (microbe) and then differentiate into effector antibody-secreting plasma cells, some of which are long lived. Effector CD4$^+$ (or CD8$^+$ [not shown]) T cells may recognize a foreign antigen (microbe), proliferate, and then differentiate into cytokine-secreting cells. Some of the antigen-activated B and T cells become long-lived memory cells. (**B**) Some salient features of naïve, effector, and memory B and T cells are listed. Ig, immunoglobulin. Adapted with permission from Abbas and Lichtman, *Basic Immunology: Functions and Disorders of the Immune System*, 3rd ed. (Saunders Elsevier, Philadelphia, PA, 2010). doi:10.1128/9781555818890.ch2.f2.12

removed, but some effector cells may migrate to particular anatomic sites and live for long periods. This prolonged survival of effector cells is best documented for antibody-producing plasma cells, which develop in response to microbes in the peripheral lymphoid organs but may then migrate to the bone marrow and continue to produce small amounts of antibody long after the infection is eradicated. Interestingly, memory cells, which also are generated from the progeny of antigen-stimulated lymphocytes, do survive for long periods in the absence of antigen. Therefore, the frequency of memory cells increases with age, presumably because of exposure to environmental microbes. In fact, memory cells make up less than 5% of peripheral blood T cells in a newborn but 50% or more in an adult. Memory cells are functionally inactive and do not function as effector cells unless stimulated by an antigen. When memory cells encounter the same antigen as induced their development, the cells rapidly respond to give rise to secondary immune responses. Currently, we have relatively little knowledge about the signals that generate memory cells, the factors that determine whether the progeny of antigen-stimulated cells will develop into effector or memory cells, or the mechanisms that keep memory cells alive in the absence of antigen or innate immunity.

Antigen-Presenting Cells

The common border crossings for microbes, i.e., the skin, gastrointestinal tract, and respiratory tract, contain specialized antigen-presenting cells located in the epithelium that capture antigens, transport them to peripheral lymphoid tissues, where immune responses are initiated, and display them to T and B cells. This function of antigen capture and presentation is best understood for dendritic cells, because of their long processes. Dendritic cells bind protein antigens of microbes that enter through the epithelia and transport the antigens to regional lymph nodes. Here, the antigen-bearing dendritic cells display fragments of the antigens for recognition by T cells. Following invasion of the epithelium by a microbe, it may be phagocytosed by macrophages that reside in tissues and in various organs. Macrophages can also present protein antigens to T cells. Note that cells that present antigens to T cells can respond to microbes by producing surface and secreted proteins that are required, together with the antigen, to activate naïve T cells to proliferate and differentiate into effector cells. Cells that display antigens to T cells and provide additional activating signals are sometimes called **professional antigen-presenting cells**. The prototypical professional antigen-presenting cells that have received considerable attention are dendritic cells, but macrophages and other cell types (e.g., B cells) may function similarly.

Less is known about cells that capture antigens for display to B cells, which may directly recognize the antigens of microbes either released or on the microbial surface. Alternatively, macrophages lining lymphatic channels may capture and display antigens to B cells. Follicular dendritic cells reside in the germinal centers of lymphoid follicles in peripheral lymphoid organs, and they display antigens that stimulate the differentiation of B cells in the follicles. Follicular dendritic cells do not present antigens to T cells and differ from the dendritic cells that function as antigen-presenting cells for T cells.

Effector Cells

Lymphocytes (T and B cells) and other nonlymphoid leukocytes (granulocytes and macrophages) that eliminate microbes are termed effector cells. The latter types of leukocytes may function as effector cells in both innate immunity and adaptive immunity (Fig. 2.12). In innate immunity, macrophages and some granulocytes directly recognize microbes and eliminate them. In adaptive immunity, the secreted products (cytokines and chemokines) of T and B cells recruit other leukocytes and activate them to destroy microbes.

Tissues of the Immune System

The tissues of the immune system consist of two types of organs: **generative (or primary) lymphoid organs**, in which T and B cells mature and become competent to respond to antigens, and **peripheral (or secondary) lymphoid organs**, in which adaptive immune responses to microbes are initiated (Fig. 2.13). Properties of peripheral lymphoid organs are featured here, as these are the important organs in which adaptive immune responses develop.

Peripheral Lymphoid Organs

The peripheral lymphoid organs, which consist of the lymph nodes, spleen, and the mucosal and cutaneous immune systems, are organized to optimize interactions between antigens, antigen-presenting cells, and lymphocytes

Figure 2.13 Lymphocyte maturation. The development of lymphocytes from precursors occurs in the primary lymphoid organs (bone marrow and thymus). Subsequently, mature lymphocytes migrate to the peripheral lymphoid organs (lymph nodes, spleen, and mucosa), where they respond to foreign antigens and recirculate to the blood and lymph. Adapted with permission from Abbas and Lichtman, *Basic Immunology: Functions and Disorders of the Immune System*, 3rd ed. (Saunders Elsevier, Philadelphia, PA, 2010). doi:10.1128/9781555818890.ch2.f2.13

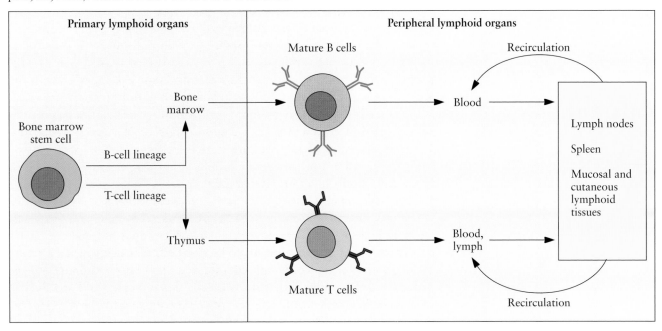

and thereby elicit adaptive immune responses. The immune system searches for microbes that enter at many sites distributed throughout the body and then responds to these microbes and eliminates them. This task may be more difficult than it seems, as only a rather low frequency (1 in 100,000) of T and B cells in the immune system are specific for any one antigen. The anatomic organization of peripheral lymphoid organs enables antigen-presenting cells to concentrate antigens in these organs and lymphocytes to locate and respond to the antigens. This localization of lymphocytes depends on their ability to circulate throughout the body; naïve lymphocytes preferentially go to the organs that serve as depots for antigens, and effector cells go to sites of infection, from which microbes have to be eliminated. Furthermore, different types of lymphocytes often need to communicate to generate effective immune responses. For example, Th cells, specific for a particular antigen, collaborate with and help B cells specific for the same antigen, resulting in antibody production. Thus, an important function of lymphoid organs is to provide a milieu in which these rare antigen-specific cells encounter each other and lead to productive interactions and responses.

Lymph nodes are nodular aggregates of lymphoid tissues located along lymphatic channels throughout the body (Fig. 2.14). A fluid known

Figure 2.14 Schematic of a lymph node. The structural organization and blood flow in a lymph node are shown. Adapted with permission from Abbas and Lichtman, *Basic Immunology: Functions and Disorders of the Immune System*, 3rd ed. (Saunders Elsevier, Philadelphia, PA, 2010).
doi:10.1128/9781555818890.ch2.f2.14

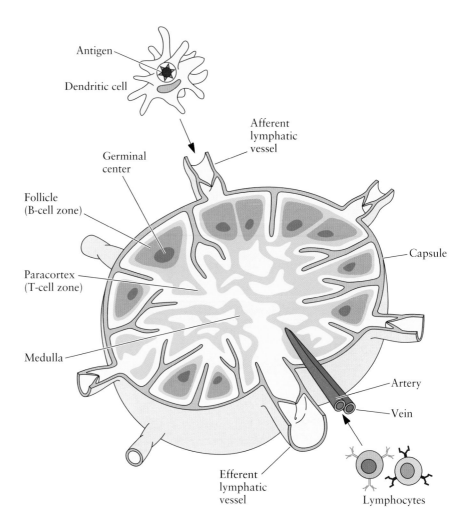

as lymph is drained by lymphatic vessels from all epithelia and connective tissues and most parenchymal organs, and these organs transport this fluid from the tissues to the lymph nodes. As the lymph passes through lymph nodes, antigen-presenting cells in the nodes sample the antigens of microbes that may enter through epithelia into tissues. In addition, dendritic cells pick up antigens of microbes from epithelia and transport these antigens to the lymph nodes. Collectively, these processes of antigen capture and transport enable the antigens of microbes that enter through epithelia or colonize tissues to be concentrated in draining lymph nodes.

The **spleen** (Fig. 2.15) is an abdominal organ that plays the same role in immune responses to blood-borne antigens as that of lymph nodes in responses to lymph-borne antigens. Blood entering the spleen flows through a network of channels (sinusoids), and blood-borne antigens are trapped and concentrated by splenic dendritic cells and macrophages. The spleen contains abundant phagocytes, which ingest and destroy microbes in the blood.

The **cutaneous and mucosal lymphoid systems** are located under the skin epithelia and the gastrointestinal and respiratory tracts, respectively. Pharyngeal tonsils and Peyer's patches of the intestine are mucosal lymphoid tissues. More than half of the body's lymphocytes are in the mucosal tissues, reflecting their large size, and many of these are memory cells. Cutaneous and mucosal lymphoid tissues are sites of immune responses to antigens that breach epithelia.

Within peripheral lymphoid organs, T cells and B cells are segregated into different anatomic compartments (Fig. 2.16). In lymph nodes, B cells localize to discrete **follicles** located around the periphery, or cortex, of each node. If the B cells in a follicle have recently responded to an antigen,

Figure 2.15 Schematic of a spleen. A splenic arteriole surrounded by the periarteriolar lymphoid sheath, which represents the T-cell zone, and an attached follicle containing a germinal center, representing the B-cell zone, are shown. Adapted with permission from Abbas and Lichtman, *Basic Immunology: Functions and Disorders of the Immune System*, 3rd ed. (Saunders Elsevier, Philadelphia, PA, 2010). doi:10.1128/9781555818890.ch2.f2.15

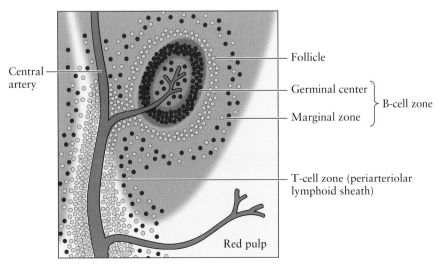

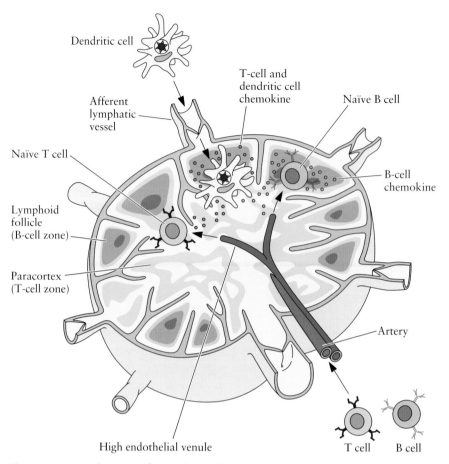

Figure 2.16 Localization of T and B cells in different regions of peripheral lymphoid tissues. T and B cells migrate to different areas of a lymph node. Lymphocytes enter through a high endothelial venule and are recruited to different areas of the node by chemokines produced in these areas, and they bind selectively to either cell type. Dendritic cells, which pick up antigens from epithelia, enter through afferent lymphatic vessels and migrate to the T-cell-rich areas of the node. Adapted with permission from Abbas and Lichtman, *Basic Immunology: Functions and Disorders of the Immune System*, 3rd ed. (Saunders Elsevier, Philadelphia, PA, 2010). doi:10.1128/9781555818890.ch2.f2.16

this follicle may contain a central region called a **germinal center**. Germinal centers contribute significantly to the production of antibodies. T cells localize to areas outside but adjacent to the follicles in the **paracortex**. The follicles contain the follicular dendritic cells that mediate B-cell activation, and the paracortex contains the dendritic cells that present antigens to T cells. In the spleen, T cells are found in periarteriolar lymphoid sheaths surrounding small arterioles, and B cells reside in the follicles.

The anatomic organization of peripheral lymphoid organs is tightly regulated to allow immune responses to develop. B cells are located in the follicles because follicular dendritic cells secrete proteins called chemokines or chemoattractant cytokines, and naïve B cells express receptors for several chemokines. These chemokines are produced constitutively, and they attract B cells from the blood into the follicles of lymphoid organs. Similarly, T cells are segregated in the paracortex of lymph nodes and the periarteriolar lymphoid sheaths of the spleen, because naïve T cells express a receptor, called CCR7, that recognizes chemokines produced in these regions of the lymph nodes and spleen. Consequently, cells are recruited from the blood into the parafollicular cortex region of the lymph nodes and the periarteriolar lymphoid sheaths of the spleen.

Upon activation by microbial antigens, lymphocytes alter the surface expression of their chemokine receptors. As a result, the B cells and T cells migrate toward each other and meet at the edge of follicles, where Th cells interact with and help B cells to differentiate into antibody-producing cells. The activated lymphocytes ultimately exit the node through efferent lymphatic vessels and leave the spleen through veins. These activated lymphocytes end up in the blood circulation and can migrate to distant sites of infection.

Lymphocyte Recirculation and Migration into Tissues

Naïve lymphocytes are in constant recirculation between the blood and peripheral lymphoid organs in which they are activated by antigens to become effector cells (Fig. 2.17). The effector lymphocytes migrate to sites

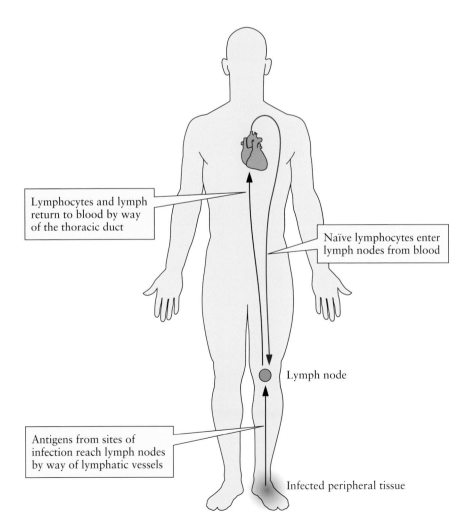

Lymphocytes and lymph return to blood by way of the thoracic duct

Naïve lymphocytes enter lymph nodes from blood

Lymph node

Antigens from sites of infection reach lymph nodes by way of lymphatic vessels

Infected peripheral tissue

Figure 2.17 Circulating lymphocytes are activated by antigens in peripheral lymphoid tissues. Naïve lymphocytes recirculate constantly through peripheral lymphoid tissues, e.g., a popliteal lymph node located behind the knee. During a foot infection, lymphocytes are activated by their antigens located in the draining lymph nodes. Activated and nonactivated lymphocytes recirculate to blood via lymphatic vessels. Adapted with permission from Murphy et al., *Janeway's Immunobiology*, 7th ed. (Garland Science, New York, NY, 2008). doi:10.1128/9781555818890.ch2.f2.17

of infection, where microbes are eliminated. Thus, lymphocytes at distinct stages of their lives migrate to the different sites where they are required for their functions. This process of lymphocyte recirculation is most relevant for T cells, as effector T cells have to locate and eliminate microbes at any site of infection. In contrast, effector B cells remain in lymphoid organs and do not need to migrate to sites of infection. Rather, B cells secrete antibodies, which enter the blood and find microbes and microbial toxins in the circulation or distant tissues.

Naïve T cells that have matured in the thymus and entered the circulation migrate to lymph nodes, where they can find antigens that enter through lymphatic vessels that drain epithelia and parenchymal organs. These naïve T cells enter lymph nodes through specialized postcapillary venules, called **high endothelial venules**, that are present in lymph nodes (Fig. 2.18). Naïve T cells express the L-selectin surface receptor, which binds to carbohydrate ligands expressed only on endothelial cells of high endothelial venules. Selectins are a family of proteins involved in cell-cell adhesion that contain conserved structural features, including a lectin or carbohydrate-binding domain. Due to the interaction of L-selectin with its ligand, naïve T cells bind loosely to high endothelial venules. In response to chemokines produced in the T-cell zones of the lymph nodes, the naïve T cells bind strongly to high endothelial venules and then migrate through the high endothelial venules into this region, where antigens are displayed by dendritic cells. Recent intravital imaging techniques have shown that in a lymph node, naïve T cells move around rapidly, scanning the surfaces of dendritic cells while searching for antigens.

After antigen recognition, T cells are transiently arrested on interacting antigen-presenting dendritic cells; they then form stable conjugates

Figure 2.18 Migration of T cells through high endothelial venules. Naïve T cells migrate from the blood through high endothelial venules into the T-cell zones of lymph nodes, where they encounter and are activated by antigens. Activated T cells exit the nodes, enter the bloodstream, and migrate preferentially to peripheral tissues at sites of infection and inflammation. Adapted with permission from Abbas and Lichtman, *Basic Immunology: Functions and Disorders of the Immune System*, 3rd ed. (Saunders Elsevier, Philadelphia, PA, 2010). doi:10.1128/9781555818890.ch2.f2.18

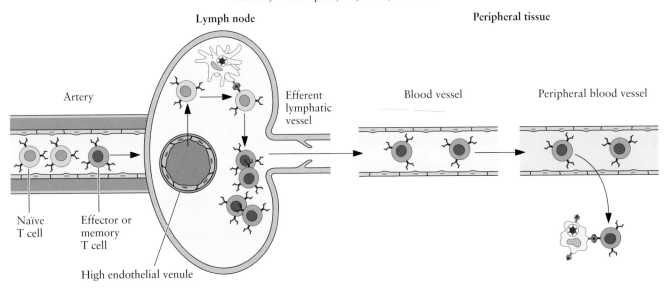

Lymph node

Peripheral tissue

Artery

Efferent lymphatic vessel

Blood vessel

Peripheral blood vessel

Naïve T cell

Effector or memory T cell

High endothelial venule

with the antigen-presenting cells and become activated. An encounter between an antigen and specific T cell occurs randomly, but most T cells circulate through some lymph nodes at least once a day. Thus, by expression of antigen-specific TCRs, some T cells have a high probability of encountering antigens. The likelihood of the correct T cell finding its antigen is increased in peripheral lymphoid organs, particularly lymph nodes, because microbial antigens are concentrated in the regions of these organs through which naïve T cells circulate. In response to a microbial antigen, naïve T cells are activated to proliferate and differentiate. During this process, naïve T cells reduce their expression of adhesion molecules and chemokine receptors and are retained in the lymph nodes. Simultaneously, T cells increase their expression of receptors for the sphingosine 1-phosphate phospholipid. Since the concentration of this phospholipid is higher in blood than in lymph nodes, activated cells are recruited from lymph nodes into the circulation. The net result of these changes is that differentiated effector cells leave the lymph nodes and enter the circulation. These effector cells preferentially migrate into the tissues that are colonized by infectious microbes, where the T cells function to eradicate infection.

Memory T-cell populations consist of some cells that recirculate through lymph nodes, where they promote secondary responses to captured antigens, and other cells that migrate to sites of infection, where they can respond rapidly to eliminate the infection. Little is known about lymphocyte circulation through the spleen or other lymphoid tissues or about the circulation pathways of naïve and activated B cells. The spleen does not contain high endothelial venules, but the general pattern of lymphocyte migration through this organ probably is similar to that of migration through lymph nodes. B cells appear to enter lymph nodes through high endothelial venules, but after they respond to an antigen, their differentiated progeny either remain in the lymph nodes or migrate, mainly to the bone marrow.

Antigen Recognition by T Cells

Certain challenges must be overcome when the immune system attempts to mount an antigen-induced immune response. First, the frequency of naïve lymphocytes specific for an antigen is extremely low, i.e., <1 in 100,000 lymphocytes. These antigen-specific lymphocytes need to locate the antigen in the body and react rapidly to it. Second, various kinds of microbes need to be eliminated by different types of adaptive immune responses elicited against the microbe at different stages of its life. Thus, if a microbe (e.g., a virus) enters the blood and circulates there, B cells are stimulated to produce antiviral antibodies that bind the virus, prevent it from infecting host cells, and help to eliminate it. In contrast, after a virus infects host cells, it is safe from antibodies, which cannot enter inside the cells, and activation of CTLs may be necessary to kill the infected cells and eliminate infection. Two important questions emerge:

1. How do low-frequency naïve lymphocytes specific for a microbial antigen find that microbe in the body?

2. How does the immune system generate effector T cells and molecules required to eradicate an infection, such as antibodies against extracellular microbes and CTLs to kill infected cells harboring intracellular microbes in their cytoplasm?

The immune system has developed an elegant way to capture and display antigens to lymphocytes. The mechanisms of how protein antigens are captured, broken down, and displayed for recognition by T cells are known. While less is known about these molecular events in B cells, B cells can recognize many more types of molecules than T cells without the need for either antigen processing or expression on the surface of host cells.

MHC Restriction

Most T cells recognize peptide antigens that are bound to and displayed by MHC molecules of antigen-presenting cells. The MHC is a genetic locus whose principal products function as the peptide display molecules of the immune system. In every individual, different clones of T cells can see peptides only when these peptides are displayed by that individual's MHC molecules. This property of T cells is called **MHC restriction** (Fig. 2.19; Milestone 2.3), a mechanism that explains how each T cell has a single TCR with dual specificity for residues of both the peptide antigen and the MHC molecule (Fig. 2.19).

Genetics and Structure of MHC Proteins

MHC molecules are transmembrane glycoproteins on antigen-presenting cells that display peptide antigens for recognition by T cells. The MHC is the genetic locus that principally determines the acceptance (MHC

Figure 2.19 Model of MHC restriction. MHC molecules expressed on the surface of antigen-presenting cells bind and present peptides enzymatically processed from protein antigens. Peptides bind to the MHC molecule by anchor residues, which attach the peptides to pockets in the binding groove of an MHC molecule. On T cells, the TCR specifically recognizes some peptide residues and some polymorphic MHC residues in a trimolecular TCR–peptide–MHC complex. The structure of this complex governs the MHC restriction of a given T-cell-mediated immune response. Adapted with permission from Abbas and Lichtman, *Basic Immunology: Functions and Disorders of the Immune System*, 3rd ed. (Saunders Elsevier, Philadelphia, PA, 2010). doi:10.1128/9781555818890.ch2.f2.19

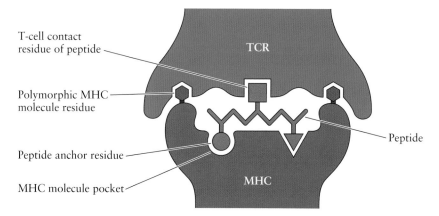

Restriction of In Vitro Cell-Mediated Cytotoxicity in Lymphocytic Choriomeningitis Within a Syngeneic or Semi-Allogeneic System
R. M. ZINKERNAGEL AND P. C. DOHERTY
Nature 248:701–702, 1974

milestone
2.3

In 1973, Rolf Zinkernagel and Peter Doherty, research fellows at the Australian National University in Canberra, investigated how T cells protect mice against infection from lymphocytic choriomeningitis virus, a virus that causes meningitis. Their seminal findings, published in 1974, showed that murine CTLs from virus-infected mice could lyse only infected target cells that express the same MHC-I antigens and not those target cells that carry a different MHC allele. They immediately recognized the important implications of this finding and proposed that in contrast to antibodies, CTLs would not recognize virus directly but only in conjunction with a self-MHC molecule. Alternatively, CTLs might recognize an MHC protein that was modified by the virus, giving rise to an "altered self" structure. This intriguing model, termed MHC restriction, stimulated intensive immunology research throughout the world that ultimately yielded the structures of MHC-I antigens, MHC-II antigens, and TCRs.

Today, we know that T cells recognize small peptides derived from intracellularly degraded proteins that are bound by a specific binding groove of MHC proteins and displayed at the cell surface for presentation to and recognition by the TCR on T cells. This discovery was made because the right people were at the right place at the right time. Indeed, the time was ripe for the discovery in the 1970s. Until then, the transplantation antigens (MHC molecules) were only known to be a major obstacle and a nuisance to transplant surgeons; their true biological role remained elusive. Independently, Hugh O. McDevitt and Baruj Benacerraf had shown that the MHC-II genes control immune responses, demonstrating a very important biological function of MHC proteins. In 1980, Benacerraf shared the Nobel Prize in Medicine with Jean Dausset and George Snell for their discoveries of the human HLA and mouse H-2 MHC gene complexes, respectively. Since these MHC genes restricted the specificity of T cells, they had been

hypothesized to encode the TCR for antigens. However, McDevitt and his team of investigators reported that murine MHC-II proteins were synthesized and expressed by B cells and not T cells, indicating that the TCR was not encoded by MHC-II genes, a surprising finding that ran counter to the dogma of the time. Owing to the work of Zinkernagel and Doherty, we know that MHC-II proteins, like MHC-I proteins, are receptors that present antigenic peptides to T cells and thereby restrict T-cell specificity. The major discovery by Zinkernagel and Doherty not only changed and shaped our concepts of T-cell immunology but also laid the foundation for our present understanding of autoimmunity and vaccine development. In 1996, the Nobel Prize in Physiology or Medicine was awarded to Rolf Zinkernagel and Peter Doherty for their discovery of how T cells of the immune system recognize foreign microbial antigens, such as viral antigens.

(Adapted from G. Hammerling, *Cell Tissue Res.* 287:1–2, 1997.)

---

identical) or rejection (MHC different) of tissue grafts exchanged between individuals. The physiological function of MHC molecules is to display peptides derived from protein antigens to antigen-specific T cells. This function of MHC molecules provides the basis for understanding the phenomenon of MHC restriction of T cells mentioned above.

The group of genes that make up the MHC locus is found in all mammals (Fig. 2.20) and includes genes that encode MHC and other proteins. Human MHC proteins are called **human leukocyte antigens (HLAs)** because these proteins were discovered as antigens of leukocytes that were identified by reactivity with specific antibodies. The HLA complex in humans maps to human chromosome 6, and the **H-2 complex** in mice maps to mouse chromosome 17. The HLA complex and H-2 complex each contain two sets of highly polymorphic genes, the MHC class I (MHC-I) and MHC class II (MHC-II) genes. These genes encode the MHC-I and MHC-II proteins that bind and present peptides to T cells. In addition, the MHC locus also contains many nonpolymorphic genes, several of which encode proteins involved in antigen presentation or another, unknown function.

HLA
human leukocyte antigen

Human (HLA)

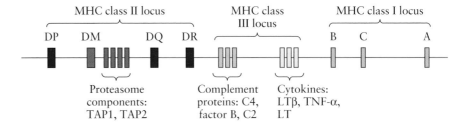

Mouse (H-2)

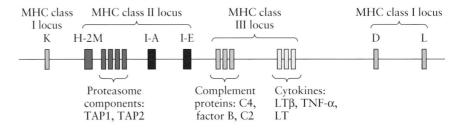

Figure 2.20 Human HLA and mouse H-2 MHC loci. Chromosomal maps of MHC and MHC-linked genes located in the human HLA and mouse H-2 complex are shown. The major genes in this complex encode molecules that regulate immune responses. MHC-II loci are shown as single blocks, but each consists of at least two genes. MHC-III loci are comprised of genes that encode molecules that do not display peptides. Several HLA- and H-2-linked MHC-I-like genes and pseudogenes are not shown. LT, lymphotoxin; TAP, transporter associated with antigen processing; TNF-α, tumor necrosis factor alpha. Adapted with permission from Abbas and Lichtman, *Basic Immunology: Functions and Disorders of the Immune System*, 3rd ed. (Saunders Elsevier, Philadelphia, PA, 2010). doi:10.1128/9781555818890.ch2.f2.20

The MHC-I and MHC-II transmembrane glycoproteins each contain a peptide-binding cleft at their amino-terminal end. Although the two classes of molecules differ in subunit composition, they are very similar in overall structure (Fig. 2.21). Each MHC-I molecule consists of an α chain noncovalently bound to the β$_2$-microglobulin protein, which is encoded by a non-MHC gene. The amino-terminal α1 and α2 domains of MHC-I form a peptide-binding cleft, or groove, that is large enough to accommodate peptides of 8 to 11 amino acids in length. The floor of the peptide-binding cleft binds peptides for display to T cells, and the sides and tops of the cleft consist of residues contacted by the peptide and TCR (Fig. 2.21). The polymorphic residues of MHC-I molecules that differ between individuals are localized in the α1 and α2 domains of the α chain. Some of these polymorphic residues contribute to variations in the floor of the peptide-binding cleft and thus in the ability of different MHC molecules to bind peptides. Other polymorphic residues contribute to variations in the tops of the clefts and thus influence TCR recognition. The α3 domain is invariant and contains the binding site for the CD8 T-cell coreceptor. T-cell activation requires recognition of an MHC-associated peptide antigen by the TCR and simultaneous recognition of MHC-I by CD8. It follows that CD8$^+$ T cells can respond only to peptides displayed by MHC-I molecules, which bind to CD8.

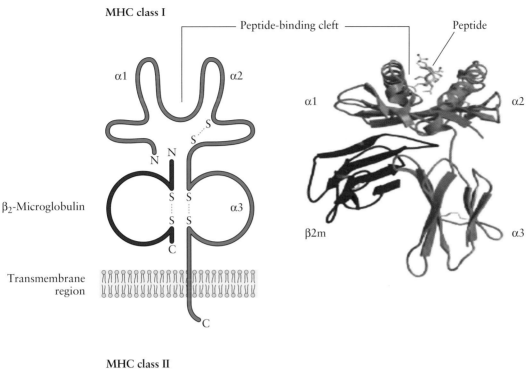

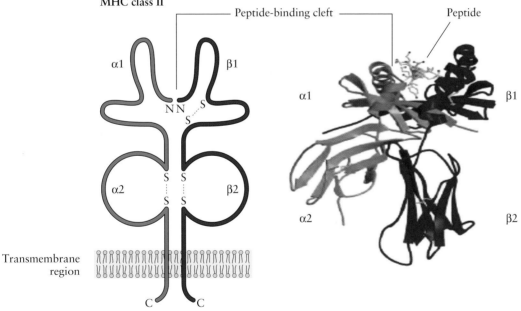

Figure 2.21 Schematic diagrams (left) and models (right) of the crystal structures of the MHC-I and MHC-II proteins showing their respective disulfide (S–S)-linked immunoglobulin (Ig) domains and similarity in overall structure. Both MHC molecules contain an antigen-binding groove and invariant portions that bind CD8 (MHC-I α3 domain) or CD4 (MHC-II β2 domain). The β_2-microglobulin protein that constitutes the light chain of MHC-I proteins is shown. Adapted with permission from Abbas and Lichtman, *Basic Immunology: Functions and Disorders of the Immune System*, 3rd ed. (Saunders Elsevier, Philadelphia, PA, 2010). doi:10.1128/9781555818890.ch2.f2.21

An MHC-II molecule consists of an α chain and β chain. The residues in the amino-terminal α1 and β1 domains of both chains are polymorphic and form a cleft that can accommodate peptides of 10 to 30 residues in length. The nonpolymorphic β2 domain contains the binding site for the CD4 T-cell coreceptor. Since CD4 binds to MHC-II molecules, CD4$^+$ T cells respond only to peptides presented by MHC-II molecules. Several features of MHC genes and proteins important for the normal function of these molecules are summarized in Fig. 2.22.

Figure 2.22 Properties of MHC genes and proteins. Some important characteristics of MHC molecules and their relevance to adaptive immunity are listed. Adapted with permission from Abbas and Lichtman, *Basic Immunology: Functions and Disorders of the Immune System*, 3rd ed. (Saunders Elsevier, Philadelphia, PA, 2010). doi:10.1128/9781555818890.ch2.f2.22

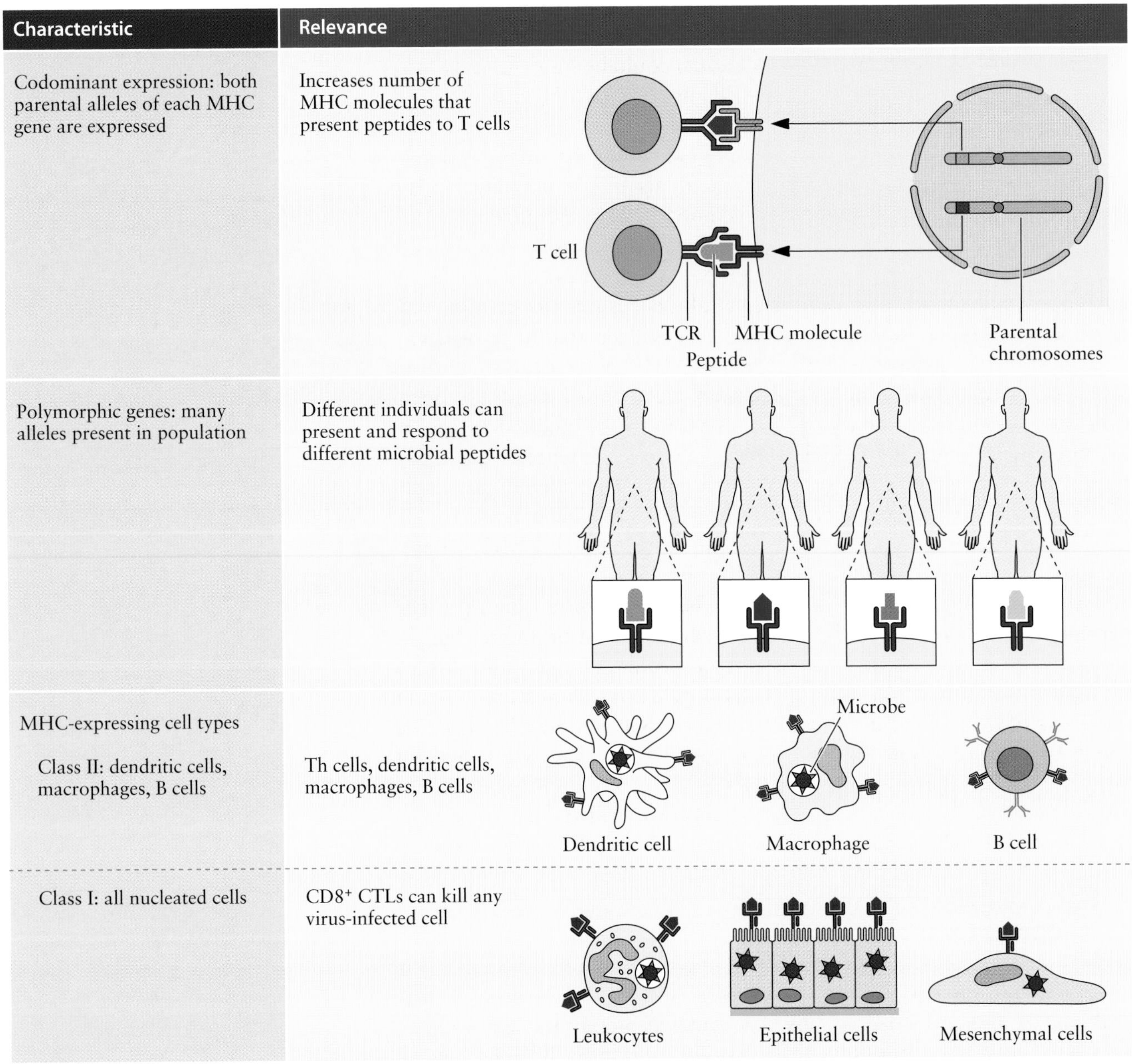

Characteristic	Relevance
Codominant expression: both parental alleles of each MHC gene are expressed	Increases number of MHC molecules that present peptides to T cells
Polymorphic genes: many alleles present in population	Different individuals can present and respond to different microbial peptides
MHC-expressing cell types Class II: dendritic cells, macrophages, B cells	Th cells, dendritic cells, macrophages, B cells
Class I: all nucleated cells	CD8$^+$ CTLs can kill any virus-infected cell

MHC genes are **codominantly expressed**, meaning that the alleles inherited from both parents are expressed equally. Since in humans there are three polymorphic MHC-I genes, *HLA-A*, *HLA-B*, and *HLA-C*, and each person inherits one set of these genes from each parent, any cell can express six different MHC-I proteins. In the MHC-II locus, every individual inherits one pair of HLA-DP genes (DPA1 and DPB1, encoding the α and β chains), one pair of HLA-DQ genes (DQA1 and DQB1, encoding the α and β chains), one HLA-DRα gene (DRA1), and one or two HLA-DRβ genes (DRB1 and DRB3, -4, or -5). Thus, a heterozygous individual can inherit six or eight MHC-II alleles, three or four from each parent (one set each of DP and DQ and one or two of DR). Due to the extra DRβ genes, and because some DQα molecules encoded on one chromosome can associate with DQβ molecules encoded from the other chromosome, the total number of expressed MHC-II proteins may be considerably more than six. The set of MHC alleles present on each chromosome is known as an **MHC haplotype**. In humans, each HLA allele is given a numerical designation; e.g., an HLA haplotype of an individual could be HLA-A2, HLA-B5, HLA-DR3, etc. All heterozygous individuals have two HLA haplotypes, one from each chromosome.

MHC genes are highly polymorphic, indicating that many different alleles are present among different individuals in the population. The polymorphism is so great that any two individuals in the population are unlikely to have exactly the same MHC genes and proteins. Because the polymorphic residues determine the peptides that are presented by selected MHC proteins, the existence of multiple alleles ensures that there are always some members of the population that will be able to present any particular microbial protein antigen. Thus, MHC polymorphism ensures that a population will be able to survive if faced with a pandemic infection and will not succumb to a newly encountered or mutated pathogen, because at least some individuals will be able to mount effective immune responses to the peptide antigens of these pathogens. Polymorphisms in MHC proteins arise from inheritance of distinct DNA sequences.

MHC-I proteins are expressed on all nucleated cells, but MHC-II proteins are expressed mainly on dendritic cells, macrophages, and B cells, and this regulates the type of T-cell-mediated response obtained. The peptide-binding clefts of MHC proteins, which consist of different pockets in the floor, bind peptides derived from protein antigens and display these peptides for recognition by T cells (Fig. 2.23). There are pockets in the floors of the peptide-binding clefts of most MHC molecules. The side chains of amino acids in the peptide antigens fit into these MHC pockets and anchor the peptides in the cleft of the MHC molecule. Peptides that are anchored in the cleft by these side chains (also called anchor residues) contain some residues that are oriented upwards to the TCR on T cells.

Antigen Capture and Presentation by Antigen-Presenting Cells

Several key features of peptide–MHC interactions are important for understanding how peptides are presented by MHC proteins to T cells (Table 2.7) and are listed in Box 2.1. Antigen-presenting cells capture microbial antigens and display them for recognition by T cells. Naïve T

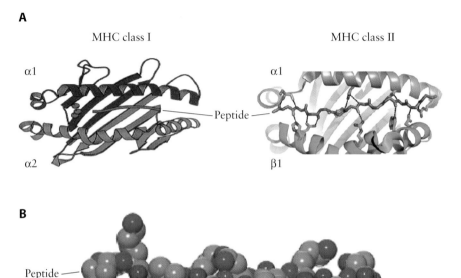

A

MHC class I   MHC class II

α1

α2   β1

Peptide

B

Peptide

Pockets in floor of peptide-
binding groove of MHC class II

Anchor residue
of peptide

Figure 2.23 Binding of peptides to MHC proteins. (**A**) Top views of the structures of MHC-I and MHC-II molecules illustrate how peptides lie on the floors of the peptide-binding grooves and are available for recognition by T cells. The amino terminus of the peptide backbone is at the left of the groove, and the carboxy terminus is at the right of the groove. Binding of the peptide to MHC-II occurs by hydrogen bonds to residues that are highly conserved in MHC-II molecules. The side chains of these residues are shown. (**B**) A side view of a peptide bound to an MHC-II protein demonstrates how anchor residues of the peptide hold it in the pockets in the floor of the peptide-binding groove of the MHC molecules. Adapted with permission from Khan et al., *J. Immunol.* **164:**6398–6405, 2000 (© 2000 The American Association of Immunologists, Inc.), for MHC-I structure and from Murphy et al., *Janeway's Immunobiology*, 7th ed. (Garland Science, New York, NY, 2008), for MHC-II structures. doi:10.1128/9781555818890.ch2.f2.23

Table 2.7 Characteristics of peptide–MHC interactions

Characteristic	Relevance
Wide specificity of peptide–MHC binding	One MHC molecule can bind to several different peptides
Each MHC molecule presents only one peptide at a time	Each TCR recognizes a single MHC-bound peptide
MHC molecules accommodate only peptides in their groove	Only protein antigens stimulate MHC-restricted T cells
During their intracellular assembly, MHC molecules bind to peptides	MHC-I and MHC-II molecules present peptides from different cellular compartments
MHC molecules are stably expressed at the cell surface only if they have a bound peptide	Only surface-associated stable peptide–MHC complexes are recognized by the TCR
The off-rate of peptide from MHC is very low	Peptides remain bound to MHC molecules sufficiently long to stimulate T-cell responses

Adapted with permission from Abbas and Lichtman, *Basic Immunology: Functions and Disorders of the Immune System*, 3rd ed. (Saunders Elsevier, Philadelphia, PA, 2011).

The principal characteristics of peptide binding to MHC proteins and their functional relevance are presented.

box 2.1
Key Features of Presentation of Peptides by MHC Proteins to T Cells

- Each MHC protein can present only one peptide at a time, because there is only one cleft, but each MHC protein can present several different peptides.

- MHC proteins bind only peptides and not other types of antigens.

This explains why MHC-restricted CD4$^+$ T cells and CD8$^+$ T cells recognize and respond to only protein antigens.

- MHC proteins acquire their peptides during their biosynthesis and assembly inside cells. MHC-I proteins acquire peptides predominantly from cytosolic proteins, and MHC-II proteins bind to peptides derived from proteins in intracellular vesicles (endosomes).

- In each individual, the MHC proteins can display peptides derived from non-self (e.g., microbial) proteins as well as peptides from self-proteins.

cells recognize protein antigens presented by dendritic cells, the most effective professional antigen-presenting cells, to initiate clonal expansion and effector cell differentiation. Differentiated effector T cells must bind to antigens presented by various antigen-presenting cells in order to activate the effector functions of the T cells in humoral and cell-mediated immune responses.

Protein antigens of microbes that enter the body are captured mainly by dendritic cells and concentrated in the peripheral lymphoid organs, where immune responses are initiated (Fig. 2.24). Microbes usually enter the body through the skin (by contact), the gastrointestinal tract (by ingestion), and the respiratory tract (by inhalation). All interfaces between the body and external environment are lined by continuous epithelia, which provide a physical barrier to infection. The epithelia and subepithelial tissues contain a network of dendritic cells; the same cells are present in the T-cell-rich areas of peripheral lymphoid organs and, in smaller numbers, in most other organs. In the skin, the epidermal dendritic cells are called Langerhans cells. Epithelial dendritic cells are immature, since they do not stimulate T cells efficiently. These immature dendritic cells express membrane receptors that bind microbes, such as receptors for terminal mannose residues on glycoproteins, a typical feature of microbial but not mammalian glycoproteins. Dendritic cells use these receptors to capture and endocytose microbial antigens. Some soluble microbial antigens may enter dendritic cells by pinocytosis. At the same time, microbes stimulate innate immune reactions by binding to TLRs and other sensors of microbes in the dendritic cells, as well as in epithelial cells and resident macrophages in the tissue. This results in production of inflammatory cytokines, such as tumor necrosis factor alpha and IL-1. The combination of the TLR signaling and cytokines activates the dendritic cells, resulting in several changes in phenotype and function.

Activated dendritic cells lose their adhesiveness for epithelia and begin to express the CCR7 chemokine receptor on their surface, which binds to chemokines produced in the T-cell zones of lymph nodes. These chemokines direct dendritic cells to exit the epithelium and migrate through lymphatic vessels to the lymph nodes draining that epithelium (Fig. 2.25). During their migration, dendritic cells mature from cells designed to

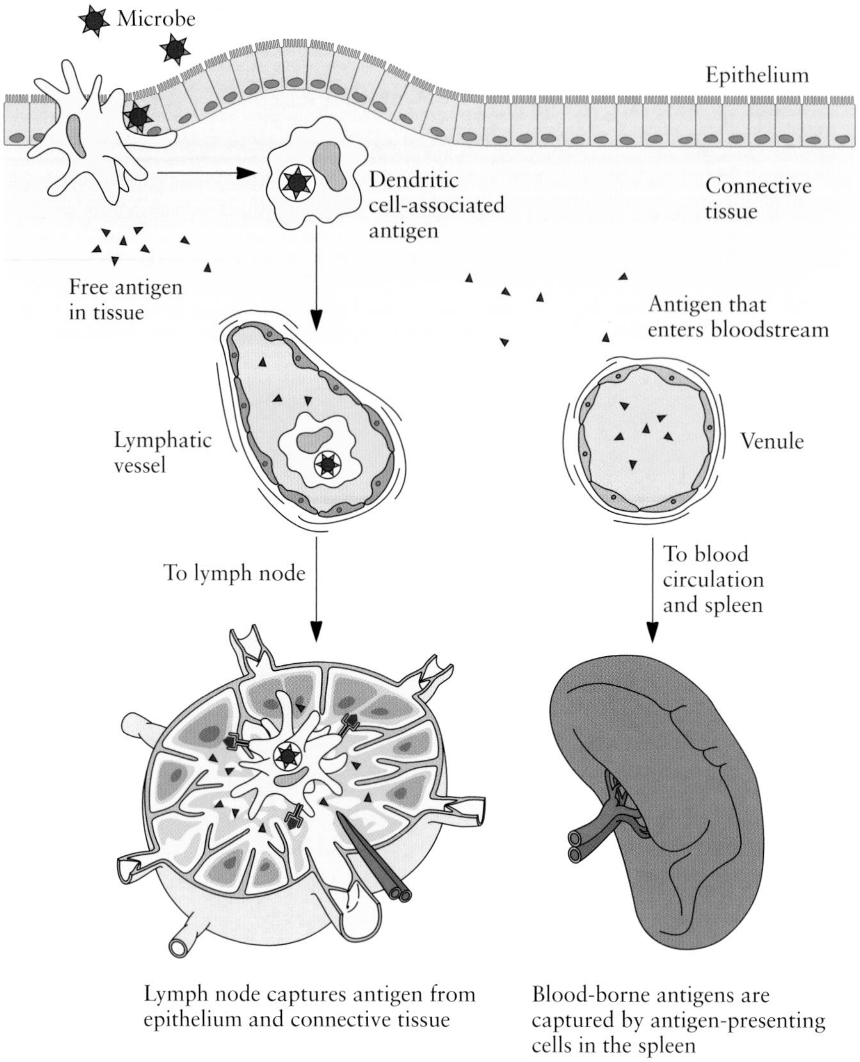

Figure 2.24 Capture and presentation of microbial antigens. Microbes pass through an epithelial cell layer and are captured by dendritic cells resident in the epithelium. Alternatively, microbes enter lymphatic vessels or blood vessels. Next, the microbes and their antigens are transported to peripheral lymphoid organs (lymph nodes and spleen), where T cells recognize these antigens and elicit immune responses. Adapted with permission from Abbas and Lichtman, *Basic Immunology: Functions and Disorders of the Immune System*, 3rd ed. (Saunders Elsevier, Philadelphia, PA, 2010).
doi:10.1128/9781555818890.ch2.f2.24

capture antigens into antigen-presenting cells capable of stimulating T cells in various immune responses. This maturation is accompanied by an increased synthesis and stable expression of MHC proteins, which present antigens to T cells via TCR recognition (signal 1), and of other costimulator proteins (e.g., CD80 and CD86) (signal 2). Both signals 1 and 2 are required for full T-cell responses. Soluble antigens in the lymph are picked up by dendritic cells that reside in the lymph nodes, and blood-borne antigens are handled similarly by dendritic cells in the spleen. The net result of this sequence of events is that the protein antigens of microbes that enter the body are transported to and concentrated in the regions of lymph nodes where the antigens are most likely to encounter T cells. Recall that naïve T cells continuously recirculate through lymph nodes and also express CCR7, which promotes their entry into the T-cell zones of lymph nodes. Therefore, dendritic cells bearing captured antigens and naïve T cells poised to recognize antigens colocalize in lymph nodes. This

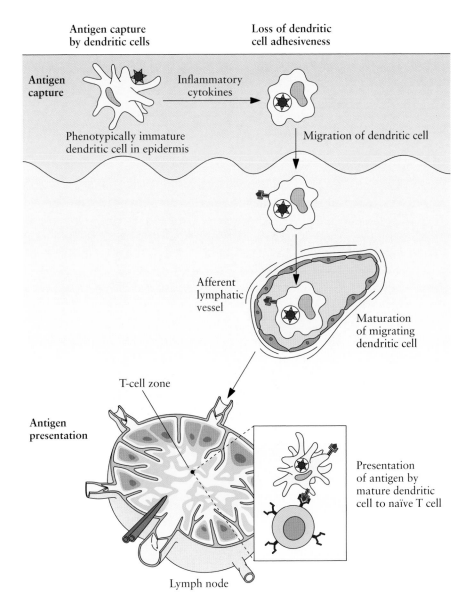

Figure 2.25 Capture and presentation of protein antigens by dendritic cells. Immature dendritic cells in the epithelium (e.g., skin, where dendritic cells are called Langerhans cells) capture microbial antigens, then leave the epithelium, and are recruited by chemokines to the draining lymph nodes. During their migration, the dendritic cells mature, likely in response to the microbe. In the lymph nodes, the dendritic cells present antigens to naïve T cells. Dendritic cells at their different stages of maturation may express different membrane proteins. Immature dendritic cells express surface receptors that capture microbial antigens, whereas mature dendritic cells express levels of MHC and other molecules that costimulate T-cell responses. Adapted with permission from Abbas and Lichtman, *Basic Immunology: Functions and Disorders of the Immune System*, 3rd ed. (Saunders Elsevier, Philadelphia, PA, 2010). doi:10.1128/9781555818890.ch2.f2.25

process is remarkably efficient: it is estimated that if microbial antigens are introduced at any site in the body, a T-cell response to these antigens begins in the draining lymph nodes within 12 to 18 h.

While different types of antigen-presenting cells serve distinct functions in T-cell-dependent immune responses, we focus here on responses stimulated by dendritic cells. Dendritic cells are the principal inducers of such T-cell-dependent responses, because they are the most potent antigen-presenting cells for activating naïve T cells. Dendritic cells not only initiate T-cell responses but also may influence the nature of the response. Importantly, various subsets of dendritic cells can direct the differentiation of naïve CD4[+] T cells into distinct populations that are equipped to defend us against infections elicited by different types of microbes. Dendritic cells can also initiate the responses of CD8$^+$ T cells to the antigens of intracellular microbes. Some microbes (e.g., viruses) rapidly infect host cells and can be eradicated only by CD8$^+$ CTLs that can respond to the antigens of these intracellular microbes and destroy the infected cells.

However, viruses may infect any typiate T-cell activation. How, then, are naïve CD8$^+$ T cells able to respond to the intracellular antigens of infected cells? It is likely that dendritic cells phagocytose infected cells and display the antigens present in the infected cells for recognition by CD8$^+$ T cells. This process is called **cross-presentation** (or cross-priming), to indicate that one cell type, the dendritic cell, can present the antigens of other cells, the infected cells, and prime (or activate) naïve T cells specific for these antigens. The dendritic cells that engulf infected cells may also present microbial antigens to CD4$^+$ Th cells. Thus, CD4$^+$ and CD8$^+$ T cells, specific for the same microbe, are activated close to one another. This process is important for the antigen-stimulated differentiation of naïve CD8$^+$ T cells to effector CTLs, which often requires help from CD4$^+$ Th cells. Once the CD8$^+$ T cells have differentiated into CTLs, they kill infected host cells without any need for dendritic cells or signals other than recognition of the antigen.

Thus, dendritic cells are essential for the regulation of microbial immunity throughout the body. Due to their migratory properties, dendritic cells can capture, process, and present protein antigens of many different microbes to different subsets of CD4$^+$ and CD8$^+$ T cells localized in peripheral lymphoid organs throughout the body. Since dendritic cells are able to orchestrate and regulate the desired antimicrobial immune responses that prevent infection, it is appropriate that dendritic cells be recognized as the most effective professional antigen-presenting cells (Milestone 2.4).

Humoral Immunity

As antibodies were the first proteins involved in specific immune recognition to be characterized, their structure and function are very well understood. An antibody molecule has two distinct functions: one is to bind specifically to regions of a pathogen that elicited the immune response, and the other is to recruit other cells and molecules to destroy the pathogen once the antibody is bound to it. For example, binding by an antibody can neutralize a virus and mark a pathogen for destruction by phagocytes (e.g., macrophages) and serum complement. Interestingly, the recognition and effector functions of an antibody molecule are controlled by distinct regions of the molecule. The antigen-binding region, which varies extensively in its amino acid sequence between antibody molecules, is located in the **variable (V) region**, which endows an antibody with exquisite specificity of binding to a particular antigenic epitope. On the other hand, the region that performs the effector functions of an antibody is termed the **constant (C) region**, which does not vary in its sequence and may be expressed in five different forms, each of which is specialized for activating different effector mechanisms. The

V region
variable region

C region
constant region

Identification of a Novel Cell Type in Peripheral Lymphoid Organs of Mice. 1. Morphology, Quantitation, Tissue Distribution

R. M. STEINMAN AND Z. A. COHN
J. Exp. Med. **137**:1142–1162, 1973

milestone
2.4

Identification of a Novel Cell Type in Peripheral Lymphoid Organs of Mice. 2. Functional Properties In Vitro

R. M. STEINMAN AND Z. A. COHN
J. Exp. Med. **139**:380–397, 1974

The clonal selection theory proposed by Sir Macfarlane Burnet in 1957 postulated that lymphocytes proliferate in response to antigens only if the antigen can be recognized by their specific receptors. However, during the next 10 years, it remained unknown how an antigen is presented to initiate an immune response. At the time, immunologists conjectured that the answer to this question was essential for the development of novel therapeutic approaches to the treatment of disease. In 1970, while working as a research fellow with Zanvil Cohn on how macrophages catabolize microbes, Ralph Steinman examined the nature of the splenic accessory cells required for immune responses in vitro. Under phase-contrast light microscopy, he found a small number of extensively branched, motile, and mitochondrion-rich cells mixed in with the expected macrophages. Because these cells had a stellate-like structure and displayed a unique morphology with many dendritic processes, Steinman and Cohn named this novel type of white blood cells dendritic cells. These investigators published on the phenotypic and functional characterization of dendritic cells in 1973 and 1974. Steinman and his colleagues then went on to show that antigen-activated dendritic cells express MHC-II

proteins on their surface, present antigens and alloantigens to T cells, and stimulate both T-cell cytotoxicity and antibody responses. Subsequent studies described dendritic cell maturation—the process by which immature dendritic cells, which capture antigens in peripheral tissues, become efficient initiators of immunity. Their high potency suggested that dendritic cells could present antigens to naïve lymphocytes, a finding that led Steinman and many other immunologists to demonstrate that dendritic cells function as the primary antigen-presenting cells of the immune system. In fact, in the steady state when the body is challenged by injury and infection, dendritic cells migrate from body surfaces to immune or lymphoid tissues, where they localize to regions rich in T cells. There, dendritic cells display antigens to T cells, and they alert these lymphocytes to the presence of injury or infection. This directs the T cells to make an immune response that is matched to the challenge at hand.

Dendritic cells are also functional in the control of immune tolerance, an immunological process that silences dangerous immune cells and prevents them from attacking innocuous materials in the body or the body's own tissues. Given these many functions of dendritic cells, these cells have commanded much current research

in medicine. During infection and cancer, microbes and tumors exploit dendritic cells to evade immunity, but dendritic cells also can capture infection- and tumor-derived protein and lipid antigens and generate resistance, including new strategies for vaccines. During allergy, autoimmunity, and transplantation, dendritic cells promote unwanted innate and adaptive responses that cause disease, but they also can suppress these conditions. Thus, since dendritic cells orchestrate innate and adaptive immune responses and provide links between antigens and all types of lymphocytes, dendritic cells are an excellent target in studying disease and in designing treatments.

More recently, techniques were developed to grow large quantities of dendritic cells in vitro, and this work set the stage for current research on the regulation of dendritic cell function and the design of dendritic-cell-based vaccines for HIV infection and cancers. Steinman shared the Nobel Prize in Physiology or Medicine in 2011 with Bruce Beutler and Jules Hoffman for his discovery of the dendritic cell and its role in adaptive immunity. Steinman was announced as a winner of the prize just 3 days after his death from pancreatic cancer in September 2011, and at the time of his death, he was being administered personalized treatments with his own dendritic cells.

membrane-bound BCR does not have these effector functions, since the C region remains inserted in the membrane of the B cell. After the V region of a BCR binds to an epitope of the antigen, a signal is transmitted via its C region and associated proteins in a signaling complex. This signaling then stimulates B-cell activation, leading to clonal expansion and specific antibody production and secretion.

IgG1
immunoglobulin G1

H chain
heavy chain

L chain
light chain

Structure of Immunoglobulins

The basic structure of an antibody monomer of the immunoglobulin G1 (IgG1) subclass is shown in Fig. 2.26. The monomer (molecular mass, 150 kDa) consists of the two **heavy (H) chains** paired via disulfide bonds and noncovalent interactions (hydrogen bonds, electrostatic forces, Van der Waals forces, and hydrophobic forces) with two **light (L) chains**. The H chains (50 kDa) are each of higher molecular mass than the L chains (25 kDa). The disulfide bonds link two H chains to each other or an H chain to an L chain. All antibodies are constructed similarly from paired H and L chains and are termed **immunoglobulins**. In each IgG molecule, the two H chains are identical and the two L chains are identical and give rise to two identical antigen-binding sites in the antigen-binding fragment (Fab) that can bind simultaneously to and cross-link two identical antigenic structures. This Fab fragment is produced upon cleavage of an antibody molecule with the protease papain, and the N-terminal half, the Fab fragment termed the **Fv fragment**, contains all of the antigen-binding activity of the intact antibody molecule.

Immunoglobulins comprise one of the most polymorphic groups of proteins in mammals, presumably because they must recognize the large array of different antigens to be encountered in nature. The mechanisms of gene recombination and gene rearrangement that juxtapose a vast array of V-gene DNA segments with the far less numerous C-gene DNA segments during B-cell development are reviewed in many articles and are not presented here.

Figure 2.26 Structure of an IgG1 molecule. The H- and L-chain variable regions (V_H and V_L) of an IgG1 antibody molecule with their CDRs (CDR1, CDR2, and CDR3) and constant regions (C_{H1}, C_{H2}, C_{H3}, and C_L) are shown. The F_V, Fab, and Fc portions of the molecule are also indicated. This IgG1 subtype molecule expresses two γ1 H chains and two covalently linked κ L chains. The hinge region consists of only one interchain disulfide bond (S–S). NH_2, amino terminal; COOH, carboxy terminal. The variable regions (Fv) contain the antigen-binding sites at one end of an antibody molecule that bind to a target epitope of an antigen. At the other end of the molecule, the Fc regions mediate complement activation and binding to phagocytic receptors during opsonization.
doi:10.1128/9781555818890.ch2.f2.26

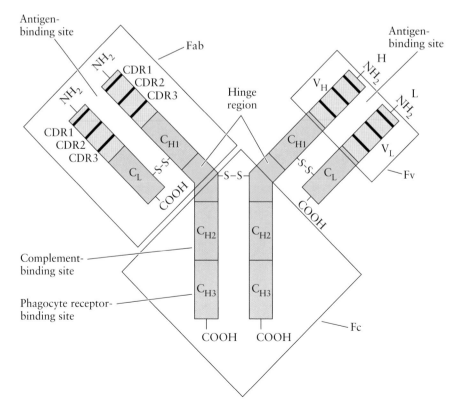

Immunoglobulin V-region and C-region genes may be easily manipulated, as the various functions of an antibody are confined to discrete domains (Fig. 2.27). Thus, antibody diversity is localized to particular parts of the V-domain sequence and, moreover, is found in a particular region on the surface of the antibody molecule. The sites that recognize and bind antigens consist of three **complementarity-determining**

Figure 2.27 Structures and properties of IgG, IgA, IgM, and IgE molecules. Selected properties of the major antibody isotypes of humans are shown. Isotypes are classified based on the structure of their H chains, and each isotype may contain either a κ or λ L chain. The distinct shapes of the secreted forms of these antibodies are presented. IgA consists of two subclasses, IgA1 and IgA2. IgG consists of four subclasses, IgG1, IgG2, IgG3, and IgG4. The concentrations in serum are average values in healthy individuals. Monomeric IgM (190 kDa) normally forms pentamers, known as macroglobulin (hence the M), of very high molecular mass. IgA dimerizes to yield a molecular mass of about 390 kDa in secretions. IgE antibody is associated with immediate-type hypersensitivity. When fixed to tissue mast cells, IgE has a much longer half-life than its half-life in plasma. Adapted with permission from Abbas and Lichtman, *Basic Immunology: Functions and Disorders of the Immune System*, 3rd ed. (Saunders Elsevier, Philadelphia, PA, 2010). doi:10.1128/9781555818890.ch2.f2.27

Antibody isotype	Subtypes	H chain	Concentration in serum (mg/mL)	Half-life in serum (days)	Secreted form	Function(s)
IgA	IgA1, IgA2	α (1 or 2)	3.5	6	IgA Monomer, dimer, trimer 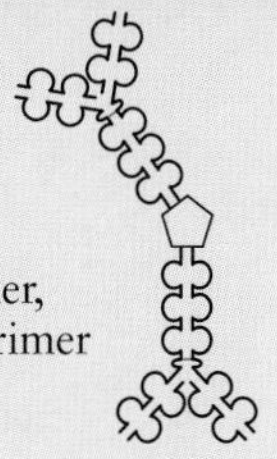	Mucosal immunity
IgD	None	δ	Trace	Trace	None	Naïve B-cell antigen receptor
IgE	None	ε	0.05	2	IgE Monomer 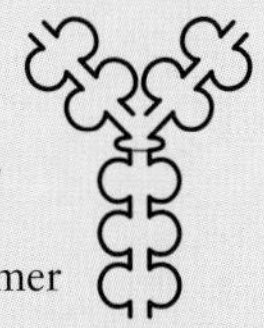	Mast cell activation (immediate hypersensitivity); Defense against helminthic parasites
IgG	IgG1–IgG4	γ (1, 2, 3, or 4)	13.5	23	IgG1 Monomer 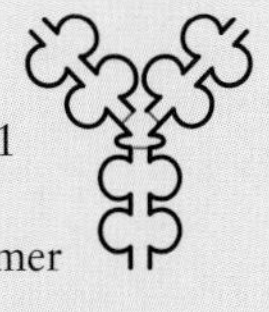	Opsonization, complement activation, antibody-dependent cell-mediated cytotoxicity, neonatal immunity, feedback inhibition of B cells
IgM	None	μ	1.5	5	IgM 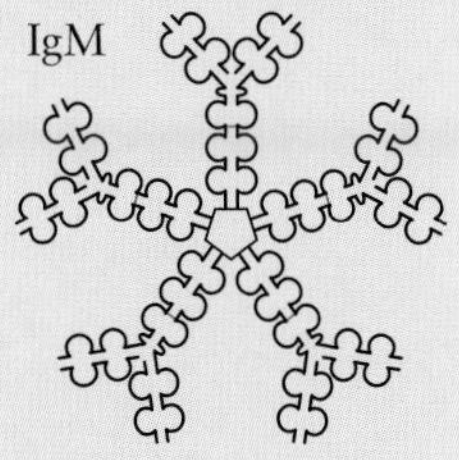 Pentamer	Naïve B-cell antigen receptor, complement activation

regions (CDRs) that lie within the variable (V_H and V_L) domains at the N-terminal ends of the two H and L chains. The CDRs display the highest variability in amino acid sequence of an antibody molecule. When the V_H and V_L domains are paired, the hypervariable loops of each domain (six loops in all) are brought close together, creating a single hypervariable site, i.e., the antigen-binding site, at the tip of each arm of the molecule (Fig. 2.26). Because CDRs from both V_H and V_L domains contribute to the antigen-binding site, the antigen specificity of an antibody is determined only by the particular combination of H chain and L chain expressed. Thus, the immune system can generate antibodies of different antigen specificities by generating different combinations of H- and L-chain regions according to a mechanism known as **combinatorial diversity**.

Functions of Immunoglobulins

In mice and humans, the effector functions of an antibody molecule are mediated by its C-region domains in the COOH-terminal part of the H chain and are not associated at all with the L chain. Each L chain contains one constant domain (C_L), and each H chain contains three constant domains (C_{H1}, C_{H2}, and C_{H3}) (Fig. 2.26). While the C_L and C_{H1} domains are situated in the Fab fragment of an antibody, the C_{H2} and C_{H3} domains are positioned in the **crystallizable fragment (Fc)** portion of the molecule. The Fc fragment of an antibody interacts with effector cells and molecules, and the functional differences between the various classes of H chains lie mainly in the Fc fragment. Five different classes of immunoglobulins (IgM, IgD, IgG, IgA, and IgE) are distinguishable by their C regions (Fig. 2.27). The class and effector function of an antibody are defined by the structure of its H chain class or **isotype**, and the H chains of the five main classes (isotypes) are designated μ, δ, γ, α, and ε, respectively. Note that IgG is the most abundant class, and it has four subclasses in humans (IgG1, IgG2, IgG3, and IgG4) and three subclasses in mice (IgG1, IgG2a, and IgG2b). The reason for such differences in subclasses between closely related species is unknown.

After antigen binding by the Fab regions of an intact antibody molecule, the Fc portion of the molecule stimulates several immune responses in humans (Fig. 2.27), including the following:

Activation of the complement cascade. The protein components of the complement system break down cell membranes, activate phagocytes, and produce signals to mobilize other components of an immunological response depending on whether a classical or alternative pathway of complement activation is elicited.

Antibody-dependent cell-mediated cytotoxicity. This response results from the binding of the Fc portion of an antibody to its Fc receptor protein on the surface of an antibody-dependent cell-mediated cytotoxicity effector cell. The bound effector cell releases substances that lyse the foreign cell to which the Fab portion of the antibody is bound.

Fc
crystallizable fragment

Phagocytosis. After the Fab portion of an antibody binds to a soluble antigen, the Fc portion of antibodies of the IgG1 and IgG3 subclasses can bind to their Fc receptor proteins (FcγR) on recruited macrophages and neutrophils. These cell types engulf and destroy (phagocytose) the antibody–antigen complex and facilitate the destruction of pathogens coated with these antibodies.

Inflammation. The Fc portion of an IgE antibody binds to a high-affinity FcεR on mast cells, basophils, and activated eosinophils, enabling these cells to respond during an allergic response to the binding of a specific antigen (or allergen) by releasing inflammatory mediators.

Transport. The Fc portion can deliver antibodies to places they would not reach without active transport. These include mucous secretions, tears, and milk (IgA) and the fetal blood circulation by transfer from the pregnant mother (IgG).

Types of Antibodies: Applications

During an immune response in mammals that protects the body from toxins and infectious pathogens, the antibodies produced and secreted by B cells bind to the pathogen-derived foreign antigens. Together with other immune system proteins that comprise the **complement system**, which bind to the Fc portion of antibody molecules, these foreign pathogens are inactivated and become neutralized. Upon stimulation of an immune response, each antibody-producing B cell synthesizes and secretes a single antibody that recognizes with high affinity a discrete epitope of the immunizing antigen. Because a target antigen generally consists of several different epitopes (Fig. 2.28), a given antigen-primed B cell produces a different antibody against only one epitope of the antigen. The classification of the different types of antibodies and some of their more frequently used applications to immune assays and immunodiagnostics are discussed below.

Polyclonal Antibodies

The set of antibodies which all react with the same antigen is termed **polyclonal antibodies**; i.e., the antibodies in this set are produced by a collection of several different B-cell clones. Thus, by definition, a polyclonal antibody reacts with different epitopes of a given antigen, and even antibodies that bind the same epitope of an antigen can be heterogeneous.

Several methods for polyclonal antibody production in laboratory animals (e.g., horses, goats, and rabbits) exist. Institutional guidelines that regulate the use of animals and related procedures are directed towards the safety of the use of adjuvants (agents that enhance the strength of an immune response to an antigen without adverse effects when administered alone). Examples of adjuvants are (i) alum and (ii) water-in-oil emulsion of heat-killed *Escherichia coli* bacteria. These guidelines include adjuvant selection, routes and sites of administration, injection volumes per site, and number of sites per animal. The primary goal of polyclonal antibody production in laboratory animals is to obtain high-titer, high-affinity antisera

Figure 2.28 Schematic of a target antigen and epitopes. The surface of the antigen shown has seven different antigenic determinants (called epitopes). When this antigen is used to immunize an animal, each epitope may elicit the synthesis of a different antibody. Taken together, the different antibodies that interact with an antigen constitute a polyclonal antibody directed against that antigen. doi:10.1128/9781555818890.ch2.f2.28

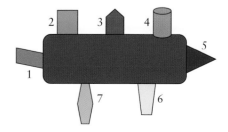

for experimental use or diagnostic testing. Most adjuvants establish an antigen depot that yields a slow release of antigen from an injection site into draining lymph nodes. Many adjuvants either contain or function as surfactants that promote concentration of protein antigen molecules over a large surface area and immunostimulatory molecules or properties. Since many antigens are weak immunogens, adjuvants are generally used with such soluble protein antigens to increase antibody titers and induce a prolonged response with accompanying immunological memory. In contrast, most complex protein antigens activate multiple B-cell clones during an immune response, and thus the response is polyclonal. Immune responses to nonprotein antigens are generally poor and are only weakly enhanced by adjuvants, and there is no immune system memory.

Antisera

Antisera are valuable tools for many biological assays, such as the blocking of infection and diagnosis of toxic substances in clinical samples. However, polyclonal antisera possess certain inherent disadvantages that relate to the heterogeneity of their antibodies. First, each antiserum differs from all other antisera, even if generated in a genetically identical animal by using the identical antigen preparation and immunization protocol. Second, antisera can be produced only in limited volumes, and it is therefore not possible to use the identical serological reagent at different stages of a complex experiment(s) or in clinical tests. Third, even antibody preparations purified by affinity chromatography by passage over an antigen column may include minor populations of antibodies that yield unexpected cross-reactions or false positives, which confound the results obtained.

Monoclonal Antibodies

To avoid the above-mentioned problems encountered with the use of polyclonal antibodies and antisera, it was desirable to make an unlimited supply of antibody molecules of homogeneous structure (derived from a single clone of B cells), high affinity, and known specificity for a specific target antigen. This was achieved by the production of MAbs from cultures of hybrid antibody-forming B cells **(hybridomas)** (Fig. 2.29; Milestone 2.5).

Recombinant Antibodies

The modular nature of antibody functions has made it possible to convert a mouse MAb into one that has some human segments in its Fc region but still retains its original antigen-binding specificity in its Fab region. This mouse–human hybrid molecule is called a chimeric antibody or, with more human sequences, a "humanized" antibody. Presently, more than 100 different humanized MAbs are in use clinically for therapeutic, diagnostic, and/or preventive applications. The methodologies developed to construct chimeric and humanized antibodies, and further details about the therapeutic applications of these types of antibodies, are discussed in chapter 9.

Spleen cells producing antibody
from mouse immunized with
antigen A

Immortal myeloma cells lacking
antibody secretion and the
enzyme HGPRT

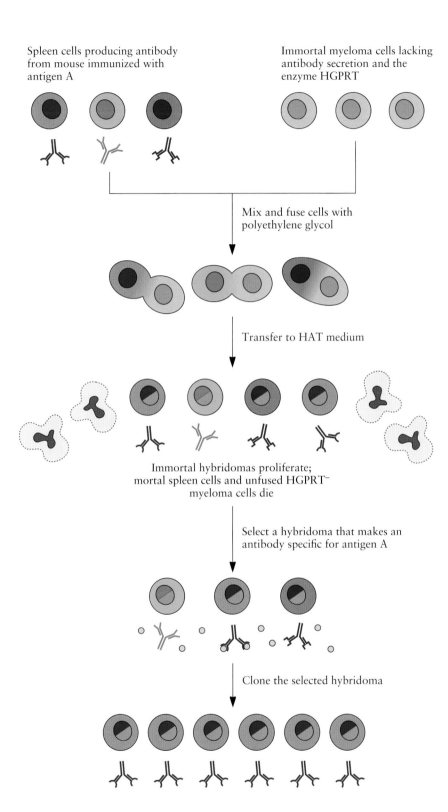

Mix and fuse cells with
polyethylene glycol

Transfer to HAT medium

Immortal hybridomas proliferate;
mortal spleen cells and unfused HGPRT⁻
myeloma cells die

Select a hybridoma that makes an
antibody specific for antigen A

Clone the selected hybridoma

Figure 2.29 Production of MAbs. Mice are immunized with antigen A and boosted intravenously 3 days before sacrifice to produce a large population of spleen cells secreting a specific antibody. Since spleen cells die after a few days in culture, to produce a continuous source of antibody they are fused with immortalized myeloma cells by using polyethylene glycol to produce a hybrid cell line called a hybridoma. The myeloma cells are selected to ensure that they are not secreting antibody themselves and that they are sensitive to the hypoxanthine–aminopterin–thymidine (HAT) medium, which is used to select hybrid cells because they lack the enzyme hypoxanthine:guanine phosphoribosyl transferase (HGPRT). The *HGPRT* gene contributed by the spleen cells allows hybrid cells to survive in HAT medium and grow continuously in culture because of the malignant potential contributed by the myeloma cells. Unfused myeloma cells and unfused spleen cells (cells with dark, irregular nuclei) die in HAT medium. Individual hybridoma cells are screened for antibody production. The cells that produce an antibody of the desired specificity are cloned by expansion from a single antibody-producing cell. The cloned hybridoma cells are grown in bulk culture to produce large amounts of antibody. Since all cells of a hybridoma line make the same antibody, this antibody is termed monoclonal. Adapted with permission from Murphy et al., *Janeway's Immunobiology*, 7th ed. (Garland Science, New York, NY, 2008). doi:10.1128/9781555818890.ch2.f2.29

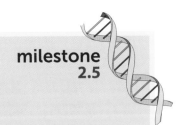

Continuous Cultures of Fused Cells Secreting Antibody of Predefined Specificity

G. KÖHLER AND C. MILSTEIN
Nature **256:**495–507, 1975

milestone
2.5

A search for homogeneous antibody preparations that could be chemically analyzed took advantage of proteins produced and secreted by patients with **multiple myeloma,** a plasma cell tumor. Knowing that antibodies are produced by plasma cells and that this disease is associated with the presence of large amounts of a homogeneous γ-globulin called a myeloma protein in a patient's serum, it seemed possible to biochemists in the 1960s and 1970s that myeloma proteins might serve as models for normal antibody molecules. Structural studies of human myeloma proteins demonstrated that MAbs could be obtained from immortalized plasma cells. However, the antigen specificity of myeloma proteins was generally unknown, which compromised their analyses and limited their application as immunological tools.

It was reasoned that B-cell hybridomas would provide a consistent and continuous source of identical antibody molecules. Unfortunately, the B cells that synthesize antibodies do not reproduce in culture. However, it was envisioned that a hybrid cell type could be created to solve this problem. This hybrid would have the B-cell genetic components for producing antibodies and the cell division functions of a compatible cell type to enable the cells to grow in culture. It was known that normal B cells sometimes become cancer cells (myelomas) that acquire the ability to grow in culture while retaining many of the attributes of B cells. Thus, myeloma cells that did not produce antibodies became candidates for fusion with antibody-producing B cells.

In the mid-1970s, these ideas were realized. Georges Köhler, Cesar Milstein, and Niels K. Jerne devised an ingenious technique for producing a homogeneous population of antibodies of known antigenic specificity. Spleen cells from a mouse immunized with a given antigen were fused to mouse myeloma cells to yield hybrid cells that both divided continuously and secreted antibody specific for the same antigen as used to immunize the spleen cell donor. The spleen B cell can make specific antibody, while the myeloma cell can grow indefinitely in culture and secrete immunoglobulin continuously. By using a myeloma cell partner that does not produce antibody proteins (i.e., a nonproducer line), the antibody produced by the hybrid cells originates only from the immune spleen B-cell partner. After fusion, the hybrid cells are selected using drugs that kill the myeloma parental cell, while the unfused parental spleen cells have a limited life span and soon die. In this way, only hybrid myeloma cell lines or hybridomas survive in culture. Those hybridomas producing antibody of the desired specificity are then identified and cloned by regrowing the cultures from single cells (Fig. 2.29). Since each hybridoma is a clone derived from fusion with a single B cell, all the antibody molecules it produces are identical in structure, including their antigen-binding site and isotype. Such antibodies are called MAbs. This technique revolutionized the use of antibodies by generating a limitless supply of antibody of a single and known specificity. As a result, Köhler, Milstein, and Jerne shared the Nobel Prize in Physiology or Medicine in 1984. Currently, both mouse and human (or humanized) MAbs are used routinely in many serological assays as diagnostic probes and as therapeutic agents (biologics) in clinical trials.

Immunological Techniques

Many immune assays are sensitive, specific, and simple. They can be used for a wide range of applications, including drug testing, assessment and monitoring of various cancers, detection of specific metabolites, pathogen identification, and monitoring of infectious agents. However, there are limitations. If the target is a protein, then the use of antibodies requires that the genes contributing to the presence of the target site be expressed and that the target site not be masked or blocked in any way that would block the binding of the antibody to this site. Conventional diagnostic assays for infectious agents rely on either a set of specific traits characteristic of the pathogen or, preferably, one unique, easily distinguishable feature. The aim is to search for the smallest number of biological characteristics that can reveal the presence and precise identity of a pathogen. For example, some infectious agents produce distinct biochemical molecules. The problem is how to identify such molecules when they are present in a biological sample. Often, such a marker molecule (biomarker) can be identified directly in a biochemical assay that is specific for the marker molecule. Nonetheless, this approach could lead to a proliferation of highly individualized detection systems for different pathogenic organisms. A standardized method of identifying any key marker molecule, regardless of its structure, is preferred. Because antibodies bind with high specificity to discrete target sites (antigens), assays based solely on identifying specific antibody–antigen complexes have eliminated the need to devise a unique identification procedure for each particular marker molecule. Many assays currently used for such applications are described briefly below. Some of these assays were developed only recently and will likely be further refined and used more frequently in the future.

Enzyme-Linked Immunosorbent Assay

There are several different ways to determine whether an antibody has bound to its target antigen. An indirect or direct enzyme-linked immunosorbent assay (ELISA) is a procedure frequently used in diagnostic immune assays. In an indirect ELISA protocol (Fig. 2.30A), the sample being tested for the presence of a specific antigen is bound to a plastic microtiter plate (e.g., a 96-well plate). The plate is washed to remove unbound molecules, and residual sticky sites on the plastic are blocked by the addition of irrelevant proteins. A marker-specific primary antibody directed against the target antigen is added to the wells, and after an appropriate time, the wells are washed to remove unbound primary antibody. A secondary antibody directed against the primary antibody is first covalently linked to an enzyme (e.g., alkaline phosphatase) that can convert a colorless substrate into a colored product. This enzyme-linked secondary antibody is added to the wells for an additional period, after which the wells are washed to remove any unbound secondary antibody–enzyme conjugate. The colorless substrate is added for a specified time, and the amount of colored product is quantitated objectively in a spectrophotometer, which speeds up the assay significantly.

> **ELISA**
> enzyme-linked immunosorbent assay

If the primary antibody does not bind to a target epitope in the sample, the second washing step removes it. Consequently, the secondary antibody–enzyme conjugate does not bind to the primary antibody and is removed during washing, with the net result that the final mixture remains colorless. Conversely, if the target epitope is present in the sample, then the primary antibody binds to it, the secondary antibody binds to the primary antibody, and the attached enzyme catalyzes the reaction to form an easily detected colored product. Since secondary antibodies that are complexed with an enzyme are usually available commercially, each new diagnostic test requires only a unique primary antibody. In addition, several secondary antibodies, each with several enzyme molecules attached, bind to one primary antibody molecule, thereby enhancing the intensity of the signal.

To further amplify the sensitivity of detection of an ELISA, advantage may be taken of a biotin–avidin detection system (Fig. 2.30B). Avidin is

A

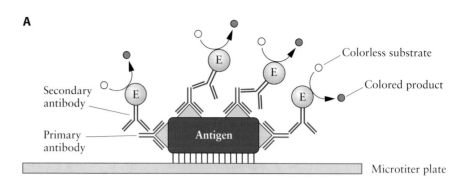

B

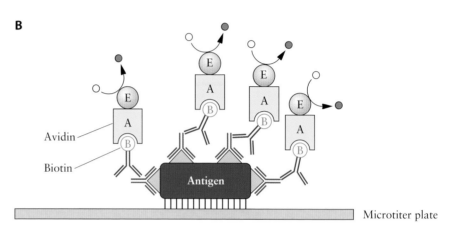

C

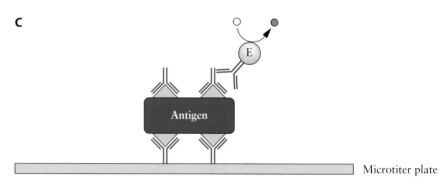

Figure 2.30 ELISA for detecting a target antigen. (**A**) Indirect ELISA. A target antigen is bound to the well of a microtiter plate. A primary antibody to this antigen is then added to the well. After a suitable time, any unbound antibody is washed away, and a secondary antibody covalently linked to an enzyme (E) and a colored substrate is added. Once this substrate is converted to a colored product, the absorbance of light by the colored product is quantified spectrophotometrically. (**B**) Biotin–avidin amplification of an indirect ELISA. The procedure is similar to that in panel A, with the exception that the secondary antibody is conjugated to biotin (B) that is bound to avidin (A). In addition, the avidin is linked to the enzyme (E). This colored reaction is amplified about 10,000-fold compared to that in panel A. (**C**) Direct ELISA. The procedure is similar to that in panel A, with the exception that after first binding a primary antibody to the well and the addition of a sample containing an antigen, an enzyme-linked secondary antibody (directed against the primary antibody) is next added to the well and the amount of colored product is measured. doi:10.1128/9781555818890.ch2.f2.30

K_d
dissociation constant

a tetrameric protein that binds to four biotin molecules. The dissociation constant (K_d) of binding of biotin to avidin is 10^{-13} L/mol, compared with 10^{-9} L/mol for the binding of an antigen to an antibody. Thus, the biotin–avidin system may yield a 10,000-fold level of amplification of an ELISA. In such an indirect ELISA, the secondary antibody is conjugated to a biotin–avidin complex, and an enzyme is linked to avidin. Otherwise, the protocol is the same as that described above for an indirect ELISA.

In a direct ELISA protocol (Fig. 2.30C), a primary antibody (polyclonal or monoclonal) specific for the target antigen is first bound to the surface of the microtiter plate. To assess the amount of a particular antigen in a sample, the sample is added to the well of the plate and allowed to interact with the bound primary antibody. This is followed by a wash to remove any unbound molecules. Then, an enzyme-linked secondary antibody is added, and the presence of bound antigen may be visualized and/or quantified spectrophotometrically.

The principal feature of an ELISA is the specific binding of the primary antibody to the target site (epitope) on the antigen. If the target antigen is a protein, then a purified preparation of this protein is generally used to generate the antibodies (polyclonal or monoclonal) that will be used to detect the epitope(s) on the target antigen. With this assay, the use of a MAb(s) generally provides for an increased affinity, specificity, sensitivity, and stability of binding of an antibody to its target antigen.

Enzyme-Linked Immunospot Assay

The ELISA was developed to detect proteins (i.e., antibodies) that are synthesized and secreted by B cells. An important adaptation of the ELISA to detect and quantitate proteins synthesized and secreted by T cells is the enzyme-linked immunospot (ELISPOT) assay (Fig. 2.31). In an ELISPOT assay, antibodies bound to the surface of a plastic well are used to

ELISPOT
enzyme-linked immunospot

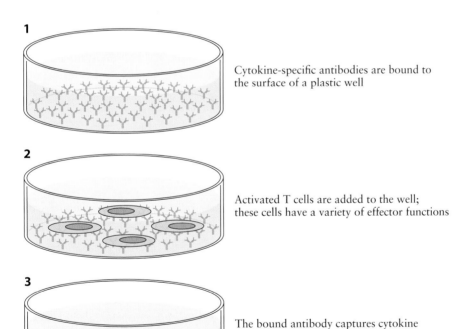

1
Cytokine-specific antibodies are bound to the surface of a plastic well

2
Activated T cells are added to the well; these cells have a variety of effector functions

3
The bound antibody captures cytokine secreted by some activated T cells

4
A second cytokine-specific antibody, coupled to an enzyme, reveals the captured cytokine, yielding a spot of colored, insoluble precipitate

Figure 2.31 Elispot assay. This assay is a modified capture ELISA that may be used to determine the frequency of T cells in a population of cells that secrete a given product, such as a cytokine(s). (**1**) Anti-cytokine capture antibody is bound to wells of a 96-well plastic plate. (**2**) The activated T cells that are added have various effector functions. (**3**) Cell-bound capture antibody may be visualized on the activated T cells that secrete cytokine. (**4**) The enzyme-coupled second cytokine-specific antibody gives rise to a spot of insoluble colored precipitate. Enumeration of the number of spots detected provides an estimate of the frequency of T cells in the mixture that secrete cytokine. Adapted from Murphy et al., *Janeway's Immunobiology*, 7th ed. (Garland Science, New York, NY, 2008). doi:10.1128/9781555818890.ch2.f2.31

capture a large array of cytokines secreted by individual T cells. Usually, cytokine-specific antibodies are bound to the surface of a well in a plastic plate and the unbound antibodies are removed (step 1). Activated T cells are then added to the well and settle onto the antibody-coated surface (step 2). If a T cell secretes a cytokine of interest, this cytokine will be captured by the plate-bound antibodies surrounding the T cell (step 3). After a suitable time, the T cells are removed, and the presence of the specific cytokine is detected using an enzyme-labeled second antibody specific for the same cytokine. Where this second antibody binds, a colored reaction product is formed (step 4). Each T cell that originally secreted a cytokine yields a single spot of color. Counting of the spots yields an estimate of the frequency of cytokine-secreting T cells in the population of cells added to the plate.

Microscopic Detection of Cellular Immune Responses In Vivo

Techniques have recently been developed to analyze various immune responses at a microscopic level. During in vitro responses, individual cells are assayed under experimental conditions. In contrast, during an immune response in vivo, the physiology and dynamics of cell populations may be determined in live animals or humans. An immune response is the sum of many complex and dynamic individual cellular responses influenced by many environmental factors. In vivo experiments maintain this natural environment, but they cannot resolve the behaviors of individual cells. In contrast, in vitro experiments provide information at the subcellular and molecular levels, but they cannot replicate adequately the full repertoire of environmental factors. Thus, techniques that allow real-time observation of single cells and molecules in intact tissues are required. Recent developments in flow cytometry and imaging technology have made such analyses of cellular and molecular immune responses in vivo possible.

Flow Cytometry

The sine qua non instrument of a cell biologist, and of a cellular immunologist in particular, is the flow cytometer, which is used frequently to define, enumerate, and isolate different types and subsets of lymphocytes. The flow cytometer detects and counts individual cells passing in a stream through a laser beam. Equipping a cytometer with the capacity to separate the identified cells enables it to function as a fluorescence-activated cell sorter (FACS), an instrument first constructed about 40 years ago and of ever-increasing importance for use by today's immunologists. These instruments are used to study the properties of cell subsets identified by using MAbs to cell surface proteins (e.g., BCR, TCR, CD4, CD8, MHC-I, and MHC-II). Individual cells in a mixed population are first tagged with fluorochrome-conjugated MAbs against some of these lymphocyte surface antigens. Together with a large volume of saline, the cell mixture is then forced through a nozzle, thereby creating a fine stream of liquid containing cells spaced singly at intervals. Each cell passes through a laser beam and scatters the laser light, and any fluorochrome molecules bound

FACS
fluorescence-activated cell sorter

to the cell surface will be excited and fluoresce. Photomultiplier tubes detect (i) the scattered light that informs us about the size and granularity of the cell and (ii) the fluorescence emissions that provide information about the binding of labeled MAbs and associated expression of the surface antigens of each cell (Fig. 2.32). FACS sorting extends flow cytometry by using electrical or mechanical means to divert and collect cells in droplets with one or more measured characteristics determined by a gate(s) set by the user. Flow cytometric data are displayed in the form of a histogram of fluorescence intensity plotted versus cell number. If two or more fluorochrome-labeled MAbs are used, the data may be displayed as a two-dimensional scatter diagram or as a contour diagram, where the fluorescence of one dye-labeled MAb is plotted against that of a second. The result is that a population of cells labeled with one MAb can be

Figure 2.32 Schematic diagram of a FACS machine. Cells suspended in a core stream (green) are carried in a sheath fluid (light gray) to the flow cell (yellow sphere), where they are interrogated by an excitation laser beam. Cells in the stream are detected by light scattered through the cells (forward scatter [FSC]) and orthogonal to the cells (side scatter [SSC]). Cells labeled with fluorescently tagged MAbs are detected by emitted fluorescent light (FL1). After detection of FSC, SSC, and FL1 signals, droplets are formed and loaded with positive or negative electrostatic charges. Droplets containing single cells are deflected to the left or right by highly charged metal plates, and sorted cells are collected into tubes. Adapted with permission from Jaye et al., *J. Immunol.* **188:** 4715–4719, 2012 (original figure from *J. Immunol.* copyright 2012, The American Association of Immunologists, Inc.). doi:10.1128/9781555818890.ch2.f2.32

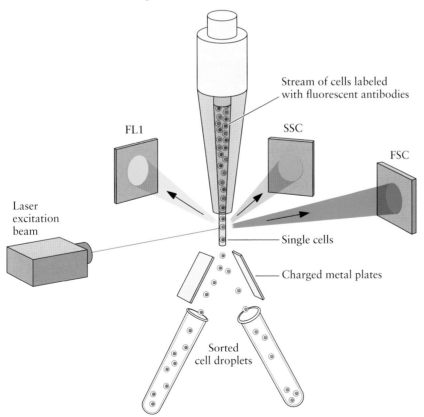

further subdivided by its labeling with the second MAb. Thus, flow cytometry can provide objective quantitative data on the percentage of cells bearing different surface antigens that mediate either early T- and B-cell development or innate and adaptive immune responses.

As the amounts of suitable fluorochrome dyes, hardware, and data analysis tools have increased significantly in recent years, it is now possible to perform as many as 20 color analyses using FACS machines that consist of several very sensitive photomultipliers. The ability to identify the cells and mechanisms that mediate the onset of many critical immune diseases, e.g., depletion of $CD4^+$ T cells in HIV, has benefitted greatly from the application of FACS analyses and sorting. This benefit is expected to continue in the future, as flow cytometry analyses currently permit the recognition of various types of cancer each with a unique pathophysiology and treatment strategy. Such analyses have enabled the isolation of tumor-free populations of hematopoietic stem cells for cancer patients undergoing stem cell transplantation. The ability of a modern-day flow cytometer to detect minute quantities of specific cells in heterogeneous cell mixtures is now used to identify residual malignant cells after therapy in common malignancies for which disease persistence predicts a worse prognosis.

Flow cytometry can also rapidly determine relative cell DNA content for acute lymphoblastic leukemia, the most common childhood blood cancer, and this helps significantly to guide treatment. Multiplex arrays of fluorescent beads that selectively capture proteins and specific DNA sequences have also been investigated by flow cytometry. The latter studies have yielded highly sensitive and rapid methods for high-throughput analyses of cytokines, antibodies, and HLA genotypes, which have already realized important applications in predicting clinical outcomes in bone marrow stem cell transplantation and cardiovascular disease. Lastly, automated analysis of very large data sets has contributed to the development of a "cytomics" field that integrates the use of flow cytometry into analyses of cellular physiology, genomics, and proteomics.

Mass Cytometry

The ability to track many genes simultaneously in a single cell is required to resolve the high diversity of cell subsets, as well as to define their function in the host. Fluorescence-based flow cytometry is the current benchmark for these functional analyses, as it is possible to quantify 18 proteins/ cell at a rate of >10,000 cells/s. Recent advances in a next-generation postfluorescence single-cell technology, termed mass cytometry, have led to a new technology that couples flow cytometry with mass spectrometry and can theoretically measure 70 to 100 parameters/cell (Fig. 2.33). Mass cytometry offers single-cell analysis of at least 45 simultaneous parameters without the use of fluorochromes or spectral overlap. In this methodology, stable nonradioactive isotopes of nonbiological rare earth metals are used as reporters to tag antibodies that may be quantified in a mass spectrophotometer detection system. By applying the resolution, sensitivity, and dynamic range of this detection system on a timescale that permits

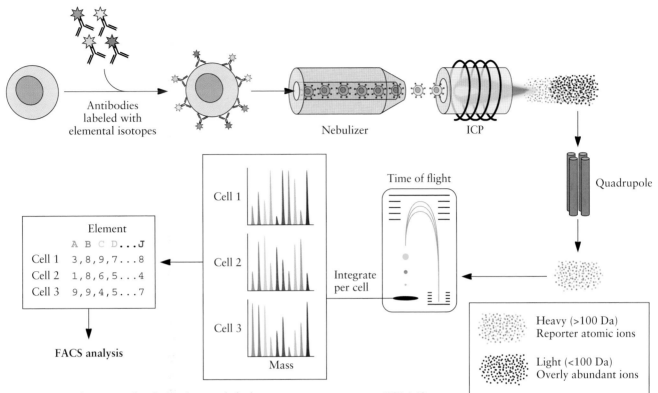

Figure 2.33 Schematic of inductively coupled plasma mass spectroscopy (ICP-MS)-based analysis of cellular markers. An affinity product (e.g., antibody) tagged with a specific element binds to the cellular epitope. The cells are introduced into the ICP by droplet nebulization. Each cell is atomized and ionized, overly abundant ions are removed, and the elemental composition of remaining heavy elements (reporters) is determined. Signals corresponding to each elemental tag are then correlated with the presence of the respective marker and analyzed with a FACS machine as described for Fig. 2.32. Da, daltons. Adapted from Bendall et al., *Trends Immunol.* **33**:323–332, 2012, with permission from Elsevier. doi:10.1128/9781555818890.ch2.f2.33

the measurement of 1,000 cells/s, this methodology offers a new approach to high-content cytometric analysis. One such type of analysis is immunophenotyping by mass spectrometry, which now provides the ability to measure >36 proteins/cell at a rate of 1,000 cells/s. Immunophenotyping is a process used to identify cells based on the types of antigens or markers on the surface of the cell. This process may also be used to diagnose specific types of leukemia and lymphoma by comparing the cancer cells to normal cells of the immune system. Hence, it is evident that further development and refinement of flow cytometry and mass cytometry will continue to provide major advances in several areas of clinical medicine, including discovery, pathophysiology, and therapy of disease.

Two-Photon Intravital Cell Imaging

T-cell interactions with antigen-bearing dendritic cells in secondary lymphoid organs mediate important adaptive immune responses to infectious microbes. These interactions trigger the T cells to proliferate, differentiate, and mature into effector T cells, which may migrate to sites of inflammation and secrete cytokines that are essential to block and/ or eliminate the infection. This targeted delivery of effector T-cell function is dependent on antigen-specific interactions with dendritic cells in the infected tissue. Knowledge of how these effector T-cell–dendritic cell interactions and the specific cytokines secreted by the effector T cells mediate productive and nonproductive adaptive immune responses is essential for the further development of novel vaccines dependent on such cell-mediated responses.

Prior to vaccine development, much more must be learned about the dynamics of effector T-cell migration through sites of infection, the frequency with which these T cells are activated to full effector function, and the precise location of effector-derived cytokine delivery during an adaptive immune response. Recent advances in dynamic imaging methods now enable the direct observation of immune cell function in complex tissues in vivo. In particular, two-photon intravital imaging techniques have been used to analyze naïve T-cell migration during antigen-specific activation by dendritic cells in primary lymphoid sites such as lymph nodes. More recently, imaging studies involving infection of several tissues (e.g., liver, brain, and skin) with various pathogens or during autoimmune processes have addressed the sequence of interactions between dendritic cells and antigen-specific effector T-cell subsets in nonlymphoid sites.

Two-photon laser scanning microscopy (intravital imaging) and flow cytometry are used to track T-cell migration and cytokine secretion consequent to T-cell–dendritic cell interactions. Two-photon fluorescence excitation uses extremely brief (<1 picosecond) and intense pulses of light to view directly into living tissues, to a greater depth and with less phototoxicity than with conventional imaging methods. Real-time imaging of fluorescently labeled cells at the single-cell level and under physiological conditions in deep-tissue environments, such as those that mediate T-cell–dendritic cell interactions in lymph nodes, has enhanced our understanding of the dynamics of T-cell–dendritic cell contacts in vivo and their mode of regulation of T-cell activation and migration. It is now possible to track the behavior of T cells, located up to 100 to 300 μm below the surface of lymph node tissue, during either the 24 h they spend, on average, in a given lymph node or during the 3 to 4 days they spend in a lymph node after exposure to an antigen. Imaging of lymphocyte behavior in lymph nodes can be performed in surgically exposed inguinal lymph nodes (located in the groin region of the abdomen) or popliteal lymph nodes (located beneath the knee joint) of live, anesthetized mice under physiological conditions of temperature and oxygen metabolism, preservation of vascular and lymphatic flow, and innervation. Thus, two-photon real-time cell imaging has extended single-cell approaches to the in vivo setting and can reveal in detail how tissue organization, extracellular factors, and cell movement combine to support the development of desirable and undesirable (infection) immune responses.

summary

Both innate and adaptive immune responses are essential for protection of humans from microbial infection and inflammation. Innate immunity mediates an initial, early antigen-nonspecific host defense mechanism of protection against microbial infections. Subsequently, adaptive immunity develops more slowly and confers specificity against a foreign antigen, and it therefore mediates the later and more vigorous defense against infections. In the absence of these defenses, individuals rapidly succumb to infection. Adaptive immunity responds more effectively against each successive exposure to a microbe, thus conferring immunological memory on the immune system.

During adaptive responses, effector cells that eliminate foreign antigens (e.g., pathogenic microbes) can be activated by T cells that express about 10^5 pathogen-specific TCRs/cell, cytokines produced and secreted by activated T cells, and phagocytic cells (e.g., macrophages) activated by their innate receptors. The activated effector T cells circulate directly to sites (primary and/or secondary lymphoid organs) of the antigen displayed on antigen-presenting cells and interact with these cells. In contrast, B cells express BCRs on their surface and receive help from Th cells to produce soluble antibodies that circulate in blood to bind specific antigens on infectious pathogens. The BCRs and circulating antibodies may have specificity for the same epitopes of an antigen.

Several phenotypically distinct subpopulations of T cells exist, each of which may have the same specificity for an antigenic epitope. However, each subpopulation may perform different functions. This is analogous to the different classes of immunoglobulin molecules, which may have identical antigenic specificities but different biological functions. The major subsets of T cells include CD4$^+$ Th cells, CD8$^+$ CTLs, and CD4$^+$ T regulatory cells. The functions of these T-cell subsets include the following.

B-cell help. Th cells cooperate with B cells to enhance antibody production. Antigen-stimulated Th cells release cytokines that provide activation, proliferation, and differentiation signals for B cells.

Inflammatory effects. On activation, certain Th cells release cytokines that induce the migration and activation of monocytes and macrophages, leading to inflammatory reactions.

Cytotoxic effects. CTLs can deliver a lethal hit on contact with their target cells, leading to their death.

Regulatory effects. Th cells can be further subdivided into different functional subsets that are commonly defined by the cytokines they release. These subsets (Th1 and Th2) have distinct regulatory properties that are mediated by the different cytokines they release. Th1 cells can negatively cross-regulate Th2 cells and vice versa. T regulatory cells coexpress CD4 and CD25 on their surface (CD25 is the IL-2 receptor α chain). Recently, the regulatory activity of these CD4$^+$ CD25$^+$ cells and their role in actively suppressing autoimmunity have been widely studied.

Cytokine effects. Cytokines produced by each of the T-cell subsets (principally Th cells) exert numerous effects on many cells, lymphoid and nonlymphoid. Thus, directly or indirectly, T cells communicate and collaborate with many cell types. Binding of an antigen to its TCR is not sufficient to activate T cells. At least two signals must be delivered to the antigen-specific T cell for activation to occur. Signal 1 involves the binding of the TCR to the antigen, which must be presented in the appropriate manner by antigen-presenting cells. Signal 2 involves costimulators (molecules that, in addition to the TCR, also stimulate the activation of T-cell signaling pathways and proliferation), including cytokines (e.g., IL-1, IL-4, and IL-6) and cell surface molecules expressed on antigen-presenting cells (e.g., CD80 and CD86). The term costimulator also includes stimuli such as microbial antigens and damaged tissue that enhance the delivery of signal 1.

Although the humoral and cellular arms are distinct components of adaptive immune responses, these two arms generally interact during a response to a given pathogen. These various interactions elicit maximal survival advantage for the host by eliminating the antigen and by protecting the host from mounting an immune response against self. Thus, the study of how the immune system works can be of great benefit for survival, best exemplified in recent times by the successful use of polio vaccines in the mid-20th century. More recently, the application of immunology, as it relates to organ (e.g., human heart and liver) transplantation, has also received much public attention, in view of the significant shortage of donor organs for transplantation.

Whereas innate and adaptive immune responses against inflammation protect from infectious diseases and are essential for human survival, an inflammatory response may also protect the host. Innate immune responses mediate the detection and rapid destruction of most infectious agents we encounter daily. These responses collaborate with adaptive immune responses to generate antigen-specific effector mechanisms that lead to the death and elimination of the invading pathogen. Thus, vaccination against infectious diseases continues to be an effective form of prophylaxis.

The application of molecular, cellular, cytometric, imaging, and bioinformatic techniques promises many significant benefits for the future of immunology, particularly in the areas of

(continued)

summary *(continued)*

vaccine development and the control of immune responses. Instead of the time-consuming empirical search for an attenuated virus or bacterium for use in immunization, one may now use pathogen-specific protein sequence data and bioinformatics to identify candidates to be tested.

The ability to modulate and control various immune responses also offers much promise for the treatment of disease. Techniques of MAb production, cell isolation and transfer, gene isolation and transfer, clonal reproduction and biosynthesis, etc., have contributed to rapid progress in the characterization and synthesis of various cytokines and chemokines that enhance and control the activation of various cells associated with immune responses. Powerful and important modulators have been synthesized using recombinant DNA technology and are being tested for their therapeutic efficacies in a variety of diseases, including many different cancers.

review questions

1. What is the raison d'être of the immune system, and what are its principal functions? Why have professionals and the lay public been so highly impacted by advances in immunology?

2. What is the most effective method of protection of an individual against infection? Cite several examples, and outline how this method proved so effective in combating infectious diseases in the 20th century.

3. How many main types of immune responses exist, and what are their salient properties, functions, and mechanisms of action? Do these responses occur independently, or can one type of response promote another type?

4. Which types of immune responses protect individuals from infections? How can inflammatory responses both elicit infection and protect against infection?

5. Which components (tissues, cells, and molecules) of the immune system are essential for the recognition and elimination of microbial pathogens?

6. What are PAMPs, pattern recognition receptors, and DAMPs, and why are they important in innate immunity? Which types of cells and cellular compartments express innate immune receptors, and why are such receptors found in these compartments?

7. How do innate and adaptive immune responses differ in specificity and diversity? What components of each type of response mediate this difference in diversity?

8. What are TLRs, how many TLRs exist, which transcription factors are activated by TLR signals, and why do we need multiple TLRs for immune responses?

9. What is an inflammasome, and how does it induce inflammatory reactions to microbes in damaged tissues?

10. How are adaptive immune responses stimulated by innate immune responses?

11. Why have two types of adaptive immunity evolved, and do these types of adaptive immune responses target the same or different microbes?

12. What are the main classes of leukocytes and lymphocytes, how can they be identified, and how do their functional roles differ?

13. What is meant by immunological memory and immune tolerance?

14. Why is it important that naïve, effector, and memory T and B cells exist, and what are the main differences in their functions?

15. Where are T and B cells found in lymph nodes before antigen stimulation, and where and how do the cells migrate in these nodes after antigen encounter? How do naïve and effector T cells differ in their patterns of migration, and which molecules mediate this differential migration?

16. What is the clonal selection hypothesis, and what are its main postulates?

17. Where are the precursor cells for the lymphoid and myeloid cell lineages found, and how do these precursors as well as their mature cell forms circulate through the body into specific tissues?

18. What are the cells of the myeloid lineage, and how do they control immune responses?

19. What are the cells of the lymphoid lineage, and how do they control immune responses?

20. What is MHC restriction?

21. What are professional antigen-presenting cells, and why are they important for MHC restriction and T-cell activation?

review questions *(continued)*

22. What are the key features of presentation of peptides by MHC proteins to T cells?

23. How is antibody diversity generated? What are the structural features of an antibody molecule that enable it to carry out its various functions? Describe these functions.

24. What different classes, subclasses, and types of antibody molecules exist? What is the functional relevance of having this collection of molecules?

25. How are polyclonal and monoclonal antibodies produced, and how may their structures be modified to permit their application to immune therapy?

26. How are the ELISA and the ELISPOT assay performed, and under which circumstances?

27. What are some of the recent advances in flow cytometry, mass cytometry, and two-photon intravital cell imaging that significantly advance our understanding of the molecular and cellular interactions that mediate immune responses?

references

Abbas, A., and A. Lichtman. 2011. *Basic Immunology: Functions and Disorders of the Immune System*, 3rd ed. Saunders Elsevier, Philadelphia, PA.

Bendall, S. C., G. P. Nolan, M. Roederer, and P. K. Chattopadhyay. 2012. A deep profiler's guide to cytometry. *Trends Immunol.* **33**:323–332.

Bendall, S. C., E. F. Simonds, P. Qiu, E. D. Amir, P. O. Krutzik, et al. 2011. Single cell mass cytometry of differential immune and drug responses across a human hematopoietic continuum. *Science* **332**:687–696.

Beutler, B. 2004. Innate immunity: an overview. *Mol. Immunol.* **40**:845–859.

Bousso, P. 2008. T cell activation by dendritic cells in the lymph node: lessons from the movies. *Nat. Rev. Immunol.* **8**:675–684.

Burnet, F. M. 1957. A modification of Jerne's theory of antibody production using the concept of clonal selection. *Aust. J. Sci.* **20**:67–69.

Cahalan, M. D., I. Parker, S. H. Wei, and M. J. Miller. 2002. Two-photon tissue imaging: seeing the immune system in a fresh light. *Nat. Rev. Immunol.* **2**:872–880.

Centers for Disease Control and Prevention. 2012. Notifiable diseases and mortality tables. *MMWR Morb. Mortal. Wkly. Rep.* **61**:184.

Germain, R. N., M. J. Miller, M. L. Dustin, and M. C. Nussenzweig. 2006. Dynamic imaging of the immune system: progress, pitfalls and promise. *Nat. Rev. Immunol.* **6**:497–507.

Glick, B. R., J. J. Pasternak, and C. L. Patten. 2010. *Molecular Biotechnology: Principles and Applications of Recombinant DNA*, 4th ed. ASM Press, Washington, DC.

Hammerling, G. 1997. The 1996 Nobel Prize to Rolf Zinkernagel and Peter Doherty. *Cell Tissue Res.* **287**:1–2.

Hozumi, N., and S. Tonegawa. 1976. Evidence for somatic rearrangement of immunoglobulin genes coding for variable and constant regions. *Proc. Natl. Acad. Sci. USA* **73**:3628–3632.

Hulett, H. R., W. A. Bonner, J. Barrett, and L. A. Herzenberg. 1969. Cell sorting: automated separation of mammalian cells as a function of intracellular fluorescence. *Science* **166**:747–749.

Jaye, D. L., R. A. Bray, H. M. Gebel, W. A. C. Harris, and E. K. Waller. 2012. Translational applications of flow cytometry in clinical practice. *J. Immunol.* **188**:4715–4719.

Khan, A. R., B. M. Baker, P. Ghosh, W. E. Biddison, and D. C. Wiley. 2000. The structure and stability of an HLA-A*0201/octameric tax peptide complex with an empty conserved peptide-N-terminal binding site. *J. Immunol.* **164**:6398–6405.

Köhler, G., and C. Milstein. 1975. Continuous cultures of fused cells secreting antibody of predefined specificity. *Nature* **256**:495–497.

Kono, H., and K. L. Rock. 2008. How dying cells alert the immune system to danger. *Nat. Rev. Immunol.* **8**:279–289.

Murphy, K., P. Travers, and M. Walport. 2008. *Janeway's Immunobiology*, 7th ed. Garland Science Publishing Inc., New York, NY.

Nielsen, L. S., A. Baer, C. Müller, K Gregersen, N. T. Mønster, S. K. Rasmussen, D. Weilguny, and A. B. Tolstrup. 2010. Single-batch production of recombinant human polyclonal antibodies. *Mol. Biotechnol.* **3**:257–266.

Steinman, R. M., and Z. A. Cohn. 1973. Identification of a novel cell type in peripheral lymphoid organs of mice. 1. Morphology, quantitation, tissue distribution. *J. Exp. Med.* **137**:1142–1162.

Steinman, R. M., and Z. A. Cohn. 1974. Identification of a novel cell type in peripheral lymphoid organs of mice. 2. Functional properties in vitro. *J. Exp. Med.* **139**:380–397.

Wikipedia. 2012. List of monoclonal antibodies. http://en.wikipedia.org/wiki/List_of_monoclonal_antibodies.

Wilson, B. A., A. A. Salyers, D. D. Whitt, and M. E. Winkler. 2011. *Bacterial Pathogenesis: A Molecular Approach*, 3rd ed. ASM Press, Washington, DC.

Zinkernagel, R. M., and P. C. Doherty. 1974. Restriction of in vitro cell-mediated cytotoxicity in lymphocytic choriomeningitis within a syngeneic or semi-allogeneic system. *Nature* **248**:701–702.

Zinkernagel, R. M., and P. C. Doherty. 1997. The discovery of MHC restriction. *Immunol. Today* **18**:14–17.

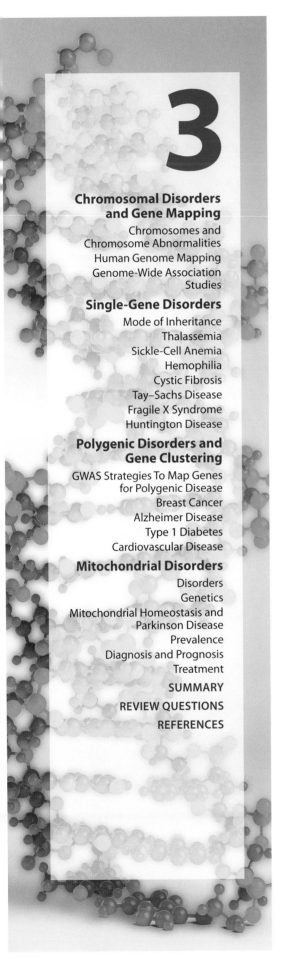

3

The Genetic Basis of Disease

D URING THE PAST DECADE, DNA-based technologies have developed dramatically. Initial research in this area began with chromosomal karyotyping, which then evolved to analyses of chromosomal structural and numerical abnormalities. These analyses were followed by the application of microarray-based and **genome-wide association study (GWAS)**-based projects of common and rare **structural variants (SVs)** in genes, and they more recently have been rapidly extended to a variety of human genetic mapping studies. The latter studies have included several projects, such as the Human Genome Project (Milestone 3.1), HapMap Project, and 1000 Genomes Project. These and other studies have significantly expanded our knowledge of chromosome structure, gene structure, and gene variation in both health and disease. As a result, genetic mapping has significantly advanced our understanding of general biology and the pathogenesis of disease. Relatively recent evidence demonstrates that genetic association studies can identify new chromosomal loci associated with a disease. Such studies have enabled us to reach much closer to our ultimate goal, i.e., to identify the major cellular pathways in which genetic variation contributes to the susceptibility and inheritance of common diseases. This chapter presents a discussion of the rapid progress made from our initial limited knowledge of chromosomes and chromosome abnormalities to the more recent developments in our understanding of the extent and mechanisms of genetic variation (Milestones 3.2 and 3.3) and their important role in the genetic control of disease susceptibility.

Chromosomal Disorders and Gene Mapping

Chromosomes and Chromosome Abnormalities

All nucleated cells contain chromosomes that consist of DNA and proteins (histones) in a compact structure. Chromosomes carry all of our genes and therefore all of our genetic information. DNA with its associated

THE HUMAN GENOME PROJECT

Initial Sequencing and Analysis of the Human Genome
E. S. LANDER ET AL.
Nature **409**:860–921, 2001

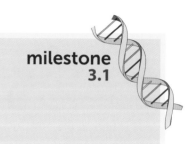

The Sequence of the Human Genome
J. C. VENTER ET AL.
Science **291**:1304–1351, 2001

The sequencing of the human genome was the centerpiece of the Human Genome Project. It was truly an international effort. Over 20 laboratories in six countries were directly involved in the sequencing itself. Many others contributed to the mapping of the human genome and to disseminating the information to the scientific community. The National Human Genome Research Institute's human sequencing program began with a set of pilot projects in 1996 and scaled up to full production levels in 1999. A draft version of the sequence was published in *Nature* in February 2001. The sequencing effort continued at full strength and is now complete. In an article entitled "Our Genome Unveiled" (Baltimore, 2001), David Baltimore commented that "the draft sequences of the human genome are remarkable achievements. . . . They provide an outline of the information needed to create a human being and show, for the first time, the overall organization of a vertebrate's DNA. . . . These papers launched the era of post-genomic science. . . . It reflects the scientific community at its best: working collaboratively, pooling its resources and skills, keeping its focus on the goal, and making its results available to all as they were acquired."

The completion of the finished sequence coincided with the April 2003 50th anniversary of the discovery of the DNA double helix by James Watson and Francis Crick and was announced at a meeting entitled "50 Years of DNA: from Double Helix to Health, a Celebration of the Genome."

COPY NUMBER VARIATIONS

Detection of Large-Scale Variation in the Human Genome
A. IAFRATE ET AL.
Nat. Genet. **36**:949–951, 2004

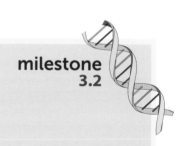

Large-Scale Copy Number Polymorphism in the Human Genome
J. SEBAT ET AL.
Science **305**:525–528, 2004

In two seminal 2004 papers in *Nature Genetics* and *Science*, researchers reported on another previously unknown layer of DNA sequence variation. Until 2004, the genetic community regarded SNPs to be the prime source of genetic variation between individuals that is associated with different phenotypes. However, it was found that larger structural changes between individuals were also associated with population differences. CNVs are regions of DNA, generally >1,000 bases long, whose chromosomal copy number differs between individuals. Larger than SNPs, but not large enough to be detected visually, these regions may be deleted, duplicated, triplicated, etc. Similar to SNPs, CNVs may be causative of or associated with disease, disease susceptibility, and other phenotypes. Alternatively, CNVs may serve only as markers for GWASs. Since 2004, geneticists have included CNVs in their analyses, and the numbers and importance of these genetic variations continue to grow. The Database of Genomic Variants, which records CNVs, currently lists more than 38,000 entries.

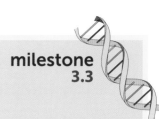

UNRAVELING THE HUMAN GENOME

ENCODE Project Writes Eulogy for Junk DNA
E. Pennisi
Science 337:1159–1161, 2012

An Integrated Encyclopedia of DNA Elements in the Human Genome
The ENCODE Project Consortium
Nature 489:57–74, 2012

Evidence of Abundant Purifying Selection in Humans for Recently Acquired Regulatory Functions
L. D. Ward and M. Kellis
Science 337:1675–1678, 2012

Systematic Localization of Common Disease-Associated Variation in Regulatory DNA
M. T. Maurano et al.
Science 337:1190–1195, 2012

milestone 3.3

In a milestone for the understanding of human genetics, in 30 research papers (including 6 in *Nature* and additional papers published online by *Science*), scientists recently announced the results of 5 to 10 years of work in unraveling the secrets of how the genome operates. Some of these papers are listed here. The Encyclopedia of DNA Elements (ENCODE) project dispensed with the idea that our DNA is largely "junk," i.e., repeating sequences with no function. Instead, The ENCODE Project Consortium found that at least 80% of the genome is important. In addition to encoding proteins, the DNA bases highlighted by ENCODE identify sites for proteins that influence gene activity, RNA strands that perform various roles, or positions in DNA where chemical modifications silence stretches of our chromosomes. The new findings are the most recent in a series of increasingly in-depth looks at the properties of the human genome. Some of the major scientific milestones reached thus far in this continuing endeavor are listed below.

1. Heredity, 1866. The realization that traits and certain diseases can be passed from parent to offspring dates back to the time of Hippocrates, who theorized that "seeds" from different parts of the body were transmitted to newly conceived embryos. Charles Darwin later proposed similar ideas. What these seeds were remained unknown until Gregor Mendel systematically tracked dominant and recessive traits in his studies of pea plants.

2. Chromosomes, 1902. In 1869, Johannes Friedrich Miescher was the first to isolate DNA. During the next 30 years, scientists discovered mitosis, meiosis, and chromosomes. In 1903, Walter Sutton discovered that chromosomes occur in pairs and separate during meiosis. He proposed that "the associations of paternal and maternal chromosomes in pairs and their subsequent separation . . . may constitute the physical basis of the Mendelian law of heredity."

3. Genes, 1941. In 1941, Edward Tatum and George Beadle discovered that genes code for proteins, explaining for the first time how genes direct metabolism in cells. Tatum and Beadle would share half of the 1958 Nobel Prize in Physiology or Medicine for their discovery, which they made by mutating bread mold with X rays.

4. DNA structure, 1953. In 1950, Erwin Chargaff identified that the nucleotides of DNA occur in specific patterns. These nucleotides are represented by A, T, G, and C, and Chargaff was the first to discover that A and T always appeared in equal measures, as did G and C. This discovery was crucial to James Watson and Francis Crick, the scientists who determined the structure of DNA in 1953. Combining Chargaff's work with studies by Maurice Wilkins and Rosalind Franklin and other scientists, Watson and Crick identified the double-helix structure of DNA.

5. Recombinant DNA, 1970s. Recombinant DNA was discovered. This enabled the first animal gene (a segment of DNA containing a gene from the African clawed frog [*Xenopus*] fused with DNA from the bacterium *Escherichia coli*) to be cloned in 1973. In addition, a method of DNA sequencing was developed by Fred Sanger, who in 1980 received his second Nobel Prize in Chemistry.

6. PCR, positional gene cloning, and identification of the first human disease gene, 1980s. PCRs simplified DNA amplification and discovery, techniques to clone and tag DNA were developed, and the first disease gene was cloned in 1986. The first human disease gene identified by positional cloning was one for chronic granulomatous disease on chromosome Xp21. The genes for Duchenne muscular dystrophy and retinoblastoma followed quickly.

7. Human genome catalogued, 2001. In 1977, the complete genome of the bacteriophage φX174 was sequenced. By 1990, a complete cataloguing of the human genome had begun. The Human Genome Project emerged and proved to be a 13-year international effort that resulted in the complete sequencing of the human genome in 2001. The project revealed that humans have about 23,000 protein-coding genes, which represented ~1.5% of the genome. It was

(continued)

milestone 3.3 *(continued)*

thought that the rest of the genome was composed of what was previously called "junk DNA," including fragments of DNA that do not encode any proteins and groups of genes that regulate other portions of the genome.

8. Junk DNA dejunked, 2012. On 5 September 2012, several scientific articles reported on the results of a decade-long ENCODE project showing that at least 80% of the genome is biologically active and that most of the non-protein-coding DNA can regulate the expression and function of nearby genes that encode proteins. Importantly, these findings reveal that the genetic basis of many diseases may be not in protein-coding genes but, rather, in their "regulatory neighbors." For example, genetic variants related to metabolic diseases are located in genetic regions that are activated only in liver cells. Similarly, regions activated in lymphocytes contain variants that are associated with autoimmune disorders, such as systemic lupus erythematosus. These studies are the first to illustrate the chromosomal locations of DNA switches that control human genes in health and disease.

GWAS
genome-wide association study

SV
structural variant

packaging proteins is referred to as **chromatin**. Some regions of chromosomes are tightly packed and are called heterochromatin, while other regions are less condensed and are called euchromatin. Less condensed packing of chromatin generally increases the transcription of genes in the region. Each species has a characteristic number and form of chromosomes, referred to as the **karyotype**. A **karyogram** is a photographic representation of stained chromosomes arranged in order of size, i.e., decreasing length (Fig. 3.1). Each chromosome is paired with its matched or **homologous chromosome**. The matched chromosomes are identical in size and structure but may carry different versions (known as alleles) of

Figure 3.1 Normal human male 46,XY karyogram. Humans have a total of 46 chromosomes that consist of two identical sets of 22 chromosomes (autosomal chromosomes) and 2 sex chromosomes (XX for female and XY for male). Because of the two identical sets of chromosomes, human cells are called diploid. Adapted from http://humandna.co.in/chromosomes.php. doi:10.1128/9781555818890.ch3.f3.1

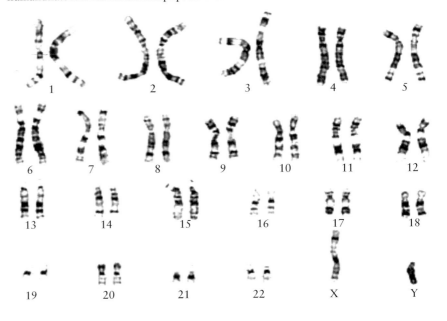

the same gene. Humans have 46 chromosomes, or 23 chromosome pairs, to carry our approximately 25,000 genes. Cells in our body that contain 46 chromosomes are **diploid** ($n = 2$); 23 chromosomes are derived from the mother's egg cell, and the other 23 are from the father's sperm. Egg cells and spermatozoa each contain only 23 chromosomes (Fig. 3.2) and therefore are **haploid** ($n = 1$). In diploid cells, the 46 chromosomes appear as 22 homologous pairs of **autosomes** (nonsex chromosomes) and one pair of sex chromosomes, XX in females and XY in males.

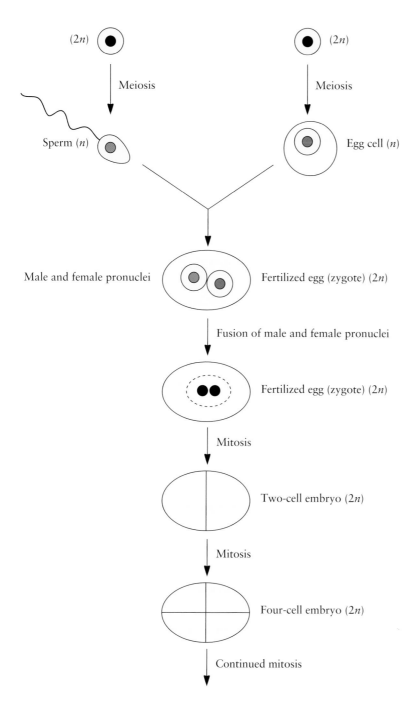

Figure 3.2 Generation of a diploid ($2n$) zygote. A diploid $2n$ fertilized egg (zygote) is produced by the fusion of a haploid (n) sperm and egg. Successive mitotic divisions generate many types of diploid cells in the body (somatic cells) during development and cell turnover (replacement of old cells with new ones) in an embryo. Adapted from Strachan and Read, *Human Molecular Genetics* (Bios Scientific Publishers, Oxford, United Kingdom, 1996), with permission. doi:10.1128/9781555818890.ch3.f3.2

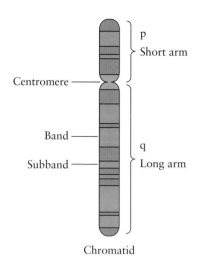

Figure 3.3 Schematic diagram of the structures of a chromosome. The p and q arms, centromere, chromatid, bands, and subbands (described in the text) are shown. A chromatid is an individual chromosome that is paired with a replicated copy of the identical chromosome. This pair of chromosomes is held together at the centromere for the process of cell division. The light and dark bands originate from the differential staining of the regions of the chromosome with the Giemsa stain, as explained in the text. The short arm of each chromosome is denoted with a "p" and the long arm with a "q." Thus, 7q refers to the long arm of chromosome 7. Each arm may be further divided into regions, depending on the size. Adapted from Wallis, *Genetic Basis of Human Disease* (The Biochemical Society, London, United Kingdom, 1999), with permission. doi:10.1128/9781555818890.ch3.f3.3

Human chromosomes are usually studied in rapidly dividing cells, such as peripheral blood lymphocytes. Cell mitosis can be arrested in the metaphase stage of the cell cycle, and the chromosomes can be differentially stained to allow their identification microscopically. This microscopic analysis of chromosomes is known as **cytogenetics**. For routine karyotyping, Giemsa staining is preferred, as this procedure produces alternating light and dark bands (**G banding**) that reflect differential chromosomal structures characteristic of each chromosomal pair (Fig. 3.1). These light and dark bands of chromosomes result from the specificity of binding of the Giemsa stain for the phosphate groups of DNA, as the stain attaches to regions of DNA where there are large amounts of A-T bonding. Thus, Giemsa staining can identify different types of changes in chromosomal structure as gene rearrangements.

Examination of a karyotype enables one to determine either if there is gain or loss of a chromosome(s) or if the structure of a given chromosome(s) is altered. The **centromere** of each chromosome separates the short arm (p) from the long arm (q). Most arms are divided into two or more regions by distinct bands, and each region is further subdivided into subbands (Fig. 3.3). For example, band Xp21.2 is found on the p arm of the X chromosome in region 2, band 1, subband 2.

Chromosome disorders are caused by abnormalities in the number (increase or decrease of genes) or the structure of chromosomes. An individual's physical characteristics are called a **phenotype**, which is the combination of all of that individual's expressed traits, including morphology, development, behavior, and biochemical and physiological properties. Phenotypes result from the expression of genes and environmental factors that interact with these genes. Thus, the phenotype of a person with a chromosomal disorder may vary with the type of chromosomal defect.

Numerical Chromosome Abnormalities

An abnormality in which the chromosome number is an exact multiple of the haploid number ($n = 23$) and is larger than the diploid number ($n = 46$) is called **polyploidy**. Polyploidy arises from fertilization of an egg by two sperm (total number of chromosomes increases to 69) or the failure in one of the divisions of either the egg or the sperm so that a diploid gamete is produced. The survival of a fetus to full term of pregnancy is rare in the instance of polyploidy. **Aneuploidy** occurs when the chromosome number is not an exact multiple of the haploid number and results from the failure of paired chromosomes (at first meiosis) or sister chromatids (at second meiosis) to separate at anaphase. Thus, two cells are produced, one with a missing copy of a chromosome and the other with an extra copy of that chromosome (Fig. 3.4). Examples of numerical chromosomal abnormalities are listed in Table 3.1.

Trisomy 21, the first human chromosomal disorder discovered (in 1959), is an abnormality that displays an extra copy (total of 3 copies) of chromosome 21 (Fig. 3.5) and causes **Down syndrome**. The genes on all three copies of chromosome 21 are normal. However, not all individuals with Down syndrome show the same physical characteristics, indicating that their phenotypes can vary. People with Down syndrome have a typical facial appearance (the face is flat and broad) that includes an

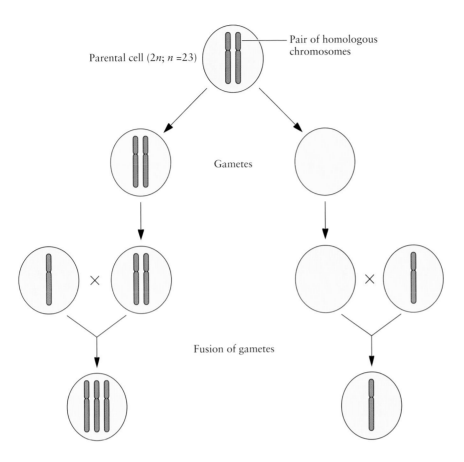

Parental cell (2*n*; *n* =23)

Pair of homologous chromosomes

Gametes

Fusion of gametes

Figure 3.4 Nondisjunction in gamete cell formation and fusion of abnormal gametes with a normal haploid gamete. Aneuploidy occurs when paired chromosomes (at first meiosis) or sister chromatids (at second meiosis) do not separate from each other at anaphase, a stage of meiosis at which sister chromosomes move to opposite sides of the cell. This failure of paired chromosomes to separate, with the chromosomes instead moving to the same side of the cell, is termed nondisjunction. As a result, two cells are produced, one with a missing copy of a chromosome and one with an extra copy of that chromosome, as shown. Redrawn from Dewhurst, *Biol. Sci. Rev.* **10**(5):11–15, 1998, by permission of Philip Allan for Hodder Education.
doi:10.1128/9781555818890.ch3.f3.4

abnormally small chin, skin folds on the inner corners of the eyes, poor muscle tone, a flat nasal bridge, a protruding tongue due to a small oral cavity, an enlarged tongue near the tonsils, a short neck, and white spots on the iris. Growth parameters such as height, weight, and head circumference are smaller in children with Down syndrome than in typical individuals of the same age.

All Down syndrome patients have some degree of mental retardation, albeit moderate. Despite this condition, many persons with Down syndrome can be educated and live with minimal daily assistance, while others require much attention and care. There are several possible health concerns, including cardiac failure and hearing loss. Individuals with

Table 3.1 Numerical chromosomal aberration syndromes

Aneuploidy condition	No. of chromosomes	Karyotype
Tetraploidy	92	XXYY
Triploidy	69	XXY
Trisomy 21 (Down syndrome)	47	XX+21
Trisomy 18 (Edward syndrome)	47	XY+18
Trisomy 13 (Patau syndrome)	47	XX+13
Klinefelter syndrome	47	XXY
Trisomy X	47	XXX
Turner syndrome	45	X

Adapted from Wallis, *Genetic Basis of Human Disease* (The Biochemical Society, London, United Kingdom, 1999), with permission.

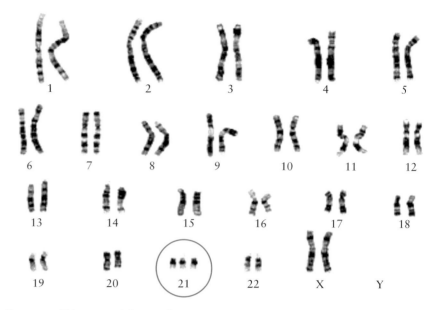

Figure 3.5 Trisomy 21. This syndrome appears when an individual inherits three copies of chromosome 21. The extra copy of chromosome 21 results in Down syndrome. Adapted from http://www.genome.gov/pages/education/modules/chromosomeanalysis .pdf. doi:10.1128/9781555818890.ch3.f3.5

Down syndrome are now living longer than they used to and can survive into their 50s and 60s. However, those individuals who survive to that age are at very high risk of developing Alzheimer disease.

Trisomy 13, the presence of three copies of chromosome 13, causes **Patau syndrome**. Only about 5% of infants with this disorder survive past their first year, and most pregnancies involving trisomy 13 end in miscarriage. Children with trisomy 13 usually have a lot of trouble breathing, especially when they sleep, and many have seizures. All individuals with Patau syndrome have severe mental retardation, and other common characteristics include a small head, extra fingers and/or toes, and a cleft lip or cleft palate.

Trisomy 18, the presence of 3 copies of chromosome 18, elicits **Edward syndrome**. Only about 10% of babies with this syndrome survive past their first year, and the majority of survivors are female, indicating a prenatal selection against males with trisomy 18 after the time of amniocentesis. Children with trisomy 18 usually have problems with breathing and eating, and many have seizures or serious heart conditions. All individuals with trisomy 18 have severe mental retardation. Most babies with trisomy 18 are very small and have certain recognizable facial features. They also tend to overlap their fingers in a very distinct pattern.

Prader–Willi syndrome results from the absence or nonexpression of a group of genes on chromosome 15. A specific form of blood cancer, **chronic myeloid leukemia**, may be caused by a chromosomal translocation, in which portions of two chromosomes (chromosomes 9 and 22) are exchanged. No chromosomal material is gained or lost, but a new, abnormal gene that leads to the development of cancer is formed.

Sex Chromosome Abnormalities

The phenotype of chromosomal disorders can vary depending on whether the chromosomal abnormality occurs on the maternally or paternally derived chromosomes (Table 3.1).

Turner syndrome (45,X) is an example of monosomy, in which a girl is born with only one sex chromosome, an X chromosome.

Klinefelter syndrome occurs in males with the genotype 47,XXY. Such males have 47 chromosomes and are classified as having a sex chromosome trisomy (three sex chromosomes), since they carry two X chromosomes and one Y chromosome. This syndrome affects about 1 in 1,000 males. Most affected males are taller than average, and they may have more body fat in the hips or chest as well as little facial and body hair. Some Klinefelter males are mentally retarded, while many others have normal intelligence. The most common feature of this syndrome is infertility; about 2% of infertile men have Klinefelter syndrome.

Structural Chromosome Abnormalities

Genomic rearrangements may alter genome architecture and yield clinical consequences. Several genomic disorders caused by structural variation of chromosomes were discovered by cytogenetics. Recent advances in molecular cytogenetic techniques have enabled the rapid and precise detection of structural rearrangements on a whole-genome scale. This high resolution illustrates the role of SVs and **single-nucleotide polymorphisms (SNPs)** in normal genetic variation.

SNP
single-nucleotide polymorphism

In analyzing the role of structural gene variants in cell function, it is important to consider the two types of such variants, i.e., **gain-of-function variants** and **loss-of-function variants**. A gain-of-function variant results from a mutation that confers new or enhanced activity on a protein. Most mutations of this type are not heritable (germ line) but, rather, are somatic mutations. A change in the structure of a gene that may arise during DNA replication and is not inherited from a parent, and also is not passed to offspring, is called a **somatic mutation**. Such a mutation that results in a single base substitution in DNA is known as a somatic point mutation. A loss-of-function variant results from a point mutation that leads to reduced or abolished protein function. Most loss-of-function mutations are recessive, indicating that clinical signs are typically observed only when both chromosomal copies of a gene (one being inherited from each parent) carry such a mutation.

In a population, any two unrelated individuals are identical at about 99.5% of their DNA sequence. At a given chromosomal site, one individual may have the A nucleotide and the other individual may have the G nucleotide. This type of site in DNA is known as an SNP (Fig. 3.6). Each of the two different DNA sequences at this site (or gene) is called an **allele**.

Molecular techniques such as array-based **comparative genomic hybridization (CGH)**, SNP arrays, array painting, and **next-generation sequencing** have facilitated and expedited the characterization of chromosome rearrangements in human genomes. These various genomic rearrangements can arise by several mechanisms, including deletions, amplifications, translocations, and inversions of DNA fragments (Fig. 3.7).

CGH
comparative genomic hybridization

1   **2**

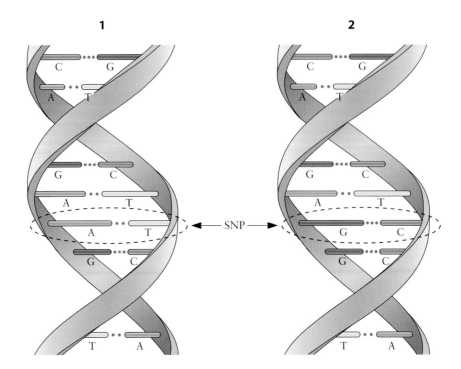

Figure 3.6 Illustration of an SNP. Two sequenced DNA fragments (1 and 2) from different individuals, TGAAA to TGAGA, contain a difference in a single nucleotide (an A/G polymorphism). In this case, the two alleles are called A and G (circled in black dashed line). Watson-Crick bonds between the bases are shown in red. Almost all common SNPs have only two alleles. The distribution of SNPs is not homogeneous, as SNPs usually occur in noncoding regions more frequently than in coding regions of DNA.
doi:10.1128/9781555818890.ch3.f3.6

Deletions occur when a portion of a chromosome is missing or removed. **Duplications** result from the copying of a portion of a chromosome that results in extra genetic material. During a **translocation**, a portion of one chromosome is transferred to another chromosome. In a reciprocal translocation, DNA segments from two different chromosomes are exchanged. A DNA **inversion** results when a portion of a chromosome is broken off, turned upside down, and reattached. When a portion of a chromosome is disrupted, the chromosome may form a circle, or ring, without any loss of DNA.

Figure 3.7 Types of common chromosomal rearrangements. Each box illustrates a healthy chromosome (left) and its derivative altered chromosome (right). The black lines indicate the regions involved. Adapted from Wijchers and de Laat, *Trends Genet.* 27:63–71, 2011, with permission from Elsevier. doi:10.1128/9781555818890.ch3.f3.7

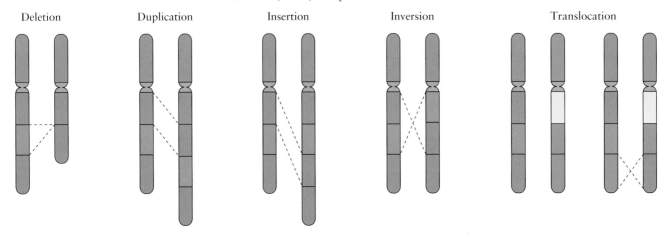

Deletion   Duplication   Insertion   Inversion   Translocation

As most chromosome abnormalities occur accidentally in the ovum or sperm, the abnormality is present in every cell of the body. However, some abnormalities arise after birth, resulting in a condition where a few cells have the abnormality and others do not. Chromosome abnormalities can either be inherited from a parent (e.g. translocation) or develop spontaneously for the first time. This is why chromosome studies are performed on parents when a child is found to have an abnormality.

How do chromosome abnormalities happen? Chromosome abnormalities usually occur when there is an error in cell division (Fig. 3.8). There are two kinds of cell division. **Meiosis** results in cells with half the number of usual chromosomes, 23 instead of the normal 46. These are

Figure 3.8 Nondisjunction of sex chromosomes during first meiosis and mitosis. The first meiotic and mitotic events, with the resultant karyotypes and syndromes (or other outcomes) that result from nondisjunction, are shown. Adapted from Connor and Ferguson-Smith, *Essential Medical Genetics*, 3rd ed. (Blackwell Scientific Publications, Oxford, United Kingdom, 1991) with permission; original figure © 1991 Blackwell Scientific Publications. doi:10.1128/9781555818890.ch3.f3.8

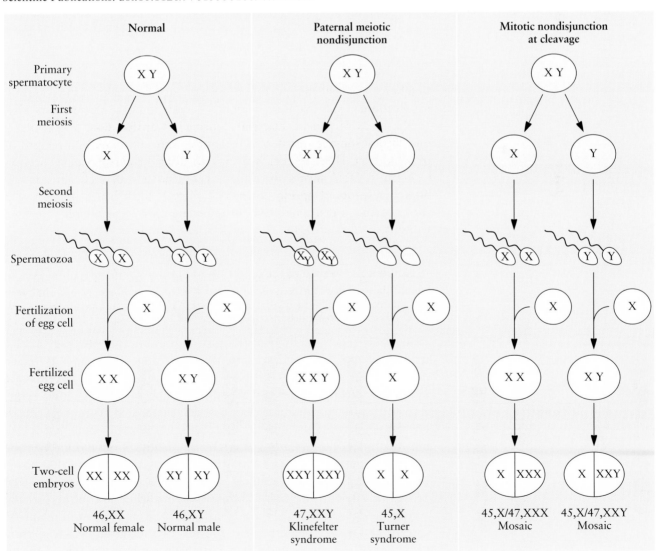

the eggs and sperm. **Mitosis** produces two cells that are duplicates (46 chromosomes each) of the original cell. This kind of cell division occurs throughout the body, except in the reproductive organs. In both processes, the correct number of chromosomes appears in the daughter cells. However, errors in cell division can result in cells with too few or too many copies of a chromosome.

Additional factors can increase the risk of chromosome abnormalities. One factor is maternal age. Women are born with all of the eggs they will ever have. Therefore, when a woman is 30 years old, so are her eggs. Chromosomal errors can appear in eggs as they age. Thus, older women are at greater risk of giving birth to babies with chromosome abnormalities than younger women. Since men produce new sperm throughout their life, paternal age does not increase the risk of chromosome abnormalities. A second factor may be environment, although conclusive evidence is currently lacking.

Genome mapping is the localization of genes that express phenotypes that correlate with DNA variation. Chromosome abnormalities arise from **genomic variants**. Advances in molecular biology and cytogenetic techniques permit the identification of many diverse types of SVs, which contribute to human disease, phenotypic variation, and karyotypic evolution. SVs in individual genomes result from chromosomal rearrangements affecting at least 50 kilobase pairs (kb) and include deletions and duplications known as **copy number variants (CNVs)**, inversions, and translocations (see above). Rearrangements are elicited by multiple events, including external factors such as cellular stress and incorrect DNA repair or recombination.

CNV
copy number variant

Human Genome Mapping

Genomic Variants

Conventional cytogenetic methods, such as chromosome banding and karyotyping, are informative and still commonly used. However, these techniques are limited to the detection of numerical chromosomal aberrations (aneuploidy and polyploidy) and microscopic SVs a few **megabases** in size (Table 3.2). Molecular cytogenetic approaches enable the detection of submicroscopic SVs and have been crucial for studying complex rearrangements generated by more than two chromosomal breakage events, refining breakpoints, and performing cross-species comparisons. These newer approaches have relied predominantly on the use of **fluorescence in situ hybridization (FISH)**, a technique in which fluorescence microscopy reveals the presence and localization of defined labeled DNA probes binding to complementary sequences on targets, traditionally metaphase chromosome spreads. FISH allows the precise identification and localization of chromosomal aberrations within a DNA stretch of about 4,000 kb.

FISH
fluorescence in situ hybridization

To detect translocations in chromosomes, chromosome-specific DNA probes or "paints" are used. **Chromosome painting** is a technique that allows the specific visualization of an entire chromosome in metaphase spreads and in interphase nuclei by in situ hybridization with a mixture of sequences generated from that particular chromosome (Fig. 3.9). Increased resolution is achieved by using oligonucleotide probes rather than

Table 3.2 Methods of identification of classes of chromosomal rearrangements

| Method (date) | Detection | | | | | Resolution | Sensitivity |
	Deletions and duplications	Insertions	Unbalanced translocations	Balanced translocations	Inversions		
G banding (early 1970s)	Yes	Yes	Yes	Yes	Yes	Low (>many Mb)	Low
CGH (early 1990s)	Yes	No	Yes	No	No	Low (>many Mb)	High
M-FISH/SKY (mid-1990s)	Yes	Yes	Yes	Yes	No	Low (>many Mb)	High
BAC array CGH (early 2000s)	Yes	No	Yes	No	No	Average (>1 Mb)	High
Tiling-path BAC array CGH (early 2000s)	Yes	No	Yes	No	No	High (>50–100 kb)	High
Oligonucleotide array CGH (early 2000s)	Yes	No	Yes	No	No	High (0.4–1 kb)	Very high
SNP arrays (late 2000s)	Yes	No	Yes	No	No	High (>5–10 kb)	High
NGS based (late 2000s)	Yes	Yes	Yes	Yes	Yes	Very high (bp level)	Very high

Adapted from Le Scouarnec and Gribble, *Heredity* **108**:75–85, 2012.

Abbreviations: M-FISH, multiple FISH; NGS, next-generation sequencing; SKY, spectral karyotyping.

whole-chromosome probes to bind to chromatin fibers. During mitosis, chromatin DNA fibers become coiled into chromosomes, with each chromosome having two chromatids joined at a centromere. The technique of FISH to chromatin fibers is known as **fiber-FISH**. Alternative targeted approaches have simplified CNV detection. For example, real-time quantitative polymerase chain reaction (PCR) and multiplex ligation-dependent probe amplification (see chapter 6) are frequently used to resolve genetic analyses of clinical material. While these different approaches are restricted to specific regions, some FISH-based techniques were developed

PCR
polymerase chain reaction

Figure 3.9 FISH of painted chromosomes. Appropriately selected painting probes can uniformly decorate all 23 chromosomes. Courtesy of S. M. Carr, Genetix (2008). Reprinted with permission from Genetix. doi:10.1128/9781555818890.ch3.f3.9

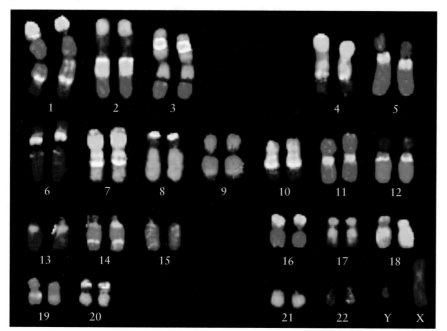

in the 1990s to detect genomic aberrations at the whole-genome level (Table 3.2). As an example, copy number differences between two genomes are now detectable by CGH. Specific translocations and complex rearrangements are characterized by techniques adapted from chromosome painting, e.g., **multiplex FISH** and **spectral karyotyping**, in which all chromosomes are differentially colored in a single experiment. Despite the introduction of these improved methods of chromosome mapping, they remain experimentally demanding and labor-intensive, and the resolution is still limited by the use of chromosomes as targets (Table 3.2).

Identification of SV boundaries is essential for accurate genotype–phenotype correlations, which depend on the extent of genes or regulatory regions that are disrupted or vary in copy number. Completion of the human genome sequence in the early 2000s and progress in molecular biology techniques have generated new genome-wide screening methods that have revolutionized our understanding of the genomes of healthy and diseased individuals. In particular, **microarray** and next-generation sequencing technologies have been very useful in the characterization of chromosome rearrangements.

Array-Based Techniques

Originally developed for gene expression profiling, DNA microarrays or chips are currently used to establish copy number changes (**array-based CGH** and **genotype single-nucleotide polymorphisms**) and analyze **DNA methylation**, alternative splicing, **microRNAs** (sequences of about 22 nucleotides that function as posttranscriptional regulators that result in translational repression or target degradation and gene silencing), and protein–DNA interactions (array-based chromatin immunoprecipitation). Each array consists of thousands of immobilized oligonucleotide probes or cloned sequences. Labeled DNA or RNA fragments are applied to the array surface, allowing the hybridization of complementary sequences between probes and targets (see chapter 1). The chief advantages of this technology are its sensitivity, specificity, and scale, as it enables the relatively rapid assay of thousands of relevant genomic regions of interest in a single experiment. Furthermore, the amount of input sample material required is generally less than 1 µg, which facilitates the assay of precious clinical samples.

Both CGH arrays and SNP arrays can detect CNVs in genomes. The genome-wide coverage of these arrays now permits the discovery of CNVs without any prior knowledge of the DNA sequence. Some arrays may identify recurrent rearrangements more easily or may genotype CNVs present in about 41% of the general population (**copy number polymorphisms**). Currently, array vendors can custom design the content of arrays to increase the resolution in a given chromosomal region(s) of interest where higher resolution is required.

Array CGH

The first whole-genome array, developed in 2004, consisted of about 430,000 overlapping fragments (cloned into bacterium-derived artificial chromosomes) that covered the whole genome. This facilitated the ability to detect changes in the copy number of genes or gene segments. CGH with FISH is a method for analyzing genomic DNA for unbalanced genetic changes. Genomic DNA from the test sample (e.g., tumor cells) is labeled orange or red and mixed with normal genomic DNA labeled another color (e.g., green). The mixture is hybridized (FISH) to a normal human metaphase spread or other reference standard. Regions of imbalance (increased or decreased copy number) in the tumor are located or mapped relative to the normal metaphase chromosomes as increases or decreases in the ratio of green to orange or red fluorescence. The array technology used allows the detection of genetic imbalances as small as just a few kilobase pairs in size, which permits the boundaries of a genetic change to be better defined.

In an array CGH analysis, test (e.g., from a tumor) and reference (e.g., from a healthy person) DNA is labeled with different fluorophores (e.g., Cy5 and Cy3) and then simultaneously hybridized onto arrays in the presence of Cot-1 DNA (enriched for repetitive sequences) to reduce the binding of repetitive sequences (Fig. 3.10). If only small amounts of DNA are available (e.g., in prenatal diagnosis or tumor analysis), amplification

Figure 3.10 Flowcharts of cytogenetics oligonucleotide arrays. (**Left**) An array CGH analysis; (**right**) a cytogenetics array analysis. White and blue boxes, sample preparation stage; orange boxes, microarray stage; green boxes, data processing stage. Methods used for array CGH labeling are enzymatic, restriction digestion, and the Universal Linkage System and can require a fragmentation step (dashed-line box). Hybridization mixtures contain blocking agents and DNA enriched for repetitive sequences to block nonspecific hybridization and reduce background signal. Hybridization times vary depending on the array format. Cy5, cyanine-5; Cy3, cyanine-3; WGA, whole-genome amplification. Adapted from Le Scouarnec and Gribble, *Heredity* **108**: 75–85, 2012. doi:10.1128/9781555818890.ch3.f3.10

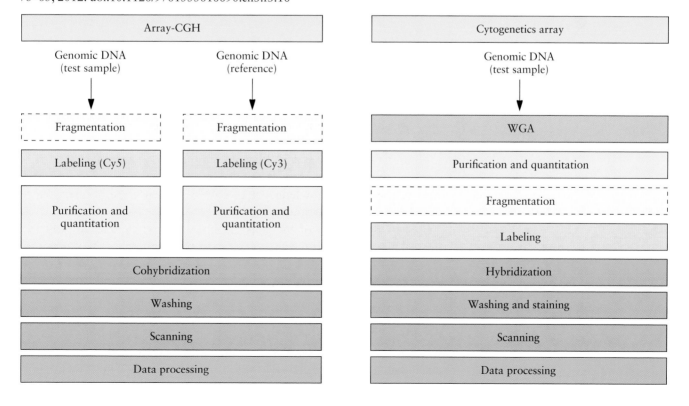

methods can be used before labeling. After hybridization, washing, and scanning, Cy5 and Cy3 fluorescence intensities are measured for each feature on the array and normalized, and $\log_2$ ratios of the test DNA (e.g., Cy5) divided by the reference DNA (e.g., Cy3) are then plotted against chromosome position (Fig. 3.11). For each position, a value of 0 indicates a normal copy number [$\log_2$ (2/2) = 0] result. A $\log_2$ ratio of 0.58 [$\log_2$ (3/2) = 0.58] indicates a gain of one copy in the test sample compared with the reference sample, and a $\log_2$ ratio of −1 ($\log_2$ (1/2) = −1) indicates a loss of one copy in the test sample compared with the reference sample. To identify CNVs in the test DNA, one attempts to minimize the

Figure 3.11 Methods for the association of copy number polymorphisms with disease. (**A**) For a copy number polymorphism with distinct copy number genotypes, the counts of each genotype are compared to those of cases (individuals with the disease) and controls (individuals without the disease). (**B**) For a copy number polymorphism where copy number genotypes are not assignable, the distribution of copy numbers is compared between cases and controls. (**C**) Copy number polymorphisms associated with disease may be identified indirectly via the association of an SNP (A → C) in linkage disequilibrium. doi:10.1128/9781555818890.ch3.f3.11

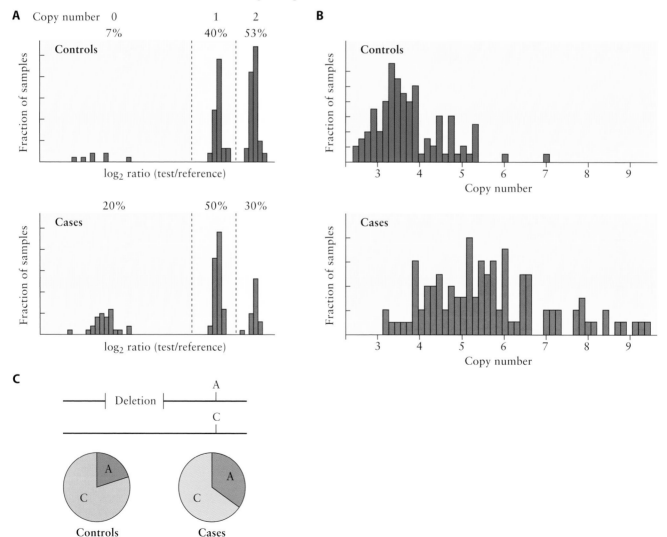

influence of CNVs in the reference DNA using a pool (about 4,100) of normal DNA samples as a reference. Several algorithms developed to detect CNVs from array CGH data search for intervals in which the average $\log_2$ ratio is greater than the assigned threshold. If probe response is good and background noise is low, a few (generally 3 to 10) probes may be enough to detect regions with different copy numbers. Algorithms detect CNVs more accurately and produce fewer false-positive results if data are corrected for artifacts such as high GC content.

Custom-designed arrays may be used to focus on chromosomal regions of interest. For example, one study reported the use of a set of 20 ultrahigh-resolution oligonucleotide arrays comprising 42 million probes in total, with a median probe spacing of just 56 base pairs (bp) across the entire genome. Such high resolution enabled the identification of 11,700 CNVs greater than 443 bp in length in the genomes of 40 healthy individuals.

Synthetic 60-mer oligonucleotide arrays are commonly used, as they yield highly reproducible results and excellent signal-to-noise ratios that ensure maximum sensitivity and specificity. These 60-mers span exon, intron, intergene, and pseudoautosomal regions (genes in these regions are inherited like an autosomal gene); duplicated segments; and CNV regions in DNA. In addition to sequences in a gene database under study, custom oligonucleotide sequences (25 to 60 bp) can be used. For every oligonucleotide on the array, array manufacturers can provide scores that predict the performance of each nucleotide on a genomic array and interpret derivative $\log_2$ ratio values in breakpoint regions. Scores are based on several parameters, including melting temperature, SNP content, sequence complexity, and uniqueness of the oligonucleotide sequence.

SNP Arrays

SNP arrays can now detect millions of different SNPs. In addition to the advances in resolution, these arrays now incorporate more SNPs linked with a given disease(s) due to large-scale studies like the **HapMap Project** and the **1000 Genomes Project**. A haplotype refers to the combination of alleles (DNA sequences) at adjacent locations (one locus, several loci, or a whole chromosome) on a chromosome that are genetically transmitted together. The HapMap Project is an international project designed to generate a haplotype map (HapMap) of the human genome that describes the common patterns of genetic variation in human health, disease, responses to drugs, and environmental factors. The 1000 Genomes Project, launched in January 2008, constitutes an international research effort to establish a comprehensive catalogue of human genetic variation, which can then be applied to association studies that relate genetic variation to disease as well as to advance our understanding of mutation and recombination in the human genome.

The initial plan for the 1000 Genomes Project was to collect whole-genome sequences for 1,000 individuals, each at 2× coverage and representing ~6 gigabase pairs of sequence per individual and ~6 terabase pairs (Tbp) of sequence in total. The term 2× coverage means that on average, individual loci in the human genome will be spanned by two independent bacterial artificial chromosome (BAC) clones. This also means that some regions will be covered by more than two clones, and some regions will be covered by a single clone or no clone at all. During recent years, this plan was revised several times, and as of March 2012, the still-growing project had generated more than 260 Tbp of data. The ongoing aims of the 1000 Genomes Project are (i) to discover >95% of the variants with minor allele frequencies of <1% across the genome and 0.1 to 0.5% in gene regions and (ii) to estimate the population frequencies, haplotype backgrounds, and linkage disequilibrium patterns of variant alleles. A so-called "Toronto Agreement" was reached to describe a set of best practices for prepublication data sharing, which were adopted in 2009 and have since facilitated the global sharing of the available data. To analyze all of the data accumulated, the Ensembl Variant Effect Predictor is a flexible and regularly updated bioinformatics method now being used to annotate all newly discovered variants and to provide information about how such variants impact genes, regulatory regions, and other features of the genome outlined below.

SNPs comprise a major part of genetic variation and play an essential role in the development of disease, evolution, and tumorigenesis. Thus, SNP arrays are now used to identify SNPs and to detect rare and common genomic rearrangements, amplifications, and deletions. SNP arrays can also detect extended regions of loss of heterozygosity or uniparental disomy (inheritance of two copies of a chromosome, or of part of a chromosome, from one parent and no copies from the other parent), provide more accurate calculation of copy numbers, and determine the parental origin of new CNVs. To facilitate the identification of CNVs, only the test sample is required to be hybridized onto each SNP array. In addition, the SNP arrays used are now manufactured so as to increase the density of SNP markers in regions of DNA that contain CNVs (Fig. 3.12).

Should SNP arrays replace CGH arrays? Despite the variety of information obtained in a single experiment and greater potential for automation and scalability, SNP arrays generally do not perform as well as CGH arrays for CNV discovery, in terms of sensitivity and resolution. If searching for very small deletion (<50 kb) or gain variants, array CGH may be the best option. However, for cancer genetics or human diseases linked to uniparental disomy, SNP arrays are more appropriate. Hybrid arrays (e.g., CGH plus SNP arrays) are now available to both analyze copy number and detect chromosomal mosaicism (chromosomes of different structure), loss of heterozygosity, uniparental disomy, or regions that have identical copies of an identical ancestral allele (Fig. 3.12).

Fine Mapping of Translocation Breakpoints Using Array Painting

CGH arrays detect deletions and amplifications, including chromosome imbalances (changes in copy number) associated with balanced (no change in copy number) translocation. However, they do not detect balanced rearrangements such as inversions and balanced reciprocal translocations. Balanced reciprocal translocations are carried by 1 in 500 individuals and also occur frequently in cancer cells. Disruption of regulatory regions,

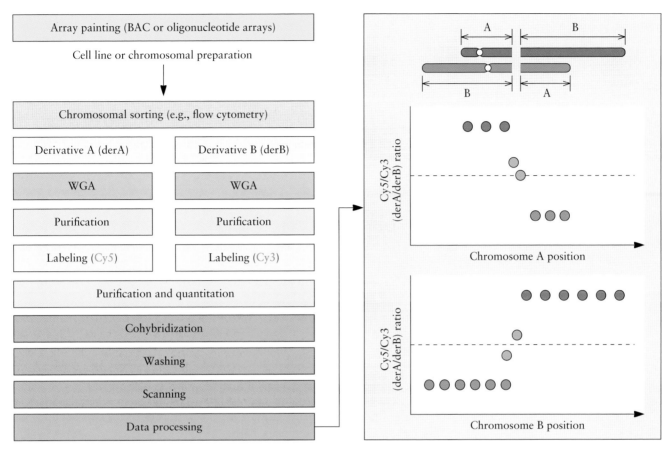

Figure 3.12 Flowchart of array painting. White and blue boxes, sample preparation stage; orange boxes, microarray stage; green box, data processing stage. BAC, bacterial artificial chromosome; WGA, whole-genome amplification. The steps of the technique of array painting and the method of data analysis are as described in the text. Adapted from Le Scouarnec and Gribble, *Heredity* **108**:75–85, 2012. doi:10.1128/9781555818890.ch3.f3.12

such as enhancers or genes, and creation of new gene fusions by a chromosome translocation can have deleterious phenotypic consequences.

Array painting, derived from reverse chromosome painting and array CGH technologies, was developed to characterize reciprocal chromosome translocation breakpoints (Fig 3.11). In reverse chromosome painting, probes are generated by degenerate-oligonucleotide-primed PCR from isolated aberrant chromosomes and are hybridized onto normal metaphase spreads using FISH. This permits the identification of aberrant chromosomal regions and the location of approximate positions of the breakpoints. However, with metaphase chromosomes as targets, breakpoints can be localized only at a resolution of 5 to 10 megabases (Mb).

To enhance the accuracy of breakpoint mapping, arrays have replaced the use of metaphase chromosomes. Two chromosomes that carry a reciprocal translocation are labeled and isolated by sorting on a flow cytometer (Fig. 3.13). Each chromosome is then amplified using oligonucleotide-primed PCR or whole-genome amplification kits. To distinguish the amplified products, they are labeled with different fluorescent

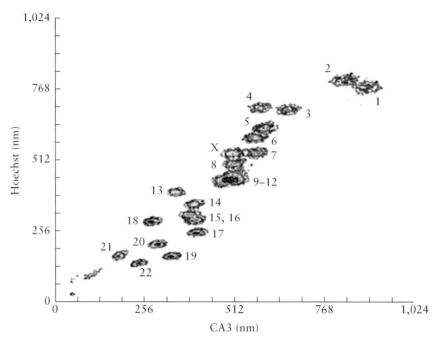

Figure 3.13 Flow sorter karyotype analysis of chromosomes from a healthy human female. A dual stain is used to take advantage of variations in the GC-to-AT ratio in different chromosomes. The dyes selected are Hoechst 33258, which shows a preference for AT-rich regions, and chromomycin A3 (CA3), which binds to GC-rich regions. To sort the chromosomes according to their intensity of fluorescence, a flow cytometer with two high-powered argon–ion lasers was used. One laser of the cytometer was tuned to provide UV light to excite the Hoechst dye (blue fluorescence), and the other laser was tuned to 458 nm to excite the CA3 dye (green fluorescence). The different chromosomes are numbered in the cytogram. Note that suitable separation between the different chromosomes is achieved, with the exception of chromosomes 9 to 12, which can be separated by other cytogenetic methods. Adapted from Ormerod (ed.), *Flow Cytometry—A Basic Introduction—Wiki Version* (2008). The cytogram was provided by C. Langford and N. Carter, The Sanger Centre, Cambridge, United Kingdom. doi:10.1128/9781555818890.ch3.f3.13

dyes (Cy5 and Cy3) and cohybridized onto an array, which is then washed to remove any excess probe and scanned (Fig. 3.11). Log_2 ratios for Cy5 and Cy3 intensities are plotted against chromosome position for each feature. Because the chromosomal regions that flank each side of the breakpoint are labeled with different dyes, the position where log_2 ratios change from high to low ratios (or vice versa) defines the breakpoint (Fig. 3.11). Fine mapping of breakpoints depends only on the resolution of the array, which has been improved significantly by the use of array painting, array CGH technology, and region-specific oligonucleotide arrays. Precise breakpoint mapping of balanced translocations is informative about associated phenotypes in patients. A 244K human genome microarray kit, which contains a platform of approximately 240,000 distinct 60-mer oligonucleotide probes spanning the entire human genome, is commonly used for clinical application. Thus, array painting performed with a 244K CGH array for a t(10;13)(q22;p13) balanced translocation indicated that the *C10orf11* gene, which was disrupted by the translocation, contributes to the mental retardation phenotype in 10q22 deletion patients. Breakpoints identified by array technologies can be independently validated by FISH assays to visually demonstrate the rearrangements in individual cells.

Array painting can also be used to explain complex chromosome rearrangements between two or more chromosomes, determine cross-species homology, and provide insight into karyotype evolution. An alternative technique to array painting is **chromatin conformation capture on chip (4C)**, in which many fragments across the breakpoints are captured by cross-linking of physically close parts of the genome, followed by restriction enzyme digestion, locus-specific PCR, and hybridization to 4C-tailored microarrays (Fig. 3.14). Clustering of positive signals displaying increased intensities predicts the positions of the breakpoints.

4C
chromatin conformation capture on chip

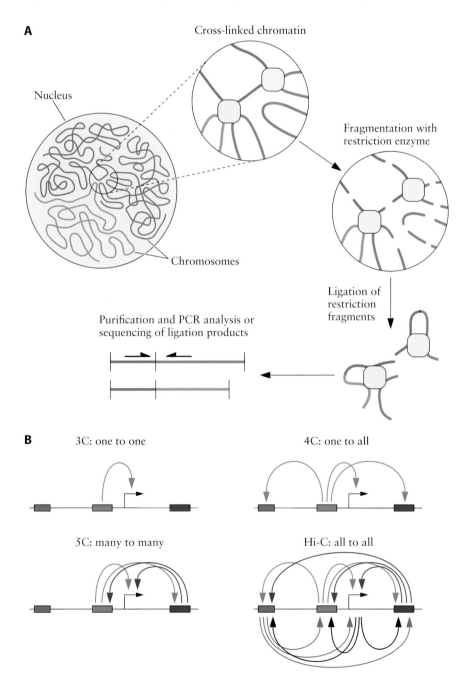

A

Nucleus

Cross-linked chromatin

Chromosomes

Fragmentation with restriction enzyme

Ligation of restriction fragments

Purification and PCR analysis or sequencing of ligation products

B

3C: one to one

4C: one to all

5C: many to many

Hi-C: all to all

Figure 3.14 Identification of DNA interaction using 3C and derivative-based technology. (**A**) The interactions or mere proximity of proteins can be studied by the clever use of cross-linking agents. For example, protein A and protein B may be very close to each other in a cell, and a chemical cross-linker could be used to probe the protein–protein interaction between these two proteins by linking them together, disrupting the cell, and identifying the cross-linked proteins. Alternatively, interaction between two DNA fragments located close to each other in the nucleus of a cell could be analyzed by chemical cross-linking. Analyses of this type of DNA–DNA interaction can be informative about the regions on different chromosomes that may undergo rearrangement and recombination between genes. 3C technology depends on the cross-linking (grey blocks), restriction digestion, and subsequent ligation of DNA fragments that are positioned close together in the nucleus. The process during which cross-linking induced DNA–DNA interactions are chemically disrupted is known as reverse cross-linking. Reverse cross-linking and purification are followed by PCR-based analysis using locus-specific primers to confirm the presence of a DNA interaction. (**B**) The different strategies for the analysis of positions at which different DNA fragments are joined together (ligation junctions), including 3C and its high-throughput derivatives 4C, 5C, and Hi-C. Colored boxes indicate hypothetical regulatory DNA elements, black arrows indicate transcriptional start sites, and curved colored arrows indicate the interactions analyzed in each of the different strategies. Reproduced from Wijchers and de Laat, *Trends Genet.* 27:63–71, 2011, with permission from Elsevier.
doi:10.1128/9781555818890.ch3.f3.14

Next-Generation Sequencing-Based Techniques

Next-generation sequencing technologies developed in 2005 currently enable the sequencing of a whole human genome to be completed in a few days and at a significantly reduced cost (see chapter 1). These technologies facilitate the sequencing of millions of DNA molecules simultaneously after library preparation of fragments. Sequence reads are aligned to the reference genome, and base variants such as small insertions and deletions (**indels**) and SVs (450 bp) can be detected. Next-generation sequencing technology can be applied for high-throughput resequencing to understand human genome variation and diseases, large-scale gene expression studies using cDNA sequencing, and whole-genome sequencing of many organisms to enhance our knowledge of evolution (see chapter 1). This technology is still under development, and third-generation platforms can yield sequence reads as large as a few kilobase pairs, whereas read lengths presently range from 30 to 400 bp depending on the platform. Until whole-genome sequencing becomes more economical, specific genomic regions may be isolated for sequencing. Chromosome suspensions may be tagged with fluorescent stains and isolated according to their fluorescence intensity by sorting on a fluorescence-activated cell sorter machine (see chapter 2). A dual stain is generally selected to take advantage of variations in the GC-to-AT ratio in different chromosomes. The dyes commonly selected are Hoechst 33258, which shows a preference for AT-rich regions, and chromomycin A3, which binds to GC-rich regions. A fluorescence-activated cell sorter with two high-powered argon–ion lasers is required; one laser provides ultraviolet (UV) light to excite the Hoechst dye (blue fluorescence), and the other laser is tuned to 458 nm to excite chromomycin A3 (green fluorescence). As shown in a typical cytogram (Fig. 3.13), isolated chromosomes of interest may be selected for DNA sequencing. When working with small genomes or specific chromosomal regions, a unique oligonucleotide tag may be added to samples before sequencing to facilitate and normalize the identification of chromosomes.

Next-generation sequencing is an attractive alternative to array-based assays in molecular cytogenetics. Information obtained by sequence analysis permits the detection of SVs of all types and sizes. In addition, breakpoints can be mapped with high resolution down to the base pair level, and complex rearrangements can be characterized to analyze multiple breakpoints in a single experiment. Four different approaches can characterize SVs: (i) read depth analysis, which can only detect chromosomal gains and losses; (ii) readpair analysis (paired-end mapping); (iii) split-read analysis; and (iv) methods of assembly that can detect all types of rearrangements, including rearrangements that do not change the copy number (inversions and translocations) (Fig. 3.15). Several tools based on one or more of these methods have been developed to analyze chromosomal rearrangements according to the genomic regions affected, size range, and breakpoint precision.

Read depth next-generation sequencing data provide information similar to that obtained from array CGH, by determining copy number gains or losses. **Sequence read depth** (the number of reads mapping at each chromosomal position) is randomly distributed over the regions.

UV
ultraviolet

Significant divergence from a normal Poisson distribution indicates copy number variation (Fig. 3.15). Duplications and amplifications are indicated by the presence of regions showing excessive read depth. In contrast, low read depth indicates heterozygous deletion, and absence of coverage suggests homozygous deletion. Statistical power is limited for smaller CNVs, but an increase in sequence coverage can improve sensitivity. Factors such as GC content, homopolymeric stretches of DNA, or preferential PCR amplification at the library preparation stage can introduce biases. Repetitive DNA regions may also create problems, as reads are aligned with low confidence—this yields little information on copy number status. Longer reads will increase mapping specificity. The range of variation in copy number achieved by read depth analysis is greater

Figure 3.15 Methods to identify SVs from next-generation sequencing data. These methods are used in combination to detect chromosomal rearrangements and characterize breakpoints (red arrows). De novo assembly methods can accurately and rapidly characterize all classes of rearrangements. MEI, mobile element insertion; RP, read pair. Reproduced from Le Scouarnec and Gribble, *Heredity* **108**:75–85, 2012. doi:10.1128/9781555818890.ch3.f3.15

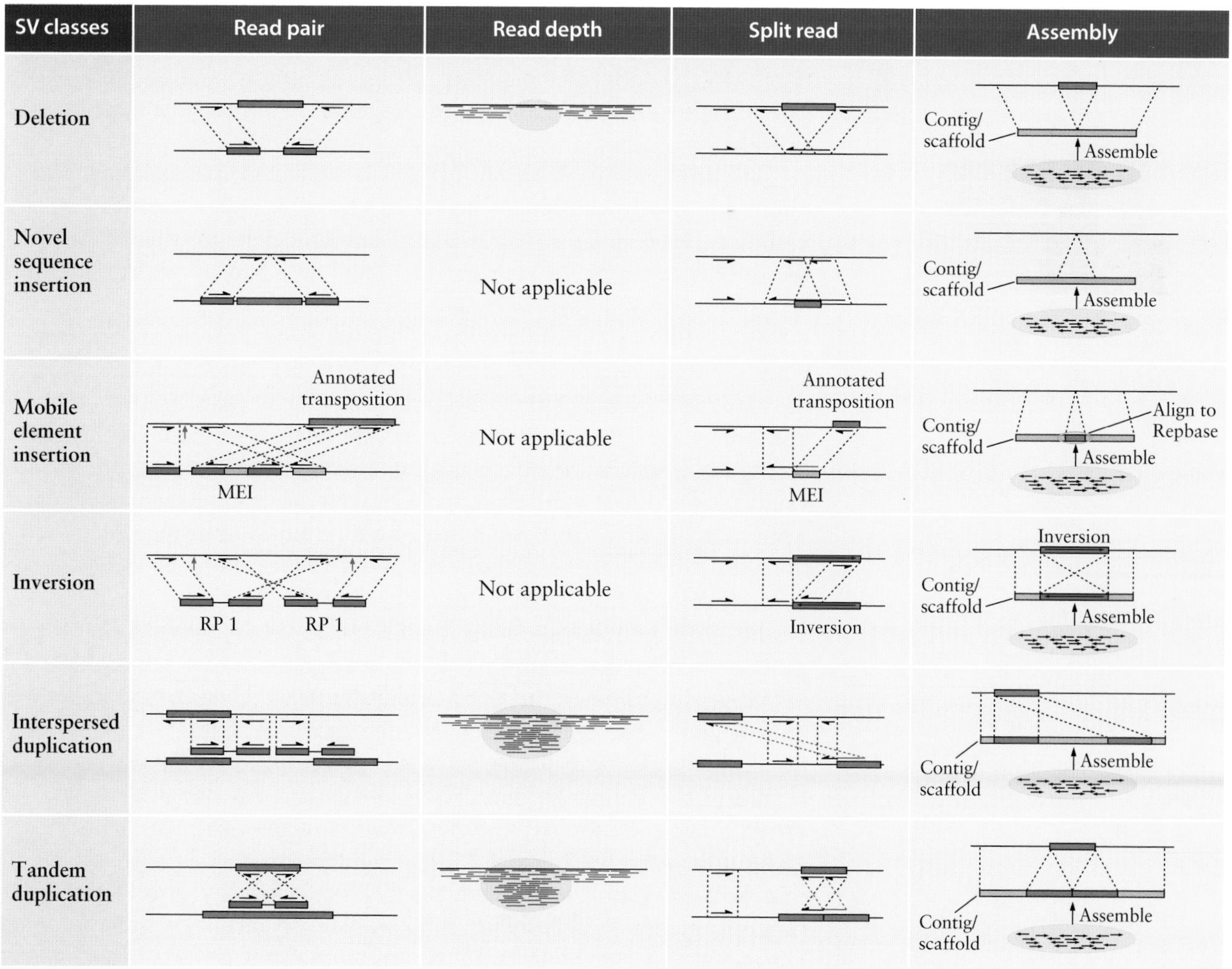

SV classes	Read pair	Read depth	Split read	Assembly
Deletion				
Novel sequence insertion		Not applicable		
Mobile element insertion		Not applicable		
Inversion		Not applicable		
Interspersed duplication				
Tandem duplication				

than that detected by SNP arrays. This greater range of next-generation sequencing may be more informative about DNA segmental duplications and multicopy gene families.

Currently, the most powerful method to study chromosome rearrangements is the **paired-end read mapping** technique (Fig. 3.15). Sequence read pairs are short sequences from both ends of each of the millions of DNA fragments generated during preparation of the library. Clustering of at least two pairs of reads that differ in either size or orientation suggests a chromosome rearrangement. When aligned to the reference genome, read pairs map at a distance corresponding to an average library insert size of 200 to 500 bp and up to 5 kb for large-insert libraries. A spanning distance significantly different from the average insert size may indicate the presence of SVs. Deletions are identified by read pairs that cover a shorter region of the genome than the reference DNA region. The latter region does not carry the deletion. In contrast, insertions or tandem duplications in the sequenced sample will cause the reads to map further, as they are absent from the reference genome. In addition to the expected span distance of a sequence read pair, aberrant mapping orientation can identify inversions and tandem duplications (Fig. 3.15). Novel insertions can be identified when only one read of a pair maps the relevant sequence. To obtain higher physical coverage at breakpoints and to facilitate the detection of SVs, data from short-insert libraries (fragments of 200 to 500 bp) can be supplemented by data from large-insert libraries (fragments of 2 to 5 kb).

The **split-read method** (Fig. 3.15) is preferred for mapping breakpoints for small deletions (1 bp to 10 kb) in unique regions of the genome and read lengths as low as 36 bp. All reads are first mapped to the reference genome. For each read pair, the location and orientation of the mapped read are used, and an algorithm is applied to search for the unmapped pair read (split read). For deletions, candidate unmapped reads are split into two fragments that map separately, and analysis of the alignment identifies the breakpoint at the base pair level. Another algorithm is used to identify exact breakpoints for tandem duplications, inversions, and complex events. Thus, this method has a significant advantage over others applied to array or next-generation sequencing data, which can identify breakpoints with high resolution but require an additional PCR or high-throughput capture step followed by conventional or next-generation sequencing to reach base pair resolution.

For the **fine mapping** of translocation breakpoints using next-generation sequencing, whole-genome sequencing (more affordable) and paired-end technology (now available) are used. These technologies are linked with large-insert paired-end libraries of about 3 kb to increase physical coverage and to maximize the detection of read pairs that span a breakpoint. If high sequence coverage is reached and reads span the breakpoint (split reads), the exact breakpoint may be identified directly without the need for an extra PCR or sequencing step.

Next-generation sequencing has revolutionized our understanding of cancer genomes by identifying the complete spectrum of somatic point mutations and providing more information about gene rearrangements in the whole genome. The latter type of mutation may occur frequently in living organisms, but it is difficult to measure the rate. Measuring this rate is important in predicting the rate at which people may develop cancer. Next-generation sequencing analyses of cancer genomes have also revealed that intrachromosomal and interchromosomal somatic gene rearrangements can be detected. The latter rearrangements are not inherited from a parent but occur somatically (see above) and result from ordered rearrangements of gene regions by DNA recombination similar to that which occurs for the joining of immunoglobulin gene segments during B-cell development (see chapter 2). These somatic rearrangements represent small genetic changes that can be detected by next-generation sequencing technology but are too small to be detected by molecular cytogenetic methods. Thus, the application of next-generation sequencing technology to the discovery of fusion genes that result from these rearrangements and have functional consequences has further explained how cancer genomes are generated.

High-throughput whole-genome sequencing reveals both point mutations (insertions and deletions) and all types of chromosome rearrangements that can be used to reconstruct genome architecture. Due to the relatively high level of sequencing error in next-generation sequencing technology, current analytical methods rely mainly on sequence alignment against a unique reference genome. Nonspecific mapping of short reads to repetitive regions may pose problems. However, third-generation sequencing technologies will provide longer reads more cheaply, enabling accurate de novo assembly, and will help to solve these issues. Taken together, the increased resolution and larger number of SVs detectable in each genome should expedite analyses of the functional significance of these SVs in health and disease.

Genome-Wide Association Studies

Comparison of human genome sequences has demonstrated that humans have limited genetic variation. About 90% of heterozygous DNA sites (e.g., SNPs) in each individual are common variants. This realization led to the hypothesis that common polymorphisms (minor allele frequency of >1%) may contribute to susceptibility to common diseases. It was then proposed that GWASs of common variants would facilitate the mapping of loci that contribute to common diseases in humans. The prediction was not that all causal mutations in these genes would be common—instead, a full spectrum of alleles was expected. This led to the hypothesis that only some common variants can localize relevant loci for further study.

The testing of this common disease–common variant hypothesis began in 2006, and since then, many publications have reported the localization of common SNPs associated with a wide range of common diseases and clinical conditions (e.g., age-related macular degeneration, type 1 and type 2 diabetes, obesity, inflammatory bowel disease, prostate cancer, breast cancer, colorectal cancer, rheumatoid arthritis, systemic lupus erythematosus, celiac disease, multiple sclerosis, atrial fibrillation, coronary

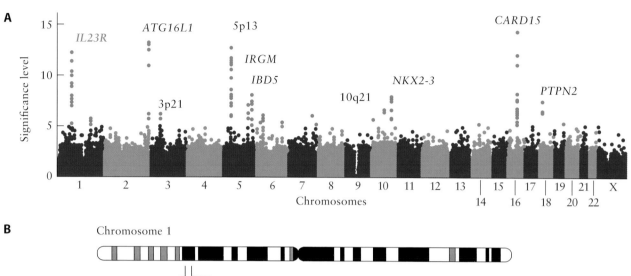

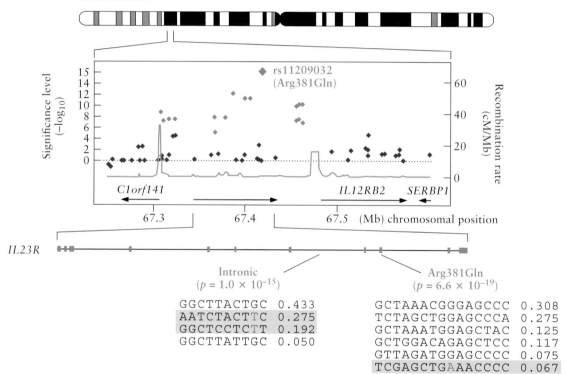

Figure 3.16 GWASs for Crohn disease. Data were obtained from the study of Crohn disease performed by the Wellcome Trust Case Control Consortium. (**A**) Significance level (*P* value on $\log_{10}$ scale) for each of the 500,000 SNPs tested across the genome. SNP locations reflect their positions across the 23 human chromosomes. SNPs with significance levels greater than 10^{-5} are in red, and the remaining SNPs are in blue. Ten regions with multiple significant SNPs are shown, labeled by their location or by the likely disease-related gene, e.g., the IL-23 receptor (*IL-23R*) gene on chromosome 21. (**B**) Enlargement of the region around the *IL-23R* locus on chromosome 21. The first part shows the significance levels for SNPs in a region of ~400 kb with colors as in panel A. The highest significance occurs at an SNP in the coding region of the *IL-23R* gene (causing

an Arg-381 → Gln change). The blue curve shows the inferred local rate of recombination across the region. There are two hot spots of recombination, with SNPs positioned between these hot spots strongly correlated with disease in a few haplotypes. The second part shows that the *IL-23R* locus contains two or more distinct, highly significant disease-associated alleles. The first site is the Arg-381 → Gln polymorphism, which has a single disease-associated haplotype (shaded in blue) with a frequency of 6.7%. The second site is in the intron between exons 7 and 8, and it tags two disease-associated haplotypes with frequencies of 27.5 and 19.2%. Adapted from Altshuler et al., *Science* **322:** 881–888, 2008, with permission.
doi:10.1128/9781555818890.ch3.f3.16

Table 3.3 Conclusions from genetic mapping of common variants by GWASs

No.	Message
1	The GWAS method works for mapping of common variants.
2	The size of variants has only a small effect.
3	The ability to detect significant associations is low.
4	Associations identify small regions, but not causal genes or mutations, to study.
5	A single locus can contain several independent common risk variants.
6	A single locus can contain common variants of weak effect and rare variants of strong effect.
7	Due to variability in allele frequencies in human populations, the relative roles of common susceptibility genes can differ among ethnic groups.

disease, glaucoma, gallstones, asthma, and restless leg syndrome) as well as various individual traits (height, hair color, eye color, freckles, and viral load of a person infected with human immunodeficiency virus). An example of a GWAS of the role of common variants in the genetic control of Crohn disease is shown in Fig. 3.16.

To date, genetic mapping of common variants by GWASs has yielded several important conclusions, which are summarized in Table 3.3 and are discussed below.

1. GWASs are effective for the mapping of common variants. In most diseases studied, GWASs have identified multiple independent genetic loci. Nonetheless, associated SNPs have not been revealed for some traits, which may be due to a small sample size, insufficient information about the phenotype, or a different genetic basis of the phenotypic trait.

2. A change in the DNA length of common variants has only a small effect. While common variants with two or more changes in DNA length per allele exist, the estimated changes are generally smaller (1.1 to 1.5 per allele).

3. The ability to detect associations of high statistical significance is low. Increasing the number of samples and replication of the analyses can enhance the significance of association.

4. Association signals identify small regions, but not causal genes or mutations, for study. Local association of genetic variants helps to identify a region but does not readily distinguish a causal mutation(s). GWASs typically identify regions of 10 to 100 kb. However, these regions need to be further mapped and resequenced to identify the specific gene and variants.

5. A single locus can contain multiple independent common risk variants. Multiple distinct alleles with different frequencies and risk ratios may be found.

6. A single locus can include common variants of weak effect and rare variants of large effect. Studies of common SNPs have identified 19 loci that influence the levels of low- or high-density lipoproteins (LDL or HDL) or triglycerides. Nine of these 19 loci carry rare mutations with large effects, e.g., loci for the LDL receptor and familial hypercholesterolemia.

LDL
low-density lipoprotein

HDL
high-density lipoprotein

7. Due to the variability in allele frequencies across human populations, the relative roles of common susceptibility genes can differ among ethnic groups. In the association of prostate cancer at 8q24, SNPs that map close to this chromosomal location play a role in all ethnic groups, but their contribution is greatest in African-Americans. This is because the risk alleles in African-Americans occur at higher frequencies and lead to a higher incidence among African-American men than European men.

Our knowledge about the functions and phenotypic associations of genes related to common diseases has also expanded rapidly, as summarized with a few selected examples in Table 3.4 and discussed below.

1. A subset of phenotypic associations detects genes related to the disease. Of 19 loci achieving genome-wide significance in a GWAS of LDL, HDL, or triglyceride levels, 12 loci contained genes with known functions in lipid biology. The gene for 3-hydroxy-3-methylglutarylcoenzyme A reductase (*HMGCR*), encoding the rate-limiting enzyme in cholesterol biosynthesis in the liver and the target of statin medications, was found by a GWAS to carry common genetic variation influencing LDL levels. Statins are a class of drugs that is used to lower cholesterol levels by inhibiting HMGCR activity.

2. Most phenotypic associations do not involve known disease genes. In some cases, GWAS results suggest that novel physiological pathways may be involved in the control of a given disease, e.g., the roles of complement factor H in age-related macular degeneration, fibroblast growth factor receptor 2 (*FGFR2*) in breast cancer, and cyclin-dependent kinase 4 inhibitors A and B (*CDKN2A* and *CDKN2B*) in type 2 diabetes. These inhibitors prevent the activation of the cyclin D-dependent kinases and function as cell growth regulators that block cell cycle progression during the G_1 phase. In other cases, closely linked genes have no known function.

3. Many phenotypic associations suggest control by DNA regions that do not encode proteins, i.e., noncoding regions. While some associated noncoding SNPs are linked to mutations in nearby coding regions of genes, many other noncoding SNPs are sufficiently distant from nearby exons and likely not linked to such mutations. Examples include the region at 8q24 associated with prostate,

Table 3.4 Functions and phenotypic associations of genes related to common diseases

No.	Message
1	A subset of phenotypic associations involves genes related to the disease.
2	Most phenotypic associations do not involve disease candidate genes.
3	Many phenotypic associations suggest control by regions of DNA that do not encode proteins (i.e., noncoding regions).
4	Some DNA regions contain expected associations across diseases and traits.
5	Some regions of DNA reveal surprising associations.

breast, and colon cancers, 300 kb from the nearest gene, and the region at 9q21 associated with myocardial infarction and type 2 diabetes, 150 kb from the nearest *CDKN2A* and *CDKN2B* genes. A role for noncoding sequence in disease risk is to be expected, as about 5% of the human genome is evolutionarily conserved and functional, and less than one-third of this 5% consists of genes that encode proteins. Noncoding mutations with roles in disease susceptibility have increased our understanding of genome biology and gene regulation. Modulation of levels of gene expression may prove more beneficial and easier to perform for treatment of disease than replacement of a fully defective protein or turning off of a gain-of-function allele.

4. Some DNA regions contain expected phenotypic associations across diseases and traits. An interesting example reveals that three autoimmune diseases, Crohn disease, psoriasis, and ankylosing spondylitis, share clinical features. The association of the same common polymorphisms in the *IL-23R* gene (gene for the receptor for the cytokine interleukin-23) in all three diseases suggests a shared molecular mechanism. SNPs in *STAT4* (gene for signal transducer and activator of transcription 4) are associated with rheumatoid arthritis and systemic lupus, two other autoimmune diseases that display common clinical properties. Multiple variants associated with type 2 diabetes are associated with insulin secretion defects in nondiabetic individuals, implicating a role for decreased function of islet β-cells in the pathogenesis of type 2 diabetes.

5. Some regions of DNA reveal surprising associations. Genetic associations among type 2 diabetes, inflammatory diseases (two loci), and cancer (four loci) have emerged unexpectedly. A single intron of *CDKAL1* (gene for CDK5 regulatory subunit associated protein 1-like 1, a member of the methylthiotransferase enzyme family) contains an SNP associated with type 2 diabetes and insulin secretion defects and another with Crohn disease and psoriasis. A variant in a coding region of the glucokinase regulatory protein is associated with levels of triglycerides and fasting blood glucose as well as with levels of C-reactive protein (a blood protein marker of infection and inflammation) and Crohn disease. Additional SNPs, e.g., in *TCF2* (gene for liver-specific transcription factor 2 of the homeobox-containing basic helix–turn–helix family) and in *JAZF1* (gene for a zinc finger protein that functions as a transcriptional repressor) are associated with an increased risk of type 2 diabetes and prostate cancer.

Disease Risk versus Disease Mechanism

It is currently held that the main value of genetic mapping is not to predict the genetic risk for a given disease but rather to provide novel insight into the biological mechanisms of induction and treatment of disease. This is because knowledge of disease pathways (not limited to the causal genes and mutations) can suggest strategies for disease prevention, diagnosis, and therapy.

The frequency of a genetic variant is not related to the magnitude of its effect or to its potential clinical value. The classic example is Brown and Goldstein's studies of familial hypercholesterolemia, which affects ~0.2% of the population and represents a very small fraction of the heritability of LDL levels and myocardial infarction. Nonetheless, the discovery of the LDL receptor emerged from studies of familial hypercholesterolemia and accompanied the development of HMGCR inhibitors (statins) for lowering LDL levels in carriers of familial hypercholesterolemia and other diseases. GWASs showed that common genetic variation in the LDL receptor and HMGCR influences LDL levels.

Genetic mapping has provided new information about many biological pathways that contribute to human disease. Nonetheless, the mapping studies are only in their infancy regarding our advanced understanding of biological processes and clinical application of the human genome to the treatment of disease. Recent GWASs and next-generation DNA sequencing studies have demonstrated the feasibility of identifying a large number of novel genetic loci. With the availability of this new information database, the primary aim continues to be the identification of all cellular pathways in which genetic variation elicits common diseases. The next sections of this chapter describe the current knowledge and path ahead for many genetic disorders caused by variations in either a single gene (monogenic), many genes (polygenic), or mutations in mitochondrial DNA.

Single-Gene Disorders

Mode of Inheritance

Single-gene disorders, or monogenic diseases, result from a mutation(s) in a single gene occurring in all cells of the body (somatic cells). Inheritance of these types of disorders follows a Mendelian segregation pattern. Although very rare, such disorders affect millions of people globally. Currently, it is estimated that more than 10,000 human diseases are monogenic. These diseases can occur in about 1 out of every 100 births and thus can cause a significant loss of life. During the last 20 years, monogenic diseases have accounted for up to 40% of the work of hospital-based pediatric practice in North America. So-called "pure genetic diseases" are caused by a single nucleotide change in a single gene in human DNA. The nature of disease depends on the functions performed by the modified gene.

Monogenic diseases can be classified into three main categories: **autosomal dominant**, **autosomal recessive**, and **X linked** (Fig. 3.17). All humans have two chromosomal copies of each gene or allele, one allele per each member of a chromosome pair. Dominant monogenic disorders involve a mutation in only one allele of a disease-related gene. In autosomal dominant inheritance, an affected person has at least one affected parent, and affected individuals have a 50% chance of passing the disorder on to their children. Huntington disease is an example of an autosomal dominant disorder. Recessive monogenic disorders occur due to a mutation in both alleles of a disease-related gene. In autosomal recessive inheritance, affected children are usually born to unaffected parents. Parents of

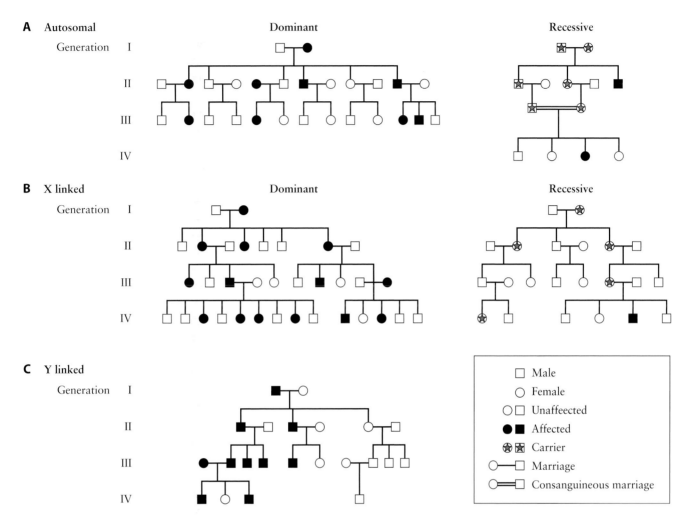

Figure 3.17 Patterns of Mendelian inheritance. (**A**) Autosomal dominant and autosomal recessive; (**B**) X-linked dominant and X-linked recessive; (**C**) Y linked. The key shows main symbols used in pedigrees. Adapted from Strachan and Read, *Human Molecular Genetics* (Bios Scientific Publishers, Oxford, United Kingdom, 1996). doi:10.1128/9781555818890.ch3.f3.17

affected children usually do not have disease symptoms but carry a single copy of the mutated gene. There is an increased incidence of autosomal recessive disorders in families in which parents are related. Children of parents who are both heterozygous for the mutated gene have a 25% chance of inheriting the disorder, and the disorder affects either sex. Cystic fibrosis and sickle-cell anemia are examples of autosomal recessive disorders. X-linked monogenic disorders are linked to mutations in genes on the X chromosome. The X-linked alleles can also be dominant or recessive. These alleles are expressed in both men and women, more so in men, as they carry only one copy of the X chromosome (XY), whereas women carry two (XX). Y-linked inheritance would affect only males, the affected males would always have an affected father, and all sons of an affected man would have the disease. However, no Y-linked diseases have ever been discovered. Apart from male infertility, mutations in Y-linked genes may not give rise to any disorders.

Table 3.5 Prevalence of some single-gene inheritable disorders

Disorder	Prevalence (approximate)
Autosomal dominant	
Familial hypercholesterolemia	1 in 500
Polycystic kidney disease	1 in 1,250
Neurofibromatosis type I	1 in 2,500
Hereditary spherocytosis	1 in 5,000
Marfan syndrome	1 in 4,000
Huntington disease	1 in 15,000
Autosomal recessive	
Sickle-cell anemia	1 in 625
Cystic fibrosis	1 in 2,000
Lysosomal acid lipase deficiency	1 in 40,000
Tay–Sachs disease	1 in 3,000
Phenylketonuria	1 in 12,000
Mucopolysaccharidoses	1 in 25,000
Glycogen storage diseases	1 in 50,000
Galactosemia	1 in 57,000
X linked	
Duchenne muscular dystrophy	1 in 7,000
Hemophilia	1 in 10,000

Adapted from http://www.news-medical.net/health/Single-Gene-Genetic-Disorder.aspx.
Approximate prevalence values are for liveborn infants.

The prevalence of some monogenic inheritable disorders is shown in Table 3.5. Some of the salient genetic and phenotypic profiles of the most common of these disorders are discussed below.

Thalassemia

Disorder and Genetics

Thalassemia is a blood-related genetic disorder that involves either the absence of or errors in genes responsible for the production of hemoglobin. Each red blood cell contains 240 million to 300 million molecules of hemoglobin. The severity of the disease depends on the gene mutations that arise and the manner in which they influence each other. Both of the hemoglobin α and β subunits are required to bind oxygen in the lungs and deliver it to other tissues. Genes on human chromosome 16 encode the α subunits, while genes on chromosome 11 encode the β subunits. A lack of expression of a particular subunit determines the type of thalassemia; e.g., a lack of an α subunit results in α-thalassemia. The lack of subunits corresponds to mutations in the genes on the appropriate chromosomes. There can be various grades of the disease depending on the gene and the type of mutations.

Prevalence

The α- and β-thalassemias are the most common inherited single-gene disorders, with the highest prevalence in areas where malaria was or still is endemic. The burden of this disorder in many regions is of such a magnitude that it represents a major public health concern. In Iran, about 8,000 fetuses may be affected by thalassemia each year. In some Mediterranean countries, control programs have achieved 80 to 100% prevention of disease in newborns.

Diagnosis and Prognosis

Diagnosis of thalassemia can be made as early as 10 to 11 weeks in pregnancy using procedures such as amniocentesis and chorionic villus sampling. Individuals can also be tested for thalassemia through routine blood counts. Thalassemic patients may have reduced fertility or even infertility. Early treatment of thalassemia has been very effective in improving the quality of life of patients. Treatments for thalassemias depend on the type and severity of the disorder. Carriers or individuals who have mild or no symptoms of α- or β-thalassemia need little or no treatment. Three types of treatments are used for moderate and severe forms of thalassemia. These include blood transfusions, iron chelation therapy (removes excess iron in the blood that increases after transfusions), and folic acid supplements (vitamin B supports red blood cell growth). Currently, genetic testing and counseling, and prenatal diagnosis, play an increasingly important role in informing individual as well as professional decisions related to the prevention, management, and treatment of this disease.

Sickle-Cell Anemia

Disorder

Sickle-cell disease is a blood-related disorder resulting from the substitution of a valine for glutamic acid in the β-globin chain of adult hemoglobin. The mutated sickle hemoglobin undergoes conformational change and polymerization upon deoxygenation, leading to red blood cell hemolysis and deformation (not doughnut shaped) and to pathology due to blockage of capillaries. Sickled red blood cells cannot migrate through small blood vessels; rather, they cluster and block these vessels, depriving organs and tissues of oxygen-carrying blood. This process leads to periodic episodes of pain and ultimately can damage tissues and vital organs and lead to other serious medical problems. Normal red blood cells live about 120 days in the bloodstream, but sickled red blood cells die much more rapidly, after about 10 to 20 days. Since they cannot be replaced fast enough, the blood is chronically short of red blood cells, leading to the condition of sickle-cell anemia.

Genetics

Hereditary persistence of fetal hemoglobin decreases the severity of sickle-cell disease. Thus, induction of fetal hemoglobin in adults has been a long-standing goal of therapies for sickle-cell disease. Recently, a GWAS for the fetal hemoglobin phenotype revealed a single strong locus, with sequence variants in the intron of a transcription factor, B-cell chronic lymphocytic leukemia/lymphoma 11A (BCL11A). BCL11A represses fetal hemoglobin expression in red blood cells. Inactivation of BCL11A in a mouse model of sickle-cell disease not only leads to the induction of fetal hemoglobin expression but also corrects the hematologic and pathological defects of the disease. Thus, BCL11A is a potential therapeutic target for sickle-cell disease.

Prevalence

Sickle-cell anemia affects millions of people throughout the world. It is particularly common among people whose ancestors originate from sub-Saharan Africa, South America, Cuba, Central America, Saudi Arabia, India, and Mediterranean countries (e.g., Turkey, Greece, and Italy). In the United States, it affects around 72,000 people, most of whose ancestors come from Africa. The disease occurs in about 1 in every 500 African-Americans and 1 in every 1,000 to 1,400 Hispanic Americans. About 2 million Americans, or 1 in 12 African-Americans, carry the sickle-cell allele.

Diagnosis and Prognosis

Sickle-cell disease can be diagnosed by a simple blood test. In many cases, sickle-cell anemia is diagnosed when newborns are screened. Vaccines, antibiotics, and folic acid supplements are administered in addition to painkillers. Blood transfusions and surgery are used in severe cases. The only known cure at present is a bone marrow transplant.

Hemophilia

Disorder and Genetics

Hemophilia is a hereditary bleeding disorder resulting from a partial or total lack of an essential blood clotting factor. It is a lifelong disorder that produces excessive bleeding, and spontaneous internal bleeding occurs very frequently. Hemophilia A, the most common form, is caused by a deficiency in clotting factor 8 (factor VIII). Hemophilia B results from a deficiency in clotting factor 9. This disorder is inherited in a sex-linked recessive manner.

Prevalence

This disorder occurs more prominently in males than in females. About one-third of new diagnoses occur without a family history. It appears globally and occurs in all racial groups. In the United Kingdom, about 6,500 people are affected with hemophilia, with a distribution of about 5,400 people (83%) with hemophilia A and about 1,100 (17%) with hemophilia B.

Diagnosis and Prognosis

Hemophilia can be diagnosed as type A or type B by blood tests in infants after 9 months of age. Administration of clotting factors helps affected individuals to live with the disease. Hemophilia, though a serious disease, can be tolerable with proper precautions and therapy, and the prospects for children with hemophilia are excellent. Recent studies have documented a greatly increased quality of life and life expectancy among hemophiliac patients in developed countries. However, it is predicted that the number of people with hemophilia in developed countries will increase gradually during the next few decades.

Cystic Fibrosis

Disorder and Genetics

Cystic fibrosis is a genetic disorder that affects the respiratory, digestive, and reproductive systems. This disorder is mediated by the production of abnormally thick mucous linings in the lungs, can lead to fatal lung infections, and can give rise to obstruction of the pancreas and impair digestion. The severity of the condition varies from the mild forms (e.g., absence of the vas deferens in men) to the more common severe forms (epithelial gland dysfunction). Since an individual must inherit two defective cystic fibrosis genes, one from each parent, to get the disease, cystic fibrosis is classified as an autosomal recessive disorder. In families in which both parents are carriers of the cystic fibrosis gene, there is a 25% chance that they will transmit cystic fibrosis to their child, a 50% chance that the child will carry the cystic fibrosis gene, and a 25% chance that the child will be a noncarrier.

The cystic fibrosis gene encodes a protein known as the cystic fibrosis transmembrane regulator (CFTR), and the *CFTR* gene contains 27 exons within 250 kb of genomic DNA. The CFTR protein consists of 1,480 amino acids distributed in two membrane-spanning domains and two ATP-binding domains. The membrane-spanning domains form a low-conductance cyclic adenosine monophosphate (cAMP)-dependent chloride channel. The ATP-binding domains control channel activity. Over 200 different mutations in the *CFTR* gene result in cystic fibrosis disease forms of various degrees of severity. About half of the known cystic fibrosis mutations are in the ATP-binding domains, which are critical for normal function. The most common mutation, a 3-bp deletion at codon 508 (ΔF508), results in the loss of a phenylalanine residue from the first ATP-binding domain and blocks the transport of the CFTR protein to the epithelial cell membrane. This mutation is present in ~70% of northern Europeans with *CFTR* mutations. The next most common mutation ($\leq$5% of *CFTR* mutations) is also present in the first ATP-binding domain and G551D (glycine at residue 551 is replaced by aspartic acid) in exon 11. Both common mutations are associated with a severe form of the disease in the homozygous state. Most of the other mutations are present at low levels and with much variation in frequency.

Prevalence

Cystic fibrosis is a common cause of death in childhood and the most common inherited disease in white populations. The incidence of cystic fibrosis varies significantly in different countries: 1 in 2,500 births in the United Kingdom, 1 in 2,000 to 3,000 births in Europe, and 1 in 3,500 births in the United States. Cystic fibrosis can arise in nonwhite populations, but only very rarely (1 in 100,000 births in African-American and East Asian populations).

Diagnosis and Prognosis

Cystic fibrosis patients have many symptoms (salty skin; persistent coughing, wheezing, or shortness of breath; and excessive appetite but poor weight gain), and their symptoms may vary due to the >200 mutations of the *CFTR* gene. A diagnostic sweat test measures the amount of salt in sweat of patients with cystic fibrosis (a high salt level indicates cystic fibrosis). This test is usually performed in babies older than 3 to 4 weeks and can also confirm the diagnosis in older children and adults. If pancreatic enzyme levels are reduced, stool analyses may reveal decreased or absent levels of the digestive enzymes (trypsin and chymotrypsin) or high levels of fat. If insulin secretion is reduced, blood sugar levels are high. Lung function tests may show that breathing is compromised. Genetic testing on a small blood sample can help determine whether an individual has a defective *CFTR* gene. During pregnancy, an accurate diagnosis of cystic fibrosis in the fetus is possible.

Life expectancy for cystic fibrosis patients has improved gradually over the past 25 years, because treatments now delay some of the changes that occur in the lungs. About 50% of cystic fibrosis patients live more than 28 years, the median age of survival of cystic fibrosis patients is ~33 years, and many cystic fibrosis patients even live until >50 to 60 years of age. Long-term survival is more common in males, individuals who do not have pancreatic problems, and individuals whose initial symptoms are restricted to the digestive system.

Due to improved treatments, about 40% of the cystic fibrosis population is currently age 18 and older. Adults may also experience cystic fibrosis-related diabetes, osteoporosis, and male sterility (in >95% of men with cystic fibrosis). The improved treatments now enable some men to become fathers; although many women with cystic fibrosis can conceive, limited lung function and other health factors make it difficult to carry a child to term.

Tay–Sachs Disease
Disorder and Genetics

Tay–Sachs disease is an autosomal recessive fatal genetic disorder caused by a genetic mutation in the *HEXA* gene on human chromosome 15, which encodes the hexosaminidase A enzyme. This mutation leads to a decrease in function of hexosaminidase A, and as a result, harmful quantities of a sphingoglycolipid, termed ganglioside GM2, accumulate in brain neurons. *HEXA* gene mutations are rare and are most frequently detected in genetically isolated populations. Tay–Sachs disease can occur from the inheritance of either two similar or two unrelated mutations in *HEXA* that cause disease. Many *HEXA* mutations have been identified, and these mutations can reach significant frequencies in certain populations (see below).

In the most common form of the disease (infantile Tay–Sachs), abnormal hexosaminidase A enzyme activity and the accompanying harmful accumulation of cell membrane-associated gangliosides in neurons lead to premature neuronal death, paralysis, dementia, blindness, psychoses, and even death of the patient. Although the degeneration of the central nervous system begins at the fetal stage, the loss of peripheral vision and motor coordination are not evident until ~6 months of age, and death usually results by 4 years of age.

Prevalence

The frequency of Tay–Sachs disease is much higher in Ashkenazi Jews (Jews of eastern European origin) than in others. Approximately 1 in 27 Jews in the United States is a carrier of the Tay–Sachs disease gene. There is also a noticeable incidence of Tay–Sachs disease in non-Jewish French Canadians, known as Acadians, who originated from France and settled in southeastern Quebec. Interestingly, while the French Canadians and Ashkenazi Jews carry different *HEXA* mutations, the Cajuns (people in southern Louisiana descended from the Acadians) carry the mutation found most frequently in Ashkenazi Jews. The ancestry of carriers from Louisiana families traces back to a single non-Jewish founder couple that lived in France in the 18th century. The Irish are also at increased risk for the Tay–Sachs gene, and among Irish-Americans the carrier rate is currently about 1 in 50. By contrast, the carrier rate in the general non-Jewish population as well as in Jews of Sephardic (Iberian or Middle Eastern) origin in the United States is about 1 in 250.

Diagnosis and Prognosis

Tay–Sachs disease may be diagnosed by a blood test that measures levels of the hexosaminidase A enzyme in serum, lymphocytes, or skin fibroblasts. During the past 25 years, carrier screening and genetic counseling in high-risk populations have greatly reduced the number of children born with Tay–Sachs disease in these groups. Thus, a high percentage of babies born with Tay–Sachs disease today are born to couples not previously considered to be at high risk. Prenatal tests of hexosaminidase A activity, such as amniocentesis (at 15 to 16 weeks of pregnancy) and chorionic villus sampling (at 10 to 12 weeks of pregnancy), can now diagnose Tay–Sachs disease in the fetus.

Fragile X Syndrome

Disorder and Genetics

Fragile X syndrome is caused by a "fragile" site at the end of the long arm of the X chromosome. This syndrome is manifested by many changes in behavior and cognitive recognition that vary widely in severity among patients. Fragile X syndrome is the most common cause of inherited mental retardation. Although it is an X-chromosome-linked recessive trait with variable expression and incomplete penetrance, 30% of all carrier women are affected. **Penetrance** is the proportion of individuals carrying a particular variant of a gene (allele or genotype) who also express an associated trait or phenotype. Full penetrance occurs when all individuals carrying a gene express the phenotype. Incomplete penetrance occurs when some individuals fail to express the phenotype, even though they carry the variant allele.

Fragile X syndrome is caused by loss-of-function mutations in the fragile X mental retardation 1 (*FMR1*) gene. *FMR1* encodes the FMRP protein found in many tissues and at particularly high levels in the brain and testes. In the brain, it may play a role in the development of neuronal synapses and cell communication. The synapses can change and adapt over time in response to experience, a characteristic called **synaptic plasticity**. The FMRP protein may help regulate synaptic plasticity and thereby control learning and memory.

Fragile X syndrome belongs to a growing class of **neurodegenerative disorders** known as **trinucleotide repeat disorders**. Among these disorders, 14 affect humans and elicit neurological dysfunction. Trinucleotide CGG repeat expansions (200 to more than 1,000 repeats) that inactivate the *FMR1* gene are the most common mutations observed at this locus. The repeat expansion mutation gives rise to high methylation in the *FMR1* promoter region that blocks transcription of *FMR1*. This expansion mutation is a null mutation, i.e., does not alter the function of the protein it codes for. Few conventional mutations occur at this locus of *FMR1*. Array-based sequence analyses showed that missense mutation (a single nucleotide change resulting in a codon that encodes a different amino acid) in *FMR1* is not a common cause of the fragile X syndrome phenotype in patients who have normal-length CGG repeat tracts. Thus, screening for small deletions of *FMR1* may be of clinical benefit.

In most people who do not have fragile X syndrome, the number of CGG repeats ranges from about 1 to 40. This CGG repeat segment is typically interrupted several times by a different trinucleotide, AGG. Having AGG scattered among the CGG trinucleotides helps to maintain the length of the long repeated segment. In patients with fragile X syndrome, the CGG trinucleotide is abnormally repeated from 200 to more than 1,000 times, which makes this region of the gene very unstable. An unstable mutation is a mutation that has a high likelihood of reverting to its original form. An unstable mutation can also be caused by the insertion of a controlling element (e.g., repeat expansion) whose subsequent deletion can result in a reversion to the original form of the gene. The inserted repeat expansion of the *FMR1* gene turns it off, and it therefore makes very little or no FMRP protein. A loss or decrease in the level of FMRP expression disrupts normal neuronal functions, causing severe learning problems, intellectual disability, and the other features of fragile X syndrome. About 1 in 3 of males with an *FMR1* gene mutation and the characteristic signs of fragile X syndrome also have features of autism spectrum disorders that affect communication and social interaction. Other changes in *FMR1* account for less than 1% of cases of fragile X syndrome.

Prevalence

Fragile X syndrome is the single most common inherited cause of mental impairment, affecting 1 in 3,600 males and 1 in 4,000 to 6,000 females worldwide. Approximately 1 in 259 women of all races carry the fragile X gene and may pass it to their children, whereas about 1 in 800 men of all races and ethnicities are carriers. Carrier females have a 30 to 40% chance of giving birth to a mentally retarded male child and a 15 to 20% chance of having a mentally retarded female child.

Diagnosis and Prognosis

The diagnosis of fragile X syndrome is made by the detection of mutations in the *FMR1* gene. Over 99% of individuals have an *FMR1* gene that expresses all of its known mutants. Tests used for diagnosis include chromosome analysis and various protein tests. Diagnosis is usually made when a child is young, and there is no current cure for this illness. Early diagnosis of the syndrome may allow for therapeutic interventions such as speech therapy, occupational therapy, psychotherapy, and special education, which can improve the quality of a patient's life considerably.

Huntington Disease

Disorder and Genetics

Huntington disease is an autosomal dominant genetic disorder; if one parent carriers the defective Huntington disease gene, his or her offspring have a 50-50 chance of inheriting the disease. Huntington disease is a neurodegenerative brain disorder in which afflicted individuals lose their ability to walk, talk, think, and reason. They easily become depressed, lose their short-term memory, and may experience a lack of concentration and focus. Every individual with the gene for the disease eventually develops the disease.

The Huntington disease gene is located on the long arm of human chromosome 4 and encodes a protein called **huntingtin** that is quite variable in its structure. The 5′ end of the Huntington disease gene contains many repeats of the CAG trinucleotide (encodes glutamine). A highly variable number of **CAG trinucleotide repeats** accounts for the Huntington disease gene mutation, which leads to the expression of an abnormally long **polyglutamine tract** at the N terminus of the huntingtin protein beginning at residue 18. Such polyglutamine tracts increase protein aggregation, which may alter cell function. Thus, like fragile X syndrome, Huntington disease is one of 14 trinucleotide repeat disorders that cause neurological dysfunction in humans. More specifically, Huntington disease is classified as a **polyglutamine disorder**. Healthy unaffected persons have a CAG repeat count of 9 to 35 (Table 3.6). However, alleles with more than 36 CAG repeats give rise to Huntington disease (the highest reported repeat length is 250). Incomplete penetrance is found in alleles with 36 to 39 CAG repeats. People with 36 to 40 CAG repeats may or may not develop Huntington disease, while people with more than 40 CAG repeats are rather likely to develop the disorder. Alleles with more than 60 CAG repeats result in a severe form of Huntington disease known as juvenile Huntington disease, and children who get Huntington disease may range from 2 to 20 years of age.

Therefore, the number of CAG repeats influences the age of onset of the disease. No case of Huntington disease has been diagnosed with a

Table 3.6 Variation in status of Huntington disease is controlled by number of CAG repeats in the *huntingtin* gene

CAG repeat count	Disease classification	Disease status
<28	Normal	Unaffected
28–35	Intermediate	Unaffected
36–40	Incomplete penetrance	Weakly affected with Huntington disease
>40	Full penetrance	All affected with Huntington disease
>60	Full penetrance	Affected with juvenile Huntington disease

CAG repeat count of less than 36. As the altered gene is passed from one generation to the next, the size of the CAG repeat expansion can change; it often increases in size, especially when it is inherited from the father. People with 28 to 35 CAG repeats have not been reported to develop the disorder, but their children are at risk of having the disease if the repeat expansion increases.

The mass of the huntingtin protein depends mainly on the number of its glutamine residues. Wild-type (normal) huntingtin consists of 3,144 amino acids, contains 6 to 35 glutamines, and has a mass of ~350 kilodaltons (kDa). In Huntington disease patients, huntingtin contains more than 36 glutamines and has an overall higher molecular mass than wild-type huntingtin. The function of huntingtin is not known, but in neurons it appears to mediate signaling, cell transport, formation of protein complexes, and protection against programmed cell death (apoptosis). Huntingtin has no known sequence homology to other proteins, but it is required for normal development before birth. It is expressed in many tissues in the body, with the highest levels of expression seen in neurons and the testes.

Prevalence

Huntington disease affects males and females equally and crosses all ethnic and racial boundaries. Usually, Huntington disease begins at age 30 to 45, but it may occur as early as the age of 2. Children who develop juvenile Huntington disease rarely live to adulthood. Everyone who carries the gene develops the disease. In the United States, Canada, and western Europe, Huntington disease affects about 1 in 20,000 people.

Diagnosis and Prognosis

Currently, there is no treatment or cure for Huntington disease, and a Huntington disease patient eventually becomes completely dependent on others for daily functioning. Individuals may also die due to other secondary complications such as choking, infection, or heart failure. Presymptomatic genetic testing is for individuals at risk for Huntington disease who do not have symptoms and involves genetic counseling. The discovery of the Huntington disease gene in 1993 facilitated the development of specialized testing that may help to confirm the diagnosis of the disease in patients with an affected parent or characteristic symptoms of the disease. Blood samples are taken from patients, and DNA is directly analyzed by PCR for Huntington disease gene mutations to determine the number of CAG repeats in the Huntington disease gene region. Additional blood samples may be obtained from close or first-degree relatives (e.g., the mother or father) with Huntington disease to help confirm the results.

Polygenic Disorders and Gene Clustering

Genetic disorders are **polygenic** if they are causally associated with the effects of many genes. Such disorders may also be **multifactorial**, if they are influenced by several different lifestyles and environmental factors. For most polygenic disorders, while the genetic background of an individual is not sufficient to cause the disorder, it may render an individual more susceptible to the disorder. Examples of multifactorial disorders include cancer, heart disease, and diabetes. Although such disorders frequently cluster in families, these disorders do not have a simple Mendelian pattern of inheritance. Moreover, as many of the factors that cause these disorders have not yet been identified, it is rather difficult to determine the risk of inheritance or transmission of these disorders. Some of these diseases (e.g., myocardial infarction, congenital birth defects, cancer, diabetes, mental illnesses, and Alzheimer disease) cause both morbidity and premature mortality. The clustering of such diseases in families may arise from similarities in genome sequence, genome architecture, and environmental triggers that promote the similar gene–gene interactions and gene–environment interactions in family members. Some principal features of the pattern of multifactorial inheritance of a polygenic disease are summarized in Table 3.7.

An individual may not be born with a disease but may be at high risk of acquiring it, a condition referred to as **genetic predisposition** or **genetic susceptibility**. The genetic susceptibility to a particular disease due to the presence of a gene mutation(s) in an allele(s) need not lead to disease. In cancer, individuals are born with genes that may not be cancer causing by themselves but upon alteration by lifestyle habits or exposure to chemicals may elicit cancer, indicating a role for gene–environment interactions in the development of cancer. Cancer susceptibility may also involve genes that suppress the formation of tumors. If tumor suppressor genes lose their function, they promote the development of carcinomas. Cardiovascular disease is generally manifested in ways unique to various communities (e.g., strokes in African communities and heart attacks among South Asians). A greater understanding of genetic predisposition to disease, as well as a knowledge of lifestyle modifications that exacerbate the condition or reduce the potential for diseases, is required for the public to make more informed choices.

Table 3.7 Characteristics of the inheritance pattern of multifactorial (polygenic) diseases

No.	Characteristic
1	A simple pattern of inheritance of a polygenic disease does not exist within a family.
2	A lower incidence of a polygenic disease within a population is associated with a higher risk of disease in first-degree relatives.
3	The relative risk of a polygenic disease is much lower in second-degree and more distantly related family members.
4	The risk of recurrence of a polygenic disease in a family is higher when more than one family member is affected.
5	The more severe the malformation associated with a polygenic disease, the greater is the risk of disease recurrence.
6	If a polygenic disease is more frequent in one sex than the other, the risk is higher for relatives of patients of the less susceptible sex.
7	If an increased risk of recurrence of disease occurs when the parents are consanguineous (e.g., first-degree relatives), many factors with additive effects may be involved.

GWAS Strategies To Map Genes for Polygenic Disease

The genome-wide strategies currently used to map polygenic disease loci include linkage analyses, association studies, and high-throughput direct DNA sequencing. Presently, only linkage and association studies are technically and financially feasible for most research groups, but genome-wide sequencing is becoming a more viable and cost-effective strategy given advances in next-generation sequencing technologies and progress in sequence annotation, especially for regulatory regions. However, large-scale sequencing studies of genetic regions that are closely linked and/or associated with a disease are currently cost-effective and can be performed with next-generation sequencing technology. Genome-wide exon sequencing, known as **exome sequencing** (sequencing of coding DNA regions only), is also now being used to map many disease loci.

Linkage analysis is the preferred genome-wide method for mapping rare variants with relatively large effect sizes. On average, complex polygenic disorders can have hundreds of alleles of genes that are genetically linked to disease susceptibility. Using this approach, loci have been mapped for many diseases, including early-onset Alzheimer disease and early-onset breast cancer.

Until recently, linkage analysis was the only strategy that could be carried out on a genome-wide basis at an affordable cost. However, the characterization of millions of SNPs and the creation of low-cost genotyping platforms made GWASs feasible by the mid-2000s. Hundreds of loci for more than 40 polygenic diseases have been mapped using GWASs, and for a few diseases new pathogenic mechanisms have been discovered. Despite this remarkable progress, it has been estimated that common loci identified to date account for only a small percentage (approximately 2 to 10%) of the genetic variance of disease susceptibility. Moreover, most of the associated SNPs have no obvious functional effects, and the pathogenesis of most polygenic diseases remains unknown. Many, if not most, of the associated SNPs are probably in linkage disequilibrium with the actual disease-predisposing variant; linkage disequilibrium refers to the nonrandom association of alleles at two or more loci on the same chromosome or different chromosomes. Resequencing studies are needed to identify the disease allele, a strategy that has been successful in a few cases.

Since the "completion" of two reference genomes in 2001, the genomes of several individuals have been sequenced at relatively high costs. The 1000 Genomes Project (http://www.1000genomes.org), which aims to characterize human variation of all types by high-throughput and unbiased sequencing of more than 1,000 human genomes from diverse populations, was initiated to identify less common (about 0.01 to 0.05%) population variants, again at relatively high costs. To date, this project has identified about 17 million variants in 742 samples from different populations. The end goal is to sequence 2,500 more individuals of diverse populations from five geographical areas to identify most of the variation that occurs at a frequency 0.1% or greater in the population. The most recent data indicate that this project has identified 38.9 million SNP sites. Notably, many of these variants are present in just one or, at most, a very few individuals.

Anticipated advances in high-throughput sequencing may soon permit more affordable genome-wide sequencing experiments to be performed with hundreds to thousands of patients. Although some rare disease variants may be mapped by GWASs using the population SNPs characterized by the 1000 Genomes Project, genome-wide sequencing of cohorts of individuals with a given disease is required to identify all rare variants that contribute to a polygenic disease(s). Advances in the record keeping, identification, and prediction of functional variation will be needed to take full advantage of sequence data. Distinguishing gene mutations that give rise to disease from gene sequence variants that occur vary rarely in the population remains a primary challenge for the field. In this regard, it is important to consider two factors that may be causally linked to disease. The first factor is the coinheritance of such rare gene sequence variants with susceptibility to disease. The second factor is the significantly higher frequency of these rare sequence variants in patients with disease compared to that found in healthy control individuals of the same ethnic background. It is hoped that a more complete understanding of the mechanism(s) by which these two factors may alter normal cell physiology and function and consequently elicit disease may be obtained from analyses of cell function in cultured cells (in vitro) and in animal models (in vivo).

The application of high-throughput methods to capture targeted regions (1 to 30 Mb) of the genome (regions linked to and/or associated with disease) and the availability of next-generation sequencing technologies have made relatively large-scale resequencing projects (using hundreds to thousands of patients) more affordable. Several groups have completed such projects, which include large-scale resequencing studies of candidate genes and relevant genomic regions (linked or associated). Genome-wide linkage and association studies both reveal that common and rare variants underlie the genetic control of most polygenic disorders. Table 3.8

Table 3.8 Types of genetic variants associated with complex disorders: methods of detection of variants that predispose to disease

Type of variant	Method of detection	Comment
Common, small to modest effect size	GWAS (500,000–1,000,000 SNPs) with thousands of unrelated cases and controls	SNP may be in LD with a disease variant; resequencing may be required
Common, very small effect size	GWAS (500,000–1,000,000 SNPs) with tens of thousands of unrelated cases and controls	SNP may be in LD with a disease variant; resequencing may be required
Rare, moderate penetrance (~0.5)	Linkage using one large family (≥10 cases) or large sample of small multiplex families; resequencing of genes in the region of interest	Alleles that are moderately penetrant usually produce multiplex families (≥2 cases with disease)
Rare, small penetrance (≤0.1)	Case or control resequencing of candidate genes or genome-wide sequencing when financially feasible: GWASs using SNPs from the 1000 Genomes Project may detect some loci with frequencies of ~0.5–1%	Alleles with small penetrance usually detect simplex families (1 case with disease)
New CNVs	Clonal and SNP arrays, resequencing (for smaller CNVs if present)	Usually detects simplex families
Spontaneous germ line mutations	Genome-wide sequencing when financially feasible	Detects simplex families; e.g., ~50% of cases of neurofibromatosis arise from spontaneous mutations in neurofibromatosis gene
No variant (e.g., nongenetic case)	Normal genome sequence	Detects simplex families in which shared phenotypes are not common. The presence of one phenotype can mask the evidence of linkage in a given family. Case or control studies yield more robust data.

Adapted with permission from Byerley and Badner, *Psychiatr. Genet.* **21**:173–182, 2011.
Abbreviation: LD, linkage disequilibrium.

summarizes the types of variants found in complex disorders and the methods for detecting variants predisposing to illness. Although many common disease variants were mapped using GWASs, rare variants may also contribute significantly to genetic variance in disease susceptibility (e.g., developmental and psychiatric disorders, including autism and obsessive-compulsive disorder). Rare variants with "moderate" penetrance (~ 0.5) can be mapped using linkage methods. Sequencing strategies will be required to detect rare variants with low penetrance in families with just one child. The recognition that rare variants can have large effects on susceptibility to complex polygenic disorders (e.g., autism, mental retardation, and schizophrenia) has rekindled the interest in rare-variant approaches. Recent advances made using such variant approaches in the genetic analyses of a selected group of polygenic disorders are presented below.

Breast Cancer

Disorder

Cancer generally results from the sequential acquisition of mutations in genes that regulate cell multiplication, cell repair, and the ability of a cell to undergo malignant transformation. This multistep process is not an abrupt transition from normal to malignant but may take 20 years or more. The mutation of critical genes, including suppressor genes, cancer-causing genes (oncogenes), and genes involved in DNA repair, leads to genetic instability and to progressive loss of differentiation. Tumors enlarge because cancer cells (i) are unable to balance cell division by cell death (apoptosis) and (ii) form their own vascular system (angiogenesis). The transformed cells lose their ability to undergo cell contact and exhibit uncontrolled growth, invade neighboring tissues, and eventually spread through the bloodstream or the lymphatic system to distant organs (metastasis).

Genetics

This malignancy is one of the most commonly inherited cancers based on observations that (i) 20 to 30% of all patients with breast cancer have a family history of the disease and (ii) twin studies show that 25% of breast cancer cases are heritable.

Discovery of the breast cancer type 1 susceptibility (BRCA1) and BRCA2 genes more than 10 years ago has had a high impact on patient care, allowing for early detection and prevention of breast cancer. However, deleterious mutations in the BRCA1 and BRCA2 genes cause at most 3 to 8% of all breast cancer cases. Carrying a deleterious BRCA1 mutation confers an estimated lifetime risk for developing breast cancer of 65%. By the age of 40, carrying a deleterious BRCA1 mutation confers a 20% chance of developing breast cancer, and the risk increases with age, with the lifetime risk being 82% by age 80. BRCA2 mutation carriers were found to carry a breast cancer risk of 45% by age 70, consistent with the finding that BRCA2 mutations give rise to fewer cases of familial breast cancer than do BRCA1 mutations. Detection of genomic rearrangements in BRCA1 and BRCA2 may identify additional carriers of nonfunctional BRCA1 and BRCA2 genes.

The *BRCA1* gene encodes an E3 ubiquitin-protein ligase, a nuclear phosphoprotein that both determines genomic stability and functions as a tumor suppressor (blocks tumor growth). The E3 ligase specifically mediates the formation of polyubiquitin chains and plays a central role in DNA repair by facilitating cellular responses to DNA damage. The E3 ligase also combines with other tumor suppressors, DNA damage sensors, and cell signaling molecules to form a large multisubunit protein complex. This complex associates with RNA polymerase II and, through the C-terminal domain, also interacts with histone deacetylase complexes. Thus, the BRCA1 protein controls gene transcription, DNA repair of double-strand breaks, and gene recombination. The *BRCA2* gene belongs to the tumor suppressor gene family, as tumors with BRCA2 mutations generally do not exhibit heterozygosity of the wild-type allele. The BRCA2 protein is found intracellularly, and it mediates the repair of chromosomal damage, with an important role in the error-free repair of DNA double-strand breaks.

Currently, polygenic mechanisms and high-frequency low-penetrance tumor susceptibility genes are thought to account for a greater proportion of familial breast cancers. Such genes are considered to be low penetrance since only a small fraction of carriers of these genes develop cancer. Candidate low-penetrance breast cancer susceptibility genes identified to date include a common variant of the type I transforming growth factor β (TGF-β) receptor, *TGFBR1*6A* (accounts for about 5% of all breast cancer cases, like *BRCA1* and *BRCA2*); *CHEK2*1100delC*; and *BRIP1*.

TGFBR1*6A

TGF-β has a dual role in cancer development. In normal mammary epithelial and breast carcinoma cells, TGF-β inhibits cell proliferation; however, as the tumor progresses, TGF-β enhances invasion and metastasis. Thus, loss-of-function mutations in the TGF-β signaling pathway in the early stages of oncogenesis contribute to tumor growth due to the lack of growth inhibitory signals. Gain-of-function mutations contribute to the late steps in tumor metastasis. *TGFBR1*6A* is a common variant of *TGFBR1*, which carries a three-alanine deletion from a nine-alanine tract in the receptor's signal sequence. Importantly, *TGFBR1*6A* converts TGF-β growth-inhibitory signals into growth-stimulatory signals in breast cancer cells, suggesting that *TGFBR1*6A* provides a selective growth advantage to cancer cells in the TGF-β-rich tumor microenvironment. *TGFBR1*6A* homozygotes have almost a threefold-increased risk compared to noncarriers. Given the high *TGFBR1*6A* carrier frequency in the general population (14.1%), the risk of breast cancer in *TGFBR1*6A* homozygotes in the general population is ~4.9%. Several SNPs have been found in the *TGFB1* gene; however, the association between these SNPs and breast cancer remains to be clarified.

CHEK2*1100delC

CHEK2 is a cell cycle checkpoint protein that blocks cell division in the presence of ionizing radiation. Inactivating mutations in *CHEK2* would be expected to promote cancerous growth in the presence of radiation-induced DNA damage. The *CHEK2*1100delC* mutation abolishes the kinase activity of the protein and blocks signaling by *CHEK2*. The *CHEK2*1100delC* variant is present in 1.1% of the population. In comparison, 5.1% of breast cancer patients who are wild type for the *BRCA* genes carry this mutation. Female carriers of *CHEK2*1100delC* have a twofold-increased risk for breast cancer compared to noncarriers. The role of *CHEK2*1100delC* in male breast cancer is controversial.

BRIP1

The *BRIP1* gene encodes a helicase that interacts with the *BRCA1* gene and mediates DNA repair. *BRIP1* was mutated in 9 out of 1,212 individuals (0.74%) with breast cancer who had a family history of breast cancer. Five different types of truncating mutations were detected among these 9 individuals, who all carried wild-type *BRCA* genes. In a control group of 2,081 people (from a 1958 birth cohort collection in the United Kingdom), only 2 people (0.1%) had truncating mutations. Individuals with *BRIP1* truncated mutations have a twofold increase in the risk of breast cancer.

TP53 and PTEN

Rare variants that account for less than 0.1% of breast cancers occur in the *TP53* and *PTEN* genes. The *TP53* gene encodes the tumor suppressor protein p53, which inhibits cell cycle progression in the presence of DNA breaks. Although mutations in *TP53* are extremely rare in the general population, those individuals with the mutation develop cancer. In a study of 100 women with breast cancer, 4 women below 31 years of age had a mutation in *TP53*, independent of their *BRCA* gene mutation status. Another study showed that 1 in 5,000 women with breast cancer express a *TP53* mutation. Thus, in the absence of genomic rearrangements in the *BRCA1* and *BRCA2* genes, *TP53* mutation screening should be considered for women with a strong family history of breast cancer.

PTEN (phosphatase and tensin homolog) encodes a tumor suppressor that inhibits cell growth during the G_1 phase of cell cycle by activating the cyclin-dependent kinase inhibitor p27(KIP1). More than 70 mutations occur in the *PTEN* gene in people with Cowden syndrome (see below). These mutations can be changes in a small number of base pairs or, in some cases, deletions of a large number of base pairs. Most of these mutations cause the *PTEN* gene to make a protein that does not function properly or does not work at all. The defective protein is unable to stop cell division or signal abnormal cells to die, which can lead to tumor growth, particularly in the breast, thyroid, or uterus. *PTEN* mutations are associated with a high-penetrance, autosomal dominant (one copy of the altered gene in each cell is sufficient to cause the disorder) syndrome termed Cowden syndrome. Individuals with Cowden syndrome have a high congenital risk for developing cancer of the breast, thyroid, and skin. To date, three different mutations in *PTEN* have been found in families with Cowden syndrome and early-onset breast cancer. While such mutations are found at low frequencies in breast cancer, loss of heterozygosity affecting the *PTEN* chromosomal locus at 10q23 occurs in about 30 to 40% of tumors, and PTEN promoter hypermethylation occurs in approximately 50% of breast cancers. Frameshift and nonsense mutations (see chapter 1) are located throughout the *PTEN* gene in tumors, and two missense mutations in the phosphatase domain cause amino acid substitutions at codons 129 (Gly to Asp) and 134. The loss of PTEN protein in tumors results in the accumulation of phosphatidylinositol-3,4,5-trisphosphate and, consequently, increased activation of the serine/threonine kinase Akt. Despite the frequency with which PTEN function is impaired in breast cancer, it is not known whether PTEN loss by itself can activate the phosphoinositide 3-kinase pathway and contribute to tumorigenesis.

Five SNPs are also associated with breast cancer risk. *CASP8* D302H, an aspartic acid-to-histidine mutation at amino acid 302 of the caspase 8 protein, is associated with a reduced risk of breast cancer in *BRCA1* mutation carriers; *IGFB3* −202 C → A, a C-to-A SNP at position −202 in the promoter of the insulin growth factor-binding protein B3 gene, is associated with a reduced risk of breast cancer; *PGR* V660L, a valine-to-leucine mutation at residue 660 of the progesterone receptor, is associated with an increased risk of breast cancer; *SOD2* V16A, a valine-to-alanine mutation at residue 16 of the manganese superoxide dismutase, is associated with an increased risk of breast cancer; and *TGFB1* L10P is a leucine-to-proline change at residue 10 of TGF-β. The respective contributions of these susceptibility genes and candidate SNPs are the focus of several ongoing studies. These investigations are complicated by the fact that the penetrance of tumor susceptibility genes is highly influenced by other factors such as modifier genes, response to DNA damage, and environmental factors such as exposure to carcinogens, hormonal and reproductive factors, and weight. Genetic testing is currently used to determine if individuals with a personal and/or family history of breast cancer carry mutations or genomic rearrangements in high-penetrance breast cancer susceptibility genes. The main aim of such tests is to provide direction about how to monitor these high-risk individuals and prevent the occurrence of breast cancer or permit early cancer detection.

Prevalence

Cancer is a leading cause of death worldwide, accounting for 7.6 million deaths (around 13% of all deaths) in 2008. Lung, stomach, liver, colon, and breast cancers cause the most cancer deaths each year. Combined, these cancers are responsible for more than 4.2 million deaths annually. Deaths from cancer worldwide are predicted to continue rising, with a projected 13.1 million deaths in 2030. Current epidemiological evidence predicts that 1 in 8 women will be diagnosed with breast cancer in her lifetime. In 2012, it was estimated that about 227,000 new cases and 39,510 deaths from breast cancer would occur in women in the United States. Breast cancer is the second most common malignancy diagnosed in women and the second leading cause of cancer-related deaths. It is a heterogeneous disease often characterized by the presence or absence of expression of estrogen receptors and human epidermal growth factor receptor 2 (HER2) on tumor cells. These molecular markers used to classify breast cancer subtypes may also predict typical responses to targeted therapies.

Diagnosis and Prognosis

The roles that genes play differ greatly, ranging from completely determining the disease state (disease genes) to interacting with other genes and environmental factors in causing cancer (susceptibility genes). The primary determinants of most cancers are lifestyle factors (e.g., smoking, dietary and exercise habits, environmental carcinogens, and infectious agents) rather than inherited genetic factors. In fact, the proportion of cancers caused by high-penetrance genes is low, ~5% for breast cancer and less for most other cancer types, except retinoblastoma in children.

Inherited mutations of the *BRCA 1* gene account for a small proportion of all breast cancers, but affected family members have a >70% lifetime risk for developing breast cancer or ovarian cancer. Identification of germ line mutations by genetic testing allows for preventative measures, clinical management, and counseling. Since the prevalence of germ line mutations such as *BRCA1* is very low in most societies, the introduction of mass screening to identify people at risk to develop cancer is not recommended.

Differences in the metabolism of chemical carcinogens can explain differences in the susceptibility of individuals to cancer, and these differences are controlled by mutations in specific genes. A major research endeavor is under way to characterize these genetic polymorphisms. It is already clear that many such polymorphisms exist and are caused by genes of low penetrance that do not follow a Mendelian inheritance pattern. However, their potential contribution to the occurrence of cancer is large. In the future, it may be possible to identify those individuals at special risk of tobacco- or diet-associated cancers and also those people susceptible to the effects of environmental contaminants. It is also anticipated that genetic tests will be useful in determining the best course of treatment for cancers, which are presently classified as a single disease but may ultimately be classified into different types, each best managed by a different therapeutic strategy. Thus, genetics will continue to play an important role in the control of cancer, including (i) the identification of individuals at risk for a specific cancer, leading to preventative or screening strategies for an individual or family members, and (ii) the identification of the subtype of a cancer so that treatment can be tailored to target that specific disease.

Alzheimer Disease

Disorder

Alzheimer disease is a devastating neurodegenerative disorder with a relentless progression. The pathogenesis of Alzheimer disease may be triggered by the accumulation of the amyloid β peptide (Aβ), which is due to overproduction of Aβ and/or the failure of mechanisms to clear the peptide. Aβ self-aggregates into oligomers of various sizes and forms diffuse and neuritic plaques in the parenchyma and blood vessels. Aβ oligomers and plaques are potent mediators of Aβ toxicity, block proteasome function, inhibit mitochondrial activity, alter intracellular Ca^{2+} levels, and stimulate inflammatory processes. Loss of normal Aβ function also contributes to neuronal dysfunction. Aβ interacts with the signaling pathways that regulate the phosphorylation of the microtubule-associated protein tau. Hyperphosphorylation of tau disrupts its normal function in regulating axonal transport and leads to the accumulation of neurofibrillary tangles and toxic species of soluble tau. Degradation of hyperphosphorylated tau by the proteasome is inhibited by the actions of Aβ. These two proteins and their associated signaling pathways therefore represent important therapeutic targets for Alzheimer disease.

Abnormalities in Aβ and tau can be measured upon neuropathological examination, in cerebrospinal fluid or by positive emission tomography scans. Alzheimer disease is characterized by dementia that typically begins with subtle and poorly recognized failure of memory and slowly becomes more severe and, eventually, incapacitating. Other common findings include confusion, poor judgment, language disturbance, agitation, withdrawal, and hallucinations. Occasionally, seizures, Parkinsonian features, increased muscle tone, involuntary muscle twitching, incontinence, and a speech disorder occur. Death usually results from one or more of the following conditions: general starvation, malnutrition, and pneumonia. The typical clinical duration of the disease is 8 to 10 years, with a range from 1 to 25 years. The distribution of the causes of Alzheimer disease is <1% chromosomal (Down syndrome [trisomy 21]); ~25% familial (≥2 persons in a family have Alzheimer disease), of which approximately 15 to 25% is late onset (age >60 to 65 years) and <2% is early onset (age <65 years); and ~75% unknown. About 1 to 6% of all Alzheimer disease is early onset, and about 60% of early-onset Alzheimer disease runs in families.

Genetics

Susceptibility to Alzheimer disease is genetically controlled, and about 13% of Alzheimer disease is inherited as an autosomal dominant disorder. However, the precise identity of the risk genes and their relation to Alzheimer disease biomarkers (see above) remains unknown. This analysis is complicated due in part to the linkage of different types of Alzheimer disease to four loci, *AD1*, *AD2*, *AD3*, and *AD4* (Table 3.9). **Late-onset familial Alzheimer disease** is a complex polygenic disorder controlled by multiple susceptibility genes, and the strongest association is with the *APOE* e4 allele at locus *AD2*. By an unknown mechanism(s), this *AD2* locus affects the age of onset of late-onset familial Alzheimer disease by shifting it toward an earlier age.

Table 3.9 Genetic control of different types of familial Alzheimer disease

Genetic locus	Time of onset	% of early-onset familial Alzheimer disease	Gene	Protein	Type of test available
AD2	Late	NA	*APOE*	Apolipoprotein E	Clinical
AD1	Early	10–15	*APP*	Aβ A4 protein	Clinical
AD3	Early	20–70	*PSEN1*	Presenilin-1	Clinical
AD4	Early	rare	*PSEN2*	Presenilin-2	Clinical

Abbreviation: NA, not applicable.

Either 1 or 2 copies of the *APOE* allele e4 (genotypes e2/e4, e3/e4, and e4/e4) are associated with late-onset Alzheimer disease. The association between *APOE* e4 and late-onset Alzheimer disease is greatest when the individual has a positive family history of dementia. The strongest association between the *APOE* e4 allele and Alzheimer disease, relative to the healthy control population, is with the e4/e4 genotype. This genotype occurs in approximately 1% of the healthy control population and in nearly 19% of the population with familial Alzheimer disease. In individuals clinically diagnosed with late-onset Alzheimer disease, the probability that Alzheimer disease is the correct diagnosis is increased to about 97% in the presence of the *APOE* e4/e4 genotype. The increased risk of Alzheimer disease associated with one *APOE* e4 allele or two *APOE* e4 alleles is also found in African-Americans and Caribbean Hispanics. Despite these associations of *APOE* e4 with late-onset Alzheimer disease, about 42% of patients with Alzheimer disease do not have an *APOE* e4 allele. Thus, *APOE* genotyping is not specific for Alzheimer disease, and the absence of an *APOE* e4 allele does not rule out the diagnosis of Alzheimer disease. These findings suggest that other genes may be involved.

In fact, several other potential genes and their human chromosomal locations that are associated with late-onset Alzheimer disease are currently being investigated. This group of genes includes three loci that regulate susceptibility to **early-onset familial Alzheimer disease**, known as *AD1*, *AD2*, and *AD4*. The relative contribution of each locus and the causative gene of each locus are shown in Table 3.9.

Prevalence

Alzheimer disease is the most common cause of dementia in North America and Europe, with an estimate of four million affected individuals in the United States. The prevalence of Alzheimer disease increases with age: about 10% of persons older than 70 have significant memory loss, and more than half of these individuals have Alzheimer disease. An estimated 25 to 45% of persons over age 85 have dementia.

Diagnosis and Prognosis

A proper diagnosis of Alzheimer disease relies on a clinical neuropathological assessment. The formation of Aβ plaques and neurofibrillary tangles is thought to contribute to the degradation of the neurons in the brain and the subsequent symptoms of Alzheimer disease. A hallmark of Alzheimer disease is the accumulation of amyloid plaques between neurons in the brain. Amyloid is a general term for protein fragments that the body produces normally. Aβ is a protein fragment cleaved from a larger amyloid precursor protein. In a healthy brain, these Aβ protein fragments are broken down and eliminated. In Alzheimer disease, the fragments accumulate to form hard, insoluble plaques (deposits). The plaques are composed of a tangle of protein aggregates that have a fiber appearance and are called amyloid fibers. Neurofibrillary tangles consist of these insoluble twisted amyloid fibers found inside neurons. These tangles are formed primarily by a protein called tau, which forms part of a microtubule that helps to transport nutrients and other substances from one part of a neuron to another. In Alzheimer disease, however, the tau protein is abnormal and the microtubule structures collapse. As a result, neurons are deprived of certain nutrients and undergo cell death.

The clinical diagnosis of Alzheimer disease, based on signs of slowly progressive dementia and findings of gross cerebral cortical atrophy on neuroimaging, is correct 80 to 90% of the time. The association of the *APOE* e4 allele with Alzheimer disease is significant; however, *APOE* genotyping is neither specific nor sensitive. While *APOE* genotyping may have an adjunct role in the diagnosis of Alzheimer disease in symptomatic individuals, it appears to have little role at this time in predictive testing of asymptomatic individuals. Three forms of early-onset familial Alzheimer disease caused by mutations in one of three genes—Aβ (A4) precursor protein (*APP*), presenilin 1 (*PSEN1*), and presenilin 2 (*PSEN2*)—are recognized. Molecular genetic testing of the three genes is available in clinical laboratories.

Treatment

Treatment is supportive, and each symptom is managed on an individual basis. Assisted-living arrangements or care in a nursing home is usually necessary. Drugs that increase cholinergic activity by inhibiting the acetylcholinesterase enzyme produce a modest but useful behavioral or cognitive benefit in a minority of affected individuals. An N-methyl-D-aspartate (NMDA) receptor antagonist is also used. Antidepressant medication may improve associated depression.

Counseling

Because Alzheimer disease is genetically heterogeneous, genetic counseling of persons with Alzheimer disease and their family members must be tailored to be family specific. Alzheimer disease is common, and the overall lifetime risk for any individual of developing dementia is approximately 10 to 12%. First-degree relatives of a single individual with Alzheimer disease in a family have a risk of developing Alzheimer disease of about 15 to 30%, which is typically reported as a 20 to 25% risk. This risk is approximately 2.5-fold higher than the background risk (about 10.4%).

Type 1 Diabetes

Diabetes is a disease in which the body does not produce or properly use insulin. Insulin is a hormone that is needed to convert sugar, starches, and other food into energy needed for daily life. The cause of diabetes continues to be a mystery, although both genetics and environmental factors such as obesity and lack of exercise appear to play roles. There are three major classes of diabetes: type 1 diabetes, type 2 diabetes, and gestational diabetes. Only type 1 diabetes, a representative polygenic autoimmune disease, is discussed in this chapter.

Disorder

Type 1 diabetes is an autoimmune disease that presents clinically with hyperglycemia resulting from the immune-mediated progressive destruction of insulin-producing pancreatic islet β-cells and associated metabolic dysfunction. The resulting insulin deficiency requires lifelong exogenous insulin treatment for survival, and long-term complications can cause substantial disability and shorten life span. About 90 to 95% of the β-cell mass is already destroyed when symptoms of type 1 diabetes first appear, demonstrating that prediction and prevention are high priorities. This disease affects about 1 in 300 individuals in North America. More than 20 million people worldwide (mostly children and young adults) are estimated to have type 1 diabetes. A possible mechanism for the development of type 1 diabetes is shown in Fig. 3.18.

Genetics

The etiology and pathogenesis of type 1 diabetes are largely accounted for by genetic predisposition. In a family with a child afflicted with type 1 diabetes, the siblings of this child are at a 15-fold-increased risk to get the disease relative to the risk in other families in the general population who are disease free. Within a pair of monozygotic twins (originating from one placenta), if one twin has type 1 diabetes, the probability of the other twin becoming a type 1 diabetic is about 40 to 50%. In pairs of dizygotic twins (originating from two placentas) in which one twin is type 1 diabetic, the probability of the other twin becoming a type 1 diabetic is only about 5 to 6%. This lower probability of susceptibility to type 1 diabetes is the same as that for unrelated children in the general population. These findings are consistent with the observed overall global increase in type 1 diabetes incidence during the last 20 to 30 years and strongly support the idea that environmental factors are also important in the control of susceptibility to type 1 diabetes.

The majority of genetic research to date has focused on the heritability that predisposes to islet autoimmunity and type 1 diabetes. The evidence indicates that type 1 diabetes is a polygenic, common, complex disease, with major susceptibility genes located in the *HLA* complex on human chromosome 6 (see chapter 2) and with other smaller effects found in non-*HLA* loci on other chromosomes. Recent advances in DNA technology, including high-throughput SNP typing and sequencing and GWASs, have advanced our understanding of the immune pathogenesis of type 1 diabetes.

GWASs have demonstrated that while more than 40 loci are associated with type 1 diabetes, only a few of these loci have been fine mapped to a particular gene or variant. Importantly, early studies established that there is a strong linkage and association of *HLA* class I and *HLA* class II gene variants with type 1 diabetes and estimated that almost half of the risk for type 1 diabetes is determined by these variants. Among the other non-*HLA* loci that are associated with risk for type 1 diabetes, variants in the insulin (*INS*) gene confer a twofold-increased risk of the disease, suggesting that insulin may be a major autoantigen in autoimmune type 1 diabetes (see chapter 4). An SNP that alters the amino acid sequence of protein tyrosine phosphatase nonreceptor type 22 (*PTPN22*) was found in the gene encoding this protein. This SNP confers a relative risk for type 1 diabetes similar to that of the SNP in the *INS* gene. Noncoding

A T-cell immune tolerance

B T-cell effector functions

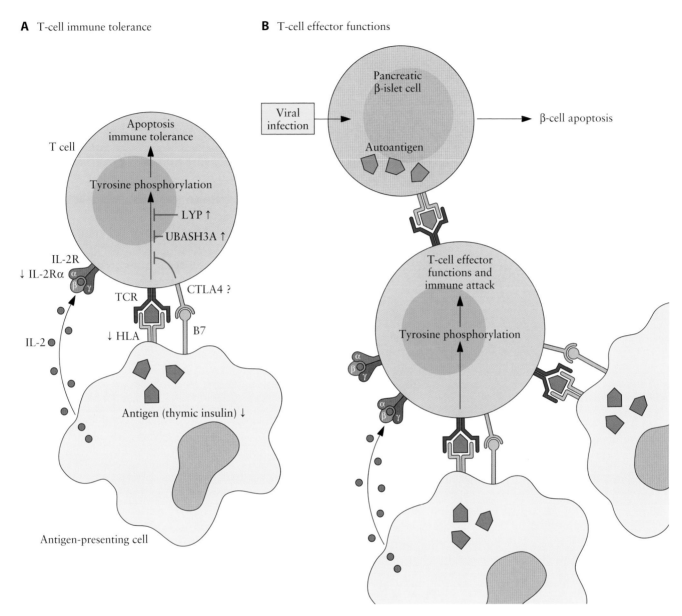

Figure 3.18 Effect of genetic variants of disease-associated loci on T-cell activation during the development of type 1 diabetes. The model presented shows the proposed mechanisms of regulation of adaptive immunity by proteins encoded by type 1 diabetes loci as well as the functional effect of known type 1 diabetes-associated variants. Molecules in green mediate T-cell activation via TCR-dependent signaling or downstream events. Molecules in red inhibit these signaling events. Arrows in grey boxes show the functional effect (increase or decrease) on the protein by the type 1 diabetes risk allele in the respective encoding genes. For HLA-encoded proteins, reduced function is likely but not yet formally proven. As TCR activation regulates both immune tolerance (in the thymus and peripheral tissues) and immune attack, type 1 diabetes-associated alleles may impair T-cell activation by a loss of function of activators or a gain of function of inhibitors. (**A**) A mild reduction in TCR activation may lead to decreased tolerance. (**B**) If T-cell signaling is impaired, T-cell effector functions can still elicit an autoimmune attack on insulin-producing pancreatic islet β cells. During inflammation (insulitis) and an increased concentration in a local autoantigen(s) in the pancreas resulting from β-cell apoptosis induced by viral infection, a partial loss of function in TCR activation may be overcome by stronger signals in T cells that previously escaped self-tolerance for the same autoantigen(s). In addition, deficient IL-2 signaling may impair the function of regulatory T cells (not shown). Only those molecules whose genetic modulation of type 1 diabetes is known or suspected are considered here. Reproduced by permission from Macmillan Publishers Ltd. (Polychronakos and Li, *Nat. Rev. Genet.* **12**:781–792, 2011). doi:10.1128/9781555818890.ch3.f3.18

variants in the genes coding for cytotoxic T-lymphocyte-associated protein 4 (*CTLA4*) and IL-2 receptor alpha chain (*IL2RA*) were also found. GWASs of several thousand SNPs revealed type 1 diabetes susceptibility to be associated with a coding allele of the interferon-induced helicase c domain-containing protein 1 (*IFIH1*). With the exception of *INS*, these loci and many others identified by GWASs are also associated with other autoimmune diseases, indicating common mechanisms and pathways in type 1 diabetes and these diseases (see chapter 4). As mentioned above, a GWAS identifies gene regions and not necessarily specific "disease causal variants" or "disease causal genes." Disease-associated markers found by GWASs may map to blocks of genes (tens to hundreds of kilobase pairs) in linkage disequilibrium. The mapping of a gene very close to a variant that is strongly associated with a disease does not prove that this gene necessarily controls disease susceptibility. Moreover, an SNP detected by a GWAS that is the most strongly disease-associated SNP will often only be a marker linked to a variant gene. This variant gene may mediate susceptibility to disease but may not actually be the disease-causing gene itself. Fine mapping to an SNP within the coding region of a disease-causing gene(s) requires that all candidate SNPs be identified, sequenced, and mapped.

For type 1 diabetes and other autoimmune diseases that share several common loci, the Immunochip, an array of 200,000 SNPs known to be associated with autoimmune diseases, was developed as a cost-efficient fine-mapping approach. It provides high-density coverage of all replicated loci for any autoimmune disease, whether derived from GWASs or from candidate gene studies. Presumably, in time, the Immunochip will have to be supplemented with SNPs identified more recently from resequencing studies. Fine-mapping studies of variants in genes that control susceptibility to type 1 diabetes have confirmed *HLA*, *INS*, *IL2RA*, and *PTPN22* to be causal genes for type 1 diabetes. A total of 144 genes have been mapped to non-HLA type 1 diabetes risk loci and remain to be analyzed.

HLA

Loci in the *HLA* gene complex (see chapter 2) are most strongly associated with type 1 diabetes (and most other autoimmune diseases). *HLA* is a 3.6-Mb segment on human chromosome 6p21. Approximately half of the 140 protein-coding genes expressed in this complex have known immune functions. Most of the genetic susceptibility to type 1 diabetes resides in the HLA class II *DRB* and *DQB* loci that encode highly polymorphic HLA class II DR and DQ β chains, respectively. The haplotypes that confer the highest risk of type 1 diabetes are *DRB*301–DQB*201* and *DRB*401–DQA*301–DQB*302* (allele numbers are shown after the asterisks). Over 90% of children in North America who develop type 1 diabetes carry at least one copy of these haplotypes. The presence of an amino acid other than aspartic acid (Asp) at position 57 of the DQB chain contributes significantly to the pathogenesis of type 1 diabetes. Conversely, most individuals who express the *DQB*602* allele are highly protected from type 1 diabetes.

INS

The type 1 diabetes locus with the second highest effect magnitude maps to chromosome 11p15.5 within and just upstream of *INS*, the gene that encodes the preproinsulin peptide. A current model proposes that genetic variants that reduce *INS* expression in a rare subset of thymic medullary epithelial cells lead to a relative loss of immune tolerance to insulin and a greater susceptibility to the development of type 1 diabetes (see chapter 4). The mechanisms of the *INS*-dependent loss of tolerance and onset of autoimmune type 1 diabetes involved are under investigation.

PTPN22

Recognition of a specific antigen by a T-cell antigen receptor (TCR) initiates an immune response that may result in the deletion of autoreactive T cells from the thymus, a loss of T-cell-mediated tolerance to self-antigens in peripheral tissues (e.g., lymph nodes or spleen), or an autoimmune T-cell-mediated attack (see chapter 4). The signaling pathway involved begins with tyrosine phosphorylation of the cytoplasmic tail of the TCR. The lymphocyte-specific tyrosine phosphatase (LYP) is a negative regulator of T-cell kinase signaling, which is essential to the balance between host defense and self-tolerance and is encoded by *PTPN22* (Fig. 3.18). The third strongest association with type 1 diabetes maps to a common *PTPN22* variant, a mutation of arginine to tryptophan at position 620. This allele confers a relative risk of about twofold for type 1 diabetes.

Other Genes Involved in the Adaptive Immune Response

Several other genes that encode proteins that regulate T-cell-mediated adaptive immune responses and tolerance have also been associated with type 1 diabetes.

IL2RA. The IL-2 receptor α chain (IL-2Rα; also known as CD25) is encoded by the *IL2RA* gene on chromosome 10p15-p14. IL-2 signaling through IL-2R is crucial for the function of regulatory T cells and effector T cells (Fig. 3.18). IL-2Rα expression is upregulated on the surfaces of newly activated effector T cells, but IL-2Rα is constitutively expressed at high levels in regulatory CD4$^+$ T cells that maintain self-tolerance (see chapter 4). The protective allele of *IL2RA* is associated with higher expression of *IL2RA* mRNA in CD4$^+$ T cells that sustain a memory immune response to an antigen (see chapter 4). Although IL-2 signaling is important for both regulatory T-cell and effector T-cell functions, its deficiency can lead to a loss of regulation of self-tolerance.

CTLA4. The *CTLA4* gene encodes a T-cell-specific transmembrane coreceptor which, like LYP, is also an important negative regulator of T-cell activation (Fig. 3.18). CTLA4 transduces its inhibitory effect through cytoplasmic phosphatases. A type 1 diabetes-associated *CTLA4* Ala → Thr polymorphism at residue 17 affects glycosylation of the mature CTLA4 protein. However, its role in the control of susceptibility to type 1 diabetes is unclear, as its genetic effect can be entirely accounted for by more

TCR
T-cell antigen receptor

LYP
lymphocyte-specific tyrosine phosphatase

strongly associated SNPs in the 3′ flanking region. It has proven difficult to replicate this finding, and therefore, the status of *CTLA4* as a causal gene for type 1 diabetes remains controversial.

UBASH3A. Ubiquitin-associated and SH3 domain-containing A (*UBASH3A*) is the only gene in linkage disequilibrium with a block of genes that map to a type 1 diabetes risk locus on human chromosome 21q22.3. *UBASH3A* is specifically expressed in lymphocytes. Like PTPN22, UBASH3A is a tyrosine-specific phosphatase that decreases the response of T cells activated through their TCR (Fig. 3.18). The *UBASH3A* allele confers risk to type 1 diabetes by the increased expression of *UBASH3A* in dividing lymphoid cells, suggesting that *UBASH3A* may cause the predisposition to autoimmune disease seen with chromosome 21 trisomy (Down syndrome). This predisposition includes a three- to eightfold-increased risk of type 1 diabetes and must be due to a 50% increase in the dosage of one or more of the estimated 300 genes on chromosome 21.

The effects of these other type 1 diabetes loci on the control of an adaptive immune response are consistent with a model in which events triggered by TCR signaling are attenuated in type 1 diabetes (Fig. 3.18). This attenuation may occur through either a loss-of-function of activating signals (e.g., variants in *IL2RA*) or a gain-of-function of inhibitory signals (e.g., variants in *PTPN22*, *UBASH3A*, and possibly *CTLA4*), which would result in a loss of self-tolerance and potential increase in autoimmune disease.

IFIH1 and EBI2 (innate immunity genes). Activation of the TCR in T cells in peripheral lymphoid organs may result in either effector (aggressive) responses or regulatory (protective) responses, depending on the presence or absence of inflammatory signals in the vicinity (see chapter 4). These signals are often provided by the innate immune system (see chapter 4). Among genes mapping to type 1 diabetes loci, the gene for interferon induced with helicase C domain 1 (*IFIH1*) is the best studied in this category. IFIH1 belongs to a family of helicases that can elicit an interferon response upon sensing viral double-stranded RNA, consistent with the notion that enteroviruses have an initiating role in type 1 diabetes autoimmunity.

A Thr-to-Ala SNP at position 946 in *IFIH1* (*IFIH1* T946A) is associated with type 1 diabetes. None of the other three genes that map to the block of genes in linkage disequilibrium with this SNP has a known SNP or has its expression level affected by the *IFIH1* polymorphism. Resequencing of both a large number of cases and controls has revealed the presence of several low-frequency loss-of-function *IFIH1* variants that generally lead to protection from type 1 diabetes. This result suggests that the predisposing allele of the common *IFIH1* SNP is associated with a gain of function. Alternatively, the *IFIH1* T946A SNP may have a role in the regulation of specific viral RNAs or the stability, targeting, and other functions of IFIH1. In this regard, a type 1 diabetes risk locus related to antiviral immunity is linked to the Epstein–Barr virus-induced gene 2 (*EBI2*).

Cardiovascular Disease

Disorder

Major cardiovascular diseases include several subtypes, such as coronary heart disease, cerebrovascular disease, heart failure, rheumatic heart disease, and congenital heart disease. Cardiovascular disease is a leading health problem, affecting more than 80 million individuals in the United States. In Canada, 1 out of 3 deaths is attributable to cardiovascular disease. The major risk factors associated with cardiovascular diseases are cigarette smoking, unhealthy diet, physical inactivity, hypertension, diabetes, and high blood cholesterol. Cardiovascular diseases may also result from a variety of genetic causes, including single-gene mutations, interactions between multiple genes, and gene–environment interactions. Economic shifts, urbanization, industrialization, and globalization bring about lifestyle changes that promote heart disease. In developing countries, life expectancy is increasing rapidly and is expected to expose people to these environmental risk factors for longer periods.

Genetics

For most cardiovascular disease disorders, the relative risk of disease is attributable to inherited DNA sequence variants. However, the role of inheritance and the magnitude of its effect vary by disease and other factors such as age of disease onset and subtype of cardiovascular disease. Most cardiovascular disease traits, such as myocardial infarction and concentrations of plasma LDL cholesterol, show complex inheritance, suggestive of an interaction between multiple genes and environmental factors. GWASs, DNA sequencing, and genotyping technology have enabled analyses of less frequent variants.

Variants associated with polygenic disorders are heterogeneous, as they appear with a wide range of frequencies extending from high (greater than 1 in 20) to low (1 in 1,000) to very low (less than 1 in 1,000). Based on DNA sequence variants, about 50% of variability between individuals occurs in their serum or plasma triglyceride levels. This variability tracks triglyceride-rich lipoproteins as well as very-low-density lipoprotein particles (chylomicrons) and their metabolic products. GWASs demonstrated that common variants at seven loci control the serum or plasma triglyceride levels. Resequencing identified an excess of rare variants across four genes in individuals with high levels of plasma triglycerides compared to individuals with low levels of plasma triglycerides.

GWASs for gene variants that control plasma levels of LDL cholesterol, HDL cholesterol, and triglycerides have evaluated more than 100,000 individuals and have mapped 95 distinct loci genetically associated with these variants. About one-third of the loci carry genes that regulate lipoprotein metabolism, including five targets of lipid-modifying therapies: statins for HMGCR, ezetimibe for Niemann–Pick C1-like (NPC1L1) protein that mediates cholesterol absorption, mipomersen for apolipoprotein B (APOB), anacetrapib and dalcetrapib for cholesterol ester transfer protein (CETP), and evacetrapib for PCSK9 (regulates LDL receptor). Two common variants are found in the introns of HMGCR and

NPC1L1, and they confer a small effect on plasma LDL cholesterol levels. However, targeting of these genes with statins or ezetimibe, respectively, has a much more dramatic effect on plasma LDL cholesterol. Thus, some disease genes may be discovered only by analyses of common variants that have small effects.

About two-thirds of the 95 loci discovered for plasma lipid traits harbor genes not previously considered to influence the biology of lipoproteins. Manipulation of several such genes at novel loci in mice have led to plasma lipid changes similar to those revealed by human genetics. For example, the 8q24 locus contains the tribbles homologue 1 (TRIB1) gene. TRIB1 is a lipid- and myocardial infarction-associated gene that regulates lipogenesis in the liver and very-low-density lipoprotein production in mice. DNA sequence variants downstream of the TRIB1 gene were initially associated with plasma lipid levels, with the minor allele linked to lower plasma triglycerides, lower plasma LDL cholesterol, and higher plasma HDL cholesterol. In addition, the minor alleles of a few other genes are involved in lipoprotein regulation and possess lower risk for coronary heart disease.

At each disease-linked locus identified by GWASs, our objective is to understand the causal variant, causal gene, mechanism of how the variant affects the function of the gene, and mechanism by which the gene affects the phenotype. These key issues have been addressed for the 1p13 causal locus for plasma LDL cholesterol levels and myocardial infarction using fine mapping, resequencing, and manipulation of positional candidate genes in cell culture and in animal models. However, at most loci, the precise variants, genes, and mechanisms are currently unknown due to linkage disequilibrium, weak effect size of common variants, small sample size, and noncoding variants. The example of the chromosome 9p21.3 locus and risk for myocardial infarction is informative about noncoding variants. About 5 years ago, several independent GWASs identified SNPs on chromosome 9p21 as associated with myocardial infarction or coronary artery disease, with ~50% of the population carrying a risk allele and each copy of the risk allele conferring an ~29% increase in risk for myocardial infarction and coronary artery disease. The highest associations were with noncoding SNPs that were located >100 kb downstream of the nearest protein-coding genes, CDKN2B (cyclin dependent kinase inhibitor 2B) and CDKN2A (cyclin-dependent kinase inhibitor 2A). Resequencing and fine-mapping studies in the gene region subsequently identified a set of SNPs, but not one of these SNPs was a causal variant. The mechanism that enables the risk allele to alter the relevant gene at the locus remains to be discovered.

Efforts to identify low-frequency coding variants that associate with cardiovascular traits and diseases coupled with the sequencing and resequencing of very rare variants identified three candidate genes that contribute to plasma HDL cholesterol variation in the population. In particular, rare mutations in ABCA1 (ATP-binding cassette, subfamily A, member 1) contribute to low HDL cholesterol in the population. Continued analyses are being performed to sequence and discover new genes that control additional complex traits in cardiovascular disease. These analyses involve the sequencing of about 20,000 genes in exomes and identification of single-nucleotide substitutions. The main obstacle to overcome may be a statistical one, as rare DNA sequence variants cannot be tested individually for association with phenotype. Instead, they need to be clustered with similar rare variants to be tested collectively for association with phenotype.

Prevalence

Heart diseases can cross over geographical, gender, and socioeconomic differences. Among the estimated 16.6 million deaths attributed to cardiovascular disease worldwide, 80% of these deaths occur in developing countries. Cardiovascular disease is now a leading cause of death in developing countries. Patients with established coronary heart disease are at high risk for subsequent coronary and cerebral events. Survivors of myocardial infarction are at increased risk of recurrent infarction and have an annual death rate five to six times higher than that of people of the same age who do not have coronary disease.

Diagnosis and Prognosis

Treatments for prevention and control of the cardiovascular disease epidemic involve addressing the major risk factors, including high blood pressure, high cholesterol, cigarette smoking, diabetes, poor diet, physical inactivity, and overweight and obesity. If effectively addressed and a reduction of such factors is achieved, these treatments may reduce the cardiovascular disease burden by 50% within 5 years. The mapping of the human genome has led to a better understanding of the genetic causal factors associated with cardiovascular disease. It is anticipated that identification of these factors will facilitate the development of more effective treatments and management of cardiovascular disease in the future.

Mitochondrial Disorders

Disorders

For cells to function normally, they must generate energy in the form of ATP. In many types of cells, mitochondrial activity is the prime source of ATP. Mitochondria also regulate several other cellular processes, including production of heat in response to changes in temperature and diet, ion homeostasis (maintenance of an internal steady state of ions), innate immune responses (see chapter 2), production of reactive oxygen species, and programmed cell death (apoptosis). It follows that if mitochondrial malfunction occurs and leads to deficient energy production and the impairment of other cellular functions, this may result in a mitochondrial disease(s).

The knowledge that mitochondrial impairment may be involved in diseases is relatively new; it was first recognized in an adult in the late 1960s and then in children in the late 1980s. Mitochondrial dysfunction can occur in the cells of many organs and systems in the body; more than 200 inherited diseases of metabolism can affect mitochondria, and more than 40 types of mitochondrial disorders have been reported. The brain, heart, muscles, and lungs are the organs that require the most energy, and when deprived of energy the functions of these organs begin to fail. The most severe effects of mitochondrial disease occur in the brain and muscles because they are heaviest users of energy. Other commonly affected organs include the liver, nervous system, eyes, ears, and kidneys.

Mitochondrial dysfunction is observed in single-gene mitochondrial disorders and may also be associated with the pathogenesis of polygenic diseases such as Alzheimer disease, Parkinson disease, cancer, cardiac disease, diabetes, epilepsy, Huntington disease, and obesity. In addition, a progressive decline in the expression of mitochondrial genes is a main feature of human aging, raising the possibility that an increased understanding of the aging process may provide more insight into one or more of the above-mentioned polygenic diseases.

Genetics

Mitochondrial dysfunction generally results from a genetic mutation(s). Interestingly, mitochondria have their own DNA for 37 genes, 13 mitochondrial proteins, 2 rRNAs, and 22 tRNAs. One mitochondrion contains dozens of copies of its mitochondrial genome. In addition, each cell contains numerous mitochondria. Therefore, a given cell can contain several thousand copies of its mitochondrial genome but only one copy of its nuclear genome. Mitochondrial DNA copy number is regulated in a tissue-specific manner by DNA methylation of the nucleus-encoded DNA polymerase gamma A and may vary from 150,000 copies in mature oocytes to about 15,000 copies in most somatic cells and about 100 copies in sperm. Although mitochondrial DNA is independent of nuclear DNA, the replication, transcription, translation, and repair of mitochondrial DNA is controlled by proteins encoded by nuclear DNA.

A mutation that induces a mitochondrial disease may occur in nuclear or mitochondrial DNA, and it may either be inherited or arise from a spontaneous mutation. Mitochondrial disorders that appear in children are usually inheritable. Mitochondrial DNA is less complex than nuclear DNA and is not inherited according to Mendelian genetics (inheritance of DNA from both parents). Rather, because egg cells but not sperm contribute mitochondria to a developing embryo, only mothers can transmit mitochondrial traits to their children. Thus, mitochondrial DNA is inherited only maternally.

Mutations in mitochondrial DNA differ from those in nuclear DNA, and many more mutations occur in mitochondrial DNA than nuclear DNA. The mutation rate in mitochondrial DNA is 10 times higher than in nuclear DNA because mitochondrial DNA is subject to damage from reactive oxygen molecules released as a by-product during oxidative phosphorylation in cells. In addition, mitochondrial DNA also lacks the DNA repair mechanisms found in the nucleus. Mutations in mitochondrial DNA are not distributed equally among all mitochondria, such that not all the hundreds of mitochondria in a particular cell contain the mutation. Commonly, a mutation in mitochondrial DNA occurs in 0 to 100% of mitochondria in an organ or body system. Mitochondrial dysfunction may also be caused by environmental factors (e.g., viral infection or drug treatment) that interfere with mitochondrial activity.

Mitochondrial Homeostasis and Parkinson Disease

After Alzheimer disease, Parkinson disease is the second most common neurodegenerative disease. Parkinson disease affects about 1% of the population that is 60 years of age or older and is characterized by the progressive reduced capacity to initiate voluntary movements arising mainly from the loss of neurons that synthesize the neurotransmitter dopamine. Among the many factors that mediate the development of Parkinson disease, mitochondrial dysfunction is considered to be a major factor in its etiology and pathogenesis. The dysfunction of mitochondria in Parkinson disease patients may result from one or more of the following conditions: deletion of mitochondrial DNA, accumulation of mitochondrial DNA mutations, increase in oxidative stress from reactive oxygen species, deficient expression and function of the mitochondrial respiratory chain, and abnormal morphology of mitochondria.

Because neurons require considerable energy to maintain their metabolically active state, they rely heavily on the maintenance of mitochondria in a functional state, a property known as **mitochondrial homeostasis**. Mitochondria constitute a population of organelles that require a careful balance and integration of many cellular processes, including the regulation of biogenesis, migration throughout the cell, shape remodeling, and autophagy (catabolic degradation of unnecessary cellular components by lysosomes). These dynamic processes prompt mitochondrial recruitment to critical subcellular compartments, content exchange between mitochondria, mitochondrial shape control, and mitochondrial communication with the cytosol. Mitochondrial homeostasis is the term used to refer to a well-maintained balance of these processes. The structure and function of the mitochondrial network are dependent on mitochondrial homeostasis, which is essential for maintaining the signaling, plasticity, and transmitter release of neurons. There are different pathways to maintain mitochondrial homeostasis. Thus, neurodegenerative diseases such as Parkinson disease are often closely associated with an imbalance of mitochondrial homeostasis (Fig. 3.19). The levels of many mitochondrial proteins are altered in postmortem samples of brains from persons with Parkinson disease. Genetic mutations in the *PTEN-induced putative kinase 1* (*PINK1*), *parkin* (*PARK2*), *DJ-1*, *alpha-synuclein* (*SNCA*), and *leucine-rich repeat kinase 2* (*LRRK2*) genes are closely linked to the recessive Parkinson disease genes. Moreover, these mutations are associated with important functions that maintain mitochondrial homeostasis, which include membrane potential, calcium homeostasis, structure of cristae (internal compartments formed by the inner membrane of mitochondria), respiratory activity, mitochondrial DNA integrity, and clearance of dysfunctional mitochondria from cells. The role of genes related to Parkinson disease in maintaining mitochondrial homeostasis is addressed below.

PINK1

Recessive mutations in PINK1, a serine/threonine-type protein kinase encoded by a nuclear gene, lead to a familial form of early-onset Parkinson disease. PINK1 possesses two specialized regions: the first contains

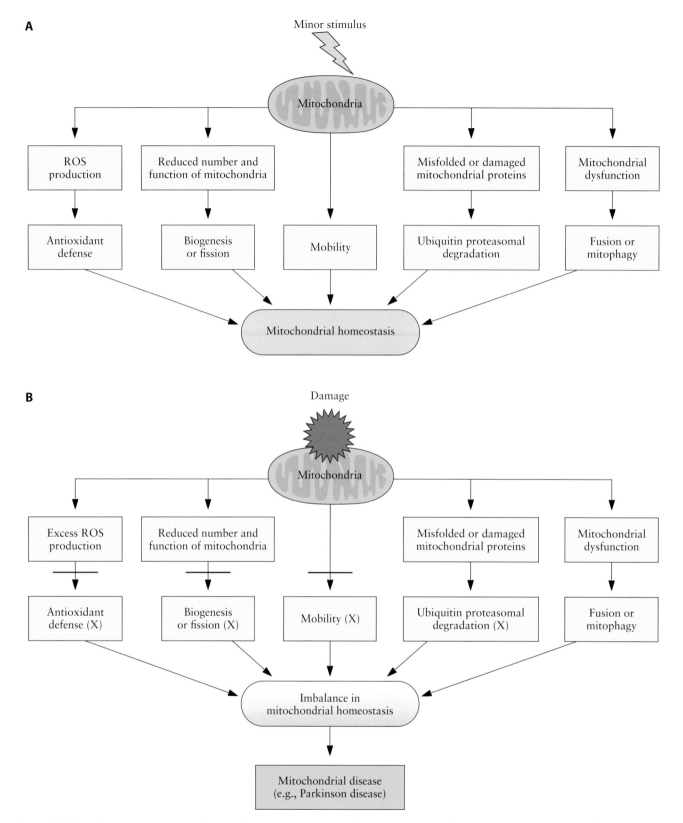

Figure 3.19 Parkinson disease is associated with an imbalance in mitochondrial homeostasis. Since mitochondria regulate energy metabolism, they are essential for the control of neuronal function. Mitochondria respond to exogenous and endogenous stimuli and maintain their homeostasis by undergoing continuous fusion, fission, mobility, mitophagy, and biogenesis. (**A**) Normally, upon exposure of mitochondria to a minor stimulus, changes in mitochondrial function and morphology occur to activate different cellular pathways that help to maintain homeostasis. (**B**) After mitochondria incur damage, changes in mitochondria occur that lead to excess reactive oxygen species (ROS) production, reduced number and mass, misfolding and/or aggregation of damaged proteins, and mitochondrial dysfunction. As a result, an imbalance in mitochondrial homeostasis ensues that may lead to the onset of Parkinson disease. Mitophagy is a catabolic process of self-degradation of mitochondria in a cell. X indicates a mitochondrial dysfunction. Adapted from Sai et al., *Neurosci. Biobehav. Rev.* **36:**2034–2043, 2012. doi:10.1128/9781555818890.ch3.f3.19

a domain highly homologous to the serine/threonine protein kinases of the calcium–calmodulin family, and the second contains a mitochondrial targeting motif at its N terminus that is essential for its function. PINK1 is located in both of the outer and inner mitochondrial membranes. Due to its subcellular localization and association with mitochondrial proteins involved in mitochondrial morphogenesis, PINK1 is thought to play a role in maintaining mitochondrial homeostasis. PINK1 may promote mitochondrial fission and protect neurons against mitochondrial malfunction under conditions of cell stress. Under stressful conditions, PINK1 promotes mitochondrial homeostasis by the phosphorylation of mitochondrial proteins that regulate cell respiratory activity and cell signaling pathways. Decreased PINK1 production that results from some *PINKI* mutations elicits neuron degeneration accompanied by mitochondrion-related structural alterations.

PARK2

PARK2 was identified in Japanese patients with autosomal recessive–juvenile Parkinson disease. The *PARK2* gene encodes parkin, an E3 ubiquitin ligase that adds ubiquitin to aberrant proteins destined for destruction by the proteasome or lysosome. The *PARK2* gene is ubiquitously transcribed in both neuronal and nonneuronal tissues. The intracellular localization of parkin is controversial, and parkin may be found in various cellular localizations, including the Golgi apparatus, synaptic vesicle, endoplasmic reticulum, nucleus, and mitochondria. Parkin is found in mitochondria of proliferating cells but migrates to the cytoplasm in quiescent cells. Parkin can translocate specifically to mitochondria to induce the clearance of damaged mitochondria from the cell. After the translocation of parkin to damaged mitochondria, a *PINK1*-dependent process, parkin can ubiquitinate some substrates destined for degradation by lysosomes and export from the cell clearance. Thus, parkin may function as a key mediator of the clearance of damaged mitochondria, which may be causal to neurodegeneration in the brain of a person with Parkinson disease.

Parkin may also mediate mitochondrial biogenesis via the transcription and replication of mitochondrial DNA. Parkin-induced mitochondrial biogenesis occurs preferentially in proliferating cells. Thus, Parkin appears to be a multifunctional protein that is involved in the entire spectrum of mitochondrial homeostasis from biogenesis to export from a cell.

DJ-1

The DJ-1 protein, also known as Parkinson disease autosomal recessive early-onset 7 (PARK7), is encoded by the *PARK7* gene on human chromosome 1. DJ-1 is a small multifunctional protein that belongs to the peptidase C56 family of proteins. It regulates transcription of androgen receptor-dependent genes, cell transformation, molecular chaperone activity, and protection of neurons against oxidative stress and cell death. Deletions and mutations of DJ-1 mediate the onset of familial Parkinson disease. While DJ-1 normally exists as a dimer protein in vivo, the stability of dimer formation is decreased in some DJ-1 mutant proteins.

Generally, DJ-1 is found in the cytosol; however, it is located occasionally in the nucleus and mitochondria, where it preferentially partitions to the matrix and intermembrane space. Under conditions of oxidative stress, DJ-1 is redistributed preferentially to mitochondria, where it acts as a neuroprotective intracellular redox sensor.

DJ-1 colocalizes with HSP70 and other proteins in the cytosol and is associated with the mitochondrial chaperone Grp75 in neurons exposed to oxidative stress, indicating that the translocation of DJ-1 to mitochondria may occur by binding to mitochondrial chaperones. This finding raises the possibility that DJ-1 influences mitochondrial homeostasis. Additional studies suggest that DJ-1 regulates mitochondrial homeostasis in an oxidative environment in a manner parallel to that of the PINK1/ parkin pathway. Thus, parkin, PINK1, and DJ-1 may cooperate at multiple levels in the maintenance of mitochondrial integrity and function.

SNCA

The alpha-synuclein protein is encoded by the *SNCA* gene on human chromosome 4. An alpha-synuclein fragment, known as the non-Aβ component of Alzheimer disease amyloid, is a fragment of the amyloid precursor protein. Alpha-synuclein is a small acidic protein (140 amino acids) that has seven incomplete repeats of 11 amino acids with a core of KTKEGV at the amino terminus, whereas the carboxyl terminus has no known structural elements. Although the function of alpha-synuclein is not known, alpha-synuclein is the main structural component of the insoluble filaments that form around a dense core of protein, known as the Lewy bodies (abnormal spherical aggregates of protein that develop inside and displace other cell components from neurons) of Parkinson disease and dementia. Point mutations in the *alpha-synuclein* gene, such as those yielding an Ala-to-Thr change at position 53 (A53T) and Glu to Lys at residue 46 (E46K), are implicated in the pathogenesis of Parkinson disease. As the E46K mutation increases the ability of alpha-synuclein to aggregate and form fibers, this mutant may be considered to be a gain-of-function mutant. However, the effect of the E46K mutation on alpha-synuclein fiber formation is weaker than that of the A53T mutation. Nonetheless, the E46K mutant yields widespread brain pathology and early-onset Parkinson disease. The substitution of Glu-46 for an Ala also increased the assembly of alpha-synuclein, but the polymers formed can have different ultrastructures, further indicating that this amino acid position has a significant effect on the assembly process. Thus, alpha-synuclein is associated closely with the development of Parkinson disease and a variety of related neurodegenerative disorders.

Alpha-synuclein is specifically upregulated in a discrete population of presynaptic terminals of the brain during the rearrangement of synapses. This protein interacts with tubulin, suggesting that alpha-synuclein may function as a microtubule-associated protein. Alpha-synuclein may also function as a molecular chaperone in the formation of protein complexes that mediate vesicle fusion in cells, in support of recent evidence that alpha-synuclein is active in the neuronal Golgi apparatus and during vesicle trafficking. The activities of mitochondrial proteins induced by alpha-synuclein overexpression can be rescued by parkin, PINK1, or DJ-1 470 coexpression. Thus, there may exist a biological pathway that may be coordinately regulated by the different Parkinson disease-linked genes.

LRRK2

LRRK2 is a member of the ROCO family of proteins, which are large multidomain proteins that contain a ROC (Ras of complex proteins)/guanosine triphosphatase (GTPase) domain followed by a COR (C-terminal of ROC) kinase domain. In a cell, damaged organelles and unused long-lived proteins localize in a double-membrane vesicle, called an autophagosome or autophagic vesicle, which is ultimately exported from the cell. Autosomal dominant mutations in LRRK2 are associated with both familial and late-onset Parkinson disease, and the Gly-to-Ser (G2019S) mutation at position 2019 is the most common of 577 mutations in LRRK2. *LRRK2* is the first gene found to be mutated frequently in late-onset autosomal dominant Parkinson disease. The evidence that G2019S is pathogenic is very strong. This mutation is very frequent in Parkinson disease and is extremely rare in healthy individuals: the mutation cosegregates with Parkinson disease in large families. The G2019S residue is extremely conserved in LRRK2 homologues, and the mutation increases the kinase activity of the protein.

> **GTPase**
> guanosine triphosphatase

LRRK2 is present mainly in the cytoplasm but may also localize to mitochondria. It is associated with membranes, such as mitochondria, endoplasmic reticulum, and synaptic vesicles, which is similar to alpha-synuclein. Parkinson disease patients with the G2019S mutation in LRRK2 exhibit reduced mitochondrial membrane potential and total intracellular ATP levels accompanied by increased mitochondrial elongation and interconnectivity.

Current evidence suggests that several Parkinson disease-linked proteins may all directly or indirectly influence mitochondrial homeostasis (Table 3.10). PINK1 and parkin are linked in a common pathway involved in the protection of mitochondrial homeostasis. Other Parkinson disease-associated proteins, such as DJ-1, LRRK2, and alpha-synuclein, interact with PINK1 and/or parkin in mammalian cells. These proteins may interact in a complex network in mitochondria. Thus, the activities of parkin, PINK1, DJ-1, alpha-synuclein, and LRRK2 all have a significant impact on mitochondrial homeostasis and function. These results further support the notion that mitochondrial dysfunction is a key event that underlies the pathogenesis of Parkinson disease.

Prevalence

In the United States, 1 in 4,000 children develops a mitochondrial disease by age 10 each year, and since mitochondrial diseases are underrecognized, a more accurate frequency may be 1 in 2,000 children per year. For similar reasons, it is not known how many people are affected by adult onset.

Table 3.10 Control of mitochondrial homeostasis by Parkinson disease-related genes

Parkinson disease-related protein	Pathways that maintain mitochondrial homeostasis	Direct action on mitochondria
PINK1	Mitochondrial biogenesis, mitophagy, mitochondrial fission and fusion, mitochondrial transport in axon, antioxidant defense	Interacts with proteins that control mitochondrial morphogenesis (HtrA2, Drp1, and Opa1/Mfn2), mitochondrial complex respiration, and calcium homeostasis
Parkin	Mitochondrial biogenesis, UPS, mitophagy, antioxidant defense	Controls replication, transcription, and repair of mitochondrial DNA by mitochondrial transcription factor A
Alpha-synuclein	Mitophagy, ROS formation, mitochondrial fusion	Binds to mitochondrial membranes and mitochondrial complex 1 to control mitochondrial membrane potential and ATP synthesis
DJ-1	Antioxidant defense	After association with mitochondrial chaperones HSP70 and Grp75 and binding to mitochondrial complex 1, DJ-1 functions as a redox-regulated chaperone that prevents misfolding and aggregation of proteins toxic to mitochondria
LRRK2		Reduces mitochondrial membrane potential and total intracellular ATP levels

Adapted from Sai et al., *Neurosci. Biobehav. Rev.* **36**:2034–2043, 2012.

Abbreviations: UPS, ubiquitin proteasome system; ROS, reactive oxygen species; HtrA2, a mitochondrial serine protease; Drp1, dynamin-related protein 1 (hydrolyzes GTP and pinches off membrane vesicles); Opa1, optic atrophy 1 (a nucleus-encoded mitochondrial protein with similarity to dynamin-related GTPases); Mfn2, mitofusin-2 (a mitochondrial membrane protein that mediates mitochondrial fusion); HSP70, heat shock protein 70 (a chaperone protein that regulates protein folding); Grp75, glucose-related protein 75 (a member of the HSP70 family of chaperones that control protein folding).

Diagnosis and Prognosis

Diagnosis is very difficult because many diseases can result from mitochondrial dysfunction in different organs or symptoms, mitochondrial diseases can be confused with other disorders or conditions, mitochondrial diseases are underrecognized, and there is a lack of screening procedures and diagnostic biomarkers that are both sensitive and specific. Diagnosis can be a lengthy process that begins with a general clinical evaluation, followed by metabolic screening, brain magnetic resonance imaging, and ultimately genetic testing and invasive biochemical and histological analysis. While the identification of some known mitochondrial mutations may assist with diagnosis, in many cases the underlying genetic mutation can elude detection.

The prognosis for patients with mitochondrial diseases is generally poor, due to their progressive deterioration and weakness. Half of those affected by mitochondrial disease are children who show characteristic systems (weak heart, seizures, failing kidneys, and/or respiratory complications) before age 5. Children may have strokes, seizures, gastrointestinal problems, swallowing difficulties, failure to thrive, blindness, deafness, respiratory difficulties, lactic acidosis, immune system problems, and liver disease. Many children are misdiagnosed with atypical cerebral palsy (a group of nonprogressive disorders of movement and posture caused by the abnormal development of, or damage to, motor control centers of the brain), various seizure disorders, or other childhood diseases. Voluntary movement (e.g., walking and grasping) is accomplished using skeletal muscles (attached to bones), whose motion is controlled by the cerebral cortex, the largest portion of the brain. "Palsy" means paralysis but may also be used to describe uncontrolled muscle movement. Therefore, cerebral palsy encompasses any disorder of abnormal movement and paralysis caused by abnormal function of the cerebral cortex.

In the absence of a suitable treatment or cure, about 80% of children with mitochondrial diseases die before the age of 20. Defects in mitochondrial function are linked to diseases of aging, including Alzheimer disease, Parkinson disease, cardiovascular disease, diabetes, cancer and stroke, and neurological conditions such as autism spectrum disorder and cerebral palsy. About 50 million people in the United States suffer from these degenerative disorders.

Treatment

Currently, a specific treatment for any mitochondrial disease does not exist. Rather, treatment is individualized by patient and is aimed at reducing symptoms or at delaying or preventing the progression of the disease. Certain vitamin and enzyme therapies, along with occupational and physical therapy, may help some patients to realize some improvement in fatigue and energy levels. Given the central role of imbalanced mitochondrial homeostasis in the pathology of Parkinson disease, molecules that help to restore a balanced mitochondrial homeostasis may represent possible targets for drug therapy in Parkinson disease.

summary

During the last 10 to 15 years, the development of DNA-based technologies has transitioned from chromosomal karyotyping to analyses of chromosomal structural and numerical abnormalities, the application of array-based and GWAS-based projects of common and rare SVs, and human genetic mapping studies. These human genomic projects have included the Human Genome Project, HapMap Project, sequencing of personal genomes, the 1000 Genomes Project, and population studies of genome-wide CNVs. Collectively, these studies have rapidly advanced our knowledge of chromosome and gene structure as well as gene variation under conditions of health and disease. The success of genetic mapping in providing new insights into biology and disease etiology and the recent evidence that systematic association studies can identify novel chromosomal loci have brought us a lot closer to our ultimate goal: the identification of all cellular pathways in which genetic variation contributes to the susceptibility and inheritance of common diseases.

Nonetheless, we still face the challenge that none of the currently available sequencing technologies can detect and assay all of the SVs in a given genome. To overcome this problem, the following advances in technology must be realized. First, the length of reads of DNA sequence produced by next-generation sequencing technologies must be increased. Second, the algorithms for detection of CNVs (deletion, duplication, or inversion of a DNA sequence longer than 1 kb) need to be improved. Third, the assembly of personal genomes is required to permit the accurate detection of SVs of all sizes and types and with sequence level resolution of DNA breakpoints, insertions, and deletions.

Since the first application of exome sequencing to validate a disease gene identification approach in 2009 using array-based targeted-capture methodologies, the application of whole-exome sequencing has received much greater emphasis (Table 3.11). The exome comprises the coding sequences of all annotated protein-coding genes (about 23,000) and is equivalent to about 1% of the total haploid genomic sequence (approximately 30 Mb). Despite the recent increase in the useful applications of whole-exome sequencing, this methodology assesses nucleotide variation in only about 2% of the genome (i.e., 20,000 to 25,000 variants), which may be sufficient to detect highly penetrant disease genes inherited in a Mendelian pattern. However, in these cases, variation in about 98% of the human genome is not assayed. This large amount of undetected variation might prove important in analyses of genomic variants associated with complex, heterogeneous, or more subtle phenotypes than the fully penetrant Mendelian diseases analyzed to date.

(continued)

summary *(continued)*

Note that a "healthy" individual is a heterozygous carrier of about 40 to 100 highly penetrant deleterious variants that can potentially cause a Mendelian disease. Many of these represent recessive carrier states. However, this estimate is based only on the exons and results from the 5 to 10% of genes and diseases are currently understood. Thus, it is possible that we all carry many more deleterious changes or potentially pathogenic variants than is now predictable.

The above-mentioned limitations of whole-genome and -exome sequencing facing us today underscore that one of our main objectives in the future is to better understand not only the functional impact but also, more importantly, the medical significance of genetic variants. The functional impact of variants can be easily determined for fully penetrant mutations in known disease genes. In contrast, much more work is required to determine the functional and medical significance of the estimated 20,000 genes for which function has not been assigned and phenotypes or associated traits have not been elucidated. It is anticipated that whole-genome sequencing may guide us to achieve personalized medicine more readily in the near future.

Table 3.11 Comparison of whole-genome sequencing and exome sequencing approaches for disease gene identification

Parameter	Whole-genome sequencing	Exome sequencing
Cost	Still expensive but declining rapidly	Cost reduced to 10–33% of that of whole-genome sequencing
Technical	No capture step, automatable	Capture step, technical bias
Variation	Detects all genetic and genomic variation (SNVs and CNVs)	Analyzes about 1% of the genome
	Determines functional gene coding and nonfunctional variation	Restricted to coding and splice site variants in annotated genes
	About 3.5 million variants	About 20,000 variants
Disease	Suitable for Mendelian and complex trait gene identification, as well as sporadic phenotypes caused by de novo SNVs or CNVs	Suitable for highly penetrant Mendelian disease gene identification

Adapted with permission from Gonzaga-Jauregui et al., *Annu. Rev. Med.* 63:35–61, 2012.

review questions

1. What are the definitions and genetic significance of karyotype, karyogram, and cytogenetics?

2. What does the term band Xp21.2 represent?

3. What role does a centromere play in the function of a chromosome?

4. Describe the known types of chromosome disorders, and then indicate the types of chromosomal abnormalities and syndromes that arise with each type of chromosomal disorder.

5. What type of chromosomal disorders may arise when a chromosomal abnormality occurs on maternally derived chromosomes?

6. What types of genomic rearrangements arise that may alter genome architecture and yield clinical consequences?

7. Define the terms SVs and SNPs, and describe how these terms have proven important for the demonstration of genetic variation in health and disease.

8. In a population, what is the predicted level of genetic identity between any two unrelated individuals?

9. How have the techniques of array-based CGH, SNP arrays, array painting, and next-generation sequencing analytical methods advanced the characterization of chromosome rearrangements in human genomes?

10. How do chromosome abnormalities happen?

review questions *(continued)*

11. What are the chromosome array painting, split-read, and fine-mapping methods, and how may these methods be used advantageously to determine chromosomal breakpoints?

12. What is the relevance of CNVs, and how would you use various array-based techniques, including CGH, to identify genomic rearrangements and disease susceptibility genes?

13. What are the steps you would follow to perform an array CGH analysis?

14. What are the HapMap Project and 1000 Genomes Project?

15. What are the advantages and current disadvantages of next-generation sequencing technologies?

16. Describe the paired-end read mapping technique, and explain why it is currently the best method to analyze chromosome rearrangements.

17. Define GWASs, and explain how they have greatly facilitated the mapping of loci that contribute to common diseases in humans. What are the important messages we have learned by using GWASs for the genetic mapping of common variants?

18. What are the main features of the functions and phenotypic associations of genes related to common diseases?

19. It is claimed that the main value of genetic mapping is not to predict the genetic risk for a given disease, but to provide novel insight into the biological mechanisms of induction and treatment of disease. Describe and defend the findings that either do or do not support this claim.

20. Genetic disorders may be classified as either monogenic or polygenic or result from mutations in mitochondrial DNA. Cite disease-related examples and describe the primary features of (i) monogenic disorders, (ii) polygenic disorders, and (iii) mitochodrial disorders.

21. What is a trinucleotide repeat disorder? How does it arise, and which diseases result from this type of disorder?

22. Discuss the two types of genetic variants that are thought to generate the architecture of complex disorders, and indicate whether current evidence supports one or both of these complex (polygenic) disorders.

23. What is meant by linkage disequilibrium, and why is linkage disequilibrium important in genetic mapping studies?

24. What are the main characteristics of mitochondrial disorders?

25. How might different factors stimulate an imbalance in mitochondrial homeostasis and elicit mitochondrial diseases, including Parkinson disease?

references

Alkan, C., B. P. Coe, and E. E. Eichler. 2011. Genome structural variation discovery and genotyping. *Nat. Rev. Genet.* **12**:363–376.

Altshuler, D., M. J. Daly, and E. S. Lander. 2008. Genetic mapping in human disease. *Science* **322**:881–888.

Baker, P. R., II, and A. K. Steck. 2011. The past, present, and future of genetic associations in type 1 diabetes. *Curr. Diab. Rep.* **11**:445–453.

Baltimore, D. 2001. Our genome unveiled. *Nature* **409**:814–816.

Bauman, J. G., J. Wiegant, P. Borst, and P. van Duijn. 1980. A new method for fluorescence microscopical localization of specific DNA sequences by in situ hybridization of fluorochrome labeled RNA. *Exp. Cell Res.* **128**:485–490.

Beckmann, J. S., X. Estivill, and S. E. Antonarakis. 2007. Copy number variants and genetic traits: closer to the resolution of phenotypic to genotypic variability. *Nat. Rev. Genet.* **8**:639–646.

Bird, T. D. 2008. Genetic aspects of Alzheimer disease. *Genet. Med.* **10**: 231–239.

Bonifati, V. 2006. Parkinson's disease: the LRRK2-G2019S mutation: opening a novel era in Parkinson's disease genetics. *Eur. J. Hum. Genet.* **14**:1061–1062.

Byerley, W., and J. A. Badner. 2011. Strategies to identify genes for complex disorders: a focus on bipolar disorder and chromosome 16p. *Psychiatr. Genet.* **21**:173–182.

Chan, E. Y. 2005. Advances in sequencing technology. *Mutat. Res.* **573**:13–40.

Cirulli, E. T., and D. B. Goldstein. 2010. Uncovering the roles of rare variants in common disease through whole-genome sequencing. *Nat. Rev. Genet.* **11**: 415–425.

Clarke, L., X. Zheng-Bradley, R. Smith, E. Kulesha, C. Xiao, I. Toneva, B. Vaughan, D. Preuss, R. Leinonen, M. Shumway, S. Sherry, P. Flicek, and The 1000 Genomes Project Consortium. 2012. The 1000 Genomes Project: data management and community access. *Nat. Methods* **9**:459–462.

Connor, J. M., and M. A. Ferguson-Smith. 1991. *Essential Medical Genetics*, 3rd ed. Blackwell Scientific Publications, Oxford, United Kingdom.

Conrad, D. F., D. Pinto, R. Redon, L. Feuk, O. Gokcumen, Y. Zhang, J. Aerts, T. D. Andrews, C. Barnes, P. Campbell, T. Fitzgerald, M. Hu, C. H. Ihm, K. Kristiansson, D. G. MacArthur, J. R. MacDonald, I. Onyiah, A. Wing Chun Pang, S. Robson, K. Stirrups, A. Valsesia, K. Walter, J. Wei, The Wellcome Trust Case Control Consortium, C. Tyler-Smith, N. P. Carter, C. Lee, S. W. Scherer, and M. E. Hurles. 2010. Origins and functional impact of copy number variation in the human genome. *Nature* **464**:704–712.

Crotwell, P. L., and H. E. Hoyme. 2012. Advances in whole-genome genetic testing: from chromosomes to microarrays. *Curr. Probl. Pediatr. Adolesc. Health Care* **42**:47–73.

Dewhurst, S. 1998. Down's syndrome. *Biol. Sci. Rev.* **10**(5):11–15.

Elias-Sonnenschein, L. S., L. Bertram, and P. J. Visser. 2012. Relationship between genetic risk factors and markers for Alzheimer's disease pathology. *Biomark. Med.* **6**:477–495.

The ENCODE Project Consortium. 2012. An integrated encyclopedia of DNA elements in the human genome. *Nature* **489**:57–74.

Fasching, P. A., et al. 2012. The role of genetic breast cancer susceptibility variants as prognostic factors. *Hum. Mol. Genet.* **21**:3926–3939.

Feuk, L., A. R. Carson, and S. W. Scherer. 2006. Structural variation in the human genome. *Nat. Rev. Genet.* **7**:85–97.

Fitzgerald, T. W., L. D. Larcombe, S. Le Scouarnec, S. Clayton, D. Rajan, N. P. Carter, and R. Redon. 2011. aCGH.Spline—an R package for aCGH dye bias normalization. *Bioinformatics* **27**:1195–1200.

Girirajan, S., C. D. Campbell, and E. E. Eichler. 2011. Human copy number, variation and complex genetic disease. *Annu. Rev. Genet.* **45**:203–226.

Gonzaga-Jauregui, C., J. R. Lupski, and R. A. Gibbs. 2012. Human genome sequencing in health and disease. *Annu. Rev. Med.* **63**:35–61.

Harold, D., et al. 2009. Genome-wide association study identifies variants at *CLU* and *PICALM* associated with Alzheimer disease, and shows evidence for additional susceptibility genes. *Nat. Genet.* **41**:1088–1093.

Haydar, T. F., and R. H. Reeves. 2012. Trisomy 21 and early brain development. *Trends Neurosci.* **35**:81–91.

Hirschhorn, J. N., and K. Z. Gajdos. 2011. Genome-wide association studies: results from the first few years and potential implications for clinical medicine. *Annu. Rev. Med.* **62**:11–24.

Huang, N., I. Lee, E. M. Marcotte, and M. E. Hurles. 2010. Characterizing and predicting haploinsufficiency in the human genome. *PLoS Genet.* **6**:e1001154.

The Huntington's Disease Collaborative Research Group. 1993. A novel gene containing a trinucleotide repeat that is expanded and unstable on Huntington's disease chromosomes. *Cell* **72**:971–983.

Iafrate, A. J., J. L. Feuk, M. N. Rivera, M. L. Listewnik, P. K. Donahoe, Y. Qi, S. W. Scherer, and C. Lee. 2004. Detection of large-scale variation in the human genome. *Nat. Genet.* **36**:949–951.

International HapMap Consortium. 2005. A haplotype map of the human genome. *Nature* **437**:1299–1320.

Kallioniemi, A., O. P. Kallioniemi, D. Sudar, D. Rutovitz, J. W. Gray, F. Waldman, and D. Pinkel. 1992. Comparative genomic hybridization for molecular cytogenetic analysis of solid tumors. *Science* **258**:818–821.

Kathiresan, S., and D. Srivastava. 2012. Genetics of human cardiovascular disease. *Cell* **148**:1242–1257.

Koopman, W. J. H., G. M. Willems, and J. A. M. Smeitink. 2012. Monogenic mitochondrial disorders. *N. Engl. J. Med.* **366**:1132–1141.

Ku, C. S., N. Naidoo, S. M. Teo, and Y. Pawitan. 2011. Regions of homozygosity and their impact on complex diseases and traits. *Hum. Genet.* **129**:1–15.

Lander, E. S., et al. 2001. Initial sequencing and analysis of the human genome. *Nature* **409**:860–921.

Le Scouarnec, S., and S. M. Gribble. 2012. Characterizing chromosome rearrangements: recent technical advances in molecular cytogenetics. *Heredity* **108**:75–85.

Mani, R. S., and A. M. Chinnaiyan. 2010. Triggers for genomic rearrangements: insights into genomic, cellular and environmental influences. *Nat. Rev. Genet.* **11**:819–829.

Maurano, R., E. Humbert, R. E. Rynes, E. Thurman, E. Haugen, H. Wang, A. P. Reynolds, R. Sandstrom, H. Qu, J. Brody, A. Shafer, F. Neri, K. Lee, T. Kutyavin, S. Stehling-Sun, A. K. Johnson, T. K. Canfield, E. Giste, M. Diegel, D. Bates, R. S. Hansen, S. Neph, P. J. Sabo, S. Heimfeld, A. Raubitschek, S. Ziegler, C. Cotsapas, N. Sotoodehnia, I. Glass, S. R. Sunyaev, R. Kaul, and J. A. Stamatoyannopoulos. 2012. Systematic localization of common disease-associated variation in regulatory DNA. *Science* **337**:1190–1195.

McLaren, W., B. Pritchard, D. Rios, Y. Chen, P. Flicek, and F. Cunningham. 2010. Deriving the consequences of genomic variants with the Ensembl API and SNP Effect Predictor. *Bioinformatics* **26**:2069–2070.

Mefford, H. C., and E. E. Eichler. 2009. Duplication hotspots, rare genomic disorders, and common disease. *Curr. Opin. Genet. Dev.* **19**:196–204.

Michalet, X., R. Ekong, F. Fougerousse, S. Rousseaux, C. Schurra, N. Hornigold, M. van Slegtenhorst, J. Wolfe, S. Povey, J. S. Beckmann, and A. Bensimon. 1997. Dynamic molecular combing: stretching the whole human genome for high-resolution studies. *Science* **277**:1518–1523.

Mills, R. E., et al. 2011. Mapping copy number variation by population-scale genome sequencing. *Nature* **470**:59–65.

The 1000 Genomes Project Consortium. 2010. A map of human genome variation from population-scale sequencing. *Nature* **467**:1061–1073.

Ormerod, M. G. 2008. *Flow Cytometry—A Basic Introduction—Wiki Version.* http://flowbook-wiki.denovosoftware.com/.

Ou, Z., P. Stankiewicz, Z. Xia, A. M. Breman, B. Dawson, J. Wiszniewska, P. Szafranski, M. L. Cooper, M. Rao, L. Shao, S. T. South, K. Coleman, P. M. Fernhoff, M. J. Deray, S. Rosengren, E. R. Roeder, V. B. Enciso, A. C. Chinault, A. Patel, S. H. Kang, C. A. Shaw, J. R. Lupski, and S. W. Cheung. 2011. Observation and prediction of recurrent human translocations mediated by NAHR between nonhomologous chromosomes. *Genome Res.* **21**:33–46.

Pennisi, E. 2012. ENCODE Project writes eulogy for junk DNA. *Science* **337**:1159–1161.

Pinkel, D., R. Segraves, D. Sudar, S. Clark, I. Poole, D. Kowbel, C. Collins, W. L. Kuo, C. Chen, Y. Zhai, S. H. Dairkee, B. M. Ljung, J. W. Gray, and D. G. Albertson. 1998. High resolution analysis of DNA copy number variation using comparative genomic hybridization to microarrays. *Nat. Genet.* 20: 207–211.

Polychronakos, C., and Q. Li. 2011. Understanding type 1 diabetes through genetics: advances and prospects. *Nat. Rev. Genet.* 12:781–792.

Rausch, T., T. Zichner, A. Schlattl, A. M. Stütz, V. Benes, and J. O. Korbel. 2012. DELLY: structural variant discovery by integrated paired-end and split-read analysis. *Bioinformatics* 28:i333–i339.

Rosman, D. S., V. Kaklaman, and B. Pasche. 2007. New insights into breast cancer genetics and impact on patient management. *Curr. Treat. Options Oncol.* 8:61–73.

Sai, Y., Z. Zou, K. Peng, and Z. Dong. 2012. The Parkinson's disease-related genes act in mitochondrial homeostasis. *Neurosci. Biobehav. Rev.* 36: 2034–2043.

Scherer, S. W., C. Lee, E. Birney, D. M. Altshuler, E. E. Eichler, N. P. Carter, M. E. Hurles, and L. Feuk. 2007. Challenges and standards in integrating surveys of structural variation. *Nat. Genet.* 39:S7–S15.

Schouten, J. P., C. J. McElgunn, R. Waaijer, D. Zwijnenburg, F. Diepvens, and G. Pals. 2002. Relative quantification of 40 nucleic acid sequences by multiplex ligation-dependent probe amplification. *Nucleic Acids Res.* 30:e57.

Scriver, C. 1995. Whatever happened to PKU? *Clin. Biochem.* 28:137–144.

Sebat, J., B. Lakshmi, J. Troge, A. J. Young, P. Lundin, S. Månér, H. Massa, M. Walker, M. Chi, N. Navin, R. Lucito, J. Healy, J. Hicks, K. Ye, A. Reiner, T. C. Gilliam, B. Trask, N. Patterson, A. Zetterberg, and M. Wigler. 2004. Large-scale copy number polymorphism in the human genome. *Science* 305: 525–528.

Smith, M. 2011. Genome architecture and sequence variation in health and disease, p. 1–22. *In* M. Smith (ed.), *Investigating the Human Genome: Insights into Human Variation and Disease Susceptibility*. FT Press, Upper Saddle River, NJ.

Speicher, M. R., and N. P. Carter. 2005. The new cytogenetics: blurring the boundaries with molecular biology. *Nat. Rev. Genet.* 6:782–792.

Stankiewicz, P., and J. R. Lupski. 2010. Structural variation in the human genome and its role in disease. *Annu. Rev. Med.* 61:437–455.

Stephens, P. J., C. D. Greenman, B. Fu, F. Yang, G. R. Bignell, L. J. Mudie, E. D. Pleasance, K. W. Lau, D. Beare, L. A. Stebbings, S. McLaren, M. L. Lin, D. J. McBride, I. Varela, S. Nik-Zainal, C. Leroy, M. Jia, A. Menzies, A. P. Butler, J. W. Teague, M. A. Quail, J. Burton, H. Swerdlow, N. P. Carter, L. A. Morsberger, C. Iacobuzio-Donahue, G. A. Follows, A. R. Green, A. M. Flanagan, M. R. Stratton, P. A. Futreal, and P. J. Campbell. 2011. Massive genomic rearrangement acquired in a single catastrophic event during cancer development. *Cell* 144:27–40.

Strachan, T., and A. Read. 1996. *Human Molecular Genetics*. Bios Scientific Publishers, Oxford, United Kingdom.

Sudmant, P. H., J. O. Kitzman, F. Antonacci, C. Alkan, M. Malig, A. Tsalenko, N. Sampas, L. Bruhn, and J. Shendure, 1000 Genomes Project, and E. E. Eichler. 2010. Diversity of human copy number variation and multicopy genes. *Science* 330:641–646.

Teo, S. M., Y. Pawitan, C. S. Ku, K. S. Chia, and A. Salim. 2012. Statistical challenges associated with detecting copy number variations with next-generation sequencing. *Bioinformatics* 28:2711–2718.

Toronto International Data Release Workshop Authors. 2009. Prepublication data sharing. *Nature* 461:168–170.

Venter, J. C., et al. 2001. The sequence of the human genome. *Science* 29: 1304–1351.

Wain, L. V., J. A. L. Armour, and M. D. Tobin. 2009. Genomic copy number variation, human health, and disease. *Lancet* 374:340–350.

Walker, F. O. 2007. Huntington's disease. *Lancet* 369:218–228.

Wallis, G. 1999. *Genetic Basis of Human Disease*. The Biochemical Society, London, United Kingdom.

Ward, L. D., and M. Kellis. 2012. Evidence of abundant purifying selection in humans for recently acquired regulatory functions. *Science* 337:1675–1678.

Wexler, N. S. 2012. Huntington's disease: advocacy driving science. *Annu. Rev. Med.* 63:1–22.

Wexler, N. S., A. B. Young, R. E. Tanzi, H. Travers, S. Starosta-Rubinstein, J. B. Penney, S. R. Snodgrass, I. Shoulson, F. Gomez, M. A. Ramos Arroyo, G. K. Penchaszadeh, H. Moreno, K. Gibbons, A. Faryniarz, W. Hobbs, M. A. Anderson, E. Bonilla, P. M. Conneally, and J. F. Gusella. 1987. Homozygotes for Huntington's disease. *Nature* 326: 194–197.

Wijchers, P. J., and W. de Laat. 2011. Genome organization influences partner selection for chromosomal rearrangements. *Trends Genet.* 27:63–71.

Yamada, N. A., L. S. Rector, P. Tsang, E. Carr, A. Scheffer, M. C. Sederberg, M. E. Aston, R. A. Ach, A. Tsalenko, N. Sampas, B. Peter, L. Bruhn, and A. R. Brothman. 2011. Visualization of fine-scale genomic structure by oligonucleotide-based high-resolution FISH. *Cytogenet. Genome Res.* 132: 248–254.

Zheng-Bradley, X., and P. Flicek. 2012. Maps for the world of genomic medicine: the 2011 CSHL Personal Genomes meeting. *Hum. Mutat.* 33:1016–1019.

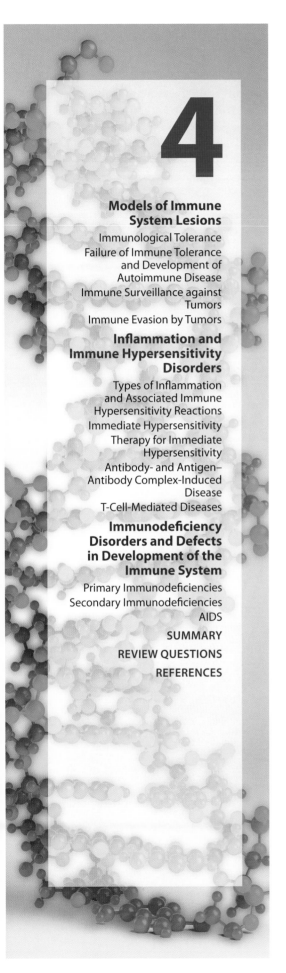

4

Immune Pathogenesis

Models of Immune System Lesions

Immunological Tolerance

Our immune system has evolved to respond to a plethora of microbes, but remarkably, it generally does not react against our own (self-)antigens. This unresponsiveness to self-antigens, termed **immunological tolerance**, is maintained even though the repertoire of lymphocyte receptors expressed is generally not skewed towards receptors that are reactive with non-self-antigens. This means that lymphocytes with receptors that recognize self-antigens are generated continuously during lymphocyte development and maturation. Since many self-antigens are accessible to the immune system, unresponsiveness to these antigens is not maintained simply by interfering with their recognition by lymphocytes and their receptors. Rather, mechanisms exist that suppress immune responses to self-antigens. These mechanisms underscore a primary function of the immune system—an ability to discriminate between self- and non-self-antigens (e.g., microbial antigens) under conditions of health, infection, and disease. If these mechanisms fail, the immune system may attack an individual's own cells and tissues. Such anti-self-responses can lead to a state of **autoimmunity**. Further, the inability to downregulate such autoimmune responses may result in **autoimmune disease**. Thus, appropriate regulation of autoimmune responses does not lead to a harmful situation and usually does not elicit autoimmune disease.

The two main objectives in studying immune tolerance versus autoimmune disease in this chapter are (i) to describe the mechanisms which permit the immune system to remain unresponsive (tolerant) to self-antigens and (ii) to identify the molecular and cellular factors which promote the breakdown of tolerance and the development of autoimmune disease.

Immunological tolerance results from a lack of response to an antigen(s) that is induced by exposure of lymphocytes to this antigen(s). Activation of lymphocytes with specific receptors for a given antigen by this antigen can

doi:10.1128/9781555818890.ch4

elicit several different outcomes. First, these lymphocytes may proliferate and differentiate into effector cells, which leads to a productive immune response. Antigens that stimulate such productive responses are considered to be **immunogenic**. Second, after exposure to antigen, the lymphocytes may be functionally inactivated or killed, resulting in tolerance. Antigens that induce tolerance are considered to be **tolerogenic**. Third, the responses of antigen-specific lymphocytes may not be detectable. This mechanism of **immunological ignorance** is mediated by the capacity of the lymphocytes to ignore (i.e., not recognize) the presence of the antigen. Normally, foreign microbes are immunogenic and self-antigens are tolerogenic. The choice between lymphocyte activation and tolerance is determined by whether the antigen-specific lymphocytes were previously exposed **(antigen-primed lymphocytes)** or not exposed **(naïve lymphocytes)** to antigen, as well as by the properties of the antigen and how it is presented to the immune system. Interestingly, the same antigen may be administered in different forms and/ or by different routes to induce either an immune response or tolerance. This finding enables an analysis of the factors that determine whether activation or tolerance develops after an encounter with an antigen.

Self-antigens normally induce tolerance. By learning how tolerance for a particular antigen is induced in lymphocytes, we may develop protocols to prevent or downregulate undesirable immune reactions. Strategies for inducing tolerance have wide application and are currently being tested to treat allergic and autoimmune diseases and to prevent the rejection of organ transplants. The same strategies may be valuable in gene therapy, to prevent responses against products of newly expressed genes or vectors, and even for stem cell transplantation if the stem cell donor and recipient differ genetically (e.g., at relevant human leukocyte antigen [HLA] loci; see chapters 2 and 3).

Two types of immunological tolerance to various self-antigens may be induced when developing lymphocytes encounter these antigens at different anatomical sites: first, if this encounter occurs in primary (central) lymphoid organs, **central tolerance** is induced; second, if mature lymphocytes encounter self-antigens in peripheral tissues, **peripheral tolerance** is induced (Fig. 4.1). Central tolerance arises only to self-antigens present in primary lymphoid organs (e.g., bone marrow and thymus). Tolerance to self-antigens that are absent from these organs must be induced and maintained by various mechanisms in peripheral tissues (e.g., spleen and lymph nodes). Our knowledge of the number and type of self-antigens that induce central tolerance or peripheral tolerance or are ignored by the immune system is currently limited.

Th cell
T helper cell

How does a failure to achieve immunological tolerance result in autoimmunity? Tolerance in CD4$^+$ T helper (Th) cells is described first here, as more is known about how CD4$^+$ T cells mediate tolerance than about CD8$^+$ T cells. CD4$^+$ Th cells control most, if not all, immune responses to protein antigens. When these cells are made unresponsive to self-protein antigens, this is sufficient to prevent both cell-mediated and humoral immune responses against these antigens. Conversely, failure of tolerance in Th cells may result in autoimmunity manifested by T-cell-mediated attack against self-antigens or by the production of autoantibodies against self-proteins.

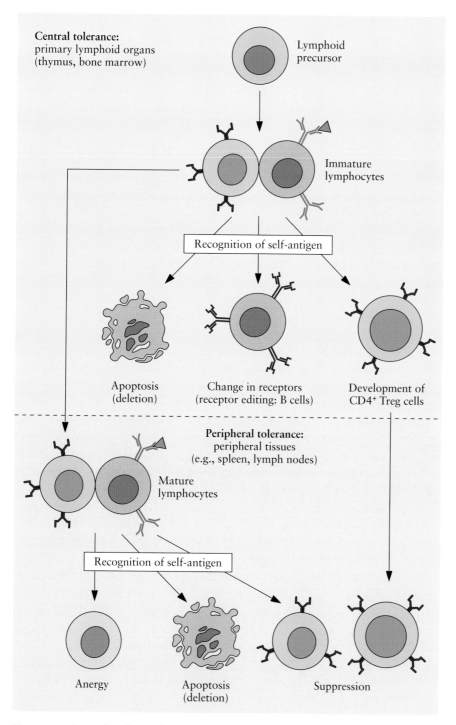

Figure 4.1 Central and peripheral immunological tolerance to self-antigens. In central tolerance, when immature lymphocytes specific for self-antigens recognize these antigens in the primary lymphoid organs (thymus and bone marrow), they may be deleted by negative selection. B cells change their specificity (receptor editing), and some T cells develop into Treg cells. Some lymphocytes that are reactive to self-antigens may undergo further maturation and enter peripheral tissues. In peripheral tolerance, mature lymphocytes reactive to self-antigens may be inactivated (anergized), deleted, or suppressed upon encounter with these self-antigens in peripheral tissues (spleen, lymph nodes, etc.). Adapted with permission from Abbas and Lichtman (ed.), *Basic Immunology: Functions and Disorders of the Immune System*, 3rd ed. (Saunders Elsevier, Philadelphia, PA, 2011). doi:10.1128/9781555818890.ch4.f4.1

Treg cell
T regulatory cell

MHC
major histocompatibility complex

Central Tolerance

The main mechanisms of central tolerance in T cells are apoptosis (cell death) and, for $CD4^+$ cells, the generation of T regulatory (Treg) cells (Fig. 4.2). The lymphocytes that develop in the thymus consist of cells with receptors that recognize both self-antigens and foreign antigens. If the receptors on an immature T cell in the thymus (thymocyte) bind with high affinity to a self-antigen displayed as a peptide–major histocompatibility complex (MHC) ligand on an antigen-presenting cell (see chapters 2 and 3), this T cell will receive a signal(s) that triggers apoptosis and will die before it is fully mature. This process is termed **negative selection** (see chapter 2), and it is a major mechanism of central T-cell tolerance.

Immature lymphocytes may interact strongly with an antigen if the antigen is present at high concentrations in the thymus and if the lymphocytes express receptors that recognize the antigen with high affinity. Antigens that induce negative selection generally include proteins that are abundant in the body, such as plasma proteins and common cellular proteins. Interestingly, many self-proteins previously thought to be expressed mainly or exclusively in peripheral tissues (e.g., expression of insulin in the pancreas) can also be expressed in some epithelial cells of the thymus. The **AIRE (auto̲immune regulator)** protein regulates the thymic expression of many of these peripheral tissue-restricted protein antigens. Mutations in the *AIRE* gene cause a rare autoimmune disorder, termed **autoimmune polyendocrine syndrome**, which is a recessive genetic disorder in which

Figure 4.2 Central T-cell tolerance. High-affinity binding of self-antigens by immature T cells in the thymus may cause apoptosis (death) of these cells (deletion). Self-antigen recognition in the thymus may also elicit the development of Treg cells that exit the thymus and enter into peripheral lymphoid tissues. Adapted with permission from Abbas and Lichtman (ed.), *Basic Immunology: Functions and Disorders of the Immune System*, 3rd ed. (Saunders Elsevier, Philadelphia, PA, 2011). doi:10.1128/9781555818890.ch4.f4.2

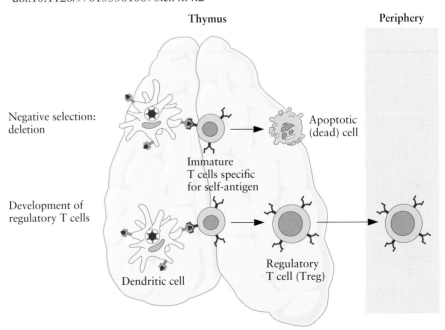

the function of many endocrine glands is altered. Autoimmune polyendocrine syndrome manifests as a mild immunodeficiency that yields persistent mucosal and cutaneous yeast infections. In addition, the function of the spleen, parathyroid gland, and adrenal gland as well as several other more minor glands is diminished. Negative selection affects self-reactive CD4$^+$ T cells and CD8$^+$ T cells, which recognize self-peptides presented by MHC class II and MHC class I molecules, respectively. The signals that induce apoptosis in immature lymphocytes following the recognition of antigens with high affinity in the thymus remain largely unknown. Current reasoning suggests that defective negative selection may explain why some autoimmunity-prone inbred strains of mice contain abnormally large numbers of mature T cells specific for various self-antigens in peripheral organs. However, why such negative selection is defective in these mice is not known.

Some immature CD4$^+$ T cells that recognize self-antigens in the thymus do not die but rather develop into Treg cells and enter peripheral tissues (Fig. 4.2). What determines whether a T cell in the thymus that recognizes a self-antigen will die or become a Treg cell is not known.

Peripheral T-Cell Tolerance

Peripheral tolerance is induced when mature T cells recognize self-antigens in peripheral tissues, leading to functional inactivation (termed **anergy**) or death when the function of self-reactive T cells is downregulated by Treg cells (Fig. 4.1). Peripheral tolerance is important for preventing T-cell responses to self-antigens that are present mainly in peripheral tissues and not in the thymus. Peripheral tolerance also may provide "backup" mechanisms for preventing autoimmunity in situations where central tolerance is incomplete.

Anergy

Anergy is defined as a state of long-term hyporesponsiveness in T cells that is characterized by an inhibition of T-cell antigen receptor (TCR) signaling and interleukin-2 (IL-2) expression. Physiologically, anergy is the functional inactivation of T cells that occurs when these cells recognize antigens in the absence of an adequate level of surface expression of **costimulator** molecules (second signals) required for maximal T-cell activation (Fig. 4.3). As discussed previously (see chapter 2), at least two signals are required for optimal naïve T-cell proliferation and differentiation into effector T cells. Signal 1 is delivered by the binding of an antigen to the TCR and is considered to be the "antigen-specific" signal. Signal 2 is provided by the binding of CD28 on T cells to the B7 family costimulator proteins CD80 and CD86 expressed on antigen-presenting cells in response to microbes and is considered to be an "antigen-nonspecific" signal. Normally, antigen-presenting cells in tissues and peripheral lymphoid organs, including dendritic cells, are in a resting state, in which they express only very low levels of CD80 and CD86 on their surfaces (see chapter 2). These antigen-presenting cells are constantly processing and displaying the self-antigens present in the tissues. T cells with specific receptors for these self-antigens can recognize the antigens and receive

TCR
T-cell antigen receptor

IL-2
interleukin-2

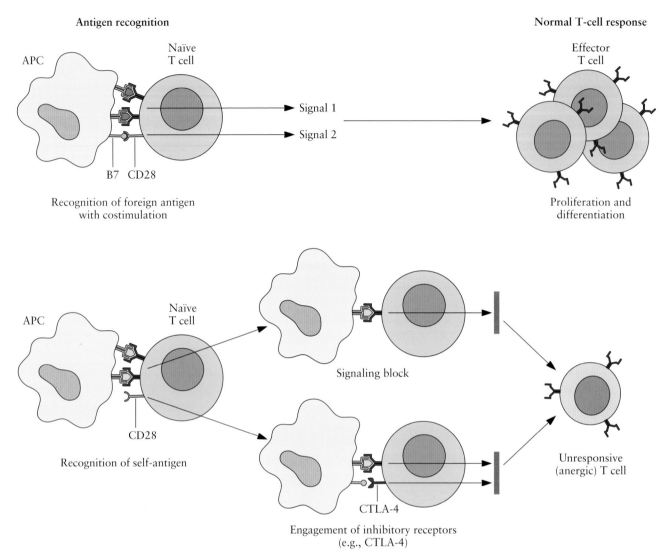

Antigen recognition

Normal T-cell response

APC

Naïve
T cell

Signal 1

Signal 2

B7   CD28

Recognition of foreign antigen
with costimulation

Effector
T cell

Proliferation and
differentiation

APC

Naïve
T cell

CD28

Recognition of self-antigen

Signaling block

CTLA-4

Engagement of inhibitory receptors
(e.g., CTLA-4)

Unresponsive
(anergic) T cell

Figure 4.3 T-cell anergy. An antigen presented by antigen-presenting cells, which express sufficient numbers of costimulator molecules (e.g., B7) on their surfaces, induces a normal T-cell response. If T cells recognize an antigen without strong costimulation or innate immunity, the TCRs may lose their ability to deliver activating signals. Alternatively, the T cells may engage inhibitory receptors, such as CTLA-4, that block activation. Adapted with permission from Abbas and Lichtman (ed.), *Basic Immunology: Functions and Disorders of the Immune System*, 3rd ed. (Saunders Elsevier, Philadelphia, PA, 2011). doi:10.1128/9781555818890.ch4.f4.3

prolonged signals from their antigen receptors (signal 1). Nevertheless, the T cells do not receive strong costimulation due to the relative lack of surface expression of B7 in the absence of an accompanying innate immune response. Under these conditions, the TCRs may lose their ability to transmit activating signals, or the T cells may preferentially engage an inhibitory receptor(s) of the CD28 family, termed CTLA-4 (cytotoxic T-lymphocyte-associated antigen 4 or CD152) or PD-1 (programmed [cell] death protein 1). The net result is long-lasting T-cell anergy (Fig. 4.3). It is intriguing that CTLA-4, which is involved in blocking or downregulating

T-cell responses, recognizes the B7 costimulators that bind to CD28 and initiate T-cell activation. How T cells choose to use CD28 or CTLA-4, with these very different outcomes, is incompletely understood.

Several experimental animal models support the importance of T-cell anergy in the maintenance of self-tolerance. Forced expression of high levels of the CD80 and CD86 costimulators in tissues from transgenic mice results in autoimmune reactions against antigens in that tissue. Thus, by artificially providing signal 2, anergy is overcome and autoreactive T cells are activated. If CTLA-4 molecules are blocked (by treatment with antibodies) or deleted (by gene knockout) in a mouse, that mouse develops vigorous autoimmune reactions against its own tissues that can lead to autoimmune disease. These results suggest that the inhibitory receptors CTLA-4 and PD-1 function to maintain autoreactive T cells in an inactive state. Importantly, polymorphisms in the *CTLA4* gene are associated with some autoimmune diseases in humans (e.g., type 1 diabetes [T1D]).

Several "factors" that induce and/or maintain the state of anergy in lymphocytes were recently identified. The **mTOR** (mammalian target of rapamycin) protein and other related metabolic sensors and regulators have expanded our view of anergy-inducing signals (Fig. 4.4). mTOR is

T1D
type 1 diabetes

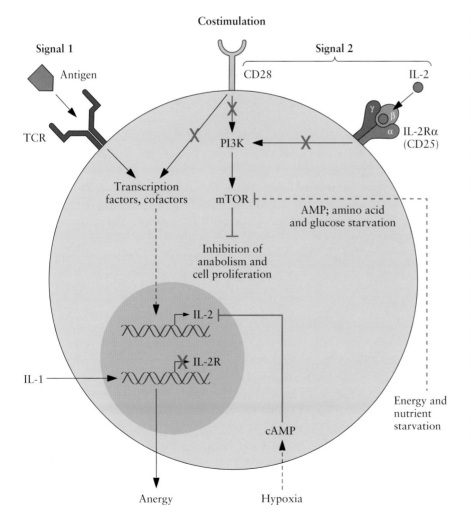

Figure 4.4 Costimulation and T-cell anergy. The signal 2 paradigm induction of anergy in T cells was initially described as the result of a TCR-dependent antigen-specific response (signal 1) without concomitant CD28 and IL-2R signaling (signal 2). Recent studies have further demonstrated that the T cell actively senses its microenvironment, through mTOR-dependent and independent mechanisms (see text), for available energy sources and nutrients, as well as additional negative cues resulting from hypoxia, e.g., cAMP. This regulates the T-cell commitment to switch its metabolic machinery and enter the S phase of the cell cycle. An inability of a T cell to become fully activated in such cases induces anergy (inactivation) and long-term tolerance in the T cell. Solid lines represent active pathways and dashed lines represent blocked pathways during the induction of T-cell anergy in an anergic environment. An **X** in red indicates a functional block in the specific pathway. cAMP, cyclic AMP; PI3K, phosphatidylinositol 3-kinase. Adapted with permission from Abbas and Lichtman (ed.), *Basic Immunology: Functions and Disorders of the Immune System*, 3rd ed. (Saunders Elsevier, Philadelphia, PA, 2011). doi:10.1128/9781555818890.ch4.f4.4

a serine/threonine protein kinase enzyme that regulates cell growth, cell proliferation, cell motility, cell survival, protein synthesis, and gene transcription. It belongs to the **phosphatidylinositol 3-kinase**-related kinase protein family and also integrates the input from upstream pathways, including insulin, insulin growth factors (IGF-1 and IGF-2), and amino acids. mTOR also senses cellular nutrient and energy levels and redox status. The mTOR pathway may be dysregulated in some human diseases, especially certain cancers. Rapamycin is a bacterial product that can inhibit the action of mTOR by associating with its intracellular receptor FKBP12.

TCR/CD28 costimulation-induced signaling in T cells results in IL-2 gene transcription and IL-2 mRNA stability and prevents induction of anergy. Costimulation also stimulates increases in Ras signaling, cellular metabolism, glucose uptake, and cell survival. Further, IL-2 signaling through the IL-2 receptor (IL-2R) complex fully activates the downstream phosphatidylinositol 3-kinase–mTOR pathway leading to entry into the S phase of the cell cycle and consequent cell proliferation. In contrast, T-cell activation involving only signal 1 (TCR and antigen) leads to deficient IL-2 transcription and induction that impairs TCR signaling in anergic T cells (Fig. 4.4). Activation of the mTOR-dependent pathway is required to induce a switch in T-cell metabolic pathways from catabolism to anabolism, via the stimulation of glycolysis and upregulation of key nutrient transporters. Anabolism comprises the various metabolic pathways that construct molecules from smaller units and is fueled by catabolism. During catabolism, large molecules are broken down, releasing energy that is used to synthesize **adenosine triphosphate (ATP)** required for many anabolic processes. This switch from catabolism to anabolism establishes the basis for the optimal activation, proliferation, and differentiation of a T cell. Anergic T cells are blocked at the G_1/S checkpoint of the cell cycle and are in a metabolically anergic state characterized by a failure to upregulate several nutrient transporters, including glucose transporter 1 (Glut1), and to switch to an anabolic state of metabolism. Thus, mTOR functions downstream of the IL-2R pathway to regulate tolerance (anergy) versus activation in a T cell. In addition, mTOR acts downstream of several energy and nutrient-sensing pathways to integrate these pathways in eukaryotic cells.

Immunosuppression by Treg Cells

Treg cells develop in the thymus or peripheral lymphoid organs on recognition of self-antigens and inhibit the activation and effector functions of potentially harmful lymphocytes specific for these self-antigens (Fig. 4.5). Most Treg cells are CD4$^+$ and express high levels of CD25, the IL-2R α chain (IL-2Rα). The development and function of these cells are dependent on the activity of the Foxp3 transcription factor. Mutations of Foxp3 in humans or knockout of the gene in mice causes a systemic, multiorgan autoimmune disease, demonstrating the importance of Treg cells for the maintenance of self-tolerance. The survival and function

IL-2R
IL-2 receptor

ATP
adenosine triphosphate

IL-2Rα
IL-2R α chain

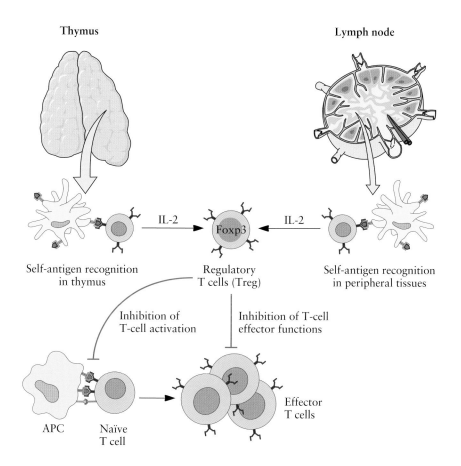

Thymus

Lymph node

IL-2

Foxp3

IL-2

Self-antigen recognition
in thymus

Regulatory
T cells (Treg)

Self-antigen recognition
in peripheral tissues

Inhibition of
T-cell activation

Inhibition of T-cell
effector functions

APC   Naïve
T cell

Effector
T cells

Figure 4.5 Suppression of immune responses by T cells. CD4$^+$ T cells that recognize self-antigens may differentiate into Treg cells in the thymus or peripheral tissues. This process is dependent on the transcription factor Foxp3 and requires the activity of IL-2. The larger arrow from the thymus, compared to the one from peripheral tissues, suggests that most of these cells presumably originate in the thymus. These Treg cells inhibit the activation of naïve T cells and their differentiation into effector T cells, by cell contact-dependent mechanisms or by the secretion of cytokines that inhibit T-cell responses. Adapted with permission from Abbas and Lichtman (ed.), *Basic Immunology: Functions and Disorders of the Immune System*, 3rd ed. (Saunders Elsevier, Philadelphia, PA, 2011).
doi:10.1128/9781555818890.ch4.f4.5

of Treg cells are dependent on IL-2, and this role of IL-2 accounts for the severe autoimmune disease that develops in mice in which the gene encoding either IL-2, IL-2Rα, or IL-2Rβ is deleted. The transforming growth factor β (TGF-β) cytokine also controls the generation of Treg cells, in part by stimulating Foxp3 expression. The source of TGF-β for inducing these cells in the thymus or peripheral tissues is not defined. Little is known about the mechanisms of inhibition of immune responses in vivo by Treg cells. Some Treg cells produce the IL-10 and TGF-β cytokines that block the activation of lymphocytes and macrophages. Treg cells may also interact directly with and suppress other lymphocytes or antigen-presenting cells by cell–cell contact-dependent mechanisms. Finally, there may also exist regulatory cell populations in addition to the CD25$^+$ Foxp3$^+$ CD4$^+$ T cells that have been the prime focus of most recent investigations.

Current evidence suggests that in some autoimmune diseases in humans, defective Treg cell function or defective resistance of pathogenic T cells to regulation may be an underlying cause of disease. However, definitive proof for these hypotheses is still lacking, mainly because it has proved difficult to define the maintenance, heterogeneity, and functions of Treg cells in humans.

TGF-β
transforming growth factor β

Deletion: Activation-Induced Cell Death

Recognition of self-antigens may trigger pathways of apoptosis that result in the deletion of self-reactive lymphocytes (Fig. 4.6). This process is known as activation-induced cell death, as it occurs after antigen recognition and activation. Self-antigens may induce T-cell death by two mechanisms. First, antigen recognition induces the production of proapoptotic (mediating apoptosis) proteins in T cells that induce cell death by the "mitochondrial pathway," in which various mitochondrial proteins leak out and activate caspases, cytosolic enzymes that induce apoptosis.

Figure 4.6 T-cell deletion. Under normal conditions, T cells respond to antigen presented by antigen-presenting cells by secreting IL-2, expressing equal amounts of proapoptotic and antiapoptotic proteins, and then undergoing proliferation and differentiation. However, under conditions in which costimulation or innate immunity is absent, self-antigen recognition by T cells may lead to an excess of intracellular proapoptotic proteins that cause cell death. Alternatively, self-antigen recognition may lead to expression of death receptors and their ligands, such as Fas and Fas ligand (FasL), on lymphocytes. Engagement of the death receptor leads to apoptotic death of the cells. Adapted with permission from Abbas and Lichtman (ed.), *Basic Immunology: Functions and Disorders of the Immune System*, 3rd ed. (Saunders Elsevier, Philadelphia, PA, 2011). doi:10.1128/9781555818890.ch4.f4.6

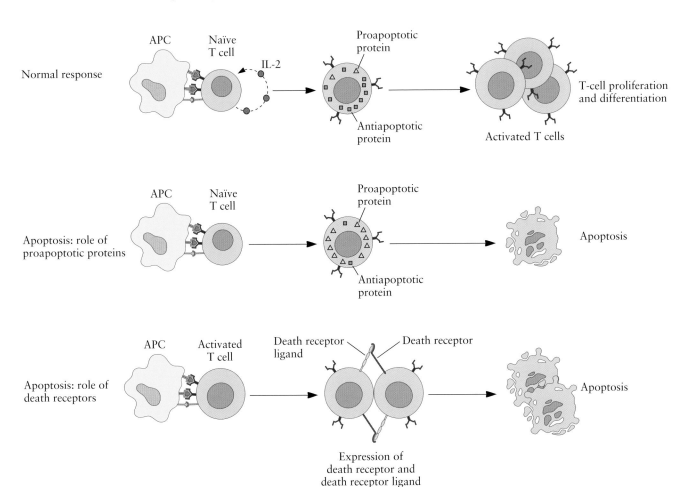

During immune responses to microbes, the activities of proteins that induce apoptosis may be blocked by other proteins activated by costimulation and growth factors. However, self-antigens recognized in the absence of strong costimulation do not stimulate the production of antiapoptotic proteins, and they stimulate the death of the cells that recognize these antigens. Second, recognition of self-antigens may lead to the coexpression of death receptors and their ligands. This ligand–receptor interaction generates signals through the death receptor that culminate in the activation of caspases and apoptosis by the death receptor pathway.

The best-understood death receptor–ligand pair involved in self-tolerance is comprised of a protein called **Fas** (CD95) that is expressed on many cell types and **Fas ligand** (FasL), which is expressed mainly on activated T cells. Binding of FasL to its Fas death receptor induces the apoptosis of T and B cells exposed to self-antigens and to mimics of self-antigens in experimental animals. Whether Fas has additional functions remains to be determined.

Evidence for a role of apoptosis in self-tolerance has emerged from genetic studies. Blocking the mitochondrial pathway of apoptosis in mice inhibits the deletion of self-reactive T cells in the thymus and peripheral lymphoid organs. Mice with mutations in the *fas* and *fasL* genes and children with mutations in the *FAS* gene all develop autoimmune diseases associated with lymphocyte accumulation. The human disease, called the **autoimmune lymphoproliferative syndrome**, is rare and is the only known example of a defect in apoptosis that causes a complex autoimmune phenotype in humans.

Analyses of the mechanisms of T-cell tolerance have shown that self-antigens differ from foreign microbial antigens in several ways. First, whereas self-antigens induce T-cell tolerance, microbial antigens stimulate T-cell activation (Table 4.1). Second, self-antigens are expressed in the thymus, where they induce deletion and generate Treg cells. In contrast, microbial antigens are actively transported to and concentrated in peripheral lymphoid organs. Third, self-antigens are displayed by resting antigen-presenting cells in the absence of innate immunity and costimulation signals, thus favoring the induction of T-cell anergy or death. On the other hand, microbes elicit innate immune reactions, leading to the expression of costimulators and cytokines that function as second signals

Table 4.1 Properties of self-antigens and foreign antigens that may determine the outcome of T-cell tolerance versus T-cell activation

Property of an antigen	Self-antigens (tissue) that induce tolerance (tolerogenic antigens)	Foreign antigens (microbe) that induce an immune response (immunogenic antigens)
Presence in primary lymphoid organs	Yes: high concentrations of self-antigens induce T-cell deletion and Treg cells (central tolerance)	No: microbial antigens are localized to peripheral lymphoid organs
Presentation with second signals (costimulation and innate immunity)	No: lack of second signals may lead to T-cell anergy or apoptosis	Yes: microbial antigens are presented. Second signals stimulate the survival and activation of lymphocytes.
Persistence	Long-lived: prolonged TCR engagement may induce anergy and apoptosis	Usually short-lived: immune responses generally eliminate microbial antigens

Certain properties of self-protein antigens and foreign (e.g., microbial) protein antigens that determine which self-antigens induce tolerance and which microbial antigens stimulate T-cell-mediated immune responses are summarized.

and promote T-cell proliferation and differentiation into effector T cells. Self-antigens are present throughout life and may therefore cause prolonged or repeated TCR engagement, again promoting anergy and apoptosis. Thus, much of our understanding of the mechanisms of T-cell tolerance, and their roles in preventing autoimmunity, originates from studies of experimental animal models. Extension of these studies to understanding T-cell tolerance versus anergy in humans presents an ongoing challenge.

Similar to T cells, B cells also undergo mechanisms of central tolerance and peripheral tolerance to protein antigens, mediated by receptor editing (expression of new antigen receptors) or negative selection (deletion via apoptosis). For the sake of brevity and to avoid duplication, the details of the mechanisms of B-cell tolerance are not discussed here.

Failure of Immune Tolerance and Development of Autoimmune Disease

Autoimmunity results from an immune response against one's own (self-) antigens. It is an important cause of disease, estimated to affect at least 1 to 2% of persons in developed countries, and with an apparently increasing prevalence. Nonetheless, note that in many cases, diseases associated with uncontrolled immune responses are called autoimmune without formal evidence that the responses are directed against self-antigens.

The principal factors that control the development of autoimmunity are the inheritance of susceptibility genes and environmental triggers, such as viral and bacterial infections (Fig. 4.7). Autoimmunity may elicit the production of both T cells and antibodies reactive against self-antigens. Using experimental animal models, we have learned much about how self-tolerance may fail and how self-reactive lymphocytes may become pathogenic. Susceptibility genes seem to both block and activate pathways of self-tolerance and promote the persistence of self-reactive T and B cells. Working in concert with such susceptibility genes, environmental stimuli and tissue injury may further activate these self-reactive lymphocytes.

Despite these findings about autoimmunity, we still do not know the etiology of any human autoimmune disease. This lack of understanding originates mainly from the following three factors. First, autoimmune diseases in humans usually are heterogeneous and multifactorial. Second, the target self-antigens that induce autoimmune diseases are frequently unknown. Third, the diseases may present in the clinic long after the initiation of the autoimmune reactions. Recent advances in the identification of disease-associated genes, methods of analysis of antigen-specific immune responses in humans, and availability of informative animal models that can be applied to clinical settings may lead to novel and improved treatments of autoimmune disease and to personalized medicine.

Genetic Control of Autoimmunity

Most autoimmune diseases are polygenic (controlled by many genes; see chapter 2) and are associated with multiple gene loci, the most critical of which are the MHC genes. A significant advance in our knowledge of the

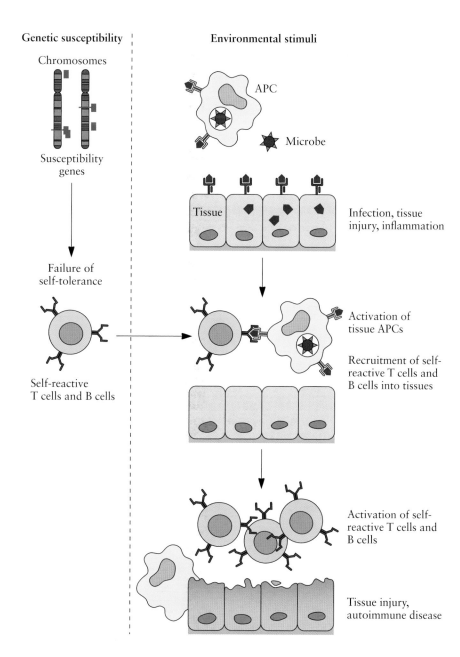

Genetic susceptibility

Chromosomes

Susceptibility genes

Failure of self-tolerance

Self-reactive T cells and B cells

Environmental stimuli

APC

Microbe

Tissue

Infection, tissue injury, inflammation

Activation of tissue APCs

Recruitment of self-reactive T cells and B cells into tissues

Activation of self-reactive T cells and B cells

Tissue injury, autoimmune disease

Figure 4.7 Possible mechanisms of induction of autoimmune disease. (**Left**) In this model of an organ-specific T-cell-mediated autoimmune disease, various genetic loci (blue bars) on either the same or different chromosomes (see chromosome bands) may confer susceptibility to an autoimmune disease(s). This control of susceptibility may occur by the influence of certain genes in these genetic loci on the maintenance of self-tolerance. (**Right**) Environmental triggers, such as infections and other inflammatory stimuli, promote the recruitment of T cells and B cells into tissues. This recruitment is accompanied by the activation of self-reactive T cells and results in tissue injury that may be followed by the onset of an autoimmune disease. doi:10.1128/9781555818890.ch4.f4.7

genetic control of autoimmune disease emerged when it was noted that if an autoimmune disease develops in one of two twins, the probability that the same disease develops in the other twin is greater than the probability of occurrence of this disease in an unrelated member of the general population. Moreover, this increased incidence is greater among monozygotic (identical) twins than among dizygotic (nonidentical, or fraternal) twins.

Genome-wide association studies coupled with breeding studies with animals have identified some of the genes that may regulate susceptibility to different autoimmune diseases. As initially mentioned in chapter 2, many autoimmune diseases in humans and in inbred mice are linked to particular MHC alleles (Table 4.2). The association between HLA alleles

Table 4.2 MHC-linked control of autoimmune diseases

Supportive evidence	Examples		
	Disease	**MHC allele**	**Relative risk**
Relative risk (ratio) of developing an autoimmune disease in individuals who inherit specific HLA alleles compared to that in individuals who do not express these alleles	Ankylosing spondylitis	HLA-B27	90
	Rheumatoid arthritis	HLA-DR4	4
	T1D	HLA-DR3/DR4	25
	Pemphigus vulgaris	HLA-DR4	14
In animal models of autoimmune diseases, susceptibility to a disease is associated with specific MHC alleles	T1D (in the NOD mouse strain)	I-A^{g7}	

Genetic susceptibility to certain autoimmune diseases is controlled by specific MHC alleles. In humans, family and linkage studies show a greater risk of developing certain autoimmune diseases in individuals who inherit particular HLA alleles than in individuals who lack these alleles. Individuals who express these specific HLA alleles have a greater "relative risk" of susceptibility to a given disease(s). For example, in people who express the HLA-B27 allele, the risk of development of ankylosing spondylitis is about 90 to 100 times higher than in B27-negative people. In the other diseases listed, various degrees of association (i.e., relative risk) are linked to other HLA alleles. Animal breeding studies have shown that the incidence of some autoimmune diseases correlates strongly with the inheritance of particular MHC alleles. An example is the association of T1D in NOD mice that express the mouse MHC class II allele called I-A^{g7}.

and autoimmune diseases in humans was first appreciated about 40 to 50 years ago and was one of the first indications that these diseases are T-cell mediated. This is because we now know that the function of MHC molecules is to present peptide antigens to TCRs on T cells. The incidence of a particular autoimmune disease in individuals who inherit a particular HLA allele(s) generally exceeds that found in the general population. This signifies an HLA-linked disease association, and this increased incidence is called the "relative risk" of an HLA-linked disease association. Note that while an HLA allele may increase the genetic risk of developing a given autoimmune disease, the HLA allele itself is not the cause of the disease. Importantly, the disease never develops in the vast majority of people who inherit an HLA allele that is frequently associated with such a disease. Particular MHC alleles may contribute to the development of autoimmunity because they present self-antigens inefficiently, which can lead to a decrease in negative selection of T cells and an increase in autoreactive T cells in the periphery. Alternatively, peptide antigens presented by these MHC alleles may fail to stimulate a minimal threshold of Treg cell activity required for active immunosuppression. The net outcome would be the trigger of immune responses against self-antigens.

Several non-HLA-linked genes also regulate susceptibility to various autoimmune diseases, as shown in Table 4.3. Recent linkage and genome-wide association studies have significantly increased the number and diversity of genetic loci associated with various autoimmune diseases. In particular, two genes were recently linked to the control of autoimmune diseases in humans. One gene encodes the tyrosine phosphatase PTPN22 (protein tyrosine phosphatase N22), which may regulate T-cell activation and is associated with numerous autoimmune diseases, including T1D. The other gene is the cytoplasmic microbial sensor NOD-2 (nucleotide-binding oligomerization domain-containing protein 2), which may lower resistance to intestinal microbes and mediate the onset of inflammatory bowel

Table 4.3 Role of some non-MHC genes in genetic susceptibility to autoimmune disease

Gene(s)	Disease association(s)	Mechanism(s)
AIRE	Autoimmune polyendocrine syndrome	Deficient expression of tissue antigens and deletion of self-reactive T cells in the thymus
Complement proteins (C2, C4)	Lupus-like disease	Impaired elimination of immune complexes; reduced B-cell tolerance
Fas, FasL	*lpr* and *gld* mouse strains; human ALPS	Deficient elimination of self-reactive T cells and B cells
FcγRIIb	Lupus-like disease	Reduced feedback inhibition of B-cell activation
Foxp3	X-linked polyendocrinopathy and enteropathy (IPEX)	Deficiency of Treg cells
IL-2; IL-2Rα/β	Many autoimmune diseases	Deficiency of Treg cells
NOD-2	Crohn's disease (inflammatory bowel disease)	Decreased resistance or reduced responses to intestinal microbes
PTPN22	Many autoimmune diseases (e.g., T1D)	Impaired tyrosine phosphatase regulation of lymphocyte activation

Adapted with permission from Abbas and Lichtman (ed.), *Basic Immunology: Functions and Disorders of the Immune System*, 3rd ed. (Saunders Elsevier, Philadelphia, PA, 2011).

Examples of some non-MHC genes that may contribute to the development of autoimmune diseases are shown. The roles of some of these genes are inferred from the autoimmune diseases that develop either in humans who carry certain mutations or in mice engineered to express specific gene knockouts. However, autoimmune diseases caused by single gene mutations are rare, and most human autoimmune diseases are complex multigenic traits (see chapter 3). *lpr* and *gld* are the mouse gene mutations for lymphoproliferation and generalized lymphoproliferative disease, respectively. ALPS, autoimmune lymphoproliferative syndrome; IPEX, immunodysregulation–polyendocrinopathy–enteropathy X-linked syndrome.

disease. Other polymorphisms associated with multiple autoimmune diseases are found in genes encoding the IL-2Rα chain CD25, believed to influence the balance of effector T cells and Treg cells, and the receptor for the cytokine IL-23 (IL-23R), which promotes the development of proinflammatory Th17-producing T cells. Ongoing attempts to elucidate these genetic associations may prove more informative about not only the etiology and pathogenesis of such autoimmune diseases but also the prediction and treatment of these diseases.

Role of Infections in Autoimmunity

The progressive increase in the incidence of autoimmune T1D in developed countries during the last 30 years is a major public health concern. This increase in incidence has inspired epidemiologists to determine the factors that may explain this disturbing trend. The fact that similar epidemiological features have been observed for other autoimmune diseases (e.g., multiple sclerosis) and allergic diseases (e.g., asthma) suggests that a common causal factor(s) underlies these diseases. Epidemiological evidence supports a causal role for both a decrease and increase of infection in the higher incidence of T1D. The outcome of disease depends on the nature of an infectious agent. A reciprocal relationship between the decline in the incidence of major infectious diseases and an increase in T1D implies, but does not formally prove, a causal relationship. Similarly, the wide but disparate geographical distribution of T1D presents interesting but indirect evidence for a link between increased infection and reduced T1D. Socioeconomic status, climate, and diet may also represent important factors that impact this relationship.

The best evidence for a causal relationship between decreased infection and increased autoimmune disease may be found in animal models.

NOD
nonobese diabetic

Both the nonobese diabetic (NOD) mouse and the biobreeding rat, which spontaneously develop a form of T1D that shares many features with T1D in humans, display a much higher incidence of T1D when they are bred in specific-pathogen-free facilities than in conventional facilities. In addition, infection of NOD mice housed in conventional facilities augments the incidence of T1D. In contrast, infection by various pathogens prevents T1D in NOD mice and biobreeding rats. For example, administration of complete Freund's adjuvant (containing heat-killed gram-negative *Escherichia coli* bacteria) to young NOD mice protects against the onset of T1D in these mice. Similarly, treatment of young NOD mice with either *Mycobacterium bovis* BCG, *Mycobacterium avium*, lymphocytic choriomeningitis virus, lactate dehydrogenase-elevating virus, murine hepatitis virus, or parasites (e.g., schistosomes) protects NOD mice from T1D.

These and other sets of studies raise the possibility that infections may activate and expand self-reactive lymphocytes and thereby trigger the development of autoimmune diseases. Clinical manifestations of autoimmunity are often preceded by infection or inflammation that might trigger the onset of disease before specific symptoms are manifested. This association between infection/inflammation and autoimmune tissue injury has been formally demonstrated in animal models (see above). Infections may contribute to autoimmunity in several ways (Fig. 4.8). An infection of a tissue may induce a local innate immune response, which may result in the increased production of costimulators and cytokines by tissue antigen-presenting cells. These activated tissue antigen-presenting cells may then stimulate self-reactive T cells that encounter self-antigens in the tissue. Thus, infection may "break" T-cell anergy and promote the activation of self-reactive lymphocytes.

Some infectious microbes may produce peptide antigens that are similar to self-antigens. Recognition of these microbial peptides by self-reactive T cells may result in stimulation of an immune response against self-antigens. Thus, cross-reactions between T cells reactive with microbial antigens or self-antigens can occur via a process known as **molecular mimicry**. Although the contribution of molecular mimicry to autoimmunity has been extensively investigated, its actual significance in the development of most autoimmune diseases, including T1D, remains unknown. Epidemiological data suggest that cross-reactivity of islet autoantigen-reactive T cells with enteroviruses (coxsackie B4 virus) may be causal to T1D. However, this causality relationship remains uncertain, perhaps due to the long lag time between the triggering infection and clinical onset of T1D. On the other hand, some rare disorders exist in which antibodies produced against a microbial protein bind to self-proteins. One example is rheumatic fever, in which antibodies against streptococcal bacteria cross-react with a myocardial antigen and elicit heart disease.

Infections also may injure tissues and release antigens that normally are sequestered from the immune system. For example, some sequestered antigens (e.g., in immune-privileged sites such as the testis and eye) normally are not "seen" by the immune system and are ignored. Release of these antigens (e.g., by trauma, infection, or inflammation) may initiate an autoimmune reaction against the tissue.

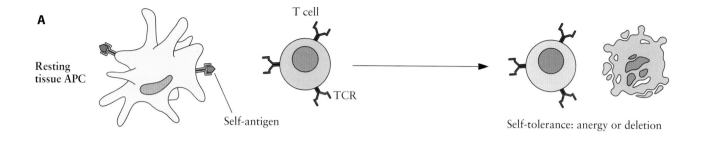

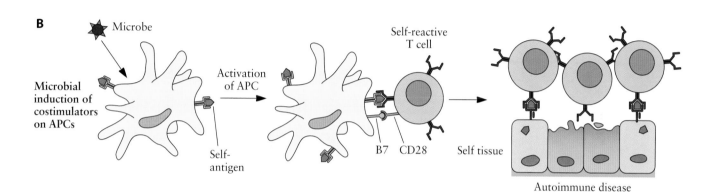

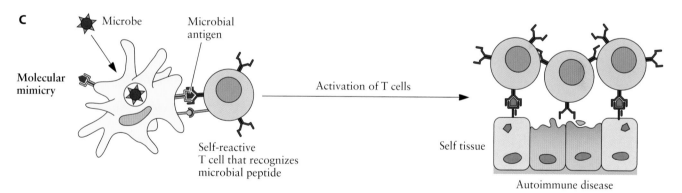

Figure 4.8 Association of infection with autoimmune disease. (**A**) Exposure of mature T cells to self-antigens presented by resting tissue antigen-presenting cells typically results in peripheral tolerance by either anergy or deletion. (**B**) Microbial antigens may activate the antigen-presenting cells to express costimulator molecules (e.g., B7). When these antigen-presenting cells present self-antigens, the specific T cells are activated in lieu of becoming tolerant. (**C**) Some microbial antigens may cross-react immunologically with self-antigens via a process known as molecular mimicry. In this manner, immune responses initiated by the microbes may become targeted toward one's own cells and tissues. While this figure illustrates concepts that apply to T cells, molecular mimicry may also apply to self-reactive B cells. doi:10.1128/9781555818890.ch4.f4.8

As discussed above, some infectious agents may suppress autoimmune responses. However, the mechanisms of suppression of such responses may not result from the induction of an immune response against their antigenic constituents. Rather, the stimulation of Toll-like receptors (TLRs) may be involved. TLRs belong to a family of proteins (nine TLRs exist) that play a key role in the innate immune system and the digestive

TLR
Toll-like receptor

system. They are single, membrane-spanning, noncatalytic receptors that mediate infection by recognition of different structurally conserved molecules derived from microbes. Once these microbes have breached physical barriers such as the skin or intestinal tract mucosa, they are recognized by TLRs, which activate immune cell responses. TLR stimulation activates the innate immune system and enhances the production of proinflammatory cytokines that can accelerate the development of T1D. The development of T1D in mice may be significantly reduced in NOD mice previously treated with various TLR4 agonists (e.g., bacterial lipopolysaccharide [LPS]) and TLR9 agonists (e.g., the CpG oligonucleotide). These observations in experimental T1D are similar to those made in experimental colitis, where the protective effect of probiotics is not observed in TLR9$^{-/-}$ knockout mice.

LPS
lipopolysaccharide

Other mechanisms may be operable here, as suggested by the appearance of immunosuppression observed upon infection with the hepatitis A virus, measles virus protein, or various parasites. Thus, some infections appear to paradoxically confer protection from autoimmune diseases.

Immune Surveillance against Tumors

During immune responses against tumors, the immune system responds to cells perceived to be foreign. The antigens that designate tumors as foreign may be expressed in malignant transformed cells. Thus, special mechanisms exist to induce immune responses against various tumor cell types. An important mechanism of tumor destruction is killing by cytotoxic T lymphocytes (CTLs). In this section, the major concepts to be discussed are as follows:

CTL
cytotoxic T lymphocyte

- Recognition of tumor antigens as foreign by the immune system
- Response to tumors by the immune system
- Manipulation of immune responses to tumors to enhance tumor rejection

Since the 1950s, it has been thought that a physiological function of the adaptive immune system is to prevent the growth of transformed cells and/or to destroy these cells before they become harmful tumors. This recognition of tumors by the immune system is called **immune surveillance**. The immune system patrols the body to recognize and destroy both invading pathogens and host cells that give rise to cancer. Current evidence suggests that cancer cells develop frequently throughout life, but these cells are killed by the immune system rapidly after they appear. There is also evidence that the immune system mounts an attack against established cancers, although it often fails.

Considerable evidence indicates that immune surveillance against tumors is important for the prevention of tumor growth (Table 4.4). However, since common malignant tumors develop in individuals with functioning immune systems, immune responses against tumors are considered to be much weaker than immune responses against infectious microbes. In fact, antimicrobial responses may be reduced in the face of rapidly growing tumors. A rational approach to enhance tumor immunity

Table 4.4 Immune system responses to tumor antigens

Findings	Consequence(s)
Histopathological and clinical data: lymphocytic infiltrates around some tumors and enlargement of draining lymph nodes correlate with better prognosis	Immune responses against tumor antigens inhibit tumor growth.
Experimental data: tumor transplants are rejected by animals previously sensitized to that tumor; immunity to tumor transplants may be transferred by lymphocytes from a tumor-bearing animal	Tumor rejection displays features of adaptive immunity (specificity and memory) and is mediated by lymphocytes (see chapter 2).
Clinical and experimental data: individuals who are immunodeficient have an increased incidence of certain types of tumors.	The immune system protects against the growth of tumors. This property is known as immune surveillance.

Adapted with permission from Abbas and Lichtman (ed.), *Basic Immunology: Functions and Disorders of the Immune System*, 3rd ed. (Saunders Elsevier, Philadelphia, PA, 2011).

Much experimental and clinical evidence supports the notion that the immune system reacts against tumors and that tumor immunity is a mechanism that is pivotal to the eradication of tumors. In particular, defense against tumors involves T-cell recognition of tumor antigens by activated antigen-presenting cells.

and suppress the rapid growth of tumors is to identify the structure of tumor antigens recognized by the immune system and then devise novel strategies to enhance antitumor immunity.

Characteristics of Tumor Antigens

Various types of proteins expressed by malignant tumors may be recognized as foreign antigens by T cells of the immune system (Table 4.5). If a person's immune system can react against a tumor in that individual, then the tumor must express antigens seen as foreign by that individual's immune system. In experimental tumors, e.g., those induced by chemical

Table 4.5 T-cell recognition of various types of tumor antigens

Type of tumor antigen presented by self-MHC	T-cell response	Examples of tumor antigens
Normal self-protein	None	NA
Mutated self-protein	Tumor-specific CD8$^+$ T cells	Mutant proteins in carcinogen- or radiation-induced animal tumors; mutated proteins in human melanoma tumors
Product of oncogene or mutated tumor suppressor gene	Tumor-specific CD8$^+$ CTLs	Oncogene products: mutated Ras, Bcr/Abl fusion proteins Tumor suppressor gene products: mutated p53 protein
Overexpressed or aberrantly expressed self-protein	Tumor-specific CD8$^+$ CTLs	Tyrosinase, gp100, and cancer/testis antigens in various tumors
Oncogenic virus	Virus-antigen specific CD8$^+$ CTLs	Human papillomavirus E6 and E7 proteins in cervical carcinoma; EBNA proteins in EBV-induced lymphomas

Tumor antigens that are recognized by tumor-specific CD8$^+$ T cells may comprise either mutated forms of normal self-proteins, products of oncogenes or tumor suppressor genes, overexpressed or aberrantly expressed self-proteins, or products of oncogenic viruses. Tumor antigens may also be recognized by CD4$^+$ T cells, since CD4$^+$ T cells provide help to tumor-specific CD8$^+$ T cells that have CTL activity against the tumors. EBNA, Epstein–Barr virus nuclear antigen; EBV, Epstein–Barr virus; gp100, glycoprotein of 100 kDa; NA, not applicable.

carcinogens or radiation, a self-protein may be mutated to produce an altered tumor antigen. Essentially, while any gene may be mutated randomly in different tumors, it is important to note that most of the mutated genes do not play any role in the generation of a tumor (**tumorigenesis**). Such mutants of diverse cellular self-proteins arise much more infrequently in spontaneous human tumors than in experimentally induced tumors.

Some tumor antigens are protein products of mutated or translocated cancer-causing genes (**oncogenes**) or tumor suppressor genes that mediate malignant transformation. Interestingly, in several human tumors, the antigens that promote immune responses appear to be normal proteins that are overexpressed. Alternatively, normal proteins whose expression is restricted to certain tissues or to particular stages of development may be dysregulated in tumors. While these normal self-antigens generally do not stimulate immune responses, their aberrant expression may trigger such responses. For example, self-proteins that are expressed only in embryonic tissues (e.g., carcinoembryonic antigen) may not induce tolerance in adults, so the same proteins expressed in tumors may be recognized as foreign by the immune system. In tumors derived by oncogenic virus transformation, tumor antigens are commonly products of viral genes (e.g., E6 and E7 proteins of the human papillomavirus that causes cervical cancer; see chapter 5).

Immune Mechanisms of Tumor Rejection

The main mechanism of immune-mediated tumor destruction is the killing of tumor cells by CTLs specific for tumor antigens. A majority of tumor antigens that induce immune responses in tumor-bearing individuals are endogenously synthesized cytosolic proteins presented on the surfaces of tumor cells as peptide–MHC class I complexes. These complexes are recognized by MHC class I-restricted CD8$^+$ CTLs that kill the antigen-producing target tumor cells. The role of CTLs in tumor rejection has been convincingly demonstrated in mouse models in which transplanted tumors were destroyed upon transfer of tumor-specific CD8$^+$ T cells to the tumor-bearing mice.

CTL responses against tumors frequently result from the recognition of tumor antigens on host antigen-presenting cells, which phagocytose (internalize) tumor cells or their antigens and present the antigens to T cells (Fig. 4.9). Any nucleated cell type may be transformed into a tumor cell. Such tumor cells display MHC class I-associated peptides, as all nucleated cells in the body express MHC class I molecules. However, the tumor cells frequently do not express costimulators or MHC class II molecules. Nonetheless, the activation of naïve CD8$^+$ T cells to proliferate and differentiate into active CTLs requires the recognition of antigen (MHC class I-bound peptide) and costimulation and/or help from MHC class II-restricted CD4$^+$ T cells (see chapter 2).

How do tumors of different cell types stimulate CTL responses? In experimental mouse models, following the phagocytosis (ingestion) of tumor cells by host dendritic cells, tumor antigens are enzymatically processed and presented by MHC class I and MHC class II molecules on the surfaces of host dendritic cells. In this manner, tumor antigens, like other protein

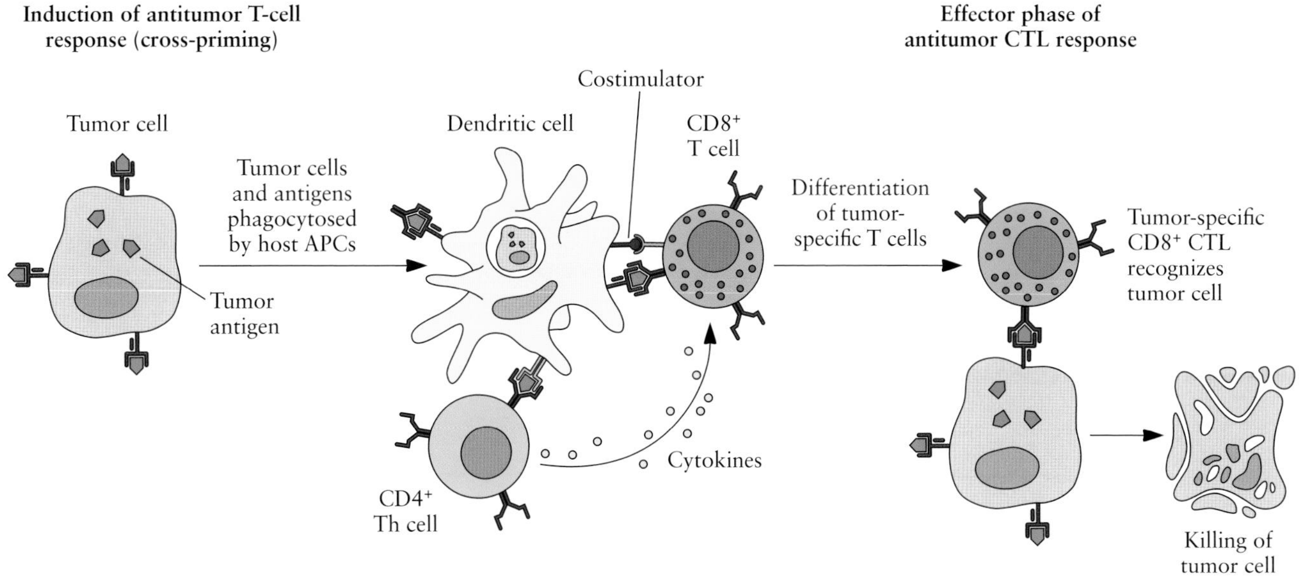

Figure 4.9 Induction of CD8$^+$ CTLs against tumors. CD8$^+$ T-cell responses to tumors may be induced by cross-presentation of tumor antigens that are taken up by dendritic cells, processed, and presented by self-MHC molecules to T cells. In certain cases, B7 costimulators expressed by these antigen-presenting cells provide the second signals for the differentiation of the activated CD8$^+$ T cells into tumor-specific CD8$^+$ CTLs. The antigen-presenting cells may also stimulate CD4$^+$ Th cells, which provide signals and cytokines for CTL development. Differentiated CTLs kill tumor cells without a requirement for costimulation of CD4$^+$ Th cells. Adapted with permission from Abbas and Lichtman (ed.), *Basic Immunology: Functions and Disorders of the Immune System*, 3rd ed. (Saunders Elsevier, Philadelphia, PA, 2011). doi:10.1128/9781555818890.ch4.f4.9

antigens, may activate both CD8$^+$ T cells and CD4$^+$ T cells. During T-cell activation and interaction with dendritic cells, these dendritic cells express the costimulator molecules (CD80 and CD86, etc.) that provide "second signals" for T-cell activation. This process is called cross-presentation or cross-priming, because one cell type (dendritic cells) presents antigens of another cell (tumor cell) and activates (or primes) T cells specific for the second cell type (tumor cell). Following the differentiation of naïve CD8$^+$ T cells into effector CTLs, the CTLs can kill target tumor cells that express the relevant antigens in the absence of costimulation or T-cell help. Thus, CTL differentiation may be induced by cross-presentation of tumor antigens by host antigen-presenting cells, but importantly, the CTLs react only against the tumor itself. The latter specificity of activated CTLs for only tumor cells is a primary objective of tumor immunity.

Antitumor CD4$^+$ T-cell responses and antitumor antibodies have also been detected in patients. However, it is still questionable whether these responses actually protect individuals against tumor growth. Experimental animal model studies have shown that activated macrophages and **natural killer (NK) cells** can destroy tumor cells in vitro, but again, the protective role of these effector mechanisms in tumor-bearing individuals remains to be determined. Further experimentation is required to solve these issues.

NK cell
natural killer cell

Antitumor immunity

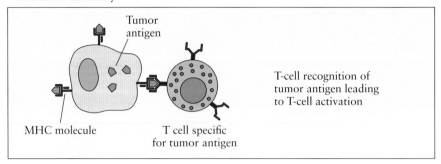

Immune evasion by tumors

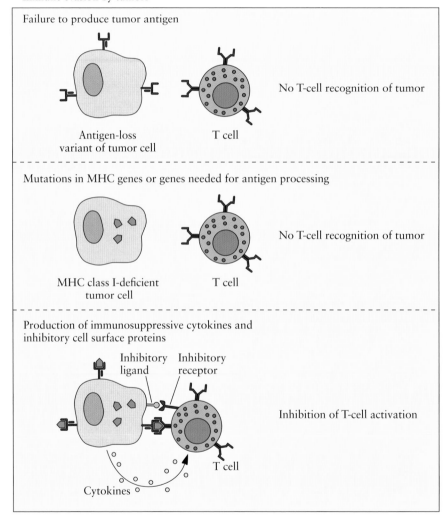

Figure 4.10 Mechanisms of immune evasion by tumors. Immune responses against tumors (antitumor immunity) develop when T cells recognize tumor antigens and are activated. Tumor cells may evade immune responses by reducing the expression of tumor antigens or MHC molecules, expressing mutated MHC molecules, or producing cytokines and inhibitory cell surface proteins (ligands and receptors) that suppress an immune response against the tumor cells. Adapted with permission from Abbas and Lichtman (ed.), *Basic Immunology: Functions and Disorders of the Immune System*, 3rd ed. (Saunders Elsevier, Philadelphia, PA, 2011). doi:10.1128/9781555818890.ch4.f4.10

Immune Evasion by Tumors

Two outcomes might impair immune surveillance and tumor destruction. First, immune responses are frequently ineffective or too weak to perform immune surveillance. Second, tumors have evolved effective mechanisms to evade immune attack. This presents a huge challenge to the immune system to effectively destroy malignant tumors, because these tumors can grow rapidly and all tumor cells must be destroyed to prevent the formation and spread of new tumors (metastasis). Often, our immune defense mechanisms are overwhelmed by the growth of the tumor. In addition, immune responses against tumors may be ineffective because many tumor antigens are only weakly immunogenic, possibly due to their structural similarities to self-antigens.

Growing tumors also develop several mechanisms for evading immune responses (Fig. 4.10). In some tumors, called "antigen loss variants," the expression of antigens targeted by the immune attack is arrested. If the lost antigens are not required to maintain the malignant properties of the tumor, the variant tumor cells may continue to grow and metastasize. Other tumors terminate expression of MHC class I molecules and/or molecules involved in antigen processing, blocking their ability to display antigens to CD8$^+$ T cells. NK cell receptors recognize their ligands expressed by tumor cells, but not by normal cells, and are activated when their target cells lack MHC class I molecules. Therefore, NK cells may provide a mechanism for killing MHC class I-negative tumors. Still other tumors may secrete immunosuppressive cytokines, such as TGF-β, which inhibits tumor responses. Finally, some tumors engage normal T-cell-inhibitory pathways, such as those mediated by CTLA-4 or PD-1, and thus suppress antitumor responses.

Inflammation and Immune Hypersensitivity Disorders

Although immune responses generally defend a host against infection and inflammation, immune responses may also inflict tissue injury and disease. Such pathogenic immune responses are termed **immune hypersensitivity reactions**, a term derived from the concept that an immune response to an antigen elicits a sensitivity to a secondary challenge with that antigen. Thus, hypersensitivity reactions represent excessive or aberrant immune responses, and they may occur in two main situations. First, responses to foreign antigens may be dysregulated (uncontrolled) and lead to inflammation and tissue injury. Second, immune responses against self-antigens, such as those that occur in autoimmune disease, may result from a failure of self-tolerance.

This section highlights the important features of hypersensitivity reactions; the inflammation, tissue injury, and diseases they cause; and their pathogenesis. The main points discussed are mechanisms of different types of hypersensitivity reactions; major inflammatory, pathological, and clinical features of diseases caused by these reactions; and treatment of such diseases.

Types of Inflammation and Associated Immune Hypersensitivity Reactions

Hypersensitivity reactions are classified based on the primary immune mechanism responsible for tissue injury and disease (Table 4.6). Immediate hypersensitivity, or type I hypersensitivity, is caused by the release of mediators from mast cells. The type I reaction involves the production of immunoglobulin E (IgE) antibody against environmental antigens and the binding of IgE to mast cells in various tissues. IgM and IgG antibodies may cause diseases in two ways: such antibodies directed against cell or tissue antigens can damage these cells or tissues or impair their functions. These diseases are antibody mediated and represent type II hypersensitivity. When complexes are formed between IgM and IgG antibodies and target soluble antigens, these immune complexes may deposit in blood vessels in various tissues, causing inflammation and tissue injury. Such diseases are called immune complex diseases and represent type III hypersensitivity. Finally, some diseases result from T-cell reactivity to self-antigens in tissues, such as $CD4^+$ delayed-type hypersensitivity and $CD8^+$ T-cell-mediated cytolysis. These T-cell-mediated diseases represent type IV hypersensitivity.

Immediate Hypersensitivity

Immediate hypersensitivity is a rapid, immunoglobulin E (IgE) antibody- and mast cell-mediated vascular and smooth muscle reaction. This reaction is often followed by inflammation that occurs in some individuals who encounter certain foreign antigens to which they have been exposed previously. Immediate hypersensitivity reactions are also called **allergy**, or **atopy**, and individuals with a strong propensity to develop such reactions are "atopic." These reactions may affect various tissues and occur with variable severities in different individuals. Common types of immediate hypersensitivity reactions include hay fever, food allergies, bronchial asthma, and anaphylaxis.

IgE
immunoglobulin E

Table 4.6 Types of hypersensitivity reactions

Type of hypersensitivity	Mediators of immune pathology	Mechanisms of tissue injury and disease
Type I (immediate hypersensitivity)	Th2 cells, IgE antibody, mast cells, eosinophils	Mast cell-derived mediators (vasoactive amines, lipid mediators, and cytokines)
Type II (antibody-mediated diseases)	Antibodies (IgM and IgG) against target cell surface antigens or extracellular matrix antigens	Antibodies bind to Fc receptors on inflammatory cells (neutrophils and macrophages) and then fix complement; complement and Fc receptors mediate recruitment and activation of inflammatory cells Phagocytosis of inflammatory cells Deficient cell function (e.g., hormone receptor signaling)
Type III (immune complex-mediated diseases)	Neutrophils, immune complexes of circulating antigens and IgM or IgG antibodies; deposition of these complexes in the basement membrane of a blood vessel wall	Complement and Fc receptor mediated recruitment and activation of inflammatory cells (e.g., mast cells and neutrophils)
Type IV (T-cell-mediated diseases)	DTH: $CD4^+$ T cells, macrophages, cytokines T cell-mediated cytolysis: $CD8^+$ CTLs, cytokines	DTH: macrophage activation and cytokine-mediated inflammation Direct target cell lysis, cytokine-mediated inflammation

The four main types of immune hypersensitivity reactions and their different immune effector molecules and mechanisms of tissue injury and disease are summarized. Ig, immunoglobulin; DTH, delayed-type hypersensitivity.

Allergies are the most frequent disorders of the immune system, estimated to affect about 20% of the population. The sequence of events in the development of immediate hypersensitivity reactions consists of the production of IgE antibodies in response to an antigen, binding of IgE to Fc receptors of mast cells, antigen-induced cross-linking of mast cell-bound IgE, and release of mast cell mediators (Fig. 4.11). Some mast

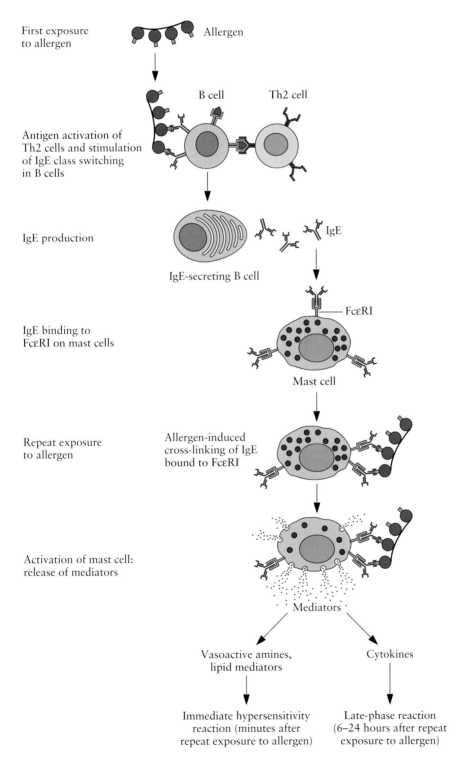

Figure 4.11 Development of an immediate hypersensitivity reaction. Exposure of an individual to an allergen stimulates Th2 cell activation that results in IgE antiallergen antibody production. These events catalyze immediate hypersensitivity reactions, during which the IgE antibodies bind to Fc receptors (FcεRI) on mast cells, a process known as sensitization. Subsequent exposure to the allergen activates mast cells to secrete several mediators, such as amines, lipid mediators, and cytokines. The actions of these various mediators and cytokines induce the immune pathology of immediate hypersensitivity reactions that occur either early (minutes) or later (6 to 24 h) after exposure to the allergen. These reactions may result in upper respiratory infections, hay fever, food allergies, bronchial asthma, and anaphylaxis. doi:10.1128/9781555818890.ch4.f4.11

cell mediators elicit a rapid increase in vascular permeability and smooth muscle contraction, leading to many symptoms of these reactions. This vascular and smooth muscle reaction generally occurs within minutes of exposure of a previously sensitized individual to antigen—i.e., immediate hypersensitivity. Other mast cell mediators are cytokines that recruit neutrophils and eosinophils to the site of inflammation over several hours. The latter inflammatory component is called a **late-phase reaction**, which accounts for the tissue injury consequent to repeated episodes of immediate hypersensitivity.

Production of IgE Antibody

In allergic but not healthy individuals, exposure to some antigens can activate Th2 cells and IgE antibody production (Fig. 4.11). Upon encounter of antigens such as proteins in pollen, certain foods, insect venoms, or animal dander or exposure to certain drugs such as penicillin, a vigorous Th2 cell response ensues. An atopic individual may be allergic to one or more of these antigens. Immediate hypersensitivity follows Th2 cell activation in response to protein antigens or chemicals that bind to proteins. Antigens that stimulate immediate hypersensitivity (allergic) reactions are usually called **allergens**. Two cytokines secreted by Th2 cells, IL-4 and IL-13, stimulate allergen-specific B cells to develop further into IgE-producing plasma cells. Therefore, atopic persons produce large amounts of IgE antibodies in response to antigens that do not elicit IgE responses in most (nonallergic) people. The preference toward Th2 development, IgE production, and immediate hypersensitivity is genetically controlled, and many different genes (polygenic control) contribute to these outcomes.

Activation of Mast Cells and Secretion of Mediators

IgE antibody produced in response to an allergen binds to high-affinity Fc receptors specific for the ε H chain expressed on mast cells (Fig. 4.12). In an atopic individual, mast cells are coated with IgE antibody specific for the antigen(s) to which the individual is allergic. This coating of mast cells

Figure 4.12 Mast cell activation. (**A**) The binding of antiallergen IgE antibodies to FcεRI receptors on mast cells sensitizes the mast cells to become activated upon repeat exposure to the allergen. (**B**) The binding of the allergen to IgE cross-links the FcεRI receptors and activates the mast cells to become degranulated. doi:10.1128/9781555818890.ch4.f4.12

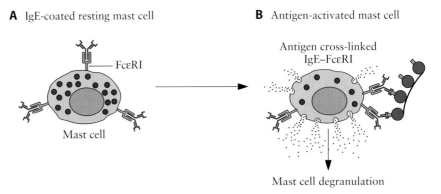

A IgE-coated resting mast cell

FcεRI

Mast cell

B Antigen-activated mast cell

Antigen cross-linked IgE–FcεRI

Mast cell degranulation

with IgE is known as "sensitization," because coating with IgE specific for an antigen renders mast cells sensitive to subsequent stimulation by that antigen. In healthy individuals, mast cells may carry IgE molecules of several different specificities, because many antigens may elicit weak IgE responses that do not provoke immediate hypersensitivity reactions. Mast cells are present in all connective tissues, and selection of those mast cells activated by cross-linking of allergen-specific IgE often depends on the route of entry of the allergen. Inhaled allergens activate mast cells in the submucosal tissues of the bronchus in the lung, whereas orally ingested allergens activate mast cells in the wall of the intestine. The high-affinity Fcε receptor, FcεRI, consists of three chains, one of which binds the Fc portion of the ε H chain very strongly, with a K_d (dissociation constant) of $\sim 10^{-11}$ M. Note that the IgE concentration in plasma is $\sim 10^{-9}$ M, so even in healthy individuals, mast cells are always coated with IgE bound to FcεRI. The other two chains of FcεRI are signaling proteins. The same FcεRI also is present on basophils, the circulating counterpart of mast cells, but the role of basophils in immediate hypersensitivity is not as well understood as that of mast cells.

Immediate hypersensitivity reactions occur after initial exposure to an allergen elicits specific IgE production and repeat exposure activates sensitized mast cells. Thus, after IgE-sensitized mast cells are exposed to an allergen, the mast cells are activated to secrete their mediators (Fig. 4.11 and 4.12). Mast cell activation results from binding of an allergen to two or more IgE antibodies on the mast cell. IgE and bound FcεRI molecules are cross-linked, and this triggers many signals downstream of FcεRI. The signals lead to three types of responses in the mast cell: rapid release of granule contents (degranulation), synthesis and secretion of lipid mediators, and synthesis and secretion of cytokines.

The key mediators produced by mast cells are vasoactive amines and proteases released from granules, products of arachidonic acid metabolism, and cytokines (Fig. 4.13). The major amine, histamine, dilates small blood vessels, increases vascular permeability, and stimulates transient smooth muscle contraction. Proteases may cause damage to local tissues. Arachidonic acid metabolites include prostaglandins that lead to dilation of blood vessels and leukotrienes that stimulate prolonged contraction of smooth muscle. Cytokines induce local inflammation (the late-phase reaction). As mast cell mediators elicit acute blood vessel and smooth muscle reactions and inflammation, they comprise the primary biomarkers of immediate hypersensitivity.

The late-phase reaction is mediated by cytokines that are secreted by mast cells and recruit leukocytes, including eosinophils, neutrophils, and Th2 cells (Fig. 4.11). Mast cell-derived tumor necrosis factor (TNF) and IL-4 promote neutrophil- and eosinophil-dependent inflammation, respectively (Fig. 4.13). Chemokines released from mast cells and epithelial cells in the tissues recruit leukocytes to these sites of inflammation. The recruited eosinophils and neutrophils secrete proteases that cause tissue damage, which may be further enhanced by the elevated production of cytokines by the recruited Th2 cells. Eosinophils are activated by IL-5 produced by Th2 cells and mast cells.

TNF
tumor necrosis factor

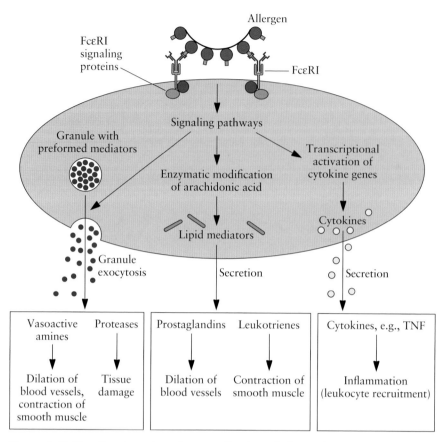

Figure 4.13 Signaling in activated mast cells. Cross-linking of IgE on mast cells by an allergen activates (phosphorylates) proteins of the FcεRI signaling complex. As a result of this activation, many downstream signaling pathways are stimulated. These pathways further promote the release of various components from mast cell granules (vasoactive amines and proteases) as well as the synthesis of metabolites of arachidonic acid, such as lipid mediators that stimulate the secretion of prostaglandins and leukotrienes. The net result of these pathways is the dilation of blood vessels (vasoactive) and contraction of smooth muscles. An additional pathway elevates the synthesis of several cytokines, such as TNF, which mediates the recruitment of neutrophils and macrophages to sites of inflammation in target tissues and cells. doi:10.1128/9781555818890.ch4.f4.13

Therapy for Immediate Hypersensitivity

The pathology of immediate hypersensitivity may vary widely in tissues according to the levels of mediators released by mast cells (Table 4.7). Mild reactions commonly seen in hay fever, e.g., allergic rhinitis and sinusitis, are reactions to inhaled allergens, such as ragweed pollen protein, dust, animal dander, and insect venom. Sinusitis results from an inflammation of the sinuses that occurs with an infection from a virus, bacterium, or fungus. Histamine is synthesized by mast cells in the nasal mucosa, and IL-13 is a product of Th2 cells. These two mediators stimulate the elevated production of mucus, which may lead to prolonged inflammation. In food allergies, ingested allergens trigger mast cell degranulation, and the released histamine causes increased peristalsis. Bronchial asthma is a form of respiratory allergy in which inhaled allergens stimulate bronchial

Table 4.7 Clinical syndromes of immediate hypersensitivity

Clinical syndrome(s)	Clinical symptoms and pathology
Allergic rhinitis, sinusitis (hay fever)	Increased secretion of mucus; inflammation of upper airways and sinuses
Food allergy	Increased involuntary constriction and relaxation (peristalsis) of muscles in the intestine due to the contraction of these muscles
Bronchial asthma	Bronchial hyperresponsiveness due to the contraction of smooth muscles and restricted breathing; inflammation and tissue injury elicited by a late-phase reaction
Anaphylaxis (may result from an administered drug, bee sting, or food ingestion)	Low blood pressure stimulated by the dilation of blood vessels; obstruction of airways due to swelling of the larynx caused by accumulation of body fluids at or near the site of inflammation

The clinical symptoms of some common immediate hypersensitivity responses are summarized. Immediate hypersensitivity may also be manifested by a skin allergy that leads to the development of skin inflammation and lesions, such as urticaria (hives or rash) and eczema (scaly and itchy rashes).

mast cells to release mediators, including leukotrienes, which result in recurrent bronchial constriction and airway obstruction. In chronic asthma, there are large numbers of eosinophils in the bronchial mucosa and excessive secretion of mucus in the airways, and the bronchial smooth muscle becomes hyperreactive to various stimuli. Some cases of asthma are not associated with IgE production, although all are caused by mast cell activation. In some affected persons, asthma may be triggered by cold or exercise, but the associated mechanisms of mast cell activation are unknown.

The most severe form of immediate hypersensitivity is **anaphylaxis**, a systemic reaction characterized by **edema** (abnormal accumulation of fluid in locations beneath the skin or in one or more cavities of the body; presents clinically as swelling) in many tissues, including the larynx, accompanied by a fall in blood pressure. This reaction is caused by widespread mast cell degranulation in response to a systemic antigen, and it is life threatening because of the sudden fall in blood pressure and airway obstruction.

Therapies for immediate hypersensitivity reactions target the inhibition of mast cell degranulation, block the effects of mast cell mediators, and reduce inflammation (Table 4.8). The most commonly used drugs are antihistamines for hay fever, drugs that relax bronchial smooth muscles for asthma, and epinephrine for anaphylaxis. Corticosteroids are used to inhibit the inflammation associated with asthma. Repeated treatment with small doses of allergens is referred to as **desensitization**. Many patients

Table 4.8 Treatment of immediate hypersensitivity

Disorder	Therapy	Mechanism(s)
Anaphylaxis	Epinephrine	Elicits contraction of smooth muscles surrounding blood vessels; stimulates cardiac output (to counter shock); blocks mast cell degranulation
Bronchial asthma	Corticosteroids	Decrease inflammation
	Phosphodiesterase inhibitors	Relax bronchial smooth muscles
Different allergic diseases	Desensitization (repeated treatment with low doses of allergens)	Unknown; may block IgE production and increase synthesis of different subclasses of antibodies (e.g., IgG and IgM)
	Anti-IgE antibody (clinical trials)	Neutralizes and eliminates IgE
	Antihistamines	Inhibit the action of histamine on blood vessels and smooth muscles
	Cromolyn	Inhibits mast cell degranulation

Cardiac output is the rate at which blood is pumped out by the heart into the circulation. Inhalation of the drug cromolyn prevents the release of mediators from inflammatory mast cells in the lungs that cause asthma symptoms.

can derive benefit from this treatment, which may work by reducing Th2 responses and/or by inducing tolerance (anergy) in allergen-specific T cells. Despite this potential benefit, additional studies are required to determine why common environmental antigens elicit the activation of Th2 cells and mast cells that can cause considerable tissue damage.

Antibody- and Antigen–Antibody Complex-Induced Disease

Chronic immunological diseases in humans frequently result from antibody-mediated hypersensitivity reactions mediated by IgM and IgG but not IgE antibodies (Fig. 4.14). IgM and IgG antibodies against cells or extracellular matrix components deposit in all tissues that express the relevant target antigen(s). Tissue resident antibody-bound activated neutrophils and macrophages may elicit tissue injury. IgG1 and IgG3 antibodies bind to neutrophil and macrophage Fc receptors and activate these leukocytes to elicit inflammation (see chapter 2). Both IgG and IgM antibodies

Figure 4.14 Types of antibody-mediated diseases. (**A**) Tissue injury results from action of antitissue antibodies. The binding of IgM and IgG antibodies to their target antigens on cells and extracellular matrix may cause type II hypersensitivity reactions that lead to tissue injury and disease. (**B**) Tissue injury results from the deposition of antibody–antigen immune complexes. IgM and IgG antibodies to target antigens on cells and extracellular matrix may form immune complexes in the blood circulation that deposit primarily in blood vessels. These deposits cause local type III hypersensitivity reactions at sites of inflammation in the vessels, an outcome known as vasculitis. doi:10.1128/9781555818890.ch4.f4.14

A

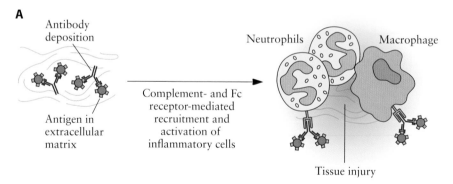

B

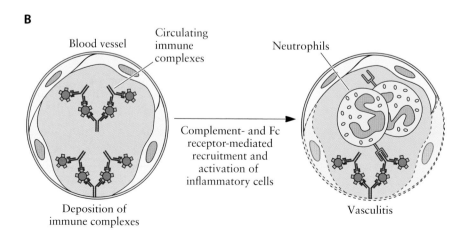

activate the complement system and yield complement by-products that recruit leukocytes and induce inflammation. When leukocytes are activated at sites of antibody deposition, these cells release substances such as reactive oxygen species and lysosomal enzymes that damage the adjacent tissues. If antibodies bind to cells, such as erythrocytes and platelets, the cells may be ingested and destroyed by host macrophages.

Antibodies bound to cells may also cause disease by forming immune complexes that deposit in blood vessels at sites of inflammation (branches of vessels) or high pressure (kidney glomeruli and synovium). Thus, immune complex diseases can affect the whole body and function in widespread inflammation, such as vasculitis, arthritis, and nephritis.

The antibodies that cause these types of inflammation and disease frequently are autoantibodies against self-antigens. Autoantibodies may bind to self-antigens in tissues or form immune complexes with circulating self-antigens. On the other hand, antibodies may also be produced against foreign (e.g., bacterial) antigens, and these antibodies usually appear during the late stages of infections. Some individuals produce antibodies against streptococcal bacterial antigens that cross-react with an antigen in heart muscle. Accumulation of these antibodies in the heart may trigger an inflammatory disease called rheumatic fever. Other individuals make antistreptococcal antibodies that accumulate in kidney glomeruli and give rise to poststreptococcal glomerulonephritis. Some immune complex diseases arise from complexes formed between antimicrobial antibodies and microbial antigens, e.g., in patients with chronic viral (Epstein–Barr virus) or parasitic (malaria) infections.

Some antibodies may cause disease without directly inducing tissue injury. Antibodies against hormone receptors may inhibit receptor function. An example of this occurs in myasthenia gravis, in which antibodies against the acetylcholine receptor inhibit neuromuscular transmission and lead to paralysis. Other antibodies may directly activate receptors. In Graves disease, a form of hyperthyroidism, antibodies against the thyroid-stimulating hormone receptor stimulate thyroid cells even in the absence of thyroid-stimulating hormone.

Therapy for Antibody-Induced Hypersensitivity-Mediated Diseases

Antitissue antibodies (Table 4.9) and immune complexes (Table 4.10) may be causal to several hypersensitivity-mediated diseases in humans. Two animal models have been particularly informative about the pathogenesis of immune complex diseases. The first model is **serum sickness**, which in mice is induced by systemic administration of a protein antigen that triggers an antibody response followed by the formation of circulating immune complexes. In humans, serum sickness may occur in a person who is injected with serum from another individual or animal, e.g., after treatment of snakebite or exposure to rabies virus. The second model is induced by subcutaneous administration of a protein antigen to a previously immunized animal. This leads to the formation of immune complexes at the site of antigen injection and a local inflammation of blood vessels (vasculitis).

Table 4.9 Human diseases caused by antibodies

Disease	Target antigen(s)	Mechanism(s)	Clinical outcome(s)
Acute rheumatic fever	Streptococcal cell wall antigen; antibody cross-reacts with heart muscle antigen	Inflammation; macrophage activation	Myocarditis, arthritis
Autoimmune hemolytic anemia	Erythrocyte membrane proteins (e.g., Rh blood group antigens)	Opsonization and phagocytosis of erythrocytes	Hemolysis, anemia
Autoimmune thrombocytopenic purpura	Platelet membrane proteins (gpIIb/IIa integrin)	Opsonization and phagocytosis of platelets	Bleeding
Graves disease (hyperthyroidism)	TSH receptor	Antibody-mediated stimulation of TSH receptors	Hyperthyroidism
Myasthenia gravis	Acetylcholine receptor	Antibody blocks binding of acetylcholine to its receptor	Muscle weakness, paralysis
Pemphigus vulgaris	Proteins in intercellular junctions of epidermal cells (e.g., cadherin)	Antibody-mediated activation of proteases, disruption of intercellular adhesins	Skin vesicles
Pernicious anemia	Intrinsic factor of gastric parietal calls	Neutralization of intrinsic factor; decreased absorption of vitamin B_{12}	Abnormal erythropoiesis, anemia

Adapted with permission from Abbas and Lichtman (ed.), *Basic Immunology: Functions and Disorders of the Immune System*, 3rd ed. (Saunders Elsevier, Philadelphia, PA, 2011).

Some human diseases in which antibodies are causal to disease onset are presented. In most of these diseases, the role of antibodies is implicated by the detection of antibodies in the blood or tissue lesions. Alternatively, the role of antibodies is based on data obtained by the ability of antibodies to transfer disease from one animal to another in experimental animal model studies. TSH, thyroid-stimulating hormone.

Therapy for these diseases is directed at blocking inflammation and injury with drugs such as corticosteroids. In severe cases, plasmapheresis (plasma removal and exchange) is used to reduce levels of circulating antibodies or immune complexes. Treatment of patients with an antibody specific for CD20, a protein antigen expressed on the surface of mature B cells, depletes B cells and may be applied clinically to treat antibody- and immune complex-mediated diseases. Current efforts in clinical trials are being devoted to (i) the inhibition of production of autoantibodies by treatment with anti-CD40 antibodies that block CD40–CD154 interaction and inhibit Th-cell-dependent B-cell activation (blockade of the CD40–CD154 signaling pathway is an effective strategy to induce immunosuppression and tolerance) and (ii) the induction of tolerance to autoantigens in diseases in which their structure is known.

Table 4.10 Human diseases caused by immune complexes

Disease	Antibody specificity	Clinical outcome(s)
Arthus reaction (experimental)	Different protein antigens	Cutaneous vasculitis
Polyarteritis nodosa	Hepatitis B virus surface antigen	Vasculitis
Poststreptococcal glomerulonephritis	Streptococcal cell wall antigen(s)	Nephritis
Serum sickness (clinical and experimental)	Different protein antigens	Systemic vasculitis, nephritis, arthritis
Systemic lupus erythematosus	DNA, nucleoproteins, etc.	Nephritis, arthritis, vasculitis

Some human diseases and two diseases detectable in experimental animal models in which antigen–antibody immune complexes are causal to disease onset are presented. Immune complexes are detected in the blood circulation or locally at sites of injury in various tissues. Complement- and Fc receptor-mediated inflammation is causal to injury encountered in all of these diseases.

T-Cell-Mediated Diseases

The pathogenic roles of T cells in human tissue injury and immunological diseases are now better understood. This is due mainly to the fact that methods to identify and isolate these T cells from lesions have improved, and animal models of human diseases have been developed in which a pathogenic role of T cells may be established experimentally.

T-cell-mediated hypersensitivity reactions are caused by autoimmune reactions and by responses to environmental antigens. The T-cell-mediated autoimmune reactions usually are directed against cellular antigens that possess a restricted tissue distribution. Contact sensitivity to chemicals found in plants (e.g., poison ivy) is T-cell mediated. Tissue injury also may accompany T-cell responses to microbes. In tuberculosis, a T-cell-mediated chronic immune response against *Mycobacterium tuberculosis* develops and leads to an infection that is difficult to overcome. The resultant inflammation causes injury to normal tissues at the site of infection. In hepatitis virus infection, the virus itself may not be toxic to cells, but the CTL response to infected hepatocytes may cause liver injury. The activation of many T-cell clones by toxins produced by some bacteria and viruses can lead to the production of large amounts of inflammatory cytokines, causing a syndrome similar to septic shock. These toxins are called **superantigens**, because they stimulate the proliferation of a high percentage (about 10 to 30%) of T cells. Superantigens yield these high responses because they bind to invariant regions of both MHC class II antigens on antigen-presenting cells and TCRs on many different clones of T cells, independent of antigen specificity.

In different T-cell-mediated diseases, tissue injury is caused either by a delayed-type hypersensitivity reaction mediated by CD4$^+$ T cells or by the killing of host cells by CD8$^+$ CTLs (Fig. 4.15). The mechanisms of tissue injury are identical to those used by T cells to eliminate cell-associated microbes. CD4$^+$ T cells may react against cell or tissue antigens and secrete cytokines that induce local inflammation and activate macrophages. Different diseases may be associated with the activation of pathogenic Th1 and Th17 cells. Th1 cells secrete γ-interferon (IFN-γ), which activates macrophages, and Th17 cells recruit leukocytes, including neutrophils. The actual tissue injury in these diseases is caused by the macrophages and neutrophils. CD8$^+$ T cells specific for antigens on host cells may directly kill these cells. In T-cell-mediated autoimmune diseases, CD4$^+$ T cells and CD8$^+$ T cells specific for self-antigens may both contribute to tissue injury.

IFN-γ
γ-interferon

Therapy for T-Cell-Mediated Hypersensitivity Disorders

Many organ-specific autoimmune diseases in humans are caused by T cells, based on the identification of these cells in lesions and similarities with animal models in which the diseases are T-cell mediated (Table 4.11). These disorders are typically chronic and progressive, in part because T-cell–macrophage interactions amplify the reaction. Tissue injury accompanied by the release and modification of self-proteins may result in

A Delayed-type hypersensitivity

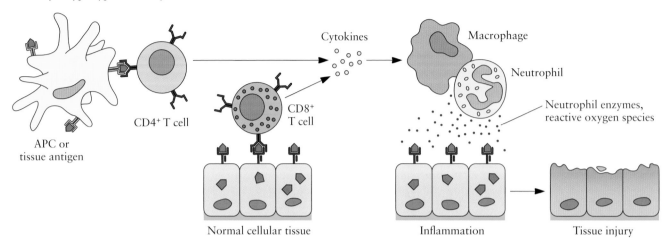

B T-cell-mediated cytolysis

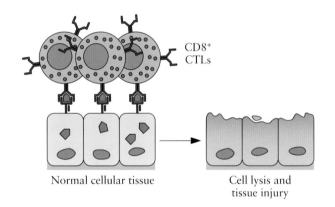

Figure 4.15 T-cell-mediated tissue injury. There are two mechanisms by which T cells may elicit tissue injury. (**A**) The first mechanism is mediated by delayed-type hypersensitivity reactions, which may be stimulated by CD4$^+$ and CD8$^+$ T cells. In this case, activated macrophages and inflammatory cells promote the onset of tissue injury. (**B**) The second mechanism is mediated by CD8$^+$ CTL-directed killing of target cells in tissues. doi:10.1128/9781555818890.ch4.f4.15

reactions against these proteins by a process known as **epitope spreading**. During such reactions, an initial immune response against one or a few epitopes (e.g., epitopes 1 to 3) on a self-protein antigen may expand to include responses against many more epitopes (e.g., epitopes 4 to 10) on this self-antigen. Thus, as an autoimmune disease develops, immune responses targeted against epitopes 1 to 3 may spread to responses against epitopes 4 to 10.

Therapies for T-cell-mediated hypersensitivity disorders are designed to reduce inflammation by the use of corticosteroids and antagonists against cytokines, such as TNF. The development of rheumatoid arthritis and inflammatory bowel disease has been blocked or reduced in individuals treated with TNF antagonists. Intervention of such T-cell-mediated hypersensitivity disorders may also be achieved with immunosuppressive drugs, e.g., cyclosporine. Many novel agents are now being developed to inhibit T-cell responses, including those that block costimulators (e.g., B7) and antagonists against receptors for cytokines (e.g., IL-2). Future clinical trials will use such agents in attempts to induce immune tolerance in pathogenic T cells.

Table 4.11 T-cell-mediated diseases

Disease	T-cell specificity	Susceptibility locus	Clinical outcome(s)
Chronic infections (e.g., tuberculosis)	Microbial proteins		Chronic inflammation (granulomatous)
Contact sensitivity (e.g., poison ivy reaction)	Modified skin proteins		DTH reaction in skin, rash
Inflammatory bowel disease	Unknown antigens in intestinal microbes	NOD-2	Inflammation of bowel wall; abdominal pain, diarrhea, hemorrhage
Multiple sclerosis	Myelin proteins	CD25	Demyelination of neurons in the central nervous system, sensory and motor neuron dysfunction
Rheumatoid arthritis	Unknown antigens in joints	PTPN22	Inflammation of synovium and destruction of cartilage and bone in joints
Superantigen-mediated disease (e.g., toxic shock syndrome)	Polyclonal–microbial superantigens activate a high percentage of T cells of many different antigen specificities		Fever, shock elicited by systemic inflammatory cytokine release
T1D	Pancreatic islet antigens	Insulin, PTPN22	Altered glucose metabolism, vascular (blood vessel) disease
Viral hepatitis (HBV, HCV)	Virally encoded proteins (EBNA)		CTL-mediated lysis of hepatocytes; liver dysfunction; fibrosis

Adapted with permission from Abbas and Lichtman (ed.), *Basic Immunology: Functions and Disorders of the Immune System*, 3rd ed. (Saunders Elsevier, Philadelphia, PA, 2011).

Several diseases in which the onset of injury is mediated by either T cells, antibodies, or immune complexes are listed. Most of these diseases are T-cell mediated. Multiple sclerosis, rheumatoid arthritis, T1D, and inflammatory bowel disease are autoimmune diseases. Inflammatory bowel disease is mediated by inflammatory responses to self-antigens in intestinal microbes. The other diseases shown are elicited by inflammatory responses against foreign antigens (microbial or environmental). The respective roles of pathogenic T cells and their specificities in the various diseases were detected in humans by the activity of isolated T cells against microbial and tissue antigens or by adoptive cell transfer studies in some experimental animal models of disease. The genetic susceptibility loci in the autoimmune diseases shown are derived from genome-wide associated studies of genetic linkage (see chapter 3). CD25, IL-2Rα; DTH, delayed-type hypersensitivity; EBNA, Epstein–Barr virus nuclear antigen; HBV, hepatitis B virus; HCV, hepatitis C virus.

Immunodeficiency Disorders and Defects in Development of the Immune System

Developmental and functional defects in the immune system may severely impair the health of an individual. Such defects may increase susceptibility to infections and reactivate latent infections (e.g., cytomegalovirus and Epstein–Barr virus infections and tuberculosis) that may be suppressed but not eradicated during normal immune responses. The incidence of certain cancers may also be increased. These consequences of defective immunity occur because the immune system normally defends individuals against infections and some cancers. Disorders that arise from defective immunity are called **immunodeficiency diseases**. Those diseases that result from genetic abnormalities in one or more components of the immune system are called **primary immunodeficiencies**. Other defects in the immune system may result from infections, nutritional abnormalities, or medical treatments that cause loss of or inadequate function of various components of the immune system. These defects are called **secondary immunodeficiencies**.

In this section, the following points are addressed: mechanisms by which immune defects give rise to primary immunodeficiency diseases; mechanisms by which secondary immunodeficiencies, such as **human immunodeficiency virus (HIV)** infection, cause **acquired immunodeficiency syndrome (AIDS)**; and therapeutic treatments of immunodeficiency diseases.

HIV
human immunodeficiency virus

AIDS
acquired immunodeficiency syndrome

Primary Immunodeficiencies

Primary immunodeficiencies are caused by genetic defects that lead to blocks in the maturation or function of different components of the immune system. About 1 in 500 individuals in North America and Europe suffers from primary immunodeficiencies (Table 4.12). Some of these deficiencies may result in increased susceptibility to infections that may occur soon after birth and may be fatal unless the immunological defects are corrected. Other disorders give rise to mild infections and may be detected and more easily resolved in adult life.

Defects in Lymphocyte Maturation

Many primary immunodeficiencies result from genetic changes that impair the maturation of B cells, T cells, or both B cells and T cells (Fig. 4.16 and Table 4.13). When defects in both B-cell and T-cell development occur, these disorders are termed **severe combined immunodeficiency (SCID)** disorders. SCID patients do not have any T cells or B cells, and thus, the function of their immune system is severely compromised.

SCID
severe combined immunodeficiency

About half of the genetic changes that cause SCID disorders affect only male children; i.e., these changes are X linked. About 50% of the cases of X-linked SCID are caused by mutations in the subunit of a receptor for cytokines that mediates cell signaling. This subunit is called the recombinase enzyme (γc), because it is a component commonly found in the receptors of several cytokines, including IL-2, IL-4, IL-7, IL-9, and IL-15. When the function of the γc chain is defective, immature lymphocytes are unable to proliferate in response to IL-7, the major growth factor for these cells. Thus, the survival and maturation of these lymphocyte precursors are significantly impaired. In humans, the defect blocks T-cell maturation primarily, resulting in a significant decrease in the numbers of mature T cells, deficiency in cell-mediated immunity, and a reduction in antibody production (a decrease in Th cells reduces the amount of antibody production in B cells).

Table 4.12 Immunodeficiency diseases

Immunodeficiency	Histology and pathology observation(s)	Consequent infections
B-cell deficiency	Decreased number (or absence) of follicles and germinal centers in lymphoid tissues (see chapter 2)	Pyrogenic bacterial infections
	Reduced serum Ig concentration	
T-cell deficiency	Reduced T-cell zones in lymphoid tissues	Viral and other intracellular microbial infections (e.g., mycobacterial and fungal)
	Fewer DTH reactions to common antigens	
	Lower T-cell proliferative responses to protein antigens	Virus-associated malignancies (e.g., EBV-associated lymphomas)
Innate immune deficiencies	Variable, depending on the innate immunity component that is defective	Variable; pyrogenic bacterial infections

Adapted with permission from Abbas and Lichtman (ed.), *Basic Immunology: Functions and Disorders of the Immune System*, 3rd ed. (Saunders Elsevier, Philadelphia, PA, 2011).

The relevant diagnostic features and consequent infections of various groups of immune deficiencies that are detectable in the clinic are summarized. In each group, different diseases and, moreover, different patients with the same disease may show much variation in clinical outcome. Reduced numbers of circulating B cells or T cells are frequently detected in some of these diseases. A pyrogenic reaction is a side effect caused by the infusion of a solution that is contaminated with a virus and commonly manifested by cold, chills, and fever. DTH, delayed-type hypersensitivity; EBV, Epstein–Barr virus; Ig, immunoglobulin.

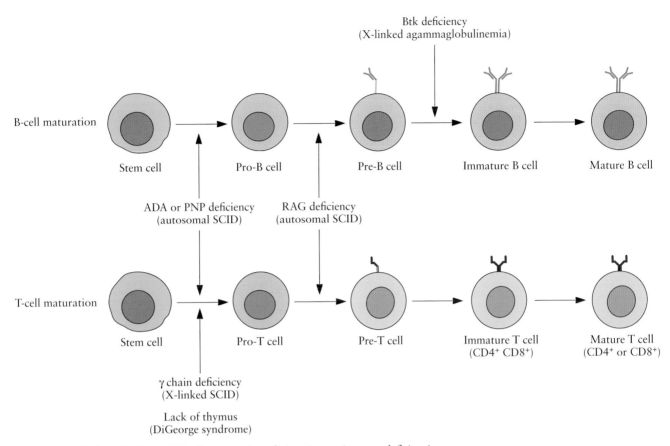

Figure 4.16 Defects in B- and T-cell maturation elicit primary immunodeficiencies. The general pathways of B- and T-cell maturation are illustrated. Note that at the pre-B and pre-T-cell stages, the IgM μ chain on a pre-B cell and the TCR β chain on a pre-T cell, respectively, are first expressed. PNP, purine nucleotide phosphorylase; RAG, recombination-activating gene; γ chain, common γ chain of many cytokine receptors; X-linked SCID, X-chromosome-linked SCID disorders that affect only male children; autosomal SCID, non-X-linked SCID disorders. DiGeorge syndrome is caused by an incomplete development of the thymus and a block in T-cell maturation. Pro-B cells are precursors of pre-B cells, pre-B cells are precursors of immature and mature B cells, pro-T cells are precursors of pre-T cells, and pre-T cells are precursors of immature and mature T cells. Adapted with permission from Abbas and Lichtman (ed.), *Basic Immunology: Functions and Disorders of the Immune System*, 3rd ed. (Saunders Elsevier, Philadelphia, PA, 2011). doi:10.1128/9781555818890.ch4.f4.16

In those cases (about 50%) in which SCID is not X linked, i.e., **autosomal SCID**, mutations occur in the gene that encodes the **adenosine deaminase (ADA)** enzyme, which catalyzes the breakdown of purines. A deficiency in ADA activity can lead to the accumulation of toxic purine metabolites in cells (e.g., lymphocytes) that are dividing and actively synthesizing DNA. As a result, these activated lymphocytes lose their function. ADA deficiency blocks T-cell maturation more effectively than B-cell maturation.

Autosomal SCID may also be generated by mutations in the purine nucleotide phosphorylase enzyme, which regulates cell signaling by the cytokine receptor γc chain, and in either of the *RAG1* and *RAG2* genes,

ADA
adenosine deaminase

Table 4.13 Characteristics of primary immunodeficiencies

Disease	Functional deficiency(ies)	Mechanism(s) of deficiency
SCID		
X-linked SCID	Decreased number of T cells; normal or increased number of B cells; reduced level of serum Ig	Cytokine receptor common γc chain gene mutations; defective T-cell maturation due to impaired IL-7 signaling
Autosomal recessive SCID due to ADA and PNP deficiency	Progressive decrease in number of T cells and B cells; reduced serum Ig in ADA deficiency, normal B cells and serum Ig in PNP deficiency	ADA or PNP deficiency elicits accumulation of toxic metabolites in lymphocytes.
Autosomal recessive SCID	Decreased number of T cells and B cells, reduced serum Ig concentration	Defective maturation of T and B cells; genetic basis unknown in most cases; may be mutations in *RAG* genes
B-cell immunodeficiency		
X-linked agammaglobulinemia	Reduced concentration of all serum Igs; reduced numbers of B cells	Block in maturation beyond pre-B cells due to mutation in *Btk* gene
T-cell immunodeficiency		
DiGeorge syndrome	Deceased number of T cells; normal number of B cells; normal or decreased levels of serum Ig	Impaired development of 3rd and 4th branchial pouches, leading to thymic hypoplasia

Shown are the most common primary immunodeficiencies in which the genetic blocks and their net effects are known. Ig, immunoglobulin; PNP, purine nucleotide phosphorylase; RAG, recombination-activating gene; branchial pouch, a pouch of embryonic endodermal tissue that develops into epithelial tissues and organs (e.g., the thymus and thyroid glands); hypoplasia, abnormal deficiency of cells.

which encode the VDJ recombinase enzyme required for Ig and TCR gene recombinations and maturation in lymphocytes (see chapter 2). The cause of about 50% of both X-linked and autosomal cases of SCID is not known.

The most common clinical disorder caused by a block in B-cell maturation is **X-linked agammaglobulinemia**. In this disorder, precursor B (pre-B) cells in the bone marrow do not mature beyond this stage, giving rise to the relative absence of mature B cells and serum immunoglobulins. Mutations in the B-cell tyrosine kinase (*BTK*) gene result in a loss of function in this enzyme and consequent disease. In pre-B cells, Btk is activated by the pre-B-cell receptor and delivers biochemical signals that promote B-cell maturation. Because the *BTK* gene is located on the X chromosome, women who carry a *BTK* mutant allele on one X chromosome are disease carriers. Male children who inherit the abnormal X chromosome are affected.

Defects in T-cell maturation are rare. The most frequent of these is **DiGeorge syndrome**, which results from incomplete development of the thymus and a block in T-cell maturation. Patients with this disease may improve with age, as the small amount of thymic tissue that does develop can support some T-cell maturation.

Treatment of primary immunodeficiencies that affect lymphocyte maturation varies with the disease. SCID is fatal in early life unless the patient's immune system is reconstituted by bone marrow transplantation. In this scenario, careful matching of donor and recipient is important to avoid rejection of the transplanted marrow cells. For B-cell deficiencies, patients may be administered pooled serum immunoglobulin from healthy donors to provide passive immunity. Serum immunoglobulin replacement therapy is quite beneficial in patients with X-linked agammaglobulinemia.

A more desirable treatment for primary immunodeficiencies is replacement gene therapy (see chapter 10). To date, successful gene therapy has been reported for X-linked SCID patients, who received their own bone marrow cells after transfection with a normal γc chain gene to achieve B-cell reconstitution. However, in some of these patients, T-cell leukemia developed because the transfected gene was inserted near an oncogene that was then activated. In all patients with immunodeficiencies, infections can occur frequently; therefore, treatment with antibiotics is generally required.

Secondary Immunodeficiencies

Secondary immunodeficiencies acquired during life can become quite serious, such as in the case of HIV infection (Table 4.14). The most frequent causes of secondary immunodeficiencies are cancers arising from bone marrow and other therapies. Cancer treatment with chemotherapeutic drugs and irradiation may damage proliferating cells, including bone marrow cell precursors and mature lymphocytes, and result in immunodeficiency. Treatments to prevent rejection of transplants and inflammatory diseases are designed to suppress immune responses. Thus, immunodeficiency is a frequent complication of such therapies.

AIDS

Since the detection of AIDS in the 1980s, it has become a devastating disease. AIDS is caused by infection with HIV. Globally, more than 42 million people (about 70% in Africa and 20% in Asia) are HIV infected, and more than 22 million deaths are attributable to this disease at a rate of about 3 million deaths per year. The infection continues to spread, and in some countries in Africa, more than 30% of the people are HIV infected.

Human Immunodeficiency Virus

HIV is a retrovirus that infects cells of the immune system, mainly CD4$^+$ T cells, and causes the progressive destruction of these cells. An infectious HIV particle consists of two RNA strands in a protein core, surrounded by a lipid envelope derived from infected host cells but containing viral

Table 4.14 The most common causes of several acquired secondary immunodeficiencies and how they may modulate normal immune responses

Cause	Modulation of immune response
HIV infection	Depletion of CD4$^+$ Th cells
Irradiation and chemotherapy treatments for cancer	Reduced bone marrow precursors for all leukocytes
Transformed bone marrow cells in cancer (e.g., leukemia)	Reduced leukocyte development
Reduced protein and calorie intake, malnutrition	Metabolic changes block lymphocyte maturation and function
Removal of spleen	Decreased phagocytosis of microbes

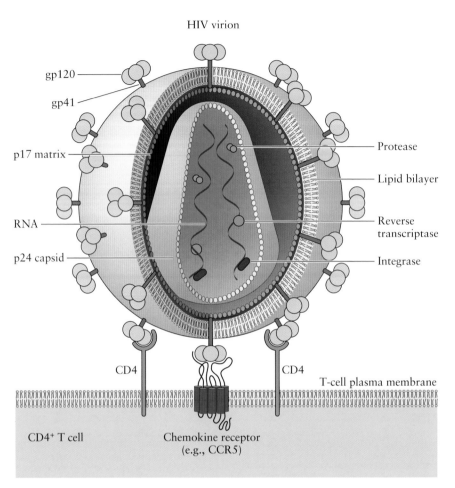

HIV virion

gp120

gp41

p17 matrix

RNA

p24 capsid

Protease

Lipid bilayer

Reverse
transcriptase

Integrase

CD4

CD4

T-cell plasma membrane

CD4⁺ T cell

Chemokine receptor
(e.g., CCR5)

Figure 4.17 HIV. An HIV-1 virion is positioned near a T-cell surface. HIV-1 consists of two identical RNA strands (viral genome) and associated enzymes. The enzymes include reverse transcriptase, integrase, and protease packaged in a cone-shaped core. The core is composed of a p24 capsid protein encompassed by a p17 protein matrix, and is surrounded by a phospholipid membrane envelope derived from the host cell (see chapter 5). HIV-1-encoded membrane proteins gp41 and gp120 are bound to the envelope. CD4 and chemokine receptors (e.g., CCR5) on the cell surface are receptors for HIV-1. Adapted with permission from Abbas and Lichtman (ed.), *Basic Immunology: Functions and Disorders of the Immune System*, 3rd ed. (Saunders Elsevier, Philadelphia, PA, 2011). For the structure of the HIV-1 genome, see chapter 5. doi:10.1128/9781555818890.ch4.f4.17

proteins (Fig. 4.17). The viral RNA encodes structural proteins, various enzymes, and proteins that regulate transcription of viral genes and the viral life cycle (see chapter 5).

The life cycle of HIV consists of the following sequential steps: infection of cells, production of viral DNA and its integration into the host genome, expression of viral genes, assembly of viral particles, and release of infectious virions (Fig. 4.18). HIV infects cells using its major envelope glycoprotein, gp120 (for 120-kilodalton [kDa] glycoprotein), to bind to CD4 and some chemokine receptors (CXCR4 on T cells and CCR5 on macrophages; see chapter 2) on human cells. Therefore, HIV infects only cells that express CD4 and these chemokine receptors.

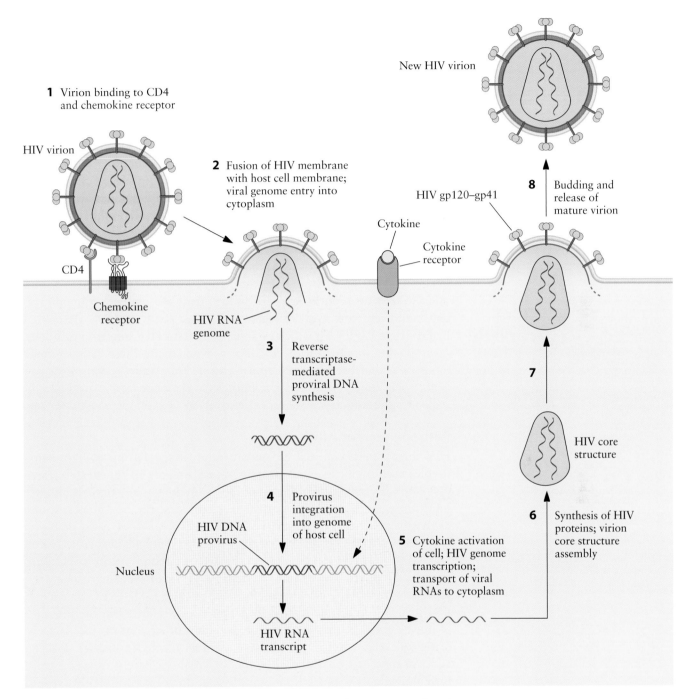

Figure 4.18 Life cycle of HIV-1. The various steps of HIV reproduction are numbered 1 to 8. These steps begin with the infection of a host cell by HIV and progress to the assembly and release of a new HIV particle (virion). Although numerous virions are synthesized and released by an infected host cell, note that the steps involved in the production and release of a single virion are shown here. The virions released by an infected cell can each infect a neighboring host cell and in this way elicit the rapid spread of an HIV-1 infection. doi:10.1128/9781555818890.ch4.f4.18

The major cell types that are infected by HIV are CD4$^+$ T cells, macrophages, and dendritic cells. After binding to cellular receptors, the viral membrane fuses with the host cell membrane, and the virus enters the cell's cytoplasm. Here the virus is uncoated by viral protease and its RNA is released. A DNA copy of the viral RNA is synthesized by the virus's reverse transcriptase enzyme (characteristic of all retroviruses), and the DNA integrates into the host cell's DNA by the action of the integrase enzyme. The integrated viral DNA is called a **provirus**. When infected T cells, macrophages, and dendritic cells are infected by a microbe such as HIV, they respond by transcribing specific genes, including several that lead to cytokine production. An undesirable caveat is that cytokines and cellular activation may also activate the provirus, which stimulates the production of viral RNAs and viral proteins. The viral core structure is assembled and migrates to the cell membrane, acquires a lipid envelope from the host, and is shed as an infectious viral particle, ready to infect another cell. The integrated HIV provirus may remain latent (dormant or inactive) within infected cells for months or years, hidden from the patient's immune system (and even from antiviral therapies). Most cases of AIDS are caused by HIV type 1 (HIV-1), which establishes a latent HIV infection in T cells, macrophages, or dendritic cells that may reactivate these cells to make more virus that becomes infectious. This increase in infectious virus can lead to the death of infected and uninfected lymphocytes, as well as subsequent immunodeficiencies and clinical AIDS (Fig. 4.19).

HIV-1
human immunodeficiency virus type 1

Pathogenesis of AIDS

HIV infection is acquired by sexual intercourse, sharing contaminated needles used by intravenous drug users, transplacental transfer, or transfusion of infected blood or blood products. Acute viremia may result after infection when the virus is present in blood, and the host may respond as in any mild viral infection. The virus infects CD4$^+$ T cells, dendritic cells, and macrophages at sites of entry through epithelia in various lymph nodes and in the blood circulation. In mucosal tissues at the sites of entry, many infected T cells may be destroyed. Because most lymphocytes, especially memory T cells, reside in these tissues, local T-cell destruction may give rise to a functional deficit that is not reflected in the presence of infected cells in the blood or the depletion of circulating T cells. Dendritic cells may capture the virus as it enters through mucosal epithelia and transport it to peripheral lymphoid organs, where it infects T cells. Rare individuals with *CCR5* mutations that do not permit HIV entry into macrophages can remain disease free for years after HIV infection, indicating the importance of macrophage infection in the progression toward AIDS. The integrated provirus may be activated in infected cells and lead to the production of viral particles and spread of the infection. During the course of HIV infection, the major source of infectious viral particles is activated CD4$^+$ T cells; dendritic cells and macrophages are reservoirs of infection.

The number of T cells lost during the progression to AIDS is much greater than the number of infected cells. The mechanism of this T-cell loss

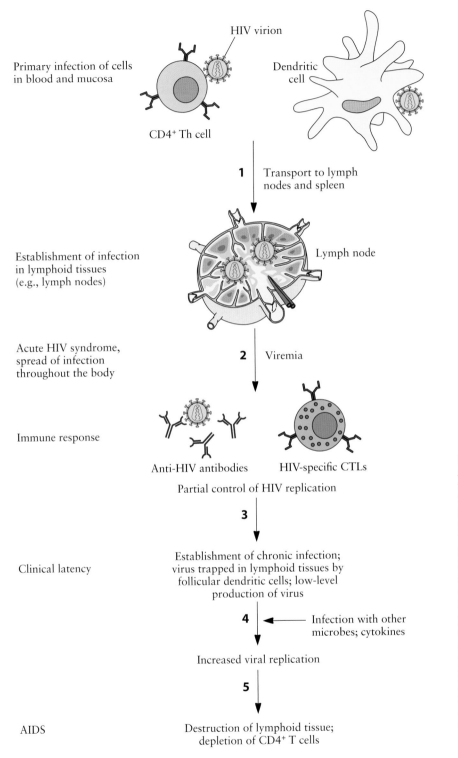

Figure 4.19 Pathology of HIV disease. A widespread HIV infection throughout the body develops (steps 1 to 5) from the progression of an infection at an initial site of infection to many sites distributed in peripheral lymphoid tissues. The host immune response mediated by CD4+ T-cell–dendritic cell interaction in lymphoid tissues initially contains and regulates the first acute infection by providing partial control of viral replication. Unfortunately, this response does not prevent the subsequent establishment and transmission of a chronic infection of cells in lymphoid tissues. Cytokines produced in response to HIV and other microbial infections stimulate HIV replication and progression to the stage of full-blown AIDS, which manifests by destruction of lymphoid tissues and significant depletion of CD4+ T cells.
doi:10.1128/9781555818890.ch4.f4.19

is not well defined. One possibility is that T cells are chronically activated, perhaps by infections that are common in these patients, and the chronic stimulation elicits apoptosis by the pathway called activation-induced cell death. Other infected cells, such as dendritic cells and macrophages, may also die, resulting in destruction of the architecture of lymphoid organs. In addition to T-cell depletion, immunodeficiency results from functional defects in T cells, dendritic cells, and macrophages. The significance of these defects is not known; however, the loss of T cells coupled with the blood CD4$^+$ T-cell count remains the most reliable indicator of disease progression.

Clinical Features of HIV Infection and AIDS

The clinical course of HIV infection is characterized by several phases that result in immunodeficiency (Fig. 4.20). Early after HIV infection, patients may experience a mild acute illness with fever and malaise, correlating with the initial viremia (spread of virus in the blood). This illness subsides in a few days, and the disease becomes latent, during which time there is a progressive loss of CD4$^+$ T cells in lymphoid tissues and destruction of the architecture of the lymphoid tissues. The blood CD4$^+$ T-cell count declines, and when it is less than 200 per mm^3 (the normal level being about 1,500 cells per mm^3), patients become susceptible to infections and suffer from AIDS.

Figure 4.20 Clinical course of HIV disease. The presence of HIV in the blood (plasma viremia) may be detected soon after infection (about 6 weeks) and may give rise to local symptoms typical of acute HIV infection. The virus spreads to lymphoid organs, but plasma viremia decreases to very low levels (detectable by reverse transcriptase PCR) and remains at these levels for many years. During this "clinical latency" period, CD4$^+$ T-cell counts decline gradually due to elevated viral replication as well as T-cell destruction (apoptosis) and loss of architecture of lymphoid tissues. As the count of CD4$^+$ T cells decreases further, increases in the risk of opportunistic infection and other clinical facets of AIDS are manifested and culminate in severe immunodeficiency and ultimately death. Adapted with permission from Abbas and Lichtman (ed.), *Basic Immunology: Functions and Disorders of the Immune System*, 3rd ed. (Saunders Elsevier, Philadelphia, PA, 2011). doi:10.1128/9781555818890.ch4.f4.20

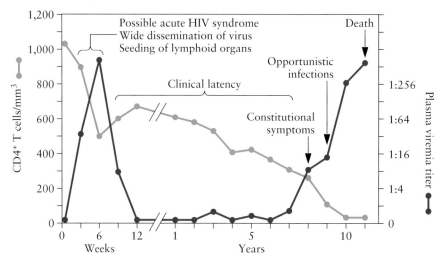

The pathology and clinical profile of full-blown AIDS result mainly from an increased susceptibility to infections and some cancers, due to immunodeficiency. Patients are frequently infected by intracellular microbes, including viruses, the *Pneumocystis jirovecii* fungus, and atypical mycobacteria. Normally, these microbes are eliminated by T-cell-mediated immunity. Many of these microbes are present in the environment, but they do not infect healthy persons with intact immune systems. Because these infections are seen in immunodeficient individuals, in whom the microbes have an opportunity to establish infection, these types of infections are regarded as "opportunistic." Many such opportunistic infections are caused by viruses, e.g., cytomegalovirus. Patients with AIDS display defective CTL responses to viruses, even though HIV does not infect $CD8^+$ T cells. CTL responses are defective likely because $CD4^+$ Th cells (the main targets of HIV) are required to help $CD8^+$ T cells to achieve maximal CTL responses against many viral antigens (see chapter 2). Patients who have AIDS are at increased risk for infections by extracellular bacteria, probably because of impaired Th-cell-dependent antibody responses to bacterial antigens. Patients also become susceptible to cancers that are caused by oncogenic viruses (Epstein–Barr virus, which causes B-cell lymphomas, and a herpesvirus that causes Kaposi's sarcoma [tumor of small blood vessels]). The clinical course of the disease has been dramatically changed by effective antiretroviral drug therapy. With appropriate treatment, patients exhibit much lower progression of the disease, fewer opportunistic infections, and greatly reduced incidence of cancers.

The immune response to HIV does not control viral spread and its pathological effects very effectively. Infected individuals produce antibodies and CTLs against viral antigens, and these responses limit the early acute HIV syndrome. But these immune responses usually do not prevent chronic progression of the disease. Antibodies against gp120 may be ineffective because the virus rapidly mutates the region of gp120 that is the target of most antibodies. CTLs often are ineffective in killing infected cells because the virus inhibits the expression of MHC class I molecules by the infected cells. Immune responses to HIV may actually promote spread of the infection. Antibody-coated viral particles may bind to Fc receptors on macrophages and follicular dendritic cells in lymphoid organs, thus increasing virus entry into these cells and creating additional reservoirs of infection. If CTLs are able to kill infected cells, the dead cells may be cleared by phagocytosis, resulting in spread of the virus to macrophages. By infecting and interfering with the function of immune cells, the virus is able to evade immune attack and prevent its own eradication.

summary

The control of immune pathogenesis by various immune mechanisms, including immunological tolerance and autoimmunity, immune responses against tumors, hypersensitivity, and immunodeficiency diseases, is presented in this chapter.

Immunological tolerance is the result of specific unresponsiveness to an antigen induced by previous exposure of lymphocytes to that antigen. All individuals are generally unresponsive to their own self-antigens. This tolerance may be induced by treatment with an antigen in a way that may prevent or reduce the frequency and/or severity of immunological diseases and the rejection of transplanted organs or tissues. Central tolerance may be induced by the death of immature lymphocytes that encounter antigens in primary lymphoid organs (thymus and bone marrow). Central tolerance of T cells is the result of high-affinity recognition of antigens in the thymus. Some of these self-reactive T cells die (negative selection), thereby eliminating potentially dangerous T cells that express high-affinity TCRs for self-antigens. This skews the repertoire of TCR specificities of T cells that exit the thymus into peripheral lymphoid tissues (periphery) towards determinants present on foreign (e.g., microbial) antigens. Peripheral tolerance results from the recognition of antigens by mature T and B cells in the periphery (e.g., lymph nodes and spleen). In T cells, peripheral tolerance is induced by several mechanisms, including the activity of $CD4^+$ Treg cells that suppress self-reactivity in peripheral lymphoid tissues. Self-reactive Treg cells suppress potentially pathogenic T cells. Deletion (death by apoptosis) may occur when T cells encounter self-antigens. Another mechanism known as functional inactivation or anergy arises when antigens are recognized in the absence of innate immunity and expression of costimulator molecules (second signals). Anergy is elicited by a block in TCR signaling that is accompanied by the engagement of inhibitory receptors, such as CTLA-4 and PD-1.

In B cells, central tolerance occurs when immature cells recognize self-antigens in the bone marrow. During maturation, some developing B cells change their receptors (receptor editing), and others die by apoptosis (negative selection, or deletion). Peripheral tolerance is induced when mature B cells recognize self-antigens in the absence of T-cell help, and this results in the anergy and death of the B cells.

Autoimmune diseases result from a failure of self-tolerance. Multiple factors contribute to these diseases, including the inheritance of susceptibility genes and environmental triggers such as infections. In humans, the strongest associations between genetic susceptibility and the development of autoimmune disease are those between HLA genes and various T-cell-mediated autoimmune diseases. Particular HLA class II alleles may elicit autoimmune disease because they present self-antigens inefficiently. This deficient presentation decreases the negative selection of T cells and increases the number and activity of autoreactive T cells in the periphery. Alternatively, antigens presented by these HLA alleles may not stimulate sufficient Treg cell activity to induce immunosuppression, enabling immune responses against self-antigens to occur. Infections may predispose to autoimmune disease by causing inflammation and stimulating the expression of costimulator molecules (e.g., CD28, CD80, and CD86), or because of T-cell recognition of cross-reactive microbial and self-antigens.

Eradication of tumors and prevention of new tumor growth are two essential functions of the immune system. Tumor antigens may be products of oncogenes or tumor suppressor genes, mutated cellular proteins, or overexpressed or aberrantly expressed molecules or products of oncogenic viruses. Tumor rejection is mediated primarily by CTLs that recognize peptides derived from tumor antigens. Induced CTL responses against tumor antigens require the internalization of tumor cells or their antigens by dendritic cells and the processing and presentation of these antigens to T cells.

Tumors may evade immune responses by reducing expression of their antigens, downregulating the expression of MHC (class I and class II) molecules or molecules involved in antigen processing, and secreting cytokines that suppress immune responses. The object of immunotherapy for cancer is to enhance antitumor immunity by transferring immune effector T cells to patients or by actively boosting the patient's own effector T cells. This boosting may include vaccination with tumor antigens or with tumor cells engineered to express certain costimulator molecules and cytokines that will result in the lysis and elimination of tumors.

Hypersensitivity reactions emerge from immune responses that elicit tissue injury and give rise to hypersensitivity diseases or immune-mediated inflammatory diseases. Such reactions may arise from dysregulated or abnormal responses to foreign antigens or autoimmune responses against self-antigens.

Hypersensitivity reactions are classified according to the mechanism of tissue injury. Immediate hypersensitivity (type I, or allergy) is caused by the sequential production of IgE antibodies against environmental antigens or drugs (allergens), binding of IgE antibodies to mast cells (sensitization), and degranulation of the mast cells on subsequent encounter with the allergen. The clinical and pathological outcomes of immediate hypersensitivity are due to the actions of mediators secreted by mast cells. These include amines, which dilate blood vessels and contract smooth muscles; arachidonic acid metabolites, which also contract muscles; and cytokines, which induce inflammation, the hallmark of the late-phase reaction. The primary aim of various treatments of allergies is to antagonize the actions of mediators and to counteract their effects on target organs.

summary *(continued)*

Antibodies against cell and tissue antigens may provoke tissue injury and disease (type II hypersensitivity). IgM and IgG antibodies may promote the phagocytosis of cells to which they bind, induce inflammation by complement- and Fc receptor-mediated leukocyte recruitment, and interfere with the functions of cells by binding to essential molecules and receptors. Antibodies may bind to circulating antigens to form immune complexes, which deposit in vessels and cause tissue injury (type III hypersensitivity). Injury is due mainly to leukocyte recruitment and inflammation. T-cell-mediated diseases (type IV hypersensitivity) are caused by Th1-mediated delayed-type hypersensitivity reactions or Th17-mediated inflammatory reactions, or by the killing of host cells by CD8$^+$ CTLs.

Immunodeficiency diseases are caused by defects in the immune system that give rise to increased susceptibility to infections and some cancers. Primary immunodeficiency diseases are caused by inherited genetic abnormalities. Secondary immunodeficiencies result from infections, cancers, malnutrition, or treatments for other conditions that adversely affect the cells of the immune system. Some primary immunodeficiency diseases are due to mutations that block lymphocyte maturation. SCID may be caused by mutations in the cytokine receptor common γc chain that reduce the IL-7-driven proliferation of immature lymphocytes, by mutations in enzymes involved in purine metabolism, or by

other defects in lymphocyte maturation. Selective B-cell maturation defects are seen in X-linked agammaglobulinemia, caused by abnormalities in the Btk enzyme involved in B-cell maturation. Certain T-cell maturation defects are seen in DiGeorge syndrome-bearing patients that do not develop a normal thymus.

Some immunodeficiency diseases are caused by defects in lymphocyte activation and function, although lymphocyte maturation occurs normally. The X-linked hyper-IgM syndrome is caused by mutations in the CD40 ligand, and as a result, Th-dependent B-cell responses (e.g., immunoglobulin heavy-chain class switching) and T-cell-dependent macrophage activation are defective. AIDS is caused by HIV, a retrovirus. HIV infects CD4$^+$ T cells, macrophages, and dendritic cells by using an envelope protein (gp120) to bind to CD4 and chemokine receptors (e.g., CCR5). The viral DNA integrates into the host genome, where it may be activated to produce more infectious virus. HIV-induced death of cells of the immune system is the principal mechanism by which the virus causes immunodeficiency. The clinical course of HIV infection typically consists of an acute viremia, a period of clinical latency during which there is a progressive destruction of CD4$^+$ T cells and lymphoid tissues. Ultimately AIDS develops, with severe immunodeficiency, opportunistic infections, some cancers, weight loss, and occasionally dementia. Treatment of HIV infection is designed to interfere with the life cycle of the virus. This approach is being adopted in many ongoing clinical trials in attempts to develop a much more effective HIV vaccine.

review questions

1. What is immunological tolerance, and why is it important for the regulation of immune responses?

2. What is central tolerance, and how is it induced in T cells and B cells?

3. What is the AIRE protein, and how does it regulate immunological tolerance?

4. What is the definition of anergy, and how is induced in T cells?

5. Explain how the escape of T cells from a state of anergy can mediate the development of autoimmune disease.

6. What are the molecular and cellular factors that promote the breakdown of tolerance and the development of autoimmune disease?

7. Discuss why and how in some autoimmune diseases in humans, defective Treg cell function or defective resistance of

pathogenic T cells to regulation may be an underlying cause of disease.

8. Indicate some of the genes that contribute to autoimmunity. Discuss, using examples, such as autoimmune type 1 diabetes, the role of MHC genes in the control of susceptibility to autoimmune diseases.

9. Discuss an important ligand–receptor interaction that generates signals through the death receptor and activates caspases and apoptosis by the death receptor pathway.

10. What are some possible mechanisms by which infections promote the development of autoimmunity?

11. What are the types of tumor antigens that the immune system reacts against? Explain why surveillance against tumors is considered to be an immunological phenomenon.

12. How do CD8$^+$ T cells recognize tumor antigens, and how are these cells activated to differentiate into effector CTLs?

(continued)

review questions *(continued)*

13. Discuss the mechanisms by which tumors may evade an immune response.

14. What are some strategies for enhancing host immune responses to tumor antigens that result in increased tumor immunity?

15. What types of antigens induce immune responses that elicit hypersensitivity reactions? What types of hypersensitivity reactions can arise?

16. What events mediate a typical immediate hypersensitivity reaction? Describe how a late-phase reaction occurs and what cells and mediators cause such a reaction.

17. Describe some examples of immediate hypersensitivity disorders, and include a discussion of the basis of their pathogenesis and forms of treatment.

18. What mechanisms induce asthma, allergies, and anaphylaxis? Which therapies and drugs are commonly used to treat these immune reactions?

19. How do antibodies cause tissue injury and disease? What are some of the differences in the manifestations of diseases caused by antibodies against extracellular matrix proteins and by immune complexes that deposit in tissues?

20. Which diseases may be caused by IgG or IgM antibodies or immune complexes? Describe their pathogenesis and their principal clinical and pathological manifestations.

21. Which diseases may be caused by T cells? Describe their pathogenesis and their principal clinical and pathological manifestations.

22. What are the main clinical and pathological features of primary and secondary immunodeficiency diseases?

23. How do immune defects elicit primary immunodeficiency diseases?

24. What are features of the SCID and DiGeorge syndrome disorders?

25. Describe the mutations and their clinical and pathological features that block T- and B-cell maturation. Also describe the mutations and their clinical and pathological features that inhibit the activation or effector functions of CD4+ T cells.

26. How does HIV infect cells and replicate inside infected cells?

27. What are the principal clinical manifestations of HIV infection, and what is the pathogenesis of these manifestations?

references

Abbas, A. K., and A. H. Lichtman (ed.). 2011. *Basic Immunology: Functions and Disorders of the Immune System*, 3rd ed. Saunders Elsevier, Philadelphia, PA.

Alderton, G. K. 2012. Tumor immunology: TIM3 suppresses antitumour DCs. *Nat. Rev. Immunol.* **12**:620–621.

Ayres, J. S., and D. S. Schneider. 2012. Tolerance of infections. *Annu. Rev. Immunol.* **30**:271–294.

Bach, J. F. 2002. The effects of infection on susceptibility to autoimmune and allergic diseases. *N. Engl. J. Med.* **347**: 911–920.

Benoist, C., and D. Mathis. 2012. Treg cells, life history, and diversity. *Cold Spring Harb. Perspect. Biol.* **4**:a007021.

Bilate, A. M., and J. J. Lafaille. 2012. Induced CD4+ Foxp3+ regulatory T cells in immune tolerance. *Annu. Rev. Immunol.* **30**:733–758.

Bordon, Y. 2013. HIV: a dead calm. *Nat. Rev. Immunol.* **13**:4–5.

Boyman, O., and J. Sprent. 2012. The role of interleukin-2 during homeostasis and activation of the immune system. *Nat. Rev. Immunol.* **12**:180–190.

Chappert, P., and R. H. Schwartz. 2010. Induction of T cell anergy: integration of environmental cues and infectious tolerance. *Curr. Opin. Immunol.* **22**: 552–559.

Chavele, K. M., and M. R. Ehrenstein. 2011. Regulatory T-cells in systemic lupus erythematosus and rheumatoid arthritis. *FEBS Lett.* **585**:3603–3610.

Cheng, M. H., and M. S. Anderson. 2012. Monogenic autoimmunity. *Annu. Rev. Immunol.* **30**:393–427.

Chevalier, M. F., and L. Weiss. 2013. The split personality of regulatory T cells in HIV infection. *Blood* **121**: 29–37.

Deeks, S. G., et al., for the International AIDS Society Scientific Working Group on HIV Cure. 2012. Towards an HIV cure: a global scientific strategy. *Nat. Rev. Immunol.* **12**:607–614.

Faivre, V., A. C. Lukaszewicz, A. Alves, D. Charron, D. Payen, and A. Haziot. 2012. Human monocytes differentiate into dendritic cell subsets that induce anergic and regulatory T cells in sepsis. *PLoS One* **7**:e47209.

Fazekas de St. Groth, B. 2012. Regulatory T-cell abnormalities and the global epidemic of immuno-inflammatory disease. *Immunol. Cell Biol.* **90**:256–259.

Feillet, H., and J. F. Bach. 2004. On the mechanisms of the protective effect of infections on type 1 diabetes. *Clin. Dev. Immunol.* **11**:191–194.

Finn, O. J. 2012. Immuno-oncology: understanding the function and dysfunction of the immune system in cancer. *Ann. Oncol. Suppl.* **8**:6–9.

Gao, J., C. Bernatchez, P. Sharma, L. G. Radvanyi, and P. Hwu. 2012. Advances in the development of cancer immunotherapies. *Trends Immunol.* **34**:90–98.

Goris, A., and A. Liston. 2012. The immunogenetic architecture of autoimmune disease. *Cold Spring Harb. Perspect. Biol.* **4**:a007260.

Gros, E., and N. Novak. 2012. Cutaneous dendritic cells in allergic inflammation. *Clin. Exp. Allergy* **42**:1161–1175.

Honda, K., and D. R. Littman. 2012. The microbiome in infectious disease and inflammation. *Annu. Rev. Immunol.* **30**:759–795.

Josefowicz, S. Z., L. F. Lu, and A. Y. Rudensky. 2012. Regulatory T cells: mechanisms of differentiation and function. *Annu. Rev. Immunol.* **30**:531–564.

Kåhrström, C. T. 2012. Viral pathogenesis: HIV hitchhikes on migratory T cells. *Nat. Rev. Immunol.* **12**:620–621.

Kaifu, T., B. Escalière, L. N. Gastinel, E. Vivier, and M. Baratin. 2011. B7-H6/NKp30 interaction: a mechanism of alerting NK cells against tumors. *Cell. Mol. Life Sci.* **68**:3531–3539.

Kroemer, G., L. Galluzzi, O. Kepp, and L. Zitvogel. 2012. Immunogenic cell death in cancer therapy. *Annu. Rev. Immunol.* **31**:51–72.

Lambrecht. B. N., and H. Hammad. 2012. Lung dendritic cells in respiratory viral infection and asthma: from protection to immunopathology. *Annu. Rev. Immunol.* **30**:243–270.

Mathis, D., and C. Benoist. 2012. The influence of the microbiota on type 1 diabetes: on the threshold of a leap forward in our understanding. *Immunol. Rev.* **245**:239–249.

Menendez, D., M. Shatz, and M. A. Resnick. 2013. Interactions between the tumor suppressor p53 and immune responses. *Curr. Opin. Oncol.* **25**:85–92.

Minton, K. 2013. Regulatory T cells: the role of PTPN22 in T cell homeostasis. *Nat. Rev. Immunol.* **13**:6.

Mohan, J. F., and E. R. Unanue. 2012. Unconventional recognition of peptides by T cells and the implications for autoimmunity. *Nat. Rev. Immunol.* **12**:721–728.

Nussbaum, J. C., and R. M. Locksley. 2012. Commensalism gone awry? Infectious (non)tolerance—frustrated. *Cold Spring Harb. Perspect. Biol.* **4**:a007328.

Oleinika, K., R. J. Nibbs, G. J. Graham, and A. R. Fraser. 2013. Suppression, subversion and escape: the role of regulatory T cells in cancer progression. *Clin. Exp. Immunol.* **171**:36–45.

Pace, L., A. Tempez, C. Arnold-Schrauf, F. Lemaitre, P. Bousso, L. Fetler, T. Sparwasser, and S. Amigorena. 2012. Regulatory T cells increase the avidity of primary CD8$^+$ T cell responses and promote memory. *Science* **338**:532–536.

Pelanda, R., and R. M. Torres. 2012. Central B-cell tolerance: where selection begins. *Cold Spring Harb. Perspect. Biol.* **4**:a007146.

Polychronakos, C., and Q. Li. 2011. Understanding type 1 diabetes through genetics: advances and prospects. *Nat. Rev. Genet.* **12**:781–792.

Powell, J. D., K. N. Pollizzi, E. B. Heikamp, and M. R. Horton. 2012. Regulation of immune responses by mTOR. *Annu. Rev. Immunol.* **30**:39–68.

Restifo, N. P., M. E. Dudley, and S. A. Rosenberg. 2012. Adoptive immunotherapy for cancer: harnessing the T cell response. *Nat. Rev. Immunol.* **12**:269–281.

Roberti, M. P., J. Mordoh, and E. M. Levy. 2012. Biological role of NK cells and immunotherapeutic approaches in breast cancer. *Front. Immunol.* **3**:375–386.

Savage, P. A., S. Malchow, and D. S. Leventhal. 2013. Basic principles of tumor-associated regulatory T cell biology. *Trends Immunol.* **34**:33–40.

Schwartz, R. H. 2012. Historical overview of immunological tolerance. *Cold Spring Harb. Perspect. Biol.* **4**:a006908.

Shultz, L. D., M. A. Brehm, J. V. Garcia-Martinez, and D. L. Greiner. 2012. Humanized mice for immune system investigation: progress, promise and challenges. *Nat. Rev. Immunol.* **12**:786–798.

Steinert, E. M., R. H. Schwartz, and N. J. Singh. 2012. At low precursor frequencies, the T-cell response to chronic self-antigen results in anergy without deletion. *Eur. J. Immunol.* **42**:2875–2880.

Steinman, R. M. 2012. Decisions about dendritic cells: past, present, and future. *Annu. Rev. Immunol.* **30**:1–22.

Stritesky, G. L., S. C. Jameson, and K. A. Hogquist. 2012. Selection of self-reactive T cells in the thymus. *Annu. Rev. Immunol.* **30**:95–114.

Sujit, V. J., K. Praveen, R. Marks, and T. F. Gajewski. 2011. Evidence implicating the Ras pathway in multiple CD28 costimulatory functions in CD4$^+$ T cells. *PLoS One* **6**:1–10.

Swain, S. L., K. K. McKinstry, and T. M. Strutt. 2012. Expanding roles for CD4$^+$ T cells in immunity to viruses. *Nat. Rev. Immunol.* **12**:136–148.

Trinchieri, G. 2012. Cancer and inflammation: an old intuition with rapidly evolving new concepts. *Annu. Rev. Immunol.* **30**:677–706.

Tse, M. T. 2013. Autoimmunity: particulate promotion of tolerance. *Nat. Rev. Immunol.* **13**:6–7.

Uzzaman, A., and S. H. Cho. 2012. Classification of hypersensitivity reactions. *Allergy Asthma Proc. Suppl.* **1**:S96–S99.

Vivier, E., S. Ugolini, D. Blaise, C. Chabannon, and L. Brossay. 2012. Targeting natural killer cells and natural killer T cells in cancer. *Nat. Rev. Immunol.* **12**:239–252.

Weiner, H. W., A. P. da Cunha, F. Quintana, and H. Wu. 2011. Oral tolerance. *Immunol. Rev.* **241**:241–259.

Xing, Y., and K. A. Hogquist. 2012. T-cell tolerance: central and peripheral. *Cold Spring Harb. Perspect. Biol.* **4**:a006957.

Ye, J., R. S. Livergood, and G. Peng. 2013. The role and regulation of human Th17 cells in tumor immunity. *Am. J. Pathol.* **182**:10–20.

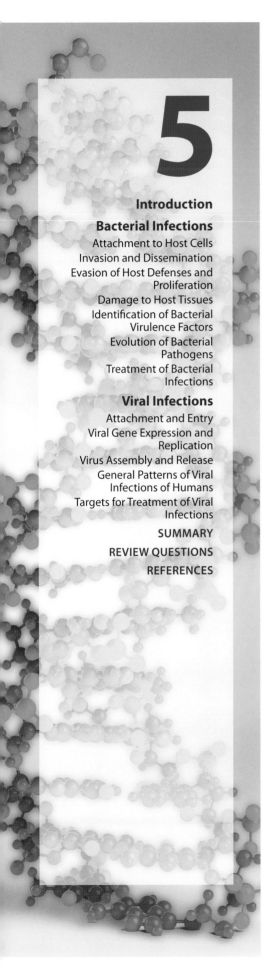

5

Microbial Pathogenesis

Introduction

We are constantly exposed to microorganisms in water, air, soil, and the food we ingest and from other animals, especially other humans, that we come in contact with. Fortunately, most are harmless. However, some bacteria, viruses, and eukaryotic microorganisms (fungi and protists) are human **pathogens**, that is, they are capable of colonizing (infecting) humans and causing disease. Moreover, new pathogens can emerge when normally benign strains acquire genes that confer **pathogenicity**, the ability to cause disease. Some microorganisms are considered **opportunistic pathogens** because they cause disease only when a host's defenses are compromised. In addition, pathogenic microorganisms differ in their **virulence** (degree of pathogenicity). Fewer cells of the highly virulent bacterium *Streptococcus pneumoniae* are required to produce a lethal infection when injected into mice than of the less virulent pathogen *Salmonella enterica* serovar Typhimurium. Although infections with most human pathogens are relatively rare, some cause a great deal of suffering. For example, the protist *Plasmodium* caused over 200 million cases of malaria in 2010, and 665,000 deaths, mostly in young children. In the same year, 2.7 million people were newly infected with human immunodeficiency virus (HIV), which can lead to acquired immunodeficiency syndrome (AIDS), and 8.8 million with *Mycobacterium tuberculosis*, the bacterium that causes the respiratory illness tuberculosis.

Contact with microorganisms occurs through our tissues that are exposed to the external environment, that is, our skin, eyes, ears, and respiratory, digestive, and urogenital tracts. In fact, our skin, nose, mouth, colon, urethra, and vagina are normally colonized by large numbers of microbes that are usually harmless and may actually provide some benefit. Problems can arise, however, when internal tissues that are normally sterile are inadvertently exposed to microorganisms, for example, through wounds and surgery.

doi:10.1128/9781555818890.ch5

We present a formidable, hostile barrier to pathogen infection of internal tissues. As discussed in chapter 2, our innate and adaptive immune systems recognize and respond to pathogens and protect us from infection. Nevertheless, pathogens can successfully breach these protective barriers and establish infections. How do they do this? In this chapter, the mechanisms of bacterial and viral infection are summarized, with an emphasis on the molecular interactions (molecules and processes) that are potential targets for biotechnological strategies to prevent or treat infectious diseases.

Bacterial Infections

Why do bacteria infect humans? What do they gain from it? Like all other organisms, bacteria must survive and reproduce. Despite our elaborate and efficient defense systems, in many ways, we present an optimal environment for some bacteria. We can provide a steady and fairly abundant supply of nutrients, for example, dead skin, dietary nutrients, and mucosal secretions such as glycogen and glycoproteins (e.g., **mucin**). We can also provide a protective environment, especially for intracellular pathogens, with a constant temperature and reduced competition.

Although many human tissues support large numbers of bacteria, for most healthy individuals, the colonizing bacteria rarely cause extensive tissue damage. Pathogens may damage tissues directly by secreting toxins that disrupt vital host cell processes, such as protein synthesis, or by growing to high cell densities that may disrupt normal host tissue function. Pathogens may also damage tissues indirectly by stimulating an inflammatory response. Especially serious is a systemic inflammatory response that may occur when cell wall fragments from lysed gram-negative bacteria induce systemic release of inflammatory cytokines. Molecules produced by pathogens that contribute to their ability to cause disease are known as **virulence factors**. Virulence factors facilitate colonization, proliferation, and dissemination in a host; they include toxins and mechanisms used by pathogens to evade host defenses.

Attachment to Host Cells

Many of the tissues that are potential sites for initial colonization of pathogens are vulnerable to flushing activities that can remove loosely attached bacteria. For example, the force of urine moving through the urethra and the flow of the food through the small intestines help to prevent urinary and intestinal tract infections, respectively. To avoid expulsion, bacteria must attach firmly to the epithelia that line these tissues. Attachment usually requires very specific molecular interactions between surface molecules present on the bacterium known as **adhesins** and specific host cell surface receptors or components of the host extracellular matrix. The specificity of this interaction determines the host and tissues that the pathogen can infect, a phenomenon referred to as tissue or **host tropism**. Attachment can also involve less specific interactions with the sticky polysaccharide **capsule** that surrounds some bacterial cells.

Fimbriae and **pili** are protein appendages that extend from the bacterial cell envelope and bind to specific receptors on the surface of host cells. Although the terms are often used interchangeably, pili (singular, pilus) and fimbriae (singular, fimbria) can be differentiated by the size of the protrusion, with pili being longer and thicker than fimbriae. Thousands of subunits of a single polypeptide known as **pilin** polymerize to form a helical filament that can be as much as 10 times the length of the cell. The filaments of most fimbriae and pili are anchored in the outer membrane of gram-negative bacteria during synthesis by a specific secretion apparatus known as the **chaperone–usher system** (Fig. 5.1). Pilin subunits are transported to the periplasm via a general secretion pathway where each subunit interacts with a **chaperone** protein. Interaction with chaperones in the periplasm ensures correct folding of the pilin polypeptides and prevents aggregation and degradation of the pilin subunits in the periplasm. The chaperones deliver the pilin subunits to a protein complex in the outer membrane known as the usher system for assembly and export. The pili of gram-positive bacteria, which lack an outer membrane, are attached to the **peptidoglycan** cell wall. The cell membrane-bound enzyme **sortase** catalyzes the polymerization of the secreted pilin subunits and the formation of a covalent bond between the pilus and peptidoglycan that anchors the assembled pilus in the cell wall (Fig. 5.2).

The specificity of the interaction between pilus and host receptor resides in the protein at the tip of the filament. The tip protein recognizes specific carbohydrate polymers on host membrane glycoproteins or glycolipids. The remarkable specificity is well illustrated by the preferred adhesion sites for different strains of **uropathogenic *Escherichia coli* (UPEC)**.

UPEC
uropathogenic *Escherichia coli*

Figure 5.1 Chaperone–usher system for pilus synthesis in gram-negative bacteria. doi:10.1128/9781555818890.ch5.f5.1

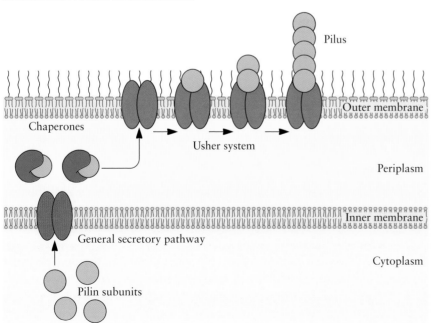

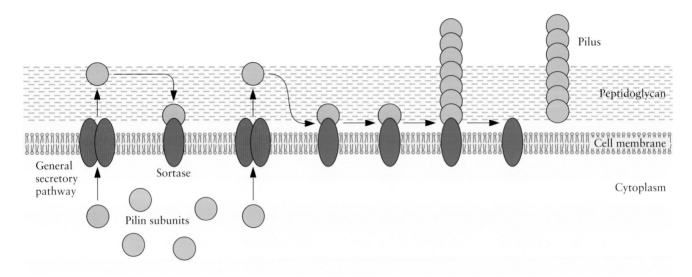

Figure 5.2 Sortase-catalyzed pilus assembly in gram-positive bacteria.
doi:10.1128/9781555818890.ch5.f5.2

Some UPEC strains produce a type I pilus made up of a helical polymer of the pilin FimA and the FimH tip protein. The FimH adhesin specifically binds to mannose residues on the glycoprotein uroplakin on the surface of human bladder epithelial cells. This strain of *E. coli* is therefore responsible for bladder infections. Another strain of UPEC produces a P pilus made up of the PapA protein that is essential for establishing more serious kidney infections. The PapG tip protein on the P pilus specifically attaches to a galactose-containing glycolipid on kidney epithelia. The presence of a receptor for a specific pilus tip protein therefore contributes to the susceptibility of a host to a particular pathogen.

Many bacteria possess surface proteins that bind to proteins of the host **extracellular matrix** such as fibronectin, fibrinogen, and collagen. *Streptococcus pyogenes*, a causative agent of pharyngitis (strep throat) and other more serious human diseases, produces **M proteins** that form thin hair-like filaments (fibrils) on the cell surface. The M proteins bind to the glycoprotein fibronectin found on the surface and in the extracellular matrix of many host cells. Infection with *S. pyogenes* can lead to rheumatic fever, an autoimmune disease, because the M protein contains an epitope that is similar in structure to an epitope of a human heart protein. In some cases, this can stimulate the production of self-reactive antibodies (autoantibodies) that damage heart tissue.

Adhesins and other bacterial surface structures are known as **pathogen-associated molecular patterns (PAMPs)** and are recognized by host phagocytes (macrophage, dendritic cells, and neutrophils) that elicit an adaptive immune response (see chapter 2). However, some bacteria can change the structure of their adhesins to avoid recognition by the host immune system. This change is known as **antigenic variation**. An example is the ability of the sexually transmitted pathogen *Neisseria gonorrhoeae* to alter the protein composition of its pilus through genetic recombination. Pilin is produced from the *pilE* gene (Fig. 5.3); however, the genome of

PAMP
pathogen-associated molecular pattern

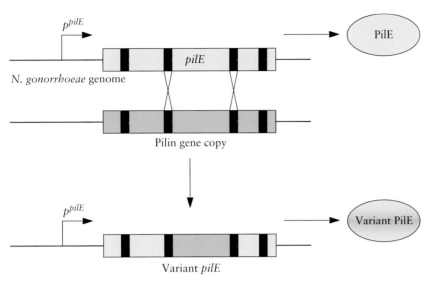

Figure 5.3 Recombination (crossed lines) between conserved regions (black boxes) in *pilE* and pilin gene copies in the N. *gonorrhoeae* genome results in production of a pilus composed of variant PilE protein subunits that are not recognized by host antibodies. p^{pilE}, promoter for *pilE* gene expression; variable regions of *pilE* and the pilin gene copy are shown in color. Adapted with permission from Hill and Davies, *FEMS Microbiol. Rev.* **33**:521–530, 2009. doi:10.1128/9781555818890.ch5.f5.3

N. *gonorrhoeae* carries several additional copies of the pilin gene that are not expressed. Rather, the pilin gene copies, which differ from the *pilE* sequence in some variable regions, serve as a reservoir of genetic variation. Although the sequences of *pilE* and the pilin gene copies are not identical, they contain regions of sufficient sequence homology that they can undergo recombination (Fig. 5.3). This process, known as **gene conversion**, results in a pilus composed of a variant pilin protein that is not recognized by existing host antibodies.

The importance of adhesins in the initial stages of infection and their immunogenicity makes them good candidates for subunit vaccines (chapter 11). Antibodies directed against adhesins could block pathogen attachment and thereby prevent bacterial colonization. Purified FimH adhesin from the type I pili of UPEC induced production of specific anti-FimH antibodies when injected into mice and protected immunized mice and monkeys from subsequent infection with the pathogen. However, the approach was not successful in preventing urinary tract infections in humans. Moreover, FimH contains an epitope that is similar to an epitope of a human protein found in the membrane of many different cells and therefore can trigger the production of autoantibodies. Antigenic variation may also reduce the effectiveness of adhesin vaccines that elicit production of antibodies that do not recognize a variant protein.

Invasion and Dissemination

Although attachment to a host cell does not necessarily result in disease, in some cases, pathogens attached to the epithelium cause serious disease without further invasion. Noninvasive pathogens can damage tissues by secreting toxins or by stimulating an inflammatory response that disrupts tissue function. Other pathogens can penetrate the protective epithelium and proliferate within host cells or disseminate to tissues beyond the site of initial infection. How do bacteria penetrate the barriers that protect us from invasion? Of course, some pathogens are introduced opportunistically by wounds, medical procedures, and insect bites; however, other pathogens have evolved mechanisms to gain access to deeper tissues.

Tissues exposed to the external environment such as the respiratory, urogenital, and intestinal tracts are protected from bacterial invasion by a mucous layer consisting of a mixture of heavily glycosylated proteins (mucins), a tightly packed layer of epithelial cells that are bound to each other by adhesion proteins (mucosal epithelium), and an underlying layer of proteinaceous connective tissue (basal lamina) (Fig. 5.4). This protective surface is known as the **mucosa**. The viscous mucin layer traps bacteria and impedes their penetration; however, pathogens may contact the underlying epithelium in regions where the mucin is thin or may produce proteases that degrade mucin components. For example, proteases produced by *S. pneumoniae* digest **secretory immunoglobulin A** molecules that are secreted into mucus by nearby B lymphocytes to trap the bacteria (see chapter 2). Once they reach the epithelium, successful invaders use two general strategies to breach this formidable barrier: (i) they produce enzymes that degrade the connections that hold the cells together, or (ii) they induce phagocytosis by host cells.

Bacterial proteins that degrade host defenses are generally referred to as **invasins**. Production of enzymes that degrade components of the host extracellular matrix enables some pathogens to migrate between cells of the epithelial layer and contributes to dissemination to deeper tissues. The enzyme hyaluronidase, produced by streptococci, staphylococci, and some clostridia, degrades hyaluronic acid, a polysaccharide in the

Figure 5.4 Mucous membrane. Tightly packed epithelial cells present a barrier to microbial invasion of deeper tissues. Heavily glycosylated proteins (mucin) secreted by goblet cells (blue) form a viscous mucous layer on the surface of the epithelium that traps microorganisms. Secretory immunoglobulin A ("Y" shapes) is secreted by B lymphocytes to help trap bacteria in the mucous layer. doi:10.1128/9781555818890.ch5.f5.4

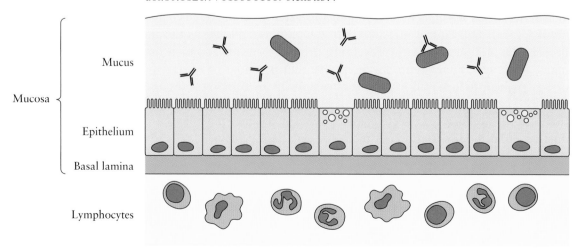

Mucosa
Mucus
Epithelium
Basal lamina
Lymphocytes

connective tissue between cells. *Clostridium perfringens* produces collagenase, which digests collagen, a fibrous protein found in the basal lamina and other connective tissues. Fibrin clots that form following wounding isolate microorganisms at the wound site. However, streptococci produce the enzyme streptokinase, which digests fibrin, enabling dissemination of the bacteria.

Attachment to host cell receptors via pili often triggers changes in both host and attached bacterial cells that lead to uptake of the bacterium by the host cell. For example, *S. enterica* serovar Typhimurium, a common food-borne pathogen, binds to specialized **M (microfold) cells** in the epithelium of the small intestine by interaction of pili and other surface proteins with host receptors (Fig. 5.5A). Attachment induces expression of bacterial virulence proteins, such as those that make up a **type III secretion system**. The secretion apparatus includes a protein needle that

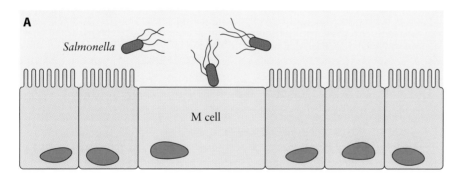

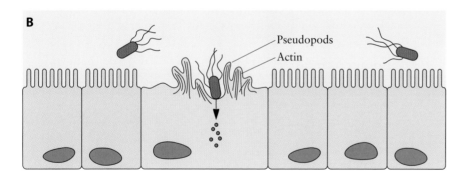

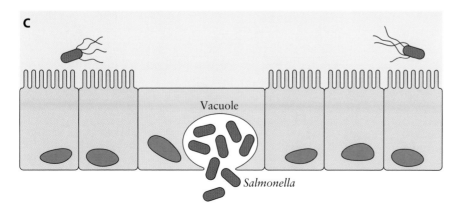

Figure 5.5 *S. enterica* serovar Typhimurium invades the intestinal epithelial barrier via M (microfold) cells. (**A**) *S. enterica* serovar Typhimurium attaches to M cell. (**B**) Effector proteins (red) secreted through a type III secretion system (Fig. 5.6) induce formation of pseudopods (membrane ruffling) that eventually engulf the bacterium. (**C**) Bacterial cells are released from the vacuole to the underlying tissues.
doi:10.1128/9781555818890.ch5.f5.5

inserts directly into the host cell membrane, connecting the cytoplasms of the pathogen and host cells (Fig. 5.6). Several bacterial proteins, referred to as effector proteins, are secreted through this structure into the host cytoplasm in an ATP-dependent manner. *Salmonella* effector proteins alter M cell gene expression and ion movement and induce rearrangement of the M-cell cytoskeleton. Recruitment and assembly of actin at the site of bacterial attachment leads to the formation of pseudopods that eventually engulf the bacterial cell (Fig. 5.5B). The pseudopods give the appearance of a membrane ruffle around the attached bacterium. Engulfment takes up the pathogen into the host cell, where it replicates within a protected vacuole. Survival of *Salmonella* within the vacuole depends on blocking fusion of the vacuole with a **lysosome** that would release toxic

Figure 5.6 *Salmonella* type III secretion system. The type III secretion system is made up of about 20 different proteins that form a continuous channel through the inner and outer membranes of gram-negative bacteria and the host cell membrane. A similar apparatus is used by many different bacterial pathogens to secrete effector proteins into host cells. Adapted with permission from Marlovits and Stebbins, *Curr. Opin. Microbiol.* **13:**47–52, 2010. doi:10.1128/9781555818890.ch5.f5.6

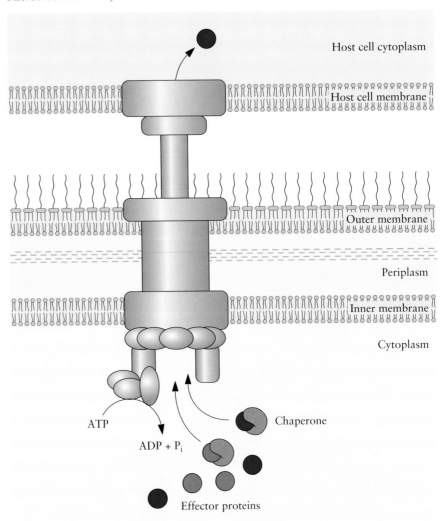

compounds to destroy the bacteria. Secreted *Salmonella* effector proteins prevent **phagosome**–lysosome fusion. From the infected M cell, the bacteria are released to the underlying tissues (Fig. 5.5C). Many other intestinal pathogens, such as *Listeria monocytogenes* and *Shigella flexneri*, and respiratory pathogens, such as *Legionella pneumophila* and *M. tuberculosis*, use similar invasion strategies. *L. monocytogenes* and *S. flexneri* can move laterally within the intestinal epithelium by acquiring actin tails. Host actin filaments assemble on a surface protein (ActA) at one end of *L. monocytogenes* cells (Fig. 5.7). This tail is used to propel the pathogen into adjacent epithelial cells.

Production of a type III secretion system and effector proteins is typical of many bacterial pathogens. Not only is the type III secretion system an efficient protein delivery mechanism, but also it avoids eliciting antibodies against the secreted bacterial proteins. Although some strains of *E. coli* are not invasive pathogens, they may produce a type III secretion system and effector proteins that induce changes in the host cell cytoskeleton. One of the proteins secreted into the host cell is the translocated intimin receptor (Tir), which is uploaded into the host cell membrane and functions as a receptor for the bacterial adhesion protein intimin (Fig. 5.8). In effect, pathogenic *E. coli* provides its own receptor, which it uses to establish an intimate interaction with the host epithelial cell. Induced actin rearrangement results in the formation of a pedestal beneath the attached pathogen (Fig. 5.8). Pedestal formation disrupts nutrient absorption by the small intestine and triggers an inflammatory response that further erodes intestinal function.

Once a pathogen breaches the protective epithelium, access to other regions of the body is facilitated. The cells of the endothelia that line the external surface of the blood circulatory system and **lymphatic system** are not as tightly packed and therefore are easier to penetrate. These systems, then, provide a highway for pathogen transport that can lead to **systemic infections**. Puerperal sepsis (commonly known as childbirth sepsis) is a severe invasive infection caused by group A *Streptococcus* (*Streptococcus* species are grouped according to their cell wall antigens). Historically a

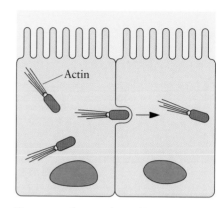

Figure 5.7 Actin tails enable some pathogens to move laterally between epithelial cells. Host cell actin polymerizes on a bacterial surface protein such as ActA in *L. monocytogenes*. doi:10.1128/9781555818890.ch5.f5.7

Figure 5.8 Pathogenic *E. coli* secretes the Tir protein into host epithelial cells via a type III secretion system. Tir acts as a receptor for the bacterial adhesin intimin. Tir also induces rearrangement of the host cytoskeleton that leads to formation of a pedestal beneath the bacterium. doi:10.1128/9781555818890.ch5.f5.8

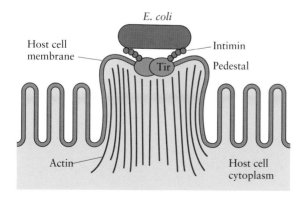

much-feared disease among pregnant women, it is still the most frequent cause of postpartum death (75,000 deaths per year worldwide). Group A *Streptococcus* can be introduced into the uterus from the vagina during delivery or through an incision during Cesarean section. The initial infection may occur in the uterine wall after separation of the placenta; however, the bacterium produces several invasins that enable rapid progression, within hours or days, to **sepsis**, the presence of the bacterium in the blood. In addition to hyaluronidase and streptokinase, group A *Streptococcus* produces proteases (exotoxin B and C5a peptidase), deoxyribonucleases (DNases), and several cytolytic toxins (streptolysins) that contribute to its rapid spread and tissue destruction.

DNase
deoxyribonuclease

Evasion of Host Defenses and Proliferation

The human body can be a dangerous place for a bacterium. Cells of the innate immune system (phagocytes and natural killer cells) constantly survey for foreign intruders by recognizing PAMPS and mount a defensive response (see chapter 2). To survive and proliferate, bacteria must evade, or neutralize, the host immune system.

Intracellular pathogens may avoid the immune system by replicating within endocytic vesicles; however, they must prevent recognition by natural killer cells that attack infected host cells and fusion with lysosomes. Lysosomes produce enzymes, such as lysozyme, proteases, and nucleases, and toxic oxygen and nitrogen species, such as hydrogen peroxide and nitric oxide, that kill bacterial cells. Following phagocytosis by alveolar macrophages, *L. pneumophila* proliferates intracellularly within a vacuole that is disguised as rough endoplasmic reticulum (Fig. 5.9). From within the phagosomal vacuole, this pathogen secretes about 300 different

Figure 5.9 *L. pneumophila* replicates within a vacuole in a host macrophage. The vacuole membrane is studded with ribosomes that were recruited from the rough endoplasmic reticulum (ER). Adapted with permission from Isberg et al., *Nat. Rev. Microbiol.* 7:13–24, 2009. doi:10.1128/9781555818890.ch5.f5.9

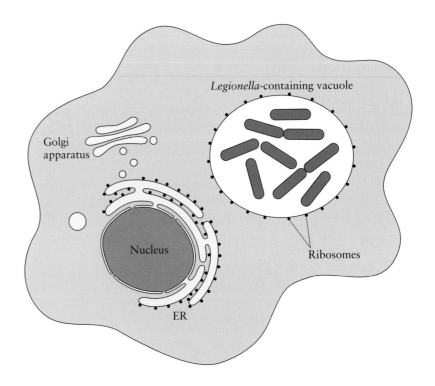

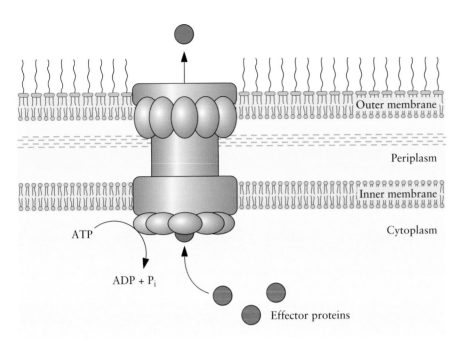

Figure 5.10 *L. pneumophila* type IV secretion system for translocation of effector proteins into host cells. The components of the type IV secretion system are homologues of proteins in the conjugative DNA transfer apparatus. Adapted with permission from Isberg et al., *Nat. Rev. Microbiol.* 7:13–24, 2009. doi:10.1128/9781555818890.ch5.f5.10

proteins through a **type IV secretion system** to manipulate the host cell (Fig. 5.10). Type IV secretion systems are found in wide range of bacteria, in which they mediate cell-to-cell transfer of proteins or nucleoprotein complexes. This includes conjugative transfer of **plasmid** DNA between bacteria and protein transfer from a bacterium to a eukaryotic host. In addition to preventing lysosomal fusion, the secreted *L. pneumophila* effector proteins recruit vesicles containing proteins from the endoplasmic reticulum to the vacuole that provide nutrients for the replicating bacterial cells (Fig. 5.9). Other intracellular pathogens manage to escape from the phagosome before lysosome fusion. *L. monocytogenes*, for example, produces a protein (listeriolysin) that forms a pore in the phagosome. Listeriolysin disrupts the phagosome membrane, and the bacterium is released into the host cytosol, where it replicates (Fig. 5.11).

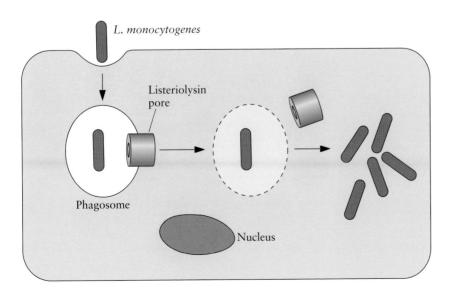

Figure 5.11 *L. monocytogenes* is taken up into a host cell by phagocytosis. The bacterium escapes the phagosome before lysosome fusion by secreting the protein listeriolysin, which forms a pore in the phagosome membrane. As a consequence, the phagosome lyses and the bacterium escapes into the cytosol, where it replicates. doi:10.1128/9781555818890.ch5.f5.11

Some extracellular pathogens have evolved effective mechanisms to evade immune surveillance. Several strategies are used by different pathogens to achieve this. Some pathogens alter the composition of their surface proteins to avoid detection by antibodies. Others, such as *Staphylococcus aureus*, coat their cells with the Fc portion of host antibodies to prevent recognition by phagocytic cells. *S. aureus* also produces coagulase, which causes fibrin clots to form around the site of infection. The clots protect the bacteria from attack by host cells. The respiratory pathogen *M. tuberculosis* is inhaled in aerosols in the environment and proliferates in lung tissue within nodules (granulomas) called tubercles. This bacterium produces mycolic acid, a waxy cell wall material that cloaks the cell as it moves beyond the respiratory epithelium. *M. tuberculosis* infects alveolar macrophages, which induce an inflammatory response that recruits other phagocytes and lymphocytes. However, rather than destroy infected macrophages, host immune cells aggregate around infected cells to form a tubercle (Fig. 5.12). Release of cytokines by the immune cells induces chronic inflammation that damages lung tissue. *M. tuberculosis* proliferates slowly within the tubercle, which may eventually rupture, disseminating the pathogen.

Many bacteria produce an extracellular capsule composed of a network of polysaccharides and/or proteins that are loosely attached to the cell wall (Fig. 5.13). Capsule composition varies among bacteria. Encasement within this protective layer prevents uptake by phagocytes, inhibits binding of host complement proteins that lead to bacterial cell lysis, and, in some cases, disguises the bacterium from host immune cells by mimicking the host. For example, the capsule of *S. pyogenes* consists of the

Figure 5.12 *M. tuberculosis* can persist for extended periods within a protective tubercle. Infected macrophages are surrounded by uninfected macrophages, lymphocytes, and a fibrous cuff of proteins such as collagen. Adapted with permission from Russell, *Nat. Rev. Microbiol.* 5:39–47, 2007. doi:10.1128/9781555818890.ch5.f5.12

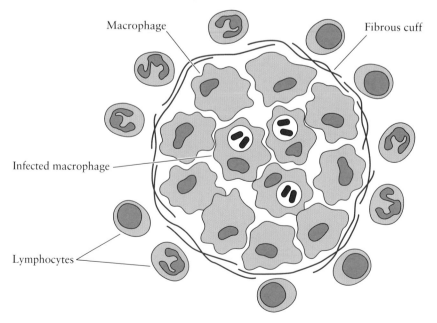

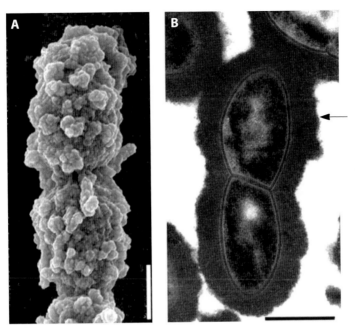

Figure 5.13 Capsule of *S. pneumoniae* viewed by scanning (**A**) and transmission (**B**) electron microscopy. Arrow indicates capsule. Bars = 0.5 μm. Reprinted with permission from Hammerschmidt et al., *Infect. Immun.* **73**:4653–4667, 2005. doi:10.1128/9781555818890.ch5.f5.13

polysaccharide hyaluronic acid, which is also a component of the host extracellular matrix. Despite the protective role of capsules, capsular polysaccharides are immunogenic. Vaccines consisting of capsular polysaccharides from *Haemophilus influenzae*, *Neisseria meningitidis*, and *S. pneuomoniae* are administered to children to prevent **meningitis**, an infection of the protective membranes covering the brain and spinal cord (meninges) commonly caused by these strains.

Bacteria that adhere to the surface of host tissues often form **biofilms** that are protected from immune attack by an extracellular matrix (Fig. 5.14). The matrix is a thick, viscous composition of polysaccharides,

Figure 5.14 Biofilms are communities of bacteria attached to a surface and embedded in a dense extracellular matrix made up of polysaccharides, proteins, and DNA. Fluid channels provide access to nutrients and removal of waste. doi:10.1128/9781555818890.ch5.f5.14

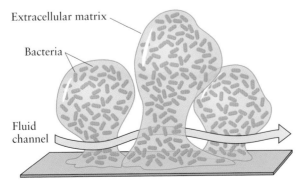

proteins, and DNA that is difficult for immune cells, antibiotics, and other antimicrobial agents to penetrate. Thus, biofilms are significant reservoirs for recurring infections. It is estimated that 65% of human infections involve biofilms. These are often chronic, intractable infections such as otitis media, prostatitis, chronic obstructive pulmonary disease, and respiratory infections in individuals with cystic fibrosis. Biofilms form on surfaces that are bathed in fluids and are particularly notable on medical implants such as catheters, heart valves, prosthetic joints, pacemakers, contact lenses, and intrauterine devices. For example, there is a >50% risk of a biofilm forming on a heart assist device (a mechanical device that helps the heart to pump blood) within 2 to 3 days of implantation. Common implant colonizers include *Staphylococcus*, *Enterococcus*, *Streptococcus*, and *E. coli*. The implant biofilms are responsible for persistent infections and can lead to malfunctioning of the implant and to systemic infections due to "planktonic showering" when large numbers of cells leave the biofilm and disseminate. *Pseudomonas aeruginosa* biofilms in the lungs of patients with cystic fibrosis are well studied. Production of copious amounts of the exopolysaccharide (extracellular polysaccharide) alginate by *P. aeruginosa* contributes not only to persistent respiratory infections but also to the blockage of airways that makes breathing difficult for individuals with cystic fibrosis.

Quorum sensing systems regulate the formation of bacterial biofilms and other cell density-dependent activities. They enable cells to monitor the environment for other bacteria and to alter their behavior in response to changes in bacterial cell density. For example, they coordinate the production of virulence factors that would be ineffective when the pathogen population is low. In gram-negative bacteria, quorum sensing signals (also known as autoinducers) are acyl-homoserine lactones, while in gram-positive bacteria, they are small linear or cyclic peptides (Fig. 5.15). These molecules are secreted, and therefore, the external concentration increases as cell density increases. When extracellular levels reach a critical threshold, a response is triggered in neighboring bacteria that leads to changes in gene expression. *P. aeruginosa* has two quorum sensing systems called Las and Rhl. LasI and RhlI are acyl-homoserine lactone synthases that produce the signal (Fig. 5.16). Acyl-homoserine lactones diffuse across the cell membrane, and within a cell, bind to and activate transcription factors LasR and RhlR. The LasR and RhlR regulons (sets of genes controlled by a transcription factor) include genes involved in biofilm formation and production of extracellular proteins (elastases and exotoxins) that are important virulence factors. Because of their central role in regulating virulence factor production, quorum sensing systems are attractive targets for antimicrobial therapies.

Although most biofilms form on tissue surfaces, in some cases, they can form inside cells. For example, UPEC forms intracellular biofilm-like communities (IBCs). UPEC binds via type I pili to mannose residues on the glycoprotein uroplakin on the surface of bladder epithelial cells (facet cells). Binding promotes exfoliation of the bladder epithelium, which is three to four cells thick. This enables the bacteria to invade more deeply

IBC
intracellular biofilm-like community

A Acyl-homoserine lactones

V. fischeri LuxI

P. aeruginosa LasI

P. aeruginosa RhlI

A. tumefaciens TraI

B Autoinducing peptides (AIP)

S. aureus AIP-I

S. aureus AIP-II

B. subtilis CSF

Glu–Arg–Gly–Met–Thr

B. subtilis ComX

Ala–Asp–Pro–Ile–Thr–Arg–Gln–Trp–Gly–Asp

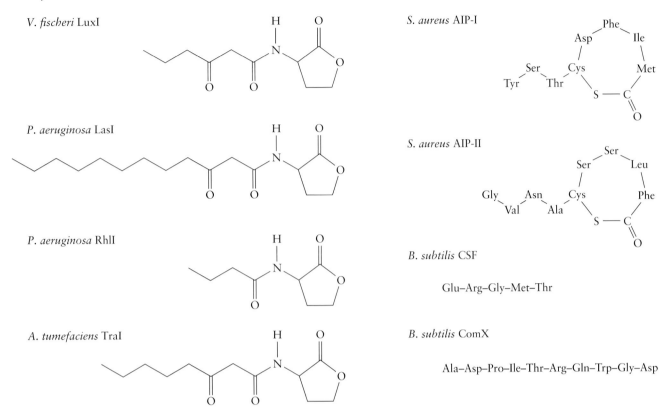

Figure 5.15 Some quorum sensing signals produced by gram-negative (**A**) and gram-positive (**B**) bacteria. Adapted from Schauder and Bassler, *Genes Dev.* **15:**1468–1480, 2001. doi:10.1128/9781555818890.ch5.f5.15

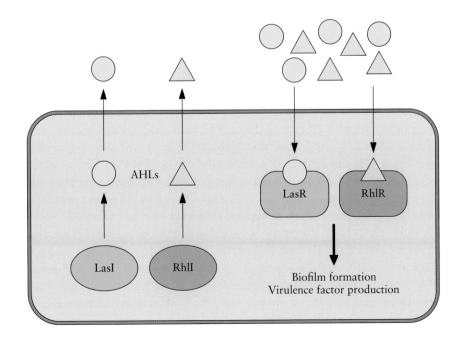

Figure 5.16 Las and Rhl quorum sensing systems of *P. aeruginosa*. Acyl-homoserine lactone (AHL) synthases LasI and RhlI catalyze the synthesis of the AHLs that diffuse out of the cell. When the extracellular concentration reaches a certain level, some of the signaling molecules enter cells and activate transcription factors LasR and RhlR. Genes required for biofilm formation and production of virulence factors are expressed.
doi:10.1128/9781555818890.ch5.f5.16

into the tissue. Binding also stimulates host actin rearrangement and engulfment of the attached bacterial cells. Within host bladder epithelial cells, *E. coli* proliferates rapidly and forms a biofilm in a polysaccharide-rich matrix. As the IBC grows, the bacterial mass pushes against the host cell membrane and produces a characteristic pod-like structure protruding from the surface of the bladder. The pathogen is protected within the cell by the impermeable uroplakin protein. The IBCs are a reservoir for recurring urinary tract infections.

Within a host, pathogens, which are mainly **chemoheterotrophs**, must acquire nutrients for growth. Damaged host cells provide organic molecules that serve as carbon and nitrogen sources and vitamins and other growth factors for bacterial biosynthetic processes. Iron is also an essential nutrient for bacterial (and host) growth, and mechanisms used by pathogens to acquire iron are considered to be virulence factors. Acquisition of iron is a competitive process. Host tissues produce iron-chelating molecules such as lactoferrin and transferrin to obtain iron for their own use and to limit iron availability for infecting microbes. However, many bacteria produce structurally diverse, iron-chelating molecules known as **siderophores** (Fig. 5.17A). When the external concentration of iron is low, as it is in host tissues, bacteria secrete siderophores that bind iron with a very high affinity. Enterobactin, the siderophore produced by *E. coli* and *S. enterica* serovar Typhimurium, has the highest affinity for ferric iron among known siderophores, with an association constant of 10^{52} M^{-1} (Fig. 5.17A). The iron–siderophore complex is taken up into the cell via specific receptors in the outer membrane (Fig. 5.17B). An ABC transporter then transports the iron–siderophore complex across the inner membrane, and then iron is released from the siderophore following reduction to the ferrous state or cleavage of the siderophore. Other bacteria compete directly with transferrin and lactoferrin for iron. *N. gonorrhoeae* does not produce siderophores, but receptors present in the outer membrane bind and remove iron complexed with transferrin or lactoferrin. The surface exposure and essential role in *N. gonorrhoeae* growth make the iron receptor an attractive candidate for a subunit vaccine.

Damage to Host Tissues

Infectious disease manifests as damage to host tissues that disrupts their function. Tissue damage can result from the host immune response to the presence of the pathogen or from secretion of toxins by the pathogen that manipulate and/or inhibit normal host cell function. The effector proteins secreted through type III and IV secretion systems, discussed in the previous section, are examples of toxins that are secreted directly into host cells. Other toxins are secreted into the extracellular fluid, from which they can damage host cells. Often toxins target particular types of host cells by interaction with specific cell surface receptors. By killing host cells, toxins release nutrients that provide the pathogen with substrates for survival and growth or protect bacteria from attack by immune cells.

The bacterial cell envelope is composed of peptidoglycan and, in the case of gram-negative bacteria, an outer membrane that contains

A

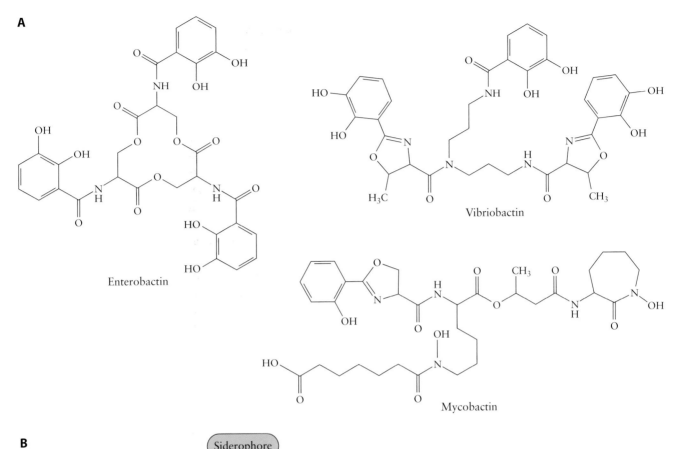

Enterobactin

Vibriobactin

Mycobactin

B

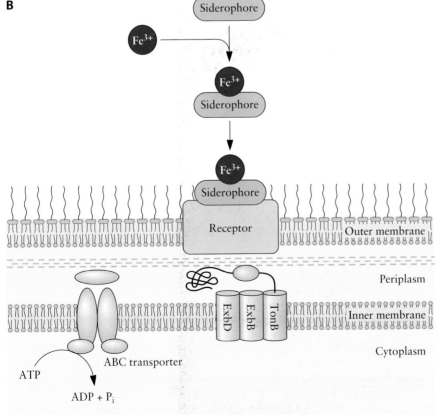

Figure 5.17 Bacterial siderophores. (**A**) Structures of the siderophores enterobactin produced by *E. coli* and *S. enterica* serovar Typhimurium, vibriobactin produced by *V. cholerae*, and mycobactin produced by *M. tuberculosis*; (**B**) acquisition of iron via siderophores. Adapted with permission from Raymond et al., *Proc. Natl. Acad. Sci. USA* **100**:3584–3588, 2003.
doi:10.1128/9781555818890.ch5.f5.17

lipopolysaccharides. One of the components of the lipopolysaccharide membrane is lipid A (Fig. 5.18). Lipid A is known as an **endotoxin** because it is a component of the cell and exerts a toxic effect when fragments of the cell envelope are released following cell lysis in the host. Lipid A binds to host phagocytes via Toll-like receptors and stimulates the release of inflammatory cytokines. Inflammatory cytokines (such as interleukin-1 and -6 and tumor necrosis factor alpha) induce fever, increase the permeability of blood vessels (which reduces blood pressure), and activate blood clotting proteins that obstruct capillaries and decrease blood flow to tissues. Similarly, fragments of cell wall peptidoglycan are released from lysed bacterial cells and can stimulate an immune response. When there is a large number of infecting bacterial cells (i.e., a high bacterial load), lysis by phagocytes or following antibiotic treatment releases large amounts of lipid A and peptidoglycan into the host. This stimulates a massive inflammatory response known as **endotoxic shock**. For example, release of cell wall fragments following treatment of septic (blood) *N. meningitidis* infections with bacteriolytic antibiotics stimulates endotoxic shock that can lead rapidly to patient death. Successful antibiotic treatment of septic infections, especially those caused by gram-negative bacteria, must be accompanied by treatments to suppress an inflammatory response.

Protein toxins synthesized by some pathogens are either secreted directly into a host cell by a type III or IV secretion system, as described above, or released into the extracellular fluid. Toxins released extracellularly are known as **exotoxins** and are of three general types: (i)

Figure 5.18 The cell wall of a gram-negative bacterium consists of a thin layer of peptidoglycan and an outer membrane containing lipopolysaccharides. The lipid A component of the lipopolysaccharide is a disaccharide of *N*-acetylglucosamine phosphate covalently linked to short fatty acid chains. Lipid A is an endotoxin that is toxic to animal cells when fragments are released during bacterial cell lysis. doi:10.1128/9781555818890.ch5.f5.18

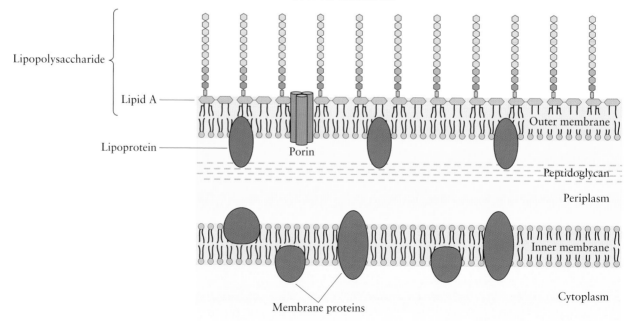

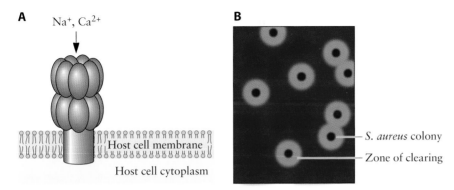

Figure 5.19 Pore-forming alpha-hemolysin toxin produced by *S. aureus*. **(A)** Structure of the heptameric pore; **(B)** zone of clearing around hemolysin-producing colonies of *S. aureus* on sheep blood agar. Panel B is reprinted with permission from Herbert et al., *Infect. Immun.* 78:2877–2889, 2010. doi:10.1128/9781555818890.ch5.f5.19

pore-forming toxins, (ii) A–B toxins, and (iii) **superantigen** toxins. These may act locally, at the site of infection, or may disseminate to damage distant tissues.

Pore-forming toxins are proteins that insert into the host cell membrane and form a channel that disrupts osmoregulation. Ions are transported through the pore and water enters the cell, rupturing the weakened membrane. An example of a pore-forming toxin is the alpha toxin produced by *S. aureus*. Seven subunits of the secreted alpha toxin form a transmembrane channel in the target cell (Fig. 5.19A). Sodium and calcium ions and other small molecules rapidly enter the cell through the pore. The consequent influx of water causes the cells to lyse. The toxin is a hemolysin that lyses red blood cells, which can be visualized as zones of clearing around *S. aureus* colonies on blood agar plates (Fig. 5.19B).

The **A–B exotoxins** are so called because they have two different components, the A domain, which has toxin activity, and the B domain, which binds to a host cell receptor. The A and B domains may be part of a single polypeptide or may be distinct polypeptides. While the general structures are similar among A–B toxins, the toxic A components differ greatly in their activities. The respiratory pathogen *Corynebacterium diphtheriae* produces a single polypeptide A–B toxin that is cleaved to separate the A and B domains, which initially are connected by a disulfide bond. The B domain interacts specifically with a transmembrane receptor (the heparin-binding epidermal growth factor-like growth factor) on the host cell (Fig. 5.20A). Binding of the toxin to the host receptor stimulates **endocytosis** and internalization of the toxin within an endocytic vesicle (Fig. 5.20B). The B domain remains bound to the vesicle; however, the A domain is released into the host cytosol by reduction of the disulfide bond that joins it to the B domain (Fig. 5.20C). The A domain of the diphtheria toxin is an **adenosine diphosphate (ADP)-ribosyltransferase** that adds an ADP-ribosyl group specifically to elongation factor 2 (EF2), a component of eukaryotic ribosomes (Fig. 5.20D). Modification of the ribosome inhibits protein synthesis, and the cell dies. Death of tissues in the throat, the primary site of *C. diphtheriae* infection, results in production of a

ADP
adenosine diphosphate

EF2
elongation factor 2

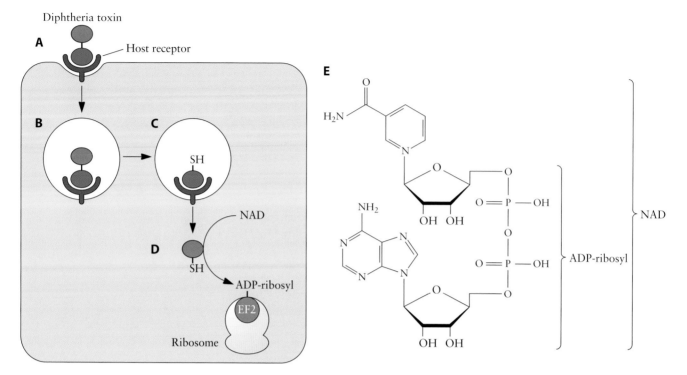

NAD
nicotinamide adenine dinucleotide

Figure 5.20 A–B exotoxin produced by *C. diphtheriae*. (**A**) The secreted toxin binds to a specific receptor on the host cell via the toxin B domain (blue oval). (**B**) Toxin binding stimulates endocytosis of the toxin. (**C**) The B domain remains bound to the endocytotic vesicle, and the A domain (red oval) is released into the host cytoplasm by reduction of disulfide bonds between the A and B domains (-SH, sulfhydryl group). (**D**) The A domain is an ADP-ribosyltransferase that catalyzes the transfer of ADP-ribosyl from nicotinamide adenine dinucleotide (NAD) (**E**) to EF2 on the host ribosome. ADP-ribosylation disrupts protein synthesis and destroys the host cell. doi:10.1128/9781555818890.ch5.f5.20

pseudomembrane consisting of host and bacterial cells. The condition can be lethal when the pseudomembrane obstructs the airways, and the disease was often referred to as "the strangler" during epidemics that swept through Europe in the 17th century.

The **enterotoxin** (an exotoxin that acts specifically on the intestines) produced by the intestinal pathogen *Vibrio cholerae* is an example of a multisubunit A–B toxin with ADP-ribosylation activity. Five B subunits of the secreted AB_5 cholera toxin form a binding site that is specific for the GM1 glycolipid in the membrane of host intestinal epithelial cells (Fig. 5.21). After endocytosis, the B subunits remain in the endocytotic vesicle; however, the A subunit is released into the cytosol. Similar to the diphtheria toxin, the cholera toxin A subunit is an ADP-ribosyltransferase; however, in this case, the ADP-ribosyl group is transferred to the host enzyme adenylate cyclase, thereby activating it (Fig. 5.21). Adenylate cyclase catalyzes the synthesis of the signal molecule cyclic adenosine monophosphate (cAMP), which, among other activities, controls ion transport systems in the small intestine. Increasing adenylate cyclase activity and therefore cAMP levels results in secretion of abnormally elevated levels of ions into the lumen of the intestine. Water follows the flow of ions, disrupting nutrient absorption and causing diarrhea.

cAMP
cyclic adenosine monophosphate

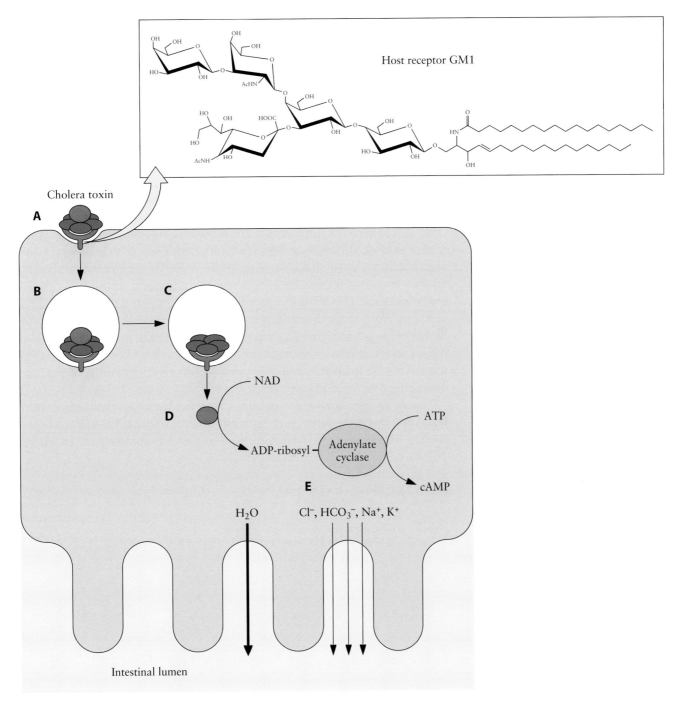

Figure 5.21 *V. cholerae* enterotoxin. The AB₅ cholera toxin binds to the GM1 glyco-lipid in the membrane of host intestinal epithelial cells (**A**) and is taken up into the cell by endocytosis (**B**). The B domain (blue ovals) remains bound to the endocytotic vesicle, and the toxin A domain is released into the host cytosol (**C**). The A domain catalyzes the transfer of ADP-ribosyl from NAD to the host enzyme adenylate cyclase (**D**). Activation of adenylate cyclase by the enterotoxin increases levels of cAMP, which stimulates ion secretion and loss of large amounts of water into the intestinal lumen (**E**). Diarrhea and dehydration ensue. doi:10.1128/9781555818890.ch5.f5.21

Not all A–B toxins have ADP-ribosyltransferase activity. Enterohemorrhagic *E. coli* is not an invasive pathogen; however, it produces an invasive A–B toxin. The Shiga-like toxin (so named because it is similar to the toxin produced by the pathogen *Shigella*) is released into the intestinal epithelium and travels through the bloodstream to the kidney. The A subunit is a glycosidase that specifically cleaves an adenine base from the 28S rRNA. Disruption of protein synthesis in endothelial cells of the kidney's blood vessels can lead to failure of kidney function (hemolytic-uremic syndrome), a condition that can be fatal. The tetanus neurotoxin released by *Clostridium tetani* is an A–B toxin that acts specifically on inhibitory interneurons. These neurons normally release glycine, which stops acetylcholine release from motor neurons, thereby allowing muscles to relax (Fig. 5.22A). The A component of the tetanus toxin, released following receptor binding and endocytosis, is an endopeptidase that cleaves proteins required for release of neurotransmitters at nerve synapses. Specifically, tetanus toxin blocks release of glycine, thereby preventing inhibition of motor neurons. This leads to uncontrolled muscle contraction (Fig. 5.22B). If respiratory muscles are affected, asphyxiation results. *C. tetani* is introduced by puncture wounds, and as a precautionary measure, a tetanus toxoid is injected to stimulate production of antibodies that neutralize toxin activity. **Toxoids**, which are purified, inactivated toxin proteins, are often effective vaccines (see chapter 11).

Superantigen toxins are peptides secreted by some pathogenic bacteria that activate large numbers of T cells. Normally, macrophages present antigens on their surfaces in a membrane complex called major

Figure 5.22 Normal release of inhibitory neurotransmitters (glycine) from inhibitory interneurons (**A**) is prevented by the tetanus neurotoxin secreted by *C. tetani* (**B**). As a consequence, motor neurons continue to release acetylcholine into the neuromuscular synapse, which stimulates muscle contraction. doi:10.1128/9781555818890.ch5.f5.22

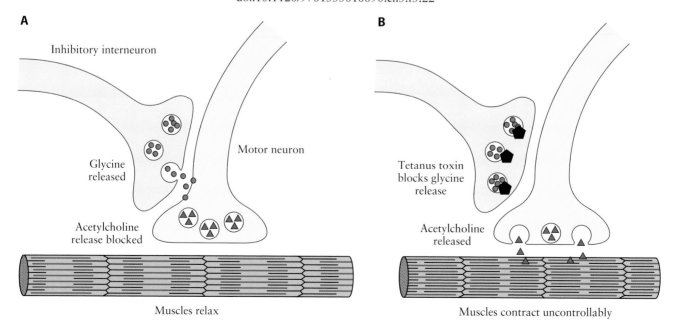

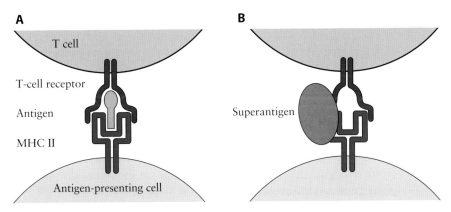

Figure 5.23 T cells are normally stimulated to produce cytokines by antigens presented by antigen-presenting cells such as macrophages (**A**). Superantigen toxins can activate large numbers of T cells by binding to both T cells and antigen-presenting cells, even in the absence of an antigen (**B**). doi:10.1128/9781555818890.ch5.f5.23

histocompatibility complex class II (MHC-II). Recognition of the antigen and the MHC-II complex stimulates a small number of T helper cells to produce cytokines that alert the host to the presence of a foreign invader (Fig. 5.23A). However, superantigen toxins bind to both T-cell receptors and MHC-II even in the absence of an antigen (Fig. 5.23B). Loss of specific recognition of antigen-presenting macrophages leads to indiscriminate activation of large numbers of T helper cells (a 2,000-fold increase) and therefore release of massive amounts of cytokines that stimulate an extensive systemic inflammatory response. Fever, diarrhea, vomiting, host tissue damage, and systemic shock result; hence, this disease is referred to as toxic shock syndrome. Highly virulent strains of *S. pyogenes* and *S. aureus* carry genes for superantigen toxins on prophage (integrated **bacteriophage**) in their genomes.

MHC-II
major histocompatibility complex class II

Identification of Bacterial Virulence Factors

Using traditional biochemical approaches, protein exotoxins were among the first virulence factors to be identified. Firstly, they are secreted proteins and therefore relatively easy to isolate and purify from bacterial cultures. Secondly, many toxins obviously damage or alter the function of host tissue, and therefore, isolated bacterial proteins with toxin activity can be assayed by introducing the toxin into animal models or cultured tissues and observing expected disease symptoms. The diphtheria A–B toxin is an example of an exotoxin that was characterized using this approach. Genetic approaches such as molecular cloning and mutagenesis have also contributed to our understanding of virulence factors. Cloning genes isolated from pathogens and introducing them into avirulent strains that consequently become virulent have led to the identification of genes encoding adhesins, invasins, and toxins. Similarly, targeted mutagenesis has revealed pathogen genes that are essential for host colonization and production of virulence factors.

The traditional approaches have contributed a wealth of knowledge about the mechanisms used by bacteria to cause disease. However, they require a priori identification of candidate gene or protein targets. On the other hand, whole-genome approaches have greatly facilitated the identification of novel factors that contribute to virulence. The genomes of most human-pathogenic bacteria have been sequenced. As new variants of the pathogens arise, their genomes are often sequenced and compared to reference sequences for identification of novel virulence traits. Comparative genomics has also been used to identify genes that are present in pathogenic but not nonpathogenic strains of a bacterial species and therefore may contribute to virulence.

Definitively, virulence factors are produced during infection, and therefore, functional genomic approaches can be used to identify the genes that are expressed when the bacterium has infected a host. Gene expression levels during an infection may be quantified relative to expression levels in vitro. Some of the functional genomic technologies were introduced in chapter 1. For example, microarray analysis and RNA sequencing are used to compare RNA produced by bacteria growing in vivo and in vitro. Proteomic approaches such as two-dimensional polyacrylamide gel electrophoresis and protein microarrays have been used to qualitatively and quantitatively compare proteins expressed by bacteria in a host and in culture media. Other whole-genome approaches that have led to identification of genes in several human pathogens that are expressed only during infection include in vivo-induced-antigen technology, signature-tagged mutagenesis, and in vivo expression technology.

In vivo-induced-antigen technology identifies proteins produced by microbial pathogens specifically during animal or human infections by screening genomic expression libraries with antibodies from infected individuals (Fig. 5.24). These proteins may be immunogenic, and therefore, the serum from patients infected with the pathogen contains antibodies that are directed against antigens produced by the pathogen. The serum is the source of the primary antibodies that are used to screen a genomic expression library. However, also present in the serum are antibodies that bind to proteins that are produced by the pathogen during growth both in vivo in the patient and in vitro. These are removed by incubating (adsorbing) the serum with pathogen cells (intact and lysed) that were cultured in vitro in laboratory medium. Antibodies that bind to proteins that are only produced during infection remain in the serum and are used to screen the library. Positive clones reveal genes encoding virulence factors that are produced in vivo or antigens that can be used for vaccine development. This approach has been used to identify vaccine candidates to protect against infections by uropathogenic and enterohemorrhagic strains of *E. coli*, *M. tuberculosis*, and *S. pyogenes* (Fig. 5.25). Potential antigens are further tested by immunizing animals and subsequently challenging them with the live pathogen.

Transposon mutagenesis of a pathogenic bacterium generates a library of random mutants that can be screened for loss of virulence. **Transposons** are small mobile DNA elements that are excised from a donor site at inverted terminal repeat sequences by the enzyme **transposase** and inserted

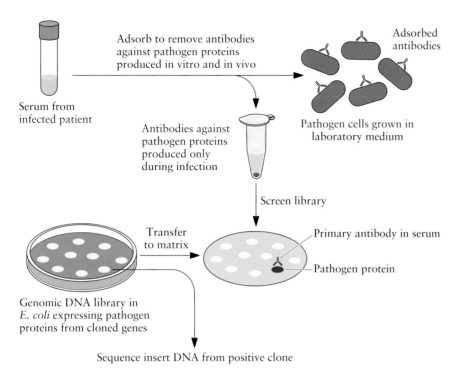

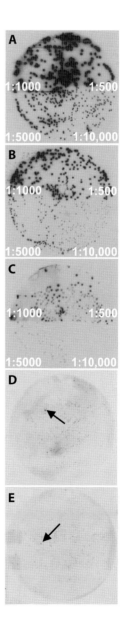

Figure 5.24 In vivo-induced-antigen technology to identify pathogen proteins produced during infection. doi:10.1128/9781555818890.ch5.f5.24

into a new recipient site in a DNA molecule (Fig. 5.26). For random transposon mutagenesis, the transposon is introduced into the bacteria on a plasmid that carries the transposase gene outside of the transposon sequence and that cannot replicate in the recipient bacterium (Fig. 5.27). Transposase expressed from the plasmid gene catalyzes the transfer of the transposon from the plasmid into a random site in the bacterial genome. Because the plasmid is not maintained in the culture, the transposase gene is lost and the transposon is stably maintained in the recipient genome. Inclusion of an antibiotic resistance gene between the inverted terminal repeats of the transposon allows selection of cells carrying a transposon. Each mutant in the library carries a transposon in a different sequence of the genome that likely disrupts the function of that sequence. To identify

Figure 5.25 Identification of proteins (antigens) produced by UPEC during urinary tract infections using in vivo-induced-antigen technology. Serum was collected from mice infected with UPEC and adsorbed against whole and lysed pathogen cells and library host *E. coli* cells cultured in vitro. (**A and B**) UPEC colonies probed with serum antibodies before (**A**) and after (**B**) adsorption. Serum dilutions are indicated. (**C**) Library host *E. coli* cells probed with adsorbed sera. (**D and E**) Antibodies remaining in the serum after adsorption were used to probe a UPEC genomic DNA library expressed in host *E. coli* cells. Arrows indicate colonies that expressed proteins from cloned UPEC genes that reacted with the serum antibodies collected from the infected mice. More than 400 serum-reactive clones were identified in the initial screen, and 83 of these were confirmed positive in a second screen. Mutagenesis confirmed that one of the positive proteins contributed to virulence of UPEC in bladder and kidney infections. Reprinted with permission from Vigil et al., *Infect. Immun.* 79:2335–2344, 2011. doi:10.1128/9781555818890.ch5.f5.25

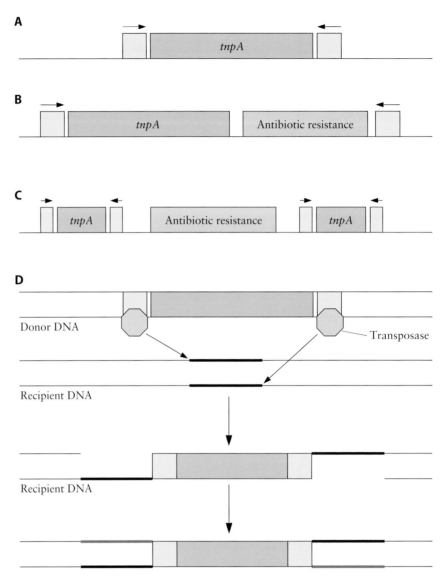

Figure 5.26 General features of transposable elements and transposition. (**A**) A simple transposable element known as an insertion sequence consists of a gene encoding transposase (*tnpA*) flanked by inverted terminal repeats (yellow boxes with black arrows). (**B**) Transposons are more complex transposable elements that carry additional genes such as genes encoding resistance to antibiotics. (**C**) Composite transposons carry additional genes between insertion sequence elements. (**D**) During transposition, the transposable element is excised from the donor DNA and inserted into a random site in the recipient DNA. Transposase excises the transposable element at the inverted terminal repeats and makes a staggered cut at a target site in the recipient DNA. After insertion of the transposable element, gaps flanking the insertion site in the recipient DNA are replicated (red lines), resulting in duplicated sequences. doi:10.1128/9781555818890.ch5.f5.26

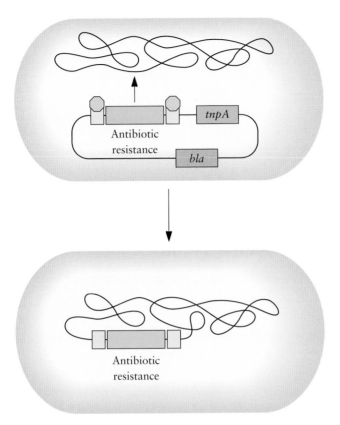

Figure 5.27 Transposon mutagenesis. The transposon is introduced into a bacterium on a plasmid that carries the transposase gene (*tnpA*) outside of the inverted terminal repeats. The *bla* gene encodes β-lactamase for selection of cells transformed with the plasmid by their resistance to ampicillin. Transposase catalyzes the transfer of the transposon into a random site in the bacterial genome. After transposition, the plasmid is not maintained in the culture because it cannot be replicated. The antibiotic resistance gene carried on the transposon enables selection of bacteria that have acquired the transposon. doi:10.1128/9781555818890.ch5.f5.27

virulence genes, thousands of mutants are tested for loss of the ability to cause disease in animal models or cultured tissues. The site of transposon insertion is determined in avirulent mutants to identify genes that are essential for virulence. Although this approach has been used extensively, screening of the library for mutants that no longer cause disease can be time-consuming.

Signature-tagged mutagenesis is a modified transposon mutagenesis approach for identification of genes that are essential for growth in a host animal. Mutants that carry a transposon in an essential gene will not be recovered from an infected host. These mutants are identified by comparing mutants that are recovered from the infected host with mutants that were present in the initial inoculum (mixture of mutants that was used to infect the host). To facilitate mutant identification, short signature tags consisting of random sequences of about 40 base pairs (bp) are inserted in the transposon (Fig. 5.28A). This generates a mixture of transposons with different signature tags that is transformed en masse into the pathogen to generate a library of mutants. Each mutant has a unique signature tag that

A

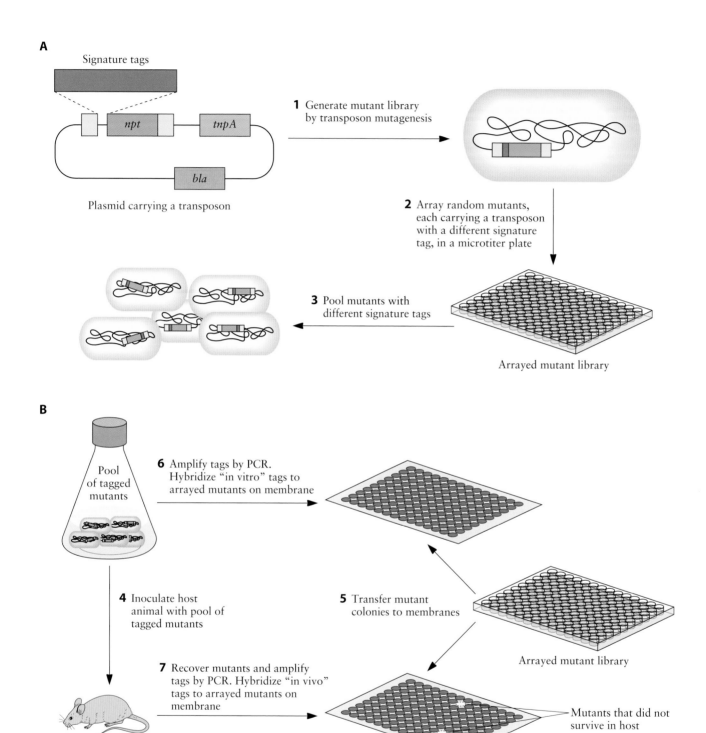

Figure 5.28 Signature-tagged mutagenesis to identify bacterial genes that are essential for survival in a host. (**A**) Generation of a library of signature-tagged transposon mutants. Signature tags are random sequences of approximately 40 bp. See Fig. 5.26 and 5.27 for descriptions of transposons and transposition. (**B**) Screening of the mutant library for genes that are essential in vivo. Mutants on the membrane that do not hybridize to the in vivo tags (amplified from mutants recovered from the host) represent strains that did not survive. The transposon in this mutant potentially disrupts a gene that is essential for survival in vivo. *bla*, β-lactamase gene that confers resistance to ampicillin; *tnpA*, transposase gene; *npt*, neomycin phosphotransferase gene for selection of transposon mutants on the antibiotic kanamycin. doi:10.1128/9781555818890.ch5.f5.28

corresponds to the site of genome insertion. The entire library is used to inoculate a host animal, and after an infection has been established, the bacteria are recovered from the host (Fig. 5.28B). Genomic DNA purified from the recovered bacteria is used as a template for PCR amplification of the signature tags. The mixture of PCR amplified tags is labeled and used as a probe to screen colonies of mutants in the initial inoculum, which are arrayed on a membrane, for signature tags that hybridize to probe. The goal is to identify mutants in the initial inoculum that do not hybridize to the probe and therefore represent mutants that are not present in the mixture of strains recovered from the infection. The region flanking the transposon in the nonhybridizing mutants is sequenced to identify genes that are essential for growth in vivo. Further testing is required to confirm the result. This procedure has been used extensively to identify virulence genes in many different human pathogens for which a suitable animal model is available (often a mouse or a pig).

Identification of virulence genes by cloning has also been improved in a procedure known as **in vivo expression technology**. In this strategy, bacterial genes that are induced during infection are identified by testing for promoters that are activated in vivo. A genomic library is generated by inserting fragments of pathogen genomic DNA into an appropriate vector upstream of a gene that is essential for bacterial growth in an animal host (for example, a gene required for purine biosynthesis) (Fig. 5.29). The promoterless copy of the essential gene on the vector replaces the chromosomal copy, which was previously deleted. The clone library is then injected into a model animal. Survival in the host animal depends on expression of the vector-borne essential gene from an active promoter on a cloned genomic fragment. Surviving bacteria are recovered from the animal, and the cloned promoters are tested for constitutive expression. A second reporter gene (e.g., *lacZ*) is included downstream from the essential gene on the vector such that it is in the same operon as the essential gene. This enables differentiation of promoters that are induced in vivo (those with low reporter gene activity on plates containing rich media) from those that are constitutively expressed (those with high reporter gene activity on plates). In vivo-induced genes are then identified by sequencing the region downstream of the promoter in the bacterial genome.

Evolution of Bacterial Pathogens

Where do pathogens come from? How do new pathogens arise? These are compelling questions that intrigue microbiology researchers. Comparative analysis of the genome sequences of pathogens and closely related nonpathogens not only has revealed genes encoding virulence factors but also has provided some insight into the evolution of pathogens. Changes in the nucleotide sequence of a bacterial genome may occur by mutations such as single nucleotide changes (point mutations) or sequence deletions, duplications, or rearrangements that alter the function or regulation of genes. Slow accumulation of mutations over time can lead to exploitation of new habitats, including new hosts, or enhanced resilience, for example, against host immune attack. Genome sequence analysis has also revealed

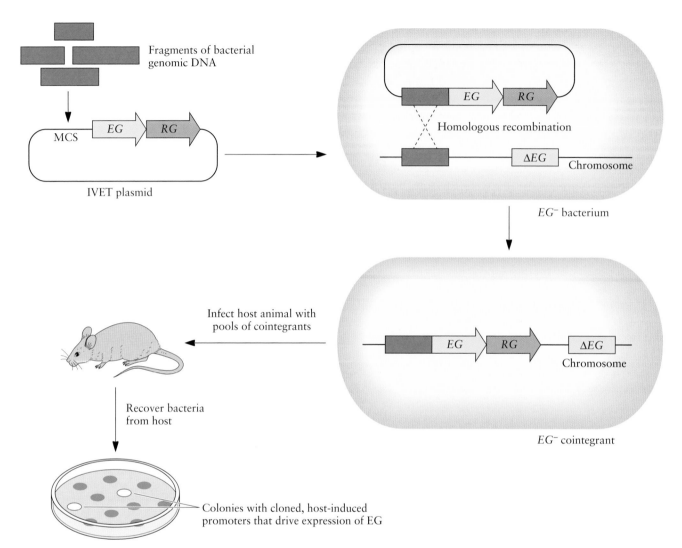

Figure 5.29 In vivo expression technology (IVET) to identify genes that are expressed in a host. To generate an IVET library, fragments of bacterial genomic DNA are inserted into the multiple-cloning site (MCS) of an IVET plasmid upstream of a promoterless essential gene (*EG*) that is required for growth in an animal host. The IVET library is introduced into a strain of pathogenic bacteria in which the essential gene was previously disrupted (*EG⁻*). Recombination (broken lines) between the cloned genomic DNA fragment and the homologous sequence in the bacterial genome results in integration of the IVET plasmid into the genome of the *EG⁻* strain (cointegrant). The library of *EG⁻* cointegrants, each with the IVET plasmid integrated at a different site in the genome, is introduced into a host animal. Cointegrant strains that carry an active promoter upstream of the integrated IVET plasmid or within the cloned fragment express a functional EG and therefore may survive in the host. Bacteria are recovered from the animal and then plated on a rich medium (i.e., contains essential nutrients that the bacterium cannot synthesize because the *EG* has been disrupted). Recovered strains that do not express the reporter gene (*RG*) in vitro (white colonies) are of interest, as these carry a promoter upstream of the integrated *EG* that is only induced during infection. Blue colonies carry cloned fragments with promoters that drive constitutive expression of *EG* and *RG*. doi:10.1128/9781555818890.ch5.f5.29

that pathogen genomes often carry clusters of genes, sometimes several gene clusters, that are not present in avirulent strains. These gene clusters are known **pathogenicity islands** because they carry virulence genes encoding adhesins, toxins, capsules, and other virulence-associated proteins. Pathogenicity islands have some characteristic features that aid in their identification and suggest that they have been acquired from other organisms (Fig. 5.30). Although rare, acquisition of foreign DNA sequences, also known as horizontal (or lateral) gene transfer, can result in rapid attainment of new traits.

Integration of foreign DNA into a genome is suggested by the differences in GC/AT ratio between the pathogenicity island and the rest of the genome sequence. In addition, pathogenicity islands contain sequences that indicate horizontal transmission via bacteriophage, transposons, or plasmids (Fig. 5.30). Bacteriophage-mediated transfer **(transduction)** is revealed by the presence of sequences for viral integration and by linkage to tRNA genes, which are often sites for bacteriophage insertion. Some bacteriophage have life cycles with two phases, a lysogenic and a lytic phase (Fig. 5.31A). During the **lysogenic phase**, viral DNA injected into the bacterium is integrated into the bacterial genome (referred to as a prophage) by the enzyme **integrase** (Fig. 5.31A). Integrase coding sequences are often apparent in pathogenicity islands. The bacteriophage DNA may remain in the bacterial genome for long periods, during which virulence genes may be expressed, or permanently if mutations occur that block prophage excision and progression to the **lytic phase**. Integrated bacteriophage can introduce toxin genes. For example, virulent strains of *Corynebacterium diphtheriae* carry a prophage that encodes the diphtheria toxin, while avirulent strains do not.

During the lytic phase of a bacteriophage life cycle, viral DNA is excised from the genome and packed into viral **capsids** (Fig. 5.31A). Rarely, the viral DNA is imperfectly excised, and the packaged DNA includes fragments of flanking bacterial DNA. When the packaged, aberrant phage is released following lysis of the host cell, it may infect and integrate its DNA into the genome of another bacterial cell, thereby transferring the bacterial DNA to a new host. This process is known as specialized transduction (Fig. 5.31B). During generalized transduction, fragments of bacterial DNA are mistakenly packaged into viral capsids instead of bacteriophage DNA (Fig. 5.31C). When the mispackaged phage infects a

Figure 5.30 General features of pathogenicity islands. Several features suggest that these sequences were acquired by horizontal gene transfer via bacteriophage, transposons, plasmids, or recombination. These include a G+C content that is different from that of the rest of the genome, insertion within a tRNA gene, direct repeat sequences (DR), a sequence homologous to bacteriophage integrase (*int*) genes, and insertion sequences (IS). Pathogenicity islands carry genes that encode virulence factors such as adhesins, invasins, or toxins. Adapted with permission from Schmidt and Hensel, *Clin. Microbiol. Rev.* **17:**14–56, 2004. doi:10.1128/9781555818890.ch5.f5.30

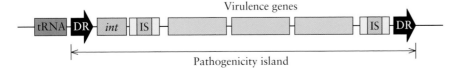

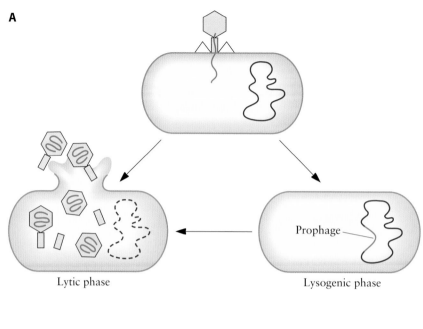

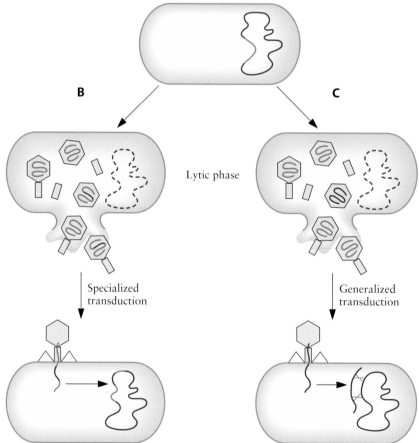

Figure 5.31 Some bacteriophages, such as bacteriophage lambda, which infects *E. coli*, have life cycles with a lytic phase and a lysogenic phase (**A**). During the lytic phase, the virus reproduces and releases new viruses by lysing the host cell. During the lysogenic phase, the genome of the virus is integrated into the host genome and is replicated with the host DNA. An integrated bacteriophage genome is referred to as a prophage. The prophage may excise from the genome and initiate the lytic cycle to release new progeny. Rarely, excision of the prophage is imprecise and host DNA flanking the prophage may be packaged into the new viruses (**B**). Following their release at the end of the lytic cycle, the aberrant viruses may infect new host cells and introduce the bacterial genes acquired from the previous host. This phenomenon, known as specialized transduction, transfers a limited number of genes adjacent to the prophage insertion site between bacterial cells. During the lytic cycle, host DNA may be erroneously packaged into the new viruses instead of viral DNA (**C**). When the aberrant viruses infect a new host, they transfer the DNA from the previous bacterial host. The transferred bacterial DNA may be integrated into the genome of the new host by recombination. This process, known as generalized transduction, can transfer any gene(s) between bacterial cells.

doi:10.1128/9781555818890.ch5.f5.31

new host, recombination between homologous regions on the transferred DNA and the host genome can result in integration of new sequences.

The presence of transposon insertion sequence elements and direct repeats flanking the pathogenicity island may indicate acquisition by **transposition** (Fig. 5.30). **Insertion sequence elements** are mobile genetic elements that are delineated by terminal inverted repeat sequences and carry a gene encoding transposase that binds to the inverted repeats and mediates transfer of the transposon to a new site (Fig. 5.26). Larger transposons often carry additional genes such as toxin or antibiotic resistance genes (Fig. 5.26). The latter contribute to the rapid spread of antibiotic resistance among organisms. The direct repeats flanking the pathogenicity island arise as a consequence of transposition. During transposition, transposase makes a staggered cut in each DNA strand at a random insertion site in the recipient genome (Fig. 5.26). After the transposon is inserted at the new site, the single-stranded gaps adjacent to the insertion site are filled in by replication, resulting in duplicated flanking sequences.

Most pathogen genomes carry one or more pathogenicity islands. The genome of the intestinal pathogen *S. enterica* carries two large pathogenicity islands. Each encodes a type III secretion system and several effector proteins that enable *Salmonella* to invade and survive within host intestinal epithelial cells (M cells). The effector proteins expressed from *Salmonella* pathogenicity island 1 (SPI-1) induce membrane ruffling and engulfment following attachment of the bacterium to M cells (Fig. 5.5). After the bacterium has been taken up in a phagosome, effector proteins encoded on SPI-2 are secreted through the type III secretion apparatus into the host cytosol to prevent phagosome–lysosome fusion. Toxin genes are often found on pathogenicity islands, for example, the genes encoding the Shiga-like toxin produced by enterohemorrhagic *E. coli*, the botulinum neurotoxin produced by *Clostridium botulinum*, and the diarrhea-inducing cholera toxin produced by *V. cholerae*. Analysis of pathogen genome sequences has revealed that horizontal transfer of these toxin genes has likely contributed to emergence of new pathogens.

Plasmids are another source of virulence traits that can be readily transferred among bacteria by **conjugation** (see chapter 1). If a newly acquired plasmid can be maintained in a recipient host, the genes encoded on the plasmids may enable the bacterium to adapt to new environments, such as the host environment. Virulence plasmids found in many human pathogens carry genes encoding virulence factors. For example, plasmid genes may confer enhanced ability to obtain nutrients or resist antibiotics, or they may encode adhesins, invasins, or toxins. *Bacillus anthracis*, the causative agent of anthrax disease, carries two plasmids that are essential for virulence. One plasmid, pXO1, carries genes encoding a potent A–B toxin, and a second plasmid, pXO2, carries capsule biosynthesis genes. Plasmids similar to pXO1 and/or pXO2 have been found in clinical isolates of the closely related but relatively benign species *Bacillus cereus*. Patients harboring the plasmid-carrying *B. cereus* strains exhibited symptoms of anthrax disease. Although pXO1 and pXO2 are not self-transmissible, they can be mobilized into new host bacteria by other, natural plasmids that carry transfer functions.

SPI-1
Salmonella pathogenicity island 1

Treatment of Bacterial Infections

Fortunately, chemotherapeutic agents have been developed to control bacterial infections that were previously often fatal (Milestone 5.1). **Antibiotics** are diverse, secondary metabolites secreted by some soil fungi and bacteria (mainly actinomycetes). They are produced naturally in soils to kill or inhibit growth of bacteria that compete for limited nutrients. Today, prescribed antibiotics are more likely to be synthetic derivatives of those original natural products that carry novel functional groups that confer broader toxicity, increased stability, or resistance to degradation. An important property of a useful antibiotic is **selective toxicity**: the antibiotic kills (bacteriocidal agents) or inhibits (bacteriostatic agents) the pathogen but does not harm the host. Selectively toxic antibiotics target specific structures or proteins that are unique to bacteria and disrupt their function.

Common bacterial cell targets that are disrupted by these antibiotics are ribosomes, cell wall synthesis enzymes, and, to a lesser extent, enzymes required for nucleic acid synthesis such as RNA polymerase and DNA gyrase. Bacterial and eukaryotic ribosomes differ in their structures, as they are comprised of different proteins and RNAs. Many antibiotics inhibit ribosomal function and thereby protein synthesis, which is essential for

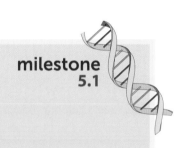

THE DISCOVERY OF PENICILLIN

milestone 5.1

On the Antibacterial Action of Cultures of a *Penicillium*, with Special Reference to Their Use in the Isolation of *B. influenzae*
A. FLEMING
Br. J. Exp. Pathol. **10**:226–236, 1929

Penicillin as a Chemotherapeutic Agent
E. CHAIN, H. W. FLOREY, A. D. GARDNER, N. G. HEATLEY, M. A. JENNINGS, J. O. EWING, AND A. G. SANDERS
Lancet ii:226–228, 1940

The first commercial antibiotic used to treat infectious disease in humans was penicillin, produced by the fungus *Penicillium*. In 1928, the Scottish physician Alexander Fleming discovered, by accident, that *Penicillium* secreted a compound that inhibited the growth of *S. aureus* in laboratory cultures. We now know that penicillin structurally resembles the terminal dipeptide in the side chain of peptidoglycan, the major component of the bacterial cell wall. As a structural analogue of the natural substrate, penicillin binds to the enzyme transpeptidase and prevents it from catalyzing the formation of peptide bonds between adjacent peptide side chains in peptidoglycan. By preventing peptide cross-linking, penicillin weakens the cell wall and renders the bacterium susceptible to lysis. Bacteria have a unique cell wall composition that is not present in humans, and therefore, penicillin meets the criterion of selective toxicity. Although Fleming is credited with the discovery of penicillin, he did not show that it was an effective treatment for human infections. In 1939, Howard Florey and Ernst Chain of Oxford University extracted penicillin from the fungus and injected it into mice that were infected with *S. pyogenes*, *S. aureus*, or *Clostridium septicum*. The mice survived. This was followed by clinical trials in the early 1940s that proved its effectiveness in humans, and penicillin began to be produced on a large scale in 1944. There was strong motivation for the rapid development of this drug because infections were the leading cause of death of soldiers in the battlefields of World War II. Since the discovery of penicillin, many other antibiotics have been discovered that are effective against human pathogens.

cell survival. For example, tetracycline binds to the 16S rRNA of the bacterial small ribosomal subunit and blocks binding of tRNA (Fig. 5.32A). Erythromycin binds to the 23S rRNA in the large ribosomal subunit and blocks translocation of peptidyl-tRNAs during translation (Fig. 5.32B). Penicillin and other structurally similar antibiotics bind to the enzyme transpeptidase and prevent cross-linking of peptidoglycan, which weakens the bacterial cell wall (Milestone 5.1). In addition to penicillin, glycopeptides, such as vancomycin, and several other antibiotics inhibit different steps in bacterial cell wall synthesis. Some peptide antibiotics target the cell membrane, for example, polymyxin produced by *Paenibacillus polymyxa*. Peptide antibiotics form holes in the membrane that permit leakage of cell contents. Unregulated ion movement across the membrane disrupts membrane potential that is essential for cell processes such as nutrient transport and ATP synthesis. Membrane-disrupting antibiotics are restricted to topical use, as they also target human cell membranes and therefore are not selectively toxic. As our understanding of virulence mechanisms increases, new targets for antibacterial agents can

Figure 5.32 Several different antibiotics target bacterial ribosomes. (**A**) Tetracycline binds to the 16S rRNA in the small (30S) ribosomal subunit and blocks binding of tRNA; (**B**) erythromycin binds the 23S rRNA in the large (50S) ribosomal subunit and blocks translocation of the growing polypeptide from the peptidyl-tRNA in the peptidyl (P) site to the aminoacyl-tRNA in the aminoacyl (A) site. doi:10.1128/9781555818890.ch5.f5.32

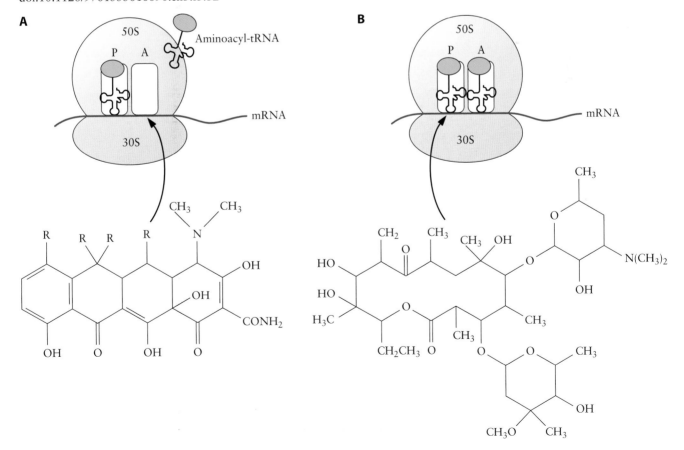

be identified. Some promising targets that are currently being explored are quorum sensing systems that control the expression of virulence factors and toxin secretion systems.

Infectious disease remains a significant cause of human mortality in underdeveloped countries. In highly developed regions (e.g., North America, Western Europe, and Australia), microbial infections are no longer a leading cause of death, due in large part to the availability of antibiotics, as well as improvements in sanitation, health care, and immunizations. The emergence of pathogen strains that are resistant to multiple antibiotics, however, is a serious threat to our ability to treat bacterial infections. Antibiotic resistance is a natural phenomenon, but it is amplified and accelerated by the overuse and misuse of antibiotics. Bacteria exposed to antibiotics must adapt or die, and therefore, the use of antibiotics selects for antibiotic-resistant strains.

There are three general mechanisms by which bacteria may acquire resistance to an antibiotic (Fig. 5.33). (i) A mutation may occur that alters the bacterial structure that is the target for the antibiotic. For example, mutations may alter the structure of RNA polymerase such that the antibiotic rifampin no longer binds but the ability of the enzyme to catalyze RNA synthesis is unaffected. (ii) The antibiotic may be enzymatically degraded or modified by addition of a functional group (a methyl, acetyl, or phosphoryl group), thereby inactivating it. β-Lactamase is an example of an enzyme secreted by some bacteria that cleaves the β-lactam ring of penicillin and cephalosporins. Synthetic derivatives of these antibiotics that

Figure 5.33 Mechanisms of antibiotic resistance.
doi:10.1128/9781555818890.ch5.f5.33

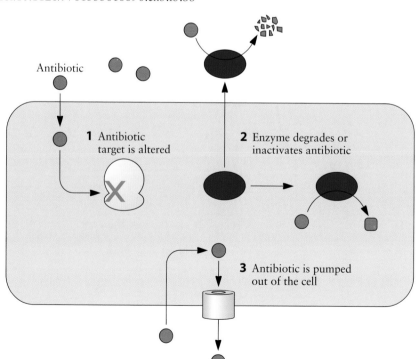

are more resistant to degradation were developed in response to the rise in β-lactamase-producing bacteria; however, strains resistant to the new antibiotics often appear within a few years of introduction. (iii) The antibiotic may be prevented from reaching its target by blocking of antibiotic transport across the cell membrane or by removal of the antibiotic from the cell by an efflux pump. **Drug efflux pumps** are specialized membrane proteins that continuously export the antibiotic from the cell. Often one pump can remove several different antibiotics. Enzymes that degrade or modify antibiotics and proteins that form drug efflux pumps are often encoded on mobile genetic elements such as plasmids or transposons that are acquired by horizontal gene transfer. This can lead to rapid spread of antibiotic resistance among bacteria, a growing public health concern.

Emerging antibiotic resistance is most evident in the bacterial pathogens that cause most human infections, for example, bacteria that cause diarrheal diseases, respiratory tract infections, meningitis, and sexually transmitted infections. Antibiotic resistance typically develops within a few years of introduction of a new antibiotic, and some strains of pathogens are resistant to all antibiotics commonly used to treat the disease. This renders first-line antibiotics, which are cheaper and less toxic, ineffective. Multidrug-resistant strains of *S. aureus* illustrate this point. In 1946, when penicillin was first in widespread use, all isolates of *S. aureus* were sensitive to the drug. By the late 1940s, hospital-acquired strains began to show resistance to penicillin, and the synthetic penicillin derivative methicillin was then introduced to treat *S. aureus* infections. Within a decade of introduction of methicillin, clinical isolates resistant to methicillin and all derivatives of penicillin began to emerge. Community-associated **methicillin-resistant *S. aureus*** is now common. Very few drugs are effective against this pathogen. So far, strains resistant to the "last resort" antibiotic (given when other treatments have failed) vancomycin are rare.

Viral Infections

Almost all organisms—plants, animals, fungi, protists, and bacteria—can be infected with viruses. Viruses are noncellular particles **(virions)** that contain genetic material but cannot replicate their genomes outside of a host cell. They carry genes but require host cell machinery to produce proteins. That is, they are obligate parasites. Although they are small and simple (the smallest known virus carries genes for two proteins), once they infect a host, they reproduce very efficiently. They have a devastating impact on human health, and a vast number of viruses can infect humans. Most of us are familiar with the symptoms of rhinoviruses and other viruses that cause the common cold and influenza virus, which causes the respiratory disease of the same name. Other common viral diseases are chicken pox (varicella-zoster virus), cold sores (herpes simplex virus 1 [HSV-1]), and warts (human papillomavirus). More serious diseases are caused by hepatitis virus (liver infections), HIV (AIDS), rabies virus, poliovirus, and many others. Viruses that infect bacteria (bacteriophage) can also have a great impact on human health, as these often introduce genes for production of toxins when integrated into the bacterial genome. Some

HSV-1
herpes simplex virus 1

human viruses can integrate their genomes into ours. The human genome contains a large number of sequences of ancient viral origin that have likely influenced human evolution (Box 5.1). Viruses that can integrate into the genomes of humans are being explored as gene delivery vehicles to replace inherited, defective genes.

The basic structure of a virus is nucleic acid contained within a protein coat (capsid). This is referred to as a **nucleocapsid**. A capsid, which is composed of protein subunits called capsomers, may be a simple helical structure that wraps around and protects the nucleic acid genomes of filamentous viruses (Fig. 5.34A), or it may take the shape of an **icosahedron**, a structure with 20 faces made up of equal-size triangles (Fig. 5.34B). Icosahedral capsids can be constructed from a minimum number of protein subunits, enabling efficient use of limited resources to synthesize many virions. **Enveloped viruses** additionally have a membrane that is derived from a host membrane as the viruses are leaving the cell and is often studded with membrane-bound **(spike)** proteins (Fig. 5.34C). The

HERVs
human endogenous retroviruses

box 5.1
Viruses: Are They All Bad?

Most of us have experienced at least the relatively mild discomfort of some viral infections, such as those that cause the common cold or chicken pox. Other viral infections are devastating. But have humans derived any benefit from viral infections? The sequence of the human genome has revealed that more than 8% of our DNA is of viral origin. About 100,000 DNA segments carry the remnants of ancient viral infections. This is likely an underestimate, as these sequences have been identified because they bear some resemblance to existing viruses. These viruses, specifically retroviruses, infected our ancient ancestors millions of years ago and became permanently integrated into the genome when mutations were acquired that prevented the virus from reproducing. The infected cells were germ cells (reproductive cells), and hence, the viral sequences were passed to subsequent generations and have persisted in our genomes. The viral sequences are known as human endogenous retroviruses (HERVs).

Both ends of the retroviral genomes carry LTRs that enable insertion of

the double-stranded viral DNA, which was generated from the single-stranded RNA retroviral genome by reverse transcriptase, into a host chromosome. In addition, the LTRs carry promoters that drive expression of viral genes. The LTRs were used to identify HERVs in human genomic sequences and to estimate when they were inserted by comparing sequences of LTRs among endogenous retroviruses in the genomes of other vertebrates (ERVs are found in every class of vertebrate but appear to be restricted to vertebrates). The promoter of an integrated LTR can also drive expression of host genes when inserted into a 5′ untranslated region. In fact, these ancient sequences have been major drivers of human evolution. They control expression of some genes (for example, genes that are expressed in reproductive tissue or genes that are controlled by the tumor suppressor protein p53), have provided alternate splice sites when inserted into protein coding sequences, and have caused chromosomal rearrangements by recombination between integrated viral sequences. Rearrangements led

to gene duplications and deletions that have affected the development of important human functions. For example, HERV-mediated duplications have contributed to the multigene families that encode components of the human immune system (MHC and T-cell receptors).

In many cases, viral proteins have been co-opted by the host and have novel functions in human reproduction and development. In fact, it seems that we owe our ability to reproduce to one such viral protein. Many enveloped viruses gain entry to host cells by fusion of viral and host cell membranes, a process that is often initiated by a viral protein. This was likely the original function of the protein syncytin, encoded on a HERV. In humans, syncytin is responsible for fusion of the placenta that surrounds a developing fetus with the mother's uterus. This fusion is essential for transport of nutrients to support the fetus. A similar protein is encoded in the genomes of other primates, suggesting that the retrovirus likely infected an ancestor that was common to all primates. Thus, while viruses are well known to cause disease, they have also contributed to important innovations in primate evolution.

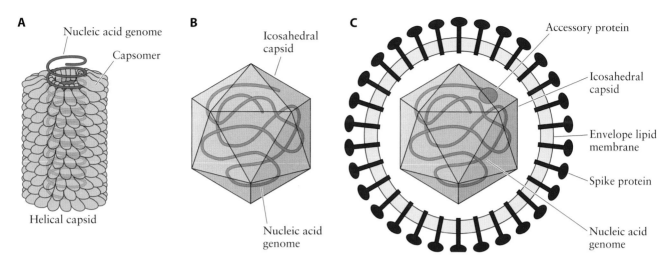

Figure 5.34 Basic viral structure. (**A**) Nucleocapsid of a filamentous virus; (**B**) nucleocapsid of an icosahedral virus; (**C**) an enveloped virus.
doi:10.1128/9781555818890.ch5.f5.34

nucleocapsids may or may not contain accessory proteins, such as specialized polymerases that initiate viral genome replication following infection.

The nucleic acid genomes may be RNA or DNA, single or double stranded, linear or circular, and whole or segmented. Viral single-stranded RNA genomes may be positive stranded or negative stranded. **Positive-sense RNA** is the same sense as mRNA and can be directly translated into proteins by host ribosomes, while **negative-sense RNA** must first be copied to produce a positive-sense strand before translation can occur. Poliovirus is an example of a virus with one linear, single-stranded, positive-sense RNA chromosome enclosed within an icosohedral capsid. Influenza virus carries eight negative-sense RNA molecules within an enveloped capsid. Human papillomavirus has a single, circular, double-stranded DNA genome. The structure of the genome determines the mechanism by which the viral genes are expressed and the viral genomes are replicated, and it is the basis for the major viral classification scheme known as the Baltimore classification system after virologist David Baltimore, who first proposed it.

All viruses require a host cell for reproduction. A basic viral infection cycle consists of recognition and attachment to a host cell, entry into the host cell, and utilization of host enzymes and resources to produce viral proteins and to replicate the viral genome (Fig. 5.35). The genetic material and capsids are then assembled to produce virions that leave the cell and infect new host cells.

Attachment and Entry

The first step in a viral infection cycle is the binding of a virion to a host cell. Virus attachment requires specific interaction between a protein on the surface of the virus and a receptor on the surface of the host cell. The interaction occurs by random collision, as viruses are nonmotile. Host

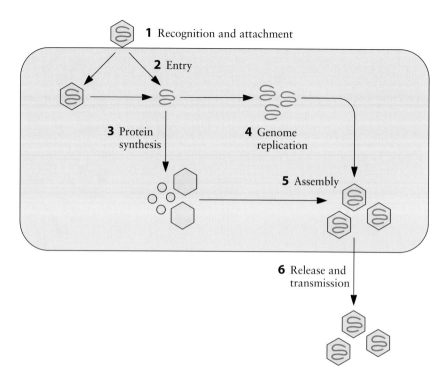

Figure 5.35 Basic steps in a viral life cycle. (**1**) Attachment of a virion to a host cell is mediated by a specific protein on the surface of the virus and a receptor on the host cell. (**2**) The nucleic acid of the virus is released from the capsid into the host cell. Uncoating of the nucleic acid may occur during or after entry of the virion. (**3**) Viral genes are expressed to produce large quantities of proteins that assist in replication of viral nucleic acids or are viral structural proteins. (**4**) Viral nucleic acids are replicated to produce many copies of the viral genome. (**5**) Viral capsids and genomes are assembled within the host cell to generate new virions. (**6**) Infectious virions are released and may be transmitted to new cells. doi:10.1128/9781555818890.ch5.f5.35

surface receptors are often carbohydrate groups on glycolipids or glycoproteins that function in important cellular processes such as signaling or adhesion. Viruses exploit these surface proteins as attachment and entry sites. The molecular interaction is very specific and determines which hosts and host cell types the virus can infect (host tropism). For example, the envelope of influenza A virus contains a spike protein known as **hemagglutinin** that interacts with a sialic acid residue on a glycoprotein in the membranes of secretory cells, especially those of the respiratory tract. Hemagglutinin in the envelope of avian influenza A virus specifically binds to $\alpha(2,3)$-linked sialic acid on bird receptors (Fig. 5.36A), whereas human influenza A virus interacts with the $\alpha(2,6)$-linked sialic acid residues on human proteins (Fig. 5.36B). Mutations that change the structure of the sialic acid-binding pocket of hemagglutinin can enable influenza viruses to infect different hosts (Box 5.2). The nonenveloped rhinovirus interacts with a host receptor via its capsid protein. The capsomer forms a depression ("canyon") into which fits the host membrane glycoprotein intercellular adhesion molecule 1 (ICAM-1).

Following attachment, the next step in the infection cycle is host cell penetration. In this step, the virus enters the host cell and the protective

ICAM-1
intercellular adhesion molecule 1

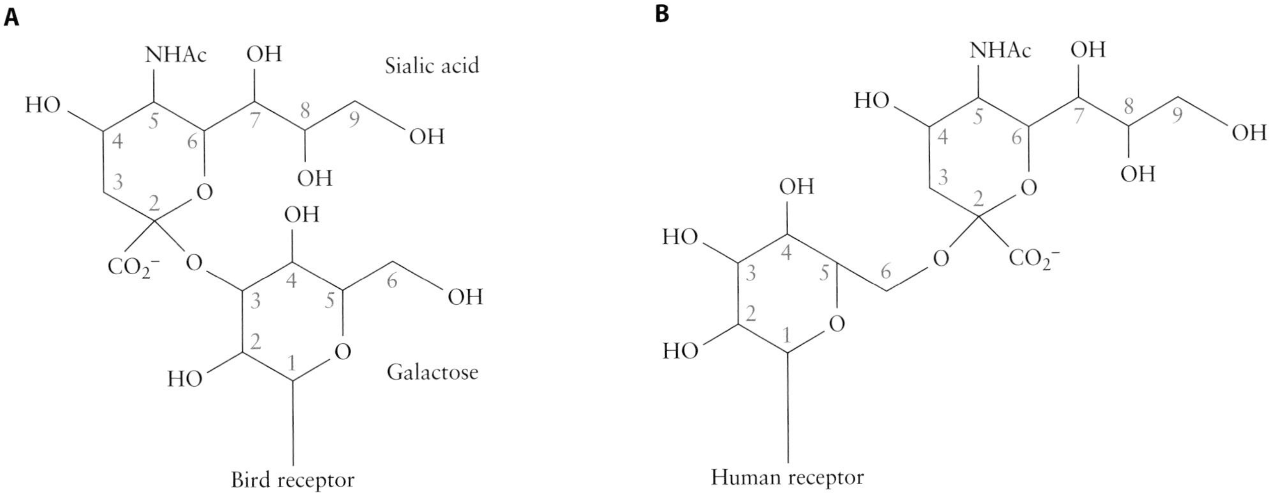

A

NHAc   OH   Sialic acid

HO

5   6   7   8   9   OH

4

3   2   O   OH

CO_2^- O   OH

4

3   5   6   OH

2   O   1

HO   Galactose

Bird receptor

B

NHAc   OH

HO

5   6   7   8   9   OH

4

OH   3   2   O   OH

HO   4

3   5   6   O CO_2^-

2   1   O

HO

Human receptor

Figure 5.36 Receptors for influenza A virus hemagglutinin spike protein on bird (**A**) and human (**B**) cells. Ac, acetate. doi:10.1128/9781555818890.ch5.f5.36

capsid is removed, exposing the viral genome for replication and expression of viral genes. Release of viral nucleic acid from the capsid is known as uncoating. Viral entry into animal cells occurs by one of two general mechanisms: (i) fusion of viral and host membranes or (ii) receptor-mediated endocytosis.

After binding to host surface receptors, some enveloped viruses may penetrate by fusion of their membranes with the membrane of the host cell (Fig. 5.37). Membrane fusion is promoted by specific proteins present in the viral envelope or in the host cell membrane and releases the

Figure 5.37 Viral penetration of host cells by membrane fusion. Nucleocapsids enter the host cell by membrane fusion and are uncoated in the host cytoplasm (**A**) or at the nucleus (**B**). doi:10.1128/9781555818890.ch5.f5.37

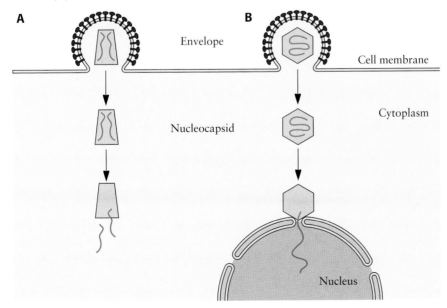

box 5.2
Why Was the Spanish Flu of 1918 So Deadly?

Many infectious diseases have had an enormous impact on human history. The bubonic plague, caused by the bacterium *Yersinia pestis*, had a great socioeconomic impact on medieval European societies. In the 1300s, one-third of Europe's population succumbed to the disease, causing serious labor shortages that resulted in unharvested fields, skyrocketing food prices, and a starving population. Smallpox virus, introduced to the indigenous peoples of North America by European explorers and settlers, wiped out much of the native population and likely facilitated European conquest and expansion in that continent. The 1918 influenza pandemic, the so-called "Spanish flu," is considered to be the most devastating event in human history.

Contemporary seasonal influenza A virus infections kill tens of thousands of people every year, mainly young children, the elderly, and immunocompromised individuals. However, tens of millions of people died during the 1918 influenza pandemic, which lasted 3 years and infected nearly 30% of the world's population. Most of the victims were young adults. The infections were also unusually severe. Within days of the first symptoms, infected individuals experienced massive bleeding of the lungs. The infection progressed rapidly, and victims typically fell to the disease within 5 days. Why was the 1918 strain of influenza virus so virulent?

Nearly a century later, researchers at the U.S. Armed Forces Institute of Pathology were able to obtain samples of this extinct virus from the lung tissues of victims of the 1918 pandemic, from preserved autopsy material and from an individual buried in the Alaskan permafrost (Taubenberger et al., 1997; Reid et al., 1999). Although the viral RNA was somewhat degraded, they determined the sequence of several key genes and eventually the entire genome sequence of the 1918 influenza virus extracted from these preserved samples (Taubenberger et al., 2005). Comparison of the gene sequences to those of contemporary strains of avian and human influenza virus revealed that the 1918 strain was probably derived from an avian influenza virus strain (H1N1; "H" and "N" indicate the alleles of the hemagglutinin and neuraminidase genes, respectively). The strain had acquired mutations in the viral surface receptor hemagglutinin which enabled it to attach to and infect human respiratory cells. After careful consideration of the public health benefits that would be gained from understanding the unusual virulence of the pathogen and instatement of stringent biosafety and containment regulations, the virus was reconstructed (Tumpey et al., 2005). In macaque infection models, the levels of the reconstructed 1918 virus were higher in both the upper and lower respiratory tracts than the levels of a contemporary human strain that was found only in the upper respiratory tract and normally replicates poorly in the lower lungs. Infection triggered a strong host inflammatory response that damaged respiratory tissue and caused severe hemorrhaging. Compared to contemporary human influenza strains, the 1918 virus also had significant differences in the sequence of its RNA-dependent RNA polymerase that likely contributed to its ability to reproduce more efficiently. Together, these mutations in hemagglutinin and the RNA polymerase complex led to human adaptation of an avian influenza virus that killed its victims rapidly and that spread effectively from human to human. Transmission was facilitated by the miserable conditions endured by the underfed and overexposed soldiers on the battlefields of World War I.

Influenza pandemics subsequent to 1918, most notably in 1957, 1968, and 2009, have been due to emergence of descendants of the 1918 strain. It appears that the human-adapted avian strain was transferred from humans to pigs and has since been maintained in pigs. The H1N1 strain that caused the global pandemic of 2009 arose following reassortment of the eight segments of the influenza RNA genome from humans, pigs, and birds that led to new combinations of viral alleles. The animal in which these were combined into a human infectious virus is unknown. Global surveillance programs for emerging influenza virus strains and an aggressive immunization campaign against the 2009 strain may have preempted a devastating pandemic in a population that otherwise lacks immunity.

nucleocapsid into the cytoplasm. Uncoating may take place in the cytoplasm, as in the case of HIV (Fig. 5.37A), or after the nucleocapsid reaches the nucleus by traveling along the actin and microtubule cytoskeleton, as for herpesvirus (Fig. 5.37B).

Both enveloped and nonenveloped viruses may be taken up via endocytosis. Viruses attach to specific receptors and migrate to regions in the host membrane that are lined on the cytoplasmic face with the protein clathrin (Fig. 5.38). Migration to clathrin-coated pits concentrates the viral particles and triggers further invagination of the membrane, which

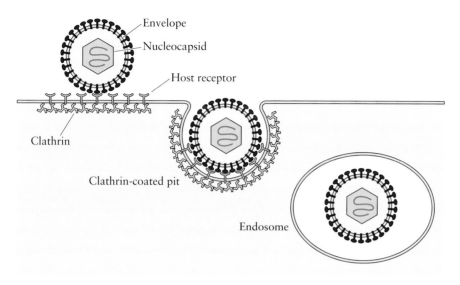

Figure 5.38 Entry of viruses by endocytosis. The virus binds to a specific receptor on the surface of a host cell and migrates to regions of the host membrane that are lined with the protein clathrin. The host membrane invaginates and eventually engulfs the virion, forming an endosome. doi:10.1128/9781555818890.ch5.f5.38

eventually pinches off to form an endocytotic vesicle (endosome). Endocytosis may also occur independently of clathrin. An advantage of entry via endocytosis is that the entire virus is internalized, and therefore, viral molecules are not left outside the cell to alert an immune response.

Most viral genomes are released from the endosome before it binds to the lysosome, which contains nucleases, among other digestive enzymes, that would otherwise destroy the viral nucleic acids. Viral genomes may be released by fusion of viral and endosomal membranes in the case of enveloped viruses or by disruption of the endosome by nonenveloped viruses (Fig. 5.39). Influenza virus, for example, enters host cells via clathrin-dependent endocytosis. Acidification of the endosome (via a proton pump) induces a conformational change in hemagglutinin which brings the viral and endosomal membranes close enough for the membranes to fuse (Fig. 5.39A). Protons also enter the interior of the virion via an ion channel formed by the influenza virus M2 protein. This destabilizes the virion and results in release of the viral RNA genome. The RNA molecules, with associated proteins (including a viral RNA polymerase), are imported into the nucleus via the **nuclear pore complex**. The nucleocapsids of other viruses that are released from the endosome via membrane fusion may be uncoated by interactions between the capsomers and cytoplasmic ribosomes that result in capsid disassembly. The RNA of Semliki Forest virus is released into the cytoplasm in this way and is then translated by the ribosomes to produce viral proteins. Nonenveloped viruses that enter the cell via endocytosis may be released by disrupting the endosome. For example, conformational changes in poliovirus capsid proteins (as a consequence of binding to host cell receptors) expose hydrophobic residues that insert in the endosomal membrane. This forms a pore through which poliovirus RNA is secreted into the host cytoplasm (Fig. 5.39B).

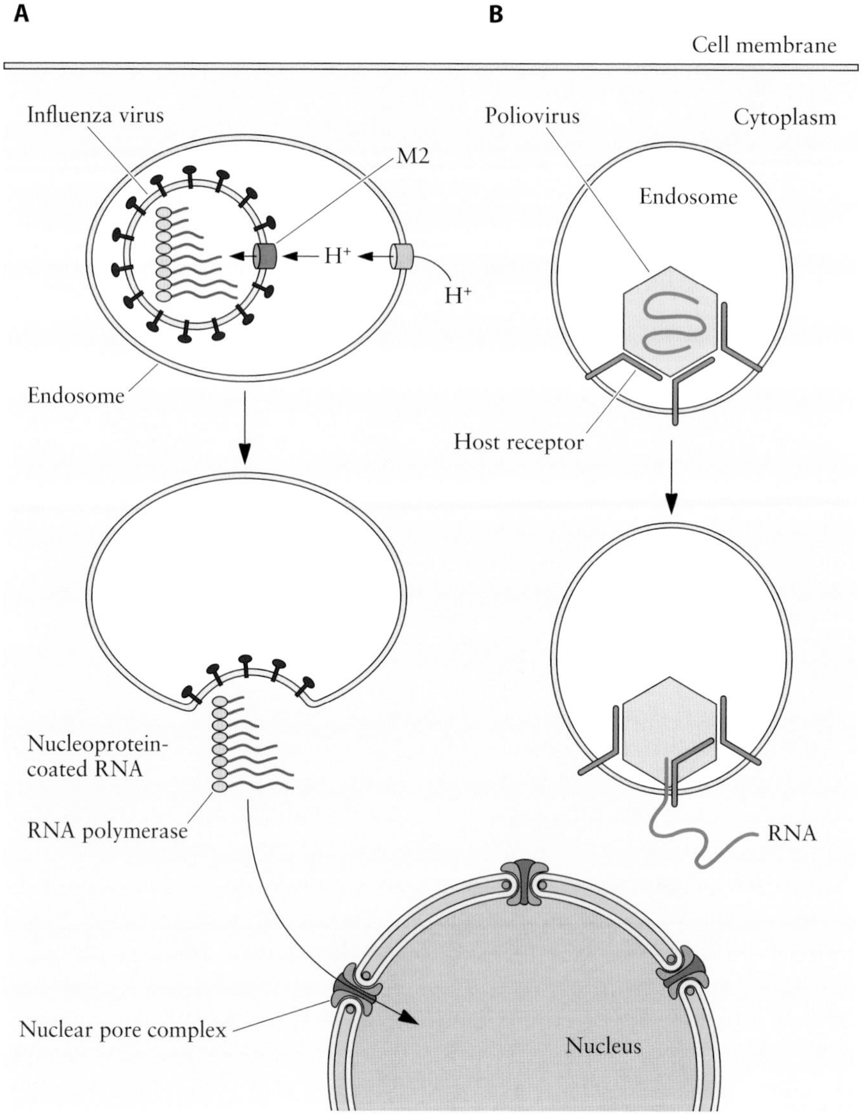

Figure 5.39 Viral genomes may be released from the endosome by fusion of viral and endosomal membranes as for influenza virus (**A**) or by disruption of the endosome as for the nonenveloped poliovirus (**B**). doi:10.1128/9781555818890.ch5.f5.39

Viral Gene Expression and Replication

Viral reproduction requires synthesis of large quantities of viral proteins, encoded by viral genes, and viral genomes, which are then assembled into new virions. Viruses use stealth and molecular mimicry to manipulate the host cell machinery to produce these. The mechanism by which this is achieved is determined by the nature of the viral genetic material, that is, whether the viral genome is made of DNA or RNA and is single or doubled stranded. After entry into the host cell, the next step in the viral infection cycle is the presentation of mRNA to the host ribosomes for viral protein synthesis. Viral proteins assist in the replication of the viral genomes, control viral and host gene expression, and are structural components of capsids and envelopes.

DNA Viruses

The genetic material of double-stranded DNA viruses is similar to that of the host, and therefore, the virus can use cellular replication and protein synthesis machinery. DNA replication and gene transcription of viral DNA genomes generally take place in the nucleus, where appropriate cellular enzymes are available. Viral mRNA produced in the nucleus is exported to the cytoplasm, where translation occurs on cytoplasmic or endoplasmic reticulum-bound ribosomes. Viral proteins that are synthesized on cytoplasmic ribosomes are transported back to the nucleus for virion assembly, while viral spike proteins are transported via the endoplasmic reticulum to membranes that will eventually comprise the viral envelope.

The nuclear membrane presents a barrier to entry of viral particles. Viral DNA may be packaged into the nucleus during cell division when the nuclear membrane disintegrates and then reforms around newly replicated chromosomes. Infections of this type occur only in cells that divide. Viral genomes may also enter the nucleus by exploiting the pathway used by cells to import proteins into the nucleus. Host proteins are targeted to the nucleus by a short sequence of mostly basic amino acids that comprise the **nuclear localization signal**. With the assistance of specific cytoplasmic receptor proteins, cellular proteins carrying a nuclear localization signal dock onto a complex of proteins that form the nuclear pore and are translocated into the nucleus. The nucleocapsids of adenoviruses and herpesviruses dock onto the nuclear pore complex via a capsid protein and release their DNA genomes (Fig. 5.40). The empty capsids of these viruses are too large to enter the nucleus and either disintegrate or remain bound to the pore complex.

Regulation of transcription initiation is a major mechanism of control of viral gene expression. RNA polymerase II, the major enzyme complex that catalyzes host mRNA synthesis, transcribes genes encoded in viral DNA genomes. Viral promoter sequences that are recognized by RNA polymerase II are therefore similar to those of host protein coding genes. Typically, genes of DNA viruses are expressed in a specific temporal order (a transcriptional cascade) that is controlled at the transcriptional level by viral and host regulatory proteins. Genes that are transcribed first ("early" genes) are those encoding proteins that participate in replication of the viral genomes and regulate viral gene expression, at the expense of host gene expression. Production of large numbers of viral genomes (often tens of thousands per cell) and early viral proteins that participate in this process places a huge demand on a host cell's resources. To avoid disruption of viral DNA replication, synthesis of viral structural proteins is delayed. Genes encoding capsid and envelope proteins are therefore transcribed later in the infection cycle ("late" genes), after initiation of viral genome replication. This ensures that all viral components are present for production of large numbers of functional virions.

The small simian virus 40 (SV40) that infects monkeys and humans provides a simple example of coordinated temporal regulation of viral genes. The DNA genome of SV40 is 5,243 bp and contains two transcriptional units, one controlled by an early promoter and the other by a late promoter (Fig. 5.41). The early mRNA encodes two proteins (the small-t

SV40
simian virus 40

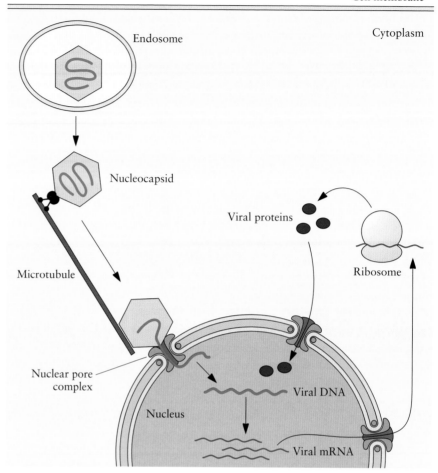

Figure 5.40 Adenovirus nucleocapsids travel along microtubules to the nuclear pore complex, where viral DNA is released into the nucleus. Viral mRNA produced by host RNA polymerase in the nucleus is exported to the cytoplasm for protein synthesis on cytoplasmic ribosomes. The viral proteins return to the nucleus for virion assembly. doi:10.1128/9781555818890.ch5.f5.40

and large-T antigens) that initiate viral DNA replication and prevent host cell **apoptosis** (programmed cell death). In addition, the large-T antigen suppresses production of early mRNA and activates transcription of late genes encoding three capsid proteins.

Although HSV-1 has a much larger genome (152,000 bp) than SV40, the temporal coordination of gene expression is generally similar. The first HSV-1 genes to be expressed immediately following infection (immediate-early, or alpha, genes) are activated by interaction between the **viral matrix** protein which entered the nucleus with the viral genome, cellular transcription factors, and an enhancer sequence adjacent to the viral immediate-early gene promoters. The HSV-1 matrix protein is a strong transcriptional activator that, by activating expression of viral and host genes, enables the virus to replicate in inactive host cells such as non-dividing cells of the central nervous system. Three of the immediate-early

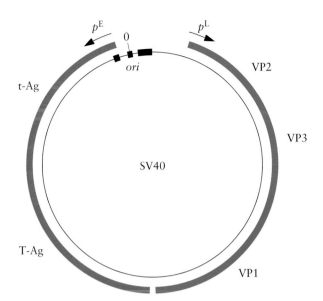

Figure 5.41 Map of the SV40 genome showing early (p^E) and late (p^L) promoters that control transcription of early (red) and late (blue) mRNA, respectively. The small-t antigen (t-Ag) encoded on the early mRNA initiates viral DNA replication at the origin of replication (*ori*; 0) on the circular SV40 genome. The large-T antigen (T-Ag), also encoded on the early mRNA, binds to the control region (black boxes) and activates transcription from the late promoter to produce capsid proteins (VP1, VP2, and VP3). doi:10.1128/9781555818890.ch5.f5.41

proteins are transcriptional activators that induce subsequent expression of early viral proteins required for viral genome replication. Proteins encoded by early genes include a replication initiation protein, DNA polymerase, a helicase–primase complex, proteins that increase levels of deoxyribonucleotides, and DNA repair enzymes. After replication of the viral genome has been initiated, expression of early genes is repressed and late genes encoding structural components are activated. The herpesvirus genome, which is circularized following introduction into the host nucleus, is replicated by **rolling-circle replication** (Fig. 5.42). Replication not only provides genomes for new virions but also increases the number of copies of viral genes from which large numbers of late viral mRNAs and proteins can be produced.

Positive-Sense RNA Viruses

The genomes of RNA viruses may be double or single stranded, and in the latter case, the RNA molecule may be in the positive or negative sense. Positive-sense RNA genomes have the same polarity (5′ to 3′) as mRNA and therefore can be presented directly to host ribosomes for translation (Fig. 5.43A). In contrast, negative-sense RNA genomes have polarity opposite (3′ to 5′) that of mRNA and must first be transcribed to produce a complementary mRNA molecule for translation by ribosomes (Fig. 5.43B). The genomes of some positive-sense RNA viruses carry a 5′ 7-methylguanosine cap and/or 3′ poly(A) tail that are characteristic of eukaryotic mRNA. For example, the single positive-sense RNA molecule of

Figure 5.42 Replication of the herpesvirus genome by rolling-circle replication. Following introduction into the host cell nucleus, the herpesvirus genome is circularized. During replication, one of the two strands of the double-stranded DNA genome is nicked. Extension of the nicked strand by a viral DNA polymerase generates a linear concatemer (solid line) that carries multiple copies of the viral genome. The other end of the same strand is displaced as DNA synthesis progresses around the circular template and serves as a template for synthesis of the complementary strand (broken line). During virion assembly, viral genomes are excised at specific sequences (*) by packaging proteins associated with a preassembled capsid, and each capsid receives a complete genome. Arrows indicate direction of DNA synthesis.
doi:10.1128/9781555818890.ch5.f5.42

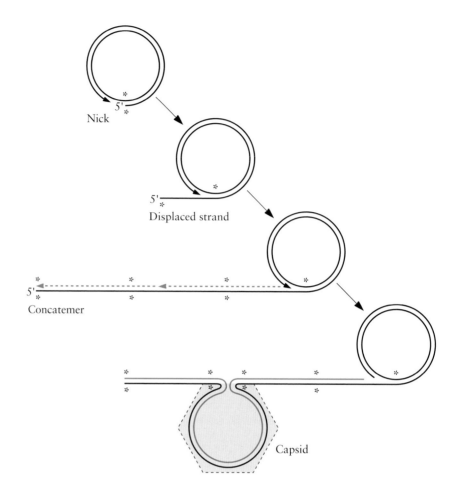

Figure 5.43 Production of mRNA and genomic RNA molecules in single-stranded positive-sense (**A**) and negative-sense (**B**) RNA viruses requires an RNA-dependent RNA polymerase (RNAP). doi:10.1128/9781555818890.ch5.f5.43

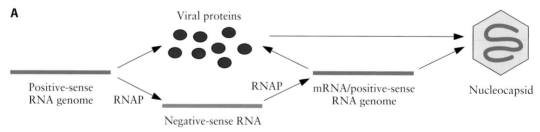

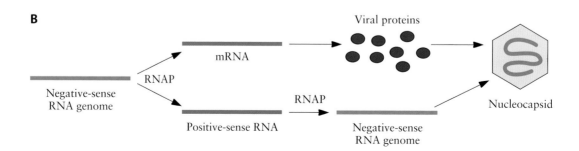

the poliovirus genome has a poly(A) tail but is capped at the 5′ end with a viral protein (VPg) (Fig. 5.44). In contrast, the yellow fever virus genome has a 5′ 7-methylguanosine cap but not a 3′ poly(A) tail.

A problem for RNA viruses is that eukaryotic cells begin mRNA translation near the 5′ end of the molecule and generally do not recognize internal translation initiation sites. Therefore, production of multiple viral proteins from a single viral genomic RNA molecule requires a mechanism to produce either monocistronic RNA molecules each encoding one protein from the polycistronic genome encoding several different proteins, or individual proteins from a single polyprotein. Poliovirus, for example, generates a single **polyprotein** from its positive-sense RNA genome. Poliovirus RNA has a long leader sequence between the 5′ end of the molecule and the translation initiation codon. Intrastrand complementary base-pairing in the leader sequence forms a structure that mediates binding of ribosomes to the RNA. Translation results in the production of a single large protein that is the precursor for all of the poliovirus proteins (Fig. 5.44). A specific region within the polyprotein has proteolytic activity, and initially, the polyprotein undergoes self-cleavage to produce three smaller polyproteins. One of the smaller polyproteins contains the capsid proteins, and the other two contain proteases for further cleavage of the polyproteins into individual proteins and an **RNA-dependent RNA polymerase** for replication of the genome.

Some other positive-sense RNA viruses are polycistronic, encoding more than one polypeptide. However, only the protein encoded at the 5′ end of the genomic RNA molecule can be translated directly because ribosomes are released from the RNA when they reach the first translation stop signal and do not recognize internal start codons. In togaviruses,

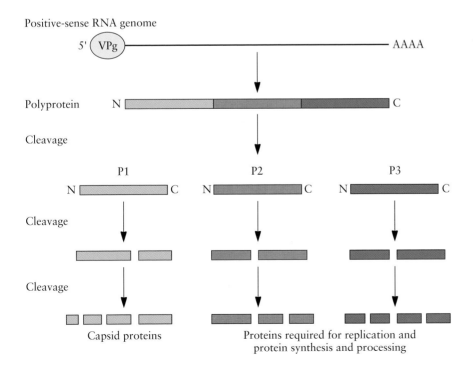

Figure 5.44 Proteolytic cleavage of the poliovirus polyprotein.
doi:10.1128/9781555818890.ch5.f5.44

the protein encoded at the 5′ end of the RNA is a replicase that can synthesize RNA from an RNA template. The enzyme catalyzes the synthesis of a negative-sense RNA molecule that is complementary to the positive-sense RNA genome (Fig. 5.45). Using the negative-sense RNA as a template, two different mRNA molecules are transcribed by the viral replicase. One is equivalent to the full-length genome, and the second is a truncated mRNA that is transcribed from an internal initiation site and represents the 3′ end of the genome (Fig. 5.45). Translation from the full-length mRNA produces more replicase, and translation from the truncated mRNA generates a polyprotein that is cleaved to produce the capsid proteins.

In addition to protein synthesis, production of progeny viruses requires production of many copies of the positive-sense RNA genome from the original RNA template. While cellular organisms have a **DNA-dependent RNA polymerase** for producing mRNA from a DNA template during transcription, animal cells are not capable of producing RNA from an RNA template. RNA-dependent RNA polymerases (RNA replicases) that recognize an RNA template are encoded in the genomes of RNA viruses. The replication of single-stranded RNA genomes first requires production of a

Figure 5.45 Protein synthesis from the polycistronic positive-sense RNA genome of togaviruses. Translation of the positive-sense RNA genome terminates at the first stop codon (red box) to produce a replicase. Replicase is an RNA-dependent RNA polymerase that catalyzes the synthesis of negative-sense RNA that is complementary to the positive-sense genome. The negative-sense RNA molecule serves as a template for the synthesis of full-length and truncated positive-sense RNA molecules. The truncated RNA is synthesized from an internal initiation site and encodes the capsid structural proteins. Arrows indicate initiation sites for transcription of the complementary RNA strand. doi:10.1128/9781555818890.ch5.f5.45

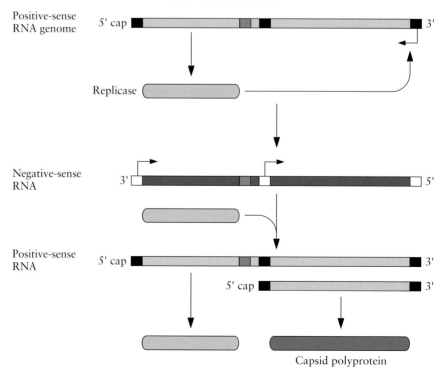

complementary RNA **(antigenomic RNA)** that serves as a template for synthesis of many molecules of the genome. In the case of positive-sense RNA viruses, the complementary copy is a negative-sense molecule generated by an RNA-dependent RNA polymerase (Fig. 5.43A). The negative-sense RNA is a template for synthesis of positive-sense RNA genomes by the RNA-dependent RNA polymerase. In addition to serving as a template for genome synthesis, the negative-sense RNA may be a template for transcription of mRNA encoding "late" (structural) viral proteins as described above for togaviruses.

Negative-Sense RNA Viruses

Many important human viruses have negative-sense RNA genomes that have polarity opposite that of mRNA. Negative-sense RNA viruses cause a variety of human diseases, such as influenza, measles, mumps, and rabies. Influenza A virus has a segmented genome made up of eight negative-sense RNA molecules encoding 11 proteins in total. The sequence of negative-sense RNA is complementary to that of mRNA, and therefore, the genomic RNAs are not directly translated by host ribosomes to produce viral proteins. A positive-sense copy (i.e., mRNA) must be synthesized before translation can occur (Fig. 5.43B). Negative-sense RNA viruses carry an RNA-dependent RNA polymerase within their capsid to transcribe mRNA once a host cell has been infected. In addition to mRNA synthesis, these enzymes catalyze the synthesis of a positive-sense RNA antigenome (complementary to the genomic RNA sequence) that serves as a template for synthesis of new negative-sense RNA genomes that are subsequently packaged into new virions (Fig. 5.43B).

In contrast to the case for most other RNA viruses that are reproduced in the host cytosol, the influenza A virus genome is transcribed and replicated in the host nucleus and, in addition to the viral RNA-dependent RNA polymerase, requires host factors for transcription. The viral RNA polymerase was prepackaged in the capsid, along with genomic RNA, in the previous host (Fig. 5.39A). To initiate transcription, 5′ 7-methylguanosine-capped RNA primers are acquired by cleaving the first 10 to 13 nucleotides with attached cap from host pre-mRNA in a process known as **cap snatching**. During transcription, the viral RNA polymerase reaches a short polyuridine tract near the 5′ end of the template RNA which causes it to slip and reread the template in this region. "Stuttering" results in the addition of a poly(A) tail to the viral mRNA. The capped and tailed mRNA is then transported to the cytoplasm for translation. Proteins destined for the viral envelope enter the endoplasmic reticulum for processing by the Golgi apparatus and are then inserted into the host cell membrane that will eventually coat newly synthesized viral capsids as they exit the cell. Other influenza virus proteins, such as RNA polymerase and nucleocapsid protein, that carry a nuclear localization signal are transported back to the nucleus, where they participate in the synthesis of viral RNA genomes. An RNA primer does not initiate synthesis of the positive-sense antigenome, nor does stuttering add a poly(A) tail. Full-length positive-sense RNA provides a template for synthesis of negative-sense genomic RNA.

Double-Stranded RNA Viruses

The reoviruses are double-stranded RNA viruses that infect humans and include rotaviruses, which are a major cause of infant gastroenteritis. Reovirus contains two concentric capsid shells, one (the outer shell) of which is degraded by lysosomal enzymes following uptake by endocytosis and fusion of the endosome with a lysosome. Transcription takes place within the intact inner capsid, which contains all of the viral enzymes required for synthesis and capping of mRNA. The RNA-dependent RNA polymerase is part of the inner capsid and is activated by proteolysis when the outer capsid is degraded. One of the two RNA strands (the negative-sense strand) serves as a template for mRNA synthesis. The mRNAs are released from the capsid for translation in the host cytoplasm (Fig. 5.46). For genome replication, the mRNAs that are assembled in new capsids are the templates for synthesis of the negative-sense strand, which remains associated with the complementary mRNA to regenerate the double-stranded RNA genome (Fig. 5.46).

Retroviruses

Retroviruses are positive-sense RNA viruses that produce a double-stranded DNA copy of their genome which is inserted into the host genome for expression of viral proteins. HIV, the causative agent of AIDS, is a well-studied example of a human retrovirus. Retroviruses release their nucleocapsids into the host cell cytoplasm following fusion of viral and host cell membranes. In addition to the RNA genome, the capsids contain

Figure 5.46 Replication cycle of a double-stranded RNA virus. doi:10.1128/9781555818890.ch5.f5.46

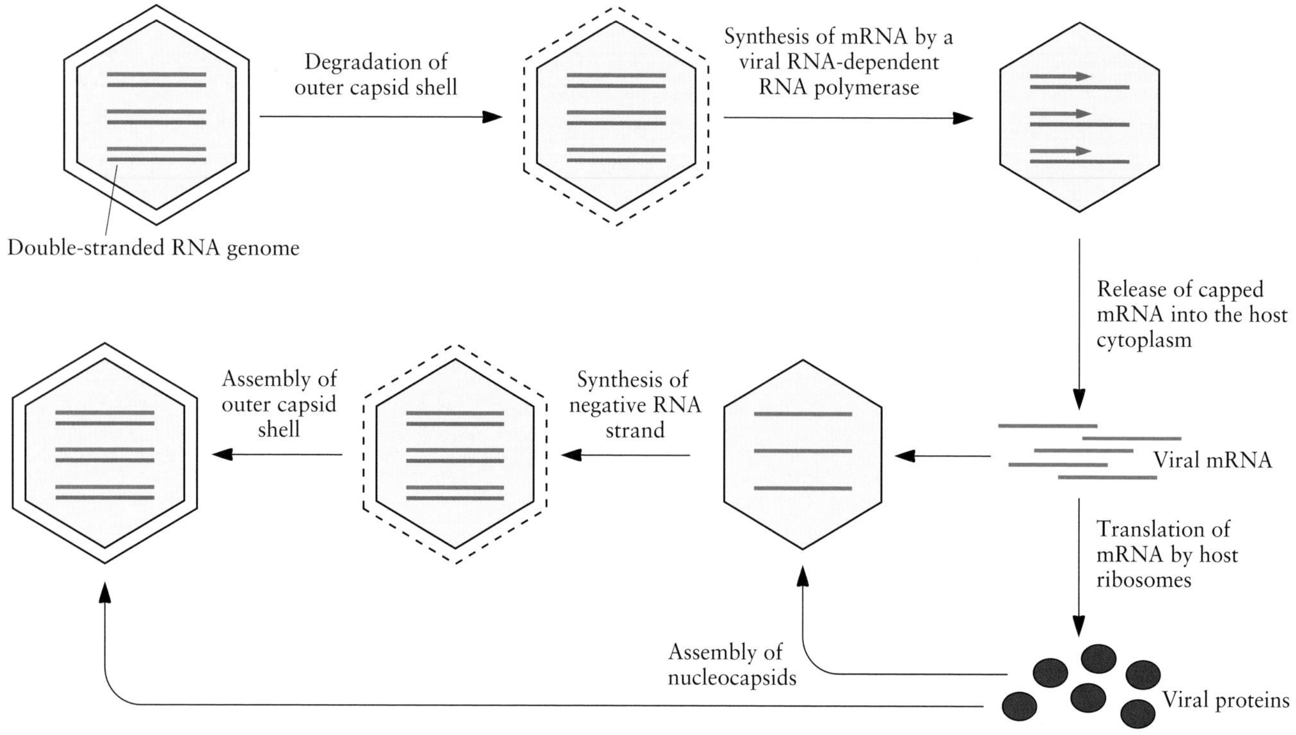

the enzymes **reverse transcriptase** and integrase. In the infected cell, reverse transcriptase, which is an RNA-dependent DNA polymerase, catalyzes the synthesis of DNA using the RNA genome as a template. Reverse transcriptase is prone to making errors and is responsible for the high mutation rate of retroviruses, which makes it difficult to develop effective vaccines. DNA synthesis is primed with a tRNA that was packaged into virions produced in the previous host. The tRNA primes the synthesis of a small fragment of DNA at the 5′ end of the RNA genome (Fig. 5.47A). The DNA fragment, which remains bound to the tRNA, then hybridizes to the 3′ end of the RNA genome (note that the 5′ and 3′ ends of the retroviral genome contain repeat sequences) and primes the synthesis of a

Figure 5.47 Retroviral DNA synthesis by reverse transcriptase. (**A**) A tRNA (blue arrow) primes the synthesis of a short DNA primer at the 5′ end of the positive-sense RNA genome. (**B**) The tRNA-DNA primer binds to a complementary sequence at the 3′ end of the viral genome and provides a 3′ OH for synthesis of a DNA strand complementary to the genomic RNA by reverse transcriptase. (**C**) Following DNA synthesis, the RNA template is degraded by RNase H, a component of the reverse transcriptase complex. (**D**) Reverse transcriptase synthesizes a second DNA strand that is complementary to the first DNA strand. (**E**) The double-stranded DNA circularizes and is translocated into the host nucleus. (**F**) The viral DNA is integrated into a host chromosome by the viral protein integrase. Three polyproteins are produced and are cleaved by proteases to yield individual proteins.
doi:10.1128/9781555818890.ch5.f5.47

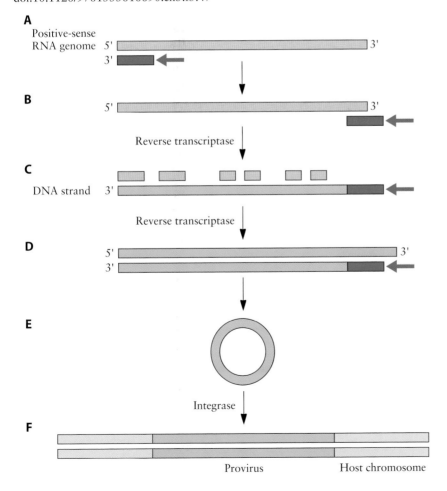

RNase
ribonuclease

DNA strand that is complementary to the genomic RNA (Fig. 5.47B). As the DNA strand is synthesized, the RNA template behind it is degraded by the ribonuclease component (RNase H) of the reverse transcriptase complex (Fig. 5.47C). Finally, reverse transcriptase synthesizes a second DNA strand that is complementary to the newly synthesized DNA template to produce double-stranded DNA (Fig. 5.47D). The DNA circularizes and is translocated into the nucleus through the nuclear pore, facilitated by accessory proteins (Fig. 5.47E). Only some retroviruses (i.e., lentiviruses) that produce specific accessory proteins for entry into the nucleus are able to replicate in nondividing cells. Other retroviruses gain entry into the nucleus during cell division when the nuclear envelope temporarily dissolves. Once in the nucleus, viral DNA is integrated into a random site in the host cell genome by recombination, which is mediated by the viral integrase protein (Fig. 5.47F). The integrated virus is referred to as a **provirus**. The host cell replication machinery does not discriminate between proviral and cellular DNAs, and the proviral DNA is replicated along with the host chromosomes.

If the integrated proviral genes are not expressed, new virions are not produced and the provirus may remain dormant in the genome for many years. However, when appropriate cellular transcription factors are available, as determined by the host cell type and external stimuli, proviral genes are activated and new virions produced. Host RNA polymerase transcribes the entire proviral sequence beginning at the left-hand long terminal repeat (LTR) and continuing to the right-hand LTR (Fig. 5.48). Some of the transcripts serve as mRNA for protein synthesis, and some are assembled (as dimers) into new virions as positive-sense RNA genomes. In addition

LTR
long terminal repeat

Figure 5.48 Retroviral protein synthesis. The entire integrated retroviral DNA sequence is transcribed beginning at the left-hand LTR and continuing to the right-hand LTR. In addition to full-length mRNA, several smaller mRNA are generated by alternate splicing of internal sequences. The mRNAs are capped at 5′ end and polyadenylated at the 3′ end (A_n), and they are then exported from the nucleus to the cytoplasm, where they are translated by host ribosomes. doi:10.1128/9781555818890.ch5.f5.48

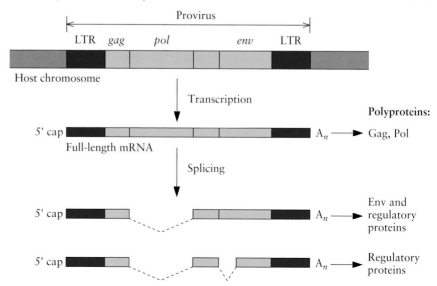

to production of a full-length transcript, some of the primary transcripts are spliced to yield several smaller mRNAs. The mRNAs are capped at the 5′ end and polyadenylated at the 3′ end, and they are then exported from the nucleus to the cytoplasm, where they are translated by host ribosomes. Three polyproteins are produced that are cleaved by proteases to yield individual proteins (Fig. 5.48). The *gag* sequence encodes capsid, matrix, and nucleocapsid proteins; *pol* encodes reverse transcriptase, integrase, and a protease; and *env* encodes envelope spike proteins. The Gag and Pol polyproteins are synthesized from the full-length mRNA, and the Env polyprotein and regulatory proteins are encoded on spliced mRNA.

Virus Assembly and Release

Viruses hijack host cell machinery to produce thousands of copies of viral genomes and proteins in a single cell. These are then assembled within the host cell to generate new virions that are released to infect new cells. Considering that virions are assembled from prefabricated parts, how are they assembled in an ordered fashion? How is the genetic material packaged? How do enveloped viruses acquire membranes containing spike proteins? Virions may be assembled in the nucleus or cytoplasm. The final assembly of enveloped viruses frequently occurs at the cell membrane, where the virus "picks up" a portion of the host membrane as it exits the cell, although viral envelopes may also be acquired from other **endomembrane** structures.

DNA viruses, exemplified by papillomavirus and adenovirus, are typically assembled in the nucleus, where viral genome replication occurs. An exception is the poxviruses, whose life cycle occurs completely in the cytoplasm. For nuclear assembly, capsid proteins that were translated in the cytoplasm are transported along the cytoskeletal filaments to the nucleus. Similar to cellular proteins that function in the nucleus, these viral proteins must carry nuclear localization signals that are recognized by the nuclear import apparatus. In some cases, capsid subunits are imported as partially assembled protein complexes; at least one of the proteins in the complex must carry the importation signal.

For many RNA viruses, the entire life cycle, including production of viral proteins and genomes and their assembly, occurs in the cytoplasm. Notable exceptions are influenza viruses, whose genomes are transcribed in the nucleus, and the retroviruses, which integrate double-stranded proviral DNA into the host genome as part of their life cycle. The mRNA and genomic RNA molecules transcribed from the influenza virus genome and from retrovirus proviral DNA are exported from the nucleus for translation of mRNA on ribosomes in the cytoplasm and endoplasmic reticulum and for packaging of genomes into virions. Influenza virus genomes are coated with nucleocapsid proteins and bind to viral RNA polymerase, matrix protein, and nonstructural proteins that carry a nuclear export signal before leaving the nucleus (Fig. 5.49). All of these viral proteins were imported into the nucleus after synthesis in the cytoplasm. Another interesting exception is poliovirus, which replicates within virus-induced vesicles derived from the host endoplasmic reticulum. The newly synthesized

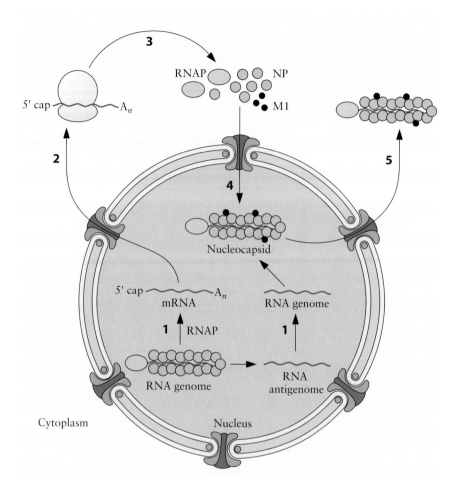

Figure 5.49 Influenza virus negative-sense RNA genomes enter the nucleus, where an RNA-dependent RNA polymerase (RNAP) produces mRNA and more copies of the RNA genome (**1**). The mRNA is exported from the nucleus (**2**), and RNAP, nucleocapsid protein (NP), and matrix proteins (M1) are synthesized on cytoplasmic ribosomes (**3**). These proteins are imported into the nucleus, where they bind to genomic RNA to form nucleocapsids (**4**). The nucleocapsids are transported back to the cytoplasm (**5**) for final assembly. A membrane and envelope proteins are acquired as the virion is released from the host cell by budding. A_n, poly(A) tail. doi:10.1128/9781555818890.ch5.f5.49

positive-sense RNA is translocated to the cytoplasm for synthesis of viral proteins and packaging into capsids.

Viral nucleocapsids self-assemble via specific protein–protein and protein–nucleic acid interactions. Many of these interactions have been demonstrated in vitro with purified viral components. The high concentration of nucleocapsid subunits typically produced within a host cell increases the likelihood that the components will interact and ensures that the assembly process proceeds efficiently. Generation of capsid proteins from polyproteins also facilitates association of structural proteins. Extensive interactions among capsomers, sometimes including nucleic acids, stabilize the structure. Final capsid assembly may involve cleavage, chemical modification, or conformational changes in capsid proteins to increase capsid stability.

The capsomers of filamentous viruses assemble around the nucleic acid, while the capsids of many icosahedral viruses are assembled first and then filled with the nucleic acids. In some cases, the nucleic acid may first interact with structural proteins to form an initiation complex. For example, in retroviruses such as HIV, interactions between genomic RNA and the viral polyprotein Gag are required for assembly. The genomes of some single-stranded RNA viruses may contain double-stranded regions due to base-pairing between complementary regions within the molecule. Intramolecular base-pairing forms secondary structures such as hairpin loops that may be recognized by proteins to initiate virion assembly. Alternatively, viral proteins may act as scaffolding proteins to ensure appropriate protein interactions during capsid assembly. HSV-1 produces a protein scaffold around which the capsid assembles. A viral protease then degrades the scaffold proteins so that the genomic DNA molecules can be packaged in the capsid.

During packaging, viral nucleic acids are distinguished from cellular DNA or RNA by the presence of unique sequences. Many viral nucleic acid packaging signals consist of short repeat sequences that are recognized by specific viral proteins. Bound proteins direct the nucleic acid to capsid entry sites. Replication of circular viral DNA by rolling-circle replication results in the formation of long concatemers containing multiple copies of genomic DNA (Fig. 5.42). Individual genomes are cleaved at specific nucleotide sequences as they are packaged such that each capsid contains a complete genome. The size of the capsid determines the size of the viral nucleic acid that can be packaged.

How are the genomes of segmented viruses packaged to ensure that each virion contains a complete complement of genes? One mechanism is serial packaging, in which the packaging of a nucleic acid molecule depends on the prior packaging of another molecule of the segmented genome. Alternatively, packaging may be random. Most of the resultant viral particles would not contain a complete genome and therefore would not be infectious. For example, the influenza virus genome is made up of eight different molecules of RNA. If eight viral RNA strands were randomly packaged, then only one in about 400 virions would be infectious. If more than eight molecules were packaged per virion, then the probability of acquiring a complete set of genes would increase. Influenza virus virions with more than eight genome molecules have been isolated. However, recent evidence indicates that influenza virus RNAs contain specific sequences required for efficient packaging, suggesting that the process is not entirely random.

Animal viruses are released to the external environment either by lysis of the host cell or by **budding** from the membrane, a special form of exocytosis. Naked viruses without membrane envelopes mainly exit by host cell lysis, as do some enveloped viruses such as poxviruses. Viruses may lyse cells directly by degrading host structures, for example, by cleaving the host cytoskeletal proteins. Viruses may also destroy host cells indirectly by inducing apoptosis or by blocking production of essential host proteins.

During budding, the nucleocapsid is wrapped in a membrane that pinches off to release an enveloped virion (Fig. 5.50). Enveloped viruses acquire their membranes from host cell membrane or from the

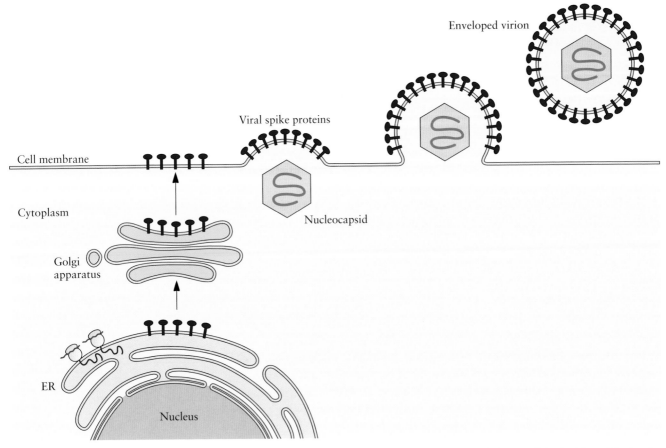

Figure 5.50 Enveloped viruses are released from the host cell by budding. Viral spike proteins are synthesized on ribosomes associated with the endoplasmic reticulum (ER) and are transported to the cell membrane via the Golgi apparatus. At the cell membrane, assembled nucleocapsids are wrapped in membrane containing the spike proteins as they leave the cell. doi:10.1128/9781555818890.ch5.f5.50

endomembrane system, i.e., from the nucleus, endoplasmic reticulum, or Golgi apparatus. The site of envelope acquisition is determined by the presence of viral envelope proteins that have been localized to a particular site via the cellular secretory pathway. Viral membrane proteins are synthesized on ribosomes associated with the endoplasmic reticulum (Fig. 5.50). Within the lumen of the endoplasmic reticulum, the proteins are properly folded, may begin to form complexes with other viral proteins, and may be chemically modified. Modifications may include formation of disulfide bonds between pairs of cysteine residues and/or addition of oligosaccharides (**glycosylation**). Note that these modifications are catalyzed by host enzymes in the endoplasmic reticulum in a manner similar to that of host protein modification. Viral spike proteins are commonly glycosylated, and the oligosaccharides are important for proper protein folding, for host recognition, and for the virus to evade host immune responses. Once they are properly folded, viral proteins are transported within membrane vesicles to the cell membrane via the Golgi apparatus (Fig. 5.50). Other viral proteins mediate binding of assembled nucleocapsids to viral integral membrane proteins. For example, the matrix proteins of influenza virus interact with nucleocapsids and direct these to glycosylated hemagglutinin and **neuraminidase** spike proteins that have been transported to the cell membrane. These protein interactions initiate the formation of a bud that eventually encloses the nucleocapsid. In this manner, many enveloped viruses acquire their membranes and spike proteins by budding from the cellular membrane (Fig. 5.50).

Once released, infectious virions may be transmitted to other cells within the same individual or to new hosts. Virions may infect cells near those from which they were released or may travel to distant sites to infect new tissues. However, extracellular virions are exposed for potential destruction by the host defense system (see chapter 2). Some viruses avoid this by moving directly from cell to cell, a process that exploits normal cell–cell junctions. For example, herpesvirus can be transferred between epithelial cells via tight junctions or at synaptic contacts between neurons. Other viruses stimulate the formation of specialized connections between cells for intercellular transfer.

General Patterns of Viral Infections of Humans

Like pathogenic bacteria, most viruses show some specificity for the types of cells that they infect. Tissue specificity (tropism) is determined by the accessibility of susceptible cells and the availability of specific receptors on a host cell. Viral infections of animal cells may be acute, with rapid reproduction that leads to the release of large numbers of infectious particles and efficient transmission. Host cells may or may not be destroyed by an **acute infection**. Recovery from the infection requires that the virus be efficiently eliminated (cleared) from the host. Alternatively, a viral infection may be chronic, persisting for long periods, sometimes the lifetime of the host, with low reproduction rates. Some viruses may remain **latent** (dormant) in a host for long periods. Initially, the infection may cause host cell lysis, but upon release, virions may infect a different cell type, in which they are latent. During the latent phase of infection, infectious particles are not produced. However, the viral genome remains inactive in the host cell and may be reactivated to produce virions that can reestablish a lytic infection. In some persistent infections, the viruses transform host cells into tumor cells. These viruses are referred to as **oncogenic viruses**.

Acute Infections

Viruses that cause acute infections typically take over the host cell machinery to reproduce rapidly. The symptoms tend to be of short duration, and the viruses are destroyed by the host adaptive immune system within a few days. Often, the host acquires immunity to subsequent infections by the same virus. Although host cell destruction may be an obvious indication of an acute infection, characteristic early changes are apparent in the infected cell as a consequence of the burden of producing massive quantities of viral nucleic acids and proteins. These may include a reduction in growth of the infected cell or morphological changes such as fusion of adjacent cells (syncytia) from coronavirus infection, shrinking of the nucleus from poliovirus infection, or formation of vacuoles in the cytoplasm from papovavirus infection. During many acute infections, the virus directly inhibits production of host proteins. Translation of host mRNAs may be inhibited by outcompetition for ribosomes by the more abundant viral mRNAs, blockage of host mRNA transport from the nucleus, host mRNA degradation, prevention of translation initiation protein complex formation, or inactivation of translation factors.

Many different viruses cause acute, lytic infections, including cold viruses (rhinoviruses, coronaviruses, and adenoviruses), influenza virus, and smallpox virus. Influenza virus causes localized infections of tissues of the upper respiratory tract, mainly the nose, throat, bronchi, and, more rarely, the lungs, and is transmitted by aerosols generated by coughing and sneezing, as well as by direct contact. Influenza virus has a rapid life cycle, releasing new infectious virions 6 to 8 h after the initial infection. Although infections are generally mild and cleared efficiently, there are three to five million severe cases annually, with 250,000 to 500,000 deaths. Influenza viruses exhibit great genetic variability. As for many RNA viruses, mutations occur frequently, in large part due to the lack of proofreading activity of the viral RNA polymerase, which makes errors in the incorporation of nucleotides into genomic RNA during replication. This has led to the generation of many different subtypes (strains) of influenza viruses with slightly different structures of the spike proteins hemagglutinin and neuraminidase (Box 5.2). As a consequence, existing antibodies do not recognize these surface protein variants (**antigenic drift**) and do not protect an individual against new infections. The packaging of multiple genome segments into a single capsid also contributes to the genetic variability of influenza virus. If a host cell is simultaneously infected with more than one strain of influenza virus, random packaging of the eight viral RNA molecules that make up the complete genome can result in release of virions containing RNA molecules from each strain. This phenomenon, known as **reassortment**, causes major genetic changes (**antigenic shift**) and can generate highly virulent strains when the new virions contain genome segments of animal and human influenza virus strains and can replicate in humans (Box 5.2). These strains can spread rapidly through a population that lacks immunity.

Viruses that cause acute infections followed by clearing and immunity will eventually become extinct when there are no more susceptible hosts. An example is the smallpox virus. A deadly and much-feared disease for thousands of years, smallpox was eradicated as a consequence of an aggressive immunization program. In contrast to the localized infections of influenza virus, smallpox caused systemic infections. The initial infection was in the lungs; however, the virus traveled via the lymphatic and blood circulatory systems to the skin epithelia, where it replicated and erupted in painful blisters often covering the whole body.

Persistent Infections

Viral infections that are not efficiently cleared may persist with no or low levels of disease. Infected cells are not destroyed but rather continuously or intermittently produce relatively low numbers of virions. Persistent infections that are eventually cleared are known as **chronic infections**, while those that persist for the lifetime of the host are known as latent infections. Some examples of viruses that establish persistent infections include adenovirus, which persists in adenoids, tonsils, and lymphocytes; hepatitis C virus, which persists in the liver; HIV, which persists in T cells and macrophages; and papillomavirus, which persists in epithelial cells. Persistent infections may occur when the immune system of a host is ineffective, for example, in tissues where there is poor immune surveillance, such as tissues of the central nervous system, or when cells of the immune system are the targets of infection, such as HIV-infected T cells.

During the latent phase of an infection, the virus stops reproducing and remains dormant. The host is asymptomatic and virions are not detectable, although antibodies against the virus may be detectable. The viral genome persists as a nonreplicating chromosome in nondividing cells such as neurons or as a replicating chromosome in dividing cells, or it is integrated into a host chromosome. The viral genome may be reactivated and direct the production of infectious virions and a subsequent lytic infection. Latency is a characteristic of the herpesvirus family of double-stranded DNA viruses. Several herpesviruses cause disease in humans. HSV-1 is a causative agent of blisters of the mouth and lips known as cold sores, HSV-2 causes lesions in the genital epithelium, and infection with Epstein–Barr virus may result in mononucleosis and, more rarely, nasopharyngeal carcinoma. Varicella–zoster virus causes chicken pox, a lytic infection common in childhood that may emerge after a lengthy dormant period to cause painful skin lesions known as shingles.

HSV-2 is transmitted via sexual contact (anal or vaginal). Initially, a lytic infection occurs in epithelial cells of genitals, with lesions appearing 6 days after exposure. The virus replicates in epithelial cells, producing infectious virions. The released virions infect the ends of nearby dorsal root neurons and travel through the cytoplasm of the axon to the cell bodies **(ganglia)** located at the base of the spine (Fig. 5.51). HSV-2 DNA enters the nucleus of the ganglion through the nuclear pore complex and the DNA circularizes, forming an **episome**. Viral latency-associated transcripts repress expression of viral genes required for a lytic infection. Intercourse, menstruation, and a variety of stresses can reactivate viral gene transcription and production of virions that travel along axons to emerge at the epithelium. The virus is transmitted to surrounding tissue, again establishing lytic infections in epithelial cells. Recurring, secondary lytic infections are generally less severe than the primary infection and diminish with age. HSV-2 can also be transmitted to new hosts, for example, by sexual contact. Transmission from mothers to infants in utero (5% of cases), during birth (85%), or after birth (10%) causes serious infections in the infants. The infants may develop lesions of the skin and eyes and may have damage in the central nervous system, lungs, and liver.

Another type of persistent infection may transform host cells into cancer cells. Most cancers arise from genetic mutations that lead to loss of regulation of cell reproduction. Abnormal new cell growth (neoplasia) forms a mass of cells (a tumor). Malignant (cancerous) tumor cells are invasive and spread to other tissues (metastases), eventually impairing the

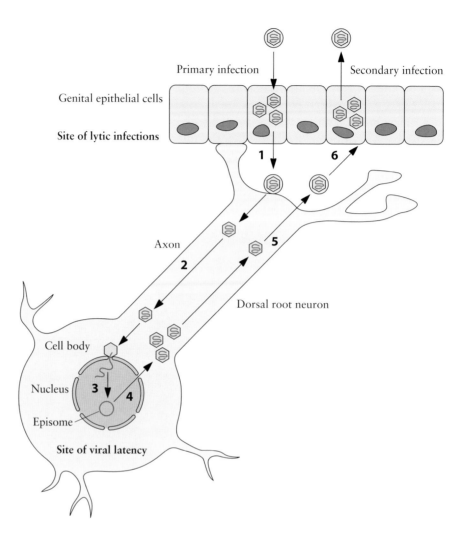

Figure 5.51 Latency in HSV-2. Virions released following a lytic cycle in genital epithelial cells infect nearby neurons (**1**) and migrate through the axonal cytoplasm to the cell body of a dorsal root neuron (**2**). The viral DNA genome is released via the nuclear pore complex into the nucleus, where it circularizes to form an episome (**3**). During latency, expression of viral genes is repressed and new virions are not produced. Stress can reactivate expression of viral genes and the production of new virions (**4**) that travel back along the axons to emerge at the epithelium (**5**). Secondary lytic infections of the epithelial cells (**6**) produce vesicular lesions that are characteristic of HSV-2 infections.
doi:10.1128/9781555818890.ch5.f5.51

normal function of those tissues or organs. Epidemiologists estimate that 15 to 20% of human cancers arise from viral infections. Members of several different virus families are known to cause cancers in humans. The RNA viruses hepatitis B virus and hepatitis C virus cause 80% of liver cancers, and the retrovirus T-cell lymphotropic virus has been implicated in some leukemias (cancer of the blood or bone marrow). Some human papillomaviruses are associated with cervical cancer, and herpesvirus 8 can cause Kaposi's sarcoma (cancer of the connective tissue), especially in immunocompromised individuals such as those with AIDS. These cancer-causing viruses, or **oncoviruses**, disrupt the normal cell cycle of the host cell. Cancer may not develop until months or years after the viral infection.

There are two general mechanisms by which a virus may induce tumorigenesis. The genomes of oncoviruses may carry **oncogenes** encoding proteins that disrupt normal cell cycle control, or the viral genomes may integrate into a host chromosome and alter expression of host proteins that regulate the cell cycle. In the latter case, the site of integration into the host genome is important, because transcriptional regulatory signals (promoters or enhancers) present on the inserted viral DNA activate the expression of adjacent cellular genes. For cell transformation, the virus must integrate adjacent to cellular genes that control cell cycling, and therefore, because integration is random, the probability of transformation is low.

When the viral genome encodes an oncogene, tumor induction can occur regardless of whether the genome is maintained in the nucleus as an episome or the viral DNA is integrated into the host chromosome. Viral oncogenes are derivatives of cellular genes that control the cell cycle. They were acquired from the genome of an infected cell by ancestral viruses during rare mispackaging events (Fig. 5.31). Acquisition of cellular genes often renders the virus unable to replicate; however, when a host cell is coinfected with a replication-competent virus that supplies missing viral proteins, the viruses can be propagated. Most characterized viral oncogenes encode proteins that activate cellular signal transduction pathways to promote cell proliferation or proteins that regulate cell cycle progression. They are homologues of cellular genes that encode growth factors or growth factor receptors, G proteins and protein kinases that transduce signals, and transcription factors that control gene expression. Human oncoviruses may also encode proteins that inactivate **tumor suppressor genes**. For example, the retinoblastoma (Rb) tumor suppressor protein normally prevents cellular transcription factors (EF2 family of transcription factors) from activating expression of proteins required for cell cycle progression (Fig. 5.52A). Interaction of Rb with a protein (E7) encoded by the oncogenic human papillomaviruses leads to inactivation of Rb (Fig. 5.52B). Consequently, proteins are expressed that enable cells to pass through the cell division checkpoint (G_1-to S-phase transition).

Rb
retinoblastoma

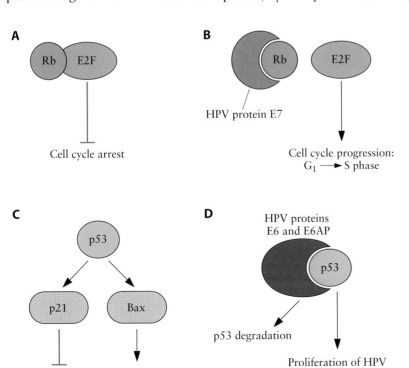

Figure 5.52 Human papillomavirus (HPV) oncogenic proteins stimulate abnormal proliferation of infected cells. In uninfected cells, retinoblastoma tumor suppressor Rb prevents transcription of cell cycle genes by blocking the activity of transcription factor E2F (**A**). Interaction with HPV protein E7 inactivates Rb (**B**). As a consequence, Rb releases E2F, which induces expression of genes that initiate DNA replication and progression of the cell cycle from G_1 phase to S phase. In response to DNA damage or viral infection, tumor suppressor protein p53 normally activates expression of proteins (e.g., p21 and Bax) that arrest the cell cycle and induce the death of damaged or infected cells (**C**). However, in HPV-infected cells, viral proteins E6 and E6AP bind to p53 and stimulate its degradation (**D**). As a result, infected cells do not undergo apoptosis, and new HPV virions are produced. doi:10.1128/9781555818890.ch5.f5.52

When cells are inappropriately stimulated to divide, such as during a viral infection, the apoptotic pathway is normally activated to destroy the damaged cells. However, papillomaviruses produce another protein (E6) that inactivates a second tumor suppressor, p53. Protein p53 is a transcription factor that normally activates expression of proteins that arrest cell cycling and initiate apoptosis in response to DNA damage (Fig. 5.52C). The gene encoding p53 is commonly mutated in human tumor cells. Infection by papillomaviruses blocks the apoptotic response and leads to loss of cell checkpoint control and propagation of damaged cells in which the virus can proliferate (Fig. 5.52D). Despite the devastation caused by oncogenic viruses, they have provided an opportunity to understand the molecular mechanisms of cell transformation.

Targets for Treatment of Viral Infections

Vaccination is the most effective method to protect individuals and populations against viral infections (see chapter 11). It is a preventative strategy that induces immunity to infection by administering a viral antigen that is not capable of causing disease. Immunized individuals produce antibodies against the antigen that prevent the subsequent development of the disease.

Once a viral infection has been established, there are two general treatment strategies. One is to stimulate the body's natural defenses against viruses, while the second directly targets a specific stage of a virus's life cycle to prevent its proliferation in the host. Viral infection, for example, the presence of double-stranded RNA, induces infected cells to produce **interferons** (a type of cytokine) that inhibit viral replication and activate an immune response (macrophages, NK cells, and cytotoxic T cells). Interferons are secreted proteins that induce a protective response in uninfected neighboring cells. Of the two major types of interferons, type I interferons generally mediate the response to viral infections. They stimulate cells to produce endonucleases to degrade mRNA and inhibit ribosome activity, thus preventing production of viral proteins. Type I interferon also induces infected cells to undergo apoptosis, which limits viral spread. Although it was initially thought that treatment with interferons could reduce viral infections, these proteins are toxic in large quantities. They are produced naturally in very small quantities and act only on cells neighboring the producing cell.

Developing antiviral drugs is more difficult than developing antibiotics to treat bacterial infections. This is because viruses proliferate within a host cell using host machinery for synthesis of viral components. Therefore, chemotherapeutic agents that inhibit viral proliferation may also be toxic for the host cell. The challenge is to develop antiviral drugs that have selective toxicity. This requires identification of antiviral drug targets that are unique to the virus and do not inhibit host cell activity. There have been some successes with this approach that were founded on a detailed understanding of the molecular events in a viral life cycle.

Some antiviral drugs prevent viral entry into or exit from host cells. Examples are drugs used to treat influenza virus infections. Recall that influenza virus infection begins with attachment of the envelope protein hemagglutinin to glycoprotein receptors on host respiratory cells. This triggers uptake of the virus by endocytosis. Acidification of the endosome and subsequently the virion leads to release of the viral RNA genome into the host cytoplasm following fusion of viral and endosomal membranes. Acidification of the virion is mediated by the influenza virus M2 ion channel protein (Fig. 5.39). The drugs amantadine and rimantadine specifically inhibit the activity of the M2 ion channel and thereby prevent uncoating of influenza A virus (Fig. 5.53). To be effective, these drugs must be administered early in the infection cycle. Drugs have also been developed to prevent influenza virus virions from leaving the host cell in which they were produced and spreading to other cells or to new hosts. Neuraminidase, another influenza virus membrane spike protein, mediates viral release, and the drugs oseltamivir (Tamiflu) and zanamivir specifically inhibit this protein and thereby prevent release of viral particles (Fig. 5.53). These drugs reduce the duration and severity of the infections. However, influenza viruses reproduce rapidly and mutate frequently, and unfortunately, misuse of these drugs has selected for resistant strains.

Other common targets for antiviral drugs are the proteins required for viral genome synthesis. Although many of the proteins that participate in this process are host proteins, some viruses require specific viral DNA

Figure 5.53 Antiviral drugs used to treat influenza virus infections. Amantadine and rimantadine inhibit the activity of the influenza virus M2 ion channel protein, thereby preventing virion acidification and release of the viral genome into the host cell cytoplasm. Oseltamivir (Tamiflu) and zanamivir are neuraminidase inhibitors that prevent release of new virions from the host cell. Ac, acetyl group. doi:10.1128/9781555818890.ch5.f5.53

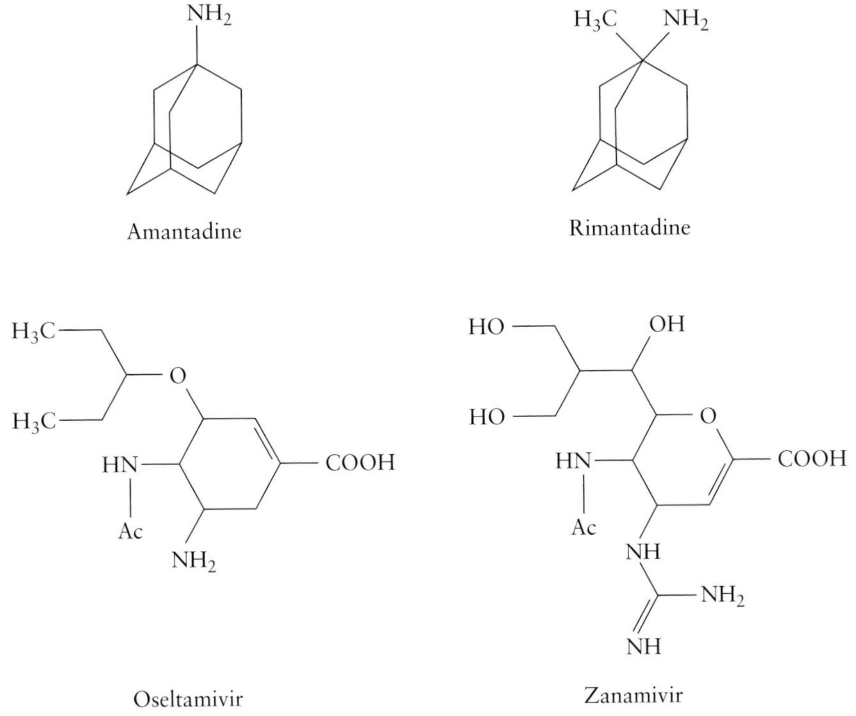

synthesis proteins. Structurally, the agents that inhibit DNA polymerases or reverse transcriptases (Fig. 5.54A) are similar to the natural substrates for these enzymes, that is, DNA nucleosides (Fig. 5.54B). However, when the triphosphorylated form of a **nucleoside analogue** is added to a growing DNA strand, DNA replication is terminated because the synthetic nucleoside lacks a 3′ OH group and therefore cannot form a phosphodiester bond with another nucleotide. Of course, these are effective only for DNA viruses or retroviruses that produce a DNA copy of their genomes. The viral polymerases are more permissive to binding the synthetic nucleosides than the normal cellular polymerases, and therefore these drugs are selectively toxic to viruses. In addition, to be used as a substrate for DNA polymerases, the nucleoside analogue must be triphosphorylated. This requires a specific viral enzyme, such as thymidine kinase produced by herpesviruses. Examples of inhibitors of viral genome synthesis are the G and T nucleoside analogues acyclovir (acycloguanosine) and zidovudine (azidothymidine [AZT]), respectively (Fig. 5.54A). Acyclovir inhibits viral DNA polymerase and is commonly used to treat herpesvirus infections, and AZT inhibits reverse transcriptase and is a treatment for HIV infections.

AZT
azidothymidine

Figure 5.54 The antiviral drugs acyclovir and zidovudine (**A**) are analogues of natural DNA nucleosides deoxyguanosine and deoxythymidine, respectively, (**B**) that prevent DNA synthesis. The synthetic nucleosides are triphosphorylated and added to a growing strand of DNA by DNA polymerase but inhibit subsequent nucleotide additions because they lack a 3′ OH group. doi:10.1128/9781555818890.ch5.f5.54

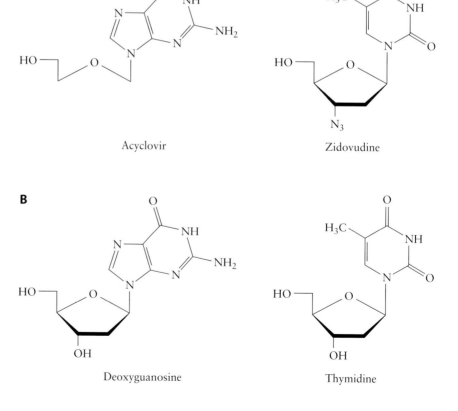

A

Acyclovir   Zidovudine

B

Deoxyguanosine   Thymidine

Many viruses make efficient use of a small genome by producing polyproteins that are cleaved by specific proteases to produce functional viral proteins. The proteases are potential antiviral targets to prevent viral proliferation. **Protease inhibitors** have been used effectively against a specific HIV protease that cleaves polyproteins. Although this is a powerful strategy for reducing infectious HIV particles, resistance often develops, as mutations occur frequently in this rapidly proliferating virus. To prevent resistant strains from arising, treatments for HIV and other viruses often consist of combinations of protease and DNA synthesis inhibitors.

summary

Many microorganisms are capable of infecting humans and causing disease. Pathogenic bacteria bind to specific human cell surface receptors via proteinaceous structures such as pili. Some bacteria may not invade any further than the epithelial barrier of surface-exposed tissues. Rather, noninvasive pathogens may secrete toxins that disrupt host cell functions such as protein synthesis or may stimulate an inflammatory response that disrupts tissue function. Invasive pathogens may secrete proteins (invasins) that disrupt connections between host cells, enabling them to disseminate to deeper tissues, while others induce phagocytosis and are taken up into epithelial cells in membrane-enclosed vesicles. Using a type III or IV protein secretion system, some intracellular pathogens secrete proteins into the host cytoplasm that manipulate the host cell and prevent fusion of the phagocytic vesicle with the lysosome that would otherwise release digestive enzymes to destroy the invading pathogen. These bacteria, such as *Legionella pneumophila*, proliferate within the endocytic vesicle and can invade neighboring cells or underlying tissues. Several features of invasive pathogens enable them to evade the host immune response.

In their quest for nutrients, pathogenic bacteria may damage host tissues, the manifestation of infectious disease. Endotoxins, comprised of the lipid A molecules of the outer membrane of gram-negative bacteria, stimulate fever and inflammation when fragments of the bacterial envelope are released following cell lysis in the host. Exotoxins, proteins secreted by a bacterium that directly block normal host cell functions, are of three general types: (i) cytolytic toxins, (ii) A–B toxins, and (iii) superantigen toxins. As the name suggests, cytolytic toxins are proteins that form a pore in the host cell membrane that causes host cell lysis. Secreted A–B toxins bind to specific host cell receptors via the B domain and are taken into the host cytoplasm within an endocytic vesicle from which the toxin, or A domain, is released into the host cytoplasm. The A domains of some toxins are glycosidases that cleave a specific nitrogenous base from rRNA, thereby disrupting host protein synthesis. The A

domains of other A–B toxins have ADP-ribosyltransferase activity which results in ADP-ribosylation of host proteins. Transfer of an ADP-ribosyl group from NAD to ribosomal proteins or adenylate cyclase disrupts protein synthesis or increases cAMP levels, respectively. In the latter case, increased ion transport into the intestinal lumen, and therefore water excretion, causes diarrheal illnesses from enterotoxigenic *E. coli* and *V. cholerae* infections. Superantigen toxins are peptides that promote interaction between T helper cells and antigen-presenting cells, even in the absence of an antigen, and stimulate massive release of cytokines and an extensive inflammatory response that damages host tissues.

Many bacteria have acquired virulence factors that promote attachment, invasion, dissemination, and damage to host tissues, as well as evasion of host defenses, by horizontal gene transfer. Evidence for this is the clustering of genes encoding virulence factors on pathogenicity islands. Pathogenicity islands typically have GC/AT contents different from the rest of the bacterial genome, coding sequences for viral proteins that mediate integration of viral sequences into the bacterial genome, and/or transposon insertion sequence elements and direct repeat sequences that indicate acquisition by transposition. Plasmid-encoded virulence traits may also be transferred among bacteria by conjugation. Virulence factors can be identified by determining the genes that are expressed in a host (in vivo expression technology), that are required for host colonization (signature-tagged mutagenesis), or that encode proteins that induce production of host antibodies (in vivo-induced-antigen technology).

Bacterial infections may be treated with selectively toxic antibiotics that target structures that are unique to bacterial cells, such as the bacterial peptidoglycan cell wall or bacterial ribosomes. However, many bacteria have acquired resistance to antibiotics by mutating antibiotic targets or by producing enzymes that inactivate the antibiotics or that pump the antibiotic from the cell. Often, the genes encoding antibiotic resistance mechanisms are carried on plasmids.

(continued)

summary (continued)

Many different viruses can infect human cells and cause a variety of mild to serious diseases. In contrast to bacteria, viruses are not cellular. Rather, they consist of genetic material, either DNA or RNA, within a capsid protein coat. These structures, known as nucleocapsids, may be enclosed within a membrane, which contains spike proteins, for enveloped viruses. Viruses are obligate parasites; they require a host cell to replicate their genomes and express proteins. Viruses attach to host cells via specific surface receptors and must enter the host cells, uncoat, and release their genomes to produce new infectious virions. The strategy a virus employs to replicate its genetic material and produce large quantities of viral proteins is determined mainly by whether its genome consists of DNA or RNA. In general, DNA viruses replicate in the host nucleus, while most RNA viruses replicate in the cytoplasm. RNA viruses carry genes encoding RNA-dependent RNA polymerases to produce copies of their genomes because host cells do not possess enzymes for producing RNA molecules from an RNA template. Viral genomes and capsid proteins are assembled into new virions and are released to infect new cells.

Viruses that cause acute infections reproduce rapidly and then are cleared by the host immune system. In contrast, viruses that cause persistent infections are not efficiently cleared and may remain in a host with no or low levels of disease for years. In one type of persistent infection, latent viruses, such as those of the herpesvirus family, do not reproduce and may remain dormant in nondividing cells, often as extrachromosomal DNA elements (episomes), for long periods. Once reactivated, they may resume a lytic cycle to produce large numbers of new virions. Retroviruses, and some DNA viruses, integrate into the host chromosome and may disrupt normal regulation of gene expression. If genes that control the host cell cycle are unregulated, oncogenesis (cancer) may result.

Because viruses utilize host cell machinery for reproduction, treatment strategies that block reproduction may also be toxic for the host. Thus, antiviral drugs target specific viral enzymes such as reverse transcriptase or those that mediate virion release. Vaccination, which is the introduction of a viral antigen to induce immunity, is an effective strategy to protect against viral infections.

review questions

1. What is a pathogen? What is an opportunistic pathogen?

2. Define and list some virulence factors.

3. What is an adhesin? Describe the molecular interactions that enable bacteria to attach to host cells. Consider the interactions that determine host specificity.

4. How are fimbriae (pili) assembled on the surface of gram-negative and gram-positive bacteria?

5. What is antigenic variation, and how does it contribute to infection? How might it reduce the efficacy of adhesin vaccines?

6. Describe the human mucosa that presents a barrier to bacterial invasion. Is bacterial invasion beyond the epithelium always required for disease?

7. Describe some of the strategies used by bacterial pathogens to invade the host epithelium.

8. What is the function of the bacterial type III secretion system in pathogenesis?

9. Outline some of the strategies used by bacterial pathogens to evade the host immune system.

10. Describe some of the ways in which bacterial pathogens manipulate host cells that enable intracellular survival.

11. Why are bacterial biofilms often responsible for persistent or recurring infections?

12. What is the function of bacterial quorum sensing systems?

13. How do bacteria compete for iron in host tissues?

14. What is an endotoxin, and how does it cause host damage?

15. Describe the general activities of the three major classes of exotoxins. Contrast the activities of different A–B toxins.

16. How has comparative genomics contributed to the identification of bacterial virulence factors?

17. Explain how each of the following approaches is used to identify new bacterial virulence factors: (i) in vivo-induced-antigen technology, (ii) transposon mutagenesis, (iii) signature-tagged mutagenesis, and (iv) in vivo expression technology.

18. In the experiment using in vivo-induced-antigen technology shown in Fig. 5.25, why did the researchers probe the UPEC (grown in vitro) and untransformed host E. coli

review questions (continued)

colonies with adsorbed serum (panels A to C) prior to probing the UPEC expression library (panels D and E)?

19. What are some of the characteristic features of pathogenicity islands found in bacterial genomes? What are the mechanisms by which these sequences may have been acquired by a bacterium?

20. What are antibiotics? What are some of the common targets for antibiotics in bacterial cells? Why is selective toxicity important?

21. Describe the three general mechanisms of bacterial antibiotic resistance.

22. Describe the basic structure of a virus, including an enveloped virus.

23. What are the basic steps in a viral infection cycle?

24. In general, what determines the host tropism of a virus?

25. What are the two general mechanisms by which viruses enter host cells?

26. Describe the diverse structures of viral genomes. Explain how the structure of a viral genome determines the process by which the viral genes are expressed and the viral genomes are replicated.

27. What are the substrates and products of DNA-dependent RNA polymerase, RNA-dependent RNA polymerase, and RNA-dependent DNA polymerase? Which organisms (i.e., human host and/or viruses) possess the genes encoding these proteins? Why are they important for that organism(s)?

28. What are the general functions of the proteins that are typically encoded by "early" and "late" viral genes? Why is the temporal regulation of these genes important?

29. How do some human viruses solve the problem of producing several different proteins from a single genomic RNA molecule?

30. Describe the infection cycle of influenza virus.

31. Describe how a DNA copy of the retroviral RNA genome is produced and integrated into the host chromosome. Following integration, how are the retroviral genomes and proteins that will be assembled into infectious virions produced?

32. Where in the host cell are DNA viruses typically assembled? How are the newly synthesized viral genomes and proteins localized to that compartment for assembly?

33. How do enveloped viruses acquire their membranes and spike proteins?

34. How are animal viruses released from host cells?

35. Define acute, persistent, chronic, and latent viral infections.

36. What are the general mechanisms by which oncogenic viruses cause cancer?

37. What is an oncogene? Describe the activity of the proteins encoded by some viral oncogenes.

38. What genetic changes in the influenza virus genome result in antigenic drift and antigenic shift?

39. How do interferons protect against viral infections?

40. Why are antiviral drugs more difficult to develop than antibacterial drugs? Describe the antiviral activity of some of the therapeutic agents that are currently available to treat viral infections.

references

Allen, H. K., J. Donato, H. H. Wang, K. A. Cloud-Hansen, J. Davies, and J. Handelsman. 2010. Call of the wild: antibiotic resistance genes in natural environments. *Nat. Rev. Microbiol.* **8:** 251–259.

Bandara, H. M. H. N., O. L. T. Lam, L. J. Jin, and L. Samaranayake. 2012. Microbial chemical signaling: a current perspective. *Crit. Rev. Microbiol.* **38:** 217–249.

Bisno, A. L., M. O. Brito, and C. M. Collins. 2003. Molecular basis of group A streptococcal virulence. *Lancet Infect. Dis.* **3:**191–200.

Buettner, D. 2012. Protein export according to schedule: architecture, assembly, and regulation of type III secretion systems from plant- and animal-pathogenic bacteria. *Microbiol. Mol. Biol. Rev.* **76:**262–310.

Camilli, A., and B. L. Bassler. 2006. Bacterial small-molecule signaling pathways. *Science* **311:**1113–1116.

Chain, E., H. W. Florey, A. D. Gardner, N. G. Heatley, M. A. Jennings, J. O. Ewing, and A. G. Sanders. 1940. Penicillin as a chemotherapeutic agent. *Lancet* ii:226–228.

Damania, B. 2007. DNA tumor viruses and human cancer. *Trends Microbiol.* **15:**38–44.

Deng, Q., and J. T. Barbieri. 2008. Molecular mechanisms of the cytotoxicity of ADP-ribosylating toxins. *Annu. Rev. Microbiol.* **62:**271–288.

Fleming, A. 1929. On the antibacterial action of cultures of a *Penicillium*, with special reference to their use in the isolation of *B. influenzae. Br. J. Exp. Pathol.* **10:**226–236.

Flemming, H.-C., and J. Wingender. 2010. The biofilm matrix. *Nat. Rev. Microbiol.* **8:**623–633.

Flint, S. J., L. W. Enquist, V. R. Racaniello, and A. M. Skalka. 2008. *Principles of Virology*, 3rd ed. ASM Press, Washington, DC.

Fronzes, R., P. J. Christie, and G. Waksman. 2009. The structural biology of type IV secretion systems. *Nat. Rev. Microbiol.* 7:703–714.

Hall-Stoodley, L., and P. Stoodley. 2009. Evolving concepts in biofilm infections. *Cell. Microbiol.* 11:1034–1043.

Hammerschmidt, S., S. Wolff, A. Hocke, S. Rosseau, E. Müller, and M. Rohde. 2005. Illustration of pneumococcal polysaccharide capsule during adherence and invasion of epithelial cells. *Infect. Immun.* 73:4653–4667.

Handfield, M., and R. C. Levesque. 1999. Strategies for isolation of *in vivo* expressed genes from bacteria. *FEMS Microbiol. Rev.* 23:69–91.

Henderson, B., S. Nair, J. Pallas, and M. A. Williams. 2011. Fibronectin: a multidomain host adhesin targeted by bacterial fibronectin-binding proteins. *FEMS Microbiol. Rev.* 35:147–200.

Herbert, S., A. K. Ziebandt, K. Ohlsen, T. Schäfer, M. Hecker, D. Albrecht, R. Novick, and F. Götz. 2010. Repair of global regulators in *Staphylococcus aureus* 8325 and comparative analysis with other clinical isolates. *Infect. Immun.* 78:2877–2889.

Hill, S. A., and J. K. Davies. 2009. Pilin gene variation in *Neisseria gonorrhoeae*: reassessing the old paradigms. *FEMS Microbiol. Rev.* 33:521–530.

Isberg, R. R., T. J. O'Connor, and M. Heidtman. 2009. The *Legionella pneumophila* replication vacuole: making a cosy niche inside host cells. *Nat. Rev. Microbiol.* 7:13–24.

Juhas, M., J. R. Van Der Meer, M. Gaillard, R. M. Harding, D. W. Hood, and D. W. Crook. 2009. Genomic islands: tools of bacterial horizontal gene transfer and evolution. *FEMS Microbiol. Rev.* 33:376–393.

Kobasa, D., S. M. Jones, K. Shinya, J. C. Kash, J. Copps, H. Ebihara, Y. Hatta, J. H. Kim, P. Halfmann, M. Hatta, F. Feldmann, J. B. Alimonti, L. Fernando, Y. Li, M. G. Katze, H. Feldmann, and Y. Kawaoka. 2007. Aberrant innate immune response in lethal infection of macaques with the 1918 influenza virus. *Nature* 445:319–323.

Kohanski, M. A., D. J. Dwyer, and J. J. Collins. 2010. How antibiotics kill bacteria: from targets to networks. *Nat. Rev. Microbiol.* 8:423–435.

Kurth, R., and N. Bannert. 2010. Beneficial and detrimental effects of human endogenous retroviruses. *Int. J. Cancer* 126:306–314.

Longworth, M. S., and A. Laimins. 2004. Pathogenesis of human papillomaviruses in differentiating epithelia. *Microbiol. Mol. Biol. Rev.* 68:362–372.

Mahan, M. J., J. S. Slauch, and J. J. Mekalanos. 1993. Selection of bacterial virulence genes that are specifically induced in host tissues. *Science* 259:686–688.

Marlovits, T. C., and C. E. Stebbins. 2010. Type III secretion systems shape up as they ship out. *Curr. Opin. Microbiol.* 13:47–52.

Mazurkiewicz, P., C. M. Tang, C. Boone, and D. W. Holden. 2006. Signature-tagged mutagenesis: barcoding mutants for genome-wide screens. *Nat. Rev. Genet.* 7:929–939.

Olsen, R. J., S. A. Shelburne, and J. M. Musser. 2009. Molecular mechanisms underlying group A streptococcal pathogenesis. *Cell. Microbiol.* 11:1–12.

Pallen, M. J., and B. W. Wren. 2007. Bacterial pathogenomics. *Nature* 449:835–842.

Pizarro-Cerda, J., and P. S. Cossart. 2006. Bacterial adhesion and entry into host cells. *Cell* 124:715–727.

Raymond, K. N., E. A. Dertz, and S. S. Kim. 2003. Enterobactin: an archetype for microbial iron transport. *Proc. Natl. Acad. Sci. USA* 100:3584–3588.

Reid, A. H., T. G. Fanning, J. V. Hultin, and J. K. Taubenberger. 1999. Origin and evolution of the 1918 "Spanish" influenza virus hemagglutinin gene. *Proc. Natl. Acad. Sci. USA* 96:1651–1656.

Rollins, S. M., A. Peppercorn, L. Hang, J. D. Hillman, S. B. Calderwood, M. Handfield, and E. T. Ryan. 2005. *In vivo* induced antigen technology (IVIAT). *Cell. Microbiol.* 7:1–9.

Russell, D. G. 2007. Who puts the tubercle in tuberculosis? *Nat. Rev. Microbiol.* 5:39–47.

Schauder, S., and B. A. Bassler. 2001. The languages of bacteria. *Genes Dev.* 15:1468–1480.

Schmidt, H., and M. Hensel. 2004. Pathogenicity islands in bacterial pathogenesis. *Clin. Microbiol. Rev.* 17:14–56.

Scott, J. R., and D. Zähner. 2006. Pili with strong attachments: Gram-positive bacteria do it differently. *Mol. Microbiol.* 62:320–330.

Smith, A. E., and A. Helenius. 2004. How viruses enter animal cells. *Science* 304:237–242.

Taubenberger, J. K., A. H. Reid, R. M. Lourens, R. Wang, G. Jin, and T. G. Fanning. 2005. Characterization of the 1918 influenza virus polymerase genes. *Nature* 437:889–893.

Taubenberger, J. K., D. Baltimore, P. C. Doherty, H. Markel, D. M. Morens, R. G. Webster, and I. A. Wilson. 2012. Reconstruction of the 1918 influenza virus: unexpected rewards from the past. *mBio* 3:e00201-12.

Taubenberger, J. K., A. H. Reid, A. E. Krafft, K. E. Bijwaard, and T. G. Fanning. 1997. Initial genetic characterization of the 1918 "Spanish" influenza virus. *Science* 275:1793–1796.

Tumpey, T. M., C. F. Basler, P. V. Aguilar, H. Zeng, A. Solórzano, D. E. Swayne, N. J. Cox, J. M. Katz, J. K. Taubenberger, P. Palese, and A. García-Sastre. 2005. Characterization of the reconstructed 1918 Spanish influenza pandemic virus. *Science* 310:77–80.

Vigil, P. D., C. J. Alteria, and H. L. Mobley. 2011. Identification of in vivo-induced antigens including an RTX family exoprotein required for uropathogenic *Escherichia coli* virulence. *Infect. Immun.* 79:2335–2344.

Vink, C., G. Rudenko, and H. S. Seifert. 2012. Microbial antigenic variation mediated by homologous DNA recombination. *FEMS Microbiol. Rev.* 36:917–948.

Wagner, E. K., M. J. Hewlett, D. C. Bloom, and D. Camerini. 2008. *Basic Virology*, 3rd ed. Blackwell Publishing, Malden, MA.

Wilson, B. A., A. A. Salyers, D. D. Whitt, and M. E. Winkler. 2010. *Bacterial Pathogenesis: a Molecular Approach*, 3rd ed. ASM Press, Washington, DC.

Production of Therapeutic Agents

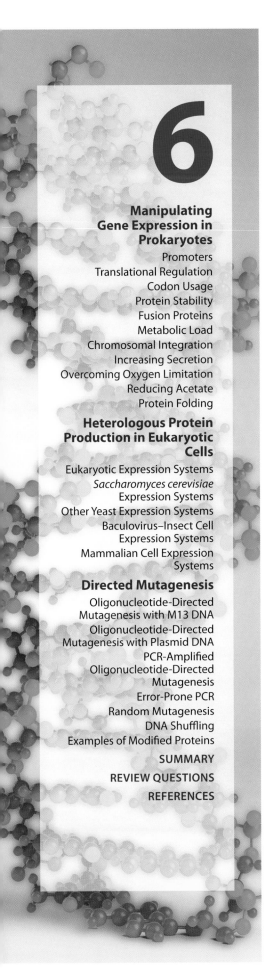

6

Modulation of Gene Expression

THE MAJOR OBJECTIVE of gene cloning for biotechnological applications is the expression of a cloned gene in a selected host organism. In addition, for many purposes, a high rate of production of the protein encoded by the cloned gene is needed. To this end, a wide range of expression vectors that provide genetic elements for controlling transcription and translation of the cloned gene as well as enhanced stability, facilitated purification, and facilitated secretion of the protein product of the cloned gene have been constructed. There is not one single strategy for obtaining maximal protein expression from every cloned gene. Rather, there are a number of different biological parameters that can be manipulated to yield an optimal level of expression.

The level of foreign gene expression also depends on the host organism. Although a wide range of both prokaryotic and eukaryotic organisms have been used to express foreign genes, initially, many of the commercially important proteins produced by recombinant DNA technology were synthesized in *Escherichia coli*. The early dependence on *E. coli* as a host organism occurred because of the extensive knowledge of its genetics, molecular biology, biochemistry, and physiology (Milestone 6.1). To date, recombinant proteins have been produced using different strains of bacteria (including *E. coli*), yeasts, and mammalian cells grown in culture and transgenic plants (Table 6.1). Each of these systems has its particular advantages and disadvantages, so, again, there is no universal optimal host for the expression of recombinant proteins, even those that will eventually be used as therapeutic agents or vaccines. Thus, for example, *E. coli* cells may be engineered to produce high levels of foreign proteins; however, these proteins are not glycosylated and are sometimes misfolded. On the other hand, with mammalian cells in culture, recombinant proteins are correctly glycosylated and folded, although the yield of proteins is much lower than in *E. coli*. Notwithstanding the very large differences between organisms, the strategies that have been elaborated for *E. coli*, in principle, are applicable to all systems.

doi:10.1128/9781555818890.ch6

Construction of Biologically Functional Bacterial Plasmids *In Vitro*

S. N. COHEN, A. C. Y. CHANG, H. W. BOYER, AND R. B. HELLING
Proc. Natl. Acad. Sci. USA **70:**3240–3244, 1973

milestone
6.1

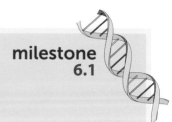

In 1972, Paul Berg and his co-workers (Jackson et al., 1972) demonstrated that fragments of bacteriophage λ DNA could be spliced into SV40. They reported, for the first time, that fragments of DNA could be covalently joined with other DNA molecules. This joining of "unrelated DNA molecules to one another" by Jackson et al. is arguably the first demonstration of the possibility of recombinant DNA technology. However, while SV40 was at the time thought to be safe in humans, the prospect of an altered form of the virus spreading unchecked, through the common bacterium *E. coli*, caused Berg to delay a portion of his research program. Thus, contrary to his original plan, Berg did not insert the recombinant virus into bacterial cells.

Soon after Berg published the results of his experiments, Stanley Cohen and Herbert Boyer and their colleagues (at Stanford University and the University of California at San Francisco) showed that a recombinant DNA molecule could be created without the use of viruses. They demonstrated that foreign DNA could be inserted into plasmid DNA and subsequently perpetuated in *E. coli*. As they state in the abstract to their research article, "The construction of new plasmid DNA species by *in vitro* joining of restriction endonuclease-generated fragments of separate plasmids is described. Newly constructed plasmids that are inserted into *Escherichia coli* by transformation are shown to be biologically functional replicons that possess genetic properties and nucleotide base sequences from both of the parent DNA molecules. Functional plasmids can be obtained by reassociation of endonuclease-generated fragments of larger replicons, as well as by joining of plamid DNA molecules of entirely different origins."

With the publishing of this article, recombinant DNA technology had truly arrived. The technology spread, first slowly to a few labs and then to dozens and, eventually, to tens of thousands of labs worldwide. In the 40 years since the groundbreaking experiments of Cohen and Boyer and their colleagues, more than 200 new drugs produced by recombinant DNA technology have been used to treat over 300 million people for a wide range of human diseases. In addition, today more than 400 additional drugs produced using this technology are in various stages of clinical trials, with many expected to be on the market within the next 5 to 10 years. Today, Cohen and Boyer are widely regarded as the founders of the scientific revolution that has become modern biotechnology.

Manipulating Gene Expression in Prokaryotes

A large number of factors affect the level of expression of a foreign gene in a prokaryotic host. These include the transcriptional promoter regulating the expression of the target gene, the translational regulatory region that is **upstream** of the coding region of every gene, the similarity of the

Table 6.1 Production of recombinant human proteins in various biological hosts

Parameter	Bacteria	Yeast	Mammalian cell culture	Transgenic plants
Glycosylation	None	Incorrect	Correct	Generally correct; small differences
Multimeric proteins assembled	Limited	Limited	Limited	Yes
Production costs	Low-medium	Medium	High	Very low
Protein folding accuracy	Low	Medium	High	High
Protein yield	High	Medium-high	Low-medium	Medium
Scale-up capacity	High	High	Low	Very high
Scale-up costs	High	High	High	Low
Time required	Low	Low-medium	Medium-high	High
Skilled workers required	Medium	Medium	High	Low
Acceptable to regulators	Yes	Yes	Yes	Not yet

codons used by the host organism and the target gene, the stability of both the recombinant protein and its mRNA, the metabolic functioning of the host cell, and the localization of the introduced foreign gene as well as the protein that it encodes.

Promoters

The minimum requirement for an effective gene expression system is the presence of a strong and regulatable **promoter** sequence upstream from a cloned gene. A strong promoter is one that has a high affinity for the enzyme **RNA polymerase**, with the consequence that the adjacent downstream region is frequently transcribed. The ability to regulate the functioning of a promoter allows the researcher to control the extent of transcription.

The rationale behind the use of strong and regulatable promoters is that the expression of a cloned gene under the control of a continuously activated (i.e., constitutive) strong promoter would likely yield a high level of continual expression of a cloned gene, which is often detrimental to the host cell because it creates an energy drain, thereby impairing essential host cell functions. In addition, all or a portion of the plasmid carrying a constitutively expressed cloned gene may be lost after several cell division cycles, since cells without a plasmid grow faster and eventually take over the culture. To overcome this potential problem, it is desirable to control transcription so that a cloned gene is expressed only at a specific stage in the host cell growth cycle and only for a specified duration. This may be achieved by using a strong regulatable promoter. The plasmids constructed to accomplish this task are called expression vectors.

For the production of foreign proteins in *E. coli* cells, a few strong and regulatable promoters are commonly used, including those from the *E. coli lac* (lactose) and *trp* (tryptophan) operons; the *tac* promoter, which is constructed from the -10 region (i.e., 10 nucleotide pairs upstream from the site of initiation of transcription) of the *lac* promoter and the -35 region of the *trp* promoter; the leftward and rightward, or p^L and p^R, promoters from **bacteriophage** λ; and the gene 10 promoter from bacteriophage T7. Each of these promoters interacts with regulatory proteins (i.e., repressors or inducers), which provide a controllable switch for either turning on or turning off the transcription of the adjacent cloned genes. Each of these promoters is recognized by the major form of the *E. coli* RNA polymerase **holoenzyme**. This holoenzyme is formed when a protein, called sigma factor, combines with the core proteins (i.e., two α, one β, and one β′ subunit) of RNA polymerase. The sigma factor directs the binding of the holoenzyme to promoter regions on the DNA.

One commonly used expression system utilizes the gene 10 promoter from bacteriophage T7 (Fig. 6.1). This promoter is not recognized by *E. coli* RNA polymerase but, rather, requires T7 **RNA polymerase** for transcription to occur. For this system to work in *E. coli*, the gene encoding T7 polymerase is often inserted into the *E. coli* chromosome under the transcriptional control of the *E. coli lac* promoter and **operator**. The *E. coli* host cells must also contain the *E. coli lacI* gene, which encodes the *lac* **repressor**. The *lac* repressor forms a tetramer that binds to the *lac*

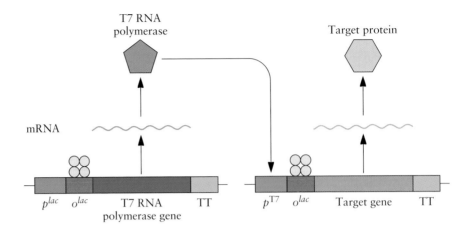

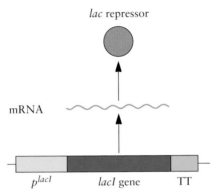

Figure 6.1 Regulation of gene expression controlled by the promoter for gene 10 from bacteriophage T7 (p^{T7}). In the absence of the inducer IPTG, the constitutively produced *lac* repressor, the product of the *lacI* gene, which is under the control of the *lacI* promoter, p^{lacI}, represses the transcription of the T7 RNA polymerase and the target gene, which are both negatively regulated by the binding of four molecules of the *lac* repressor to the *lac* operator (o^{lac}). In the absence of T7 RNA polymerase, the target gene, which is under the transcriptional control of p^{T7}, is not transcribed. When lactose or IPTG is added to the medium, it binds to the *lac* repressor, thereby preventing it from repressing the transcription of the T7 RNA polymerase gene directed by the *lac* promoter (p^{lac}) and the target gene directed by p^{T7}. In the presence of T7 RNA polymerase, the target gene is transcribed. TT, transcription termination sequence. doi:10.1128/9781555818890.ch6.f6.1

IPTG
isopropyl-β-D-thiogalactopyranoside

operator and prevents T7 RNA polymerase from being made before it is needed. Following transformation of the host *E. coli* cells with a plasmid containing the cloned target gene under the transcriptional control of the T7 promoter, the compound isopropyl-β-D-thiogalactopyranoside (IPTG) is added to the growth medium. Under these conditions, IPTG binds to the *lac* repressor, thereby removing it from the *lac* operator and permitting the *lac* promoter to be transcribed so that T7 RNA polymerase can be synthesized. The T7 RNA polymerase binds to the T7 promoter and transcribes the gene of interest. To ensure that synthesis of the target protein does not interfere with cellular metabolism, potentially depleting cellular resources, in the absence of IPTG, the *lac* operator is often inserted between the T7 promoter and the target gene so that expression of the target gene is also negatively controlled by the *lac* repressor. With this arrangement of regulatory regions, there is no synthesis of the target gene. However, following the addition of IPTG, and after a delay of about 1 h, a large burst of synthesis of the target protein occurs. While the features of the DNA constructs may vary when other strong and regulatable promoters are used, the experimental design is similar and includes the entry of the host cells into a growth phase, when the target protein is not expressed, which is followed by an induction phase when the target protein is expressed at a high level.

A problem with the T7 expression system is that with the very high rate of **transcription** of the T7 RNA polymerase compared to that of the native *E. coli* enzyme (the T7 enzyme transcribes DNA into mRNA eight times faster), the recombinant protein may not fold properly and hence form insoluble **inclusion bodies**. In addition, when the target protein is a membrane protein, very high levels may be toxic to the cell. To avoid these problems, it is possible to downregulate (attenuate) the activity of the T7 RNA polymerase by producing the inhibitor T7 lysozyme. Moreover, the amount of T7 lysozyme in the cell can be finely regulated by placing the gene encoding this enzyme under the transcriptional control of the p^{rhaBAD} promoter (Fig. 6.2). This promoter is regulated by a wide range in the concentration of (the deoxy sugar) L-rhamnose (i.e., 10 to 2,000 μM). Thus, by decreasing the amount of T7 RNA polymerase, the maximum amount of a target protein may be synthesized without the production of inclusion bodies for soluble proteins or host toxic levels for membrane proteins.

E. coli is not necessarily the microorganism of choice for the expression of all foreign proteins. However, an understanding of the genetics and molecular biology of most other microorganisms is not nearly as well developed. Moreover, there is no one vector or promoter-repressor system that gives optimal levels of gene expression in all bacteria, or even in all gram-negative bacteria. Fortunately, most of the strategies that have been developed for *E. coli* are also useful with a wide range of microorganisms as well as other host cells.

Figure 6.2 Attenuation of the activity of T7 RNA polymerase by the inhibitor T7 lysozyme. The amount of T7 lysozyme is controlled by the level of L-rhamnose added to the system (which activates the p^{rhaBAD} promoter). TT, transcription termination sequence. doi:10.1128/9781555818890.ch6.f6.2

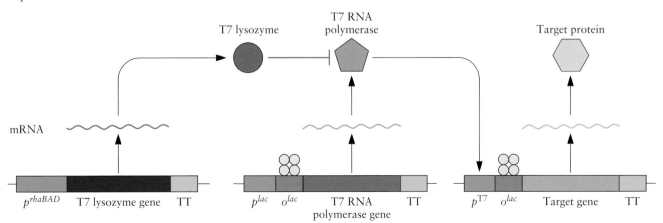

Translational Regulation

Placing a cloned gene under the control of a regulatable, strong promoter, although essential, may not be sufficient to maximize the yield of the cloned gene product. Other factors, such as the efficiency of **translation** and the stability of the newly synthesized cloned gene protein, may also affect the amount of product.

In prokaryotic cells, various proteins are not necessarily synthesized with the same efficiency. In fact, they may be produced at very different levels (up to several hundredfold) even if they are encoded within the same **polycistronic mRNA**. Differences in translational efficiency and in transcriptional regulation enable the cell to have hundreds or even thousands of copies of some proteins and only a few copies of others.

The molecular basis for differential translation of bacterial mRNAs is the presence of a translational initiation signal called a **ribosome-binding site** which precedes the protein-coding portion of the mRNA. A ribosome-binding site is a sequence of six to eight nucleotides in mRNA that can base-pair with a complementary nucleotide sequence on the 16S RNA component of the small ribosomal subunit. Generally, the stronger the binding of the mRNA to the rRNA, the greater the efficiency of translational initiation.

Thus, many *E. coli* expression vectors have been designed to ensure that the mRNA of a cloned gene contains a strong ribosome-binding site. Inclusion of an *E. coli* ribosome-binding site just upstream from the protein-coding open **reading frame** ensures that heterologous prokaryotic and eukaryotic genes can be translated readily in *E. coli*. However, certain conditions must be satisfied for this approach to function properly. First, the ribosome-binding sequence must be located within a short distance (generally 2 to 20 nucleotides) from the translational start codon of the cloned gene. At the RNA level, the translational **codon** is usually AUG (adenosine, uridine, and guanidine). In DNA, the coding strand contains the ATG sequence (where T is thymidine) that functions as a start codon, and the complementary noncoding strand is a template for transcription. Second, the DNA sequence that includes the ribosome-binding site through the first few codons of the gene of interest should not contain nucleotide sequences that have regions of complementarity and can fold back (form intrastrand loops) after transcription (Fig. 6.3), thereby blocking the interaction of the mRNA with the ribosome. The local secondary structure of the mRNA, which can either shield or expose the ribosome-binding site, determines the extent to which the mRNA can bind to the appropriate sequence on the ribosome and initiate translation. Thus, for each cloned gene, it is important to ensure that the mRNA contains a strong ribosome-binding site and that the secondary structure of the mRNA does not prevent its access to the ribosome. However, since the nucleotide sequences that encode the amino acids at the N-terminal region of the target protein vary from one gene to another, it is not possible to design a vector that will eliminate the possibility of mRNA fold-back in all instances. Therefore, no single optimized translational initiation region can guarantee a high rate of translation initiation for all cloned genes. Consequently, the optimization of translation initiation needs to be on a gene-by-gene basis.

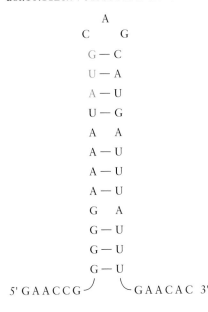

Figure 6.3 Example of secondary structure of the 5′ end of an mRNA that would prevent efficient translation. In this example, the ribosome-binding site is GGGGG, the initiator codon is AUG (shown in red), and the first few codons are CAG-CAU-GAU-UUA-UUU. The mRNA is oriented with its 5′ end to the left and its 3′ end to the right. Note that in addition to the traditional A·U and G·C base pairs in mRNA, G can also base-pair to some extent with U. doi:10.1128/9781555818890.ch6.f6.3

Codon Usage

While the **genetic code** for amino acids, on average, includes about three different codons (any particular amino acid may have from one to six codons), these codons are used to different extents in various living organisms. Any organism, e.g., *E. coli*, produces cognate tRNAs for each codon in approximately the same relative amount as that particular codon is used in the production of its proteins. Various organisms preferentially use different subsets of codons (Table 6.2) and contain various amounts of the cognate amino acyl-tRNAs for the synthesis of proteins encoded by their mRNAs. Thus, expressing a foreign gene in a particular host organism may result in a cellular incompatibility that can interfere with efficient translation when a cloned gene has codons that are rarely used by the host cell. For example, AGG, AGA, AUA, CUA, and CGA are the least-used codons in *E. coli*. When a foreign protein is expressed

Table 6.2 Genetic code and codon usage in *E. coli* and humans

| Codon | Amino acid | Frequency of use in: | | Codon | Amino acid | Frequency of use in: | |
		E. coli	Humans			*E. coli*	Humans
GGG	Glycine	0.13	0.23	UAG	Stop	0.09	0.17
GGA	Glycine	0.09	0.26	UAA	Stop	0.62	0.22
GGGU	Glycine	0.38	0.18	UAU	Tyrosine	0.53	0.42
GGC	Glycine	0.40	0.33	UAC	Tyrosine	0.47	0.58
GAG	Glutamic acid	0.30	0.59	UUU	Phenylalanine	0.51	0.43
GAA	Glutamic acid	0.70	0.41	UUC	Phenylalanine	0.49	0.57
GAU	Aspartic acid	0.59	0.44	UCG	Serine	0.13	0.06
GAC	Aspartic acid	0.41	0.56	UCA	Serine	0.12	0.15
GUG	Valine	0.34	0.48	UCU	Serine	0.19	0.17
GUA	Valine	0.17	0.10	UCC	Serine	0.17	0.23
GUU	Valine	0.29	0.17	AGU	Serine	0.13	0.14
GUC	Valine	0.20	0.25	AGC	Serine	0.27	0.25
GCG	Alanine	0.34	0.10	CGG	Arginine	0.08	0.19
GCA	Alanine	0.22	0.22	CGA	Arginine	0.05	0.10
GCU	Alanine	0.19	0.28	CGU	Arginine	0.42	0.09
GCC	Alanine	0.25	0.40	CGC	Arginine	0.37	0.19
AAG	Lysine	0.24	0.60	AGG	Arginine	0.03	0.22
AAA	Lysine	0.76	0.40	AGA	Arginine	0.04	0.21
AAU	Asparagine	0.39	0.44	CAG	Glutamine	0.69	0.73
AAC	Asparagine	0.61	0.56	CAA	Glutamine	0.31	0.27
AUG	Methionine	1.00	1.00	CAU	Histidine	0.52	0.41
AUA	Isoleucine	0.07	0.14	CAC	Histidine	0.48	0.59
AUU	Isoleucine	0.47	0.35	CUG	Leucine	0.55	0.43
AUC	Isoleucine	0.46	0.51	CUA	Leucine	0.03	0.07
ACG	Threonine	0.23	0.12	CUU	Leucine	0.10	0.12
ACA	Threonine	0.12	0.27	CUC	Leucine	0.10	0.20
ACU	Threonine	0.21	0.23	UUG	Leucine	0.11	0.12
ACC	Threonine	0.43	0.38	UUA	Leucine	0.11	0.06
UGG	Tryptophan	1.00	1.00	CCG	Proline	0.55	0.11
UGU	Cysteine	0.43	0.42	CCA	Proline	0.20	0.27
UGC	Cysteine	0.57	0.58	CCU	Proline	0.16	0.29
UGA	Stop	0.30	0.61	CCC	Proline	0.10	0.33

at high levels in *E. coli*, the host cell may not produce enough of the aminoacyl-tRNAs that recognize these rarely used codons, and either the yield of the cloned gene protein is much lower than expected or incorrect amino acids may be inserted into the protein. Any codon that is used less than 5 to 10% of the time by the host organism may cause problems. Particularly detrimental to high levels of expression are regions of mRNA where two or more rarely used codons are close or adjacent to, or appear in, the sequence encoding the N-terminal portion of the protein. Fortunately, there are several experimental approaches that can be used to alleviate this problem. First, if the target gene is eukaryotic, it may be cloned and expressed in a eukaryotic host cell. Second, a new version of the target gene containing codons more commonly used by the host cell may be chemically synthesized (i.e., **codon optimization**). Third, a host *E. coli* cell engineered to overexpress several rare tRNAs may be employed. In fact, some *E. coli* strains have been transformed with plasmids that encode genes that lead to the overproduction of some tRNAs which are specific for certain rare *E. coli* codons. These transformed *E. coli* cell lines are available commercially and can often facilitate a high level of expression of foreign proteins that use these rare *E. coli* codons (Fig. 6.4). For example, with one of the commercially available *E. coli* cell lines, it was possible to overexpress the Ara h2 protein, a peanut allergen, approximately 100-fold over the amount that was synthesized in conventional *E. coli* cells. With this approach, it should be possible to produce large quantities of a variety of heterologous proteins that are otherwise difficult to express in different hosts.

Protein Stability

The expression of some foreign proteins in *E. coli* host strains, which are typically grown at 37°C, often results in the formation of inclusion bodies of inactive protein. This occurs because the foreign protein misfolds when it cannot attain its native active conformation. A variety of strategies have been developed, albeit with limited success, to circumvent this problem. Cultivation of recombinant strains at lower temperatures sometimes facilitates slower, and hence proper, protein folding, often significantly increasing the amount of recoverable active protein. However, **mesophilic** bacteria like *E. coli* grow extremely slowly at low temperatures. In one study, the chaperonin 60 gene (*cpn60*) and the cochaperonin 10 gene (*cpn10*) from the **psychrophilic** bacterium *Oleispira antarctica* were introduced into a host strain of *E. coli*, with the result that the *E. coli* strain gained the ability to grow and to express foreign proteins at a high rate at temperatures of 4 to 10°C (Fig. 6.5). It has been suggested that at temperatures below around 20°C, *E. coli* cells are unable to grow to any appreciable extent as a consequence of the cold-induced inactivation of several *E. coli* **chaperonins** that normally facilitate protein folding in this bacterium. Thus, transforming *E. coli* with chaperonins from a cold-tolerant bacterium allowed the introduced proteins to perform the functions at low temperature that *E. coli* proteins perform at higher temperatures. Although very high levels of expression of the cloned gene were

A

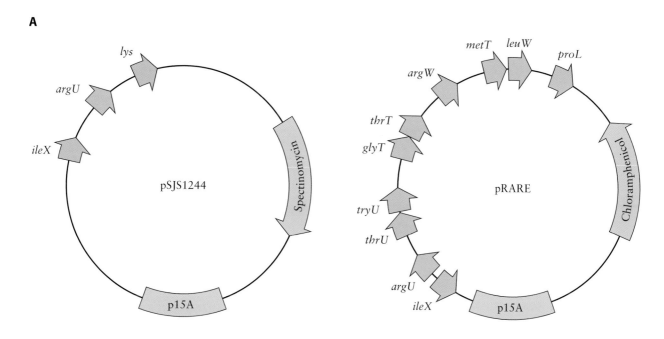

B

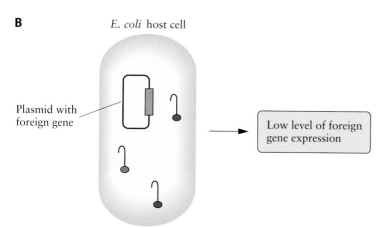

C

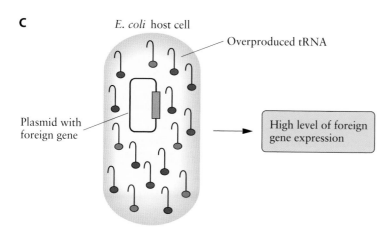

Figure 6.4 Schematic representation of two commercially available plasmids that may be used to increase the pool of certain rare tRNAs in *E. coli*. Plasmid pSJS1244 carries 3 and plasmid pRARE carries 10 rare *E. coli* tRNA genes. p15A represents the replication origins of these plasmids. Spectinomycin and chloramphenicol are the antibiotics for which resistance genes are carried within these plasmids (**A**). The expression of foreign proteins in a typical *E. coli* host cell is also shown; the concentration of rare tRNAs is shown schematically (**B**) and in an *E. coli* host cell that has been engineered (by introduction of one of the plasmids shown in panel A) to overexpress several rare tRNAs (**C**). doi:10.1128/9781555818890.ch6.f6.4

Nontransformed *E. coli*   Transformed *E. coli*

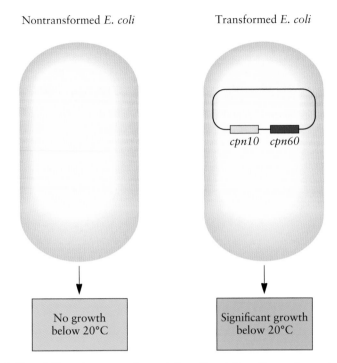

Figure 6.5 The ability of nontransformed *E. coli* and *E. coli* transformed to express plasmid-borne chaperonin genes *cpn10* and *cpn60*, which were isolated from a psychrophilic bacterium, and consequently grow at low temperatures. doi:10.1128/9781555818890.ch6.f6.5

not attained, this work is an important first step in the development of expression systems for proteins that are sensitive to high temperatures and might otherwise be difficult to produce. The next logical step in the development of this system would likely be the construction of an *E. coli* host cell that contains stably integrated copies of these chaperonin genes in the chromosome.

Fusion Proteins

Occasionally, foreign proteins are found in smaller-than-expected amounts when they are produced in heterologous host cells. This apparent low level of expression may be due to degradation of the foreign protein within the host cell. One solution is to engineer a DNA construct encoding a target protein in frame with DNA encoding a stable host protein (Fig. 6.6). The combined, single protein that is produced is called a **fusion protein**, and it protects the cloned foreign gene product from attack by host cell proteases. In general, fusion proteins are stable because the target proteins are fused with proteins that are not especially susceptible to proteolysis.

Fusion proteins are constructed by ligating a portion of the DNA coding regions of two or more genes. In its simplest form, a fusion vector system entails the insertion of a target gene into the coding region of a cloned host gene, or fusion partner (Fig. 6.6). The fusion partner may be positioned at either the N- or C-terminal end of the target gene. Knowledge of the nucleotide sequences of the various coding segments joined at

Figure 6.6 Schematic representation of a DNA construct encoding a fusion protein. A plasmid carrying such a construct would also contain a selectable marker gene. P, transcriptional promoter; RBS, ribosome-binding site; TT, transcriptional terminator. The arrow indicates the direction of transcription. doi:10.1128/9781555818890.ch6.f6.6

the DNA level is essential to ensure that the **ligation** product maintains the correct reading frame. If the combined DNA has an altered reading frame, i.e., a sequence of successive codons that yields either an incomplete or an incorrect translation product, then a functional version of the protein encoded by the cloned target gene will not be produced.

When the protein encoded by the cloned gene is intended for human use, it is generally necessary to remove the fusion partner from the final product. This is because fusion proteins require more extensive testing before being approved by regulatory agencies, such as the U.S. Food and Drug Administration (FDA). Therefore, strategies have been developed to remove the unwanted amino acid sequence from the target protein. One way to do this is to join the gene for the target protein to the gene for the stabilizing fusion partner with specific **oligonucleotides** that encode short stretches of amino acids that are recognized by a particular nonbacterial **protease**. For example, an oligonucleotide linker encoding the amino acid sequence Ile-Glu-Gly-Arg may be joined to the cloned gene. Following synthesis and purification of the fusion protein, a blood coagulation factor called X_a can be used to release the target protein from the fusion partner, because factor X_a is a specific protease that cleaves peptide bonds uniquely on the C-terminal side of the Ile-Glu-Gly-Arg sequence (Fig. 6.7). Moreover, because this peptide sequence occurs rather infrequently in native proteins, this approach can be used to recover many different cloned gene products.

The proteases most commonly used to cleave a fusion partner from a target protein interest are enterokinase, tobacco etch virus protease, thrombin, and factor X_a. However, following this cleavage, it is necessary to perform additional purification steps in order to separate both

FDA
U.S. Food and Drug Administration

Figure 6.7 (A) Proteolytic cleavage of a fusion protein by blood coagulation factor X_a. The factor X_a recognition sequence (X_a linker sequence) lies between the amino acid sequences of two different proteins. A functional cloned gene protein (with Val at its N terminus) is released after cleavage. (B) Schematic representation of a tripartite fusion protein including a stable fusion partner, a linker peptide, and the cloned target protein. doi:10.1128/9781555818890.ch6.f6.7

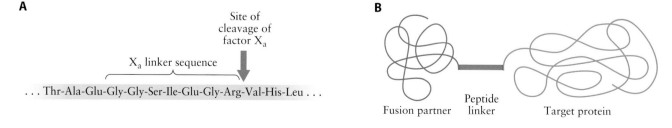

Table 6.3 Some protein fusion systems used to facilitate the purification of foreign proteins in *E. coli* and other host organisms

Fusion partner	Size	Ligand	Elution conditions
ZZ	14 kDa	Immunoglobulin G	Low pH
Histidine tail	6–10 amino acids	Ni^{2+}	Imidazole
Strep tag	10 amino acids	Streptavidin	Iminobiotin
Pinpoint	13 kDa	Streptavidin	Biotin
Maltose-binding protein	40 kDa	Amylose	Maltose
GST	26 kDa	Glutathione	Reduced glutathione
Flag	8 amino acids	Specific MAb	EDTA or low pH
Polyarginine	5–6 amino acids	SP-Sephadex	High salt at pH >8.0
c-myc	11 amino acids	Specific MAb	Low pH
S tag	15 amino acids	S fragment of RNase A	Low pH
Calmodulin-binding peptide	26 amino acids	Calmodulin	EGTA and high salt
Cellulose-binding domain	4–20 kDa	Cellulose	Urea or guanidine hydrochloride
Chitin-binding domain	51 amino acids	Chitin	SDS or guanidine hydrochloride
SBP tag	38 amino acids	Streptavidin	Biotin

ZZ, a fragment of *Staphylococcus aureus* protein A; Strep tag, a peptide with affinity for streptavidin; Pinpoint, a protein fragment that is biotinylated and binds streptavidin; GST, glutathione *S*-transferase; Flag, a peptide recognized by enterokinase; EDTA, ethylenediaminetetraacetic acid; c-myc, a peptide from a protein that is overexpressed in many cancers; S tag, a peptide fragment of ribonuclease (RNase) A; EGTA, ethylene glycol-bis(β-aminoethyl ether)-N,N,N′,N′-tetraacetic acid; SBP (streptavidin-binding protein), a peptide with affinity for streptavidin; SP-Sephadex, a cation-exchange resin composed of sulfopropyl groups covalently attached to Sephadex beads; SDS, sodium dodecyl sulfate.

IL-2
interleukin-2

the protease and the fusion protein from the protein of interest. Unfortunately, sometimes proteases also cleave the protein of interest. When this occurs to any significant extent, it is necessary to change either the linker peptide or the digestion conditions.

In addition to reducing the degradation of cloned foreign proteins, a number of fusion proteins have been developed to simplify the purification of recombinant proteins (Table 6.3). This approach is useful for purification of proteins expressed in either prokaryotic or eukaryotic host organisms. For example, a vector that contains the human **interleukin-2 (IL-2)** cytokine gene joined to DNA encoding the fusion partner (marker peptide) sequence Asp-Tyr-Lys-Asp-Asp-Asp-Asp-Lys serves the dual function of reducing the degradation of the expressed IL-2 gene product and facilitating the purification of the product. Following expression of this construct, the secreted fusion protein can be purified in a single step by immunoaffinity chromatography, in which monoclonal antibodies directed against the marker peptide have been immobilized on a solid support and act as ligands to bind the fusion protein (Fig. 6.8). Because this particular marker peptide is relatively small, it does not significantly decrease the amount of host cell resources that are available for the production of IL-2; thus, the yield of IL-2 is not compromised by the concomitant synthesis of the marker peptide. In addition, while the fusion protein has the same biological activity as native IL-2, as mentioned above, to more readily satisfy the government agencies that regulate the

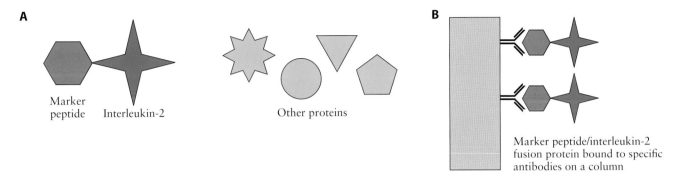

Figure 6.8 Immunoaffinity chromatographic purification of a fusion protein that includes a marker protein and IL-2. A monoclonal antibody that binds to the marker peptide of the fusion protein (anti-marker peptide antibody) is attached to a solid matrix support. The secreted proteins (**A**) are passed through the column containing the bound antibody. The marker peptide portion of the fusion protein is bound to the antibody (**B**), and the other proteins pass through. The immunopurified fusion protein can then be eluted from the column by the addition of pure marker peptide. doi:10.1128/9781555818890.ch6.f6.8

use of pharmaceuticals, it is necessary to remove the marker peptide if the product is intended for human use.

Alternatively, it has become popular to generate a fusion protein containing six or eight histidine residues attached to either the N- or C-terminal end of the target protein. The histidine-tagged protein, along with other cellular proteins, is then passed over an affinity column of nickel–nitrilotriacetic acid. The histidine-tagged protein, but not the other cellular proteins, binds tightly to the column. The bound protein may be eluted from the column by the addition of imidazole (the side chain of histidine). With this protocol, some cloned and overexpressed proteins have been purified up to 100-fold with greater than 90% recovery in a single step.

Metabolic Load

The introduction and expression of foreign DNA in a host organism often change the metabolism of the organism in ways that may impair normal cellular functioning (Fig. 6.9). This biological response is due to a **metabolic load** (metabolic burden, metabolic drain) imposed upon the host by the presence and expression of foreign DNA. A metabolic load can occur as the result of a variety of conditions, including the following.

- Increasing plasmid copy number and/or size requires increasing amounts of cellular energy for plasmid replication and maintenance.
- The limited amount of dissolved oxygen in the growth medium is often insufficient for both host cell metabolism and plasmid maintenance and expression (see the section on overcoming oxygen limitation below).
- Overproduction of both target and marker proteins may deplete the pools of certain aminoacyl-tRNAs (see the section on codon usage above) and/or drain the host cell of its energy (in the form of ATP or guanosine 5′-triphosphate [GTP]).

GTP
guanosine 5′-triphosphate

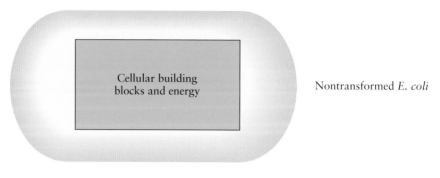

Figure 6.9 Schematic representation of nontransformed *E. coli* cells containing a high level of cellular building blocks and energy and *E. coli* cells transformed with plasmid DNA (circle) that encodes the synthesis of a foreign protein (shown in blue), thereby depleting the host cell of much of its cellular building blocks and energy. The overexpression of a foreign protein prevents the cell from obtaining sufficient energy and resources for its growth and metabolism, so it is less able to grow rapidly and attain a high cell density.
doi:10.1128/9781555818890.ch6.f6.9

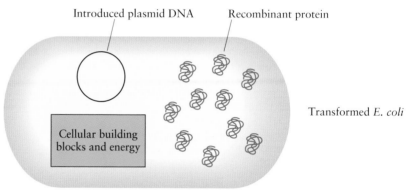

- When a foreign protein is overexpressed and then exported from the cytoplasm to either the cell membrane, the periplasm, or the external medium, it may "jam" membrane export sites and thereby prevent the proper localization of other, essential, host cell proteins.
- A foreign protein may sometimes interfere with the functioning of the host cell, for example, by converting an important and needed metabolic intermediate into a compound that is irrelevant, or even toxic, to the cell.

One of the most commonly observed consequences of a metabolic load is a decrease in the rate of cell growth after introduction of foreign DNA. Sometimes, a metabolic load may result in plasmid-containing cells losing all or a portion of the plasmid DNA. This is especially problematic during the large-scale growth of a transformed host because this step is usually carried out in the absence of selective pressure. Since cells growing in the presence of a metabolic load generally have a decreased level of energy for cellular functions, energy-intensive metabolic processes such protein synthesis are invariably adversely affected by a metabolic load. A metabolic load may also lead to changes in the host cell size and shape and to increases in the amount of extracellular polysaccharide produced by the bacterial host cell. This additional extracellular carbohydrate may cause the cells to stick together, making harvesting, e.g., by cross-flow microfiltration procedures, and protein purification more difficult.

When a particular aminoacyl-tRNA becomes limiting, as is often the case when a foreign protein is overexpressed in *E. coli*, there is an increased probability that an incorrect amino acid will be inserted in place of the limiting amino acid. In addition, translational accuracy, which depends upon the availability of GTP as part of a proofreading mechanism, is likely to be further decreased as a consequence of a metabolic load from foreign protein overexpression. In one instance, a high level of expression of mouse **epidermal growth factor** in *E. coli* caused about 10 times the normal amount of incorrect amino acids to be incorporated into the recombinant protein. This increase in error frequency can dramatically diminish the usefulness of the target protein as a therapeutic agent.

The extent of the metabolic load can be reduced by using a low-copy-number rather than a high-copy-number plasmid vector. An even better strategy might be to avoid the use of plasmid vectors and integrate the introduced foreign DNA directly into the chromosomal DNA of the host organism (see the section below on integrating foreign genes into the host chromosomal DNA). In this case, plasmid instability will not be a problem. With an integrated cloned gene, without a plasmid vector, the transformed host cell will not waste its resources synthesizing unwanted and unneeded antibiotic resistance marker gene products. The use of strong but regulatable promoters is also an effective means of reducing metabolic load (see the section on promoters above). In this case, the production-scale fermentation process may be performed in two stages. During the first, or growth, stage, the promoter controlling the transcription of the target gene is turned off, while during the second, or induction, stage, this promoter is turned on.

When the codon usage of the foreign gene is different from the codon usage of the host organism, depletion of specific aminoacyl-tRNA pools may be avoided by either completely or partially synthesizing the target gene to better reflect the codon usage of the host organism (see the section on codon usage above). In one study, it was found that levels of the protein **streptavidin** were 10-fold higher in *E. coli* when expression was directed by a synthetic gene with a G+C content of 54% than when it was directed by the natural gene with a G+C content of 69%.

Although it may at first seem counterintuitive, one way to increase the amount of foreign protein produced during the fermentation process is to accept a modest level of foreign gene expression—perhaps 5% of the total cell protein—and instead focus on attaining a high host cell density. Thus, for example, an organism with a 5% foreign protein expression level and a low level of metabolic load that can be grown to a density of 40 g (dry weight) per liter produces more of the target protein than one with a 15% expression level for the same protein and a cell density of only 5 to 10 g (dry weight) per liter.

Chromosomal Integration

As a consequence of metabolic load, a fraction of the cell population often loses its plasmids during cell growth. In addition, cells that lack plasmids grow faster than those that retain them, so plasmidless cells eventually dominate the culture. After a number of generations of cell growth, the loss of plasmid-containing cells diminishes the yield of the cloned gene product. Plasmid-containing cells may be maintained by growing the cells in the presence of either an antibiotic or an essential metabolite that enables only plasmid-bearing cells to thrive. But the addition of either antibiotics or metabolites to industrial-scale fermentations can be extremely costly, and it is imperative that anything that is added to the fermentation, such as an antibiotic or a metabolite, be completely removed prior to certifying the product fit for human use. For this reason, cloned DNA is often introduced directly into the chromosomal DNA of the host organism. When DNA is part of the host chromosomal DNA, it is quite stable and can be maintained indefinitely in the absence of selective agents.

For integration of DNA into a chromosomal site, the input DNA must share some sequence similarity, usually at least 50 nucleotides, with the chromosomal DNA, and there must be a physical exchange (recombination) between the two DNA molecules. To integrate DNA into the host chromosome, it is first necessary to identify a desired chromosomal integration site, i.e., a segment of DNA on the host chromosome that can be disrupted without affecting the normal functions of the cell. Once the chromosomal integration site has been isolated and spliced onto a plasmid, a marker gene is inserted in the middle of the cloned chromosomal integration site (Fig. 6.10). The DNA on the plasmid can base-pair with identical sequences on the host chromosome and subsequently integrate into the host chromosome as a result of a host enzyme-catalyzed double

Figure 6.10 Insertion of a foreign gene into a unique predetermined site on a bacterial chromosome. In step 1, a marker gene is integrated into the host cell chromosomal DNA by homologous recombination. In step 2, the selectable marker gene is replaced by the target gene. doi:10.1128/9781555818890.ch6.f6.10

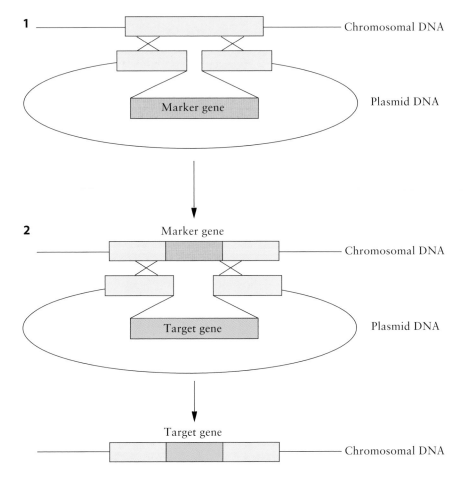

crossover. In this case, transformants are selected for the acquisition of the marker gene (often an antibiotic resistance gene). Then, the target gene, under the control of a regulatable promoter, is inserted in the middle of the cloned chromosomal integration site on a different plasmid. This plasmid construct is used to transform host cells that contain the marker gene integrated into its chromosome, and following a host enzyme-catalyzed double crossover, the target gene and its transcriptional regulatory region are inserted into the chromosome in place of the marker gene. The final construct is selected for the loss of the marker gene.

Several other methods can also be used to integrate foreign genes into host chromosomal DNA. For example, when a marker gene is flanked by certain short specific DNA sequences and then inserted into either a plasmid or chromosomal DNA, the gene may be excised by treatment of the construct with an enzyme that recognizes the flanking DNA sequences and removes them (Fig. 6.11). One combination of an enzyme and DNA sequence that is useful for this sort of manipulation is the Cre–*loxP* recombination system, which consists of the Cre recombinase enzyme and two 34-bp *loxP* recombination sites. The marker gene to be removed is flanked by *loxP* sites, and after integration of the plasmid into the chromosomal DNA, the marker gene is removed by the Cre enzyme. A gene

Figure 6.11 Removal of a selectable marker gene following integration of plasmid DNA into a bacterial chromosome. A single crossover event (×) occurs between chromosomal DNA and a homologous DNA fragment (hatched) on a plasmid, resulting in the integration of the entire plasmid into the chromosomal DNA. The selectable marker gene, which is flanked by *loxP* sites, is excised by the action of the Cre enzyme, leaving one *loxP* site on the integrated plasmid. The Cre enzyme is on a separate plasmid within the same cell under the transcriptional control of the *E. coli lac* promoter, so excision is induced when IPTG is added to the growth medium. doi:10.1128/9781555818890.ch6.f6.11

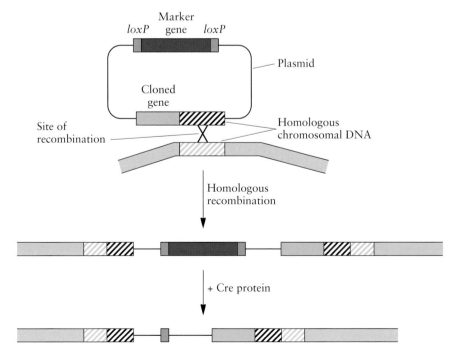

encoding the Cre enzyme is located on its own plasmid, which can be introduced into the chromosomally transformed host cells. Marker gene excision is triggered by the addition of IPTG to the growth medium; this derepresses the *lacI* gene (encoding the *lac* repressor), which turns on the *E. coli lac* promoter–operator, which was present upstream of the *Cre* gene, and causes the Cre enzyme to be synthesized. Once there is no longer any need for the Cre enzyme, the plasmid that contains the gene for this enzyme under the control of the *lac* promoter may be removed from the host cells merely by raising the temperature. This plasmid has a temperature-sensitive replicon that allows it to be maintained in the cell at 30°C but not above 37°C.

Increasing Secretion

For most *E. coli* proteins, secretion entails transit through the inner (cytoplasmic) cell membrane to the **periplasm**. Directing a foreign protein to the periplasm, rather than the cytoplasm, makes its purification easier and less costly, as many fewer proteins are present here than in the cytoplasm. Moreover, the stability of a cloned protein depends on its cellular location in *E. coli*. For example, recombinant **proinsulin** is approximately 10 times more stable if it is secreted (exported) into the periplasm than if it is localized in the cytoplasm. In addition, secretion of proteins to the periplasm often facilitates the correct formation of disulfide bonds because the periplasm provides an oxidative environment, as opposed to the more reducing environment of the cytoplasm. Table 6.4 provides some examples of the amounts of secreted recombinant pharmaceutical proteins attainable with various bacterial host cells.

Normally, an amino acid sequence called a **signal peptide** (also called a signal sequence or leader peptide), located at the N-terminal end of a newly synthesized protein, facilitates its export by enabling the protein to pass through the cell membrane (Fig. 6.12). It is sometimes possible to engineer a protein for secretion to the periplasm by adding the DNA

Table 6.4 Yields of several secreted recombinant proteins produced in different bacteria

Protein	Yield	Host bacterium
Hirudin	>3 g/L	*Escherichia coli*
Human antibody fragment	1–2 g/L	*Escherichia coli*
Human insulin-like growth factor	8.5 g/L	*Escherichia coli*
Monoclonal antibody 5T4	700 mg/L	*Escherichia coli*
Humanized anti-CD18 F(ab′)$_2$	2.5 g/L	*Escherichia coli*
Human epidermal growth factor	325 mg/L	*Escherichia coli*
Alkaline phosphatase	5.2 g/L	*Escherichia coli*
Staphylokinase	340 mg/L	*Bacillus subtilis*
Human proinsulin	1 g/L	*Bacillus subtilis*
Human calcitonin precursor	2 g/L	*Staphylococcus carnosus*
Organophosphohydrolase	1.2 g/L	*Ralstonia eutropha*
Human CD4 receptor	200 mg/L	*Streptomyces lividans*
Human insulin	100 mg/L	*Streptomyces lividans*

sequence encoding a signal peptide to the cloned gene. When the recombinant protein is secreted to the periplasm, the signal peptide is precisely removed by the cell's secretion apparatus, so the N-terminal end of the target protein is identical to the natural protein.

Unfortunately, the fusion of a target gene to a DNA fragment encoding a signal peptide sequence does not necessarily guarantee a high rate of secretion. When this simple strategy is found to be ineffective in producing a secreted protein product, alternative strategies need to be employed. One approach that was found to be successful for the secretion of the IL-2 cytokine was the fusion of the IL-2 gene downstream from the gene for the entire **propeptide** maltose-binding protein, rather than just the maltose-binding protein signal sequence, with DNA encoding the factor X_a recognition site as a linker peptide separating these two genes (Fig. 6.13). When this genetic fusion, on a plasmid vector, was used to transform *E. coli* cells, as expected, a large fraction of the fusion protein was found localized in the host cell periplasm. Functional IL-2 could then be released from the fusion protein by digestion with factor X_a.

Sometimes too high a level of translation of a foreign protein can overload the cell's secretion machinery and inhibit the secretion of that protein. Thus, to ensure that secretion of a target protein occurs most efficiently, it is necessary to lower the level of expression of that protein.

E. coli and other gram-negative microorganisms generally cannot secrete proteins into the surrounding medium because of the presence of an outer membrane (in addition to the inner or cytoplasmic membrane) that restricts this process. Of course, it is possible to use as host organisms gram-positive prokaryotes or eukaryotic cells, both of which lack an outer membrane and therefore can secrete proteins directly into the medium. Alternatively, it is possible to take advantage of the fact that some gram-negative bacteria can secrete a bacteriocidal protein called a **bacteriocin** into the medium. A cascade mechanism is responsible for this specific secretion. A bacteriocin release protein activates **phospholipase A**, which is present in the bacterial inner membrane, and cleaves membrane phosopholipids so that both the inner and outer membranes are **permeabilized** (Fig. 6.14A). This results in some cytoplasmic and periplasmic

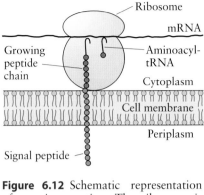

Figure 6.12 Schematic representation of protein secretion. The ribosome is attached to a cellular membrane, and the signal peptide at the N terminus is transported, by the secretion apparatus, across the cytoplasmic membrane, followed by the rest of the amino acids that constitute the mature protein. Once the signal peptide has crossed the membrane, it is cleaved from the remainder of the protein by an enzyme associated with the membrane called a signal peptidase. Membrane proteins as well as secreted proteins generally contain a signal peptide (prior to removal by processing). doi:10.1128/9781555818890.ch6.f6.12

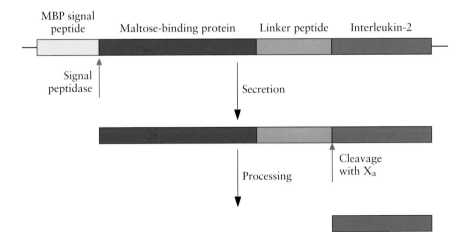

Figure 6.13 Engineering the secretion of IL-2. When IL-2 is fused to the *E. coli* maltose-binding protein and its signal peptide, with the two proteins joined by a linker peptide, secretion to the periplasm occurs. Following the purification of the secreted fusion protein, the maltose-binding protein and the linker peptide are removed by digestion with factor X_a. doi:10.1128/9781555818890.ch6.f6.13

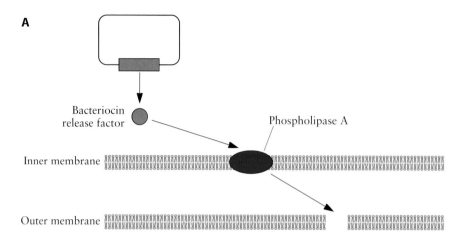

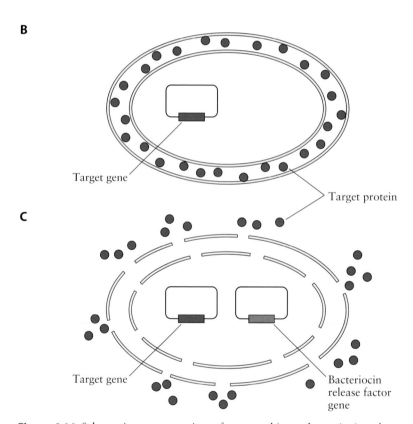

Figure 6.14 Schematic representation of a recombinant bacteriocin release protein activating phospholipase A (present within the *E. coli* inner membrane) to permeabilize the cell membranes (**A**). Also shown are schematics of *E. coli* cells engineered to secrete a foreign protein to the periplasm by fusing the gene of interest (green) to a secretion signal (**B**) and to the growth medium by permeabilizing cell membranes with a bacteriocin release protein encoded on another plasmid (red) (**C**). doi:10.1128/9781555818890.ch6.f6.14

proteins being released into the culture medium. Thus, by putting the bacteriocin release protein gene onto a plasmid under the control of a regulatable promoter, *E. coli* cells may be permeabilized at will. *E. coli* cells that carry the bacteriocin release protein gene on a plasmid are transformed with another plasmid carrying a cloned gene that has been fused to a secretion signal peptide sequence that causes the target protein to be secreted into the periplasm. The cloned gene is placed under the same transcriptional regulatory control as the bacteriocin release protein gene so that the two genes can be induced simultaneously, with the cloned gene protein being secreted into the medium (Fig. 6.14B and C).

Overcoming Oxygen Limitation

E. coli and most other microorganisms that are used to express foreign proteins generally require oxygen for optimal growth. Unfortunately, oxygen has only a limited solubility in aqueous media. Thus, as the cell density of a growing culture increases, the cells rapidly deplete the growth medium of dissolved oxygen. When cells become oxygen limited, exponential growth slows and the culture rapidly enters a stationary phase during which cellular metabolism changes. One consequence of the stationary phase is the production by the host cells of proteases that can degrade foreign proteins. Oxygen dissolves into the growth medium very slowly, so the amount of dissolved oxygen available to growing cells is often not increased fast enough when large amounts of air or oxygen are added to the growth medium, even with high stirring rates. Modification of the fermenter (**bioreactor**) configuration to optimize the aeration and agitation of cells and addition of chemicals to the growth medium to increase the solubility of oxygen have been tried in an effort to deal with the limited amount of dissolved oxygen. However, these efforts have met with only limited success.

Some strains of the bacterium *Vitreoscilla*, a gram-negative obligate aerobe, normally live in oxygen-poor environments such as stagnant ponds. To obtain a sufficient amount of oxygen for their growth and metabolism, these organisms synthesize a hemoglobin-like molecule that tightly binds oxygen from the environment and subsequently increases the level of available oxygen inside cells (Fig. 6.15). When the gene for this protein is expressed in *E. coli*, the transformants display higher levels of protein synthesis of both cellular and recombinant proteins, higher levels of cellular respiration, a higher ATP production rate, and higher ATP contents, especially at low levels of dissolved oxygen (0.25 to 1.0%) in the growth medium, than do nontransformed cells. In these transformants, the *Vitreoscilla* hemoglobin increases the intracellular oxygen concentration, which raises the activities of both cytochromes *d* and *o*. This causes an increase in proton pumping, with the subsequent generation of ATP, thereby providing additional energy for cellular metabolic processes (Fig. 6.15). For this strategy to be effective in different host cells, not only must the *Vitreoscilla* sp. hemoglobin gene be efficiently expressed but also the host cells must be able to synthesize the heme portion of the hemoglobin molecule. Once these conditions have been met, this strategy

A

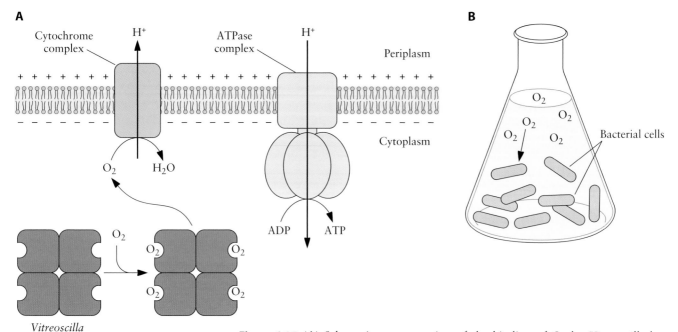

B

Figure 6.15 (A) Schematic representation of the binding of O_2 by *Vitreoscilla* hemoglobin, the utilization of this O_2 in pumping (by proteins such as cytochromes) H^+ from the cytoplasm to the periplasm, and the subsequent coupling of H^+ uptake (by ATPase) to ATP generation. **(B)** Host cells engineered to express *Vitreoscilla* hemoglobin are more efficient at taking up oxygen from the growth medium. doi:10.1128/9781555818890.ch6.f6.15

can be used to improve growth as well as foreign gene expression in a range of different industrially important bacteria, including *E. coli*. Thus, host cells expressing the *Vitreoscilla* sp. hemoglobin gene often undergo several additional doublings before entering stationary phase, producing a much higher cell density and a much greater yield of the target recombinant protein.

Reducing Acetate

It is often difficult to achieve high levels of foreign-gene expression and a high host cell density at the same time because of the accumulation of harmful waste products, especially acetate, which inhibits both cell growth and protein production and also wastes available carbon and energy resources. Since acetate is often associated with the use of glucose as a carbon source, lower levels of acetate, and hence higher yields of protein, are generally obtained when fructose or mannose is used as a carbon source. In addition, several different types of genetically manipulated *E. coli* host cells that produce lower levels of acetate have been developed. One of these modified strains was produced by introducing a gene (from *B. subtilis*) encoding the enzyme acetolactate synthase into *E. coli* host cells. This enzyme catalyzes the formation of acetolactate from pyruvate, thereby decreasing the flux through acetyl coenzyme A to acetate (Fig. 6.16). In practice, the acetolactate synthase genes are introduced into the cell on one plasmid, while the target gene (encoding the protein

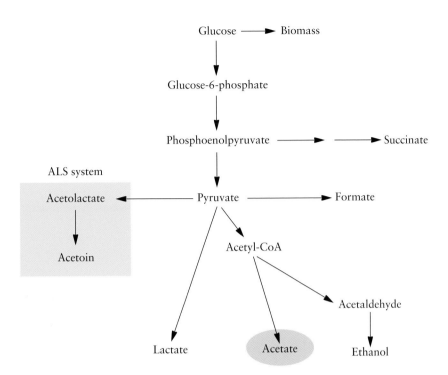

Figure 6.16 Schematic representation of the pathways for glucose metabolism in an *E. coli* strain that has been transformed with a plasmid carrying the genes for the protein subunits of acetolactate synthase (ALS). Introduction of this pathway results in the synthesis of acetoin. Note that the conversion of glucose to biomass is a multistep process. Acetyl-CoA, acetyl coenzyme A. doi:10.1128/9781555818890.ch6.f6.16

that is to be overexpressed in *E. coli*) is introduced on a second plasmid from a different incompatibility group. The cells that were transformed with the acetolactate synthase genes produced 75% less acetate than the nontransformed cells and instead synthesized acetoin, which is approximately 50-fold less toxic to cells than acetate. The recombinant protein yield was also doubled.

In another approach to lowering the level of acetate, researchers transformed *E. coli* host cells with a bacterial gene for the enzyme pyruvate carboxylase, which converts pyruvate directly to oxaloacetate and is not present in *E. coli* (Fig. 6.17). With the introduction of pyruvate carboxylase, acetate levels were decreased to less than 50% of their normal level, the cell yield was increased by more than 40%, and the amount of foreign protein synthesized was increased by nearly 70%. This result reflects the fact that the addition of pyruvate carboxylase allows *E. coli* cells to use the available carbon more efficiently, directing it away from the production of acetate toward biomass and protein formation.

Similar to the strategy discussed above, the tricarboxylic acid (TCA) cycle may also be replenished by converting aspartate to fumarate (Fig. 6.17). To do this, *E. coli* host cells were transformed with the gene for L-aspartate ammonia lyase (aspartase) under the control of the strong *tac* promoter on a stable low-copy-number plasmid. The target recombinant protein was introduced on a separate plasmid. Using this system, aspartase activity was induced by the addition of IPTG at the middle to

TCA
tricarboxylic acid

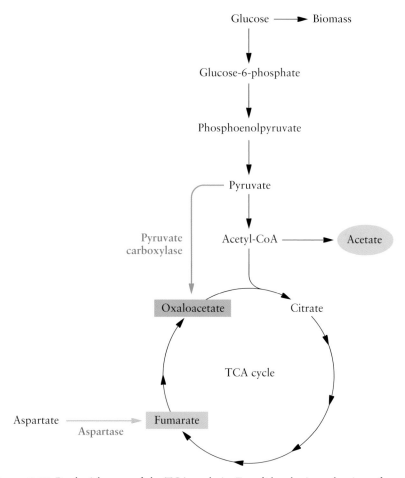

Figure 6.17 Replenishment of the TCA cycle in *E. coli* by the introduction of a gene encoding pyruvate carboxylase. This avoids the conversion of pyruvate to acetate. The TCA cycle may also be replenished by the introduction of a gene encoding aspartase, converting aspartate in the medium to fumarate. Note that the conversion of glucose to biomass is a multistep process. Acetyl-CoA, acetyl coenzyme A. doi:10.1128/9781555818890.ch6.f6.17

late log phase of growth. When the recombinant *E. coli* cells were grown in minimal medium containing aspartate, the production of different recombinant proteins could be increased up to fivefold, with 30 to 40% more biomass production.

Protein Folding

The use of conditions that result in very high rates of foreign gene expression in *E. coli* often also lead to the production of misfolded proteins that can aggregate and form insoluble inclusion bodies within the host cell. While it is possible to solubilize inclusion bodies and subsequently establish conditions that allow at least a portion of the recombinant protein to fold correctly, this is typically a tedious, inefficient, expensive, and time-consuming process, one that is best avoided if possible. A simple strategy to avoid the formation of misfolded proteins and hence inclusion bodies involves reducing the rate of synthesis of the target gene product so that it has more time to fold properly. This may be achieved by various means, including using weaker promoters, decreasing the concentration of inducers (such as IPTG), or lowering the growth temperature to 20 to 30°C. These strategies are sometimes, but not always, effective in preventing the formation of inclusion bodies.

An alternative strategy to improve the yield of properly folded (and therefore active) recombinant proteins in *E. coli* involves the coexpression of one or more molecular **chaperones** (proteins that facilitate the correct folding of other proteins) by the host *E. coli* strain (Table 6.5). The "folding chaperones" utilize ATP cleavage to promote conformational changes to mediate the refolding of their substrates. The "holding chaperones" bind to partially folded proteins until the folding chaperones have done their job. The "disaggregating chaperone" promotes the solubilization of proteins that have become aggregated. Protein folding also involves the "trigger factor," which binds to nascent polypeptide chains, acting as a holding chaperone. Although there are a large number of chaperone molecules that are involved in the proper folding and secretion of proteins

Table 6.5 *E. coli* proteins that facilitate the correct folding of recombinant proteins

Localization	Function	Name
Cytoplasm	Holding chaperone	Hsp31
		Hsp33
		IbpA
		IbpB
		Trigger factor
	Folding chaperone	GroEL (Hsp60)
		DnaK (Hsp70)
		HscA
		HscC
		HtpG (Hsp90)
	Disaggregase	ClpB
	Secretory chaperone	SecB
Periplasm	Generic chaperones	Skp (OmpH)
		FkpA
	Specialized chaperones	SurA
		LolA
		PapD
		FimC
	PPIases	SurA
		PpiD
		FkpA
		PpiA (RotA)
	Disulfide bond formation	DsbA
		DsbB
		DsbC
		DsbD
		DsbE
		DsbG
		CcmH

Adapted from Baneyx and Mujacic, *Nat. Biotechnol.* **22**:1399–1408, 2004.

in *E. coli*, a detailed understanding of the roles of many of these proteins has begun to emerge, and the proper folding of a number of recombinant proteins has been facilitated by coexpressing some of these chaperone proteins. Thus, for example, significantly enhanced correct folding of periplasmic proteins, as well as reduced recombinant protein degradation and inclusion body formation, is observed when the chaperone Skp and the peptidyl-prolyl *cis/trans* isomerase (PPIase) FkpA are coexpressed along with the recombinant (secreted) protein. Finally, it is important to note that the periplasm provides an oxidizing environment (compared to the reducing environment of the cytoplasm), so in nontransformed *E. coli*, disulfide bond-containing proteins are found only in the cell envelope, where disulfide formation and isomerization are catalyzed by a set of thiol-disulfide oxidoreductases known as the Dsb proteins. In practice, this means that disulfide bond-containing recombinant proteins are unlikely to be effectively synthesized in the cell cytoplasm.

In a study in which the chaperones DnaK and GroEL (and their co-chaperonin protein molecules) were overexpressed, the yields of 7 of 10 eukaryotic **kinases** expressed at the same time as target proteins were increased up to fivefold. Moreover, the overexpression of additional chaperones did not affect the detected levels of these kinases. This result was interpreted as indicating that the ability of overexpressed chaperones to facilitate the proper folding of recombinant proteins is a protein-specific phenomenon effective in some, but not all, instances.

While some researchers now routinely coexpress molecular chaperones along with the recombinant protein of interest, others have sought technically simpler solutions to the problem of protein folding. Thus, it has been observed that high levels of certain osmolytes (substances that contribute to the osmotic pressure in cells), such as sorbitol and betaine, along with a high level of salt in the growth medium, can enhance the correct folding and solubility of several recombinant proteins produced in *E. coli*.

Heterologous Protein Production in Eukaryotic Cells

Eukaryotic Expression Systems

The **eukaryotic** proteins produced in bacteria do not always have the desired biological activity or stability. In addition, despite careful purification procedures, bacterial compounds that are toxic or that cause a rise in body temperature in humans and animals may contaminate the final product. Moreover, any human protein intended for medical use must be identical to the natural protein in all its properties. The inability of prokaryotic organisms to produce authentic versions of eukaryotic proteins is, for the most part, due to improper posttranslational protein processing and to the absence of appropriate mechanisms that add chemical groups to specific amino acid acceptor sites. To avoid these problems, investigators have developed eukaryotic expression systems in fungal/yeast, insect, and mammalian cells for the production of therapeutic agents (Milestone 6.2).

Synthesis of Rabbit β-Globin in Cultured Monkey Kidney Cells Following Infection with a SV40 β-Globin Recombinant Genome

R. C. MULLIGAN, B. H. HOWARD, AND P. BERG
Nature 277:108–114, 1979

milestone
6.2

Conceptually, the development of a eukaryotic expression system appears to be a relatively simple matter of assembling the appropriate regulatory sequences, cloning them in the correct order into a vector, and then putting the gene of interest into the precise location that enables it to be expressed. In reality, the development of the first generation of eukaryotic expression vectors was a painstaking process following a trial-and-error approach. Before the study of Mulligan et al., a number of genes had been cloned into the mammalian SV40 vectors, but mature, functional mRNAs were never detected after infection of host cells. This problem was overcome by inserting the rabbit cDNA for β-globin into an SV40 gene that had nearly all of its coding region deleted but retained "all the regions implicated in transcriptional initiation and termination, splicing and polyadenylation. . . ." Both rabbit β-globin mRNA and protein were synthesized in cells that were transfected with this β-globin cDNA–SV40 construct. Mulligan et al. concluded, "The principal conceptual innovation is the decision to leave intact the regions of the vector implicated in . . . mRNA processing." This study established that an effective eukaryotic expression system could be created by placing the cloned gene under the control of transcription and translation regulatory sequences. It also stimulated additional research that pinpointed in detail the structural prerequisites for the next generation of eukaryotic expression vectors.

The basic requirements for expression of a target protein in a eukaryotic host are similar to those in prokaryotes. Vectors into which the target gene is cloned for delivery into the host cell can be specialized plasmids designed to be maintained in the eukaryotic host, such as the yeast 2μm plasmid; host-specific viruses, such as the insect **baculovirus**; or artificial chromosomes, such as the yeast artificial chromosome (YAC). The vector must have a eukaryotic promoter that drives the transcription of the target gene, eukaryotic transcriptional and translational stop signals, a sequence that enables **polyadenylation** of the mRNA, and a selectable eukaryotic marker gene (Fig. 6.18). Because recombinant DNA procedures are technically difficult to carry out with eukaryotic cells, most eukaryotic vectors are shuttle vectors with two origins of replication and two selectable marker genes. One set functions in *E. coli*, and the other set functions in the eukaryotic host cell.

Many eukaryotic proteins undergo posttranslational processing that is required for protein activity and stability. Some proteins are produced as inactive precursor polypeptides that must be cleaved by proteases at specific sites to produce the active form of the protein. In addition, ~50% of all human proteins are glycosylated (i.e., certain amino acids are modified by adding specific sugars), often providing stability and distinctive binding properties to a protein, including protein folding, targeting a protein to a particular location, or protecting it from proteases. In the cell, sugars are attached to newly synthesized proteins in the endoplasmic reticulum and in the Golgi apparatus by enzymes known as glycosylases and glycosyltransferases.

The most common **glycosylations** entail the attachment of specific sugars to the hydroxyl group of either serine or threonine (O-linked glycosylation) and to the amide group of asparagine (N-linked glycosylation).

YAC
yeast artificial chromosome

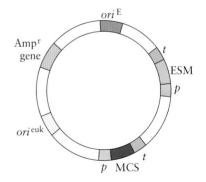

Figure 6.18 Generalized eukaryotic expression vector. The major features of a eukaryotic expression vector are a eukaryotic transcription unit with a promoter (*p*), a multiple-cloning site (MCS) for a gene of interest, and a DNA segment with termination and polyadenylation signals (*t*); a eukaryotic selectable marker (ESM) gene system; an origin of replication that functions in the eukaryotic cell (*ori*^{euk}); an origin of replication that functions in *E. coli* (*ori*^{E}); and an *E. coli* selectable marker (Amp^{r}) gene. doi:10.1128/9781555818890.ch6.f6.18

Other amino acid modifications include phosphorylation, acetylation, sulfation, acylation, γ-carboxylation, and the addition of C_{14} and C_{16} fatty acids, i.e., **myristoylation** (or myristylation) and **palmitoylation** (or palmitylation), respectively. Unfortunately, there is no universally effective eukaryotic host cell that performs the correct modifications on every protein.

Saccharomyces cerevisiae Expression Systems

Yeasts, like prokaryotes, grow rapidly in low-cost medium, generally do not require the addition of growth factors to the medium, can correctly process eukaryotic proteins, and can secrete large amounts of heterologous proteins. The yeast *S. cerevisiae*, traditionally employed in baking and brewing, has been used extensively as a host cell for the expression of cloned eukaryotic genes.

High levels of recombinant protein production can be achieved using *S. cerevisiae*. This is because of the following. (i) The detailed biochemistry, genetics, and cell biology of this single-celled yeast are well known; its **genome** sequence was completely determined in 1996. (ii) It can be grown rapidly to high cell densities on relatively simple media. (iii) Several strong promoters have been isolated from *S. cerevisiae*, and a naturally occurring plasmid, called the 2μm plasmid, can be used as part of an endogenous yeast expression vector system. (iv) *S. cerevisiae* is capable of carrying out many posttranslational modifications. (v) *S. cerevisiae* normally secretes so few proteins that, when it is engineered for extracellular release of a recombinant protein, the product can be easily purified. (vi) Because of its use in the baking and brewing industries, *S. cerevisiae* has been listed by the FDA as a "generally recognized as safe" organism. Therefore, the use of the organism for the production of human therapeutic agents does not require the extensive experimentation demanded for unapproved host cells.

S. cerevisiae Vectors

YEp
yeast episomal plasmid

YIp
yeast integrating plasmid

There are three main classes of *S. cerevisiae* expression vectors: **episomal**, or plasmid, vectors (yeast episomal plasmids [YEps]), integrating vectors (yeast integrating plasmids [YIps]), and YACs. Of these, episomal vectors have been used extensively for the production of either intra- or extracellular heterologous proteins. Typically, the vectors contain features that allow them to function in both bacteria and *S. cerevisiae*. An *E. coli* origin of replication and bacterial antibiotic resistance genes are usually included on the vector, enabling all manipulations to first be performed in *E. coli* before the vector is transferred to *S. cerevisiae* for expression.

The YEp vectors are based on the 2μm plasmid, a small, independently replicating circular plasmid found in about 30 copies per cell in most natural strains of *S. cerevisiae*. Many *S. cerevisiae* selection schemes rely on **mutant** host strains that require a particular amino acid (e.g., histidine, tryptophan, or leucine) or nucleotide (e.g., uracil) for growth. Such auxotrophic strains cannot grow on minimal growth medium unless it is supplemented with a specific nutrient. In practice, the vector is equipped with a functional version of a gene that complements the mutated gene in the host strain. For example, when a YEp with a wild-type *LEU2* gene is transformed into a mutant *leu2* host cell and plated onto medium that lacks leucine, only cells that carry the plasmid will grow.

Generally, regulatable, inducible promoters are preferred for producing large amounts of recombinant protein during large-scale growth. In this context, the galactose-regulated promoters respond rapidly to the addition of galactose with a 1,000-fold increase in transcription. Repressible, constitutive, and hybrid promoters that combine the features of different promoters are also available. Maximal expression also depends on efficient termination of transcription.

Plasmid-based yeast expression systems are often unstable under large-scale ($\geq$10 L) growth conditions, even in the presence of selection pressure. To remedy this problem, a heterologous gene is integrated into the host genome to provide a more reliable production system. The major drawback of this strategy is the low yield of recombinant protein from a single gene copy.

To increase the number of copies of an integrated heterologous gene and thereby increase the overall yield of the recombinant protein, the heterologous gene can be integrated into nonessential portions of the *S. cerevisiae* genome. In one study, 10 copies of a target gene were inserted into the yeast genome and produced a significant amount of the recombinant protein.

A YAC is designed to clone a large segment of DNA (100 kilobase pairs [kb]), which is then maintained as a separate chromosome in the host yeast cell. A YAC vector mimics a chromosome because it has a sequence that acts as an origin of DNA replication, a yeast **centromere** sequence to ensure that after cell division each daughter cell receives a copy of the YAC, and **telomere** sequences that are present at both ends after linearization of the YAC DNA for stability (Fig. 6.19). However, to date, YACs have not been used as expression systems for the commercial production of heterologous proteins.

Secretion of Heterologous Proteins by *S. cerevisiae*

All glycosylated proteins of *S. cerevisiae* are secreted, and each must have a leader sequence to pass through the secretory system. Consequently, the coding sequences of recombinant proteins that require either O-linked or N-linked sugars for biological activity must be equipped with a leader sequence. Under these conditions, correct disulfide bond formation, proteolytic removal of the leader sequence, and appropriate posttranslational modifications can occur, and an active recombinant protein is secreted.

In recent years, the amount of heterologous protein that can be produced per liter of yeast culture has increased 100-fold (from about 0.02 to 2 g/L). This increase is mainly due to improvements in growing cultured cells to high cell densities; the level of protein produced per cell has remained largely unchanged.

One of the major reasons for producing a recombinant protein for use in human therapeutics in yeasts rather than in bacteria is to ensure that the protein is processed correctly following synthesis. Correct

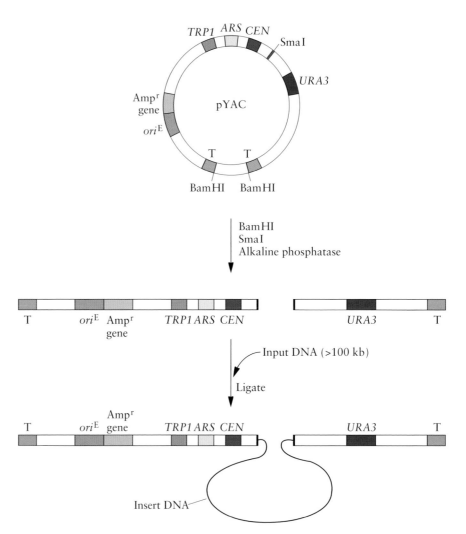

Figure 6.19 YAC cloning system. The YAC plasmid (pYAC) has an *E. coli* selectable marker (Amp^r) gene; an origin of replication that functions in *E. coli* (*ori*^E); and yeast DNA sequences, including *URA3*, *CEN*, *TRP1*, and *ARS*. *CEN* provides centromere function, *ARS* is a yeast autonomous replicating sequence that is equivalent to a yeast origin of replication, *URA3* is a functional gene of the uracil biosynthesis pathway, and *TRP1* is a functional gene of the tryptophan biosynthesis pathway. The T regions are yeast chromosome telomeric sequences. The SmaI site is the cloning insertion site. pYAC is first treated with SmaI, BamHI, and alkaline phosphatase and then ligated with size-fractionated (100-kb) input DNA. The final construct carries cloned DNA and can be stably maintained in double-mutant *ura3* and *trp1* cells. doi:10.1128/9781555818890.ch6.f6.19

protein folding occurs in the endoplasmic reticulum in eukaryotes and is facilitated by a number of different proteins, including molecular chaperones, enzymes for disulfide bond formation, signal transduction proteins that monitor the demand and capacity of the protein-folding machinery, and proteases that clear away improperly folded or aggregated proteins (Fig. 6.20). The enzyme protein disulfide isomerase is instrumental in forming the correct disulfide bonds within a protein. Poor yields of overexpressed proteins often occur because the capacity of the cell to properly fold and secrete proteins has been exceeded. One possible way

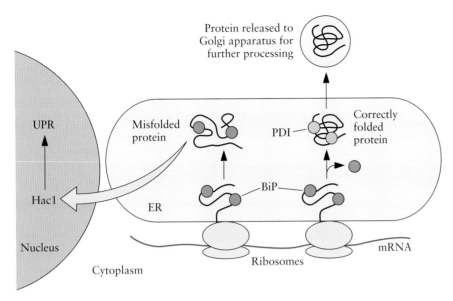

Figure 6.20 Summary of protein folding in the endoplasmic reticulum (ER) of yeast cells. During synthesis on ribosomes associated with the ER, nascent proteins are bound by the chaperones BiP and calnexin, which aid in the correct folding of the protein. Protein disulfide isomerases (PDI) catalyze the formation of disulfide bonds between cysteine amino acids that are nearby in the folded protein. Quality control systems ensure that only correctly folded proteins are released from the ER. Proteins released from the ER are transported to the Golgi apparatus for further processing. Prolonged binding of BiP to misfolded proteins leads to activation of the *S. cerevisiae* transcription factor Hac1, which controls the expression of several proteins that mediate the unfolded-protein response (UPR). Adapted from Gasser et al., *Microb. Cell Fact.* 7:11–29, 2008. doi:10.1128/9781555818890.ch6.f6.20

around this problem includes the overproduction of molecular chaperones and protein disulfide isomerases. Thus, when the yeast protein disulfide isomerase gene was cloned between the constitutive glyceraldehyde phosphate dehydrogenase promoter and a transcription **terminator** sequence in a YIp vector, and the entire construct was integrated into a chromosomal site, the modified strain showed a 16-fold increase in protein disulfide isomerase production compared with that of the wild-type strain. When protein disulfide isomerase-overproducing cells were transformed with a YEp vector carrying the gene for human platelet-derived growth factor B, there was a 10-fold increase in the secretion of recombinant protein over that of transformed cells with normal levels of protein disulfide isomerase. Higher levels of secreted products are also obtained for some recombinant proteins in *S. cerevisiae* cells that overexpress the chaperone BiP.

Overexpression of the molecular chaperone BiP or protein disulfide isomerase increases the secretion of some heterologous proteins; however, overexpression of a single chaperone does not always have the desired outcome. This is because proper protein folding requires the coordinated efforts of many interacting factors (Fig. 6.20). Even when levels of one chaperone are adequate, the levels of cochaperones or cofactors may be limiting.

Other Yeast Expression Systems

One of the major drawbacks of using *S. cerevisiae* is the tendency for the yeast to hyperglycosylate heterologous proteins by adding 50 to 150 mannose residues in N-linked oligosaccharide side chains that can alter protein function. Also, proteins that are designed for secretion frequently are retained in the periplasmic space, increasing the time and cost of purification. Moreover, *S. cerevisiae* produces ethanol at high cell densities, which is toxic to the cells and, as a consequence, lowers the quantity of secreted protein. For these reasons, researchers have examined other yeast species that could act as effective host cells for recombinant protein production.

Pichia pastoris is a yeast that is able to utilize methanol as a source of energy and carbon. It is an attractive host for recombinant protein production because glycosylation occurs to a lesser extent than in *S. cerevisiae* and the linkages between sugar residues are of the α-1,2 type, which are not allergenic to humans. With these characteristics as a starting point, a *P. pastoris* strain was extensively engineered so that it glycosylates proteins in a manner identical to that of human cells. Both human and yeast cells add the same small (10-residue) branched oligosaccharide to nascent proteins in the endoplasmic reticulum (Fig. 6.21). However, this is the last common precursor between the two cell types, because once the protein is transported to the Golgi apparatus, further processing is different. To create a "humanized" strain, the enzyme responsible for addition of the α-1,6-mannose was eliminated from *P. pastoris* to prevent hypermannosylation. Next, the gene encoding a mannose-trimming enzyme (a mannosidase) from the filamentous fungus *Trichoderma reesei* was inserted into the yeast genome and was found to trim the oligosaccharide to a human-like precursor. Genes encoding enzymes for the sequential addition of sugar residues that terminate the oligosaccharide chains in galactose were also added. It should be noted that the coding sequences for all engineered genes contained a secretion signal for localization of the encoded protein to the Golgi apparatus. Finally, several genes for proteins that catalyze the synthesis, transport to the Golgi apparatus, and addition of **sialic acid** to the terminal galactose on the protein precursor were inserted into the *P. pastoris* genome. Several properly sialylated recombinant proteins that can be used as human therapeutic agents have been produced by the humanized *P. pastoris*.

During growth on methanol, enzymes required for catabolism of this substrate are expressed at very high levels with alcohol oxidase, the first enzyme in the methanol utilization pathway, encoded by the gene *AOX1*, representing as much as 30% of the cellular protein. Transcription of *AOX1* is tightly regulated; in the absence of methanol, the *AOX1* gene is completely turned off, but it responds rapidly to the addition of methanol to the medium. Therefore, the *AOX1* promoter is an excellent candidate for producing large amounts of recombinant protein under controlled conditions. Moreover, the expression of the cloned target gene can be timed to maximize recombinant protein production during large-scale fermentations. In contrast to *S. cerevisiae*, *P. pastoris* does not synthesize ethanol, which can limit cell yields; therefore, very high cell densities of *P. pastoris* are attained, with the concomitant secretion of large quantities of

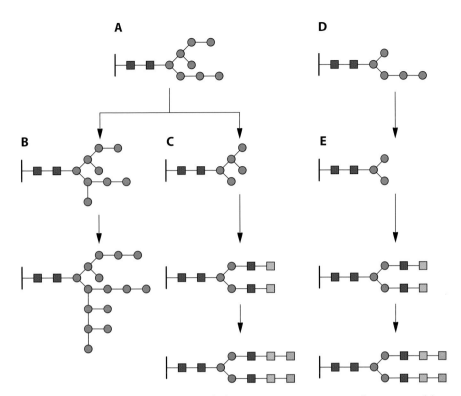

Figure 6.21 Differential processing of glycoproteins in *P. pastoris*, humans, and humanized *P. pastoris*. Initial additions of sugar residues to glycoproteins in the endoplasmic reticulum are similar in humans and *P. pastoris* cells (**A**). However, further N glycosylation in the Golgi apparatus differs significantly between the two cell types. N-glycans are hypermannosylated in *P. pastoris* (**B**), while in humans, mannose residues are trimmed and specific sugars are added, leading to termination of the oligosaccharide in sialic acid (**C**). *P. pastoris* cells have been engineered to produce enzymes that process glycoproteins in a manner similar to that of human cells. In humanized *P. pastoris*, a recombinant glycoprotein produced in the endoplasmic reticulum (**D**) is transported to the Golgi apparatus, where it is further processed to yield a properly sialylated glycoprotein (**E**). Blue squares, *N*-acetylglucosamine; red circles, mannose; green squares, galactose; orange squares, sialic acid. Adapted from Hamilton and Gerngross, *Curr. Opin. Biotechnol.* **18**:387–392, 2007.
doi:10.1128/9781555818890.ch6.f6.21

recombinant protein. *P. pastoris* normally secretes very few proteins, thus simplifying the purification of secreted recombinant proteins.

To avoid the problems of plasmid instability during long-term growth, most *P. pastoris* vectors are designed to be integrated into the host genome, usually within the *AOX1* gene, the *HIS4* gene for histidine biosynthesis, or ribosomal DNA (rDNA). The *P. pastoris* expression system has been used to produce more than 100 different biologically active proteins from bacteria, fungi, invertebrates, plants, and mammals, including humans.

Authentic heterologous proteins for industrial and pharmaceutical uses have also been generated in other yeasts, including the methanol-utilizing (methylotrophic) yeast *Hansenula polymorpha* and the thermotolerant **dimorphic** yeasts *Arxula adeninivorans* and *Yarrowia lipolytica*. It is often necessary to try several host types in order to find the one that produces the highest levels of a biologically active recombinant protein. Differences in the processing and productivity of a particular protein can occur among

rDNA
ribosomal DNA

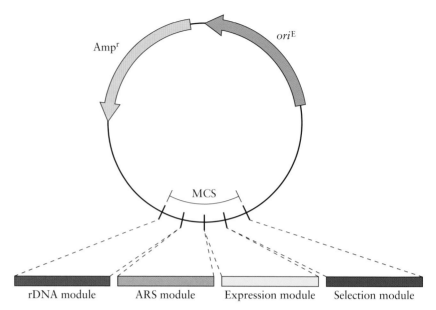

rDNA module   ARS module   Expression module   Selection module

Figure 6.22 A wide-range yeast vector system for expression of heterologous genes in several different yeast hosts. The basic vector contains a multiple-cloning site (MCS) for insertion of selected modules containing appropriate sequences for chromosomal integration (rDNA module), replication (ARS module), selection (selection module), and expression (expression module) of a target gene in a variety of yeast host cells. Sequences for maintenance (ori^E) and selection (Amp^r) of the vector in *E. coli* are also included. doi:10.1128/9781555818890.ch6.f6.22

different yeast strains. The construction of a wide-range yeast vector for expression in several fungal species has facilitated this trial-and-error process (Fig. 6.22). The basic vector contains features for propagation and selection in *E. coli* and a multiple-cloning site for insertion of interchangeable modules that are chosen for a particular yeast host, including a sequence for vector integration into the fungal genome, a suitable origin of replication, a promoter to drive expression of the heterologous gene, and selectable markers. By selecting from a range of available modules, customized vectors can be rapidly and easily constructed for expression of the same gene in several different yeast cells to determine which host is optimal for heterologous-protein production.

Baculovirus–Insect Cell Expression Systems

Baculoviruses are a large, diverse group of viruses that specifically infect arthropods, including many insect species, and are not infectious to other animals. During the infection cycle, two forms of baculovirus are produced: **budded** and **occluded** (Fig. 6.23). The infection is initiated by the occluded form of the virus. In this form, the viral nucleocapsids (virions) are clustered in a matrix that is made up of the protein polyhedrin. The occluded virions packaged in this protein matrix are referred to as a polyhedron and are protected from inactivation by environmental agents. Once the virus is taken up into the midgut of the insect, usually through ingestion of contaminated plant material, the polyhedrin matrix dissolves due to the alkaline gut environment, and the virions enter midgut cells to begin the infection cycle in the nucleus. Within the insect midgut, the infection can spread from cell to cell as viral particles (single nucleocapsids) bud off from an infected cell and infect other midgut cells. The budding form is not embedded in a polyhedrin matrix and is not infectious to other individual insect hosts, although it can infect cultured insect cells. **Plaques** produced in insect cell cultures by the budding form of baculovirus have a morphology different from that of the occluded form. During the late stages of the infection cycle in the insect host, about 36 to 48 h after infection, the polyhedrin protein is produced in massive quantities and production continues for 4 to 5 days, until the infected cells rupture and the host organism dies. Occluded virions are released and can infect new hosts.

The promoter for the polyhedrin (*polyh*) gene can account for as much as 25% of the mRNA produced in cells infected with the virus. However, the polyhedrin protein is not required for virus production, so replacement of the polyhedrin gene with a coding sequence for a heterologous protein, followed by infection of cultured insect cells, results in the production of large amounts of the heterologous protein. Furthermore, because of the similarity of posttranslational modification systems between insects and mammals, it was thought that the recombinant protein would mimic closely the authentic form of the original protein. Baculoviruses have been highly successful as delivery systems for introducing target genes for production of heterologous proteins in insect cells. More than a thousand different proteins have been produced using this system, including enzymes, transport proteins, receptors, and secreted proteins.

The specific baculovirus that has been used extensively as an expression vector is *Autographa californica* multiple nucleopolyhedrovirus (AcMNPV). *A. californica* (the alfalfa looper) and over 30 other insect species are infected by AcMNPV. This virus also grows well on many insect cell lines. The most commonly used cell line for genetically engineered AcMNPV is derived from the fall armyworm, *Spodoptera frugiperda*. In these cells, the polyhedrin promoter is exceptionally active, and during infections with wild-type baculovirus, high levels of polyhedrin are synthesized.

Baculovirus Expression Vectors

The first step in the production of a recombinant AcMNPV to deliver the gene of interest into the insect host cell is to create a transfer vector. The transfer vector is an *E. coli*-based plasmid that carries a segment of DNA from AcMNPV (Fig. 6.24A) consisting of the polyhedrin promoter region (without the polyhedrin gene) and an adjacent portion of upstream AcMNPV DNA, a multiple-cloning site, the polyhedrin termination and polyadenylation signal regions, and an adjacent portion of downstream AcMNPV DNA. The upstream and downstream AcMNPV DNA segments included on the transfer vector provide regions for homologous recombination with AcMNPV. A target gene is inserted into the multiple-cloning site between the polyhedrin promoter and termination sequences, and the transfer vector is propagated in *E. coli*.

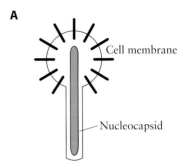

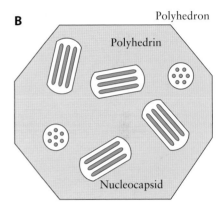

Figure 6.23 Budded (**A**) and occluded (**B**) forms of AcMNPV. During budding, a nucleocapsid becomes enveloped by the membrane of an infected cell. A polyhedron consists of clusters of nucleocapsids (occluded virions) embedded in various orientations in a polyhedrin matrix.
doi:10.1128/9781555818890.ch6.f6.23

AcMNPV
Autographa californica multiple nucleopolyhedrosis virus

A

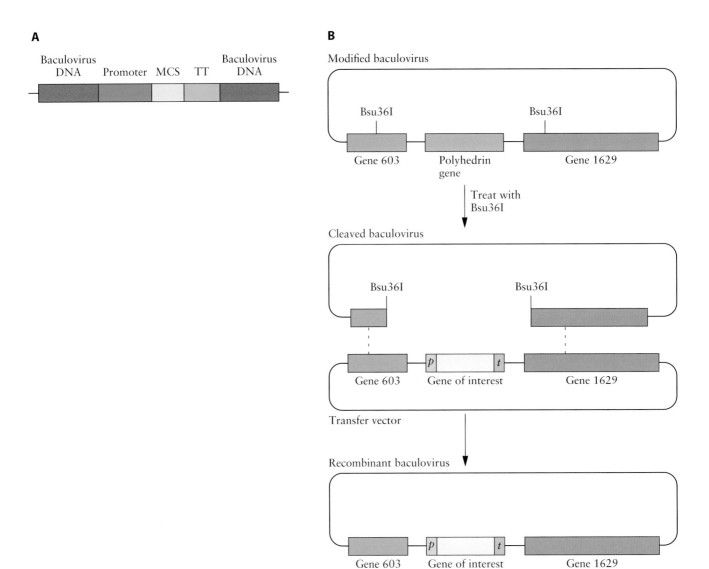

B

Figure 6.24 Organization of the expression unit of a baculovirus (AcMNPV) transfer vector. (**A**) The target gene of interest is inserted into the multiple-cloning site (MCS), which lies between the polyhedrin gene promoter and polyhedrin gene transcription termination sequences (TT). The baculovirus DNA upstream from the polyhedrin promoter and downstream from the polyhedrin TT provides sequences for integration of the expression unit by homologous recombination into the baculovirus genome. (**B**) Production of recombinant baculovirus. Single Bsu36I sites are engineered into gene 603 and a gene (1629) that is essential for AcMNPV replication. These genes flank the polyhedrin gene in the AcMNPV genome. After a baculovirus with two engineered Bsu36I sites is treated with Bsu36I, the segment between the Bsu36I sites is deleted. Insect cells are cotransfected with a Bsu36I-treated baculovirus DNA and a transfer vector with a gene of interest under the control of the promoter (*p*) and terminator (*t*) elements of the polyhedrin gene and the complete sequences of both genes 603 and 1629. A double-crossover event (dashed lines) generates a recombinant baculovirus with a functional gene 1629. With this system, almost all of the progeny baculoviruses are recombinant. doi:10.1128/9781555818890.ch6.f6.24

Next, insect cells in culture are cotransfected with AcMNPV DNA and the transfer vector carrying the cloned gene. Within some of the doubly transfected cells, a double-crossover recombination event occurs at homologous polyhedrin gene sequences on the transfer vector and in the AcMNPV genome, and the cloned gene with polyhedrin promoter and termination regions becomes integrated into the AcMNPV DNA, with the concomitant loss of the polyhedrin gene.

Unfortunately, the identification of occlusion-negative plaques is subjective, and purification of recombinant baculovirus is tedious due to the low frequency of recombination (~0.1%) between the AcMNPV DNA and the transfer plasmid. However, linearization of the AcMNPV genome before **transfection** into insect cells can dramatically increase the frequency of recombinant plaques. The AcMNPV genome was engineered with two Bsu36I sites that were placed on either side of the polyhedrin gene (Fig. 6.24B). One is in gene 603, and the other is in gene 1629, which is essential for viral replication. When DNA from this modified baculovirus is treated with Bsu36I and transfected into insect cells, no viral replication occurs, because a segment of gene 1629 is missing. As part of this system, a transfer vector is constructed with the gene of interest between intact versions of gene 603 and gene 1629. This transfer vector is introduced into insect cells that were previously transfected with linearized, replication-defective AcMNPV DNA that is missing the segment between the two Bsu36I sites. A double-crossover event both reestablishes a functional version of gene 1629 and incorporates the cloned gene into the AcMNPV genome (Fig. 6.24B). With this system, over 90% of the baculovirus plaques are recombinant.

Integration of Target Genes into Baculovirus by Site-Specific Recombination

To eliminate the need to use plaque assays to identify and purify recombinant viruses, several methods have been developed that introduce the target gene into the baculovirus genome at a specific nucleotide sequence by recombination. Transfection of insect cells is required only for the production of the heterologous protein. AcMNPV DNA can be maintained in *E. coli* as a plasmid known as a bacmid, which is a baculovirus–plasmid hybrid molecule. In addition to AcMNPV genes, the bacmid contains an *E. coli* origin of replication, a kanamycin resistance gene, and an integration site (attachment site) that is inserted into the *lacZ′* gene without impairing its function (Fig. 6.25A). Another component of this system is the transfer vector that carries the gene of interest cloned between the polyhedrin promoter and a terminator sequence. In the transfer vector, the target gene expression unit (expression cassette) and a gentamicin resistance gene are flanked by DNA attachment sequences that can bind to the attachment site in the bacmid (Fig. 6.25B). An ampicillin resistance (Amp^r) gene lies outside the expression cassette for selection of the transfer vector.

Amp^r
ampicillin resistance

Bacterial cells carrying a **bacmid** are cotransformed with the transfer vector and a helper plasmid that encodes the specific proteins (transposition proteins) that mediate recombination between the attachment sites on the transfer vector and on the bacmid and that carries a tetracycline

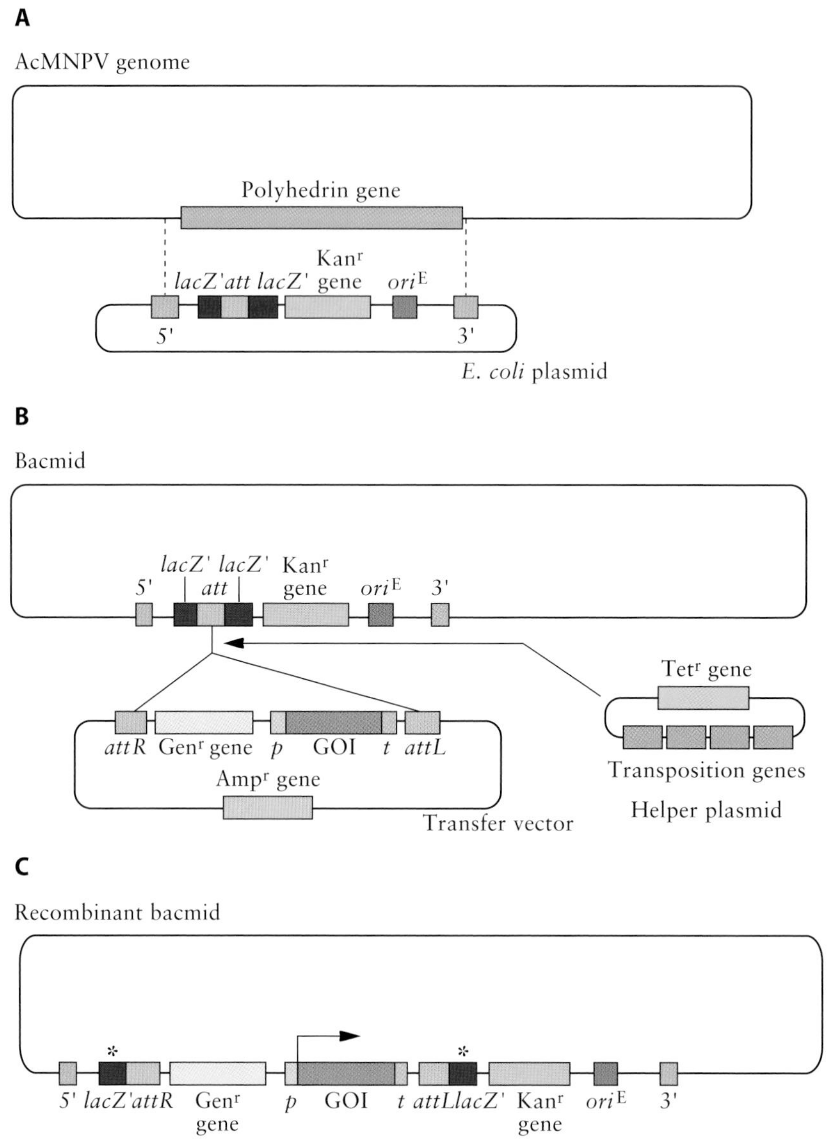

A

AcMNPV genome

B

Bacmid

C

Recombinant bacmid

Figure 6.25 Construction of a recombinant bacmid. (**A**) An *E. coli* plasmid is incorporated into the AcMNPV genome by a double-crossover event (dashed lines) between DNA segments (5′ and 3′) that flank the polyhedrin gene to create a shuttle vector (bacmid) that replicates in both *E. coli* and insect cells. The gene for resistance to kanamycin (Kan^r), an attachment site (*att*) that is inserted in frame in the *lacZ′* sequence, and an *E. coli* origin of replication (*ori*^E) are introduced as part of the plasmid DNA. (**B**) The transposition proteins encoded by genes of the helper plasmid facilitate the integration (transposition) of the DNA segment of the transfer vector that is bounded by two attachment sequences (*attR* and *attL*). The gene for resistance to gentamicin (Gen^r) and a gene of interest (GOI) that is under the control of the promoter (*p*) and transcription terminator (*t*) elements of the polyhedrin gene are inserted into the attachment site (*att*) of the bacmid. The helper plasmid and transfer vector carry the genes for resistance to tetracycline (Tet^r) and ampicillin (Amp^r), respectively. (**C**) The recombinant bacmid has a disrupted *lacZ′* gene (*). The right-angled arrow denotes the site of initiation of transcription of the cloned gene after transfection of the recombinant bacmid into an insect cell. Cells that are transfected with a recombinant bacmid are not able to produce functional β-galactosidase.
doi:10.1128/9781555818890.ch6.f6.25

resistance gene. After recombination, the DNA segment that is bounded by the two attachment sites on the transfer vector (the expression cassette carrying the target gene) is transposed into the attachment site on the bacmid, destroying the reading frame of the *lacZ'* gene (Fig. 6.25C). Consequently, bacteria with recombinant bacmids produce white colonies in the presence of IPTG and the chromogenic substrate 5-bromo-4-chloro-3-indolyl-β-D-galactopyranoside (X-Gal). Moreover, white colonies that are resistant to kanamycin and gentamicin and sensitive to both ampicillin and tetracycline carry only a recombinant bacmid and no transfer or helper plasmids. After all of these manipulations, the integrity of the cloned gene can be confirmed by PCR. Finally, recombinant bacmid DNA can be transfected into insect cells, where the cloned gene is transcribed and the heterologous protein is produced.

Mammalian Glycosylation and Processing of Precursor Proteins in Insect Cells

Although insect cells can process proteins in a manner similar to that of higher eukaryotes, some mammalian proteins produced in *S. frugiperda* cell lines are not authentically glycosylated. For example, insect cells do not normally add galactose and terminal sialic acid residues to N-linked glycoproteins. Where these residues are normally added to mannose residues during the processing of some proteins in mammalian cells, insect cells will trim the oligosaccharide to produce paucimannose (Fig. 6.26). Consequently, the baculovirus system cannot be used for the production of several important mammalian glycoproteins. To ensure the production of humanized glycoproteins with accurate glycosylation patterns, an insect cell line was constructed to express five different mammalian glycosyltransferases.

X-Gal
5-bromo-4-chloro-3-indolyl-
β-D-galactopyranoside

Figure 6.26 N glycosylation of proteins in the Golgi apparatus of insect, human, and humanized insect cells. While the sugar residues added to N-glycoproteins in the endoplasmic reticulum are similar in insect and human cells, further processing in the Golgi apparatus yields a trimmed oligosaccharide (paucimannose) in insect cells and an oligosaccharide that terminates in sialic acid in human cells. To produce recombinant proteins for use as human therapeutic agents, humanized insect cells have been engineered to express several enzymes that process human glycoproteins accurately. Blue squares, *N*-acetylglucosamine; red circles, mannose; green squares, galactose; orange squares, sialic acid. doi:10.1128/9781555818890.ch6.f6.26

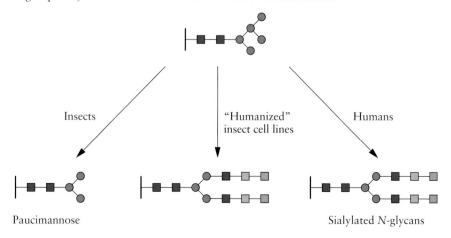

CHO
Chinese hamster ovary

SV40
simian virus 40

Mammalian Cell Expression Systems

Currently, about half of the commercially available therapeutic proteins are produced in mammalian cells. However, these cells are slow growing, have more fastidious growth requirements than bacteria or yeast cells, and can become contaminated with animal viruses. Chinese hamster ovary (CHO) cells and mouse myeloma cells are most commonly used for long-term (stable) gene expression and when high yields of heterologous proteins are required. About 140 recombinant proteins are currently approved for human therapeutic use, most produced in CHO cells that have been adapted for growth in high-density suspension cultures, and many more are in clinical trials. Although mammalian cells have been used for some time to produce therapeutic proteins, current efforts are aimed at improving productivity through the development of high-production cell lines, increasing the stability of production over time, and increasing expression by manipulating the chromosomal environment in which the recombinant genes are integrated.

Vector Design

Many cloning vectors for the expression of heterologous genes in mammalian cells are based on simian virus 40 (SV40) DNA (Table 6.6) that can replicate in several mammalian species. However, its use is restricted to small inserts because only a limited amount of DNA can be packaged into the viral capsid. The genome of this virus is a double-stranded DNA molecule of 5.2 kb that carries genes expressed early in the infection cycle that function in the replication of viral DNA (early genes) and genes expressed later in the infection cycle that function in the production of viral capsid proteins (late genes). Other vectors are derived from adenovirus, which can accommodate relatively large inserts; bovine papillomavirus, which can be maintained as a multicopy plasmid in some mammalian cells; and adeno-associated virus, which can integrate into specific sites in the host chromosome.

All mammalian expression vectors tend to have similar features and are not very different in design from other eukaryotic expression vectors. A typical mammalian expression vector (Fig. 6.27) contains a eukaryotic origin of replication, usually from an animal virus. The promoter sequences that drive expression of the cloned gene(s) and the selectable marker gene(s), and the transcription termination sequences (**polyadenylation** signals), must be eukaryotic and are frequently taken from either

Table 6.6 Genomes of some animal viruses that are used as cloning vectors in mammalian cells in culture

Virus	Genome	Genome size (kb)
SV40	Double-stranded DNA	5.2
Adenovirus	Double-stranded DNA	26–45
Bovine papillomavirus	Double-stranded DNA	7.3–8.0
Adeno-associated virus	Single-stranded DNA	4.8
Epstein–Barr virus	Double-stranded DNA	170

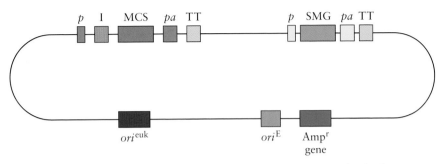

Figure 6.27 Generalized mammalian expression vector. The multiple-cloning site (MCS) and selectable marker gene (SMG) are under the control of eukaryotic promoter (*p*), polyadenylation (*pa*), and termination-of-transcription (TT) sequences. An intron (I) enhances the production of heterologous protein. Propagation of the vector in *E. coli* and mammalian cells depends on the origins of replication *ori*^{E} and *ori*^{euk}, respectively. The Amp^r gene is used for selecting transformed *E. coli*. doi:10.1128/9781555818890.ch6.f6.27

human viruses or mammalian genes. Strong constitutive promoters and efficient polyadenylation signals are preferred. Inducible promoters are often used when continuous synthesis of the heterologous protein may be toxic to the host cell. Expression of a target gene is often increased by placing the sequence for an intron between the promoter and the multiple-cloning site within the transcribed region. Sequences required for selection and propagation of a mammalian expression vector in *E. coli* are derived from a standard *E. coli* cloning vector.

For the best results, a gene of interest must be equipped with translation control sequences (Fig. 6.28). Initiation of translation in higher eukaryotic organisms depends on a specific sequence of nucleotides surrounding the start (AUG) codon in the mRNA called the Kozak sequence, i.e., GCCGCC(A or G)CC<u>AUG</u>G in vertebrates. The corresponding DNA sequence for the Kozak sequence, which is often followed by a signal sequence to facilitate secretion, a protein sequence (tag) to enhance the purification of the heterologous protein, and a proteolytic cleavage sequence that enables the tag to be removed from the recombinant protein, is placed at the 5′ end of the gene of interest. A stop codon is added to ensure that translation ceases at the correct location. Finally, the sequence content of the 5′ and 3′ untranslated regions (UTRs) is important for efficient translation and mRNA stability. Either synthetic 5′ and 3′ UTRs or those from the human β-globin gene are used in most

UTR
untranslated region

Figure 6.28 Translation control elements. A target gene can be fitted with various sequences that enhance translation and facilitate both secretion and purification, such as a Kozak sequence (K), signal sequence (S), protein affinity tag (T), proteolytic cleavage site (P), and stop codon (SC). The 5′ and 3′ UTRs increase the efficiency of translation and contribute to mRNA stability. doi:10.1128/9781555818890.ch6.f6.28

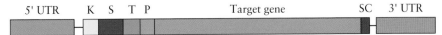

mammalian expression vectors. The codon content of the target gene may also require modification to suit the codon usage of the host cell.

The majority of mammalian cell expression vectors carry a single gene of interest. However, the active form of some important proteins consists of two different protein chains. The in vivo assembly of dimeric and tetrameric proteins is quite efficient. Consequently, various strategies have been devised for the production of two different recombinant proteins within the same cell.

Single vectors that carry two cloned genes provide the most efficient means of producing heterodimeric or tetrameric protein. The two genes may be placed under the control of independent promoters and polyadenylation signals (double-cassette vectors) (Fig. 6.29A), or, to ensure that equal amounts of the proteins are synthesized, vectors (bicistronic vectors) can be constructed with the two cloned genes separated from each other by a DNA sequence that contains an internal ribosomal entry site (IRES) (Fig. 6.29B). IRESs are found in mammalian virus genomes, and after transcription, they allow simultaneous translation of different proteins from a polycistronic mRNA molecule. Transcription of a "gene α–IRES–gene β" construct is controlled by one promoter and polyadenylation signal. Under these conditions, a single "two-gene" (bicistronic) transcript is synthesized, and translation proceeds from the 5′ end of the mRNA to produce chain α and internally from the IRES element to produce chain β (Fig. 6.29B).

IRES
internal ribosomal entry site

Selectable Markers for Mammalian Expression Vectors

For the most part, the systems that are used to select transfected mammalian cells are the same as those for other eukaryotic host cells. In addition, a number of bacterial marker genes have been adapted for eukaryotic cells. For example, the bacterial *neo* gene, which encodes neomycin phosphotransferase, is often used to select transfected mammalian cells. However, in eukaryotic cells, G-418 (Geneticin), which is phosphorylated by neomycin phosphotransferase, replaces neomycin as the selective agent because neomycin is not an effective inhibitor of eukaryotic protein synthesis.

Some selection schemes are designed not only to identify transfected cells but also to increase heterologous-protein production by amplifying the copy number of the expression vector. For example, dihydrofolate reductase catalyzes the reduction of dihydrofolate to tetrahydrofolate, which is required for the production of purines. Sensitivity to methotrexate, a competitive inhibitor of dihydrofolate reductase, can be overcome if the cell produces excess dihydrofolate reductase. As the methotrexate concentration is increased over time, the dihydrofolate reductase gene in cultured cells is amplified. In fact, methotrexate-resistant cells can have hundreds of dihydrofolate reductase genes. The standard dihydrofolate reductase–methotrexate protocol entails transfecting dihydrofolate reductase-deficient cells with a vector carrying a dihydrofolate reductase gene as the selectable marker gene and treating the cells with methotrexate. After the initial selection of transfected cells, the concentration of methotrexate is gradually increased, and eventually cells with very high copy numbers of the expression vector are selected.

A

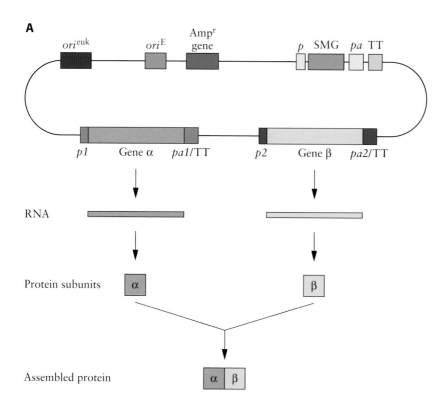

B

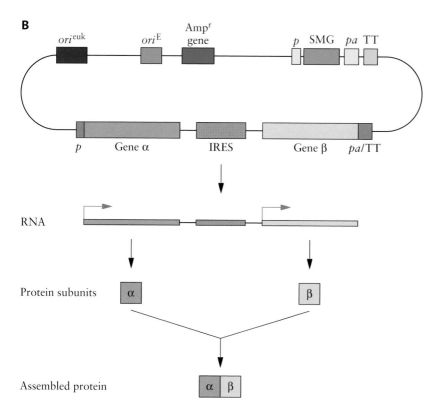

Figure 6.29 (**A**) A two-gene expression vector. The cloned genes (gene α and gene β) encode subunits of a protein dimer (αβ). The cloned genes are inserted into a vector and are under the control of different eukaryotic promoter (*p*), polyadenylation (*pa*), and termination-of-transcription (TT) sequences. Each subunit is translated from a separate mRNA, and a functional protein dimer is assembled. The vector has origins of replication for *E. coli* (*ori*^E) and mammalian cells (*ori*^euk), a marker gene (Amp^r) for selecting transformed *E. coli*, and a selectable marker gene (SMG) that is under the control of eukaryotic promoter (*p*), polyadenylation (*pa*), and TT sequences. (**B**) A bicistronic expression vector. Each cloned gene is inserted into a vector on either side of an IRES sequence; together they form a transcription unit under the control of a single eukaryotic promoter (*p*), polyadenylation (*pa*), and TT sequences. Translation of the mRNA occurs from the 5′ end and internally (right-angled arrows). Both subunits are synthesized and assembled into a functional protein dimer. The vector carries origins of replication for *E. coli* and mammalian cells, a selectable marker for selecting transformed *E. coli*, and a selectable marker gene that is under the control of eukaryotic promoter, polyadenylation, and TT sequences. doi:10.1128/9781555818890.ch6.f6.29

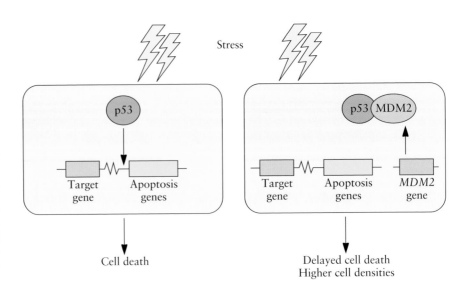

Figure 6.30 Strategy to increase yields of recombinant mammalian cells. Cell death (apoptosis), stimulated by the transcription factor p53, can lead to decreased yields of recombinant mammalian cells grown under stressful conditions in large bioreactors. To prevent cell death, the gene encoding MDM2 (the mouse double-mutant 2 protein) is introduced into mammalian cells. The MDM2 protein binds to p53 and prevents it from inducing expression of proteins required for apoptosis. Engineered cells not only showed delayed cell death but also achieved higher cell densities in bioreactors. doi:10.1128/9781555818890.ch6.f6.30

MDM2
mouse double-mutant 2 protein

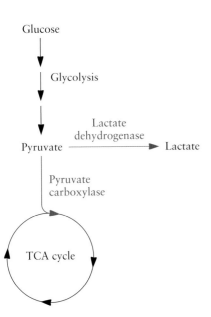

Engineering Mammalian Cell Hosts for Enhanced Productivity

In large-scale bioreactors, depleted nutrients and accumulation of toxic cell waste can limit the viability and density of cells as they respond to stress by inducing cell death, also known as apoptosis. One method to improve cell growth and viability under culture conditions in bioreactors is to prevent the tumor suppressor protein p53, which is a transcription factor, from activating the cell death response pathway. The mouse double-mutant 2 protein (MDM2) binds to protein p53 and prevents it from acting as a transcription factor (Fig. 6.30). MDM2 also marks p53 for degradation. When mammalian cells were transfected with plasmids containing a regulatable *MDM2* gene and cultured under conditions that mimicked the late stages of cell culture and in nutrient-limited medium, cultures expressing *MDM2* had higher cell densities and delayed cell death compared to those of nontransfected cells, especially in nutrient-deprived medium.

Many cultured mammalian cells are unable to achieve high cell densities in cultures because toxic metabolic products accumulate in the culture medium and inhibit cell growth. Many cells secrete the acidic waste product lactate as they struggle to obtain energy from glucose. Under these conditions, pyruvate is converted to lactate by lactate dehydrogenase rather than entering into the TCA cycle, where it is further oxidized through the activity of pyruvate carboxylase (Fig. 6.31). To counteract the acidification of the medium from lactate secretion, the human pyruvate carboxylase gene was cloned into an expression vector under the control of the cytomegalovirus promoter and the SV40 polyadenylation signals and transfected into CHO cells. When the pyruvate carboxylase gene was

Figure 6.31 When oxygen is present, pyruvate, which is formed from glucose during glycolysis, is converted by the enzyme pyruvate carboxylase to an intermediate compound in the TCA cycle. This metabolic pathway is important for the generation of cellular energy and for the synthesis of biomolecules required for cell proliferation. However, under low-oxygen conditions, such as those found in large bioreactors, pyruvate carboxylase has a low level of activity. Under these conditions, lactate dehydrogenase converts pyruvate into lactate, which yields a lower level of energy. Cultured cells secrete lactate, thereby acidifying the medium. doi:10.1128/9781555818890.ch6.f6.31

stably integrated into the CHO genome and expressed, the enzyme was detected in the mitochondria, where glucose is degraded. After 7 days in culture, the rate of lactate production decreased by up to 40% in the engineered cells.

Many eukaryotic DNA viruses from which the vectors used in mammalian cells are derived maintain their genomes as multicopy episomal DNA (plasmids) in the host cell nucleus. These viruses produce proteins, such as the large-T antigen in SV40 and the nuclear antigen 1 protein in Epstein–Barr virus, that help to maintain the plasmids in the host nucleus and to ensure that each host cell produced after cell division receives a copy of the plasmid. To increase the copy number of the target gene by increasing the plasmid copy number, HEK 293 cells have been engineered to express the SV40 large-T antigen or Epstein–Barr virus nuclear antigen 1.

Many proteins of therapeutic value are secreted. However, the high levels of these proteins that are desirable from a commercial standpoint can overwhelm the capacity of the cell secretory system. Thus, protein processing is a major limiting step in the achievement of high recombinant protein yields. Researchers have therefore engineered cell lines with enhanced production of components of the secretion apparatus. In this regard, an effective strategy may be to simultaneously overexpress several, if not all, of the proteins that make up the secretory mechanism. This can be achieved through the enhanced production of the transcription factor X box protein 1 (Xbp-1), a key regulator of the secretory pathway. Normally, full-length, unspliced *xbp-1* mRNA is found in nonstressed cells and is not translated into a stable, functional protein (Fig. 6.32). However,

Xbp-1
X box protein 1

Figure 6.32 Strategy to increase yields of secreted recombinant proteins from mammalian cells by simultaneously upregulating the expression of several proteins in the secretion apparatus. The expression of chaperones and other proteins of the secretion apparatus is controlled by the transcription factor Xbp-1. In nonstressed cells, the intron is not cleaved from the *xbp-1* transcript, and therefore, functional Xbp-1 transcription factor is not produced. In stressed cells with accumulated misfolded proteins, an endoribonuclease cleaves the transcript to remove the intron and yield mature *xbp-1* mRNA that is translated into transcription factor Xbp-1. Recombinant CHO cells transfected with a gene including only the *xbp-1* exons overproduced a functional Xbp-1 transcription factor that directed the production of high levels of proteins required for protein secretion. doi:10.1128/9781555818890.ch6.f6.32

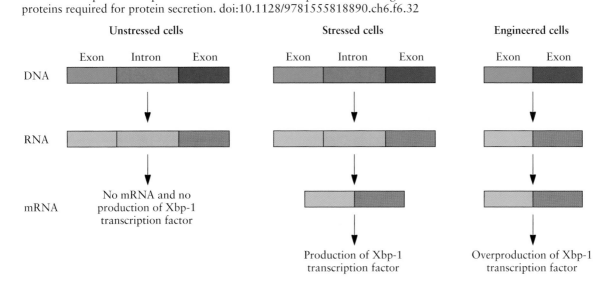

when unfolded or misfolded proteins accumulate in the endoplasmic reticulum, a ribonuclease is activated that specifically cleaves *xbp-1* mRNA (Fig. 6.32). This results in the production of a functional transcription factor that activates the expression of a number of proteins of the secretion apparatus. An *xbp-1* gene without its intron encodes an active form of *xbp-1* mRNA (Fig. 6.32), so the Xbp-1 protein is overproduced. The overexpression of Xbp-1 facilitates the secretion of human erythropoietin, human γ-interferon, and human monoclonal antibodies, especially in cell lines engineered to express the target proteins transiently.

Directed Mutagenesis

It is possible with recombinant DNA technology to isolate the gene (or **cDNA**) for any protein that exists in nature, to express it in a specific host organism, and to produce a purified product. Unfortunately, the properties of some of these "naturally occurring" proteins are sometimes not well suited for a particular end use. On the other hand, it is sometimes possible, using traditional **mutagenesis** (often ionizing radiation or DNA-altering chemicals) and selection schemes, to create a mutant form of a gene that encodes a protein with the desired properties. However, in practice, the mutagenesis–selection strategy only very rarely results in any significant beneficial changes to the targeted protein, because most amino acid changes decrease the activity of a target protein.

By using a variety of different directed mutagenesis techniques that change the amino acids encoded by a cloned gene, proteins with properties that are better suited than naturally occurring counterparts can be created. For example, using directed mutagenesis techniques, it is possible to change the specificity, stability, or regulation of target proteins.

Determining which amino acids of a protein should be changed to attain a specific property is much easier if the three-dimensional structure of the protein, or a similar protein, has been characterized by X-ray crystallographic analysis. But for many proteins, such detailed information is often lacking, so directed mutagenesis becomes a trial-and-error strategy in which changes are made to those nucleotides that are most likely to yield a particular change in a protein property. Moreover, it is not always possible to know in advance which individual amino acid(s) contributes to a particular physical, biological, or chemical property. Regardless of what types of alterations are made to a target gene, the protein encoded by each mutated gene has to be tested to ascertain whether the mutagenesis process has indeed generated the desired activity change.

Oligonucleotide-Directed Mutagenesis with M13 DNA

Oligonucleotide-directed mutagenesis **(site-specific mutagenesis)** is a straightforward method for producing defined point **mutations** in a cloned gene (Fig. 6.33). For this procedure, the investigator must know the precise nucleotide sequence in the region of DNA that is to be changed and the amino acid changes that are being introduced. In the original version of this method, the cloned gene was inserted into the

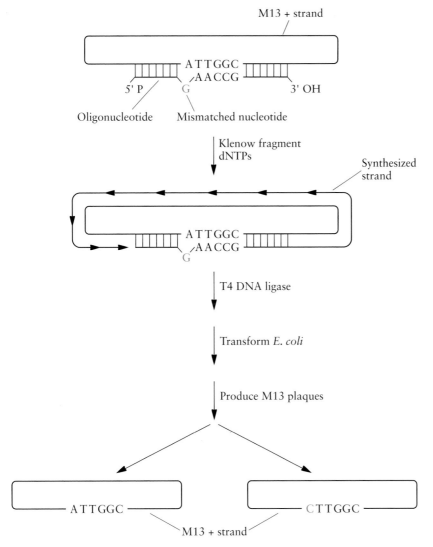

Figure 6.33 Oligonucleotide-directed mutagenesis. Single-stranded bacteriophage M13 (M13 + strand), carrying a cloned gene, is annealed with a complementary synthetic oligonucleotide containing one mismatched base, i.e., one base that is not complementary to its counterpart in the target DNA. With the oligonucleotide as the primer, DNA synthesis is catalyzed by the Klenow fragment of *E. coli* DNA polymerase I; the cloned gene and the M13 vector are the template. Synthesis continues until the entire strand is copied. The newly synthesized DNA strand is circularized by T4 DNA ligase. The ligation reaction mixture is used to transform *E. coli*. Both the target DNA with its original sequence and the mutated sequence are present in the progeny M13 phage. dNTPs, deoxynucleoside triphosphates.
doi:10.1128/9781555818890.ch6.f6.33

double-stranded form of an M13 bacteriophage vector (Milestone 6.3). The single-stranded form **(M13 + strand)** of the recombinant vector was isolated and mixed with a synthetic oligonucleotide. The oligonucleotide had, except for one nucleotide, a sequence exactly complementary to a segment of the cloned gene. The nucleotide difference (i.e., mismatch) coincided precisely with the nucleotide of the mRNA codon that was targeted for change. In Fig. 6.33, the sequence ATT, which encodes the

Oligonucleotide-Directed Mutagenesis Using M13-Derived Vectors: an Efficient and General Procedure for the Production of Point Mutations in Any Fragment of DNA

M. J. Zoller and M. Smith

Nucleic Acids Res. **10**:6487–6500, 1982

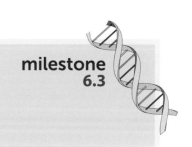

milestone
6.3

The technique of oligonucleotide-directed mutagenesis (site-specific mutagenesis) was developed mainly in the laboratory of Michael Smith as incremental changes to the technique of "marker rescue." In the marker rescue procedure, a mutation in a bacteriophage genomic DNA is corrected after the mutant DNA is annealed to a fragment of complementary wild-type DNA. Subsequently, it was demonstrated that a chemically synthesized oligonucleotide annealed to bacteriophage genomic DNA could produce a specific mutation. Unfortunately, these and other early procedures for oligonucleotide-directed mutagenesis required specialized skills and initially could be performed only in a few research laboratories. However, the procedure using bacteriophage M13 described by Zoller and Smith made it a relatively straightforward matter for thousands of laboratories throughout the world to specifically and rapidly alter the DNA sequence of any cloned gene. The key to the success of the protocol developed by Zoller and Smith lay in the use of *E. coli* bacteriophage M13. It was possible to clone foreign DNA into the double-stranded form of the virus, add an oligonucleotide with a specified change to the single-stranded form to produce a mutated DNA copy, and then recover the mutated double-stranded form in a relatively high yield. Since it was originally described, this procedure has been enhanced, simplified, and optimized and has been used by a large number of researchers to specifically modify thousands of different genes.

isoleucine codon AUU, is to be changed to CTT, which encodes the leucine codon CUU. Under a specific set of conditions, the oligonucleotide hybridizes to the complementary region of the cloned gene. The 3′ end of the hybridized oligonucleotide acts as a primer site for the initiation of DNA synthesis that uses the intact M13 strand as the template. DNA replication is catalyzed by the Klenow fragment of *E. coli* DNA polymerase I (which retains both the 5′-3′ polymerase and the 3′ exonuclease activities, but not the 5′ exonuclease activity, of the complete *E. coli* DNA polymerase I protein). T4 DNA ligase is added to ensure that the last nucleotide of the synthesized strand is joined to the 5′ end of the primer.

Each complete double-stranded M13 molecule, now containing the mismatched nucleotide, is introduced into *E. coli* cells by transformation. Infected cells produce M13 virus particles, which eventually lyse the cells and form plaques. Theoretically, because plasmid DNA is replicated semiconservatively, half of the phage that are formed carry the wild-type sequence and the other half contain the mutated sequence that has the specified nucleotide change. Phage produced in the initial transformation step are propagated in *E. coli*, and particles that contain only the mutated gene are identified by DNA hybridization under highly stringent conditions. The original oligonucleotide, containing the mismatched nucleotide, is the probe in these hybridization experiments and will bind only to the mutated gene under these conditions. After the double-stranded form of M13 is isolated, the mutated gene is excised by digestion with restriction enzymes and then spliced onto an *E. coli* plasmid expression vector for production of the altered protein in *E. coli*.

In practice, when oligonucleotide-directed mutagenesis is used, the expected 50% of the M13 viruses carrying the mutated form of the target gene is not recovered. Rather, for a variety of technical reasons, only <1% of the plaques actually contain phage carrying the mutated gene. Consequently, the oligonucleotide-directed mutagenesis method has been modified to enrich for the number of mutant phage plaques that can be obtained.

Oligonucleotide-Directed Mutagenesis with Plasmid DNA

As an alternative to the M13 system and its many time-consuming steps, a number of protocols that allow oligonucleotide-directed mutagenesis to be performed with plasmid rather than M13 DNA have been developed. With this approach, the need to subclone a target gene from a plasmid into M13 and then, after mutagenesis, clone it back into a plasmid is avoided. In one plasmid-based mutagenesis protocol, the target DNA is inserted into a multiple-cloning site on a plasmid vector that contains a functional tetracycline resistance gene and a nonfunctional Amp^r gene as the result of a single nucleotide substitution in the middle of the Amp^r gene (Fig. 6.34). The plasmid vector carrying the target DNA is transformed into *E. coli* host cells to increase the amount of DNA through plasmid replication. Following growth of the transformed cells, the double-stranded plasmid DNA is extracted and then denatured by treatment with an alkaline solution to form single-stranded circular DNA molecules. Three different oligonucleotides that anneal to one of the single-stranded circular DNA molecules are added to the sample of denatured plasmid DNA. One oligonucleotide is designed specifically to alter the target DNA, another is designed to correct the substituted nucleotide in the nonfunctional ampicillin resistance gene, and the third is designed to change a single nucleotide in the tetracycline resistance gene so that this gene will become nonfunctional. Following DNA synthesis, the nicks in the synthesized strand are sealed by the enzyme T4 DNA ligase and then the reaction mixture is used to transform *E. coli* cells. Transformants are selected for ampicillin resistance and tetracycline sensitivity. With this procedure, about 90% of the selected transformants have the specified mutation in the target gene. The cells with the specified mutation in the target gene may be identified by DNA hybridization or, where possible, by the activity of the altered target protein. All of the plasmids, host bacterial strains, enzymes, oligonucleotides (other than the one needed to alter the target gene), and buffers for this method are sold commercially as a kit, facilitating its widespread use.

PCR-Amplified Oligonucleotide-Directed Mutagenesis

The **polymerase chain reaction (PCR)** can also be used to introduce a specific mutation. In fact, there are commercially available kits to which a researcher merely adds the target plasmid carrying the gene of interest, along with forward and reverse PCR primers that are typically 24 to 30 nucleotides in length, and, following PCR, a high percentage of

PCR
polymerase chain reaction

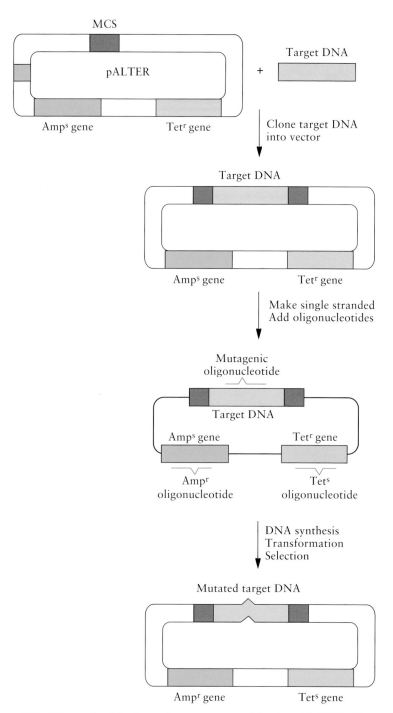

Figure 6.34 Oligonucleotide-directed mutagenesis with plasmid DNA. The target DNA is inserted into the multiple-cloning site (MCS) on the vector pALTER. Plasmid DNA isolated from the *E. coli* cells is alkaline denatured before the mutagenic oligonucleotide, the Amp^r oligonucleotide, and the tetracycline sensitivity (Tet^s) oligonucleotide are annealed. The oligonucleotides act as primers for DNA synthesis by T4 DNA polymerase with the original strand as the template. The gaps between the synthesized pieces of DNA are sealed by T4 DNA ligase. The final reaction mixture is used to transform *E. coli* host cells, and cells that are Amp^r and Tet^s are selected. doi:10.1128/9781555818890.ch6.f6.34

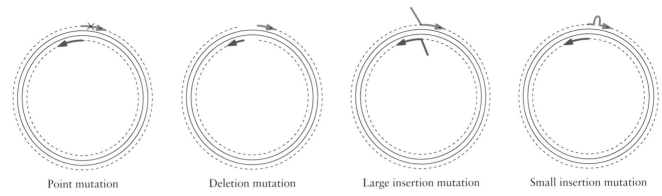

| Point mutation | Deletion mutation | Large insertion mutation | Small insertion mutation |

Figure 6.35 Overview of the basic methodology to introduce point mutations, insertions, or deletions into DNA cloned into a plasmid. The forward and reverse primers are shown in red and green, respectively. Solid circles represent template DNA. Dashed lines represent newly synthesized DNA. The × indicates an altered nucleotide(s). doi:10.1128/9781555818890.ch6.f6.35

the plasmids produced will have the desired mutation. In this case, no special plasmid vectors are required; any plasmid up to approximately 10 kb in length is effective. For PCR-based mutagenesis point mutations, nucleotide changes are introduced in the middle of the primer sequence (Fig. 6.35). To create deletion mutations, primers must border the region of target DNA to be deleted on both sides and be perfectly matched to their annealing (or template) sequence. To create mutations with long insertions, a stretch of mismatched nucleotides is added to the 5′ end of one or both primers, while for mutations with short insertions, a stretch of nucleotides is designed in the middle of one of the primers. In all of these procedures, the only absolute requirements are that the nucleotide sequence of the target DNA must be known and the 5′ ends of the primers must be phosphorylated. Following PCR amplification, the linear DNA is circularized by ligation with T4 DNA ligase. The circularized plasmid DNA is then used to transform *E. coli* host cells. Since this protocol yields a very high frequency of plasmids with the desired mutation, it is not necessary to utilize any enrichment procedures. Rather, screening three or four clones by sequencing the target DNA is generally sufficient to find the desired mutation. Given its simplicity and effectiveness, this procedure has come to be widely used.

Error-Prone PCR

Unfortunately, investigators seldom know which specific nucleotide changes are needed to be introduced into a cloned gene to modify the properties of the target protein. Consequently, they have to use methods that generate a wide range of possible amino acid changes at a particular site. This approach has two advantages. First, detailed information regarding the role of particular amino acid residues in the functioning of the protein is not required. Second, unexpected mutants encoding proteins with a range of interesting and useful properties may be generated because the introduced changes are not limited to one amino acid.

Random-mutagenesis procedures (including conventional mutagenesis, **error-prone PCR**, or one of a large number of published protocols), whether they cause nucleotide deletions, insertions, or alterations, can result in mutations that affect the activity of the target protein in one of three ways. The activity of the target protein may be decreased or abolished, it may be increased, or it may remain unchanged (silent mutation). To find the small number of potentially useful mutations from a very much larger number of negative and silent mutations, it is necessary either to have a very good selection scheme or to undertake the time-consuming task of testing very large numbers of mutants.

Some of the temperature-stable DNA polymerases that are used to amplify target DNA by PCR occasionally insert incorrect nucleotides into the replicating DNA. If one is attempting to amplify a DNA fragment with high fidelity, this will obviously be a problem. On the other hand, if the construction of a library of mutants of the target gene is the objective, then this approach is a very powerful method of random mutagenesis (Fig. 6.36). Moreover, with DNA of up to 10 kb in size, it is possible to vary the number of nucleotide changes per gene from about 1 to about 20. The number of altered (incorrect or mutated) nucleotides is a function of the DNA concentration, the particular temperature-stable DNA polymerase

Figure 6.36 (A) Error-prone PCR of a target gene yields a variety of mutated forms of the gene. Introduced mutations (typically 1 or 2 per gene) are shown in red. The horizontal arrows represent PCR primers. (B) Schematic representation of the changes (indicated by stars) introduced into a protein by either random mutagenesis or error-prone PCR, both of which cause single-amino-acid substitutions at one or more sites throughout the protein molecule.
doi:10.1128/9781555818890.ch6.f6.36

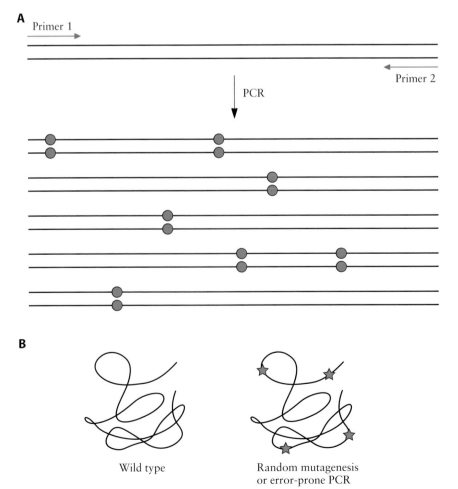

employed, the presence of Mn^{2+}, and the concentration of Mg^{2+}. Following error-prone PCR, the randomly mutagenized DNA is cloned into expression vectors and screened (tested) for altered or improved protein activity. The DNAs from those clones that encode the desired activity are isolated and sequenced so that the changes to the target DNA may be elaborated. The advantage of this approach is that it produces a wide range of mutations throughout the target protein, some of which may be beneficial.

Random Mutagenesis

Altered (mutant) oligonucleotides may be incorporated into a target gene by a variety of procedures. One strategy entails inserting a target gene into a plasmid between two unique restriction endonuclease sites and using PCR, in separate reactions, to amplify overlapping fragments (Fig. 6.37). The primer pair that is used to amplify the left fragment consists of mismatched oligonucleotides that were synthesized to contain degenerate

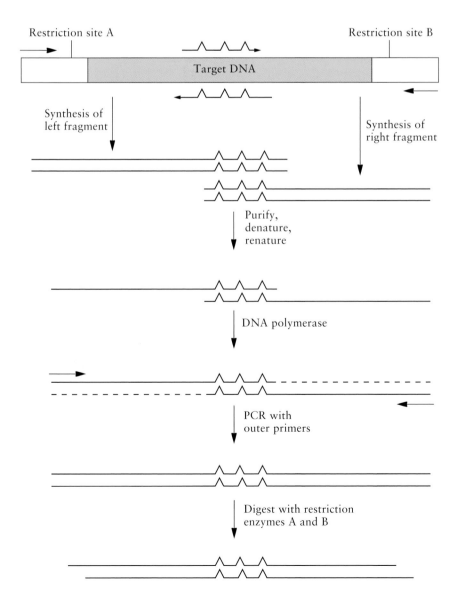

Figure 6.37 Random mutagenesis of a target DNA by using degenerate oligonucleotides and PCR. The left and right portions of the target DNA are amplified separately by PCR. The primer pairs are shown by horizontal arrows. A mutation-producing oligonucleotide is shown as a line with three spikes; each spike indicates a position that contains a nucleotide that is altered from what is found in the wild-type gene (or cDNA). The amplified fragments are purified, denatured to make them single stranded, and then reannealed. Complementary regions of overlap are formed between complementary mutation-producing oligonucleotides. The single-stranded regions are made double stranded with DNA polymerase, and then the entire fragment is amplified by PCR. The resultant product is digested with restriction endonucleases A and B and then cloned into a plasmid vector that has been digested with the same enzymes. doi:10.1128/9781555818890.ch6.f6.37

oligonucleotides and that bind to the lower strand of the target DNA along with a regular, completely complementary primer that hybridizes to a region of the upper strand that flanks the left unique restriction endonuclease site. For the right fragment, the PCR primers are the mismatched oligonucleotides that were synthesized to contain degenerate oligonucleotides and that bind to the upper strand of the target DNA along with a primer that is complementary to a region of the lower strand that lies outside the second (right) unique restriction endonuclease site. After PCR amplification, the products are purified and combined. Denaturation and reannealing of the DNA in the mixture produce some DNA molecules that overlap in the target region. DNA polymerase is then used to form complete double-stranded DNA molecules. These molecules are amplified by PCR with a pair of primers that bind to opposite ends of the DNA molecule. The amplified DNA is then treated with the two restriction enzymes for which there are unique sites at the ends of the fragment, and the DNA is cloned into a suitable plasmid vector. This procedure results in the production of an altered gene that has mutated sites (altered nucleotides) in the region of the overlap of the original oligonucleotides.

Using error-prone PCR and/or random mutagenesis, it is possible to change several different properties of a target protein. One way to do this is to separately introduce amino acid changes that alter particular protein properties and then, at the DNA level, combine these mutations into a single modified protein (Fig. 6.38).

Figure 6.38 Schematic representation of the introduction of several changes into a single protein.
doi:10.1128/9781555818890.ch6.f6.38

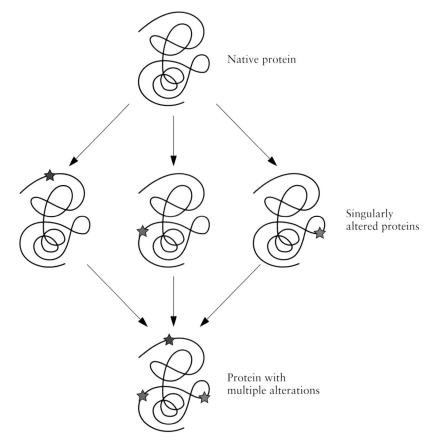

Native protein

Singularly
altered proteins

Protein with
multiple alterations

DNA Shuffling

Some medically important proteins such as α-interferon are encoded by a family of several related genes, with each protein having slightly different biological activity. If all, or at least several, of the genes or cDNAs for a particular protein have been isolated, it is possible to recombine portions of these genes or cDNAs to produce hybrid or chimeric forms (Fig. 6.39A). This **"DNA shuffling"** is done in an effort to find hybrid proteins with unique properties or activities that were not encoded in any of the original sequences. As well, some of the hybrid proteins may combine important attributes of two or more of the original proteins, e.g., high activity and thermostability.

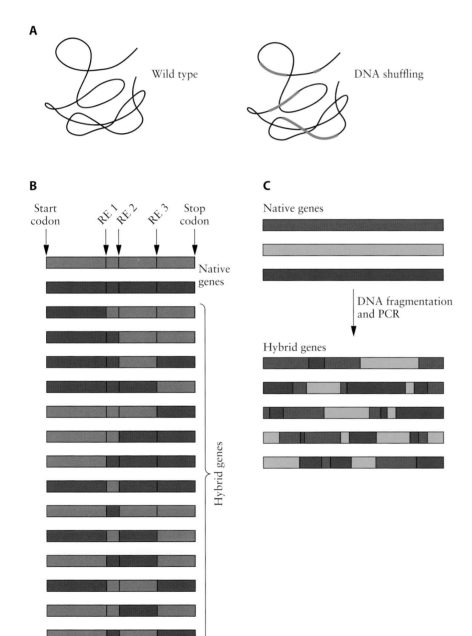

A

Wild type

DNA shuffling

B

Start codon   RE 1   RE 2   RE 3   Stop codon

Native genes

Hybrid genes

C

Native genes

DNA fragmentation and PCR

Hybrid genes

Figure 6.39 (A) Schematic representation of the changes introduced into a protein by DNA shuffling, in which genes are formed with large regions from different sources. (B) The 14 different hybrid genes that can be generated by combining restriction enzyme fragments from two genes from the same gene family that have three different restriction sites in common. RE, restriction enzyme. (C) A few of the hybrid DNAs that can be generated during PCR amplification of three members of a gene family. doi:10.1128/9781555818890.ch6.f6.39

The simplest way to shuffle portions of similar genes is through the use of common restriction enzyme sites (Fig. 6.39B). Digestion of two or more of the DNAs that encode the native forms of similar proteins with one or more restriction enzymes that cut the DNAs in precisely the same place, followed by ligation of the mixture of the resultant DNA fragments, can potentially generate a large number of hybrids. For example, two DNAs that each have three unique identically situated restriction enzyme sites can be recombined (shuffled) to produce 14 different hybrids in addition to the original DNAs (Fig. 6.39B). Another way to shuffle DNA involves combining several members of a gene family, fragmenting the mixed DNA by partial digestion with deoxynuclease (DNase I), selecting smaller DNA fragments, and PCR amplifying these fragments. During PCR, gene fragments from different members of a gene family can cross-prime each other after DNA fragments bind to one another in regions of high homology or complementarity. The final full-length products are obtained by including "terminal primers" in the PCR. After 20 to 30 PCR cycles, a panel of hybrid (full-length) DNAs will be established (Fig. 6.39C). The hybrid DNAs are then used to create a library that can be screened for the desired activity. Although DNA shuffling works well with gene families—it is sometimes called molecular breeding—or with genes from different families that nevertheless have a high degree of homology, this technique is not especially useful when proteins have little or no homology. Thus, the DNAs must be very similar to one another or else the PCR will not proceed.

Examples of Modified Proteins

tPA
tissue plasminogen activator

The enzyme **tissue plasminogen activator (tPA)** is a multidomain serine protease that is medically useful for the dissolution of blood clots. However, like **streptokinase**, tPA is rapidly cleared from the circulation, so it must be administered by infusion. To be effective with this form of delivery, high initial concentrations of tPA must be used. Unfortunately, under these conditions, tPA can cause nonspecific internal bleeding. Thus, a (i) long-lived tPA (ii) with increased specificity for fibrin in blood clots and (iii) not prone to induce nonspecific bleeding would be desirable. It was found that these three properties could be separately introduced by directed mutagenesis of the gene for the native form of tPA (Fig. 6.40). First, changing Thr-103 to Asn causes the enzyme to persist in rabbit plasma approximately 10 times longer than the native form. Second, changing amino acids 296 to 299 from Lys-His-Arg-Arg to Ala-Ala-Ala-Ala produces an enzyme that is much more specific for fibrin than is the native form. Third, changing Asn-117 to Gln causes the enzyme to retain the level of fibrinolytic activity found in the native form. Moreover, combining these three mutations in a single construct allows all three activities to be expressed simultaneously. Based on clinical trials that indicated that this genetically modified form of tPA caused a significantly lower level of major bleeding in patients, it was approved by the FDA for treatment of acute myocardial infarction and is currently commercially available.

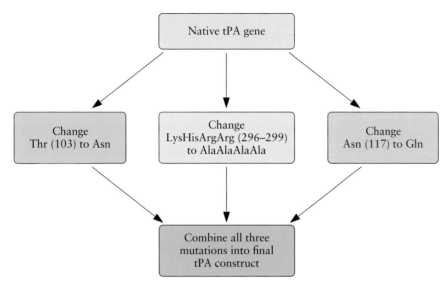

Figure 6.40 Schematic representation of the stepwise introduction of several specific amino acid changes into recombinant tPA in an effort to make it a more effective therapeutic agent. doi:10.1128/9781555818890.ch6.f6.40

All of the clinically important semisynthetic cephalosporin antibiotics are produced from either 7-aminocephalosporanic acid (7-ACA) or the related compound 7-aminodesacetoxycephalosporanic acid (7-ADCA). Cephalosporins are extremely useful antibiotics since they have few toxic effects on humans and are effective against many different pathogenic bacteria. The compound 7-ADCA is produced from penicillin G made by *Penicillium chrysogenum* and involves several polluting chemical steps followed by an enzymatic deacylation step that is catalyzed by penicillin acylase. In order to develop a simplified, more environmentally friendly means of producing 7-ADCA, a genetically modified version of *P. chrysogenum* that produces adipyl-7-ADCA was engineered. However, for the deacylation of this unique compound and its conversion to 7-ADCA, an adipyl acylase was needed. This was problematic since at the time that this work was undertaken, all known acylases showed little or no activity toward adipyl-7-ADCA. It was therefore decided to try to convert, by error-prone PCR, a known glutaryl acylase into an adipyl acylase. The starting point for these experiments was the glutaryl acylase from *Pseudomonas* strain SY-77. This enzyme has a high activity in removing the glutaryl side chain from glutaryl-7-A(D)CA as well as a low level of activity on adipyl-7-ADCA, but no activity on cephalosporin C. Since deacylation of adipyl-7-ADCA could not be used for a growth selection, researchers selected for the enzymatic hydrolysis of adipyl-leucine, which allows for growth of leucine-deficient *E. coli* host strains on minimal medium and therefore could be used as a selection. This selection is based on the notion that an enzyme that cleaves adipyl-leucine would also cleave adipyl-7-ADCA. To select mutants that could efficiently cleave adipyl-leucine, the gene encoding the β-subunit of glutaryl acylase was divided into five overlapping parts that were each mutagenized separately using error-prone

7-ACA
7-aminocephalosporanic acid

7-ADCA
7-aminodesacetoxycephalosporanic acid

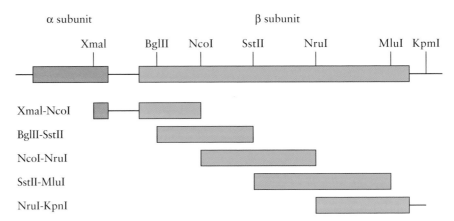

Figure 6.41 Schematic representation of the α- and β-glutaryl acylase genes, with several unique restriction enzyme sites indicated. The five fragments shown below these two genes were generated by digestion with the indicated restriction enzymes, and then each fragment was separately subjected to error-prone PCR. Mutated fragments were subsequently inserted into the native gene before being selected for the ability to cleave adipyl-leucine and utilize the released leucine to facilitate bacterial growth. doi:10.1128/9781555818890.ch6.f6.41

PCR (Fig. 6.41). The mutagenesis was done in five separate reactions since it was thought that this approach would yield the largest number of independent mutations distributed throughout the gene. In this way, a total of 24 mutants were found to have a significant level of adipyl-leucine-cleaving activity as well as an improved ratio of adipyl-7-ADCA versus glutaryl-7-ACA hydrolysis. In particular, one mutant, in which Asn-266 was changed to His-266, i.e., mutant SY-77^{N266H}, displayed a nearly 10-fold increase in catalytic efficiency (k_{cat}/K_m) compared to the native form of the enzyme toward adipyl-7-ADCA. This was the result of a 50% increase in k_{cat} combined with a sixfold decrease in K_m toward adipyl-7-ADCA without decreasing the catalytic efficiency toward glutaryl-7-ACA. In the future, it may be possible to further improve the activity of the enzyme toward adipyl-7-ADCA and decrease its activity toward glutaryl side chains by combining the most effective selected mutation with some of the other mutations that were isolated using this procedure.

summary

In prokaryotes, a promoter region is necessary for the initiation of transcription at the correct nucleotide site, and a terminator sequence at the end of the gene is essential for the cessation of transcription. To produce large amounts of protein, it is necessary to use a promoter that supports transcription at a high level (strong promoter). However, continuous transcription of a cloned gene drains the energy reserves of the host cell, so it is also necessary to use a promoter system whose activity can be regulated.

In addition to appropriate transcriptional signals, it is essential that a ribosome-binding site be located in the DNA segment that precedes the translation initiation site and a termination sequence at the end of the cloned gene to ensure that translation stops at the correct amino acid. If secretion of the protein is desired, the DNA sequence preceding the cloned gene should include a signal sequence in the same reading frame as the target gene.

Since various organisms preferentially use different subsets of codons, they contain different amounts of the cognate

summary *(continued)*

amino acyl-tRNAs for the synthesis of proteins that they encode. When a recombinant protein is overproduced, the difference in host and recombinant gene codon usage can lead to a shortage of some rare aminoacyl-tRNAs, resulting in either low yields or proteins with incorrect amino acids. To avoid these problems, host cells have been engineered to overexpress several rare tRNAs so that foreign DNAs with different codon usage profiles may be highly expressed with no deleterious consequences.

To stabilize the recombinant protein, a DNA construct that encodes the target protein that is in frame with DNA encoding a stable host protein is often utilized. The host protein typically confers increased stability on the target protein portion of the fusion protein construct. In addition, the amino acids that are added to the recombinant protein can sometimes be used for purifying the fusion protein by, for example, immunoaffinity column chromatography. In these cases, the junction point of a fusion protein is usually designed to be enzymatically cleaved in vitro.

The introduction and expression of foreign DNA in a host organism often change the metabolism of the organism and thereby impair its normal functioning. This phenomenon is called a metabolic load. A variety of strategies have been developed to minimize the extent of the perturbations caused by a metabolic load and at the same time optimize the yield of the target protein and the stability of the transformed cell.

Whole or parts of recombinant plasmid-based systems can be lost during host cell growth. To overcome this problem, researchers have developed protocols for integrating a cloned gene into a chromosomal site of the host organism. Under these conditions, the gene is maintained stably as part of the DNA of the host organism.

Most microorganisms that are used to express foreign proteins require oxygen for growth. However, oxygen has only a limited solubility in water and is rapidly depleted from the growth medium of actively growing cultures, especially when the cultures attain a high cell density. To deal with the limited amount of dissolved oxygen available for cell growth and maintenance, researchers have introduced the gene for the *Vitreoscilla* sp. hemoglobin that binds oxygen from the environment, creating a higher level of intracellular oxygen and thereby causing increases in both host and foreign protein synthesis.

Conditions that result in very high rates of foreign gene expression in *E. coli* often also lead to the production of misfolded proteins that can aggregate and form insoluble inclusion bodies. This problem may be remedied by the coexpression of one or more molecular chaperones (proteins that facilitate the correct folding of other proteins) by the host *E. coli* strain.

A number of heterologous proteins have been successfully synthesized in prokaryotic host cells. However, many proteins require eukaryote-specific posttranslational modifications, such as glycosylation, to be functional. Consequently, expression systems were devised for yeast, insect, and mammalian cells. With respect to the ease and likelihood of obtaining an authentic protein from a cloned gene, each of these systems has distinct merits and shortcomings. In other words, there is no single eukaryotic host cell that is capable of producing an authentic protein from every cloned gene.

Notwithstanding the differences in detail, the fundamental strategies that have been developed for prokaryotes may be used to create a variety of different eukaryotic expression systems. All eukaryotic expression vectors have the same basic format. The gene of interest, which may be equipped with sequences that facilitate the secretion and purification of the heterologous protein, is under the control of eukaryotic promoter and polyadenylation and transcription terminator sequences. To simplify both maintenance and recombinant DNA manipulations, eukaryotic expression vectors are routinely maintained in *E. coli*.

Several different yeast-based expression systems have been developed for the production of heterologous proteins. The yeast *S. cerevisiae*, which is well characterized genetically and can be grown in large fermenters, has been used extensively for this purpose. Both episomal and integrating expression vectors have been constructed. However, with *S. cerevisiae* as the host cell, many recombinant proteins are hyperglycosylated, and in some cases, protein yields are low because the capacity of the cell to properly fold and secrete proteins has been exceeded. Other yeast systems have been developed for the production of heterologous proteins. Of these, the methylotrophic yeast *P. pastoris* has been used successfully because of the low occurrence of hyperglycosylation, the ease of obtaining high cell densities, and the rapid and strong response of the *AOX1* promoter (usually used to drive transcription of the target gene) to methanol. A humanized strain of *P. pastoris* has been genetically altered to produce glycoproteins with glycosylation patterns that are identical to those found on the same proteins produced in human cells.

A large number of biologically active heterologous proteins have also been produced in insect cells grown in culture using baculoviruses to deliver the gene of interest into the insect host cell. This system is advantageous because posttranslational protein modifications are similar in insects and mammals, and the baculoviruses used in these systems do not infect humans or other insect cells. The baculovirus most commonly used as a vector is AcMNPV. A gene of interest

(continued)

summary *(continued)*

is inserted into the AcMNPV genome by homologous or site-specific recombination between sequences on a transfer vector carrying the target gene and the AcMNPV DNA. Once the target gene has been inserted, recombinant AcMNPV DNA is introduced into insect cells for heterologous-protein production. Improved insect host cells have been developed through genetic engineering to increase protein yields and to ensure that target proteins are properly glycosylated. In addition to production of a single protein of interest, the baculovirus–insect expression system is particularly amenable to producing functional multimeric protein complexes, such as virus-like particles, which are effective vaccines.

Many therapeutic proteins that require a full complement of posttranslational modifications are produced in cultured mammalian cells, such as CHO cells. Most of the vectors that have been developed to introduce foreign genes into mammalian cells are based on mammalian viruses. The viral genome has been altered to remove some viral genes required for replication and viral-protein production and to include suitable mammalian transcription and translation signals to drive expression of the cloned gene.

The proper functioning of a protein is due to its conformation, which is a consequence of its amino acid sequence and subunit structure. Certain amino acids in a polypeptide chain play important roles in determining the specificity, thermostability, and other properties of a protein. Changing even a single nucleotide of the gene encoding a target protein can result in the incorporation of an amino acid that can either disrupt the normal activity or enhance a specific property of the protein. With the emergence of recombinant DNA technology, it is possible to replace nucleotides of a cloned gene and produce proteins with specific amino acids at defined sites. This procedure is called directed mutagenesis, and it includes various strategies for introducing these kinds of changes into cloned genes. The choice of which amino acid to change is often based on knowledge of the role of a particular amino acid in the functional protein. This knowledge comes from genetic studies or X-ray crystallographic data for the three-dimensional organization of the protein. Specific sites or regions can be altered, or combined, to improve the thermostability, pH tolerance, specificity, allosteric regulation, cofactor requirements, and other properties of protein. In many instances, the amino acid change(s) that might enhance a particular property of a target protein is not known a priori. In these cases, random mutagenesis, error-prone PCR, or DNA shuffling rather than oligonucleotide-directed mutagenesis is preferred.

review questions

1. Suggest several ways that the expression of a cloned gene can be manipulated for optimal expression.

2. Sometimes the strategy for the expression of a target protein in a host organism involves synthesizing the protein as part of a fusion protein. Why is this approach useful? How is a fusion protein created?

3. What are inclusion bodies, and how can their formation be avoided?

4. How would you avoid some of the problems associated with the limited amount of oxygen that is available to growing *E. coli* cells when a foreign protein is overproduced?

5. How can a specific target DNA fragment be integrated into the chromosomal DNA of a host bacterium?

6. What factors are responsible for metabolic load?

7. Suggest strategies to limit the extent of metabolic load on *E. coli* cells that are designed to overproduce a recombinant protein.

8. How can *E. coli* host cells be engineered to yield high levels of expression of foreign proteins that contain significant numbers of rare *E. coli* codons?

9. What is the T7 expression system, and how does it work?

10. What strategies may be used to ensure the correct folding of overproduced recombinant proteins?

11. How can fusion proteins be utilized to facilitate the purification of a target recombinant protein?

12. What is the Cre–*lox* system, and how does it work?

13. How can *E. coli* cells be engineered to secrete proteins to the external medium?

14. How can *E. coli* be engineered so that the acetate that is produced as a by-product of its growth does not accumulate to the point where it becomes growth inhibitory?

15. What are the major posttranslational modifications of eukaryotic proteins in the endoplasmic reticulum and Golgi apparatus?

16. Describe the features of a eukaryotic expression vector.

references

Andersson, C. I. J., N. Holmberg, J. Farrés, J. E. Bailey, L. Bülow, and P. T. Kallio. 2000. Error-prone PCR of *Vitreoscilla* hemoglobin (VHb) to support the growth of microaerobic *Escherichia coli*. *Biotechnol. Bioeng.* 70:446–455.

Ansari, A., and V. C. Emery. 1998. Baculoviruses, p. 219–233. *In* R. Rabley and J. M. Walker (ed.), *Molecular Biomethods Handbook*. Humana Press Inc., Totowa, NJ.

Baneyx, F., and M. Mujacic. 2004. Recombinant protein folding and misfolding in *Escherichia coli*. *Nat. Biotechnol.* 22:1399–1408.

Barnes, L. M., and A. J. Dickson. 2006. Mammalian cell factories for efficient and stable protein expression. *Curr. Opin. Biotechnol.* 17:381–386.

Berger, I., D. J. Fitzgerald, and T. J. Richmond. 2004. Baculovirus expression system for heterologous multiprotein complexes. *Nat. Biotechnol.* 22:1583–1587.

Bowman, K. K., J. Clark, L. Yu, K. Mortara, K. Radika, J. Wang, and H. Zhan. 2000. Expression, purification, and characterization of deglycosylated human pro-prostate-specific antigen. *Protein Expr. Purif.* 20:405–413.

Chaitan, K., J. E. Curtis, J. DeModena, U. Rinas, and J. E. Bailey. 1990. Expression of intracellular hemoglobin improves protein synthesis in oxygen-limited *Escherichia coli*. *Bio/Technology* 8:849–853.

Chatterjee, R., and L. Yuan. 2006. Directed evolution of metabolic pathways. *Trends Biotechnol.* 24:28–38.

Chong, S. R., F. B. Mersha, D. G. Comb, M. E. Scott, D. Landry, L. M. Vence, F. B. Perler, J. Benner, R. B. Kucera, C. A. Hirvonen, J. J. Pelletier, H. Paulus, and M. Q. Xu. 1997. Single-column purification of free recombinant proteins using a self-cleavable affinity tag derived from a protein splicing element. *Gene* 192:271–281.

Cohen, S. N., A. C. Y. Chang, H. W. Boyer, and R. B. Helling. 1973. Construction of biologically functional bacterial plasmids in vitro. *Proc. Natl. Acad. Sci. USA* 70:3240–3244.

Cole, E. S., K. Lee, K. Lauziere, C. Kelton, S. Chappel, B. Weintraub, D. Ferrara, P. Peterson, R. Bernasconi, T. Edmunds, S. Richards, L. Dickrell, J. M. Kleeman, J. H. McPherson, and B. M. Pratt. 1993. Recombinant human thyroid stimulating hormone: development of a biotechnology product for detection of metastatic lesions of thyroid carcinoma. *Bio/Technology* 11:1014–1024.

Condreay, J. P., and T. A. Kost. 2007. Baculovirus vectors for insect and mammalian cells. *Curr. Drug Targets* 8:1126–1131.

Donovan, R. S., C. W. Robinson, and B. R. Glick. 1996. Optimizing inducer and culture conditions for expression of foreign proteins under the control of the *lac* promoter. *J. Ind. Microbiol.* 16:145–154.

Eijsink, V. G. H., A. Bjørk, S. Gåseidnes, R. Sirevåg, B. Synstad, B. van den Burg, and G. Vriend. 2004. Rational engineering of enzyme stability. *J. Biotechnol.* 113:105–120.

Eiteman, M. A., and E. Altman. 2006. Overcoming acetate in *Escherichia coli* recombinant protein fermentations. *Trends Biotechnol.* 24:530–536.

Ernst, J. F. 1988. Codon usage and gene expression. *Trends Biotechnol.* 6:196–199.

Ferrer, M., T. N. Chernikova, M. M. Yakimov, P. N. Golyshin, and K. N. Timmis. 2003. Chaperonins govern growth of *Escherichia coli* at low temperatures. *Nat. Biotechnol.* 21:1266–1267.

Gasser, B., M. Saloheimo, U. Rinas, M. Dragosits, E. Rodríguez-Carmona, K. Baumann, M. Giuliani, E. Parrilli, P. Branduardi, C. Lang, D. Porro, P. Ferrer, M. L. Tutino, D. Mattanovich, and A. Villaverde. 2008. Protein folding and conformational stress in microbial cells producing recombinant proteins: a host comparative overview. *Microb. Cell Fact.* 7:11–29.

Geisow, M. J. 1991. Both bane and blessing—inclusion bodies. *Trends Biotechnol.* **9:**368–369.

Georgiou, G., and L. Segatori. 2005. Preparative expression of secreted proteins in bacteria: status report and future prospects. *Curr. Opin. Biotechnol.* **16:**538–545.

Glick, B. R. 1995. Metabolic load and heterologous gene expression. *Biotechnol. Adv.* **13:**247–261.

Goldstein, M. A., and R. H. Doi. 1995. Prokaryotic promoters in biotechnology, p. 105–128. *In* M. R. El-Gewely (ed.), *Biotechnology Annual Review*, vol. 1. Elsevier Science B. V., Amsterdam, The Netherlands.

Gurtu, V., G. Yan, and G. Zhang. 1996. IRES bicistronic expression vectors for efficient creation of stable mammalian cell lines. *Biochem. Biophys. Res. Commun.* **229:**295–298.

Haacke, A., G. Fendrich, P. Ramage, and M. Geiser. 2009. Chaperone over-expression in *Escherichia coli*: apparent increased yields of soluble recombinant protein kinases are due mainly to soluble aggregates. *Protein Expr. Purif.* **64:**185–193.

Hamilton, S. R., and T. U. Gerngross. 2007. Glycosylation engineering in yeast: the advent of fully humanized yeast. *Curr. Opin. Biotechnol.* **18:**387–392.

Hannig, G., and S. C. Makrides. 1998. Strategies for optimizing heterologous protein expression in *Escherichia coli*. *Trends Biotechnol.* **16:**54–60.

Herlitze, S., and M. Koenen. 1990. A general and rapid mutagenesis method using polymerase chain reaction. *Gene* **91:**143–147.

Hockney, R. C. 1994. Recent developments in heterologous protein production in *Escherichia coli*. *Trends Biotechnol.* **12:**456–463.

Jackson, D. A., R. H. Symons, and P. B. Berg. 1972. Biochemical method for inserting new genetic information into DNA of simian virus 40: circular SV40 DNA molecules containing lambda phage genes and the galactose operon of Escherichia coli. *Proc. Natl. Acad. Sci. USA* **69:**2904–2909.

Jarvis, D. L. 2003. Developing baculovirus-insect cell expression systems for humanized recombinant glycoprotein production. *Virology* **310:**1–7.

Jespers, L. S., J. H. Messens, A. De Keyser, D. Eeckhout, I. Van Den Brande, Y. G. Gansemans, M. J. Lauwereys, G. P. Vlasuk, and P. E. Stanssens. Surface expression and ligand-based selection of cDNAs fused to filamentous phage gene VI. *Biotechnology* (New York) **13:**378–382.

Karnaukhova, E., Y. Ophir, L. Trinh, N. Dalal, P. J. Punt, B. Golding, and J. Shiloach. 2007. Expression of human α_1-protease inhibitor in *Aspergillus niger*. *Microb. Cell Fact.* **6:**34–44.

Kaur, J., and R. Sharma. 2006. Directed evolution: an approach to engineer enzymes. *Crit. Rev. Biotechnol.* **26:**165–199.

Kim, M.-D., S.-K. Rhee, and J.-H. Seo. 2001. Enhanced production of anticoagulant hirudin in recombinant *Saccharomyces cerevisiae* by chromosomal δ-integration. *J. Biotechnol.* **85:**41–48.

Kirchhoff, F., and R. C. Desrosiers. 1995. Random mutagenesis of short target DNA sequences via PCR with degenerate oligonucleotides. *Methods Mol. Biol.* **57:**323–333.

Kjeldsen, T., M. Hach, P. Balschmidt, S. Havelund, A. F. Pettersson, and J. Markussen. 1998. Prepro-leaders lacking N-linked glycosylation for secretory expression in the yeast *Saccharomyces cerevisiae*. *Protein Expr. Purif.* **14:**309–316.

Kjeldsen, T., A. F. Pettersson, and M. Hach. 1999. Secretory expression and characterization of insulin in *Pichia pastoris*. *Biotechnol. Appl. Biochem.* **29:**79–86.

Kost, T. A., J. P. Condreay, and D. L. Jarvis. 2005. Baculovirus as versatile vectors for protein expression in insect and mammalian cells. *Nat. Biotechnol.* **23:**567–575.

Kwaks, T. H. J., and A. P. Otte. 2006. Employing epigenetics to augment the expression of therapeutic proteins in mammalian cells. *Trends Biotechnol.* **24:**137–142.

Kwaks, T. H. J., R. G. A. B. Sewalt, R. van Blokland, T. J. Siersma, M. Kasiem, A. Kelder, and A. P. Otte. 2005. Targeting of a histone acetyltransferase domain to a promoter enhances protein expression levels in mammalian cells. *J. Biotechnol.* **115:**35–46.

Landt, O., H.-P. Grunert, and U. Hahn. 1990. A general method for rapid site-directed mutagenesis using the polymerase chain reaction. *Gene* **96:**125–128.

Leahy, D. J., C. E. Dann III, P. Longo, B. Perman, and K. X. Ramyar. 2000. A mammalian expression vector for expression and purification of secreted proteins for structural studies. *Protein Expr. Purif.* **20:**500–506.

Liang, M., S. Dübel, D. Li, I. Queitsch, W. Li, and E. K. Bautz. 2001. Baculovirus expression cassette vectors for rapid production of complete human IgG from phage display selected antibody fragments. *J. Immunol. Methods* **247:**119–130.

Liu, X., S. N. Constantinescu, Y. Sun, J. S. Bogan, D. Hirsch, R. A. Weinberg, and H. F. Lodish. 2000. Generation of mammalian cells stably expressing multiple genes at predetermined levels. *Anal. Biochem.* **280:**20–28.

Looman, A. C., J. Bodlaender, M. de Gruyter, A. Vogelaar, and P. H. van Knippenberg. 1986. Secondary structure as primary determinant of the efficiency of ribosomal binding sites in *Escherichia coli*. *Nucleic Acids Res.* **14:**5481–5497.

Majander, K., L. Anton, J. Antikainen, H. Lang, M. Brummer, T. K. Korhonen, and B. Westerlund-Wikström. 2005. Extracellular secretion of polypeptides using a modified *Escherichia coli* flagellar secretion apparatus. *Nat. Biotechnol.* **23:**475–481.

Makrides, S. C. 1999. Components of vectors for gene transfer and expression in mammalian cells. *Protein Expr. Purif.* **17:**183–202.

Malissard, M., S. Zeng, and E. G. Berger. 1999. The yeast expression system for recombinant glycosyltransferases. *Glycoconj. J.* **16:**125–139.

Martinez-Morales, F., A. C. Borges, A. Martinez, K. T. Shanmugam, and L. O. Ingram. 1999. Chromosomal integration of heterologous DNA in *Escherichia coli* with precise removal of markers and replicons used during construction. *J. Bacteriol.* **181:**7143–7148.

Mayer, A. F., K. Hellmuth, H. Schlieker, R. Lopez-Ulibarri, S. Oertel, U. Dahlems, C. Mielke, M. Tümmler, D. Schübeler, I. von Hoegen, and H. Hauser. 2000. Stabilized, long-term expression of heterodimeric proteins from tricistronic mRNA. *Gene* **254:**1–8.

Miksch, G., F. Buttenworth, K. Friehs, E. Flaschel, A. Saalbach, T. Twellmann, and T. W. Nattkemper. 2005. Libraries of synthetic stationary-phase and stress promoters as a tool for fine tuning of expression of recombinant proteins in *Escherichia coli*. *J. Biotechnol.* **120**: 25–37.

Mulligan, R. C., B. H. Howard, and P. Berg. 1979. Synthesis of rabbit β-globin in cultured monkey kidney cells following infection with a SV40 β-globin recombinant genome. *Nature* **277**: 108–114.

Nilsson, J., S. Ståhl, J. Lundeberg, M. Uhlén, and P.-Å. Nygren. 1997. Affinity fusion strategies for detection, purification, and immobilization of recombinant proteins. *Protein Expr. Purif.* **11**: 1–16.

Nygren, P. Å., S. Ståhl, and M. Uhlén. 1994. Engineering proteins to facilitate bioprocessing. *Trends Biotechnol.* **12**: 184–188.

Oganesyan, N., I. Ankoudinova, S.-H. Kim, and R. Kim. 2007. Effect of osmotic stress and heat shock in recombinant protein overexpression and crystallization. *Protein Expr. Purif.* **52**: 280–285.

Otten, L. G., C. F. Sio, J. Vrielink, R. H. Cool, and W. J. Quax. 2002. Altering the substrate specificity of cephalosporin acylase by directed evolution of the β-subunit. *J. Biol. Chem.* **277**: 42121–42127.

Pérez-Pérez, J., G. Márquez, J. L. Barbero, and J. Gutiérrez. 1994. Increasing the efficiency of protein export in *Escherichia coli*. *Bio/Technology* **12**: 178–180.

Robinson, A. S., V. Hines, and K. D. Wittrup. 1994. Protein disulfide isomerase overexpression increases secretion of foreign proteins in *Saccharomyces cerevisiae*. *Bio/Technology* **12**: 381–384.

Sassenfeld, H. M. 1990. Engineering proteins for purification. *Trends Biotechnol.* **8**:88–93.

Schmidt, F. R. 2004. Recombinant expression systems in the pharmaceutical industry. *Appl. Microbiol. Biotechnol.* **65**:363–372.

Schröder, M. 2008. Engineering eukaryotic protein factories. *Biotechnol. Lett.* **30**:187–196.

Shi, X., and D. L. Jarvis. 2007. Protein N-glycosylation in the baculovirus-insect cell system. *Curr. Drug Targets* **8**:1116–1125.

Simmons, L. C., and D. G. Yansura. 1996. Translational level is a critical factor for the secretion of heterologous proteins in *Escherichia coli*. *Nat. Biotechnol.* **14**:629–634.

Smith, M. 1985. In vitro mutagenesis. *Annu. Rev. Genet.* **19**:423–462.

Sørensen, H. P., and K. K. Mortensen. 2005. Advanced genetic strategies for recombinant protein expression in *Escherichia coli*. *J. Biotechnol.* **115**: 113–128.

Steinborn, G., T. Wartmann, G. Gellissen, and G. Kunze. 2007. Construction of an *Arxula adeninivorans* host-vector system based on *trp1* complementation. *J. Biotechnol.* **127**:392–401.

Steinborn, G., G. Gellissen, and G. Kunze. 2007. A novel vector element providing multicopy vector integration in *Arxula adeninivorans*. *FEMS Yeast Res.* **7**:1197–1205.

Stemmer, W. P. C. 1994. DNA shuffling by random fragmentation and reassembly: in vitro recombination for molecular evolution. *Proc. Natl. Acad. Sci. USA* **91**:10747–10751.

Strocchi, M., M. Ferrer, K. N. Timmis, and P. N. Golyshin. 2006. Low temperature-induced systems failure in *Escherichia coli*: insights from rescue by cold-adapted chaperones. *Proteomics* **6**: 193–206.

Sung, B. H., C. H. Lee, B. J. Yu, J. H. Lee, J. Y. Lee, M. S. Kim, F. R. Blattner, and S. C. Kim. 2006. Development of a biofilm production-deficient *Escherichia coli* strain as a host for biotechnological applications. *Appl. Environ. Microbiol.* **72**:3336–3342.

Takahashi, K., T. Yuuki, T. Takai, C. Ra, K. Okumura, T. Yokota, and Y. Okumura. 2000. Production of humanized Fab fragment against human high affinity IgE receptor in *Pichia pastoris*. *Biosci. Biotechnol. Biochem.* **64**: 2138–2144.

Talmadge, K., and W. Gilbert. 1982. Cellular location affects protein stability in *Escherichia coli*. *Proc. Natl. Acad. Sci. USA* **79**:1830–1833.

Terpe, K. 2003. Overview of tag protein fusions: from molecular and biochemical fundamentals to commercial systems. *Appl. Microbiol. Biotechnol.* **60**:523–533.

Wagner, S., M. M. Klepsch, S. Schlegel, A. Appel, R. Draheim, M. Tarry, M. Högbom, K. J. van Wijk, D. J. Slotboom, J. O. Persson, and J.-W. de Gier. 2008. Tuning *Escherichia coli* for membrane protein overexpression. *Proc. Natl. Acad. Sci. USA* **105**: 14371–14376.

Welch, M., S. Govindarajan, J. E. Ness, A. Villalobos, A. Gurney, J. Minshull, and C. Gustafsson. 2009. Design parameters to control synthetic gene expression in *Escherichia coli*. *PLoS One* **4**(9):e7002.

Whittaker, M. M., and J. W. Whittaker. 2000. Expression of recombinant galactose oxidase by *Pichia pastoris*. *Protein Expr. Purif.* **20**:105–111.

Wolff, A. M., O. C. Hansen, U. Poulsen, S. Madrid, and P. Stougaard. 2001. Optimization of the production of *Chondrus crispus* hexose oxidase in *Pichia pastoris*. *Protein Expr. Purif.* **22**: 189–199.

Zhang, L., Y. Li, Z. Wang, Y. Xia, W. Chen, and K. Tang. 2007. Recent developments and future prospects of *Vitreoscilla* hemoglobin application in metabolic engineering. *Biotechnol. Adv.* **25**:123–136.

Zoller, M. J., and M. Smith. 1982. Oligonucleotide-directed mutagenesis using M13-derived vectors: an efficient and general procedure for the production of point mutations in any fragment of DNA. *Nucleic Acids Res.* **10**: 6487–6500.

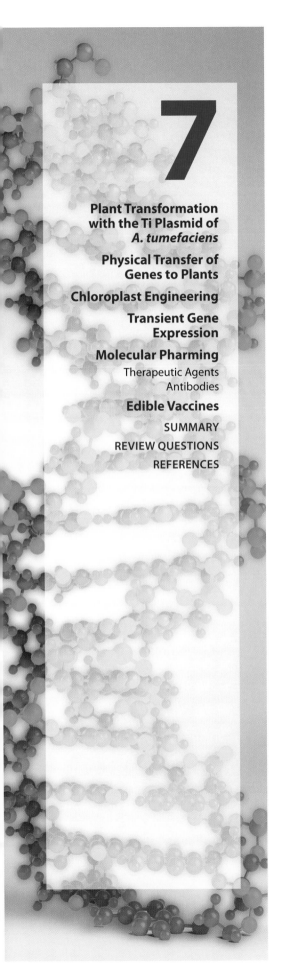

7

Genetic Engineering of Plants

THE MARKET FOR recombinant biopharmaceuticals was estimated to be around $100 billion in 2010. Moreover, this market is growing quite rapidly. Thus, with increasing demand for recombinant biopharmaceuticals, coupled with the high costs and inefficiency of the existing production systems, including bacterial, yeast, and animal cells in culture and (to a very limited extent) transgenic animals, there is currently a shortage in manufacturing capacity worldwide, a situation which is likely to worsen. As a result, transgenic plants are gaining more and more attention as a potential new generation of bioreactors.

The production of therapeutic proteins in transgenic plants has several potential advantages over their synthesis in recombinant microbial cells. Plants are easy to grow and can generate considerable biomass. Unlike recombinant bacterial, yeast, or animal cells in culture, which are grown in large bioreactors—a process that requires highly trained personnel and expensive equipment—crops can be produced relatively inexpensively by less skilled workers. In addition, when proteins that are intended for human use are produced in transgenic plants, there is a significantly reduced risk of mammalian virus contamination in comparison to that with proteins that are produced in animal cells grown in culture. The processing and assembly of foreign proteins in plants are similar to those in animal cells, whereas bacteria do not readily process, assemble, or posttranslationally modify eukaryotic proteins. Ultimately, the biggest hurdle to overcome in the production of foreign proteins in plants is the purification of the product of a transgene from the mass of plant tissue. On a laboratory scale, plants have been used to produce therapeutic agents (Table 7.1), monoclonal antibodies and antibody fragments (Table 7.2), and a number of potential and vaccine antigens (Table 7.3).

With the advance of this technology, in July 2011, regulators in the United Kingdom announced in a press conference that they had, for the first time, approved a human clinical trial of a monoclonal antibody

doi:10.1128/9781555818890.ch7

Table 7.1 Some pharmaceutical proteins that have been produced in transgenic plants

Pharmaceutical protein	Plant(s)	Application(s)
α-Tricosanthin	Tobacco	HIV therapy
Allergen-specific T-cell epitope	Rice	Pollinosis
Angiotensin-1-converting enzyme	Tobacco, tomato	Hypertension
Cyanovirin-N	Tobacco	HIV micobicide
Glucocerebrosidase	Tobacco	Gaucher disease
Human α$_1$-antitrypsin	Rice, tomato	Cystic fibrosis, liver disease, hemorrhage
Human apolipoprotein	Safflower	Plaque reduction
Human aprotinin	Corn	Trypsin inhibitor for transplantation surgery
Human enkephalins	Arabidopsis, canola	Antihyperanalgesic by opiate activity
Human epidermal growth factor	Tobacco	Wound repair, control of cell proliferation
Human erythropoietin	Tobacco	Anemia
Human granulocyte-macrophage colony-stimulating factor	Tobacco	Neutropenia
Human growth hormone	Tobacco	Dwarfism, wound healing
Human hemoglobin	Tobacco	Blood substitute
Human hirudin	Canola, tobacco	Thrombin inhibitor, anticoagulant
Human homotrimeric collagen I	Tobacco	Collagen synthesis
Human insulin	Potato, *Arabidopsis*, safflower	Diabetes
Human α interferon	Rice, turnip	Hepatitis C and B
Human β interferon	Tobacco	Antiviral
Human interleukin-2 and interleukin-4	Tobacco	Immunotherapy
Human lactoferrin	Potato, rice	Antimicrobial, diarrhea
Human muscarinic cholinergic receptors	Tobacco	Central and peripheral nervous systems
Human placental alkaline phosphatase	Tobacco	Achonodroplasia or cretinism in children
Human protein C	Tobacco	Anticoagulant
Human serum albumin	Tobacco	Liver cirrhosis, burns, surgery
Human somatotropin	Tobacco	Growth hormone
Lipase	Corn	Cystic fibrosis

HIV
human immunodeficiency virus

directed against human immunodeficiency virus (HIV) produced using genetically modified tobacco plants. The initial trials, which include 11 participants, will test the safety of the antibody, called P2G12, which is applied topically to the vaginal cavity and is thus specific to viral transmission to females. If it is proven to be safe, it will then be tested for effectiveness. The purified monoclonal antibody is produced in Germany in the first facility to be granted a license to manufacture recombinant pharmaceutical products from plants in Europe.

Recombinant DNA technology, which has been used extensively with microbial systems, is also an important tool for the genetic manipulation of plants. There are a large number of effective DNA delivery systems and expression vectors that work with a wide range of plant cells. Furthermore, most plant cells are totipotent—meaning that an entire plant can be regenerated from a single plant cell—so fertile plants that carry an introduced gene(s) in all cells (i.e., transgenic plants) can be produced from genetically engineered cells. If the transgenic plant flowers and produces viable seed, the desired trait is passed on to successive generations.

Table 7.2 Examples of potentially therapeutic antibodies and antibody fragments that have been produced in plants

Host plant	Disease or antigen
Tobacco	38C13 mouse B-cell lymphoma
Tobacco	Anthrax
Tobacco	B-cell lymphoma
Tobacco	Breast and colon cancer
Tobacco	Broad-spectrum anticancer
Tobacco	Botulism
Tobacco	CD40 (cell surface protein)
Tobacco	Cell surface protein from mouse B-cell lymphoma
Tobacco	Hepatitis
Soybean	Herpes simplex virus
Tobacco	Human carcinoembryonic antigen
Pea	Human cancer cell surface antigen
Tobacco	Human CD40 cell surface protein
Tobacco	Human creatine kinase
Alfalfa	Human IgG
Tobacco	Rabies
Tobacco	*Salmonella* surface antigen
Tobacco	*Streptococcus mutans* cell surface antigen SA I/II
Tobacco	Substance P (neuropeptide)

Table 7.3 Some potential vaccine antigens that have been expressed in plants

Disease or causative agent	Plant(s) or vector
Hepatitis B	Tobacco, potato, yellow lupin, lettuce
Malaria	Virus
Rabies	Tomato, spinach, virus
Human rhinovirus	Virus
HIV	Virus
E. coli	Tobacco, potato, corn
Norwalk virus	Tobacco, potato, corn
Diabetes	Tobacco, potato, carrot
Foot-and-mouth disease	*Arabidopsis*, alfalfa
Cholera	Potato, rice
Human cytomegalovirus	Tobacco
Dental caries	Tobacco
Respiratory syncytial virus	Tomato
Human papillomavirus	Potato, tobacco
Anthrax	Tobacco
SARS	Tomato, tobacco
Staphylococcus aureus	Cowpea
Measles	Lettuce
Influenza virus	Tobacco
Tuberculosis	*Arabidopsis*
Rotavirus	Alfalfa

Note that in some cases the antigen was cloned into a transient-expression system, such as a plant virus, often facilitating high levels of expression within a period of 1 to 2 weeks. SARS, severe acute respiratory syndrome.

T-DNA
transfer DNA

Plant Transformation with the Ti Plasmid of *A. tumefaciens*

The gram-negative soil bacterium *Agrobacterium tumefaciens* is a phytopathogen that, as a normal part of its life cycle, genetically transforms plant cells. This genetic transformation leads to the formation of **crown gall** tumors, which interfere with the normal growth of an infected plant (Fig. 7.1A).

Crown gall formation is the consequence of the transfer, integration, and expression of genes of a specific segment of bacterial plasmid DNA—called the **transfer DNA (T-DNA)**—into the plant cell genome. The T-DNA is part of the "tumor-inducing" (Ti) plasmid that is carried by most strains of *A. tumefaciens* (Fig. 7.1B). Depending on the **Ti plasmid**, the length of the T-DNA region can vary from approximately 10 to 30 kilobase pairs (kb). Strains of *A. tumefaciens* that do not possess a Ti plasmid cannot induce crown gall tumors.

After the initial attachment step, *A. tumefaciens* responds to certain plant phenolic compounds, such as acetosyringone and hydroxyacetosyringone, which are excreted by wounded plants. These wound response compounds resemble some of the products of phenylpropanoid metabolism such as lignins and flavonoids and act to induce the expression of the **virulence (*vir*) genes** located on the Ti plasmid.

The ~25 *vir* genes are located on a 35-kb region of the Ti plasmid that is outside of the T-DNA region. The products of the *vir* genes are

Figure 7.1 (**A**) Infection of a plant with *A. tumefaciens* and formation of a crown gall tumor. (**B**) Schematic representation of a Ti plasmid. The T-DNA is defined by its left and right borders and includes genes for the biosynthesis of auxin, cytokinin, and an opine; these genes are transcribed and translated only in plant cells. Outside the T-DNA region, there is a cluster of *vir* genes required for transfer of the T-DNA into the host plant genome, a gene(s) that encodes an enzyme(s) for opine catabolism, and an origin of DNA replication (*ori*) that permits the plasmid to be stably maintained in *A. tumefaciens*. None of these features is drawn to scale. doi:10.1128/9781555818890.ch7.f7.1

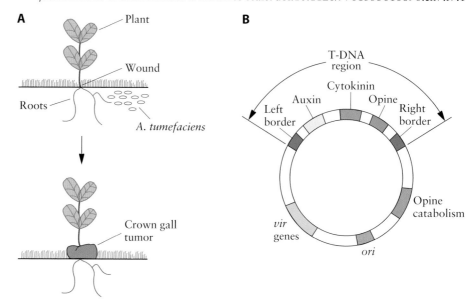

essential for the transfer and integration of the T-DNA region into the genome of a plant cell. Once the *vir* genes have been induced, the T-DNA is transferred as a linear, single-stranded molecule from the Ti plasmid and eventually becomes integrated into the plant chromosomal DNA.

The formation of the single-stranded form of T-DNA is initiated by strand-specific cutting, by an enzyme encoded by one of the *vir* genes, at both borders of the intact T-DNA region (the right border and left border in Fig. 7.1B). The 5′ end of the single-stranded T-DNA carries the right-border sequence, and the left-border sequence is at the 3′ end. Integration of the T-DNA into the plant genome depends on specific sequences (a 25-bp repeating unit) located at the right border of the T-DNA. Although the left border contains a similar 25-bp repeat, this region is not involved in the integration process.

Most of the genes that are located within the T-DNA region are activated only after the T-DNA is inserted into the plant genome. This reflects the fact that these are essentially plant genes, which cannot be expressed in bacteria because of the differences in transcriptional and translational regulatory sequences between the two types of organisms. The products of these genes are responsible for crown gall formation. The T-DNA region includes the genes *iaaM* and *iaaH*, which encode enzymes that synthesize the plant hormone **auxin** (indoleacetic acid). In addition, the T-DNA region carries the *tmr* gene (also known as *ipt*), which encodes isopentenyltransferase, an enzyme that catalyzes the synthesis of the plant hormone **cytokinin**. Both auxin and the cytokinin regulate plant cell growth and development; however, in excess, they can cause the plant to develop tumorous growths, such as crown galls.

The T-DNA region also carries a gene for the synthesis of a molecule called an opine, a unique condensation product of either an amino acid and a keto acid or an amino acid and a sugar. The opines are synthesized within the crown gall and then secreted. They can be used as a carbon source, and sometimes also as a nitrogen source, by any *A. tumefaciens* cell that carries a Ti plasmid-borne gene (not part of the T-DNA region) for the catabolism of that particular opine. Thus, each strain of *A. tumefaciens* genetically manipulates plant cells to be biological factories for the production of a carbon compound that it alone is able to use.

Although Ti plasmids are effective as natural vectors, they have several serious limitations as routine cloning vectors.

1. The production of phytohormones by genetically transformed cells growing in culture prevents the cells from being regenerated into mature plants. Therefore, the auxin and cytokinin genes must be removed from any Ti plasmid-derived cloning vector.
2. A gene encoding opine synthesis is not useful to a transgenic plant and may divert plant resources into opine production. Therefore, the opine synthesis gene should be removed.
3. Ti plasmids are large (typically ~200 to 800 kb). For recombinant DNA experiments, however, a much smaller version is preferred, so large segments of DNA that are not essential for a cloning vector must be removed.

A

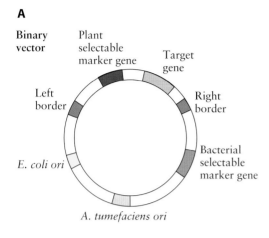

Binary vector

Plant selectable marker gene

Target gene

Left border

Right border

E. coli ori

Bacterial selectable marker gene

A. tumefaciens ori

B

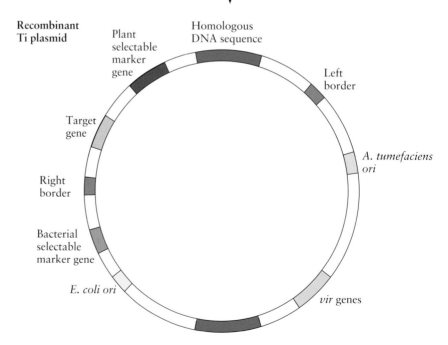

Cointegrate vector

Right border

Target gene

Bacterial selectable marker gene

E. coli ori

Plant selectable marker gene

Homologous DNA sequence

Disarmed Ti plasmid

Left border

A. tumefaciens ori

vir genes

Recombine

Recombinant Ti plasmid

Plant selectable marker gene

Homologous DNA sequence

Left border

Target gene

A. tumefaciens ori

Right border

Bacterial selectable marker gene

vir genes

E. coli ori

Figure 7.2 Two Ti plasmid-derived cloning vector systems. (**A**) The binary cloning vector has origins of DNA replication (*ori*) for both *E. coli* and *A. tumefaciens* (or a broad-host-range origin), a selectable marker gene that can be used in either *E. coli* or *A. tumefaciens*, and both a target gene and a plant selectable marker gene inserted between the T-DNA left and right borders. (**B**) The cointegrate cloning vector (top) carries only an *E. coli* origin of replication and cannot exist autonomously within *A. tumefaciens*. It also contains a selectable marker that can be used in either *E. coli* or *A. tumefaciens*, a T-DNA right border, a plant selectable marker gene, a cloned target gene, and a sequence of Ti plasmid DNA that is homologous to a segment on the disarmed Ti plasmid. The disarmed Ti plasmid (middle) contains the T-DNA left border, the *vir* gene cluster, and an *A. tumefaciens ori*. Following recombination between the cointegrate cloning vector and the disarmed Ti plasmid, the final recombinant plasmid (bottom) has the T-DNA left and right borders bracketing the cloned and plant reporter genes. doi:10.1128/9781555818890.ch7.f7.2

4. Because the Ti plasmid does not replicate in *Escherichia coli*, the convenience of perpetuating and manipulating Ti plasmids carrying inserted DNA sequences in that bacterium is not available.

To overcome these constraints, recombinant DNA technology was used to create two types of Ti plasmid-based vectors. These vectors are similarly organized and contain the following components.

- The right border sequence of the T-DNA region. This region is absolutely required for T-DNA integration into plant cell DNA, although most cloning vectors include both right- and left-border sequences.
- A multiple-cloning site to facilitate insertion of the cloned gene into the region between the T-DNA border sequences
- A genetic marker gene that readily allows a researcher to select transformed plant cells
- An origin of DNA replication that allows the plasmid to replicate in *E. coli*. In some vectors, an origin of replication that functions in *A. tumefaciens* has also been added.

Because these cloning vectors lack *vir* genes, they cannot by themselves effect the transfer and integration of the T-DNA region into recipient plant cells. Two different approaches have been used to achieve these ends. In one approach, a **binary vector system** is used (Fig. 7.2A). The binary cloning vector contains either both *E. coli* and *A. tumefaciens* origins of DNA replication or a single broad-host-range origin of DNA replication (that allows the plasmid to exist stably in either bacterium). In addition, no *vir* genes are present on a binary cloning vector. All the cloning steps are carried out in *E. coli* before the binary vector is introduced into *A. tumefaciens*. The recipient *A. tumefaciens* strain carries a modified (disarmed) Ti plasmid that contains a complete set of *vir* genes but lacks the T-DNA region, so that no T-DNA can be transferred from this *A. tumefaciens* strain and no crown gall tumors can form. With this system, the disarmed Ti plasmid synthesizes the *vir* gene products that mobilize the T-DNA region of the binary cloning vector, enabling the T-DNA from the binary cloning vector to be inserted into the plant chromosomal DNA. Since transfer of the T-DNA is initiated from the right border, the plant selectable marker, which will eventually be used to detect the presence of the T-DNA inserted into the plant chromosomal DNA, is usually placed next to the left border. If the selectable marker were adjacent to the right border, transfer of only a small portion of the T-DNA might yield plants that contained the selectable marker but not the gene of interest.

In the second approach, called the **cointegrate vector system**, the cloning (cointegrate) vector has a plant selectable marker gene, the target gene, the right border, an *E. coli* origin of DNA replication, and a bacterial selectable marker gene. The cointegrate vector recombines with a different modified (disarmed) Ti plasmid. As for the binary system, the disarmed Ti plasmid of the cointegrate system lacks both the tumor-producing genes and the right border of the T-DNA (Fig. 7.2B). The cointegrate cloning vector and the disarmed helper Ti plasmid each carry homologous DNA sequences that provide a site for in vivo homologous recombination. Following recombination, the entire cloning vector becomes integrated into the disarmed Ti plasmid, which provides the *vir* genes necessary for the transfer of the T-DNA to the host plant cells. Because the cointegrate vector lacks an *A. tumefaciens* origin of replication, the only way that this cloning vector can be maintained in *A. tumefaciens* is as part of a cointegrate structure. In this cointegrated configuration, the genetically engineered T-DNA region carrying a target gene can be transferred to plant cells.

Although *A. tumefaciens*-mediated gene transfer systems are effective in many species of plants, monocotyledonous plants (**monocots**), including the world's major cereal crops (rice, wheat, and corn), are not as readily transformed by *A. tumefaciens*. However, by refining and carefully controlling conditions, protocols have been devised for the transformation of corn and rice by *A. tumefaciens* carrying Ti plasmid vectors. Many of the early plant transformation experiments were conducted with limited-host-range strains of *Agrobacterium*. However, more recently, broad-host-range strains that infect most plants have been tested and found to be effective, so many of the plant species that previously appeared to be refractory to transformation by *A. tumefaciens* can now be transformed. Thus, when setting out to transform a plant species for the first time, it is necessary to determine which *Agrobacterium* strain and Ti plasmid are best suited to that particular plant. In addition, modification of the plant tissue culture conditions by the inclusion of antioxidants, which prevent the buildup of reactive oxygen species that can cause tissue senescence, during the transformation of a number of plants (including grape, rice, corn, and soybean) has been found to increase the transformation frequencies of those plant cells.

A systematic examination of the conditions that are used in *Agrobacterium*-mediated plant transformation revealed that gaseous ethylene, a plant stress hormone, significantly decreased the transfer of genes to plant genomes. During plant transformation, ethylene is produced when *Agrobacterium* infects the plants. To remedy this, a bacterial gene encoding the enzyme 1-aminocyclopropane-1-carboxylate (ACC) deaminase was introduced into an *A. tumefaciens* strain and, when expressed, lowered plant ethylene levels by cleaving ACC, which is the immediate precursor of ethylene (Fig. 7.3A). This genetically modified *A. tumefaciens* strain was then utilized to introduce foreign DNA into plants. When melon cotyledon segments were genetically transformed using the *A. tumefaciens* strain expressing ACC deaminase, the transformation frequency of the plants (as judged by the level of introduced marker enzyme activity) increased significantly (Fig. 7.3B). More recently, another group of researchers found that an *A. tumefaciens* strain that had been genetically engineered to express the ACC deaminase gene was more efficient at transforming commercial **canola cultivars** than the nonengineered strain. Thus, it is expected that the introduction of this ethylene-lowering gene will increase the transformation frequencies for a wide range of different plants.

ACC
1-aminocyclopropane-1-carboxylate

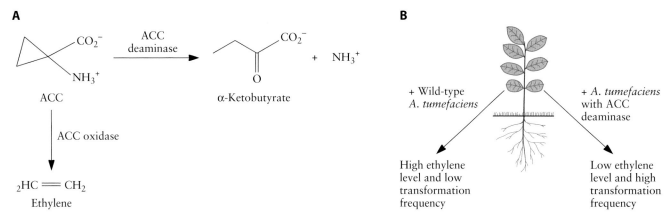

Figure 7.3 (A) Cleavage of ACC by bacterial ACC deaminase decreases the amount of ACC that can be converted into ethylene by plant ACC oxidase. (B) Improving the transformation of plants by *A. tumefaciens* by engineering the bacterium to produce the enzyme ACC deaminase lowers the amount of ethylene that forms following the interaction of the plant and bacterium. doi:10.1128/9781555818890.ch7.f7.3

Physical Transfer of Genes to Plants

When the difficulties in transforming some plant species first became apparent, a number of procedures that could act as alternatives to transformation by *A. tumefaciens* were developed (Table 7.4). Some of these methods require the removal of the plant cell wall to form **protoplasts**. Plant protoplasts can be maintained in culture as independently growing cells, or with a specific culture medium, new cell walls can be formed and whole plants can be regenerated. In addition, transformation methods have been developed that introduce cloned genes into a small number of cells of a plant tissue from which whole plants can be formed, thereby bypassing the need for regeneration from a protoplast. More than 100 different plant species have been genetically transformed with these various techniques.

Table 7.4 Methods of transforming plant cells

Method	Comment
Ti plasmid	The most commonly used method; highly effective; works better with some plants than with others
Microprojectile bombardment	Works well with a wide range of plants and plant tissues; sometimes results in transient expression of introduced DNA; the second most common method of plant cell transformation
Viral vectors	Used only to a limited extent
Direct gene transfer into plant protoplasts	Limited to plant cell protoplasts that can be regenerated into viable plants
Microinjection	Limited usefulness; only one cell at a time is injected; requires a highly skilled individual
Electroporation	Limited to plant cell protoplasts that can be regenerated into viable plants
Liposome fusion	Limited to plant cell protoplasts that can be regenerated into viable plants

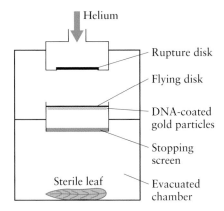

Figure 7.4 Schematic representation of a microprojectile bombardment apparatus. When the helium pressure builds up to a certain point, the plastic rupture disk bursts, and the released gas accelerates the flying disk with the DNA-coated gold particles on its lower side. The gold particles pass the stopping screen, which holds back the flying disk, and penetrate the cells of the sterile leaf (or other plant tissue).
doi:10.1128/9781555818890.ch7.f7.4

The technique of microprojectile bombardment, sometimes called **biolistics**, is the most important alternative to Ti plasmid DNA delivery systems for plants. Spherical gold or tungsten particles (approximately 0.4 to 1.2 mm in diameter, or about the size of some bacterial cells) are coated with DNA that has been precipitated with $CaCl_2$, **spermidine**, or polyethylene glycol. The coated particles are accelerated to high speed (300 to 600 m/s, or ~1,100 to 2,200 km/h) with a special apparatus called a particle gun (or gene gun). This device employs high-pressure helium as the source of particle propulsion (Fig. 7.4). The microprojectiles can penetrate plant cell walls and membranes; however, the particle density (that is, the number of particles per plant cell) used does not significantly damage the cells. The extent of particle penetration into the target plant cells may be controlled by varying the intensity of the explosive burst, altering the distance that the particles must travel before reaching the target cells, or using different-size particles.

Once inside a cell, some of the DNA is removed from the particles and, in some cells, integrates into the plant DNA. Sometimes, stable DNA that does not integrate into the plant genome may be expressed transiently for periods of up to several months. Microprojectile bombardment can be used to introduce foreign DNA into plant cell suspensions, **callus** cultures, **meristematic tissues**, immature embryos, **protocorms**, **coleoptiles**, and pollen in a very wide range of different plants, including monocots and **conifers**, plants that are somewhat less susceptible to *Agrobacterium*-mediated DNA transfer. Furthermore, this method has also been used to deliver genes into **chloroplasts** and mitochondria, thereby opening up the possibility of introducing exogenous (foreign) genes into these organelles.

Typically, plasmid DNA dissolved in buffer is precipitated onto the surfaces of the microprojectiles. Using this procedure, it is possible to increase the transformation frequency by increasing the amount of plasmid DNA; however, too much plasmid DNA can be inhibitory. It is estimated that approximately 10,000 cells are transformed per bombardment. With this technique, cells that appear to be transformed, based on the expression of a marker gene, often only transiently express the introduced DNA. Unless the DNA becomes incorporated into the genome of the plant, the foreign DNA will eventually be degraded.

The configuration of the vector that is used for biolistic delivery of foreign genes to plants influences both the integration and expression of those genes. For example, transformation is more efficient when linear rather than circular DNA is used. In addition, large plasmids (>10 kb) may become fragmented during microprojectile bombardment and therefore produce lower levels of foreign-gene expression. However, large segments of DNA may be introduced into plants using **yeast artificial chromosomes**. Yeast artificial chromosomes up to 150 kb in total size have a good chance of being transferred to plant cells, and the transferred DNA can be stably integrated into the plant cell. Thus, the production of transgenic plants that contain several foreign genes is feasible; eventually, entire biosynthetic pathways may be introduced into plant cells.

Chloroplast Engineering

Most higher plants have approximately 50 to 100 chloroplasts per leaf cell, and each chloroplast has about 10 to 100 copies of the chloroplast DNA genome (Fig. 7.5). Thus, while a single copy of a foreign gene inserted into plant chromosomal DNA is present in one copy per plant cell, a single copy of the same gene inserted into chloroplast DNA will be present in 500 to 10,000 copies per plant cell. The many copies of chloroplast DNA in each plant cell means that much higher levels of foreign gene expression should be attainable when the introduced gene is present as a part of the chloroplast DNA. Researchers using this approach have reported that the target protein may be expressed to a level of 1 to 5% of the soluble plant protein. This is 10- to 100-fold higher than protein expression levels that are typically obtained when foreign DNA is inserted into the plant genomic DNA.

While the vast majority of plant genes are found as part of the nuclear DNA, both the chloroplast and mitochondrion contain genes that encode a number of important and unique functions. Moreover, not all of the proteins that are present in these organelles are encoded by organellar DNA. For chloroplasts in particular, some proteins are encoded in the nuclear DNA, synthesized in the cell's cytoplasm, and then imported into the chloroplast by a specialized transport mechanism. Accordingly, there are two ways that a specific foreign protein can be introduced into the chloroplast. In one approach, a fusion gene encoding the foreign protein and additional amino acids that direct the transport of the protein to the organelle can be inserted into the nuclear chromosomal DNA using the *A. tumefaciens*–Ti plasmid system and a plant selectable marker (Fig. 7.6A). After the recombinant protein is synthesized, it can be transported into the chloroplast by the chloroplast targeting/transit peptide. This peptide is removed upon insertion of the target protein through the chloroplast membrane. This approach does not necessarily result in very high levels of expression of the target protein.

In the other approach, the gene for the foreign protein can be inserted directly into the chloroplast DNA (Fig. 7.6B). In this case, foreign DNA is typically introduced into the chloroplast genome by microprojectile bombardment of the genetic construct on a plasmid vector. The plasmid generally contains both the foreign DNA and a chloroplast selectable marker flanked by specific chloroplast DNA sequences. Chloroplast selectable markers are designed to be expressed only within chloroplasts and within the cytoplasm. Homologous recombination is the normal mode of DNA integration into the chloroplast genome; in this case, integration is targeted to a region of the DNA that does not encode proteins (an intergenic region).

The efficient expression of foreign genes in the chloroplast requires not only the use of an appropriate (chloroplast-specific) promoter sequence but also the presence of the correct sequences in the 5' and 3' untranslated regions of the mRNA. The promoter is fused at the DNA level with translational control sequences that function only within the chloroplast and not within the cytoplasm, followed by the gene of interest and a

Figure 7.5 Plant cells, showing multiple chloroplasts and double-stranded circular chloroplast DNA. doi:10.1128/9781555818890.ch7.f7.5

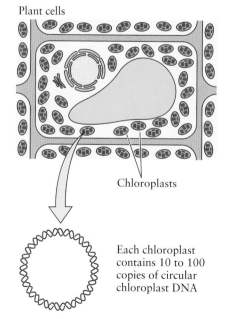

Plant cells

Chloroplasts

Each chloroplast contains 10 to 100 copies of circular chloroplast DNA

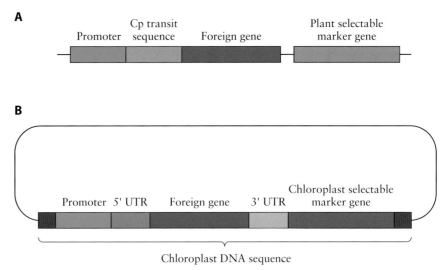

Figure 7.6 (A) A foreign gene that is introduced on a Ti plasmid and hence inserted into the chromosomal DNA. The gene encodes a fusion protein that includes a chloroplast (Cp) targeting/transit sequence that facilitates the localization of the foreign protein in the chloroplast. The Cp targeting/transit sequence is cleaved during transport into the chloroplast. (B) Plasmid vector used for integrating a foreign gene and a marker gene directly into the chloroplast genome. Homologous recombination can occur between chloroplast DNA sequences on the vector and the chloroplast genome. The regulatory sequences controlling expression of the selectable marker genes are not shown. 5′ UTR and 3′ UTR, 5′ and 3′ untranslated regions, respectively. doi:10.1128/9781555818890.ch7.f7.6

3′ chloroplast untranslated region containing a stem-and-loop structure, which may act as a transcription termination signal (Fig. 7.6B). An additional advantage of inserting a target gene into the chloroplast genome is that since chloroplast transcription is prokaryotic-like, several genes may be introduced at the same time as part of a single operon.

Chloroplast DNA is inherited in a non-Mendelian fashion. In most **angiosperm** plants, it is maintained in egg cells but not sperm (pollen) cells and is therefore transmitted uniparentally by the female. Thus, pollen cannot transmit the contents of the chloroplast genome to the zygote. This has practical importance, since this trait can prevent the spread of foreign genes, localized in the chloroplast, through pollen to neighboring plants, thereby addressing one of the concerns of some critics of the genetic engineering of plants.

Most chloroplast transformation protocols include selectable markers that confer resistance to antibiotics that inhibit protein synthesis on prokaryotic-type ribosomes. These antibiotics inhibit greening (chlorophyll production), the rate of proliferation, and shoot formation. However, plant cells with transformed chloroplasts that express the antibiotic resistance genes are readily identified in the presence of (otherwise) inhibitory antibiotics by their greening, faster proliferation, and shoot formation. Since only one, or at most a few, of the chloroplasts actually incorporates the foreign DNA, repeated rounds of growth on selective antibiotics are often required so that the chloroplast population becomes enriched and eventually dominated by the transformants. Eventually, all chloroplast genomes that do not carry the transgenic DNA with its selectable marker gene are lost.

Chloroplasts belong to a group of plant organelles called **plastids**, which contain approximately 120 to 180 kb of circular double-stranded DNA bounded by a double membrane. Plastids include **amyloplasts**, which contain starch grains; chloroplasts, which contain chlorophyll; **elaioplasts**, which contain oil; and **chromoplasts**, which contain other pigments. While photosynthetic tissues, such as green leaves, contain chloroplasts, other tissues contain other types of plastids. For example, tomato fruit contains a large number of chromoplasts. Thus, in much the same way that foreign DNA may be expressed as a part of the chloroplast genome, it can also be targeted for expression in the chromoplast.

Researchers can transform tomato plastids and obtain high-level expression of foreign proteins both in green leaves and in tomato fruit. One of the advantages of this system is that transgenic tomatoes expressing high levels of certain foreign proteins (which are normally found as part of an animal or human pathogen) may be used as edible vaccines.

Transient Gene Expression

The development of transgenic plants that can act as factories for the production of large amounts of therapeutic proteins is a time-consuming, expensive, and inefficient process. Nevertheless, it is possible to achieve high yields of recombinant proteins in plants. Unfortunately, all of the strategies currently available have some limitations. For example, high expression has been reported for nuclear transgenes in seeds; however, because the amount of seed harvested per hectare is much lower than the green biomass harvest (i.e., only a small percentage), the yield of the recombinant protein per hectare is relatively low. Similarly, recombinant proteins can be expressed in high yield in chloroplasts; however, proteins synthesized in chloroplasts are not glycosylated, a situation that is problematic for many therapeutic proteins.

To get around the above-mentioned constraints to the high-level expression of foreign proteins in plants, it is possible to transiently express foreign genes in plants in large quantities. Foreign genes that are transiently expressed in plants do not integrate into the plant nuclear or chloroplast DNA. Rather, they exist stably for a limited period (often weeks to months) in the plant cell cytoplasm, typically under the control of strong foreign regulatory signals such as those found in plant viruses. In this way, the protein may be produced rapidly (transgenic plants do not need to be selected or propagated) and in high yield. Since the foreign gene is not incorporated into the plant genome, it is not inherited. With "first-generation viral vectors," the virus expressed all of its own genes in addition to the target gene, which was spliced into its genome, either under the control of a strong viral promoter or fused to a gene encoding the viral coat protein, which is highly expressed and is present on the surface of assembled viruses. The foreign gene may be delivered to plant cells either using infectious (naked) nucleic acid versions of the vector that are not assembled into a whole virus or as mature viral particles.

When mature viral particles are used, the entire process can be performed on a large scale by spraying plants in the field with a mixture of viral particles and an abrasive compound such as **Carborundum**. Depending on how efficiently the vector can move through the plant, it may take about 2 to 3 weeks for most of the tissues of the transformed plants to become infected. This approach has been somewhat successful in that there are numerous reports of the expression of various therapeutic proteins as fusions to viral coat proteins. In addition, after purification from plants and injection into animals, potentially therapeutic proteins have been found to be immunogenic and in some instances to demonstrate some level of protection against pathogenic agents in animal models. Unfortunately, only small epitopes (i.e., fewer than 25 amino acids) have been successfully expressed as coat protein fusions. This is probably because larger epitopes can prevent the viral particle from assembling. To overcome this limitation, scientists constructed "second-generation viral vectors." In this approach, the recombinant viral DNA is delivered to the plant by *A. tumefaciens* infection. The viral genome is modified so that only the viral elements required for efficient expression of the target gene are maintained (Fig. 7.7). Since the modified virus is delivered to the plant by *A. tumefaciens*, viral genes responsible for infectivity, amplification/replication, cell-to-cell movement, assembly of viral particles, shutoff of the synthesis of plant cell components, and systemic spread of the virus are dispensable and therefore removed. In addition, by selectively modifying the vector, e.g., (i) introducing silent mutations to remove putative cryptic splice sites (mutations that do not alter the encoded amino acid), (ii) changing the codon usage, and (iii) adding multiple plant introns, it has been possible to construct more efficient viral delivery systems. The plant introns are thought to make the viral RNA more plant mRNA-like and therefore more efficiently transcribed, processed, and translated. With the improved vectors there is typically one successful infection event per 10 to 20 copies of *A. tumefaciens* that are introduced by vacuum infiltration, a process in which the aerial part of the plant is immersed in an *A. tumefaciens* suspension and a weak vacuum is then applied for 10 to 30 s. Vacuum infiltration replaces the viral functions of infection and systemic movement. Amplification of the viral genome within a plant cell and movement from one plant cell to another are performed by the viral proteins (i.e., replicase and movement protein) that make up the modified viral replicon. Starting with vacuum infiltration, the entire process takes 4

Figure 7.7 Second-generation transient-expression vector system. Shown is T-DNA carrying viral genes for replicase and movement protein as well as a target protein gene. The viral replicase is an RNA-dependent RNA polymerase. The movement protein facilitates movement from one plant cell to another. The arrow indicates the direction of transcription. The heavy vertical lines within genes represent introduced introns. In addition to the genes shown, between the left (LB) and the right (RB) borders there is a plant selectable marker gene. doi:10.1128/9781555818890.ch7.f7.7

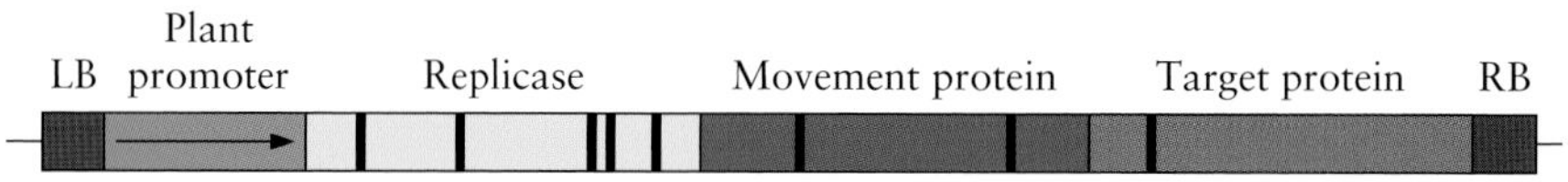

to 10 days before recombinant protein can be isolated from plant leaves. Moreover, it is effective with a range of different plants. It is estimated that in a 1-hectare growth facility, this system may be able to produce up to 400 kg of recombinant protein per year. This system does not require a large amount of land, and therefore, growth of transformed plants could be limited to a contained greenhouse facility.

It was demonstrated by one group of workers that with a transient-expression system, it was possible to dip a head of lettuce, obtained from a grocery store, into an *A. tumefaciens* cell suspension, vacuum infiltrate the bacterium (carrying the engineered viral DNA), and within about 1 week's time obtain around 20 to 100 mg of functional target protein per kilogram of fresh lettuce leaf tissue (Fig. 7.8). With this system, plant growth facilities are not required; the system is very efficient and inexpensive and is easily scaled up. This transient-expression system, based on expression from viral RNA replicons delivered into plant cells by *A. tumefaciens*, allows production of recombinant proteins at yields up to 5 g per kg of fresh leaf biomass. Such high yields are possible because plant biomass accumulation takes place prior to infection. Notwithstanding the high yields of foreign protein that can be obtained, this approach requires special equipment for vacuum infiltration and may be expensive to perform on a large scale. In addition, containment of the genetically engineered agrobacteria producing modified viral replicons that are used for

Figure 7.8 Transformation of a head of lettuce with T-DNA carrying viral DNA and a specific target gene using a strain of *A. tumefaciens* carrying a specific target gene. Following incubation and vacuum infiltration, the target protein may be harvested from the transiently tranformed lettuce within one to several weeks' time. LB and RB, left and right borders, respectively. doi:10.1128/9781555818890.ch7.f7.8

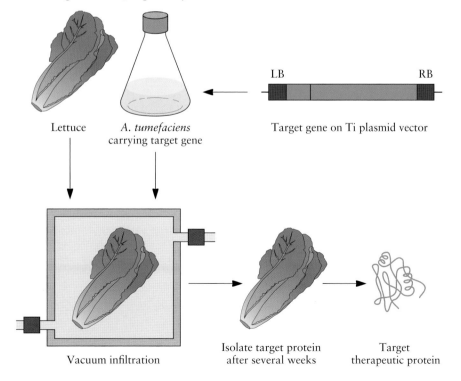

A

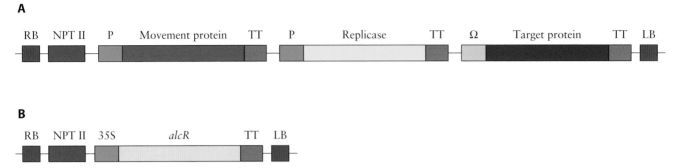

B

Figure 7.9 (A) Second-generation transient-expression vector system requiring the addition of ethanol for activation. The construct contains left (LB) and right (RB) T-DNA border sequences; a neomycin phosphotransferase gene (NPT II), whose promoter and transcription termination regions are not shown, that acts as a plant selectable marker; movement and replicase genes, each under the transcriptional control of an *alc* promoter (P) and transcription termination sequences (TT); and the target gene preceded by a sequence that enhances translation (Ω). (B) T-DNA encoding the *alcR* gene under the control of the constitutive 35S promoter. doi:10.1128/9781555818890.ch7.f7.9

transfection may also be a problem. To avoid these problems, researchers have developed an ethanol-inducible system that provides efficient release of viral RNA replicons from proreplicons contained in a stably transformed cassette (Fig. 7.9). The cassette carrying the viral vector has two components, the viral replicon and a gene encoding the movement protein, and each of these components has been placed separately under the transcriptional control of an ethanol-inducible promoter. In this system, the viral replicon is under the control of the alcohol dehydrogenase (*alc*) promoter from the fungus *Aspergillus nidulans*. This promoter is activated only when it is bound by the AlcR transcriptional activator, encoded on a separate construct, in the presence of ethanol. In practice, to activate the production of a target protein, transgenic plants carrying the two constructs shown in Fig. 7.9 are sprayed twice a day for around 5 days with a solution of 4% ethanol. Alternatively, expression of the target gene was strongly induced when the roots were soaked and then sprayed once with 4% ethanol. Ethanol was chosen as the inducer in this system because it is relatively inexpensive and is nontoxic and because the system exhibits a very low background level of expression in the absence of the inducer. This system should be readily amenable to scale-up. What remains is to determine whether this system is as effective and reliable in practice as it seems.

Molecular Pharming

Therapeutic Agents

Given the significant financial advantages of producing therapeutic agents in plants, this approach is likely to replace, wherever possible, other more expensive and labor-intensive expression systems over the next 10 to 20 years. Facilitating the move toward greater use of plants to produce foreign proteins is the promise of very high expression levels following the use of either transient-expression systems or the expression of foreign genes in the chloroplast. With this is in mind, it is worth examining in some detail how one group of researchers was able to obtain an exceptionally high level of synthesis of a therapeutic protein in transgenic tobacco plants.

The target protein for experiments to optimize the expression of therapeutic proteins in plants was a bacteriophage lytic protein that acts as an antibiotic against pathogenic bacteria (group A and group B streptococci). The PlyGBS lysin gene was isolated from a bacteriophage that infects *Streptococcus agalactiae*, a group B streptococcus. Infections caused by group A and B streptococci are the leading causes of **neonatal sepsis** and **meningitis**, and as persistent colonizers of the human genital and gastrointestinal tracts, pathogenic group B streptococci can sometimes be transmitted to a fetus during pregnancy. The PlyGBS lysin kills all known group B streptococcus strains and reduces colonization by these bacteria in both the vagina and the oropharynx (the part of the pharynx between the soft palate and the upper edge of the epiglottis). PlyGBS is also active against group A streptococci such as *Streptococcus pyogenes*, a human pathogen that colonizes the skin and the mucous membranes of the upper respiratory tract, causing a variety of different infections.

To maximize the expression of the PlyGBS lysin in transgenic plants, several high-expression strategies were combined (Fig. 7.10). Following purification of the bacteriophage, the *plyGBS* gene was isolated and its sequence was determined. To ensure that a high level of the PlyGBS lysin could be produced in chloroplasts, the *plyGBS* gene was completely resynthesized so that its codon usage corresponded to the codon usage found in the AT-rich tobacco chloroplast genome. The synthetic gene was then fused to the strongest known expression signal in chloroplasts, i.e., the constitutive rRNA operon promoter; the promoter was followed by the gene 10 leader RNA from phage T7 (a strong ribosome-binding site). Next, the *plyGBS* gene expression cassette was inserted into a chloroplast transformation vector between chloroplast DNA sequences to target the transgene to the intergenic spacer between the *trnfM* and *trnG* genes by homologous recombination. Chloroplasts were transformed by microprojectile bombardment of tobacco leaves followed by selection on spectinomycin-containing plant regeneration medium. A spectinomycin resistance gene was used as the chloroplast selectable marker gene (Fig. 7.10).

Not surprisingly, transgenic tobacco plants expressing the *plyGBS* gene cassette were somewhat debilitated in their growth compared to wild-type tobacco plants. This is a reflection of the enormous metabolic load imposed on these plants by the overproduction of the lysin protein. Notwithstanding the slower growth, the plants developed normally and produced viable seeds. When this technology is commercialized, these transgenic plants will likely be confined to a greenhouse environment to avoid release of lysin into the environment that might lead to the development of lysin-resistant bacteria.

It has been suggested that for bacteriophage lysins to be effective at killing their hosts, the lysins must be highly resistant to degradation by

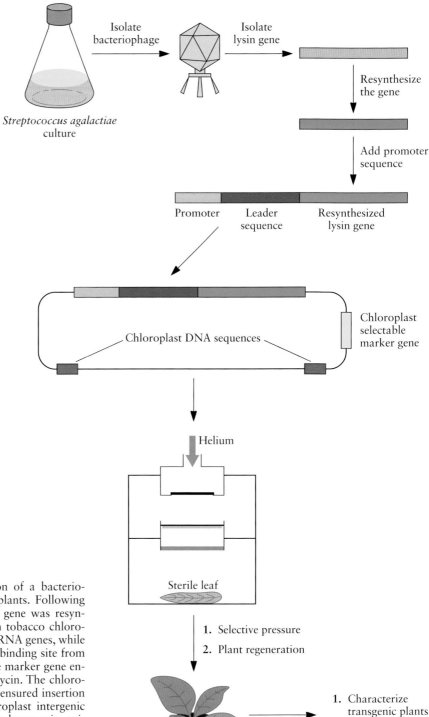

Figure 7.10 Strategy for the overproduction of a bacteriophage lysin protein in transgenic tobacco plants. Following its isolation, the native bacteriophage lysin gene was resynthesized to better reflect the codon usage in tobacco chloroplasts. The promoter was from chloroplast rRNA genes, while the leader sequence was a strong ribosomal binding site from bacteriophage T7. The chloroplast selectable marker gene encoded resistance to the antibiotic spectinomycin. The chloroplast DNA sequences on the plasmid vector ensured insertion by homologous recombination into a chloroplast intergenic region. The construct was used to transform tobacco using microprojectile bombardment.
doi:10.1128/9781555818890.ch7.f7.10

bacterial proteases. Thus, at least in part, the extraordinary stability of the lysin protein facilitates its stable accumulation in chloroplasts. Since most other proteins are not as stable as the PlyGBS lysin, it may be difficult to express other foreign proteins in chloroplasts to the extent observed for the PlyGBS lysin. Nevertheless, the strategy of (i) optimizing the codon usage of the foreign gene, (ii) fusing the foreign gene to the rRNA operon promoter and the T7 gene 10 leader peptide, and (iii) expressing the construct in plant chloroplasts should provide an effective means of obtaining relatively high levels of many different therapeutic proteins.

Antibodies

It has been estimated that it costs approximately $5,000 per gram to produce antibodies in animal (hybridoma) cells in culture, $1,000 per gram to produce antibodies in transgenic bacteria, and $10 to $100 per gram to produce antibodies in transgenic plants. However, since most harvested plant tissues cannot usually be stored for long periods, foreign proteins have sometimes been produced in seeds, where they are stable for long periods under ambient conditions and where they exist in a more concentrated form, thereby facilitating their purification. To date, a large number of antibodies, including immunoglobulin G (IgG), IgM, single-chain Fv fragments, and Fab fragments, have been produced in plants (Table 7.2). Some of these plant-produced antibodies (sometimes called plantibodies) have been purified and used in the laboratory for diagnostic and therapeutic purposes, and others have been used to protect the plant against certain pathogenic agents, such as viruses. In a small number of instances, plant-produced antibodies are being tested in clinical trials to determine whether they are essentially equivalent to antibodies produced in other host cells.

IgG
immunoglobulin G

Proper protein glycosylation of therapeutic proteins, including antibodies, is important because it contributes to protein conformation by influencing protein folding; can target a protein to a particular location, for example, through interaction with a specific receptor molecule; or can increase protein stability by protecting the protein from proteases. In mammals, both the heavy and light antibody chains are synthesized with a peptide signal, which targets the nascent proteins to the endoplasmic reticulum lumen, where the antibody is assembled. Then, glycosylation takes place in the endoplasmic reticulum and the Golgi apparatus by specific enzymes, known as glycosylases and glycosyltransferases, before the antibody is secreted. Plant and animal glycosylations are similar in the early steps but differ in the later steps. However, recently plants have been genetically engineered to glycosylate proteins in a manner that mimics the typical animal glycosylation pattern and so prevent potential side effects and rapid clearance from the bloodstream of plant-made pharmaceuticals.

The effective industrial-scale production of antibodies synthesized by plants has until recently been hampered by a very low yield, typically in the range of 1 to 10 mg/g of fresh plant tissue. Moreover, with conventional transgenic plant technology, it is estimated that it takes about 2 years from the beginning of the cloning process to produce gram quantities of

Figure 7.11 Schematic representation of a portion of the viral vectors used to produce full-size IgG antibodies in plants. In each case, the viral replicase and movement protein (MP) are cloned together with the gene (cDNA) for either a light chain or a heavy chain. The expression of each antibody gene is controlled by a promoter, a signal peptide (to ensure secretion), and a transcription termination region from the appropriate virus, none of which are shown. The viruses used included tobacco mosaic virus (TMV) and potato virus X (PVX). In both cases, the viruses were unable to replicate because of the absence of the viral coat protein. Recombinant viruses were introduced into plants as part of the T-DNA that is transferred during *A. tumefaciens* infection. doi:10.1128/9781555818890.ch7.f7.11

antibody. The use of transient-expression systems (discussed above) can significantly speed up this process, but it is still problematic to coordinate the synthesis and assembly of the two different polypeptides that are integral components of antibody molecules. These systems generally produce only very low levels of active antibody. Recently, an alternative transient-expression system has been developed. This expression system involves coinfection of plant cells with two separate plant virus vectors, one based on tobacco mosaic virus and the other on potato virus X, so that the two vectors do not compete with one another but, rather, can coexist within the same cell. The two vectors can replicate within the plant at the same time, with each vector expressing a different antibody chain; i.e., one expresses the light chain, and the other expresses the heavy chain (Fig. 7.11). This system has been used to increase the amount of IgG antibody that is typically synthesized by a plant transient-expression system about 100-fold, so that it is possible to produce up to 0.5 mg of assembled IgG antibody per gram of fresh leaf biomass. At this high level of expression, it should be possible to grow transgenic plants that produce specific IgG molecules in small areas indoors in controlled greenhouses, thereby avoiding environmental concerns regarding the inadvertent environmental "escape" of genes from antibody-producing plants. Interestingly, the potential value of this technology provided the incentive for the sale of the company by which it was originally developed to a larger company that hopes to open a clinical-grade manufacturing plant with the objective of beginning clinical trials of the antibodies produced.

Edible Vaccines

Despite the fact that vaccines were developed some time ago for a number of previously devastating childhood diseases, it is estimated that worldwide approximately 20% of infants are left unimmunized, resulting in around two million unnecessary deaths per year. This situation is a result of the fact that in many countries, either the vaccine itself is too expensive to be used on a large scale or there is a lack of physical infrastructure (e.g., roads and refrigeration) that makes it not readily feasible to disseminate the vaccine. For example, the worldwide cost of keeping current vaccines refrigerated is estimated to be about $200 million to $300 million per year. Commercial vaccines are expensive to produce and package and require trained personnel to administer injections. Moreover, for some infectious diseases, immunizations either do not exist or are unreliable or very expensive.

The majority of animal pathogens initiate disease following interaction with the mucosal surfaces lining the digestive, respiratory, or genital tract, and the primary defense of these tissues is the mucosal immune system. However, the vast majority of existing vaccines are administered by subcutaneous injection, so they generate effective antibody- and cell-mediated responses in the systemic compartment but not in mucosal sites. On the other hand, mucosal vaccines may be administered either orally or nasally and are effective in inducing antigen-specific immune responses in both the systemic and mucosal compartments. Thus, mucosal vaccination may be a good strategy for combating infectious diseases caused by these pathogens.

A mucosal immune response is triggered by an ingested antigen, which may be expressed as part of a plant. The plant cell wall protects the antigenic protein from destruction by gastric secretions, and when it reaches the intestine, the plant material is degraded, releasing the antigen. The antigen binds to and is taken up by M cells present in the lining of the intestine and then passed to other cells in the immune system, including macrophages and B cells (Fig. 7.12). The macrophages display portions of

Figure 7.12 Schematic representation of how an edible vaccine generates an immune response against an antigen from an infectious agent.
doi:10.1128/9781555818890.ch7.f7.12

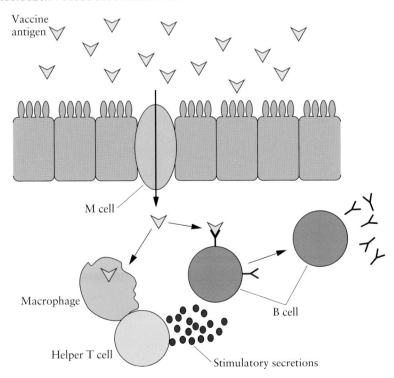

the antigen to the T helper cells, which, in turn, respond by secreting small molecules that activate B cells to synthesize and release antibodies that can neutralize the antigen. Interestingly, both the Bill and Melinda Gates Foundation and the U.S. National Institutes of Health (in an effort to develop more efficacious vaccines) have proposed that mucosal vaccines be a focus of most future vaccine development.

An edible vaccine, in contrast to traditional vaccines, would not require elaborate production facilities, purification, sterilization, or packaging or specialized delivery systems. Moreover, unlike many currently utilized recombinant protein expression systems, plants glycosylate proteins, a factor that may contribute to the immunogenicity and stability of a target protein. Much of the work on edible vaccines that has been reported so far utilizes potatoes as the delivery vehicle. Potatoes were originally chosen for this work because they were easy to manipulate genetically. However, potatoes are extremely unlikely to be a vaccine delivery plant; they require cooking to make them palatable, and cooking destroys (inactivates) most protein antigens. Other plants that are being considered for the delivery of edible vaccines include bananas (although banana trees require several years to mature), tomatoes (although tomatoes spoil readily), lettuce, carrots, peanuts, rice, and corn (mainly for "vaccinating" animals). In particular, a rice-based oral vaccine is stable at room temperature for up to several years and is also protected from digestive enzymes. To avoid having to cook the rice and likely denature the antigenic protein, the transgenic rice is ground into a fine powder that is dissolved or suspended in water and is then easily ingested.

Cholera is an infectious diarrheal disease caused by the bacterial toxin produced by *Vibrio cholerae*. Globally, there are more than 5 million cases and 200,000 deaths from cholera each year. *V. cholerae* colonizes the small intestine and secretes large amounts of a hexameric toxin, which is the actual pathogenic agent. This multimeric protein consists of one A subunit, which has ADP-ribosylation activity and stimulates adenylate cyclase, and five identical B subunits that bind specifically to an intestinal mucosal cell receptor (Fig. 7.13A). The A subunit has two functional domains: the A1 peptide, which contains the toxic activity, and the A2 peptide, which joins the A subunit to the B subunits.

Traditional cholera vaccines, in common use for many years, consisted of phenol-killed *V. cholerae*. This vaccine generated only moderate protection, generally lasting from about 3 to 6 months. More recently, an oral vaccine (Dukoral) consisting of heat-inactivated *V. cholerae* Inaba classic strain, heat-inactivated *V. cholerae* Ogawa classic strain, formalin-inactivated *V. cholerae* Inaba El Tor strain, formalin-inactivated *V. cholerae* Ogawa classic strain, and a recombinant cholera toxin B subunit has come into use. The vaccine is taken orally (two doses 1 week apart), and it is claimed that an additional booster immunization is not required for about 2 years.

Having demonstrated the efficacy of an oral vaccine against cholera, researchers turned their attention to the development of an edible vaccine against this disease. To do this, potato plants were transformed

Figure 7.13 Schematic representation of cholera toxin (**A**) and an engineered cholera vaccine (**B**). In the engineered vaccine, the cholera toxin B subunit is synthesized as a fusion protein with the rotavirus peptide, and the cholera toxin A2 subunit is fused to the enterotoxigenic *E. coli* fimbrial colonization factor. The cholera toxin B and A2 peptides are noncovalently attached to one another. Both fusion proteins are expressed in transgenic potatoes and combine spontaneously to form a cholera toxin-like protein, as shown.
doi:10.1128/9781555818890.ch7.f7.13

A

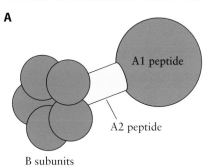

B

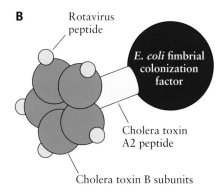

using *A. tumefaciens* to transfer the cholera toxin subunit B gene in the T-DNA into potato genomic DNA for expression. One gram of transgenic potato produced approximately 30 mg of subunit B protein. After the transgenic potatoes were cooked in boiling water until they were soft enough to be edible by humans, approximately 50% of the subunit B protein remained undenatured. The cooked potatoes were fed to mice once a week for 4 weeks before the mice were tested for the presence of antibodies against the subunit B protein and for resistance to *V. cholerae*-caused diarrhea. These tests indicated that the mice had acquired a significant level of protection against *V. cholerae*. Moreover, although mucosal antibody titers declined gradually after the last immunization, they were rapidly restored after an oral boost (an additional feeding) of transgenic potato.

To increase the immunogenicity of the cholera vaccine, the cholera toxin subunit B and A2 genes were each fused to genes encoding antigenic proteins and then used to generate transgenic potato plants. To create these two fusion proteins, a 22-amino-acid epitope from murine (mouse) rotavirus enterotoxin NSP4 was fused to the C-terminal end of the cholera toxin subunit B protein, and the enterotoxigenic *E. coli* fimbrial colonization factor CFA/I was fused to the N-terminal end of the cholera toxin subunit A2 protein (Fig. 7.13B). Normally, the A2 peptide links the A1 peptide, which has the toxic activity, with the subunit B peptide, which has the binding activity. Transgenic potatoes that expressed both of the fusion proteins were fed to mice, which generated antibodies against cholera toxin subunit B protein, murine rotavirus enterotoxin NSP4, and *E. coli* fimbrial colonization factor CFA/I and were protected against diarrhea caused by rotavirus, cholera, and enterotoxigenic *E. coli*.

More recently, the cholera toxin subunit B protein has been expressed in banana and rice. To express the cholera toxin B subunit in rice plants, first the B subunit gene was completely resynthesized using codons that are optimized for expression in plants. This gene was introduced into rice plants using a binary vector of the Ti plasmid of *A. tumefaciens* (Fig. 7.14). The construct included the codon-optimized cholera toxin B subunit gene under the control of the rice seed storage glutelin 2.3-kb *GluB-1* promoter and signal sequence to facilitate secretion. In

Figure 7.14 Construct used to express the cholera toxin B subunit in rice seeds. The left (LB) and right (RB) borders of the T-DNA construct are indicated; the promoter and transcription termination region of the marker gene are not shown. For the expression of the cholera toxin B subunit, the promoter (P), the signal sequence, and the transcription termination region (TT) are all from the rice seed storage protein gene glutelin *GluB-1*; the KDEL region is a signal that retains the protein on the endoplasmic reticulum. The arrow indicates the direction of transcription. doi:10.1128/9781555818890.ch7.f7.14

addition, a DNA sequence encoding the KDEL signal was located at the 3' end of the cholera toxin B subunit gene. The KDEL sequence (lysine, aspartic acid, glutamic acid, and leucine) is expressed as part of the protein and targets the cholera toxin subunit to the endoplasmic reticulum, where the protein is glycosylated. The KDEL sequence is also thought to increase the expression level of some proteins in plants. The glutelin 2.3-kb *GluB-1* transcription termination sequence is also included to the 3' end of the cholera toxin B subunit gene. When this construct was introduced into two different rice cell varieties, expression levels as high as 30 μg per seed (or ~2.1% of the seed protein) were observed. Moreover, mice that were fed a suspension of ground rice powder in water were protected against cholera after ingesting approximately 50 mg of rice powder containing around 75 μg of cholera toxin B subunit protein. It is expected that this type of edible vaccine may become the prototype for a large number of similar edible vaccines. However, it is first necessary for this vaccine to undergo successful clinical trials in humans.

It has been estimated that Shiga toxin-producing strains of *E. coli* cause approximately 100,000 cases of hemorrhagic colitis (an acute disease characterized by overtly bloody diarrhea) a year. About 6% of those infections produce severe complications, including kidney failure. The overall structure of the Shiga toxin is similar to that of cholera toxin in that it contains one A subunit, which encodes the toxin activity (that after entry into the cytosol inhibits protein synthesis), per five B subunits, which act (together) to bind to animal cell surface receptors. To develop an oral vaccine against type 2 Shiga toxin (type 2 is responsible for the most severe Shiga toxin-caused disease in humans), the genes for a genetically inactivated version of the Shiga toxin A and B peptides were both cloned and expressed in tobacco plant cells (Fig. 7.15). To test the ability of transformed tobacco plants that synthesized the modified Shiga toxin to protect mice against the toxin, scientists infected mice with Shiga toxin-producing strains of *E. coli*. Before the introduction of the Shiga toxin-producing bacteria, some of the mice were fed leaves from transgenic tobacco plants expressing the inactivated Shiga toxin once a week for 4 weeks, while other mice were left untreated. One week after the introduction of the toxic *E. coli* strain, all of the mice that were not fed the antigen-producing tobacco had died. In contrast, 2 weeks after treatment with the toxic *E. coli* strain, all of the orally vaccinated mice were still alive. This experiment serves as a proof of the concept that oral administration of the inactivated Shiga toxin is a highly effective means of protecting animals against Shiga toxin-producing *E. coli*. Of course, for a human oral vaccine, a more suitable host plant, such as tomato, rice, or banana, would be desirable.

To express proteins in plants that can be used to prime the immune system as part of an edible vaccine, the first option is to translate the protein in the cell cytoplasm. However, this often results in low expression levels or can cause stunted growth of the resulting plants. In the second option, the gene encoding the foreign antigen may be introduced into

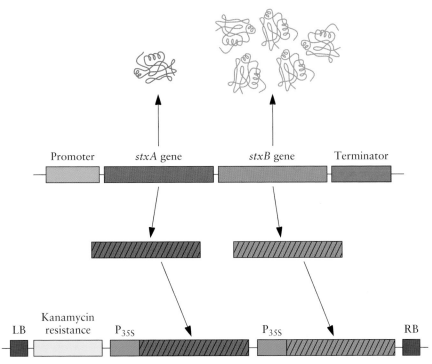

Figure 7.15 Schematic representation of Shiga toxin A and B subunits encoded in a bacterial operon (*stxA* and *stxB* genes, respectively) under the control of a single promoter and transcription terminator. The bacterium produces five copies of the B subunit protein for each copy of the A subunit protein. Following the isolation of these two genes, it was necessary to modify them (shown as diagonal lines). First, the A subunit was inactivated, and then nucleotides sequences on both genes that might adversely affect the transcription of these bacterial genes in plant cells were eliminated. The modified genes were inserted between the left and right borders (LB and RB) of the T-DNA of a binary Ti plasmid-based vector, each under the control of a separate constitutive cauliflower mosaic virus 35S promoter. The T-DNA construct also contained a gene encoding kanamycin resistance in plants under the control of its own promoter. The transcription terminator sequences of the three genes contained within the T-DNA are not shown for the sake of simplicity. The final construct was used to transform tobacco plants. doi:10.1128/9781555818890.ch7.f7.15

the genome of the chloroplast. Chloroplast transformation is an attractive alternative to nuclear transformation because it produces relatively high transgene expression. In the third option, in those instances where posttranslational modification is required, the protein of interest is targeted to the endoplasmic reticulum, where it will be N glycosylated. The protein of interest can be retained in the endoplasmic reticulum by using a KDEL sequence or allowed to move through the Golgi apparatus and secretory pathway, where N-glycan modifications occur. In the fourth option, it is possible to transiently express many proteins at high levels in plants; it is likely that this approach will be utilized to a much greater extent in the future.

summary

Strains of the soil bacterium *A. tumefaciens* can genetically engineer plants naturally. In this system, after responding to chemical signals from a surface wound, *A. tumefaciens* makes contact with an exposed plant cell membrane. A series of steps then occurs that results in the transfer of a segment (T-DNA) of a Ti plasmid from the bacterium into the nucleus of the plant cell. The T-DNA region becomes integrated into the plant genome, and subsequently, the genes on the T-DNA region are expressed.

The *A. tumefaciens*–Ti plasmid system has been modified for use as a mechanism of delivery of cloned genes into some plant cells. In these vector systems, the phytohormone and opine metabolism genes have been removed from the T-DNA region, and the modified T-DNA sequence has been cloned into a plasmid that can exist stably in *E. coli*, for genetic manipulation. A cloned gene that is inserted into this T-DNA region is part of the DNA that is transferred into the nucleus of a recipient plant cell. To achieve this transfer, *A. tumefaciens* is used as a delivery system. In one system, the shuttle vector with the T-DNA-cloned gene segment is introduced into an *A. tumefaciens* strain that carries another, compatible plasmid with genes (*vir* genes) that are essential for transferring the T-DNA region into a plant cell. In addition to this binary vector system, a cointegrate system has been designed so that after the introduction of the shuttle vector carrying the target gene into *A. tumefaciens*, it recombines with the *vir* gene-containing, disarmed Ti plasmid, yielding a single plasmid that has both *vir* gene functions and the T-DNA-cloned gene segment. Since the *A. tumefaciens* T-DNA system is not effective with all plants, microprojectile bombardment (biolistics) has been an effective procedure for delivering DNA

to a wider range of plant cells. This transferred DNA can be stably integrated into the genome of the plant cells, and it is also possible to target genes to the chloroplast DNA, where, as a consequence of the many chloroplasts per cell and the many copies of chloroplast DNA per chloroplast, it is possible to obtain much higher levels of foreign gene expression than would be otherwise possible.

To rapidly and easily obtain high-level transient expression of foreign proteins in plants, plant virally derived vectors may be delivered to the plant by *A. tumefaciens*. From the whole virus, only the viral elements required for efficient expression of the target gene are maintained. In addition, by modifying the vector to introduce silent mutations to remove cryptic splice sites, changing the codon usage, and adding plant introns, it has been possible to construct more efficient viral delivery systems.

Using a variety of production strategies, a large number of therapeutic agents, including both antibodies and antibody fragments, have been produced in transgenic plants. A number of these proteins are currently in clinical trials.

Since the majority of animal pathogens initiate disease following interaction with the mucosal surfaces lining the digestive, respiratory, or genital tract, and the primary defense of these tissues is the mucosal immune system, a logical method of delivering vaccines is the construction of edible vaccines. Thus, a number of vaccine antigens have been cloned and expressed in transgenic plants and are currently being tested both in the lab and in clinical trials for efficacy. To date, a number of these edible vaccines have been found to be effective in inducing both systemic and mucosal antigen-specific immune responses.

review questions

1. Why is the Ti plasmid from *A. tumefaciens* well suited for developing a vector to transfer foreign genes into plant chromosomal DNA?

2. How do (i) binary and (ii) cointegrate Ti plasmid-based vector systems for plant transformation differ from one another?

3. How are plants transformed by microprojectile bombardment?

4. How is foreign DNA targeted for integration into chloroplast DNA?

5. How can foreign proteins be transiently expressed in plants? What is the advantage of doing this?

6. How can transient expression of foreign genes in plants be optimized to produce a very high level of the target protein?

7. What is the advantage of introducing foreign genes into chloroplast rather than nuclear DNA?

8. How would you ensure that a foreign gene that has been inserted into the chloroplast is expressed at a high level?

9. Briefly describe a vector system that may be used to engineer plants to produce large amounts of full-size IgG molecules.

review questions *(continued)*

10. Why are plants an attractive host system, compared to bacteria and animal cells in culture, for the production of human therapeutic proteins?

11. Briefly describe how an edible vaccine against cholera might work.

12. Describe a strategy for developing a plant vaccine against type 2 Shiga toxin.

13. Briefly discuss the advantages and disadvantages of using different plants as the basis for an edible vaccine against a human disease.

references

Bock, R., and H. Warzecha. 2010. Solar-powered factories for new vaccines and antibiotics. *Trends Biotechnol.* **28**:246–252.

Christou, P. 1992. Genetic transformation of crop plants using microprojectile bombardment. *Plant J.* **2**:275–281.

Cox, K. M., J. D. Sterling, J. T. Regan, J. R. Gasdaska, K. K. Frantz, C. G. Peele, A. Black, D. Passmore, C. Moldovan-Loomis, M. Srinivasan, S. Cuison, P. M. Cardarelli, and L. F. Dickey. 2006. Glycan optimization of a human monoclonal antibody in the aquatic plant *Lemna minor*. *Nat. Biotechnol.* **24**:1591–1597.

Darbani, B., A. Eimanifar, C. N. Stewart, Jr., and W. N. Camargo. 2007. Methods to produce marker-free transgenic plants. *Biotechnol. J.* **2**: 83–90.

Desai, P. N., N. Shrivastava, and H. Padh. 2010. Production of heterologous proteins in plants: strategies for optimal expression. *Biotechnol. Adv.* **28**:427–435.

Gelvin, S. B. 2003. Agrobacterium-mediated plant transformation: the biology behind the "gene-jockeying" tool. *Microbiol. Mol. Biol. Rev.* **67**:16–37.

Gleba, Y., V. Klimyuk, and S. Mariollett. 2007. Viral vectors for the expression of proteins in plants. *Curr. Opin. Biotechnol.* **18**:134–141.

Goldsbrough, A. P., C. N. Lastrella, and J. I. Yoder. 1993. Transposition mediated re-positioning and subsequent elimination of marker genes from transgenic tomato. *Bio/Technology* **11**: 1286–1292.

Hager, M., and R. Bock. 2000. Enslaved bacteria as new hope for plant biotechnologists. *Appl. Microbiol. Biotechnol.* **54**:302–310.

Hao, Y., T. C. Charles, and B. R. Glick. 2010. ACC deaminase increases the *Agrobacterium tumefaciens*-mediated transformation of commercial canola cultivars. *FEMS Microbiol. Lett.* **307**: 185–190.

Hefferon, K. 2010. Clinical trials fuel the promise of plant-derived vaccines. *Am. J. Clin. Med.* **7**:30–37.

Heifetz, P. B., and A. M. Tuttle. 2001. Protein expression in plastids. *Curr. Opin. Plant Biol.* **4**:157–161.

Iamtham, S., and A. Day. 2000. Removal of antibiotic resistance genes from transgenic tobacco plastids. *Nat. Biotechnol.* **18**:1172–1176.

Ishida, Y., H. Saito, S. Ohta, Y. Hiei, T. Komari, and T. Kumashiro. 1996. High efficiency transformation of maize (*Zea mays* L.) mediated by *Agrobacterium tumefaciens*. *Nat. Biotechnol.* **14**: 745–750.

Jamal, A., K. Ko, H.-S. Kim, Y.-K. Choo, H. Joung, and K. Ko. 2009. Role of genetic factors and environmental conditions in recombinant protein production for molecular farming. *Biotechnol. Adv.* **27**:914-923.

Karg, S. R., and P. T. Kallio. 2009. The production of biopharmaceuticals in plant systems. *Biotechnol. Adv.* **27**: 879–894.

Maliga, P. 2004. Plasmid transformation in higher plants. *Annu. Rev. Plant Biol.* **55**:289–313.

Marillonet, S., C. Thoeringer, R. Kandzia, V. Klimyuk and Y. Gleba. 2005. Systemic *Agrobacterium tumefaciens*-mediated transfection of viral replicons for efficient transient expression in plants. *Nat. Biotechnol.* **23**:718–723.

Meyers, B., A. Zaltsman, B. Lacroix, S. V. Kozlovsky, and A. Krichevsky. 2010. Nuclear and plastid genetic engineering of plants: comparison of opportunities and challenges. *Biotechnol. Adv.* **28**:747–756.

Miki, B. L., and S. McHugh. 2004. Selectable marker genes in transgenic plants: applications, alternatives and biosafety. *J. Biotechnol.* **107**:193–232.

Mullen, J., G. Adam, A. Blowers, and E. Earle. 1998. Biolistic transfer of large DNA fragments to tobacco cells using YACs retrofitted for plant transformation. *Mol. Breed.* **4**:449–457.

Negrouk, V., G. Eisner, H.-I. Lee, K. Han, D. Taylor, and H. C. Wong. 2005. Highly efficient transient expression of functional recombinant antibodies in lettuce. *Plant Sci.* **169**:433–438.

Nochi, T., H. Takagi, Y. Yuki, L. Yang, T. Masumura, M. Mejima, U. Nakanishi, A. Matsumura, A. Uozumi, T. Hiroi, S. Morita, K. Tanaka, F. Takaiwa, and H. Kiyono. 2007. Rice-based mucosal vaccine as a global strategy for cold-chain- and needle-free vaccination. *Proc. Natl. Acad. Sci. USA* **104**: 10986–10991.

Nonaka, S., M. Sugawara, K. Minamis-awa, K. Yuhashi, and H. Ezura. 2008. 1-Aminocyclopropane-1-carboxylate deaminase enhances *Agrobacterium tumefaciens*-mediated gene transfer into plant cells. *Appl. Environ. Microbiol.* 74:2526–2528.

Obembe, O. O., J. O. Popoola, S. Leelavathi, and S. V. Reddy. 2011. Advances in molecular farming. *Biotechnol. Adv.* 29:210–222.

Oey, M., M. Lohse, B. Kreikemeyer, and R. Bock. 2009. Exhaustion of the chloroplast protein synthesis capacity by massive expression of a highly stable protein antibiotic. *Plant J.* 57:436–445.

Potrykus, I. 1991. Gene transfer to plants: assessment of published approaches and results. *Annu. Rev. Plant Physiol.* 42:205–225.

Ruf, A., M. Hermann, I. J. Berger, H. Carrer, and R. Bock. 2001. Stable genetic transformation of tomato plastids and expression of foreign protein in fruit. *Nat. Biotechnol.* 19:870–875.

Saint-Jure-Dupas, C., L. Faye, and V. Gomord. 2007. From plant to pharma with glycosylation in the toolbox. *Trends Biotechnol.* 25:317–323.

Southgate, E. M., M. R. Davey, J. B. Power, and R. Marchant. 1995. Factors affecting the genetic engineering of plants by microprojectile bombardment. *Biotechnol. Adv.* 13:631–651.

Thomas, D. R., C. A. Penney, A. Majumder, and A. M. Walmsley. 2011. Evolution of plant-made pharmaceuticals. *Int. J. Mol. Sci.* 12:3220–3226.

Tremblay, R., D. Wang, A. M. Jevnikar, and S. Ma. 2010. Tobacco, a highly efficient green bioreactor for production of therapeutic proteins. *Biotechnol. Adv.* 28:214–221.

Werner, S., O. Breus, Y. Symonenko, S. Marillonet, and Y. Gleba. 2011. High-level recombinant protein expression in transgenic plants by using a double-inducible viral vector. *Proc. Natl. Acad. Sci. USA* 108: 14061–14066.

Yoder, J. I., and A. P. Goldsbrough. 1994. Transformation systems for generating marker-free transgenic plants. *Bio/Technology* 12:263–267.

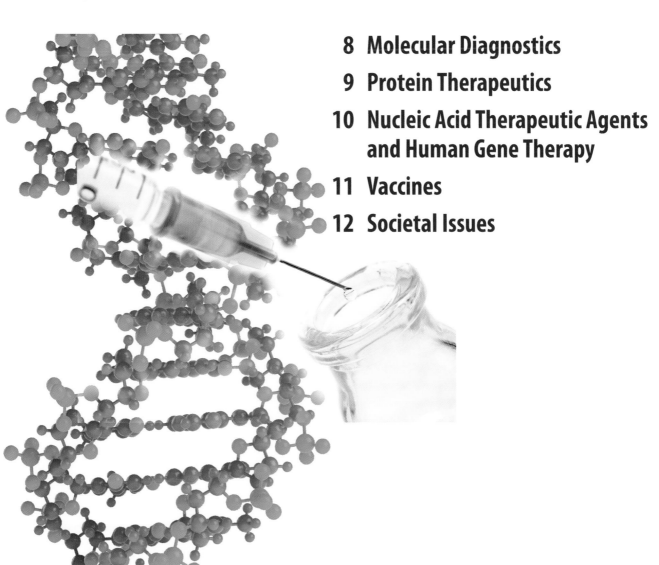

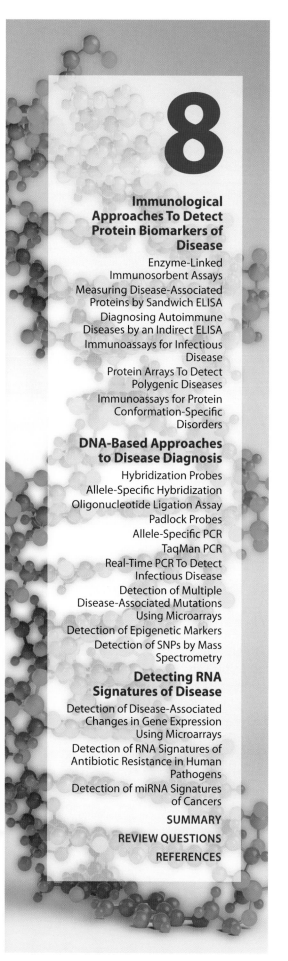

8

Molecular Diagnostics

A MAJOR GOAL OF A DIAGNOSTIC TEST is to determine the presence of a disease as early as possible before it has significantly progressed, which limits treatment efficacy. Diagnostic tests are also used to predict susceptibility to a disease or the response to a treatment, to determine disease **prognosis** (progress and outcome), and to monitor treatment efficiency. A diagnostic test must have high specificity for the target molecule or pathogen (i.e., few false-positive results), have high sensitivity to detect low levels of the target (i.e., few false-negative results), and be sufficiently rapid and inexpensive for routine, and possibly high-throughput, analysis. Noninvasive tests that can detect a target molecule or pathogen in body fluids such as blood, urine, sputum, or pus from a wound are preferable; however, many techniques are available to detect molecular targets in tissues acquired by **biopsy**.

Molecular diagnostic approaches detect molecular biomarkers of disease. In this context, a disease biomarker is a specific molecule that has been determined to be present, or present at higher or lower levels, in diseased tissues compared to normal tissue or to be an indicator of disease prognosis or a response to therapy. It can be a protein, a specific sequence of DNA or RNA, or a small metabolite. Specific molecules may be analyzed qualitatively or quantitatively in urine, blood serum, or tissue samples. Samples that contain a small amount of the target molecule, for example, small core needle biopsy samples or blood samples, may require some amplification of the target for accurate detection and quantification. The complex heterogeneity of clinical samples may also present a challenge for molecular diagnostics, and methods such as tissue microdissection may be employed to increase sample purity.

Molecular diagnostic tests are available to detect the **etiological agents** of infectious and genetic diseases. The latter may be single-gene disorders such as cystic fibrosis or complex multigene disorders such as Alzheimer disease. Although many biomarkers have been identified, clinically relevant biomarkers are not well defined for most human diseases.

doi:10.1128/9781555818890.ch8

Immunological Approaches To Detect Protein Biomarkers of Disease

Proteins play a critical role in all cellular processes, including metabolism, communication, defense, reproduction, transport, and motility. Regulatory proteins maintain tight control over protein production to ensure that cells function normally. Dysregulation of these processes in diseased tissue is reflected in alterations in protein composition or levels of specific proteins. Characteristic changes in proteins have been used extensively to diagnose disease, either by detecting the presence or measuring levels of a specific protein biomarker or by determining protein profiles in polygenic diseases.

There are several advantages to using proteins as diagnostic biomarkers of disease. Abnormal levels of gene expression as a consequence of disease-associated mutations are more accurately quantified by measuring the protein directly rather than by measuring mRNA levels, which often do not correlate with protein levels. Furthermore, proteins in diseased tissues may exhibit irregularities in posttranslational modification that cannot be detected using nucleic acids. In addition, many diseases often are a consequence of altered protein conformation (e.g., prion diseases and some neurodegenerative diseases) that may not be detected from nucleic acid sequences. On the other hand, the unique shape and often very low cellular levels of a protein are a challenge for development of a diagnostic assay. Antibodies can be produced that bind to target proteins with high affinity and specificity, and therefore immunological approaches meet the criteria of sensitivity, specificity, and simplicity for diagnostic assays.

Agglutination is a simple, inexpensive, rapid, and highly specific immunological test that is widely performed in diagnostic laboratories. For example, it is often used for human blood typing based on the presence of specific antigens on the surface of red blood cells, which vary among individuals. **Antiserum** containing antibodies against either A or B surface antigen is mixed with red blood cells. Clumping (agglutination) indicates the presence of the antigen (Fig. 8.1A). Some individuals produce only the A antigen (blood type A), which agglutinates with anti-A antiserum but not with anti-B antiserum (Fig. 8.1A). Blood type B individuals produce only the B antigen, which reacts with anti-B antiserum and not with anti-A antiserum (Fig. 8.1B). Others carry both antigens (blood type AB) or neither (blood type O). Blood samples from individuals with blood type O, the most common blood type, do not produce a positive agglutination result with either antiserum. This test, referred to as hemagglutination, is important for determining which blood to use in transfusions, as antibodies naturally produced against nonself red blood cell antigens would destroy the introduced blood. In some agglutination tests, the antigen or antibody employed to detect either a specific antibody or antigen, respectively, in patient samples is used to coat the surface of small latex beads.

Enzyme-Linked Immunosorbent Assays

Enzyme-linked immunosorbent assays (ELISAs) are widely used for diagnosing human diseases, including various cancers, autoimmune diseases, allergies, and infectious diseases. An ELISA measures antigens or

ELISA
enzyme-linked immunosorbent assay

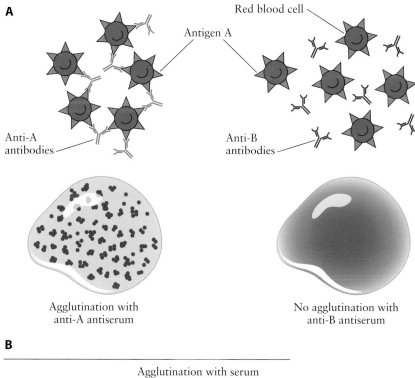

Figure 8.1 Hemagglutination test to determine blood type. (**A**) Antiserum containing antibodies against A or B surface antigens is mixed with red blood cells carrying A surface antigens. Clumping (agglutination) in the presence of anti-A antiserum but not with anti-B antiserum indicates blood type A. (**B**) Agglutination responses for different blood types. doi:10.1128/9781555818890.ch8.f8.1

antibodies produced against an antigen in a clinical sample from blood, urine, or tissues (see chapter 2). It is based on the specific and high-affinity interaction between an antibody and an antigen and is a sensitive assay that can be used for rapid detection on a large scale (i.e., high-throughput assays). Detection relies on the activity of an enzyme that is covalently bound to an antibody employed in the assay. An **indirect ELISA** can detect the presence of specific antibodies in patient serum that indicates an immune response to the presence of a particular protein or pathogen. A **sandwich ELISA** detects the presence of a specific antigen in a patient's sample and hence is sometimes referred to as an antigen capture assay.

In an indirect ELISA used for diagnostic purposes, a standardized antigen is bound to a solid support, usually the surface of a well in a microtiter plate (Fig. 8.2A; see also Fig. 2.30A in chapter 2). A patient's serum sample is applied to the well, and specific **primary antibodies** in the serum bind to the immobilized antigens. A **secondary antibody** that binds

A

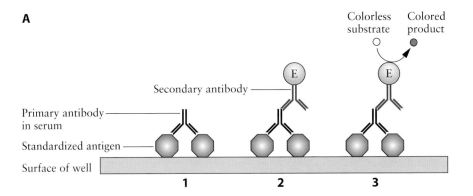

B

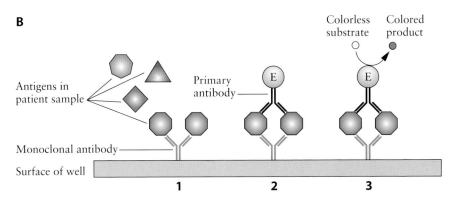

Figure 8.2 ELISAs. (**A**) Indirect ELISA. A specific antigen is immobilized on the surface of the wells in a microtiter plate. (1) Patient serum is applied, and primary antibodies present in the serum bind to the antigen. Unbound antibodies are removed by a washing step. (2) A secondary antibody that binds specifically to the primary antibody is applied. (3) The secondary antibody is conjugated to an enzyme (E) that catalyzes the conversion of a colorless substrate into a colored product. The formation of a colored product indicates that the serum sample contains antibodies directed against the antigen. (**B**) Sandwich ELISA. Monoclonal antibodies specific for a target antigen are bound to the surface of a well in a microtiter plate. (1) A patient's sample is added to the wells, and specific antigens present in the sample bind to the immobilized antibody. A wash step removes any unbound molecules. (2) A primary antibody is applied to detect the presence of bound antigen. (3) The primary antibody is conjugated to an enzyme that catalyzes the formation of a colored product that indicates the presence of the target antigen. doi:10.1128/9781555818890.ch8.f8.2

specifically to the primary antibody is applied. Because the primary antibody is present in human serum, the secondary antibody is an anti-human immunoglobulin antibody that was raised in another animal, for example, a goat, by injecting the animal with human immunoglobulin. The secondary antibody is covalently bound (conjugated) to an enzyme such as alkaline phosphatase or horseradish peroxidase that catalyzes the conversion of a colorless substrate into a colored product. A colorless substrate is applied, and the formation of a colored product by the enzyme indicates the presence of the secondary antibody and, hence, the primary antibody–antigen complex.

In contrast to an indirect ELISA, a sandwich ELISA directly detects a particular antigen in a complex clinical sample. To capture the antigen, a

monoclonal antibody (see chapter 2) that is specific for the target antigen is first bound to the surface of a microtiter plate (Fig. 8.2B). The patient's sample is then added to a well; if the specific antigen is present in the sample, it binds to the immobilized antibody. A labeled primary antibody is added to detect the presence of bound antigen. The labeled antibody may detect a different **epitope** on the same antigen.

A sandwich ELISA is the basis for many diagnostic tests, including the home pregnancy test. In one pregnancy test, the human chorionic gonadotropin (hCG) protein produced during pregnancy by the developing placenta is detected in urine. Monoclonal and **polyclonal antibodies** (see chapter 2) specific for hCG are deposited on different regions of a membrane (Fig. 8.3). Urine is applied to one end of the membrane and is drawn through the membrane by capillary action to the region containing

hCG
human chorionic gonadotropin

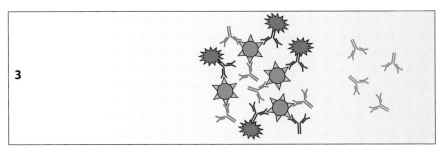
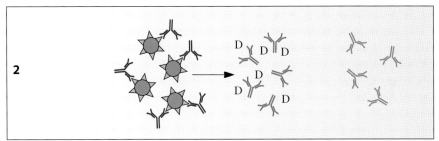
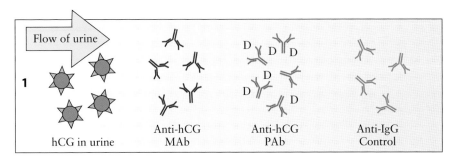
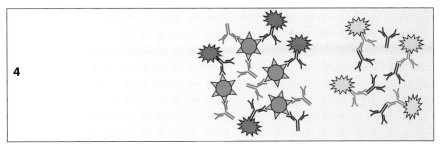

Figure 8.3 Pregnancy test using a sandwich ELISA. (**1**) Urine is applied to one end of the membrane. (**2**) Urine is drawn down the membrane by capillary action, and hCG molecules in the urine of a pregnant woman bind to the anti-hCG monoclonal antibodies (MAb). (**3**) The hCG molecules bound to the MAbs are subsequently carried with the flow of urine and are captured by the immobilized anti-hCG polyclonal antibodies (PAb). The MAb are conjugated to an enzyme that reacts with dye molecules (D) to form a colored product (red star) that indicates a positive test result. (**4**) A control region contains anti-IgG antibodies that bind to the MAb whether or not hCG is present and produce a colored product (yellow star) to indicate that the test reagents are functional. doi:10.1128/9781555818890.ch8.f8.3

the anti-hCG monoclonal antibodies. If hCG molecules are present, they bind to the monoclonal antibodies. The antibodies with bound hCG are carried with the urine through the membrane to the site of the immobilized anti-hCG polyclonal antibodies. The polyclonal antibodies capture the hCG–monoclonal antibody complexes. Monoclonal antibodies that are not bound to hCG are carried away with the urine by capillary action. The monoclonal antibodies are conjugated to an enzyme that reacts with dye molecules contained in the region with the polyclonal antibodies. Formation of a colored product indicates a positive pregnancy test result. Farther downstream in the membrane, a control region contains anti-IgG antibodies that bind to the monoclonal antibodies whether or not hCG is present. A colored product in the control region indicates that the test components are functional. Timing of the test is important for accuracy because the hCG protein is produced only after implantation of the embryo in the uterine wall, which occurs 6 to 12 days after fertilization, after which protein levels continue to rise during the first 20 weeks of pregnancy. False-negative results may occur if the test is performed too early in pregnancy.

Measuring Disease-Associated Proteins by Sandwich ELISA

Ovarian cancer is a devastating disease that kills over 15,000 women a year in the United States. More than 22,000 new cases are diagnosed each year, most at an advanced stage when the survival rate is less than 20%. An immunoassay widely used to monitor progression and recurrence of ovarian carcinoma measures levels of the protein CA125 in serum. CA125 is a high-molecular-weight glycoprotein that is present at higher levels in 50% of women with ovarian cancer compared to healthy women. Levels may also be elevated in women with lung, pancreatic, breast, cervical, and colorectal cancers and with noncancerous disorders such as pelvic inflammatory disease, hepatic disorders, and nonmalignant ovarian cysts. Because CA125 may indicate other disorders and is not elevated in all patients with ovarian cancer, especially at early stages of the disease when tumors are small and therefore secrete only low levels of the protein, it is not recommended as a screening test. Rather, it is commonly used to monitor a patient's response to treatment. Researchers have therefore sought to identify other biomarkers that are more specific for ovarian cancer.

A useful biomarker for ovarian cancer is a protein that is secreted specifically by ovarian tumors, and not by normal tissue or other tumor types, and that can be detected at low levels in the blood or urine, enabling early, and noninvasive, detection. Using cDNA microarrays and quantitative polymerase chain reaction (PCR) (see chapter 1), human epididymis protein 4 (HE4) was identified as a promising new biomarker that was expressed in ovarian carcinomas but not normal tissue or benign ovarian tumors. Mouse monoclonal antibodies were generated against two different HE4 epitopes and used to develop an ELISA. In an initial blinded study (researchers were not informed of the source of the patient samples), HE4 was found to be elevated in sera from patients with early- and late-stage ovarian cancer compared to sera from healthy women and from women with benign ovarian tumors.

PCR
polymerase chain reaction

HE4
human epididymis protein 4

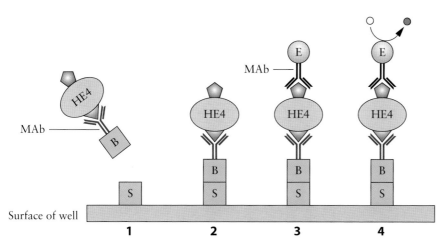

Figure 8.4 Sandwich ELISA to monitor progression or recurrence of ovarian cancer. (1) Patient samples are incubated with biotinylated (B) monoclonal antibodies (purple MAb) generated against one HE4 epitope (blue). (2) Biotin, together with the MAb–HE4 complex, binds with high affinity to streptavidin (S) that coats the wells of a microtiter plate. (3) A second antibody (black MAb) that binds to a different HE4 epitope (red) is added, and after generation of a colored product by horseradish peroxidase (E) conjugated to the antibody (4), HE4 is quantified spectrophotometrically. doi:10.1128/9781555818890.ch8.f8.4

The U.S. Food and Drug Administration (FDA) has approved the use of a HE4 sandwich ELISA to monitor the recurrence or progression of ovarian cancer in women who are being treated for the disease. Patient samples are incubated with biotinylated monoclonal antibodies generated against one HE4 epitope in streptavidin-coated wells (Fig. 8.4). A second antibody, conjugated to the enzyme horseradish peroxidase, that binds to a different HE4 epitope is added, and after generation of a colored product by the horseradish peroxidase, HE4 is quantified spectrophotometrically. Using this assay, HE4 was found to be present at elevated levels in more than 75% of patients with ovarian cancer, compared to 5% of healthy women and 13% of individuals with other nonmalignant conditions. An increase in HE4 levels of more than 25% is considered significant to suggest recurrence or disease progression, while a decrease of this magnitude suggests a positive response to treatment.

FDA
U.S. Food and Drug Administration

Diagnosing Autoimmune Diseases by an Indirect ELISA

Autoimmune diseases occur when the body's immune system does not recognize normal cellular molecules and structures as "self" but rather produces antibodies (autoantibodies) that destroy those targets (see chapter 4). Several autoimmune diseases have been identified, including celiac disease, type 1 diabetes, lupus erythematosus, and rheumatoid arthritis. Indirect ELISAs have been developed to diagnose some autoimmune diseases. In these assays, autoantibodies produced against a self-protein are detected in patient blood samples.

Rheumatoid arthritis is an autoimmune disease that results in chronic, systemic inflammation of the joints, mainly the synovial (flexible) joints. The synovial membrane that lines these joints secretes a fluid

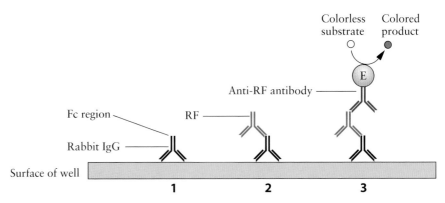

Figure 8.5 Indirect ELISA for diagnosis of the autoimmune disease rheumatoid arthritis. (**1**) Rabbit IgG molecules are bound to the surface of a microtiter plate, and diluted patient serum is applied. (**2**) If rheumatoid factor (RF) is present in the serum, it binds to the Fc region of the rabbit IgG molecules. (**3**) After a washing step to remove unbound molecules, an anti-RF antibody conjugated to an enzyme (E) is applied for colorimetric detection. doi:10.1128/9781555818890.ch8.f8.5

that lubricates the articulating bones. Initially, an inflammatory response occurs in the synovial membranes in the small joints of the hands and feet; it then occurs in larger joints, causing accumulation of excess fluid and damage to the joints. Systemic inflammation may damage organs such as the heart and lungs. The disease affects about 1% of the population of the United States.

Rheumatoid factor is an autoantibody that targets the Fc region of immunoglobulin G (IgG) antibodies and contributes to rheumatoid arthritis. It is commonly used as a diagnostic biomarker to differentiate rheumatoid arthritis from other forms of arthritis and other inflammatory conditions. An indirect ELISA has been developed to detect the presence of rheumatoid factor in patient blood samples. In this test, IgG molecules, usually from a rabbit, are the standardized antigen bound to the surface of a multiwell plate, and diluted patient sera are applied to the wells (Fig. 8.5). After incubation to allow binding of rheumatoid factor in the serum to the rabbit IgG molecules and washing to remove unbound molecules, an anti-human IgG antibody is added. The anti-human IgG antibody specifically targets rheumatoid factor (and not rabbit IgG) and is conjugated to an enzyme such as horseradish peroxidase for colorimetric detection. Other ELISAs have been developed that target rheumatoid factor IgM and IgA that can assist in accurate diagnosis of rheumatoid arthritis. Early diagnosis is important to prevent irreversible joint damage, as rheumatoid arthritis may be managed in the early stages by administering antirheumatic drugs.

| **IgG** |
| immunoglobulin G |

Immunoassays for Infectious Disease

Clinical laboratories often identify pathogenic microbes in patient samples based on their physiological or biochemical characteristics. For example, a pathogenic bacterium may ferment specific carbohydrates or produce specific enzymes, and detection of the products of these reactions is the basis for some diagnostic assays. While these methods are effective, they require growth and isolation of the pathogen and therefore are slower than immunological and other molecular methods, typically requiring more than 48 h. Some human pathogens, such as the intracellular bacterium *Chlamydia trachomatis*, are fastidious and do not grow well in laboratory cultures. Viruses can be propagated only in host cells and cannot be identified using metabolic characteristics. Moreover, some clinical isolates exhibit atypical metabolic profiles that confound accurate identification. Immunological detection methods such as agglutination assays and ELISAs eliminate the need to grow the pathogen in culture and can be used to detect specific viral, bacterial, fungal, or protozoan pathogens in body fluids and tissues. Because antibody-based approaches detect a target antigen with high specificity and sensitivity, they are well suited to distinguish a specific pathogen from the other hundreds of microbes that are normally present in some human tissues. Immunological assays for infectious disease may target proteins produced by a pathogen or may detect the presence of antibodies produced against the pathogen. The latter, however, does not differentiate between current and past infections.

To prevent transmission of infectious disease, donated blood is screened for several pathogens, including human immunodeficiency virus, hepatitis virus, human T-lymphotropic virus, and *Treponema pallidum*, the bacterium that causes syphilis. Infection with hepatitis virus is relatively common in the United States (25 per 100,000 individuals). Although five hepatitis viruses can infect the liver and cause liver inflammation, cell death, and even liver failure, most cases of chronic viral hepatitis are caused by hepatitis B and C viruses. Both viruses may be transmitted via blood transfusion. ELISAs have been developed for routine blood screening for hepatitis B surface antigen, a viral envelope lipoprotein, and for antibodies produced against both hepatitis B and C viruses. Viral surface antigens are the first markers to be detected in serum after infection, while antibodies appear up to several months later. Detection of the viral surface antigen with a sandwich ELISA utilizes specific monoclonal antibodies to capture the hepatitis B surface antigens present in blood in wells of an assay plate (Fig. 8.6A). The monoclonal antibodies were generated using recombinant hepatitis B surface protein expressed in yeast. An anti-hepatitis B surface antigen detection antibody with a conjugated enzyme and the enzyme substrate are added to measure virus

Figure 8.6 Diagnosis of hepatitis B virus infection by ELISA. (**A**) A sandwich ELISA detects the presence of hepatitis B surface antigen in blood. (**B**) An indirect ELISA detects the presence of anti-hepatitis B virus antibodies in blood.
doi:10.1128/9781555818890.ch8.f8.6

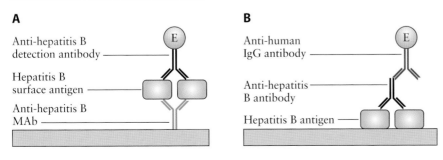

titers. An indirect ELISA that detects the presence of anti-hepatitis B virus antibodies in the blood sample is often used to confirm a positive result (Fig. 8.6B); however, blood from individuals immunized with a hepatitis B vaccine also produces a positive result with this test. Blood may additionally be screened for hepatitis B and C virus RNA. The current risk for hepatitis B virus infection through blood transfusion is between 1 in 200,000 and 1 in 500,000, and that for hepatitis C virus is 1 in 1,390,000.

Protein Arrays To Detect Polygenic Diseases

An ELISA typically measures a single target protein; however, for some diagnoses, it may be more informative to measure multiple target proteins in a single assay. Analysis of **proteomes** is commonly used in research to identify and quantify protein changes in diseased tissue versus normal tissue and is useful for diagnosis of **polygenic diseases** (resulting from mutations in more than one gene) such as breast cancer, Alzheimer disease, type 1 diabetes, and cardiovascular disease (see chapter 3). Protein microarrays are multiplex immunoassays that can detect multiple biomarkers in a clinical sample (see chapter 1). Biomarkers may be subsets of proteins in complex clinical samples such as biopsied tumor tissue. Currently, commercially available protein microarrays are most commonly used to detect antibodies in serum from patients suffering from allergies, autoimmune diseases, or infections.

Approximately 25% of the population in industrialized countries suffers from type I allergies, which are IgE-mediated immediate hypersensitivity reactions, such as asthma, hay fever, and eczema (see chapter 4). In susceptible individuals, the first exposure to an allergen elicits the production of high levels of IgE antibodies that bind to mast cells. In this way, the individual becomes sensitized to the allergen. On subsequent exposures, the allergen reacts with mast cell-bound IgE and causes bridging between adjacent IgE molecules, thereby stimulating the release of inflammatory mediators such as histamine, leukotrienes, prostaglandins, and tumor necrosis factor. These biochemicals stimulate smooth muscle contraction and dilation of capillaries that cause edema (swelling), itching, and development of a rash. Severe reactions can lead to systemic **anaphylaxis** that may be fatal if not treated quickly. Diagnostic allergen microarrays detect IgEs produced against common allergens, for example, pollen, food proteins, and molds. Purified allergens are arrayed in triplicate on a solid support and probed with sera from allergic patients (Fig. 8.7). Specific IgEs present in the serum bind to an immobilized allergen and are retained on the surface of the array. Bound IgEs are detected using an anti-human IgE antibody, often labeled with a fluorescent dye. The array format enables detection of hundreds of allergens in a single assay.

Another type of protein microarray measures levels of fifty different protein biomarkers of heart disease in a patient serum sample. Increased or decreased levels of these proteins have been associated with increased risk of heart disease. For example, adiponectin is a protein hormone secreted by adipocytes (fat cells) that regulates glucose and lipid metabolism. Reduced levels of this hormone are found in individuals with increased

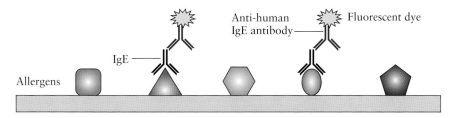

Figure 8.7 Diagnostic allergen microarray. Purified allergens are arrayed on a solid support and probed with serum from an allergic patient. Specific IgE present in the serum binds to an immobilized allergen and is detected using an anti-human IgE antibody conjugated to a fluorescent dye. doi:10.1128/9781555818890.ch8.f8.7

risk of heart attack. Elevated levels of the protein plasminogen activation inhibitor 1 have been found in patients with myocardial infarction. This protein inhibits tissue plasminogen activator, a key enzyme in fibrinolysis, synthesized by endothelial cells, platelets, and hepatocytes. A liquid bead array has been developed to quantify levels of these and other protein biomarkers in the serum samples. In this system, all of the different protein biomarkers are detected in a single well of a multiwell plate (Fig. 8.8), which enables rapid analysis of many patient samples simultaneously. Primary antibodies directed against each protein biomarker are attached to

Figure 8.8 Liquid bead assay to detect protein biomarkers associated with increased risk of cardiovascular disease. (**A**) Patient serum is applied to a well of a microtiter plate containing capture beads bound to primary antibodies directed against all of the protein biomarkers. Capture beads of different colors distinguish primary antibodies that bind to different protein biomarkers. Target proteins in the sample bind to their cognate antibodies. (**B**) For detection of the bound proteins, secondary antibodies conjugated to a fluorescent molecule are added to the well. Each bead, and any captured proteins, is analyzed by passing by a laser that measures fluorescence. The presence of specific protein biomarkers is determined by the color of the capture bead to which they are bound. doi:10.1128/9781555818890.ch8.f8.8

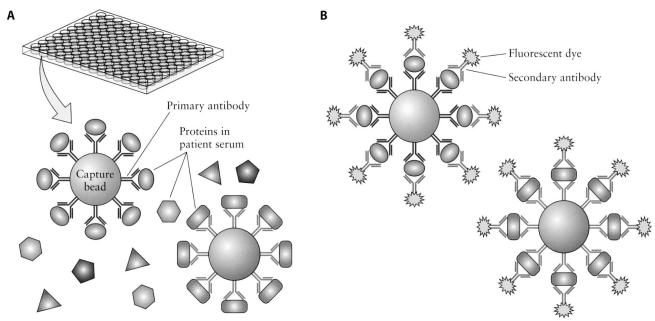

tiny capture beads that are distinguished by either size, color, or shape. A serum sample is applied to a well, and specific proteins in the sample bind to their cognate antibody (Fig. 8.8A). For detection of the bound proteins, secondary antibodies conjugated to a fluorescent molecule are added to the well (Fig. 8.8B). Each bead, and any captured proteins, is analyzed by passing by a laser that measures fluorescence emitted from bound secondary antibodies and distinguishes the different capture beads by their size, color, or shape.

Reverse-phase protein microarrays can also detect specific proteins in clinical samples; however, for this type of array, the samples, such as cell lysates or tissue slices, are immobilized in a single spot on a support (Fig. 8.9). Several samples, for example, from different patients, are spotted on the microarray, which is then probed with a single antibody to detect a specific target protein. This format contrasts with some other types of protein microarrays in which each immobilized spot contains a single, purified protein; hence, the term "reverse" is used. The advantage of the reverse-phase protein microarray is that a large number of clinical samples can be processed at one time.

One of the many possible medical diagnostic applications for reverse-phase protein microarrays is to monitor responses to treatment in clinical trials. The response to a cancer therapy can vary widely among patients, and therefore, the ability to predetermine the most effective treatment not only could increase the probability of a favorable outcome but also would reduce the high cost of the trial-and-error approach. Although tumor diagnosis (benign, noninvasive, malignant primary, or metastatic) is still determined on the basis of morphological characteristics

Figure 8.9 Reverse-phase protein microarray. Multiprotein samples, such as cell lysates, are spotted on a solid support (**1**) and incubated with a primary antibody (**2**). The primary antibody may be labeled with a biotin (B) molecule that can be detected with a streptavidin–fluorescent dye conjugate (**3**). doi:10.1128/9781555818890.ch8.f8.9

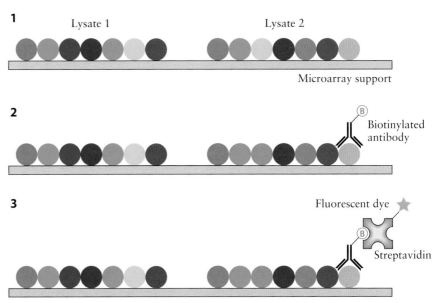

(e.g., shape and size of a tumor, tumor cells, and cell nuclei), tumors that look similar under a microscope can have quite different outcomes. A better predictor of tumor progression or regression is the molecular activities of tumor cells that determine tumor behavior, especially in response to treatment. The most accurate assessments of tumor cell activity are based on the types and quantities of proteins present in the cell.

Most new molecular cancer therapies are targeted to inhibit specific proteins in signaling pathways that control cell proliferation. Many of these targets are protein kinases and phosphatases that control the phosphorylation of proteins. Reverse-phase protein microarrays provide a rapid, inexpensive method to assess the phosphorylation state of these signaling pathways in response to treatment. Cells obtained by tumor biopsy and **laser capture microdissection** are lysed and printed onto multiple arrays. Each array is probed with a single antibody, for example, an antiphosphoprotein antibody. By analyzing multiple arrays each with a different antibody that is specific for a protein in a signaling pathway, the prognosis can be predicted and treatment individualized.

Colorectal cancer is the third leading cause of cancer-related deaths in the United States. The most common site of metastases is the liver, which significantly reduces patient survival. In a recent clinical trial, imatinib (a tyrosine kinase inhibitor) alone or in combination with panitumumab (a monoclonal antibody that blocks receptors for epidermal growth factor) was tested as a treatment for late-stage colorectal cancer that had spread to the liver. One of the goals of the study was to determine whether the response to treatment could be predicted by tumor proteome profiles. Patients with metastatic colorectal cancer were scored based on the phosphoproteome profiles of liver tumors analyzed using reverse-phase protein microarrays. Many of the tumor cells had high levels of phosphorylated tyrosine kinases c-KIT, c-Abl, and platelet-derived growth factor receptor, which suggests that these patients may respond well to imatinib treatment.

Immunoassays for Protein Conformation-Specific Disorders

Several human neurological disorders arise as a consequence of protein misfolding that leads to protein aggregation and cell death. Parkinson disease, Alzheimer disease, and prion diseases are examples of protein conformational disorders. Alzheimer disease is a degenerative brain disorder that is characterized by the progressive loss of abstract thinking and memory, personality change, language disturbances, and a slowing of physical capabilities. Clinical diagnosis of Alzheimer disease is poor, although 1% of the population between 60 and 65 years old and 30% of the population over 80 years old may develop it. Two hallmarks of Alzheimer disease found in the brain are (i) neurofibrillary tangles of the cytoskeletal protein tau that accumulate within nerve cell bodies and (ii) dense extracellular aggregates of insoluble proteins called amyloid plaques that develop at the ends of inflamed nerves (Fig. 8.10).

The principal protein of an amyloid plaque is a small protein called Aβ (β-amyloid protein). The Aβ protein ranges in length from 39 to 42

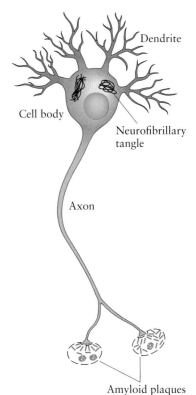

Figure 8.10 Neurofibrillary tangles and amyloid plaques associated with neurons in the brain are hallmarks of Alzheimer disease. doi:10.1128/9781555818890.ch8.f8.10

Dendrite

Cell body

Neurofibrillary tangle

Axon

Amyloid plaques

APP
β-amyloid precursor protein

amino acid residues; the Aβ40 and Aβ42 forms are the main variants. All Aβ proteins are derived from the β-amyloid precursor protein (APP) by proteolytic cleavage. Abnormal cleavage of APP results in production of Aβ40 and Aβ42 and alters protein folding, causing exposed regions of the protein to self-interact; hence, the proteins aggregate in amyloid plaques.

Diagnosis of Alzheimer disease using immunological methods may exploit the development of antibodies that differentiate between the conformations of the disease-associated Aβ proteins and normal APP. To develop conformation-specific antibodies, researchers used an approach that mimics protein aggregation during the disease process, that is, the self-interaction of the Aβ proteins. They grafted short motifs (10-mers) from the Aβ42 protein into a complementarity-determining region (CDR3, V$_H$ domain; see chapter 2) of an antibody to generate antibodies that would interact with aggregated Aβ42 via the grafted motif (Fig. 8.11A).

Figure 8.11 Development of conformation-specific antibodies to detect aggregated Aβ proteins indicative of Alzheimer disease. (**A**) Short (10-amino-acid) overlapping Aβ42 peptide segments were grafted into a complementarity-determining region (CDR3, V$_H$ domain) to generate antibodies that bind specifically to aggregated, insoluble Aβ42 proteins. (**B**) Out of 12 grafted antibodies, three (Aβ12-21, Aβ15-24, and Aβ18-27) bound to aggregated Aβ42 proteins, and not Aβ42 monomers, that were deposited on a membrane. A sequence of 4 amino acids (highlighted in red) was found to be the minimal requirement for binding. Adapted from Perchiacca et al., *Proc. Natl. Acad. Sci. USA* **109**:84–89, 2012, with permission. doi:10.1128/9781555818890.ch8.f8.11

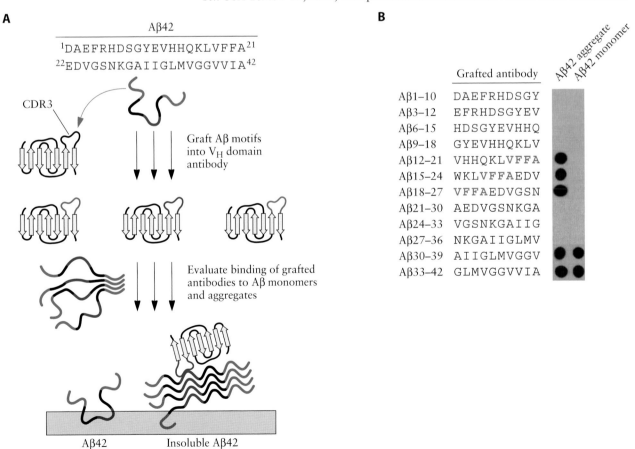

Out of twelve antibodies, carrying different but overlapping Aβ42 sequences, three bound specifically to insoluble Aβ42 aggregates but not Aβ42 monomers (Fig. 8.11B). A motif of four amino acids, common to all three positive antibodies, was found to be the minimal sequence required for binding. To determine whether binding of the grafted antibodies is mediated by interactions between the Aβ motif on the antibody and the same motif in the aggregated Aβ proteins, each grafted antibody was bound to immobilized Aβ42 aggregates and then a second antibody was applied. Binding of the second antibody was reduced only when it recognized a motif that overlapped with that grafted on the first antibody. This relatively simple approach may be used to generate conformation-specific antibodies against other proteins that not only aid in accurate disease diagnosis but also may be used in therapeutic strategies to target these proteins.

DNA-Based Approaches to Disease Diagnosis

DNA-based diagnostic tests determine the existence of specific nucleotide sequences, including human genetic mutations and sequences present in human pathogens. They are highly sensitive and specific and can detect single nucleotide mutations or copy number variations. The ability to diagnose diseases in humans at the genetic level makes it possible to determine the cause of an illness and to predict whether individuals or their offspring are predisposed to the disease. Because a DNA-based test does not require expression of a gene, in contrast to diagnostic detection of proteins, DNA analysis can be used for the identification of asymptomatic carriers of hereditary disorders, for prenatal diagnosis of serious genetic conditions, and for early diagnosis before the onset of symptoms. DNA sequence-specific diagnostic approaches include hybridization of a unique DNA probe to a complementary target sequence, target sequence amplification by PCR, microarray analysis to detect multiple sequences in a single sample, and mass spectrometry to identify single-nucleotide polymorphisms (SNPs).

SNP
single-nucleotide polymorphism

Hybridization Probes

Hybridization is the formation of hydrogen bonds between two complementary strands of nucleic acids. A diagnostic test involving DNA hybridization utilizes a DNA probe to detect a complementary target DNA sequence that is characteristic of the disease. The probe is labeled with a reporter molecule that indicates hybridization between the target and probe DNA. Typically, the cells in a clinical specimen such as infected or biopsied tissue are lysed, and the genomic DNA in the lysed cells is denatured by treatment with a strong alkali to generate single-stranded target DNA. Then, a labeled oligonucleotide probe is added under appropriate conditions of temperature and ionic strength to promote base-pairing between the probe and the target DNA. Unbound probe DNA is removed by washing, and hybridization is detected by measuring the activity of the reporter molecule attached to the probe. A hybridization probe used in a diagnostic assay must hybridize exclusively to the selected target nucleic acid sequence. False positives (i.e., detection of a signal in the absence of the target sequence) and false negatives (i.e., no detection of a signal when the target is present) severely reduce the utility of a diagnostic test and can have serious consequences for the patient. Oligonucleotide probes are usually less than 100 nucleotides in length, although they may be longer, and are labeled with an enzyme that produces a color change when it acts on certain substrates (see the discussion of the ELISA procedure above) or with a fluorescent dye.

Hybridization probes are often employed to detect the presence of microbial pathogens. Malaria, caused by the parasite *Plasmodium falciparum*, is one of the most common infectious diseases and is especially fatal in young children. The parasite infects and destroys red blood cells, leading to fever and, in severe cases, damage to the brain, kidneys, and other organs. Sensitive, simple, and inexpensive methods are required to identify the source(s) of the parasite in various localities, to assess the progress of eradication programs, and to facilitate early treatment. Currently, malarial infections are diagnosed by either microscopic examination of blood smears or immunological detection of parasite antigens, effective but labor-intensive and time-consuming processes, especially given the large numbers of samples that need to be examined. Although immunological procedures for *Plasmodium* detection, such as ELISAs, are rapid and amenable to automation, they do not always discriminate between current and past infections, because they are designed to detect anti-*Plasmodium* antibodies in the blood of affected individuals.

A DNA diagnostic test for active infections that measures the presence of the pathogen was developed by using highly repetitive DNA sequences (present in many copies) from *P. falciparum*. First, a genomic library of the parasite DNA was screened with labeled whole-genome parasite DNA. The most intensely labeled hybridizing colonies were selected because they were expected to contain repetitive DNA. The DNA from each of the selected colonies was then tested for its ability to hybridize with DNA from several other *Plasmodium* species that do not cause malaria. The DNA sequence that was chosen as a specific probe hybridized with *P. falciparum* but not with *Plasmodium vivax*, *Plasmodium cynomolgi*, or human DNA, despite the fact that *P. vivax* causes a less severe form of malaria. This probe can detect as little as 10 picograms (pg) of purified *P. falciparum* DNA or 1 nanogram (ng) of *P. falciparum* DNA in blood.

More than 100 different DNA diagnostic probes have been developed for the detection of various pathogenic strains of bacteria, viruses, and parasites. For example, probes have been developed for the diagnosis of human bacterial infections caused by *Legionella pneumophila* (causative agent of pneumonia), *Salmonella enterica* serovar Typhi (food poisoning), enterotoxigenic *Escherichia coli* (gastroenteritis), and *Neisseria gonorrhoeae* (gonorrhea, a sexually transmitted infection). In principle, nearly all pathogenic organisms can be detected by this procedure.

Allele-Specific Hybridization

In addition to infectious diseases, hybridization probes are widely used to detect specific disease-associated alleles. Monogenic diseases are caused by mutations in a single gene; however, any one of several alterations to the normal nucleotide sequence of a gene may be responsible. This is exemplified by cystic fibrosis, a common lethal autosomal recessive disorder that affects approximately 1 in every 2,500 live births (see chapter 3). Mutations in a single gene, the cystic fibrosis transmembrane conductance regulator (CFTR) gene, result in defects in chloride ion transport. As a consequence, the mucus of lung and other mucosal tissues is thick and viscous, obstructing the respiratory, digestive, and reproduction system functions. Every state in the United States now routinely screens newborns for cystic fibrosis. An ELISA is used to detect higher-than-normal levels of immunoreactive trypsinogen in blood. This protein is produced by the pancreas and is linked to cystic fibrosis. Premature babies and babies from stressful births also may have elevated levels of immunoreactive trypsinogen, and in these cases, genetic tests are employed for confirmation.

About 1,900 different mutations are reported to occur in the CFTR genes of patients with cystic fibrosis. Screening individuals who may be at risk for cystic fibrosis for such a large number of possible mutations is a daunting task. However, some of the mutations that cause cystic fibrosis are much more common than others. The most common mutation is an in-frame deletion of three nucleotides in exon 10 of the CFTR gene that leads to loss of the amino acid phenylalanine at codon 508 (ΔF508). Over 90% of cystic fibrosis patients carry at least one ΔF508 allele, and nearly 50% of cystic fibrosis patients are individuals who are homozygous for ΔF508. It is estimated that about 160 different mutations account for 96 to 97% of cystic fibrosis alleles.

Allele-specific hybridization is commonly used to screen for cystic fibrosis. With this technique, an individual's CFTR gene is amplified by PCR and then hybridized to labeled oligonucleotide probes for the mutant (e.g., ΔF508) and wild-type genes, separately (Fig. 8.12A). In this way, it is possible to distinguish between healthy individuals with two wild-type alleles, cystic fibrosis carriers with one mutant and one wild-type allele, and cystic fibrosis-affected individuals with two mutant alleles (Fig. 8.12B). Diagnostic kits are commercially available that test patient blood samples for the presence of a panel of common CFTR mutations as recommended by the American College of Medical Genetics.

CFTR
cystic fibrosis transmembrane conductance regulator

Oligonucleotide Ligation Assay

The oligonucleotide ligation assay (OLA) is also commonly used to detect SNPs known to be associated with human diseases with a high degree of accuracy. In this diagnostic assay, two short oligonucleotide probes (~50 nucleotides) are designed to anneal to adjacent sequences within a gene that encompass the polymorphic nucleotide (Fig. 8.13). Importantly, one of the probes (allele-specific probe) has as its last base at the 3' end the nucleotide that is complementary to the polymorphic nucleotide. The second probe (common probe) is complementary to the sequence

OLA
oligonucleotide ligation assay

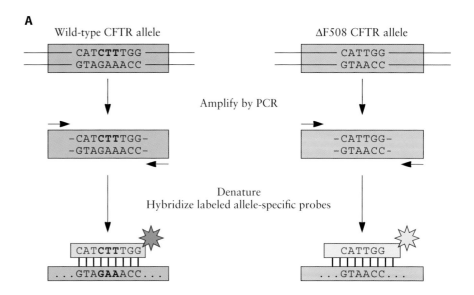

A

Wild-type CFTR allele

ΔF508 CFTR allele

Amplify by PCR

Denature
Hybridize labeled allele-specific probes

B

Source of blood sample	Hybridization test results	
Normal individual	★	★
Cystic fibrosis carrier	★	☆
Cystic fibrosis-affected individual	☆	☆

Figure 8.12 Allele-specific hybridization to screen for cystic fibrosis. (**A**) The CFTR gene is amplified by PCR, and then the PCR products are denatured and incubated with oligonucleotide probes that specifically hybridize to wild-type or mutant (usually ΔF508) alleles (note that only part of the sequence of the CFTR gene and probe are shown). After a washing step to remove unbound probe, hybridization is detected by measuring fluorescence emitted by the fluorophore attached to the probe. (**B**) Healthy individuals are homozygous for the wild-type allele, while those affected by the disease carry two mutant alleles. Cystic fibrosis carriers are heterozygous for the CFTR alleles. doi:10.1128/9781555818890.ch8.f8.12

immediately downstream of the polymorphic nucleotide. When these two probes are hybridized with the target gene (which has been amplified by PCR), base-pairing occurs between the allele-specific probe and the target sequence, including the 3′ nucleotide if the complementary polymorphic nucleotide is present, and the common probe binds immediately downstream. DNA ligase, added to the reaction, covalently joins the two probes. However, if the nucleotide at the 3′ end of the allele-specific probe is mismatched, it will not base-pair with the polymorphic nucleotide in the target DNA sequence, although the common probe will be perfectly aligned. As a consequence of the single-nucleotide misalignment, DNA ligase cannot join the two probes. In short, OLA is designed to distinguish between two possibilities: ligation if the probes are perfectly matched and no ligation if the allele-specific probe carries a mismatched nucleotide. Other oligonucleotide probes may be designed to detect different SNPs in the same gene.

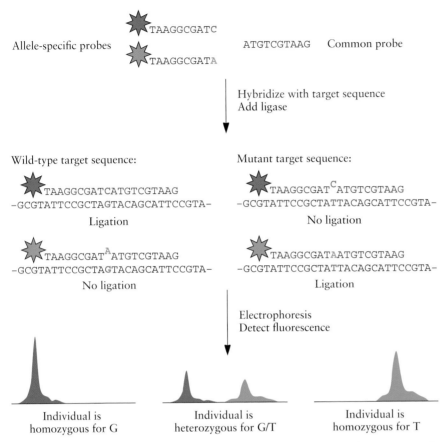

Figure 8.13 OLA. Ligation of the allele-specific and common probes occurs only when the allele-specific probe is perfectly complementary to the target sequence. Longer ligated probes may be separated from shorter unligated probes by electrophoresis, and then probe fluorescence can be detected to distinguish homozygous wild-type, homozygous mutant, and heterozygous individuals.
doi:10.1128/9781555818890.ch8.f8.13

To determine whether ligation has occurred, one of the two probes is labeled with a reporter molecule. For example, the allele-specific probe may be labeled at the 5′ end with a fluorescent dye that can be detected by laser excitation and fluorescence emission. Probes may be labeled with different fluorescent molecules corresponding to different SNPs. If an individual is homozygous for a particular allele, either two copies of a normal gene or two copies of a disease-associated SNP, a positive signal will be detected only from one of the fluorescent molecules. Heterozygous individuals will yield positive signals from two different probes. Overall, the OLA system is rapid, sensitive, highly specific, and amenable to automation.

Padlock Probes

Diagnostic assays for SNPs that use padlock probes are very similar to those that use OLA probes, except that the former utilizes only one probe rather than two as used in the OLA procedure. A padlock probe is an

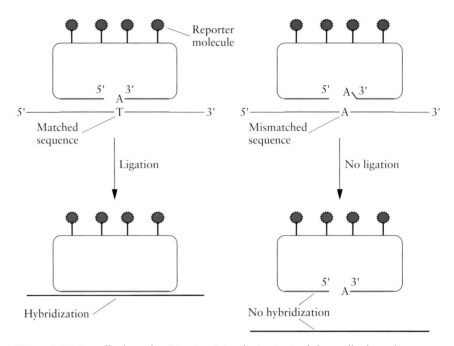

Figure 8.14 A padlock probe. Ligation (circularization) of the padlock probe occurs only when the 5′ and 3′ ends of the probe are perfectly complementary to the target sequence. When there is a single-base mismatch at the 3′ end of the probe, ligation does not occur and the probe assumes a conformation that does not allow hybridization. Under stringent conditions, the ligated probe remains bound to the target DNA, while a nonligated probe is removed in a washing step. Hybridized probe can be detected by the activity of the reporter molecule, for example, by fluorescence emitted from a fluorescent dye. doi:10.1128/9781555818890.ch8.f8.14

oligonucleotide that is complementary to a target sequence at its 5′ and 3′ ends but not in its middle region (Fig. 8.14). When a padlock probe hybridizes to its target sequence, the 5′ and 3′ ends of the probe come into close proximity to each other and the middle portion loops out. If the ends of the probe are exactly complementary to the target sequence, after hybridization they can be joined together by DNA ligase. If there is a mismatch between the target and probe, ligation does not occur. The requirement for both ends of the probe to bind perfectly to the target sequence for ligation to occur ensures a high specificity of detection and therefore the ability to easily detect SNPs. Following the ligation reaction, the probe–target hybrid can be detected by the activity of reporter molecules attached to the middle (linker) portion of the probe. Padlock probes typically have sequences approximately 15 to 20 nucleotides in length at each end that are complementary to the target sequence and a middle region of approximately 50 nucleotides.

Allele-Specific PCR

Many nucleic acid diagnostic tests are based on PCR (see chapter 1). Advantages of PCR-based tests include (i) specificity that enables detection of a particular nucleotide sequence in complex samples, (ii) sensitivity that enables detection of low-abundance targets, (iii) an amplification step that generates substantial amounts of a target sequence for additional analyses such as hybridization or sequencing, (iv) rapid analysis (usually completed in 1 to 2 h or less), (v) multiplexing that enables identification of multiple targets in a single sample, and (vi) low cost. Conventional PCR has been used extensively in diagnostic laboratories over the last 15 years; however, more recently, it has become possible not only to detect but also to quantify the pathogen in a clinical sample using quantitative real-time PCR.

Allele-specific PCR (also referred to as PCR amplification refractory mutation system) screens for known SNPs. Different forward PCR primers are used to distinguish among alleles that differ by a single nucleotide. One primer is exactly complementary to the normal DNA sequence, and another primer anneals to a variant sequence containing the disease-associated SNP (Fig. 8.15). The primers are usually designed to place the polymorphic nucleotide at the 3′-terminal end of the primer because most polymerases used for PCR do not extend 3′ mismatched primers efficiently. A third reverse primer is complementary to the opposite strand and is common to all reactions. Each reaction contains only one of the allele-specific forward primers and the common reverse primer, together with the patient DNA sample, a thermostable DNA polymerase, and all four deoxyribonucleotides. PCR amplification occurs only when a forward primer is present that is exactly complementary to the target sequence in the patient sample; mismatches between primer and template DNA prevent primer annealing and therefore primer extension during DNA synthesis. An advantage of this method, compared to allele-specific hybridization described above, is that the amplification and diagnostic steps are combined.

One variation of allele-specific PCR is known as competitive oligopriming, in which two different SNP-specific forward primers are included in a single reaction. To discriminate between the PCR products, each forward primer is labeled with a different fluorescent dye. For example, one forward primer that is exactly complementary to a normal allele

Figure 8.15 Allele-specific PCR to detect SNPs. Amplification occurs only when the forward and reverse PCR primers perfectly match the target sequence. Different allele-specific primers that carry an SNP-specific nucleotide at the 3′ end are employed to distinguish among alleles that differ by a single nucleotide. doi:10.1128/9781555818890.ch8.f8.15

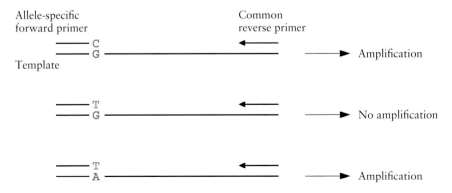

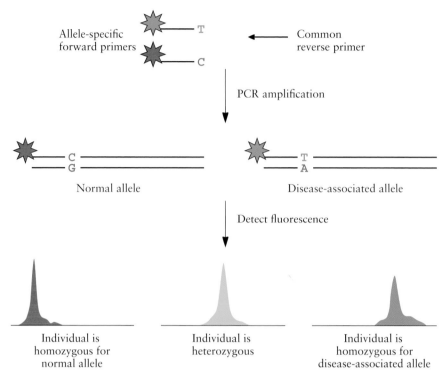

Figure 8.16 Competitive oligopriming to detect SNPs. Two allele-specific forward primers that differ in the nucleotide at the 3′ end are included in a single reaction together with a reverse primer that anneals to a sequence that is common among variant alleles. PCR amplification occurs only when a forward primer is perfectly complementary to the target DNA. To discriminate between PCR products corresponding to normal and disease-associated alleles, each forward primer is labeled with a different fluorescent dye. In the example shown here, PCR products from individuals who are homozygous for the normal allele fluoresce red, those from individuals who are homozygous for the disease-associated allele fluoresce green, and those from individuals who are heterozygous fluoresce yellow. doi:10.1128/9781555818890.ch8.f8.16

may be labeled at its 5′ end with rhodamine (which fluoresces red), while another forward primer that is complementary to the disease-associated allele may be labeled at its 5′ end with fluorescein (which fluoresces green) (Fig. 8.16). In both cases, amplification requires a third, unlabeled primer that is complementary to the opposite strand. PCR amplification occurs only when a forward primer is exactly complementary to the target DNA; therefore, the presence of these two forward primers in the same reaction mixture will result in the amplification of either the normal or disease-associated DNA sequence or both, depending on which alleles are present. If an individual is homozygous for the normal allele, after PCR and removal of unincorporated primers, the reaction mixture will fluoresce red; if he or she is homozygous for the disease-associated allele, the reaction mixture will fluoresce green; and if he or she is heterozygous, the reaction mixture will fluoresce yellow. This assay can be automated and adapted for any single-nucleotide target site of any gene that has been sequenced.

TaqMan PCR

The TaqMan PCR protocol is used to screen individuals for the presence of SNPs that are indicative of any of a variety of genetic diseases. Made popular by one particular company, it is based on the 5′ nuclease activity of *Taq* polymerase, which is commonly used to amplify DNA in PCR applications. To simultaneously detect normal and disease-associated alleles, two TaqMan probes are utilized. Each probe is exactly complementary to either the normal or the disease-associated DNA sequence, and each probe has a different fluorescent dye attached to its 5′ end (Fig. 8.17). Intact probes, whether unbound or bound to template DNA, do not fluoresce because of the presence of a quencher molecule at the 3′ end of the probe. PCR primers anneal to sequences that flank the probe hybridization site, and as PCR amplification proceeds, the TaqMan probe is displaced by the growing DNA strand. The 5′ nuclease activity of the *Taq* polymerase degrades the 5′ end of the TaqMan probe, thereby releasing the fluorescent dye and removing it from the proximity of the quencher molecule. Thus, only TaqMan probes that were previously bound to target DNA are degraded and subsequently fluoresce. Any mismatched probes

Figure 8.17 TaqMan assay. (**A**) A TaqMan probe is an oligonucleotide that hybridizes to a target sequence and contains a fluorescent dye covalently attached to its 5′ end and a quencher attached to its 3′ end. The quencher suppresses fluorescence of the fluorophore on the intact probe. (**B**) During the PCR, *Taq* polymerase extends the primer and the TaqMan probe is displaced by the growing DNA strand. (**C**) The 5′-to-3′ exonuclease activity of the *Taq* polymerase cleaves the 5′ end of the probe and releases the fluorophore. Separation from the quencher results in fluorescence of the fluorophore. The amount of fluorescence is proportional to the amount of target DNA present. doi:10.1128/9781555818890.ch8.f8.17

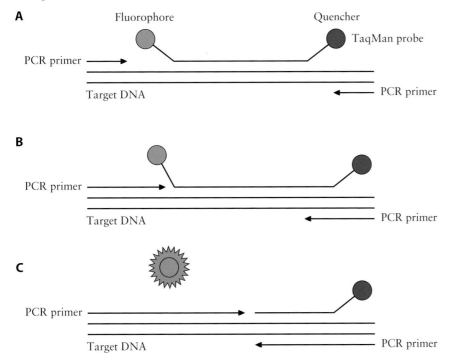

that are not complementary to the SNP in the region where the TaqMan probe binds will be displaced but not cleaved, so they will not fluoresce. By monitoring the fluorescence at two different wavelengths (one for each TaqMan probe), it is possible to distinguish individuals that are homozygous for each SNP (either normal or disease associated) and heterozygous (carrying one normal and one disease-associated allele). This technique may be used to detect more than two different SNPs at the same time as long as the fluorescent dyes attached to each SNP-specific probe have well-separated, nonoverlapping fluorescence maxima.

A TaqMan diagnostic assay has been developed to screen individuals for specific mutations in the *BRAF* gene that are associated with melanoma. Melanoma is a type of skin cancer that is caused by changes in melanocytes that produce the skin pigment melanin. More than 75,000 cases are diagnosed each year in the United States. More than half of these are associated with mutations in the *BRAF* gene, which encodes B-Raf, a serine/threonine protein kinase that regulates a signal transduction pathway (BRAF–MEK–ERK pathway) that controls cell division, differentiation, and apoptosis in response to growth factors (Fig. 8.18A). In 90% of the *BRAF* mutations associated with cutaneous melanomas, glutamate, a negatively charged amino acid, has been replaced with the hydrophobic amino acid valine in codon 600 (V600E; GTG→GAG). This change alters the conformation of the activation segment of the protein, leading

Figure 8.18 TaqMan assay to detect *BRAF* alleles associated with melanoma. (**A**) Mutations in *BRAF* may lead to constitutive kinase activity of B-Raf and consequently constitutive activation of the MEK–ERK signaling pathway, which controls cell proliferation. (**B**) A TaqMan PCR assay to detect the V600E *BRAF* allele commonly associated with cutaneous melanoma uses an allele-specific PCR primer and a reverse PCR primer that binds to a sequence present in both wild-type and mutant *BRAF*. An allele-specific TaqMan probe binds to the sequence that spans the mutation but in the DNA strand opposite to that bound to the forward primer. As PCR proceeds, the probe is displaced and cleaved by *Taq* polymerase as it extends the strand from the reverse primer. An allele-specific PCR primer and a TaqMan probe labeled with a different fluorescent dye bind specifically to the normal *BRAF* sequence. The presence of mutant and/or normal *BRAF* is determined by monitoring fluorescence of the cleaved probe(s). doi:10.1128/9781555818890.ch8.f8.18

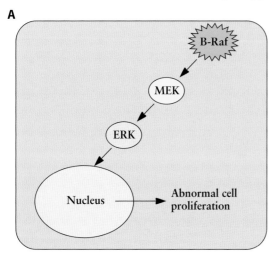

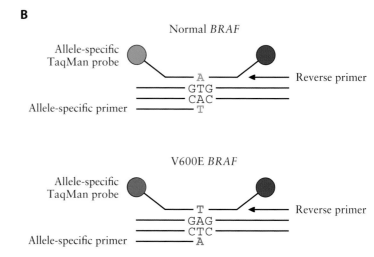

to constitutive kinase activity of B-Raf and uncontrolled proliferation of melanocytes. A TaqMan PCR assay to detect the V600E *BRAF* mutation uses a forward primer that specifically binds to the mutant *BRAF* sequence and a reverse primer that binds to a sequence present in both wild-type and mutant *BRAF* (Fig. 8.18B). A fluorescently labeled TaqMan probe also anneals to the sequence that spans the mutation but in the DNA strand opposite to that bound to the forward primer. A second probe, labeled with a different fluorescent dye, binds specifically to the wild-type sequence. As PCR proceeds, the probe is displaced and cleaved by *Taq* polymerase as it extends the strand from the reverse primer. The presence of mutant and/or wild-type *BRAF* is determined by monitoring fluorescence of the released probe(s).

The mutation can be detected both in tumor tissue samples and in DNA from blood plasma. The screen assists in identifying patients that have a high probability of responding well to treatment with vemurafenib, a B-Raf kinase inhibitor. Studies have shown that tumors with the V600E mutation are more sensitive to vemurafenib and that treatment with vemurafenib reduced the risk of death by 56% and decreased tumor size in 52% of melanoma patients with this allele compared to those who received chemotherapy. In 2011, the FDA approved the use of vemurafenib, and the test for the V600E mutation, for diagnosis and treatment of melanoma associated with this allele.

Real-Time PCR To Detect Infectious Disease

Most nucleic acid diagnostic tests for etiological agents of infectious disease are based on PCR. As for other molecular approaches, PCR circumvents the requirement to grow the pathogens, which is time-consuming for viruses and slow-growing or fastidious bacteria. Not only does this technique enable identification of a viral, bacterial, or fungal pathogen, but also it can reveal specific characteristics of that pathogen, for example, the presence of antibiotic resistance genes that can be used to determine the best course of treatment. A successful PCR assay to detect microbes in human-derived samples must be able to distinguish among the vast number of commensal species that exist in the human intestinal, genitourinary, and respiratory tracts and on skin.

In a real-time PCR, the double-stranded DNA product is bound by a fluorescent dye and the fluorescence is measured after each amplification cycle to quantify the amount of product in real time. Post-PCR processing is not required, and therefore, time to pathogen identification is shorter than for conventional PCR. Real-time PCR may be described as occurring in four phases (Fig. 8.19A). In the first, or linear, phase (generally about 10 to 15 cycles), fluorescence emission at each cycle has not yet risen above the background level. In the early exponential phase, a sufficient amount of double-stranded DNA has been produced to increase the amount of fluorescence above a threshold level that is significantly higher than the background. The cycle at which this occurs is known as the threshold cycle (C_T). The C_T value is inversely correlated with the amount of target DNA in the original sample. During the exponential phase, the amount of

C_T
threshold cycle

fluorescence doubles as the DNA products of the reaction double in each cycle under ideal conditions. In the plateau phase, the reaction components become limited and measurements of the fluorescence intensity are no longer useful.

To quantify the amount of target DNA in a sample, a standard curve is first generated by serially diluting a sample with a known number of copies of the target DNA, and assuming that all samples are amplified with equal efficiencies, the C_T values for each dilution are plotted against the starting amount of sample (Fig. 8.19B). The number of copies of a target DNA in a clinical sample can be determined by obtaining the C_T value for the sample and extrapolating the starting amount from the standard curve. In addition, since during the exponential phase the amount of DNA doubles with each cycle, a sample that has four times the number of

Figure 8.19 (A) Plot of normalized fluorescence (ΔRn) versus cycle number in a real-time PCR experiment. Four phases of PCR are shown: (1) a linear phase, where fluorescence emission is not yet above background level; (2) an early exponential phase, where the fluorescence intensity becomes significantly higher than the background (the cycle at which this occurs is generally known as C_T); (3) an exponential phase, where the amount of product doubles in each cycle; and (4) a plateau phase, where reaction components are limited and amplification slows down. (B) Plot of C_T versus the starting amount of a target nucleotide sequence. Fluorescence detection is linear over several orders of magnitude.
doi:10.1128/9781555818890.ch8.f8.19

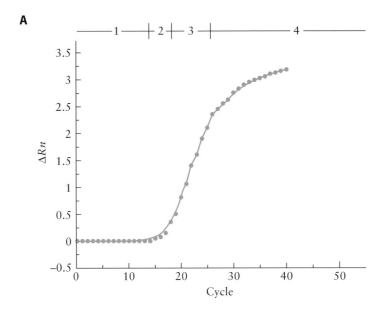

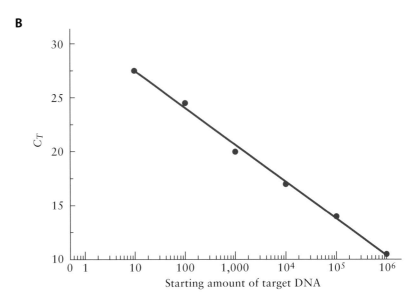

starting copies of the target sequence compared to another sample would require two fewer cycles of amplification to generate the same number of product strands. Often, a **melt curve** is generated to assess the specificity of the products.

One application for this technology in infectious disease diagnosis is the detection of methicillin-resistant *Staphylococcus aureus* (MRSA) in nasal swabs. MRSA is an increasing problem worldwide. Infections with these multidrug-resistant bacteria are usually associated with hospitals and other health care facilities, although community-associated infections are increasingly common. In the United States, health care-associated MRSA infections account for 90,000 deaths and $4.5 billion in health care costs annually; thus, there is strong motivation to develop rapid, accurate screening tests to prevent transmission of these bacterial pathogens among hospitalized patients.

As for PCR in general, the specificity of a real-time PCR assay is determined by the sequence of the oligonucleotides that are used to prime amplification of a target sequence. Thus, the primers must be designed carefully and validated to ensure that they do not anneal to DNA from other nontarget microorganisms. In one dual-target detection assay to detect MRSA, one set of PCR primers is used to amplify the *mecA* sequence (Fig. 8.20). The *mecA* gene encodes a cell wall penicillin-binding protein (PBP2a) that protects *S. aureus* from the bacteriocidal effects of methicillin and structurally related β-lactam antibiotics. The gene is encoded within a genomic region (pathogenicity island) known as the staphylococcal cassette chromosome *mec* (SCC*mec*), which was likely acquired by some *S. aureus* strains by horizontal gene transfer (see chapter 5). SCC*mec* is inserted in the *S. aureus orfX* gene, and the SCC*mec*–*orfX* junction is a target for some MRSA diagnostic tests that employ real-time PCR. A second set of primers is designed such that the forward primer anneals to the right extremity of SCC*mec* and the reverse primer anneals to a sequence within *orfX*. Seven different forward primers are employed; each binds specifically to a variant sequence of the right extremity SCC*mec*. This enables discrimination of the seven possible subtypes of MRSA that differ in the sequence of the SCC*mec* right extremity. Each primer set is labeled with a different fluorescent dye, and fluorescent signals for both target sequences indicate a positive test for MRSA. While detection of the specific

MRSA
methicillin-resistant *Staphylococcus aureus*

Figure 8.20 Dual-target real-time PCR assay to detect MRSA. One set of PCR primers (P1 and P2) primes amplification of the *mecA* sequence in SCC*mec*. A second primer pair is designed such that the forward primer (P3) anneals to the right extremity of SCC*mec* and the reverse primer (P4) anneals to a sequence within *orfX*. This confirms the presence of *S. aureus* and enables determination of the specific MRSA strain based on the variant SCC*mec*. doi:10.1128/9781555818890.ch8.f8.20

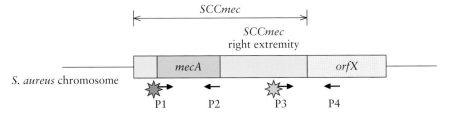

SCC*mec*–*orf*X junction is important to confirm that *S. aureus* is present and for monitoring transmission of specific MRSA strains, detection of the *mecA* gene is also required because this gene has been lost from the genome of some strains ("empty cassette variants").

When a pathogen cannot be detected using available primers, broad-range PCR may be used. This approach employs universal primers that anneal to a common nucleotide sequence, for example, the 16S rRNA gene present in all bacteria. Following amplification, products are sequenced for pathogen identification by comparison of the sequence to a database such as the Ribosomal Database Project. This method is more suited to determine infections in samples of low complexity, for example, meningitis, pneumonia, and endocarditis, which occur in tissues that are normally sterile (meninges, lower lungs, and heart, respectively).

In addition to its use in the measurement of pathogenic agents in the environment, a variant of real-time PCR may be used to quantify the levels of a variety of mRNAs in different eukaryotic tissues or prokaryotic cells. In these instances, the initial target is RNA and not DNA, and therefore, a reverse transcription (RT) step is needed to generate cDNA before the real-time PCR (see chapter 1). With quantitative RT-PCR, it is possible to measure mRNA levels that are about 10,000- to 100,000-fold lower than those measurable by traditional techniques.

RT
reverse transcription

Detection of Multiple Disease-Associated Mutations Using Microarrays

While hybridization probes and PCR are used extensively to detect mutations in single genes, microarray analysis enables simultaneous detection of mutations at multiple loci associated with complex genetic diseases (see chapter 1). Arrays have been developed to screen for more than 500,000 SNPs simultaneously in human genomic DNA. MammaPrint, a microarray to detect breast cancer, and AmpliChip CYP450, a microarray to predict the response to certain drugs, were the first diagnostic arrays to be approved by the FDA (see chapter 12).

Cytochrome P450 oxidases are a group of enzymes that regulate the metabolism of approximately 25% of prescription drugs, including antidepressants, antipsychotics, beta blockers, and opiates (Table 8.1). Different alleles of two cytochrome P450 genes, CYP2D6 and CYP2C19, are associated with differential drug metabolism. Over 80 different alleles of CYP2D6 and three different alleles of CYP2C19 have been discovered in humans and are linked to different levels of cytochrome P450 enzyme activity. Based on genotype, individuals are classified into four groups: poor, intermediate, extensive, and ultrarapid metabolizers. Individuals exhibiting ultrarapid drug metabolism may have multiple (3 to 13) copies of the CYP2D6 gene and remove the drug rapidly. Thus, the dosage of a drug may have to be increased in patients with this genetic makeup. Poor metabolizers, on the other hand, may have two inactive CYP2D6 alleles, and therefore, the drug may accumulate, leading to undesirable side effects, or drugs that require activation in vivo may be ineffective. For example, tamoxifen, a therapy for breast cancer, is administered in an inactive form and is converted into an active form (endoxifen) by several enzymes, including the CYP2D6 enzyme. This drug is unlikely to be effective in

Table 8.1 Drugs metabolized by CYP2D6 and CYP2C19

Cytochrome P450 oxidase	Antidepressants	Beta blockers	Antipsychotics	Proton pump inhibitors	Antiepileptics	Others
CYP2D6	Amitriptyline	Carvedilol	Haloperidol			Atomoxetine
	Clomipramine	Metoprolol	Risperidone			Codeine
	Desipramine	Propafenone	Thioridazine			Dextromethorphan
	Paroxetine	Timolol				Flecainide
	Imipramine					Mexiletine
	Venlafaxine					Ondansetron
						Tamoxifen
						Tramadol
CYP2C19	Amitriptyline			Omeprazole	Diazepam	Cyclophosphamide
	Clomipramine			Lansoprazole	Phenytoin	Progesterone
				Pantoprazole	Phenobarbitone	

Adapted from Li et al., *Curr. Genomics* 9:466–474, 2008.

patients who are poor or intermediate metabolizers, such as those with inactive or low activity CYP2D6 alleles. Thus, by identifying the alleles that an individual possesses, it may be possible to predict how that person will metabolize particular drugs. The AmpliChip CYP450 microarray detects 33 different CYP2D6 alleles, including deletions and duplications, and 3 different CYP2C19 alleles. Genomic DNA is extracted from blood samples or buccal (cheek) swabs, PCR amplified, fragmented, labeled, and hybridized to the array to identify the alleles present. The results facilitate drug selection, optimization of drug dosage, and prediction of adverse drug effects.

Detection of Epigenetic Markers

Epigenetic modifications refer to the covalent modification of DNA or histones associated with DNA. Specific bases in DNA may be methylated, and acetyl, ubiquitin, methyl, phosphate, or **sumoyl** groups may be added to specific amino acids in histone proteins. Epigenetic changes alter gene expression and cellular phenotype, are stable, and can be transmitted to offspring (i.e., they are heritable), although they do not change the genotype of a cell. DNA methylation almost always involves the addition of a methyl group to a cystosine residue in CpG dinucleotides, generating 5-methylcytosine (Fig. 8.21). The addition is catalyzed by a DNA methyltransferase. About 75% of the 28 million CpG dinucleotides in the mammalian genome are methylated. DNA methylation is a normal process that serves a regulatory function. It downregulates gene expression in response to environmental conditions by inhibiting binding of transcription factors or by recruiting specific methyl-binding domain proteins that induce the formation of **heterochromatin**. DNA methylation plays an important role in animal development, for example, by silencing genes on the X chromosome or on one of the two copies of some autosomal genes, depending on which parent it was inherited from.

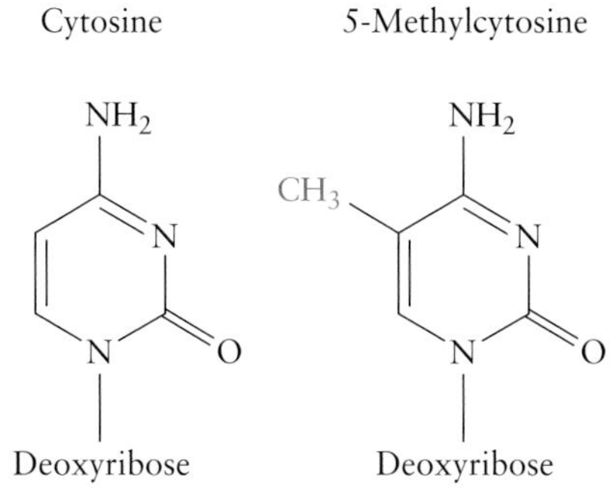

Recently, altered DNA methylation patterns have been found to be associated with a number of human diseases, especially cancers. In tumors, levels of genome methylation generally decrease; however, the promoters

Figure 8.21 DNA methylation. A methyl group is covalently bound to a cystosine residue in CpG dinucleotides, generating 5-methylcytosine.
doi:10.1128/9781555818890.ch8.f8.21

of some genes exhibit increased methylation of normally unmethylated CpG dinucleotides. Methylation of core promoter sequences inactivates expression of these genes, some of which encode tumor suppressors (Fig. 8.22). Thus, epigenetic modification can have an effect similar to that of a mutation or deletion. However, because DNA methylation is an epigenetic alteration, it is, at least in theory, reversible and therefore may be a target for demethylation drugs to treat human disease.

Aberrant methylation seems to be nonrandom, and CpG dinucleotides at different loci have different propensities for methylation. These factors, as well as the chemical stability of methylcytosine, make methylation patterns useful as biomarkers of disease. Genome-wide or locus-specific screening may be used to detect aberrant DNA methylation. Strategies to detect methylated sequences employ (i) methylation-sensitive restriction endonucleases that bind to DNA only when CpG dinucleotides in their recognition site are methylated (or only when they are unmethylated), (ii) 5-methylcytosine-specific antibodies to capture methylated fragments that can be identified using a microarray or by DNA sequencing, and (iii) chemical deamination of nonmethylated cytosines to uracils using sodium bisulfite; methylated cytosines are protected against deamination (Fig. 8.23). The last method converts differences in methylation patterns into differences in sequence (i.e., C to U) that can be identified by PCR, sequencing, or microarrays.

In one diagnostic approach to detect epigenetic biomarkers associated with specific cancers, DNA is first treated with bisulfite and then hybridized to a pool of oligonucleotide probes that specifically detect methylated or unmethylated CpG dinucleotides in target sequences of known cancer-related loci (Fig. 8.24). At each locus, a pair of allele-specific oligonucleotides discriminates between the methylated sequence and a sequence containing a

Figure 8.22 Hypermethylation of DNA in promoter regions decreases gene expression. In tumors, levels of genome methylation generally decrease; however, the promoters of some genes exhibit increased methylation of normally unmethylated CpG dinucleotides. Methylated CpG dinucleotides are associated with heterochromatin (condensed chromatin structure) and transcriptional silencing. Expression of tumor suppressor genes may be inhibited. Open circles, unmethylated CpG; closed circles, methylated CpG. Adapted from Lao and Grady, *Nat. Rev. Gastroenterol. Hepatol.* **8:** 686–700, 2011. doi:10.1128/9781555818890.ch8.f8.22

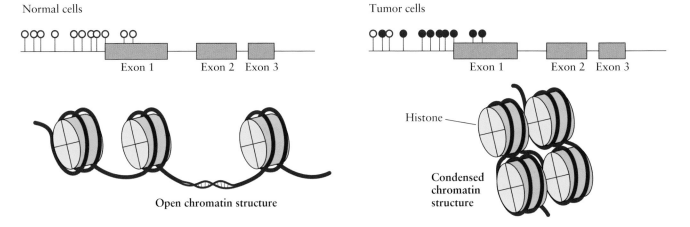

Normal cells

Exon 1   Exon 2   Exon 3

Open chromatin structure

Tumor cells

Exon 1   Exon 2   Exon 3

Histone

Condensed chromatin structure

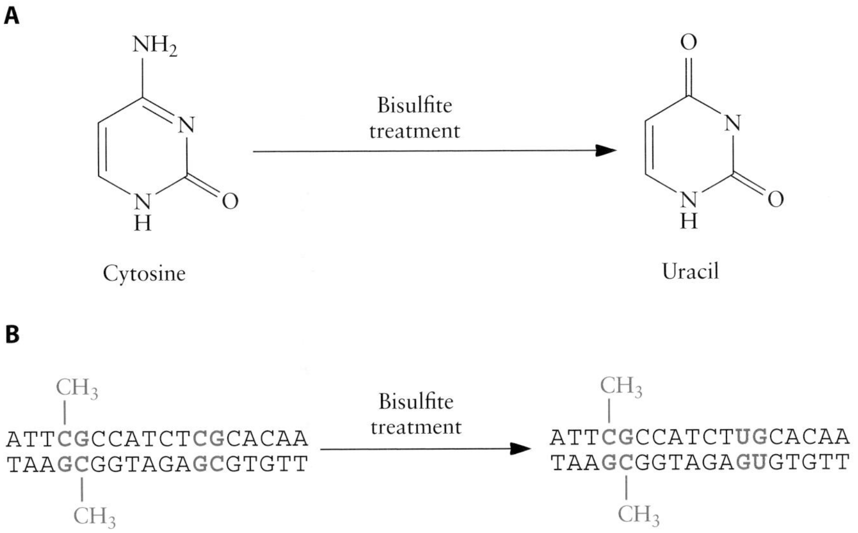

A

NH₂ → (Bisulfite treatment) → O

Cytosine   Uracil

B

CH₃
|
ATTCGCCATCTCGCACAA
TAAGCGGTAGAGCGTGTT
|
CH₃

→ Bisulfite treatment →

CH₃
|
ATTCGCCATCTUGCACAA
TAAGCGGTAGAGUGTGTT
|
CH₃

Figure 8.23 Sodium bisulfite converts cytosine to uracil (**A**). Unmethylated cytosines in CpG dinucleotides are converted to uracil, while methylated cytosines are protected from deamination (**B**). doi:10.1128/9781555818890.ch8.f8.23

Figure 8.24 Detection of epigenetic biomarkers associated with specific cancers. Genomic DNA extracted from a patient's sample is treated with sodium bisulfite that converts unmethyated cytosines in CpG dinucleotides to uracil; methylated cytosines are not converted. Allele-specific oligonucleotide probes are added that specifically anneal to either methylated or converted CpG dinucleotides due to the presence of a complementary G or A, respectively, at the 3′ end of the probe. A common oligonucleotide probe (blue line) binds one nucleotide downstream of the allele-specific probes. DNA polymerase extends the allele-specific probe by a single nucleotide to close the gap between the allele-specific and common probes. DNA ligase catalyzes the formation of a phosphodiester bond to seal the nick in the backbone. PCR primers (black arrows) bind at the ends of the probes and prime amplification of the ligated probes. The PCR primers are labeled with different fluorescent dyes (red and blue stars) to reflect the original methylation state of the CpG site in the genome. doi:10.1128/9781555818890.ch8.f8.24

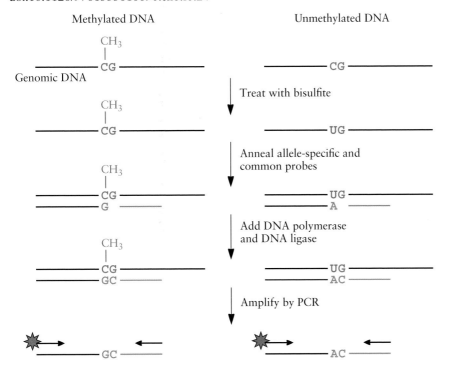

uracil converted from an unmethylated cytosine by the presence of a complementary G or A, respectively, at the 3′ end of the probe (the hybridization behavior of uracil is similar to that of thymine). A third oligonucleotide binds downstream of the allele-specific oligonucleotide-binding site such that there is a gap of one nucleotide between the bound oligonucleotides. Addition of DNA polymerase extends the allele-specific oligonucleotide by a single nucleotide to close the gap between the bound oligonucleotides, which is sealed by DNA ligase. The requirement for binding of two oligonucleotides at each locus increases the specificity of the assay. Primer-binding sites at the 5′ ends of the oligonucleotides enable amplification of the ligation products with PCR primers that are labeled with different fluorescent dyes to reflect the methylation state of the CpG site. The identity of the PCR products may be determined by sequencing or by hybridization to a microarray. It is possible to probe more than 1,500 CpG sites that are known to be susceptible to alterations in methylation status in tumors simultaneously in a single patient DNA sample.

Detection of SNPs by Mass Spectrometry

MALDI-TOF
matrix-assisted laser desorption
ionization–time of flight

Matrix-assisted laser desorption ionization–time of flight (MALDI-TOF) mass spectrometry can be used to detect SNPs known to be associated with a disease. This approach is rapid, requires only small amounts of DNA, can detect multiple SNPs simultaneously, and can be use to process many samples at the same time. In principle, the method is similar to that used for protein identification (see chapter 1); however, rather than ionizing proteins, when excited, the matrix transfers some of its energy to the DNA fragments in the sample, which are then ionized. The ionized DNA fragments are separated in an electromagnetic field and identified by their mass-to-charge ratio. Analysis of SNPs by this method requires the generation of small (<30 nucleotides), allele-specific DNA fragments. Most protocols produce these fragments by single- or multiple-base primer extension reactions (Fig. 8.25); single-base primer extension is described here. To generate large amounts of template, the target DNA is first amplified by PCR. After removal of unincorporated nucleotides and PCR primers, another primer is added that anneals to the amplification products such that its 3′ end is immediately adjacent to the polymorphic site. The primer is extended by addition of a single nucleotide that is complementary to the nucleotide at the polymorphic site. DNA polymerase catalyzes the extension, which is limited by incorporation of a dideoxynucleotide (see Fig. 1.19). The next step is identification of the incorporated nucleotide based on its mass, which differs among the four possible nucleotides. Because the mass differences among the nucleotides are small, the terminating nucleotides are often modified to enhance the differences. Care must be taken to remove all traces of metal cations (sodium and potassium) that may bind to negatively charged DNA and increase the mass of the DNA fragments, thereby distorting the results.

Diagnostic assays based on mass spectrometry are being developed for a number of different diseases. For example, mutations that predispose individuals to familial adenomatous polyposis can be detected by

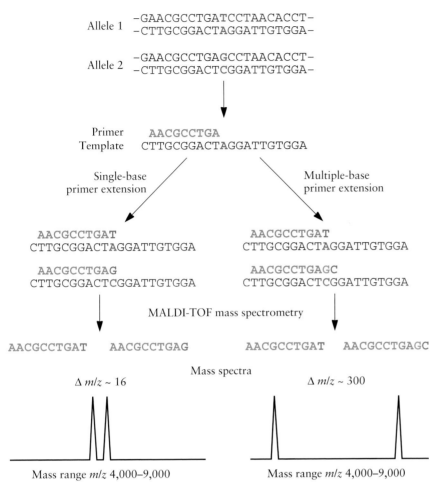

Figure 8.25 Detection of SNPs by mass spectrometry. After PCR amplification of the target sequence, a primer (brown sequence) is added that anneals to the single-stranded template such that its 3' end is immediately adjacent to the polymorphic site (red). In a single-base extension assay, the primer is extended by a single dideoxynucleotide (blue) that is complementary to the polymorphic nucleotide. In a multiple-base extension assay, the primer may be extended by addition of several deoxynucleotides (green) before termination with a dideoxynucleotide. The mass (actually the mass-to-charge ratio [m/z]) of the extension products is determined by mass spectrometry and is the basis for identification of the incorporated nucleotide(s) and hence the SNP. Adapted from Tost and Gut, *Clin. Biochem.* **38:**335–350, 2005, with permission. doi:10.1128/9781555818890.ch8.f8.25

MALDI-TOF mass spectrometry. Familial adenomatous polyposis is an inherited disorder characterized by the formation of numerous benign tumors (polyps, adenomas) in the colon. They may become malignant and lead to early development of colorectal cancer. Currently, diagnosis is by colonoscopy. This disorder is caused by germ line mutations in the adenomatous polyposis coli (*APC*) gene, which encodes a tumor suppressor. Many of the common mutations in this gene are frameshift and nonsense mutations that produce truncated proteins. The most severe disease (high number of polyps, early onset of colorectal cancer) is associated with a small deletion in the *APC* gene. Four of the most frequent mutations, including three small (<6-base pair [bp]) deletions and one nucleotide substitution, were detected by MALDI-TOF mass spectrometry. The *APC* region was amplified by PCR from DNA extracted from patient blood samples, and the purified PCR products were then used as a template in a primer extension assay (Fig. 8.26A). In this protocol, the allele-specific primers were extended by one or more bases by including a mixture of deoxynucleotides and chain-terminating dideoxynucleotides in the reaction. Purified extension products were mixed with an organic matrix and their masses determined by MALDI-TOF mass spectrometry. In the mass spectra, extension products could clearly be distinguished between healthy individuals carrying two wild-type *APC* alleles and heterozygous

Figure 8.26 Detection of a mutation frequently associated with familial adenomatous polyposis by MALDI-TOF mass spectrometry. (**A**) A 5-bp deletion (Δ1309; highlighted in blue in the sequence of the normal allele) in the *APC* gene is frequently associated with colorectal cancer. To detect this small deletion, purified *APC* PCR products (partial sequence shown), amplified from patient blood samples, were used as a template in a primer extension assay. An extension primer was extended by one or more bases by including a mixture of deoxynucleotides (deoxyadenosine triphosphate [dATP]) and chain-terminating dideoxynucleotides (dideoxyribosylthymine triphosphate [ddTTP] and dideoxycytidine triphosphate [ddCTP]) in the reaction. Extension products from the normal and Δ1309 alleles were mixed with an organic matrix and their masses determined by MALDI-TOF mass spectrometry. Extended deoxynucleotides are shown in green, and dideoxynucleotides are shown in blue in the extension products. Da, daltons. (**B**) MALDI-TOF mass spectra generated from the *APC* allele of a healthy homozygous individual (control DNA: WT) and an affected heterozygous individual (patient DNA). A synthetic oligonucleotide was used as a control for homozygous Δ1309 DNA because this genotype is lethal. The masses of the unextended (6,323 Da) and extended (6,596 and 6,924 Da for the wild-type and Δ1309 alleles, respectively) primers are shown above the peaks. Redrawn from Bonk et al., *Clin. Biochem.* 35:87–92, 2002, with permission. doi:10.1128/9781555818890.ch8.f8.26

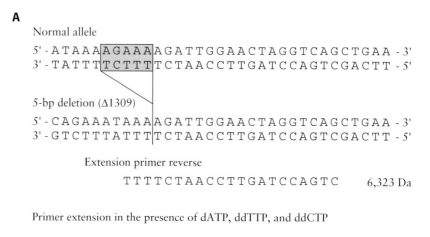

A

Normal allele

5'-ATAAA AGAAA AGATTGGAACTAGGTCAGCTGAA-3'
3'-TATTT TCTTT TCTAACCTTGATCCAGTCGACTT-5'

5-bp deletion (Δ1309)

5'-CAGAAATAAA AGATTGGAACTAGGTCAGCTGAA-3'
3'-GTCTTTATTT TCTAACCTTGATCCAGTCGACTT-5'

Extension primer reverse

TTTTCTAACCTTGATCCAGTC   6,323 Da

Primer extension in the presence of dATP, ddTTP, and ddCTP

Extension product: normal allele
CTTTTCTAACCTTGATCCAGTC   6,596 Da
Extension product: 5-bp deletion (Δ1309)
TATTTTCTAACCTTGATCCAGTC   6,924 Da

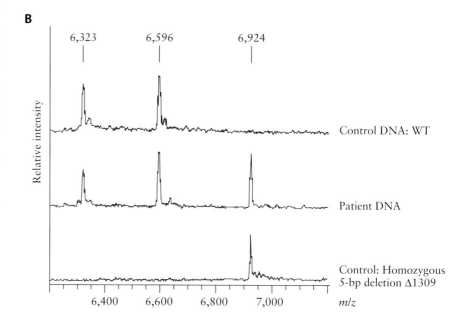

B

familial adenomatous polyposis patients carrying one wild-type allele and a deletion or nucleotide substitution in the other allele (Fig. 8.26B). Homozygous mutations in the *APC* gene are lethal, and affected fetuses do not survive; therefore, synthetic oligonucleotides were used as a control. This example shows that in addition to point mutations, deletions of several nucleotides that are associated with a genetic disease can be detected in clinical samples using mass spectrometry.

Detecting RNA Signatures of Disease

Detection of Disease-Associated Changes in Gene Expression Using Microarrays

In addition to applications for microarrays in genotyping as described above for the AmpliChip CYP450 microarray, this tool has also been used to determine and detect RNA profiles associated with disease (see chapter 1). For example, the clinical microarray MammaPrint measures expression of 70 different genes in breast tumors. Breast cancer differs among patients in terms of risk of metastasis and response to treatment. Prior to development of a diagnostic microarray, disease prognosis was based primarily on tumor size and lymph node status; however, these are not accurate predictors of long-term outcome. The goal of the Mamma-Print diagnostic array is to determine if a patient with breast cancer that has not spread to lymph nodes is at risk for recurrence or metastasis (Fig. 8.27). Individuals with low risk have a 10% chance of having the cancer return within 10 years without further treatment, while those with high risk have a 29% chance of recurrence in the same period.

To develop the MammaPrint microarray, RNA from 98 breast tumors was extracted and hybridized to microarrays carrying 25,000 human gene sequences. Two distinct gene expression profiles were apparent; these correlated with disease progression, that is, with either progressive, metastatic disease (poor prognosis) or lack of metastasis 5 years after primary diagnosis (good prognosis). Elevated expression levels of genes involved in cell cycle regulation, cell invasion, metastasis, and angiogenesis provide an RNA signature for poor prognosis, and these individuals are recommended for chemotherapy. Without this test, many patients are unnecessarily subjected to cytotoxic chemotherapy; 70% would have survived without this aggressive treatment. If the microarray results indicate a good prognosis, a less aggressive hormonal therapy such as tamoxifen may be prescribed. Tamoxifen blocks the activity of the hormone estrogen that breast tumors need to grow.

Detection of RNA Signatures of Antibiotic Resistance in Human Pathogens

Treatment of microbial infections requires identification of the etiological agent and determination of an effective course of treatment. Unfortunately, suitable treatments are often difficult to predict due to the prevalence of antibiotic-resistant strains of pathogens. For some slow-growing pathogens,

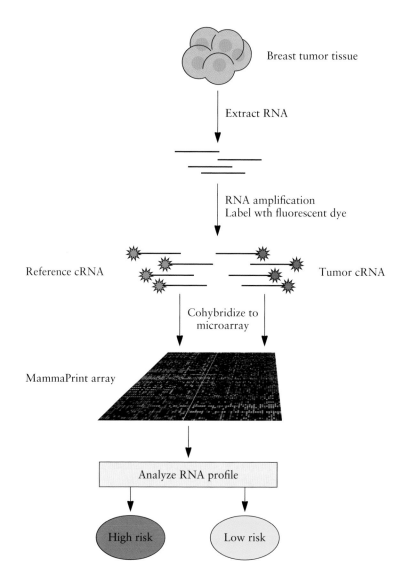

Figure 8.27 MammaPrint diagnostic array assesses risk for recurrence or metastasis of breast cancer. The microarray contains sequences that target mRNA expressed from 70 different breast cancer prognosis genes. RNA is extracted from patient tumor tissue, amplified using T7 RNA polymerase to generate complementary RNA (cRNA), and then labeled with a fluorescent dye (e.g., Cy5). Levels of expression of each gene are determined by hybridizing labeled cRNA to the microarray and measuring fluorescence intensity relative to reference cRNA; reference cRNA consists of a pool of RNA from 100 breast tumors (from equal numbers of patients with good and poor outcomes) labeled with a different fluorescent dye (e.g., Cy3). Tumor RNA expression profiles stratify women into two groups: those with a low risk for breast cancer recurrence and those with a high risk for recurrence. doi:10.1128/9781555818890.ch8.f8.27

for example, *Mycobacterium tuberculosis*, current clinical diagnostic methods that require culturing of the pathogen to determine antibiotic sensitivities can take weeks to months. In the meantime, patients may be treated with ineffective drugs. Furthermore, DNA-based approaches such as hybridization or PCR to detect specific antibiotic resistance genes in bacterial genomes are limited by the available knowledge of these gene sequences, which represents only a fraction of such genes. A possible solution is the use of specific RNA signatures not only for identification of pathogens in clinical samples but also for rapid determination of their antibiotic susceptibilities. Rather than detection of specific antibiotic resistance genes, this approach detects RNA profiles that are associated with an antibiotic susceptibility response. Exposure of susceptible strains to an antibiotic induces a characteristic transcriptional response that is antibiotic specific. Direct detection of RNA using fluorescent hybridization probes therefore offers a rapid, sensitive, and specific method for distinguishing antibiotic-resistant from antibiotic-sensitive strains in clinical samples.

RNA profiles associated with susceptibility to the antibiotics ciprofloxacin, ampicillin, and gentamicin were determined following a brief exposure of susceptible and resistant laboratory strains of *E. coli*. Crude cell lysates from cultured bacteria were hybridized to a pool of oligonucleotide probes that target specific RNA sequences; each gene-specific probe was tagged with a different fluorescent dye. While expression of some genes increased in response to several different antibiotics, a small set of genes exhibited antibiotic-specific expression profiles (Fig. 8.28A). The transcriptional responses were observed in the antibiotic-sensitive strains but not in the resistant strains. To test the efficacy of this approach on clinical samples, urine was obtained from patients suspected of having urinary tract infections. The urine specimens were exposed briefly (30 min) to ciprofloxacin, cells in the urine were lysed rapidly, and transcriptional profiles were determined by hybridization of probes directly to RNA. RNA signatures enabled identification of *E. coli* in the samples and discrimination of ciprofloxacin-susceptible and -resistant strains as confirmed by culture-dependent diagnostic methods (Fig. 8.28B). Although similar results could be achieved using PCR-based approaches, direct detection of specific RNAs associated with antibiotic susceptibility via probe hybridization avoids having to purify and amplify the nucleic acids.

Figure 8.28 RNA signatures of bacterial antibiotic susceptibility. (**A**) Antibiotic-sensitive and -resistant strains of *E. coli* were exposed to the antibiotics ciprofloxacin (CIP), ampicillin (AMP), and gentamicin (GENT) for a short time (10 min). Levels of expression of several genes were measured by hybridizing RNA from antibiotic-treated and untreated cultures to oligonucleotide probes labeled with fluorescent dyes (a subset of antibiotic-responsive genes are labeled on the *x* axis). Changes in the levels of expression of each gene were determined by comparing RNA levels in antibiotic-treated and untreated cultures. Significant changes in gene expression were observed in the antibiotic-sensitive strains but not in the antibiotic-resistant strains. (**B**) Ciprofloxacin-resistant (R) and -sensitive (S) strains of *E. coli* were differentiated in *E. coli*-positive urine samples from 13 patients based on RNA profile scores. For each patient sample, a single RNA profile score was calculated from the transcript levels of the individual antibiotic-responsive genes in *E. coli* present. A low score indicates a susceptible strain, while a high score indicates a resistant strain. Redrawn from Barczak et al., *Proc. Natl. Acad. Sci. USA* **109**:6217–6222, 2012, with permission. doi:10.1128/9781555818890.ch8.f8.28

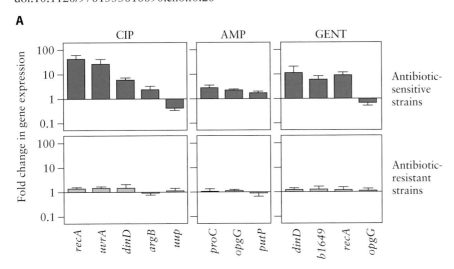
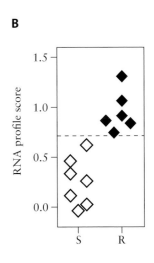

miRNA
microRNA

RISC
RNA-induced silencing complex

Detection of miRNA Signatures of Cancers

MicroRNAs (miRNAs) are small (~20-nucleotide) natural RNA molecules that regulate production of specific proteins by preventing translation of mRNA. They are produced from longer primary miRNA molecules that are folded into stem–loop structures and cleaved by the RNases Drosha and Dicer into smaller double-stranded miRNA (Fig. 8.29). The large nuclease complex RNA-induced silencing complex (RISC) separates the strands of the double-stranded miRNA, and the resultant single-stranded RNA products, together with RISC, bind to complementary sequences on mRNA molecules. The nuclease component of RISC then degrades the mRNA, thereby preventing synthesis of the encoded protein. In some cases where the miRNA is imperfectly complementary to its target mRNA, the mRNA is not destroyed but, rather, translation is repressed.

The miRNAs play a role in initiation and progression of several human cancers, and therefore, changes in their expression could provide an early indicator of these diseases. Abnormal expression of some miRNAs as a consequence of mutations, epigenetic modifications, or alterations in miRNA processing has been detected in several different tumor types. These miRNAs target mRNA involved in tumorigenesis such as Ras oncogenes, antiapoptotic genes, and transcription factors.

B-cell chronic lymphocytic leukemia is a common form of adult leukemia that can progress to a more aggressive lymphoma or leukemia.

Figure 8.29 Production and activity of miRNA. Double-stranded miRNA is produced from primary miRNA (pri-miRNA) that is folded into a stem–loop structure by intramolecular complementary base-pairing and then cleaved by RNases Drosha and Dicer. RISC mediates binding of single-stranded miRNA to complementary sequences in target mRNAs, which prevents production of the encoded proteins. doi:10.1128/9781555818890.ch8.f8.29

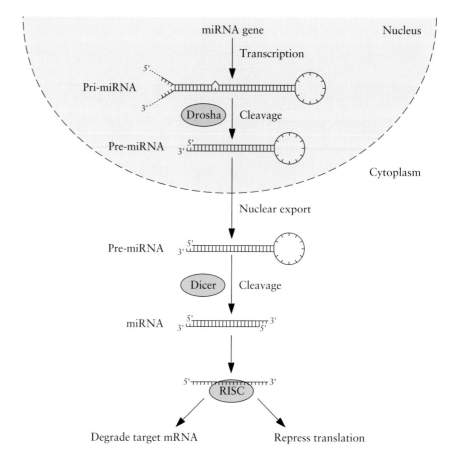

Affected individuals produce an abnormally high number of lympho-cytes (white blood cells). However, these do not function normally; that is, they do not help reduce infections. Moreover, they impair the function of other normal blood cells such as healthy lymphocytes, ery-throcytes, and platelets. A microarray containing oligonucleotide probes corresponding to 190 human miRNAs was used to determine miRNA expression profiles in 144 patients with chronic lymphocytic leukemia. Expression levels of 13 miRNAs were found to correlate with disease development (prognosis and progression). For example, levels of two miRNAs (*miR-16-1* and *miR-15a*) were found to be lower in patients with a good prognosis and slow disease progression, while lower levels of another miRNA (*miR-29*) were found to be associated with a poor prognosis. The target gene for *miR-29* is the oncogene *TCL1*, which is overexpressed in patients with more aggressive chronic lymphocytic leu-kemia. These miRNAs may be early predictors of disease development in patients with this malignancy.

summary

Molecular diagnostics aims to detect biological molecules (biomarkers) associated with infectious or genetic disease in patient samples. The biomarkers may be specific proteins, DNA or RNA sequences, or metabolites that are present, or present at altered levels, in body fluid or tissue samples of af-fected individuals compared to healthy individuals. The pur-pose of a diagnostic test is to identify or confirm the cause of a disease (etiology), to determine risk of developing the disease in asymptomatic individuals, to determine the best course of treatment, or to monitor progression of the disease or response to treatment. Therefore, the test must have a high degree of specificity (i.e., detect only the targeted molecule) and sensitivity (i.e., detect the targeted molecule even when it is present in low abundance). Importantly, for clinical appli-cations, a diagnostic test must also be rapid, inexpensive, and amenable to high throughput.

Molecular technologies have greatly facilitated accurate de-tection of biomarkers. Immunological approaches exploit antibody–antigen interactions to detect the presence of a pathogen or antibodies against a pathogen, proteins that are dysregulated or have altered conformation, or antibodies against self-proteins (autoantibodies). Three common ap-proaches to detect protein biomarkers of disease are ELISAs, protein microarrays, and mass spectroscopy. An indirect ELISA detects the presence of specific antibodies (primary antibodies) in patient serum by their binding to a purified an-tigen immobilized on a solid support. The antibody–antigen interaction is detected by addition of a labeled secondary an-tibody. This assay may be used to determine if a patient has mounted an autoimmune response or an immune response to a pathogen. A sandwich ELISA measures levels of an an-tigen, often a protein associated with a disease or particular condition, in a patient's sample. In this assay, specific (usu-ally monoclonal) antibodies are bound to a solid support to which a complex clinical sample is applied. If the cognate antigen is present in the sample, it is captured by the immo-bilized antibody and can be detected by applying a labeled antibody.

Microarrays may be used to detect multiple protein biomark-ers in clinical samples from patients affected by polygenic diseases or the presence of IgE molecules produced against specific allergens in patients who suffer from allergies. For detection of protein biomarkers, primary antibodies are bound to a solid support (e.g., small capture beads) and then probed with a patient sample. Secondary antibodies, often labeled with a fluorescent dye, are added to detect captured antigens. In a reverse-phase microarray, patient samples are immobilized on a solid support and then probed with an antibody to detect disease-associated proteins. For detection of specific antibodies in patient serum, purified proteins or other antigens (e.g., allergens) are arrayed and then probed with a serum sample.

DNA-based diagnostic tests determine the existence of spe-cific nucleotide sequences, including genetic mutations and sequences present in pathogens. In addition to being highly sensitive and specific, these tests can be used to assess risk for a disease before onset of symptoms. Specific DNA se-quences can be detected in patient samples by hybridization, PCR, microarray analysis, and mass spectrometry. Hybrid-ization utilizes a labeled oligonucleotide probe to detect

(continued)

summary (continued)

a complementary target DNA sequence that indicates the presence of a specific disease-associated allele (SNP) or a specific pathogen. Perfect complementarity between a target sequence that includes a polymorphic site and a probe is required for ligation of two tandem probes in an OLA and for ligation of two ends of a padlock probe.

Allele-specific PCR requires a perfect match between an oligonucleotide primer and a target sequence for amplification. One of the pair of PCR primers includes the nucleotide that is complementary to an SNP at its 3′ end. Mismatches prevent primer extension by the thermostable DNA polymerase. Several variations of allele-specific PCR are employed in diagnostic assays, including the TaqMan protocol, which measures the activity of a fluorescent probe in real time as it is released during primer extension. *Taq* polymerase, which catalyzes the synthesis of the DNA strand, possesses nuclease activity and cleaves the probe that is bound to the template. Cleavage removes a quencher molecule that prevents fluorescence of the dye. An alternative real-time PCR protocol measures fluorescence of a dye that specifically binds to double-stranded DNA that is produced after each

amplification cycle. Because the amount of double-stranded DNA doubles after each cycle, the target DNA can be quantified in a sample, such as a sample containing an infectious agent.

Allele-specific primer extension products may also be identified by mass spectrometry. In one protocol, a primer binds to a template DNA strand such that its 3′ end is immediately upstream of a polymorphic nucleotide and is extended by a single chain-terminating dideoxynucleotide. Differences in the masses of primer extension products from different alleles are dependent on differences in the mass of the incorporated nucleotide and can be measured by MALDI-TOF mass spectrometry.

Microarray analyses enable simultaneous detection of SNPs at multiple loci for polygenic diseases or determination of RNA signatures associated with a disease. For example, microarrays have been approved by the FDA to detect cytochrome P450 oxidase alleles associated with drug metabolism (AmpliChip CYP450 microarray) and to measure expression of specific genes in breast tumors (MammaPrint microarray). The results from these tests can be used to determine an effective course of treatment for the disease.

review questions

1. What are the goals of a molecular diagnostic test? What are some of the factors that are considered in the development of an accurate diagnostic test?

2. What is a biomarker?

3. Compare and contrast indirect and sandwich ELISAs. Describe an application for each in the diagnosis of a human disease.

4. Describe a method to identify the etiological agent of an infection.

5. What is the advantage of using a protein microarray in a diagnostic test compared to an ELISA?

6. What is a reverse-phase protein microarray? How might it be used in a diagnostic assay?

7. Why might an immunological approach be more appropriate than a DNA-based approach to diagnose Alzheimer disease?

8. What is a hybridization probe? How are hybridization probes employed to detect disease-associated SNPs?

9. Why is PCR commonly used in molecular diagnostic tests?

10. Describe an allele-specific PCR.

11. How can two different alleles be distinguished by using a TaqMan assay?

12. Why is real-time PCR quantitative? Why would this be useful for a diagnostic test?

13. Describe one application for DNA microarrays in disease diagnosis.

14. What is DNA methylation? What role does it play in human disease? How can DNA methylation patterns be detected in clinical samples?

15. How is mass spectrometry used to detect SNPs?

16. When assessing the antibiotic susceptibility of a pathogen, what are the advantages of determining RNA profiles rather than detecting the antibiotic resistance gene directly?

17. What are miRNAs? Why are they useful biomarkers of some diseases?

references

Aqeilan, R. I., G. A. Calin, and C. M. Croce. 2010. *miR-15a* and *miR-16-1* in cancer: discovery, function and future perspectives. *Cell Death Differ.* **17**: 215–220.

Barczak, A. K., J. E. Gomez, B. B. Kaufmann, E. R. Hinson, L. Cosimi, M. L. Borowsky, A. B. Onderdonk, S. A. Stanley, D. Kaur, K. F. Bryant, D. M. Knipe, A. Sloutsky, and D. T. Hung. 2012. RNA signatures allow rapid identification of pathogens and antibiotic susceptibilities. *Proc. Natl. Acad. Sci. USA* **109**:6217–6222.

Belluco, C., E. Mammano, E. Petricoin, L. Prevedello, V. Calvert, L. Liotta, D. Nitti, and M. Lise. 2005. Kinase substrate protein microarray analysis of human colon cancer and hepatic metastasis. *Clin. Chim. Acta* **357**:180–183.

Bibikova, M., Z. Lin, L. Zhou, E. Chudin, E. W. Garcia, B. Wu, D. Doucet, N. J. Thomas, Y. Wang, E. Vollmer, T. Goldmann, C. Seifart, W. Jiang, D. L. Barker, M. S. Chee, J. Floros, and J.-B. Fan. 2006. High-throughput DNA methylation profiling using universal bead arrays. *Genome Res.* **16**:383–393.

Bonk, T., A. Humeny, C. Sutter, J. Gebert, M. von Knebel Doeberitz, and C.-M. Becker. 2002. Molecular diagnosis of familial adenomatous polyposis (FAP): genotyping of adenomatous polyposis coli (*APC*) alleles by MALDI-TOF mass spectrometry. *Clin. Biochem.* **35**:87–92.

Calin, G. A., and C. M. Croce. 2006. MicroRNA signatures in human cancers. *Nat. Rev. Cancer* **6**:857–866.

Calin, G. A., M. Ferracin, A. Cimmino, G. Di Leva, M. Shimizu, S. E. Wojcik, M. V. Iorio, R. Visone, N. I. Sever, M. Fabbri, R. Iuliano, T. Palumbo, F. Pichiorri, C. Roldo, R. Garzon, C. Sevignani, L. Rassenti, H. Alder, S. Volinia, C. Liu, T. J. Kipps, M. Negrini,

and C. M. Croce. 2005. MicroRNA signature associated with prognosis and progression in chronic lymphocytic leukemia. *N. Engl. J. Med.* **353**:1793–1801.

Erickson, H. S. 2012. Measuring molecular biomarkers in epidemiologic studies: laboratory techniques and biospecimen considerations. *Stat. Med.* **31**:2400–2413.

Ferec, C., and G. R. Cutting. 2012. Assessing the disease-liability of mutations in CFTR. *Cold Spring Harb. Perspect. Med.* **2**:a009480.

Hartmann, M., J. Roeraade, D. Stoll, M. F. Templin, and T. O. Joos. 2009. Protein microarrays for diagnostic assays. *Anal. Bioanal. Chem.* **393**: 1407–1416.

Hellström, I., J. Raycraft, M. Hayden-Ledbetter, J. A. Ledbetter, M. Schummer, M. McIntosh, C. Drescher, N. Urban, and K. E. Hellström. 2003. The HE4 (WFDC2) protein is a biomarker for ovarian carcinoma. *Cancer Res.* **63**:3695–3700.

Kim, J. U., C. H. Cha, H. K. An, H. J. Lee, and M. N. Kim. 2013. Multiplex real-time PCR assay for detection of methicillin-resistant *Staphylococcus aureus* (MRSA) strains suitable in regions of high MRSA endemicity. *J. Clin. Microbiol.* **51**:1008–1013.

Kit, A. H., H. M. Nielsen, and J. Tost. 2012. DNA methylation based biomarkers: practical considerations and applications. *Biochimie* **94**: 2314–2337.

Lao, V. V., and W. M. Grady. 2011. Epigenetics and colorectal cancer. *Nat. Rev. Gastroenterol. Hepatol.* **8**:686–700.

Li, X., R. J. Quigg, J. Zhou, W. Gu, P. N. Rao, and E. F. Reed. 2008. Clinical utility of microarrays: current status, existing challenges and future outlook. *Curr. Genomics* **9**:466–474.

Mueller, C., L. A. Liotta, and V. Espina. 2010. Reverse phase protein microarrays advance to use in clinical trials. *Mol. Oncol.* **4**:461–481.

Perchiacca, J. M., A. R. A. Ladiwala, M. Bhattacharya, and P. M. Tessier. 2012. Structure-based design of conformation- and sequence-specific antibodies against amyloid β. *Proc. Natl. Acad. Sci. USA* **109**:84–89.

Pinzani, P., F. Salvianti, R. Cascella, D. Massi, V. De Giorgi, M. Pazzagli, and C. Orlando. 2010. Allele specific Taqman-based real-time PCR assay to quantify circulating BRAFV600E mutated DNA in plasma of melanoma patients. *Clin. Chim. Acta* **411**: 1319–1324.

Sibley, C. D., G. Peirano, and D. L. Church. 2012. Molecular methods for pathogen and microbial community detection and characterization: current and potential application in diagnostic microbiology. *Infect. Genet. Evol.* **12**: 505–521.

Tost, J., and I. G. Gut. 2005. Genotyping single nucleotide polymorphisms by MALDI mass spectrometry in clinical applications. *Clin. Biochem.* **38**: 335–350.

van't Veer, L. J., H. Dai, M. J. van de Vijver, Y. D. He, A. A. M. Hart, M. Mao, H. L. Peterse, K. van der Kooy, M. J. Marton, A. T. Witteveen, G. J. Schreiber, R. M. Kerkhoven, C. Roberts, P. S. Linsley, R. Bernards, and S. H. Friend. 2002. Gene expression profiling predicts clinical outcome of breast cancer. *Nature* **415**:530–536.

9

Protein Therapeutics

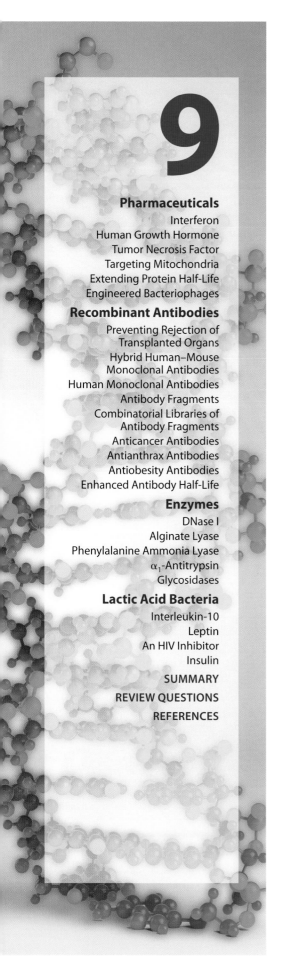

PRIOR TO THE DEVELOPMENT of recombinant DNA technology, most human protein pharmaceuticals were available in only limited quantities, they were extremely costly to produce, and in a number of cases, their biological modes of action were not well characterized. When recombinant DNA technology was first developed, it was heralded as a means of producing a whole range of potential human therapeutic agents in sufficient quantities for both efficacy testing and eventual human use. For the most part, this prediction has turned out to be true. Today, the "genes" (mostly cDNAs) for several thousand different proteins that are potential human therapeutic agents have been cloned. Most of these recombinant proteins have been produced in mammalian as well as bacterial host cells, and currently more than 500 are undergoing clinical testing with human subjects for the treatment of various diseases. More than 250 of these "biotechnology drugs" have been approved for use in the United States or the European Union (Table 9.1). However, it will be several years before many of the other proteins are commercially available because medical products must first be tested rigorously in animals and then undergo thorough human trials, which typically last for several years, before being approved for general use. Each year, only about 25 to 30 new U.S. Food and Drug Administration (FDA)-approved pharmaceuticals are introduced to the marketplace. However, the financial incentive for pharmaceutical companies is considerable. It has been estimated that in 2013 the annual global market for human recombinant protein drugs will be about $160 billion, with therapeutic monoclonal antibodies constituting about half of this market and a few "blockbuster" drugs such as insulin, erythropoietin, interferon (IFN), and blood clotting factors making up the largest part of the rest of this market.

doi:10.1128/9781555818890.ch9

Table 9.1 Examples of recombinant proteins that have been approved for human use in the United States or the European Union

Alglucosidase α	Human luteinizing hormone
Anakinra	Hyaluronidase
Antihemophilic factor	Iduronate-2-sulfatase
Antithrombin III	Insulin
α_1-Antitrypsin	Insulin analogue
Botulinum toxin A	Insulin-like growth factor
C1 esterase inhibitor	IFN-α2a
Darbepoetin α	IFN-α2b
Dibotermin	IFN-αN3
DNase I	IFN-β1a
Drotrecogin α	IFN-β1b
Erythropoietin	IFN-γ1b
Factor VIIa	IFN N
Factor VIII	IFN analogue
Factor IX	Interleukin-2
Fibrinogen	Interleukin-2 analogues
Follicle-stimulating hormone	Interleukin-11
α-Galactosidase	Interleukin-11 analogue
Galsulfase	Kallikrein inhibitor
Glucagon	Keratinocyte growth factor
β-Glucocerebrosidase analogue	Laronidase
α-Glucosidase	Novel erythropoiesis-stimulating protein
Granulocyte-macrophage colony-stimulating factor	Osteogenic protein
Hirudin	Platelet-derived growth factor
Human bone morphogenic protein 7	Stem cell factor
Human chorionic gonadotropin	Tissue plasminogen activator
Human growth hormone	Thyrotropin-α
Human growth hormone analogue	Truncated tissue plasminogen activator

FDA
U.S. Food and Drug Administration

IFN
interferon

Pharmaceuticals

Protein therapeutic agents may be divided into several different categories based on the general mode of action of the therapeutic agent. In this regard, a large number of the currently available protein therapeutic agents that are produced by recombinant DNA technology replace a deficient or abnormal protein. Other recombinant therapeutics may augment an existing pathway or provide a novel function or activity. Some classic examples of replacing a deficient protein include providing diabetics with insulin or hemophiliacs with specific blood coagulation factors.

Interferon

A number of different strategies have been used to isolate either the genes or cDNAs for human proteins. Previously, scientists isolated the target protein and determined a portion of its amino acid sequence; from the deduced DNA coding sequence, oligonucleotide probes were synthesized and used as a DNA hybridization probe to isolate the gene or cDNA from either a genomic or a cDNA library. Alternatively, antibodies raised against the purified protein were used to screen a gene or cDNA expression library. More recently, knowledge of the DNA sequence of the human genome has simplified the isolation of human genes or cDNAs.

IFNs are animal glycoproteins that are synthesized by cells in response to various pathogens, especially viruses. IFN "interferes" with viral replication within host cells. IFNs can also activate immune cells, such as natural killer cells and macrophages; increase recognition of infected or tumor cells by upregulating antigen presentation to T lymphocytes; and increase the ability of uninfected host cells to resist new infection by an infecting virus. There are several different types of IFNs, all with slightly different activities. IFNs may induce the production of hundreds of other proteins, upregulate major histocompatibility molecules, activate cells that are part of the immune system, promote apoptosis, and inhibit protein synthesis of both viral and host genes. At the present time, 12 different IFNs and IFN derivatives have been approved by the FDA for the treatment of various diseases (Milestone 9.1).

Several research groups have attempted to engineer IFNs with combined properties based on different members of the IFN-α gene family that vary in the extents and specificities of their antiviral activities. This can be achieved by splicing a portion of one IFN-α gene with a DNA sequence from a different (although similar) IFN-α gene to create, after translation, a hybrid protein that exhibits novel properties, i.e., properties different from those of the proteins encoded by either of the contributing genes.

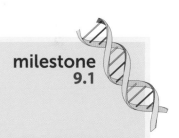

**milestone
9.1**

Synthesis in *E. coli* of a Polypeptide with Human Leukocyte Interferon Activity

S. NAGATA, H. TAIRA, A. HALL, L. JOHNSRUD, M. STREULI, J. ECSÖDI, W. BOLL, K. CANTELL, AND C. WEISSMANN
Nature **284**:316–320, 1980

The late 1970s and early 1980s were a time of tremendous excitement in molecular biotechnology. The promise of this new technology was being touted to both the public and large institutional investors. One of the products of the new biotechnology that captured the imagination of a large number of people was IFN, which at the time was seen by many as a possible miracle cure for a wide range of diseases, including viral infections and cancer. Thus, the isolation of a human IFN cDNA and its subsequent expression in *E. coli* were reported in newspapers and magazines around the world.

Several features of IFN made it particularly difficult to synthesize and isolate a cDNA encoding the polypeptide. First, although IFN had been purified more than 80,000-fold, only minuscule amounts were available, so researchers did not even have an accurate estimate of its molecular mass. Second, unlike many other proteins, IFN did not have a chemical or biological activity that was easy to monitor. At the time, its activity was measured by the reduction in the cytopathic effect of an animal virus on cells in culture, which was an extremely complex and time-consuming process. Third, unlike with insulin, researchers had no idea if there was

one particular human cell that produced high levels of IFN and therefore could serve as a source of mRNA that was enriched for IFN mRNA. These problems notwithstanding, a cDNA encoding IFN was eventually isolated and characterized. Since that time, researchers have discovered that there are several different types of IFNs. Unfortunately, IFN is not the panacea that was dreamed of by both investors and the press. However, the genes for several IFNs have been isolated, and clinical trials have shown that they are effective treatments for a variety of viral diseases.

A

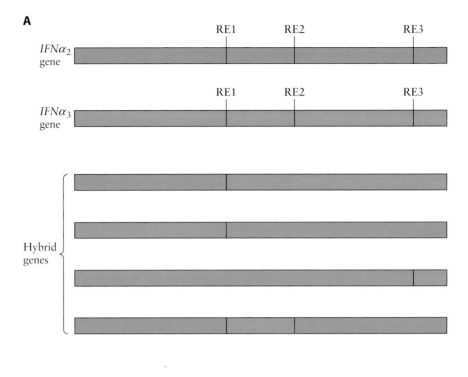

B

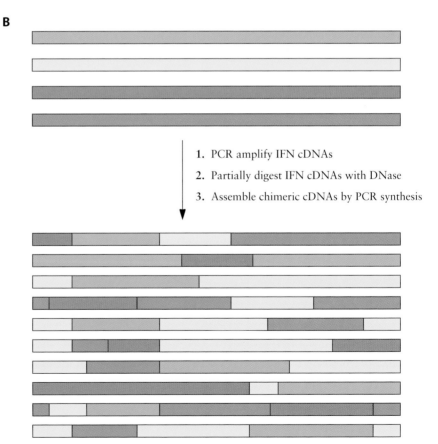

1. PCR amplify IFN cDNAs

2. Partially digest IFN cDNAs with DNase

3. Assemble chimeric cDNAs by PCR synthesis

Figure 9.1 Two strategies for constructing hybrid IFN genes. (**A**) Structures of the IFN-α2 and IFN-α3 genes and four hybrid genes. Comparison of the sequences of the IFN-α2 and IFN-α3 genes shows shared restriction enzyme sites (RE1, RE2, and RE3). Digestion of the genes at the indicated restriction sites and ligation of the resulting fragments generate a number of different hybrid IFN genes, of which four possibilities are shown. (**B**) Construction of hybrid IFN-α genes following partial DNase digestion and PCR synthesis of several different IFN-α genes. The resultant IFN-α gene-shuffled libraries are tested for antiproliferative and antiviral activities. doi:10.1128/9781555818890.ch9.f9.1

In one study, hybrid genes from IFN-α2 and IFN-α3 were constructed in an effort to create proteins with novel IFN activities. Comparison of the sequences of the two IFN-α cDNAs indicated that they had common restriction sites at nucleotide positions 60, 92, and 150. Digestion of both cDNAs at these sites and ligation of the DNA fragments yielded a number of hybrid derivatives of the original genes (Fig. 9.1A). These hybrids were expressed in *Escherichia coli*, and the resultant proteins were purified and tested for various biological functions. Some of the hybrid IFNs were found to provide a greater level of protection of mammalian cells in culture against viral infection than the parental molecules. Other hybrid IFNs had an antiproliferative activity against various human cancers that was greater than that of either of the parental molecules. In addition, another group of scientists generated other hybrid IFN molecules by a variation of the above-mentioned procedure. In this case, the entire IFN-α cDNA family was amplified by polymerase chain reaction (PCR) and then partially digested with **DNase I** into small DNA fragments (∼50 to 60 nucleotides long) before the fragments were shuffled and amplified by PCR (Fig. 9.1B). This procedure works to generate new IFN cDNAs because, following the denaturation step, the PCR mixture contains many overlapping single-stranded DNAs that can act as PCR primers. After testing of the many shuffled IFN cDNAs, it is possible to select hybrid IFNs with vastly improved antiviral or antiproliferative activities. Some of the hybrid IFNs have undergone successful clinical trials (Box 9.1) and have been approved for use as human therapeutic agents. The strategies for creating hybrid IFNs can also be applied to other gene families whose products have therapeutic potential.

Hepatitis C virus infection is one of the most common causes of liver disease, which affects nearly 200 million people worldwide. Many of these individuals eventually develop either cirrhosis of the liver or hepatocellular carcinoma (liver cancer). Therapeutic agents that maximize the early antiviral response and maintain viral suppression throughout the course of therapy have the best chance of achieving lasting eradication of the virus from an infected individual. One effective treatment for hepatitis C includes the combined use of the antiviral chemical compound ribavirin and IFN-α. However, for this sort of application, longer-acting IFNs are needed so that the side effects from IFN treatment can be minimized, e.g., by lowering the required dosage and decreasing the required frequency of the treatments. One approach to creating long-acting IFNs includes **pegylation** of the therapeutic agent. The process of pegylation entails covalently attaching polyethylene glycol to proteins (Fig. 9.2). This binding is typically achieved by incubation of a reactive derivative of polyethylene glycol with the target protein molecule. As a consequence of pegylation, the effective size of modified IFNs in solution increases, thereby prolonging the circulatory time of the therapeutic protein by reducing its renal (kidney) clearance.

Another means of generating longer-acting IFNs is to fuse an IFN gene with the gene for a stable and relatively inert protein, such as **human serum albumin**, that, after translation, produces a stable hybrid protein. This combination is an albumin–IFN hybrid molecule (Zalbin, formerly Albuferon) that retains all of the biological activity of the native IFN

PCR
polymerase chain reaction

Figure 9.2 (A) Structure of polyethylene glycol. **(B)** Schematic representation of a pegylated therapeutic protein. The polyethylene glycol residues attached to the protein (shown as zigzag lines) shield the protein from proteases and antibodies and increase the protein's effective size and solubility, thereby decreasing the rate at which it is cleared from the kidneys.
doi:10.1128/9781555818890.ch9.f9.2

A

Polyethylene glycol

B

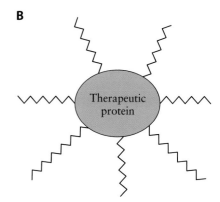

box 9.1
Clinical Trials

After the discovery of a new drug or course of treatment, and before it is made available to the public, it is essential that extensive studies and analysis of its safety and efficacy be conducted and then reviewed by an impartial agency. Although a large number of countries have developed their own approaches to test new therapeutics, the "gold standard" for clinical trials is the set of requirements established by the FDA. This process is briefly described here.

The preclinical phase of therapeutic drug development (i.e., the initial stage of the process of bringing a new therapeutic agent to market) entails thorough and extensive laboratory research on the mode of action, structure, and other biochemical and physical properties of a potential new drug. Scientists working at universities, research institutes, and drug and biotechnology companies are continually discovering and testing new molecules, as well as new uses for known compounds. However, it is impossible to know with any certainty which avenues of research will eventually bear fruit. Once a promising result has been obtained in the laboratory and has been shown to be reproducible, sufficient quantities of a highly purified version of the potential therapeutic compound must be produced so that it can be tested on small animals, such as mice. If the animal tests are positive and there is no evidence of any serious side effects, the organization seeking to commercialize the research files an "investigational new drug" application with the FDA. This is an application to begin the process of clinical trials. Based on the preclinical research data

that are provided, about 85% of these applications are approved.

Clinical trials are conducted in three distinct phases (described below), generally requiring a total of about 7 to 9 years at a cost of approximately $100 million to $400 million to complete. At each stage, various compounds are dropped from consideration based on the results obtained. Eventually, approximately 20% of the compounds that looked promising based on preclinical results will, after a careful review of all the data, finally be approved. This slow and expensive process is claimed to be "the most objective method ever devised to assess the efficacy of a treatment" (Zivin, 2000). The three phases of the FDA review process are as follows.

Phase I: With between 10 and 100 healthy people, the safety of the drug and, starting with very low doses, the highest dosages that can be administered are assessed. When there is a chance that serious side effects may result, individuals affected with the disorder that the drug is designed to alleviate may be used.

Phase II: With 50 to 500 affected patients, the optimal dosing regimen is determined. A control group is used so that it is possible to clearly distinguish between the effects of the drug and the natural remission of the disease. The use of a control group also helps to delineate real from apparent side effects of the treatment.

Phase III: Depending upon the disease, approximately 300 to 30,000 patients who have the disease are tested. After it is established that the drug is not harmful (phase I)

and the optimal dosing regimen has been determined (phase II), the effectiveness of the treatment needs to be proven.

The requirement for careful and thorough clinical trials ensures both the safety and efficacy of approved drugs. However, since the costs of both the preclinical research and the clinical trials are borne by pharmaceutical companies, this system makes it difficult for small companies that discover a new product to eventually bring that product to market without the involvement of a large corporation with significant financial resources. Furthermore, the high cost of clinical trials and the low probability of a new drug's being approved mean that it is unlikely that therapeutic agents will even be considered for clinical trials unless there is a strong possibility that there will be significant financial gains from the sale of that agent. Representatives from the pharmaceutical industry estimate that in 2007 it cost more than $1.4 billion to develop a new FDA-approved drug, including both laboratory research and clinical trials. However, a very large fraction of this estimated cost of bringing drugs to market is the so-called "lost opportunity cost." That is, many pharmaceutical companies include in their calculations what their money would have earned if it had been invested in the stock market instead of in new drug research and development. Nevertheless, regardless of how one calculates the actual cost of bringing a new pharmaceutical to market, it is quite expensive and clearly beyond the means of most small to medium-size companies. This financial disincentive may discourage research on therapeutic agents for diseases that either affect only a relatively small number of people or affect only populations in poor, underdeveloped countries.

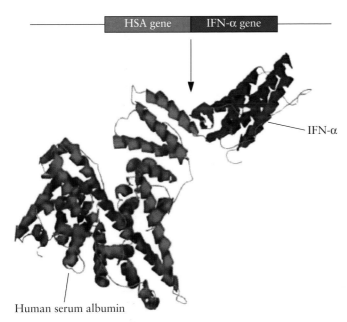

HSA gene	IFN-α gene

IFN-α

Human serum albumin

Figure 9.3 Schematic representation of the synthesis of the albumin–IFN fusion protein (Zalbin, formerly Albuferon), which includes human serum albumin (HSA) (red) at the N terminus and human IFN-α2b (blue) at the C terminus. Modified from an original image supplied by Human Genome Sciences.
doi:10.1128/9781555818890.ch9.f9.3

molecule (Fig. 9.3). Native IFN levels in the blood of a treated patient typically decrease rapidly, so 2 days after administration, they are undetectable. On the other hand, with the albumin–IFN hybrid molecule, the drug (in this case, the fusion protein) remains in serum at a therapeutically effective level for a much longer time, so it needs to be administered no more than once every 2 weeks. Based on successful phase III clinical trials of the albumin–IFN hybrid molecule, this compound has recently been approved for human use.

Human Growth Hormone

Human growth hormone (somatotropin) is a 191-amino-acid pituitary protein that stimulates the production of insulin-like growth factor 1. Insulin-like growth factor 1 is an essential component of the promotion of growth, especially in children. Infants and children who lack sufficient endogenous levels of human growth hormone, patients with chronic renal insufficiency (defective kidneys), and individuals with Turner syndrome respond to treatment with growth hormone, which stimulates tissue and bone growth, increases protein synthesis and mineral retention, and decreases body fat storage. In addition, in 2004, the FDA approved the use of recombinant human growth hormone for individuals whose short stature was caused by a variety of medical conditions other than human growth hormone deficiency.

Human growth hormone was one of the first recombinant therapeutic proteins in the world to be approved for human use. Prior to its production by recombinant DNA technology, growth hormone used to treat various deficiencies was extracted from the pituitary glands of cadavers; growth hormone from other primates was found to be inactive in humans. In 1985, cases of Creutzfeldt-Jakob disease were found in individuals that had previously received cadaver-derived human growth hormone. Based on the assumption that infectious prions causing the disease were transferred along with the cadaver-derived growth hormone, this product was removed from the market.

The first recombinant growth hormone was called somatrem (Protropin); it was produced and marketed by Genentech beginning in 1985. It had an amino acid sequence that was identical to that of human growth hormone, except that there was an extra methionine residue at the N-terminal end of the peptide chain (which was thought to prolong its half-life). It was discontinued in the late 1990s. At the present time, six different major pharmaceutical companies manufacture and sell recombinant human growth hormone. The recombinant form of the protein is produced in *E. coli* and is identical to native pituitary-derived human growth hormone.

As a consequence of its relatively short half-life in plasma, human growth hormone therapy currently requires **subcutaneous** injection once a day. This treatment is both inconvenient and expensive. Treatment of children with human growth hormone typically entails daily injections during the years when the child is growing. The cost of the treatment varies depending on the country and the size of the child but is generally approximately $10,000 to $30,000 per year. Therefore, it would be advantageous to have a long-lasting form of human growth hormone. To this end, DNA encoding the extracellular domain of the human growth hormone receptor was fused to human growth hormone cDNA using a DNA fragment encoding a 20-amino-acid-long linker peptide consisting of four repeats of the amino acids Gly_4Ser (Fig. 9.4). This construct has a very strong tendency to dimerize as the growth hormone moiety from one molecule binds with the receptor portion of another molecule. When this growth hormone construct was tested in rats, a single injection promoted growth for 10 days (compared to the usual requirement in rats for daily injections). It is thought that the dimerization of the engineered growth hormone stabilizes human growth hormone in vivo so that it is cleared from plasma approximately 300 times more slowly than free monomeric human growth hormone. Under these conditions, the active monomeric form (Fig. 9.4A) is slowly released from the inactive dimeric growth hormone (Fig. 9.4B), allowing it to bind to the growth hormone receptor (Fig. 9.4C). This experiment is certainly intriguing; however, it remains to be determined whether humans respond in a similar manner to the dimerized complex.

In an attempt to boost muscle mass, human growth hormone has been abused by competitors in sports since the 1970s, and it has been banned by the International Olympic Committee. However, because urine analysis could not detect the presence of synthetic human growth hormone, the ban was unenforceable until the early 2000s, when blood tests that could distinguish between natural and artificial human growth hormone were developed. In February 2010, a British rugby player had the dubious distinction of being the first athlete to be caught illegally using recombinant human

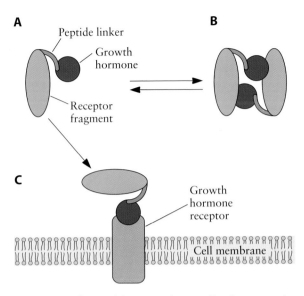

Figure 9.4 Derivatization of growth hormone by coupling it to a portion of the growth hormone receptor using a specific 20-amino-acid peptide. (**A**) Monomeric derivative; (**B**) dimeric derivative; (**C**) monomeric derivative bound to a growth hormone receptor. doi:10.1128/9781555818890.ch9.f9.4

growth hormone. In their agreement with Major League Baseball before the 2012 season, baseball players agreed to be randomly tested for the presence of excessive amounts of human growth hormone in their systems.

Tumor Necrosis Factor

Tumor necrosis factor (TNF) is a **cytokine** whose primary role is regulation of the immune system. TNF is produced as a 212-amino-acid-long transmembrane protein arranged in stable homotrimers that are subsequently cleaved by the metalloprotease TNF-converting enzyme to form a soluble homotrimeric cytokine TNF (Fig. 9.5A).

While a number of studies have clearly shown that TNF is a potent antitumor agent, it has not been widely used in this capacity because of its severe toxicity. If TNF could be delivered directly to its site of action, i.e., the tumor, then lower doses could be used and the unwanted side effects would be diminished. To develop a version of TNF with tumor specificity, DNA encoding the short peptide Cys-Asn-Gly-Arg-Cys-Gly, which targets a tumor cell surface protein, was fused to TNF DNA. The fusion protein contained the 6-amino-acid extension at its N-terminal end. In mice, the cytotoxic activities of Cys-Asn-Gly-Arg-Cys-Gly–TNF and TNF were identical, indicating that the additional amino acids did not prevent protein folding, trimer formation, or binding to receptors. However, the modified version of TNF was 12 to 15 times more effective at inhibiting tumor growth than the unmodified form (Fig. 9.5B). Moreover, a higher percentage of mice with **lymphoma** survived after treatment with the modified factor. In addition, all the mice that were treated with the modified factor and survived for 30 days survived second and third challenges with mouse lymphoma cells. These data indicate that there is a significant benefit, at least in mice, to fusing TNF with a short targeting peptide.

TNF
tumor necrosis factor

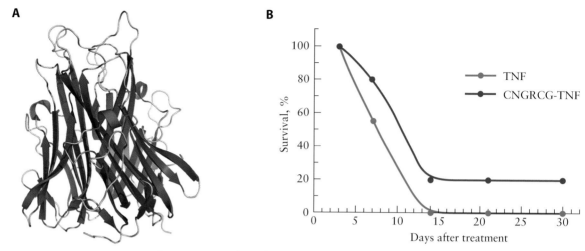

Figure 9.5 (A) Three-dimensional representation of TNF; (B) survival of lymphoma-bearing mice following treatment with 3 µg of either TNF or Cys-Asn-Gly-Arg-Cys-Gly–TNF (CNGRCG-TNF) as a function of the number of days after treatment. doi:10.1128/9781555818890.ch9.f9.5

Targeting Mitochondria

A mitochondrion is a membrane-enclosed organelle, about 0.5 to 1.0 µm in diameter (Fig. 9.6A), found in most eukaryotic cells that plays a critical role in the production of ATP and in cellular energy metabolism. The human mitochondrial genome consists of circular DNA approximately 16 kilobase pairs (kb) in size, encoding 37 separate proteins. Moreover, more than 1,000 additional proteins that function within the mitochondrion are encoded within the nuclear DNA, synthesized in the cytosol, and subsequently imported into the mitochondria, across the characteristic double membrane. The nucleus-encoded proteins are synthesized with a mitochondrial targeting sequence, which allows them to be imported into the mitochondria. Once a protein is inside, the mitochondrial targeting sequence is recognized and cleaved off, allowing the protein to function without any impediment.

Mitochondrial dysfunction can lead to a wide range of disorders. It has been estimated that mitochondrial disease affects approximately 1 in

Figure 9.6 (A) Targeting proteins to the mitochondria. Shown is a schematic view of a typical mitochondrion. (B) Amino acid sequence of the TAT peptide. (C) Domains of a therapeutic protein directed to the mitochondria, including the TAT peptide, MTS (a mitochondrial transit sequence), and LAD (the enzyme lipoamide dehydrogenase). doi:10.1128/9781555818890.ch9.f9.6

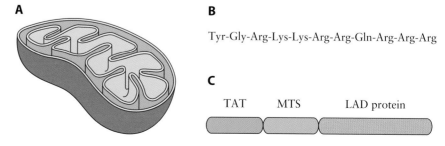

every 8,000 people in the population. Most mitochondrial disorders (encoded within the nuclear DNA) appear as neurological disorders, but they can also manifest as myopathy (muscle disease), diabetes, multiple endocrinopathy (endocrine disease and hormone imbalance), or any one of a variety of other systemic disorders. On the other hand, diseases that are caused by mutation of the mitochondrial DNA per se include Kearns–Sayre syndrome (a neuromuscular disorder), MELAS syndrome (mitochondrial myopathy, encephalopathy, lactic acidosis, and stroke; this syndrome is a form of dementia), and Leber hereditary optic neuropathy (a form of vision loss). The vast majority of the diseases that are caused by mutations to mitochondrial DNA are transmitted by a mother to her children, as the zygote derives its mitochondria from the ovum—i.e., mitochondrial DNA is maternally inherited. At the present time, there are no effective therapies for mitochondrial disorders. However, symptomatic therapy (i.e., reducing the manifestations of the disease) is typically utilized to deal with a variety of mitochondrial disorders, e.g., diabetes.

Currently, enzyme replacement therapy is used to treat several **lysosomal storage diseases** such as Gaucher disease, Fabry disease, Hunter syndrome, and Pompe disease. **Enzyme replacement therapy** might also be an effective strategy for mitochondrial disorders provided that a simple and effective means of directing enzymes to the mitochondrion can be found. In this regard, one possibility is to synthesize the target enzyme as a fusion protein with the 11-amino-acid-long arginine- and lysine-rich transcriptional activator of transcription (TAT) peptide (Fig. 9.6B). When protein transduction domains, which are short cationic peptides, are fused with other proteins, they facilitate passage through cell membranes. The most widely investigated and tested of these protein transduction domains is the TAT peptide that is equivalent to amino acid residues 47 to 57 of the Tat protein encoded by human immunodeficiency virus (HIV). TAT fusion proteins may be rapidly and efficiently introduced into cultured cells, intact tissues, and live tissues (when injected into mice) while retaining the biological activity of the target protein.

The protein lipoamide dehydrogenase (LAD), the third catalytic subunit of three multicomponent enzyme complexes in the mitochondrial matrix, including pyruvate dehydrogenase, is responsible for the metabolism of carbohydrates and amino acids. A defect in LAD results in extensive metabolic disturbances and biochemical abnormalities in an affected individual. The clinical course of LAD deficiency is variable, presenting in infancy as a neurological disease or later in life with life-threatening episodes of liver failure or myoglobinuria (i.e., the presence of myoglobin in the urine). In one series of experiments, the ability of the TAT peptide to help to deliver human LAD enzyme into human cells in culture was examined. In this case, the genetic construct that was used encoded the TAT peptide, a mitochondrial targeting sequence peptide, and the LAD enzyme (Fig. 9.6C). In fact, in these experiments, when it was attached to the TAT peptide, the enzyme LAD was efficiently delivered into cells in culture and to their mitochondria, restoring LAD activity to normal values. In a subsequent study using LAD-deficient mice, it was shown that TAT was able to mediate enzyme replacement therapy in vivo. That is, it

TAT
transcriptional activator of transcription

HIV
human immunodeficiency virus

LAD
lipoamide dehydrogenase

was possible to demonstrate the delivery of the TAT–LAD protein into deficient mouse tissues and that a single administration of the TAT–LAD protein resulted in a significant increase in the enzymatic activity of the mitochondrial multienzyme complex pyruvate dehydrogenase within the liver, heart, and brain in the treated mice. If this approach turns out to be as effective in vivo in humans as in mice, it has the potential to provide a general and highly effective means of treating individuals who have a wide range of different mitochondrial disorders.

Extending Protein Half-Life

The most common strategy for extending the in vivo half-life of therapeutic proteins is pegylation. In some instances, pegylation may also reduce the immunogenicity and increase the stability of therapeutic proteins. However, pegylation may significantly increase the costs of the therapeutic agent, and in some cases pegylated proteins or their metabolites may accumulate in the kidney, thereby interfering with normal kidney functioning. Thus, scientists sought to determine whether an unstructured synthetic protein might provide some of the benefits of pegylation without any of the drawbacks. In designing such a protein, it was decided to exclude all hydrophobic amino acids (which often lead to compact structures or protein aggregation). Amide-containing amino acids that may become unstable upon long-term storage, positively charged side chains that may cause binding to membranes, and cysteine residues that become either oxidized or cross-linked were also excluded. Thus, only six amino acids were used in the design of the unstructured protein, i.e., alanine, glutamic acid, glycine, proline, serine, and threonine. A library of randomized 36-amino-acid segments was constructed and screened for expression level. Highly expressed sequences were iteratively ligated and then rescreened for maximal activity. The five most highly expressed proteins were tested for (i) stability of the genetic construct, (ii) solubility, (iii) heat stability, (iv) aggregation resistance, and (v) the presence of contaminating host cell proteins and lipopolysaccharides, eventually resulting in the selection of a single 864-amino-acid protein called XTEN. The DNA encoding the XTEN protein was fused at a genetic level to the 3′ end of a target therapeutic protein, while a cellulose-binding domain was fused to the therapeutic protein gene at the 5′ end (Fig. 9.7). Following the translation of the construct into protein in *E. coli*, the cell lysate was purified by passage over a cellulose column, with the fusion protein binding strongly to the column. The cellulose-binding domain was subsequently removed from the fusion protein by digestion with a specific protease (the amino acid recognition site for that protease had previously been engineered into the region between the cellulose-binding domain and the therapeutic protein). The proteins (and peptides) exenatide (similar to glucagon-like peptide 1), glucagons, factor VII, and human growth hormone have all been fused to XTEN and purified as described above. In mice, rats, and monkeys, the plasma half-life of the fusion protein was at least 3 and as much as 125 times greater than in the absence of the fusion protein. It is argued that this approach is less expensive and more effective than pegylation or fusion to known proteins such as albumin. However, human testing has just begun.

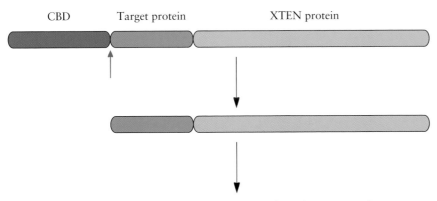

Test the physical, chemical, and biological properties of
the target protein-XTEN complex

Figure 9.7 Schematic representation of a tripartite fusion protein including a cellulose-binding domain (CBD), a target therapeutic protein, and the 864-amino-acid unstructured XTEN protein. The red arrow indicates the site where, following digestion with a specific protease, the CBD is removed.
doi:10.1128/9781555818890.ch9.f9.7

Engineered Bacteriophages

In recent years, the widespread clinical and agricultural use of antibiotics has played a significant role in the emergence of antibiotic-resistant strains of bacteria. For example, multidrug-resistant *Staphylococcus aureus* is especially troublesome in hospitals and nursing homes, where patients with open wounds, invasive devices, and weakened immune systems are at greater risk of infection than the general public. Increased rates of multidrug-resistant infections have been observed with various antibiotics, including glycopeptides, cephalosporins, and quinolones. There is some indication that antibiotic-resistant bacterial strains are emerging at a faster pace than new antibiotics are being discovered. The economic impact of multidrug-resistant infections is difficult to quantify, but it has been estimated to be between $5 billion and $24 billion per year in the United States. Should this trend continue, it is possible that the health of a large number of individuals may be put at risk. One long-term solution to this problem includes developing vaccines against different pathogenic bacteria. However, this is not a simple or inexpensive task. Therefore, scientists have sought to develop a variety of innovative approaches to deal with this problem. One approach includes utilizing lytic bacteriophages directed against various multidrug-resistant bacteria. However, just as bacteria can develop resistance to antibiotics, they can also develop resistance to infection and lysis by bacteriophages.

To avoid the development of bacteriophage-resistant bacteria, a recent and innovative approach, still in an early stage of development, utilizes a nonlytic bacteriophage to deliver a protein that increases bacterial susceptibility to specific antibiotics (Fig. 9.8). The LexA3 protein represses the bacterial SOS response, a global response to DNA damage in which DNA repair is induced. The *E. coli lexA3* gene was placed under the control of a strong inducible promoter and spliced into double-stranded bacteriophage M13, which has been previously used extensively in both DNA

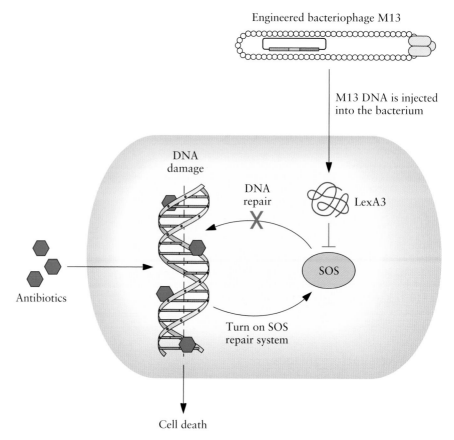

Engineered bacteriophage M13

M13 DNA is injected into the bacterium

DNA damage

DNA repair

LexA3

SOS

Antibiotics

Turn on SOS repair system

Cell death

Figure 9.8 Use of an engineered bacteriophage as an adjunct to antibiotic therapy. Bacteriophage M13 was engineered to introduce the *lexA3* gene, as part of the phage genome, into bacteria that infect an animal host. Upon induction, large amounts of the LexA3 protein are produced in the bacteria, which prevents the bacterial SOS system from repairing DNA that has been damaged by a quinolone antibiotic. The antibiotic is therefore more effective at killing the bacteria. doi:10.1128/9781555818890.ch9.f9.8

sequencing and directed mutagenesis protocols. Following synthesis of LexA3 inside a phage-transformed bacterium, the SOS system is constitutively repressed and therefore no longer able to repair DNA that is damaged by added quinolone antibiotics. Thus, the LexA3 protein enhances the bacteriocidal activity of certain antibiotics. In fact, this approach works well for a number of different antibiotics and not just quinolones. Moreover, this approach can enhance the killing of bacteria that have already acquired antibiotic resistance as well as the killing of wild-type bacteria. When the engineered bacteriophage system was tested in mice, it was found to significantly increase their survival after they were infected with a pathogenic strain of *E. coli*. Researchers are now developing engineered bacteriophage strains that have bacterial targets other than the SOS repair system, so one might eventually have engineered bacteriophages that use multiple mechanisms to debilitate pathogenic bacteria. However, a note of caution is in order, as bacteriophages have not yet gained acceptance in clinical practice. Thus, the first application of this technology is likely to be in the treatment of diseased animals.

Recombinant Antibodies

Previously, animal antisera were used, in a limited number of situations, to provide passive immunity to individuals who typically required immediate treatment for some highly dangerous situation such as a bite from a poisonous snake. However, since the antisera used in these treatments are immunogenic in humans, it is generally not possible to provide more than one such treatment without generating a severe immune response, including anaphylaxis, in the treated individual. One reason for the renewed interest in therapeutic antibodies is that it is now possible to engineer antibodies with a greatly reduced level of immunogenicity in humans. In addition, this technique can be used to maintain a continuous supply of pure monospecific recombinant antibodies. In fact, a number of monoclonal antibodies have been approved for treating human diseases, and many more are currently undergoing clinical trials (Table 9.2).

Preventing Rejection of Transplanted Organs

In the 1970s, passive immunization was reconsidered as a way of preventing immunological rejection of a transplanted organ. The rationale was to administer to patients a specific antibody that would bind to certain lymphocytes and diminish the immune response directed against the

Table 9.2 Some therapeutic monoclonal antibodies that have been approved for human use in either the United States or the European Union

Yr of approval	Antibody	Trade name	Antibody type	Therapeutic use
1986	Muromomab	Orthoclone	Murine	Prevention of acute kidney transplant rejection
1994	Abciximab	ReoPro	Chimeric	Prevention of blood clots
1997	Daclizumab	Zenapax	Humanized	Prevention of acute kidney transplant rejection
1998	Rituximab	Rituxan	Chimeric	Treatment of non-Hodgkin lymphoma
1998	Infliximab	Remicade	Chimeric	Treatment of Crohn disease, psoriasis, and rheumatoid arthritis
1998	Basiliximab	Simulect	Chimeric	Prevention of transplantation rejection
1998	Palivizumab	Synagis	Humanized	Treatment of viral infections in children
1998	Trastuzumab	Herceptin	Humanized	Treatment of metastatic breast cancer
2000	Gemtuzumab	Mylotarg	Humanized	Treatment of acute myeloid leukemia
2001	Alemtuzumab	Campath	Humanized	Treatment of chronic lymphocytic leukemia
2002	Adalimumab	Humira	Human	Treatment of rheumatoid arthritis
2002	Ibritumomab	Zevalin	Chimeric	Treatment of non-Hodgkin lymphoma
2003	Efalizumab	Raptiva	Humanized	Treatment of severe plaque psoriasis
2003	Omalizumab	Xolair	Humanized	Treatment of severe persistent asthma
2003	Tositumomab	Bexxar	Murine + iodine-131	Treatment of non-Hodgkin lymphoma
2004	Cetuximab	Erbitux	Chimeric	Treatment of various cancers
2004	Natalizumab	Tysabri	Humanized	Treatment of multiple sclerosis
2004	Bevacizumab	Avastin	Humanized	Treatment of various cancers
2006	Panitumumab	Vectibix	Human	Treatment of colorectal cancer
2009	Ofatumomab	Arzerra	Human	Treatment of chronic lymphocytic leukemia
2010	Denosumab	Prolia	Human	Treatment of osteoporosis in women
2010	Denosumab	Xgeva	Human	Treatment of bone cancer
2011	Belimumab	Benlysta	Human	Treatment of systemic lupus erythematosus

As of mid-2012, an additional 50 humanized or human monoclonal antibodies were undergoing clinical trials.

transplanted organ. In 1986, the mouse monoclonal antibody OKT3 was approved by the FDA for use as an immunosuppressive agent after organ transplantation in humans (Table 9.2). Lymphocytes that differentiate in the thymus are called T cells. Various members of the T-cell population act as immunological helper and effector cells and are responsible for organ rejection. The OKT3 monoclonal antibody binds to a cell surface receptor called CD3, which is present on all T cells. As a result, a full immunological response is blocked, and the transplanted organ is not rejected. Immunosuppression by this means was reasonably effective, although as anticipated, because the antibody was from a mouse, there were some side effects, including fever and rash formation.

Hybrid Human–Mouse Monoclonal Antibodies

The modular nature of antibody functions has made it possible to convert a mouse monoclonal antibody into one that has some human segments but still retains its original antigen-binding specificity. This hybrid molecule is called a chimeric antibody (Fig. 9.9A), or, with more human sequences, a "humanized" antibody (Fig. 9.9B). The difference between a chimeric and a humanized mouse monoclonal antibody is the portion of the mouse antibody that has been removed. In a chimeric antibody, the portion of the mouse monoclonal antibody that was targeted for replacement with a human sequence was the mouse Fc fragment. This is because the mouse Fc fragment functions poorly as an effector of immunological responses in humans, and it is also the most likely fragment to elicit the production of human antibodies. Thus, the DNA coding sequences for the Fv regions of both the light (L) and heavy (H) chains of a human immunoglobulin were substituted for the Fv DNA sequences for the L and H chains from a specific mouse monoclonal antibody (Fig. 9.9A). Chimeric antibodies are composed of approximately 70% human and 30% mouse DNA sequences.

The humanizing of mouse monoclonal antibodies has been taken one step further than the formation of chimeric molecules by substituting into human antibodies only the complementarity-determining regions (CDRs) of the mouse monoclonal antibodies (Fig. 9.9B). Humanized antibodies consist of approximately 95% human and 5% mouse DNA sequences. Because these engineered humanized antibodies have antigen-binding affinities similar to those of the original mouse monoclonal antibodies, they are more effective therapeutic agents and are less likely to generate an immune response.

The humanizing of mouse monoclonal antibodies may be performed as follows (Fig. 9.10). Starting with a mouse hybridoma cell line, cDNAs for the L and H chains are isolated. The variable regions of these cDNAs are amplified by PCR. The oligonucleotide primers that are used for this amplification are complementary to the sequences at the 5′ and 3′ ends of the DNA encoding the variable regions. From the nucleotide sequences of the cDNAs for the L and H regions (V_L and V_H), it is possible to delineate the limits of the CDRs. It is usually straightforward to determine where the CDRs begin and end, because these regions are

L chain
light chain

H chain
heavy chain

CDR
complementarity-determining region

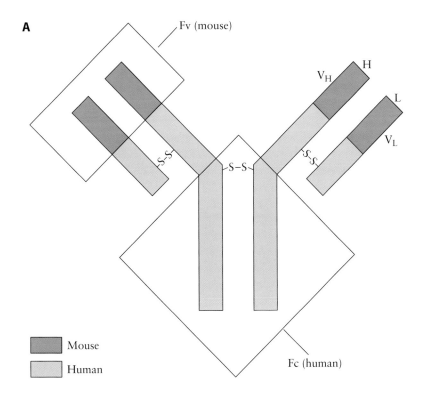

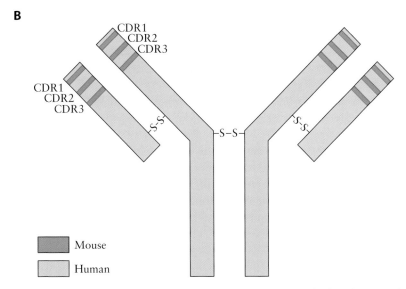

Figure 9.9 Genetically engineered antibodies. (**A**) Chimeric antibody. The V_L and V_H DNA regions from the immunoglobulin L and H genes that encode part of a mouse monoclonal antibody were substituted for the V_L and V_H DNA regions of a human immunoglobulin molecule. The product of the constructed gene is a chimeric (partially humanized) immunoglobulin with the antigen-binding specificity of the mouse monoclonal antibody and both lowered immunogenicity in humans and human Fc effector capabilities. (**B**) Humanized antibody. The CDRs (CDR1, CDR2, and CDR3) from the genes for H and L immunoglobulin chains of a mouse monoclonal antibody replace the CDRs of the genes for a human antibody. The product of this constructed gene is an immunoglobulin with the antigen-binding specificity of the mouse monoclonal antibody and all the other properties of a human antibody molecule. doi:10.1128/9781555818890.ch9.f9.9

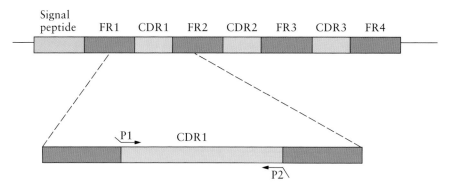

Figure 9.10 PCR amplification of CDR1 from a rodent monoclonal antibody L-chain cDNA. The PCR primers P1 and P2 contain oligonucleotides complementary to the rodent CDR1 DNA. In addition, P1 and P2 each contain 12 nucleotides at their 5′ ends that are complementary to the FRs of human monoclonal L-chain cDNAs. Using six separate pairs of oligonucleotide primers—three for the V_L region and three for the V_H region—each of the rodent CDRs is separately amplified by PCR. Then, by PCR, the amplified rodent CDRs are spliced into human antibody genes in place of the resident CDRs. This grafting is made possible by the presence of DNA complementary to the human FRs on the amplified rodent CDR DNAs. doi:10.1128/9781555818890.ch9.f9.10

FR
framework region

CHO
Chinese hamster ovary

highly variable in sequence, while the sequences of the framework regions (FRs) are relatively conserved. On the basis of the sequences of the DNAs encoding the rodent CDRs, six pairs of oligonucleotide PCR primers are synthesized. Each pair of primers is designed to initiate the synthesis of the DNA for one of the six mouse CDRs—three from the L chain and three from the H chain. In addition, each primer includes an extra 12 nucleotides at its 5′ end, complementary to the flanking regions within the human framework DNA into which the DNA for the mouse CDRs is targeted. Oligonucleotide-directed mutagenesis is then used to replace, one at a time, the complete DNA sequence for each of the human CDRs with the amplified DNA for the mouse CDRs. Thus, it is necessary to carry out six cycles of oligonucleotide-directed mutagenesis, one cycle to replace each CDR. This procedure, in effect, "grafts" the mouse CDRs onto the human antibody framework. The humanized variable-region cDNAs are then cloned into expression vectors, which are then introduced into appropriate host cells, usually Chinese hamster ovary (CHO) cells, for the production of antibodies.

To date, more than 50 different monoclonal antibodies have been humanized. While this technology is clearly effective and widely applicable, it is nevertheless time-consuming and expensive.

Human Monoclonal Antibodies

Although most of the immunotherapeutic agents that have been developed have been effective, there are drawbacks to the use of monoclonal antibodies that contain nonhuman sequences. For example, if multiple treatments are required, which is often the case, it is desirable that the antibody contain no or only a very limited amount of nonhuman sequences to prevent immunological cross-reactivity and sensitization of the patient.

Unfortunately, it is extremely difficult to create entirely human monoclonal antibodies for a number of technical and ethical reasons. Therefore, it has been necessary to devise other approaches for obtaining human monoclonal antibodies.

To address this need, researchers constructed a **XenoMouse** in which (i) the mouse antibody production machinery is inactivated and (ii) all of the human immunoglobulin loci (both L and H chains) are integrated into a mouse chromosome (Fig. 9.11). The human H-chain genes and the human κ and λ L-chain genes (κ and λ are different classes of L-chain genes) were cloned into a yeast artificial chromosome (YAC) that can carry very large amounts of foreign DNA. The YAC with the human

YAC
yeast artificial chromosome

Figure 9.11 Generation of a XenoMouse. Mouse antibody genes are inactivated by specific deletions in embryonic stem cells, which are subsequently used to generate transgenic mice unable to make antibodies. The human genes encoding immunoglobulin L and H chains are introduced on a YAC into mouse embryonic stem cells. These cells are used to generate transgenic mice able to synthesize both mouse and human antibodies. The mice generated from these two types of manipulation are cross-bred, and mice that can synthesize only human immunoglobulins are selected, immunized, and used to make hybridomas producing human antibodies. doi:10.1128/9781555818890.ch9.f9.11

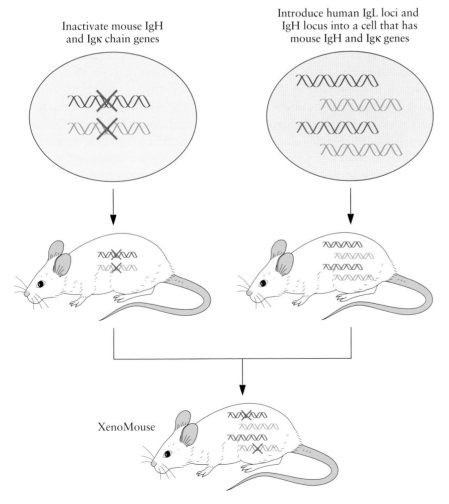

immunoglobulin genes was then introduced into mouse embryonic stem cells by fusing YAC-containing yeast spheroplasts (cells from which the cell wall has been removed) with the **embryonic stem cells**. This procedure yielded a large number of embryonic stem cells in which all of the introduced human immunoglobulin genes have become stably integrated into the chromosomal DNA. These transfected cells were used to generate mice containing human (as well as mouse) immunoglobulin gene loci. Crossbreeding of two mouse lines, one carrying both mouse and human immunoglobulin genes and the other carrying only the deleted mouse immunoglobulin genes, produced a mouse strain (called XenoMouse) that expresses only human immunoglobulins. It is now possible, after immunization of a XenoMouse with a particular antigen, to produce a fully human immunoglobulin, several of which have already been approved for human use, with many more at various stages of clinical development (Table 9.2).

Antibody Fragments

Naturally occurring antibodies provide animals with a powerful means of defending themselves against a wide range of pathogenic organisms and toxins. Immunoglobulin G (IgG) is the main antibody found in mammalian serum, and it is the form that is almost exclusively used in therapeutic antibodies (Fig. 9.12). The fact that IgG molecules have two identical sites that bind to two identical antigens (i.e., they are bivalent) generally increases their effectiveness in vivo. While the Fc portion of the IgG molecule is important in recruiting cytotoxic effector functions through complement or interaction with specific receptors, Fc-mediated effects are not necessary for all applications and may even sometimes be undesirable. By manipulating portions of the IgG L- and H-chain cDNAs, researchers have constructed a variety of IgG derivatives or fragments that may be used instead of whole antibody molecules (Fig. 9.12). Some of these

IgG
immunoglobulin G

Figure 9.12 Schematic representation of active antibodies and antibody fragments. doi:10.1128/9781555818890.ch9.f9.12

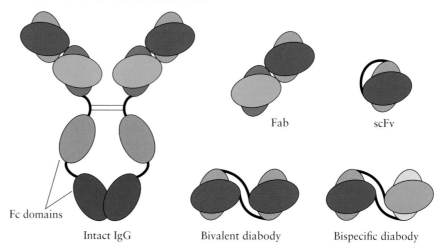

Fc domains

Intact IgG   Bivalent diabody   Bispecific diabody

Fab   scFv

molecules, because of their small size, bind more efficiently to targets that are inaccessible to conventional whole antibodies. Others have multiple sites for binding to the same antigen, while still others have binding specificities for two or more target antigens.

For example, antigen-binding single-chain antibodies (scFv), consisting of only V_L and V_H domains, may be used for a variety of therapeutic and diagnostic applications in which Fc effector functions are not required and when small size is an advantage. Single-chain antibodies have a molecular mass of approximately 27 kilodaltons (kDa), compared with approximately 150 kDa for IgG molecules. Because of their small size, single-chain antibodies can penetrate and distribute in large tumors more readily than intact antibodies. In addition, a protein-coding sequence can be linked to a single-chain antibody sequence to create a dual-function molecule that can both bind to a specific target and deliver a toxin or some other specific activity to a cell (Fig. 9.13).

Computer simulation of the three-dimensional structure of potential single-chain antibodies showed that the V_L and V_H domains have to be separated by a linker peptide to assume the correct conformation for antigen binding (Fig. 9.13A). Thus, DNA constructs of V_L and V_H sequences from a cDNA of a cloned monoclonal antibody were each ligated to a chemically synthesized DNA linker fragment in the order V_L–linker–V_H. After expression, the single-chain protein was purified, and both its affinity and specificity were found to be equivalent to those of the original intact monoclonal antibody. Moreover, instead of linking the V_H and V_L chains with a short peptide, amino acids in the FR portion of the molecule can be modified to form a disulfide linkage between the V_H and V_L chains (Fig. 9.13B). The effectiveness of this disulfide-stabilized Fv molecule (V_L–S–S–V_H) coupled to a cancer cell toxin was compared with that of an scFv molecule coupled to the same toxin. The disulfide-stabilized

Figure 9.13 (A and B) Schematic representation of the use of monoclonal antibody fragments to deliver other molecules to target cells. A toxin is coupled to an scFv molecule (**A**) and to a disulfide-stabilized Fv molecule (**B**). (**C**) A monoclonal antibody-based drug delivery system. The drug is coupled directly to a monoclonal antibody (or fragment). (**D**) An enzyme that converts an inactive prodrug to an active drug is attached to a monoclonal antibody (or fragment). The active drug is formed only in the immediate vicinity of the target cells. In all cases, the monoclonal antibody binds to a specific protein on the surface of the target cell. doi:10.1128/9781555818890.ch9.f9.13

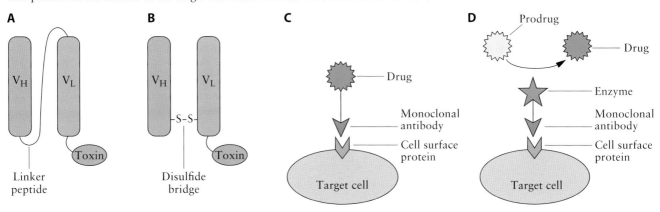

and scFv immunotoxins had the same activities and specificities. However, the former molecule was severalfold more stable than the latter. In addition, by altering the number of amino acids in the linker in an scFv molecule (usually to five or fewer amino acids), it is possible to direct the self-assembly of these molecules into either bivalent dimers (called diabodies) (Fig. 9.12), trimers (triabodies), or tetramers (tetrabodies). Shortening the linker affects not only the multimerization but also the stability of the molecule, with molecules with a shorter linker tending to be more stable. It is also possible to combine two different antigen specificities into a single bispecific diabody (Fig. 9.12). In this way, a wide range of small antibodies may be designed for different applications.

Drugs that are effective when tested in cell culture are often much less potent in a whole organism. This difference is typically due to the drug not being able to reach the targeted site in the whole animal at a concentration sufficient to be effective. Increasing the dose of a drug is usually not the answer to this problem, because high drug concentrations often have deleterious side effects. One strategy to enhance the delivery of a drug to its target site includes coupling the drug to a monoclonal antibody fragment that is specific for proteins found only on the surfaces of certain cells, e.g., tumor cells (Fig. 9.13C). Alternatively, a specific enzyme that converts an inert prodrug (precursor) to the active form of the drug may be coupled to a monoclonal antibody directed against a specific cell surface antigen (Fig. 9.13D). To ensure that the drug is released only in the vicinity of the target cells, the monoclonal antibody or single-chain antibody that is complexed with the prodrug-converting enzyme must bind to a protein that is highly specific to the target cell and be stable under physiological conditions but cleared rapidly from circulation. With this approach, only specifically targeted cells are exposed to the drug, permitting the use of a much lower concentration than if it was administered directly.

When an antibody (particularly one that is conjugated to a toxin or radiochemical) is able to destroy a tumor or pathogen cell, it is often advantageous to use antibody fragments, since the Fc portion of the molecule may impede or prevent the rest of the molecule from binding to relatively inaccessible antigens. However, despite the usefulness of antibody fragments in a variety of applications, a major limitation of using them as therapeutic agents is that, since they lack the Fc portion of the molecule, they are unable to mount a complete immune response.

One example of coupling a toxin to an antibody fragment is the fusion of a portion of *Pseudomonas* exotoxin A to an scFv. *Pseudomonas* exotoxin A is a 66-kDa protein with three separate domains; domain I is responsible for cell binding, domain II for translocation of the protein into the cell, and domain III for ADP-ribosylation (Fig. 9.14A). An immunotoxin is generally synthesized by replacing the N-terminal domain of the toxin, e.g., *Pseudomonas* exotoxin A (domain I), with the single-chain antibody sequence, thereby creating molecules very similar in size to the original toxin with the ability to bind, enter, and kill a specific cell (Fig. 9.14B). A number of immunotoxins that have antitumor activity in vitro and in animal models have been constructed. These include antibodies directed against the p55 subunit of the interleukin-2 receptor, the

A

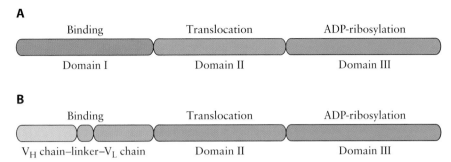

B

Figure 9.14 Domain structures of *Pseudomonas* exotoxin A (**A**) and a single-chain antibody–*Pseudomonas* exotoxin A fusion protein (**B**). The functions of the various domains are shown. doi:10.1128/9781555818890.ch9.f9.14

transferrin receptor, carbohydrate antigens, the epidermal growth factor receptor, and some cancer cell surface proteins. It is also possible to direct toxin molecules to cancer cells by using a bispecific diabody that is engineered to bind to a surface-specific tumor-associated antigen and then to a toxin molecule, thereby directing the toxin molecule to the tumor (Fig. 9.15). Several different engineered immunotoxins are currently in clinical trials.

It may be possible to create peptides that are smaller than scFvs and still retain the ability to bind to a specific antigen. The rationale for developing smaller antibody–toxin complexes is that they are more likely to penetrate a tumor and may therefore more completely stop tumor growth. It is well established that antibody-binding specificity resides within the six hypervariable loops, CDRs, three from the variable region of the H chain and three from the variable region of the L chain (Fig. 9.16). In all antibody molecules, the CDRs are flanked by FRs. Hence, it was speculated, at least for some antibodies, that the major portion of the antigen-binding site might reside primarily within two CDRs, one from the H chain and the other from the L chain, and not require significant contributions toward antigen binding from the four other CDRs. To test this possibility, starting

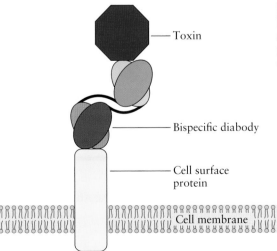

Figure 9.15 Schematic representation of the binding of a diabody to a protein molecule on the surface of a cancerous cell, as well as the binding of a toxin protein molecule to the other portion of the diabody. doi:10.1128/9781555818890.ch9.f9.15

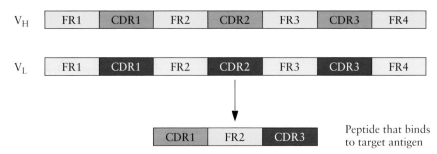

Figure 9.16 Organization of V_H and V_L regions of a monoclonal antibody and the development of a peptide, based on portions of the CDRs and FRs of the V_H and V_L regions of the antibody molecule, with a similar binding specificity. doi:10.1128/9781555818890.ch9.f9.16

with genes for the variable portion of a monoclonal antibody against a surface protein from **Epstein–Barr virus**, eight different peptide combinations were synthesized. Each peptide contained at least one CDR3 loop (known to be the major antigen-contacting segment), as well as one other CDR loop and an FR spacer (which acts as a linker peptide). All eight of these peptides were tested in vitro for the ability to compete with the parental antibody for binding to the Epstein–Barr virus (thought to be the causative agent of **Burkitt lymphoma** and other cancers) surface protein. One of the peptide combinations, V_HCDR1–V_HFR2–V_LCDR3 (Fig. 9.16), appeared to be promising. The short peptide (28 amino acids long, or ~3 kDa) apparently retained the binding specificity of the whole monoclonal antibody (~150 kDa) from which it was derived.

Next, this peptide was coupled to a toxin molecule, colicin Ia, and the combination was tested both with cells in culture and with mice. Colicin Ia has a molecular mass of approximately 69 kDa. In mice, the peptide–colicin adduct efficiently traveled through the circulatory system and then found and killed the tumor cells expressing the target antigen. Importantly, without the antibody fragment, colicin Ia by itself does not affect these tumors to any significant extent. Also, the original, full-size monoclonal antibody is unable to penetrate into the tumor. On the other hand, the peptide–colicin adduct accumulated at the cores of the targeted tumors. This very exciting work is at an early stage of development, so a large number of issues remain to be addressed before it can become an effective human therapeutic measure. Nevertheless, the demonstration that a small peptide can mimic the binding specificity of an entire antibody molecule and successfully deliver a cellular toxin to targeted cells may provide the basis for a whole new approach for treating certain tumors.

Combinatorial Libraries of Antibody Fragments

An elaborate series of manipulations makes it possible to select, as well as produce, functional antibodies in *E. coli*. Following cDNA synthesis from mRNA isolated from mouse antibody-producing cells (B lymphocytes), the variable portions of the H- and L-chain sequences in the cDNA preparation are amplified separately by PCR. The cDNA sequences of the H and L chains are each cloned, separately, into a bacteriophage λ vector that can carry relatively large fragments of DNA (Fig. 9.17A and B). The cDNAs of one H and one L chain are cloned into a single "combinatorial" vector, thereby enabling the bacteriophage-infected *E. coli* to coexpress both chains, thus forming an assembled antibody Fv fragment (Fig. 9.17C). The H and L chains are expressed in *E. coli* during the lytic cycle of bacteriophage λ, and the library of combinatorial bacteriophage clones is screened for the presence of antigen-binding activity.

The step in which L- and H-chain cDNAs are combined on one vector creates a vast array of diverse antibody genes, some of which encode unique target-binding sites whose isolation would never have been possible by standard hybridoma procedures. The mammalian antibody repertoire has the potential to produce approximately 10^6 to 10^8 different antibodies (from 10^3 to 10^4 L chains and the same number of H chains). A bacteriophage λ library contains approximately this number of clones, so one combinatorial library can be expected to produce as many different antibodies (in this case, Fv molecules) as a mammal. In addition, it is possible to obtain even greater variation by random mutagenesis of the DNAs in the combinatorial library. Moreover, because millions of bacteriophage plaques can be screened in a relatively short period, the identification of Fv molecules with the desired antigen specificity typically takes only about 7 to 14 days.

Because they lyse bacterial host cells, bacteriophage λ vectors are not particularly useful for the production of large quantities of protein. To overcome this drawback, the bacteriophage λ vector was engineered so that the H- and L-chain DNA sequences were inserted into a site that was flanked by plasmid DNA sequences. This plasmid DNA, containing an H- and L-chain DNA combination, can be excised from the bacteriophage λ vector and transformed into *E. coli*. As part of a plasmid, large numbers of Fv fragments can be produced in *E. coli* cells.

As an alternative to the use of bacteriophage λ, filamentous bacteriophages, such as M13 and fd, have also been used for the production of combinatorial libraries (Fig. 9.17D). Following infection of *E. coli* by the bacteriophage, the antibody fragment is synthesized as part of a fusion protein that is located on the outer surface of the bacteriophage. A combinatorial library of antibody fragments displayed on the surface of a filamentous bacteriophage can be screened by an enzyme-linked immunosorbent assay-like system using a multiwell plate that is coated with the target antigen. This approach is easier and faster than using plaque assays with bacteriophage λ to select and subsequently purify a bacteriophage producing an antibody fragment that binds to a specific antigen. Once a desired antibody fragment-encoding bacteriophage (either bacteriophage λ or M13) has been identified, the DNA can be isolated and subcloned into an expression vector.

Until recently, all of the combinatorial libraries of antibodies included either single-chain antibodies or Fab fragments and not full-length antibodies. However, for many applications, it is advantageous for therapeutic antibodies to be full length. With this in mind, researchers have also cloned and expressed complete antibody molecules in *E. coli*.

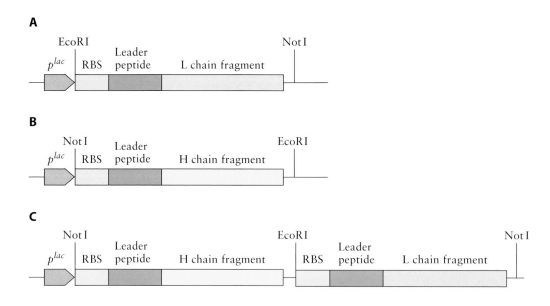

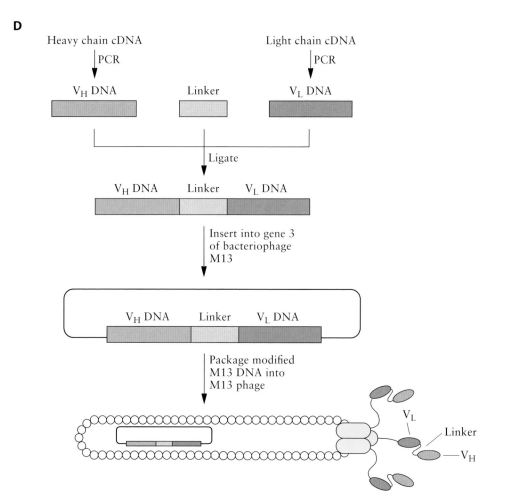

Figure 9.17 DNA constructs of an Fv combinatorial gene library cloned into bacteriophage λ DNA. (**A and B**) Portions of the cDNAs of the L (**A**) and H (**B**) chains are separately cloned into bacteriophage λ vectors. (**C**) The sequences encoding the fragments from the H-chain library (following EcoRI digestion) are ligated to the sequences encoding the fragments from the L-chain library, thereby creating a combinatorial library that contains all possible combinations of L- and H-chain fragments. (**D**) Formation of an Fv antibody combinatorial library in the filamentous bacteriophage M13. Following extraction of mRNA and its conversion to cDNA, the cDNAs for the V_L and V_H regions are amplified (separately) by PCR and then ligated to DNA that encodes a short linker peptide. Each single-chain antibody–DNA construct is cloned into gene 3 of bacteriophage M13. There are three copies of the phage gene 3 protein, which is a phage surface protein, per M13 bacteriophage; these are expressed after the modified bacteriophage is used to infect *E. coli*. p^{lac}, *E. coli lac* promoter; RBS, ribosome-binding site.
doi:10.1128/9781555818890.ch9.f9.17

Anticancer Antibodies

A number of therapeutic antibodies are directed against antigenic proteins that are overexpressed on the surfaces of cancer cells compared to noncancerous cells. For example, women with a type of breast cancer that overproduces the human epidermal growth factor receptor 2 (HER2) protein on the surface of the tumor may be effectively treated with the humanized anti-HER2 monoclonal antibody trastuzumab (Herceptin) (Box 9.2). Another example of a highly successful anticancer antibody is rituximab (Rituxan) (Box 9.3), which is used to treat patients with non-Hodgkin lymphoma. Unfortunately, with this approach, (i) antibodies directed against these proteins may also bind to some noncancerous cells expressing the same or a similar antigen, and (ii) targeting overexpressed surface proteins presents researchers with only a limited number of targets for therapeutic antibodies. One way to select for additional cell surface targets would be to identify proteins whose expression is selectively induced in tumor cells exposed to chemotherapeutic drugs. For example, when colorectal cancer cells were treated with the drug **irinotecan**, which is a topoisomerase enzyme inhibitor and is commonly used to treat this type of cancer, several newly synthesized proteins were found on the surfaces of those cells. (Irinotecan itself is a major chemotherapeutic agent, with sales of >$560 million in 2008. Topoisomerases are enzymes that unwind

HER2
human epidermal growth factor receptor 2

---

box 9.2
Trastuzumab: The First Humanized Monoclonal Antibody Approved for the Treatment of Breast Cancer

In 25 to 30% of women with aggressive metastatic breast cancer, there is a genetic alteration in the *HER2* gene that results in the production of an increased amount of HER2 protein on the surface of the tumor. Overexpression of the HER2 protein can readily be determined by using an immunohistochemistry-based assay. Some years ago, researchers at the biotechnology company Genentech isolated a mouse monoclonal antibody with high affinity for the HER2 protein and then (using procedures similar to those described in this chapter) humanized it. The humanized anti-HER2 monoclonal antibody, trastuzumab (Herceptin), contains human FRs and mouse CDRs and is produced commercially using mammalian (CHO) cells grown in suspension culture as the host for the expression of the antibody. Antibodies produced in CHO cells

are glycosylated similarly to bona fide human antibodies. After humanization, trastuzumab was found to bind to the HER2 protein with a dissociation constant of approximately 5×10^{-9} M, indicating that the monoclonal antibody's high level of specificity for the substrate had been maintained through the process of humanization.

In the laboratory, and then in initial clinical trials with more than 800 patients, trastuzumab mediated antibody-dependent cellular cytotoxicity (i.e., it signaled the immune system to target the cancerous cells) and inhibited the proliferation of human tumor cells that overexpressed HER2 (i.e., it stopped the cancerous cells from growing). Trastuzumab was most effective when it was administered together with some of the chemicals that are currently used for the treatment (chemotherapy) of breast cancer, provided that

the breast cancer was at a later stage of development. In two large clinical trials that included over 3,700 women, those who received trastuzumab and chemotherapy had a 52% higher chance that the cancer would not return than those who were treated with chemotherapy alone. Trastuzumab is provided by the manufacturer as a sterile white to pale yellow powder containing 440 mg per vial, and after reconstitution, it is typically administered intravenously over a period of 30 min and is taken weekly for 52 weeks. Since a small number of individuals treated with trastuzumab may develop heart problems, it is necessary to carefully monitor the cardiac functions of all patients on this therapy, especially older patients and those with a family history of heart problems. In the relatively short time that it has been available, trastuzumab has become a blockbuster drug, with annual sales of approximately $1.4 billion. In 2008, in the United States, trastuzumab treatment for one individual cost approximately $40,000 for the year.

box 9.3
Rituximab and Ibritumomab: Therapeutic Monoclonal Antibodies That Treat Non-Hodgkin Lymphoma

Non-Hodgkin lymphoma is a malignant growth of B or T cells of the lymph system. It has been estimated by the American Cancer Society that in 2007 alone approximately 63,000 new cases of non-Hodgkin lymphoma were diagnosed, resulting in approximately 19,000 deaths. In fact, about 5 million people worldwide have non-Hodgkin lymphoma, 5 to 10% of these people die every year, and the incidence of the disease is growing. It is the fifth most common cancer (although there are about 29 different lymphomas in this category), with an individual's chance of developing the disease in their life-time being about 1 in 50.

There are a variety of treatments for patients with non-Hodgkin lymphoma, including radiation therapy, chemotherapy, immunotherapy, bone marrow transplantation, and "watching and waiting" for slowly growing cases. In 1997, the FDA approved the use of rituximab (Rituxan) for the treatment of non-Hodgkin lymphoma. Rituximab is a genetically engineered chimeric (murine–human) monoclonal antibody directed against the CD20 antigen (a protein on the surfaces of B lymphocytes). Following binding of the antibody to CD20, the body's defenses attack and kill the antibody-marked B cells. Stem cells in bone marrow lack CD20, so they are uninhibited by this treatment. Healthy B cells can regenerate from those stem cells, after the completion of the course of rituximab treatment (given once a week for 4 to 8 weeks), and return to normal levels within several months. In 2006, the FDA approved the use of rituximab in combination with CHOP (cyclophosphamide, doxorubicin, vincristine, and prednisone) and other anthracycline-based chemotherapy regimens. In addition, the use of rituximab in combination with the chemical compound methotrexate was approved for the treatment of moderately to severely active rheumatoid arthritis in patients who had been refractory to other treatments.

Notwithstanding some severe side effects in some patients, rituximab has been enormously successful. In 2008, sales of rituximab were approximately $2.6 billion. Hundreds of thousands of people worldwide who did not respond well to conventional chemotherapy have been successfully treated with rituximab. In fact, while the incidence of non-Hodgkin lymphoma continues to increase, since the introduction of rituximab, mortality from the disease in the United States has declined at a rate of approximately 2.3% a year. In 2002, the FDA approved the use of ib-ritumomab tiuxetan (Zevalin) together with rituximab. Ibritumomab is also a monoclonal antibody that targets B cells. However, ibritumomab is linked to a chemical chelator molecule (tiux-etan) that binds tightly to radioactive indium-111 or yttrium-90. Thus, a therapeutic regimen with ibritumomab tiuxetan targets tumor cells with a high dose of radiation. In late 2007, treatment with ibritumomab tiuxetan was priced at approximately $24,000 per month, with treatments typically lasting 1 or 2 months. Treatment with ibritumomab tiuxetan is quite toxic, and around half of the treated individuals experience side effects. Therefore, ibritumomab tiuxetan is approved only for patients who have failed to respond to other treatments.

DNA during either DNA replication or mRNA transcription.) The new cell surface proteins were expressed early after exposure to the chemotherapeutic compound, prior to any major effects on the viability of treated cells. Monoclonal antibodies directed against one newly synthesized cell surface protein (called LY6D/E48) were generated, and then the antibodies were complexed with the cellular toxin **auristatin E**. The antibody–toxin conjugate was then used to treat tumor cells that were first treated with irinotecan (Fig. 9.18). Following binding to the cell surface protein, the antibody–toxin conjugate was internalized inside the tumor cell. With this strategy, in six out of eight mice, colorectal tumors disappeared entirely, while in the other two mice, the tumors were dramatically decreased in size. This exciting approach must be shown to be effective with larger numbers of animals before it can begin to be used in human clinical trials. However, provided that it is possible to identify one or more proteins that are specifically induced by chemotherapeutic agents and are not found on the surfaces of nontumor cells, this procedure could become an additional strategy that is used to treat a variety of different types of human cancer.

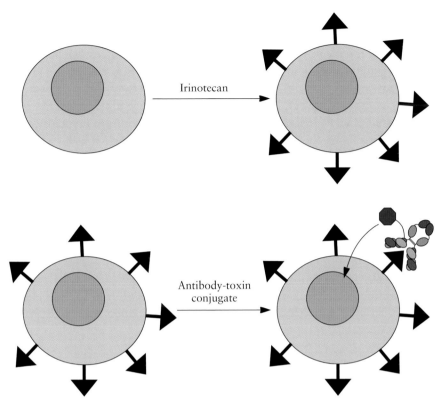

Figure 9.18 Targeting tumor cells for destruction by monoclonal antibody–toxin conjugates. Tumor cells are first treated with the chemotherapeutic agent irinotecan, which induces the synthesis of a unique cell surface protein. Then, a monoclonal antibody directed against the newly synthesized cell surface protein and conjugated to a toxin molecule is added. After binding of the antibody to the cell surface, the toxin is internalized, thereby killing the tumor cell. doi:10.1128/9781555818890.ch9.f9.18

Antianthrax Antibodies

Anthrax is a lethal disease in animals and humans that is caused by *Bacillus anthracis*, a spore-forming, gram-positive, rod-shaped bacterium. *B. anthracis* produces an exotoxin (anthrax toxin) made up of three proteins: (i) protective antigen (PA), (ii) edema factor (EF), and (iii) lethal factor (LF). These three proteins act together to impart their physiological effects. Assembled anthrax toxin complexes are taken up in an **endosome** of a target cell before they translocate into the cytoplasm. Once in the cell cytosol, the enzymatic components of the anthrax toxin disrupt cell signaling pathways that normally recruit immune cells, allowing the bacteria to evade the immune system, proliferate, and kill the host animal. PA binds to cellular receptors and forms a pore through which it delivers LF and EF into the cytosol, LF is a Zn-dependent protease that cleaves mitogen-activated protein kinase kinases (resulting in lysis of macrophages), and EF is a Ca–**calmodulin**-dependent adenylate cyclase (causing local inflammation and edema, i.e., abnormal accumulations of fluids). PA together with LF results in the formation of lethal toxin, while PA with EF forms edema toxin. PA is both the major component in the current anthrax vaccine and the target antigen for most monoclonal antibodies that are used to protect individuals against anthrax. However,

PA
protective antigen

EF
edema factor

LF
lethal factor

some researchers have expressed concern about the long-term efficacy of this approach. Thus, it was decided to develop anti-EF monoclonal antibodies with the idea of using them either alone or in concert with anti-PA monoclonal antibodies in the treatment of anthrax infection.

Following the immunization of chimpanzees with purified EF protein, a phage combinatorial cDNA library of chimpanzee antibody genes was constructed (Fig. 9.19). The library was screened immunologically, and four unique clones encoding Fab fragments specific for EF were selected. The Fab fragments were then converted to chimeric IgG molecules, with the remainder of the molecule being of human origin. One of the four selected antibodies was able to inhibit EF-mediated cyclic AMP production (adenylate cyclase activity). The same antibody also inhibited edema formation in mice treated with the anthrax toxin and protected mice against edema toxin-mediated death. In addition, the selected antibody both prevented Ca–calmodulin from binding to EF and displaced Ca–calmodulin that was already bound to EF. Thus, it is thought that the anti-EF monoclonal antibody that has been selected or constructed may be used either alone or in tandem with anti-PA antibodies to broaden the spectrum of protection against anthrax.

Figure 9.19 Flowchart of the development of a humanized monoclonal antibody against anthrax edema toxin. doi:10.1128/9781555818890.ch9.f9.19

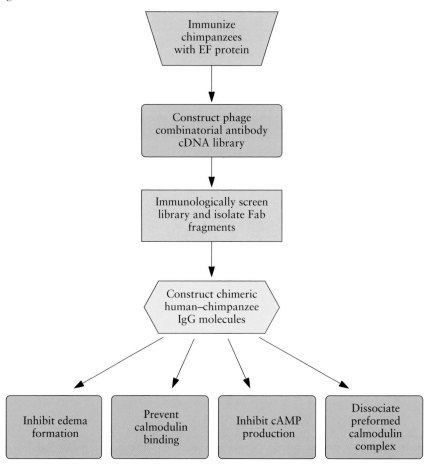

Antiobesity Antibodies

It has been estimated that as much as 15% of the world's population and more than 30% of the population of the United States are overweight or obese. Weight loss has become a major obsession, especially in many of the more affluent countries of the world, with billions of dollars spent annually on (mostly unsuccessful) attempts by individuals to lose weight. Notwithstanding the social components of carrying excess weight, overweight and obese individuals are significantly at risk for a number of potentially debilitating diseases. Consequently, there has been a considerable focus on finding a simple and painless way to deal with this epidemic.

Enzymes typically increase the rates of chemical reactions. They do this by stabilizing a high-energy transition state intermediate (between the substrate and the final product), which lowers the activation energy of the chemical reaction. Chemical analogues of the transition state intermediate (i.e., transition state analogues) mimic the high-affinity transition state intermediate but do not undergo any chemical reaction. As a result, transition state analogues can bind more tightly to an enzyme than either substrate or product analogues. Moreover, since transition state analogues do not undergo a chemical reaction, they act as enzyme inhibitors by blocking the active site.

In the mid-1980s, several investigators developed antibodies that had enzyme-like catalytic activity. They did this by first immunizing mice or rats with chemically synthesized small molecules that represented transition state analogues of enzyme-catalyzed reactions. The transition state analogues were chemically coupled to larger, generally inert, protein carrier molecules before the complex was used to immunize an animal. The transition state analogue haptens bind but do not react with the catalytic antibodies (a hapten is a small molecule that reacts with a specific antibody but generally cannot induce the formation of antibodies unless it is bound to a carrier protein). Rather, it is the substrates that resemble the transition state analogues that react with catalytic antibodies. Mice or rats that produce catalytic antibodies may then be used to produce monoclonal antibodies with the desired catalytic activity. The genes for those monoclonal antibodies may then be isolated, manipulated, and overproduced in a variety of different host organisms.

The active form of the 28-amino-acid peptide ghrelin, which stimulates food intake, is acylated on serine 3 with *n*-octanoic acid (Fig. 9.20A). In humans, ghrelin levels increase during dieting, with the unfortunate consequence that ghrelin facilitates weight regain and therefore impedes sustained weight loss. It is believed that lowering the level of this peptide could reduce food intake and promote weight loss.

To develop a catalytic antibody that is directed against ghrelin, the following approach was used (Fig. 9.20B). Several transition state analogues of ghrelin were chemically coupled to the protein keyhole limpet hemocyanin and used to inoculate mice. Keyhole limpet hemocyanin is an extremely large protein that is often used as an inert carrier protein in the production of antibodies. Its many surface lysine residues may be used to couple a large number of haptens to its surface. The inoculated mice then became the source of monoclonal antibodies that bind and then

A

Phe-Leu-Ser-Pro-Glu-His-Gln-Arg-Val-Gln-Gln-Arg-Lys-Glu-Ser-Lys-Lys-Pro-Pro-Ala-Lys-Leu-Gln-Pro-Arg

B

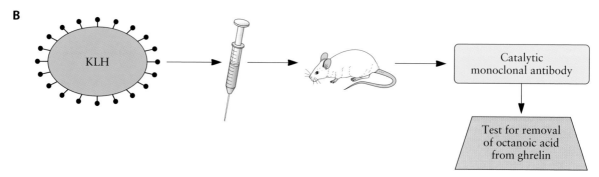

Figure 9.20 Developing a catalytic antibody designed to facilitate weight loss. (**A**) Structure of the 28-amino-acid peptide ghrelin, including acylation of the serine residue at position 3 with *n*-octanoic acid. (**B**) Chemically synthesized transition state analogues of ghrelin are chemically coupled to keyhole limpet hemocyanin (KLH) and then injected into mice. Catalytic mouse monoclonal antibodies that can both bind and cleave ghrelin are selected and tested for the ability to remove the *n*-octanoic acid moiety. This cleavage renders the ghrelin inactive.
doi:10.1128/9781555818890.ch9.f9.20

specifically remove the *n*-octanoic acid moiety from ghrelin (i.e., from the active form of the peptide molecule). The peptide in which the *n*-octanoic acid moiety is removed no longer has biological activity. Although this work is very preliminary, it is envisioned that this catalytic antibody may effectively lower ghrelin levels in humans and become an important adjunct to other weight loss strategies.

Enhanced Antibody Half-Life

It would be advantageous, for both patients and physicians, if the frequency that a patient needed to be treated with a particular therapeutic agent could be decreased as much as possible. This not only would save both patients' and physicians' time but also might decrease both the side effects and the costs of various treatments. To this end, scientists have worked to develop therapeutic antibodies with an extended in vivo half-life. One way to achieve this goal might be to engineer the Fc portion of a therapeutic antibody so that it binds more tightly to specific human Fc receptors. For example, one group of scientists constructed variants of the humanized anti-vascular endothelial growth factor (anti-VEGF) IgG antibody bevacizumab (Avastin), which has been approved by the FDA for the treatment of colorectal, lung, breast, and renal cancers. Variants of this antibody were constructed, based on a computer simulation of the interaction of the Fc region with its receptor, by directed mutagenesis of the portion of the antibody DNA encoding the Fc region (Fig. 9.21). When

anti-VEGF
anti-vascular endothelial growth factor

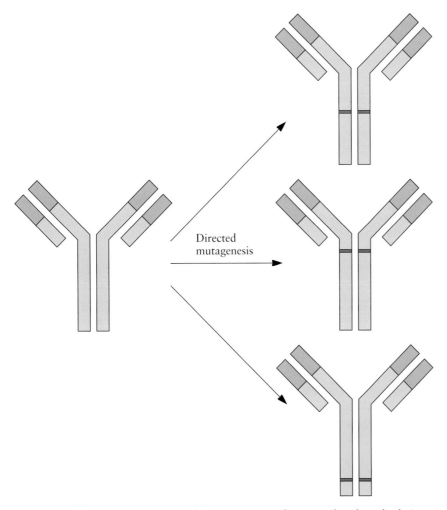

Figure 9.21 Directed mutagenesis of the Fc portion of a monoclonal antibody in order to develop more stable variants of the target antibody. The dark lines indicate amino acid changes within the Fc region. doi:10.1128/9781555818890.ch9.f9.21

the variants were tested in monkeys, not only did they bind more tightly to the receptor than the native form of the antibody, but also one had a serum half-life of 31.1 days, compared to 9.7 days for the native form of the antibody. This work has yet to be tested in humans; however, it suggests that it may be possible, for some antibody therapies, to decrease the frequency that a patient needs to be treated, thereby providing greater convenience to patients.

Enzymes

Enzymes may be used therapeutically in a variety of ways. For example, they may be used to augment an existing metabolic pathway, thereby increasing the amount of a particular compound or metabolite that is a product of that pathway. Alternatively, some enzymes may be used to relieve the disease pressure caused by a pathogen or may help to lower the level of an overproduced metabolite.

DNase I

Cystic fibrosis is one of the most common fatal hereditary diseases among Europeans and their descendants, with approximately 30,000 diagnosed cases in the United States and another 23,000 cases in Canada and Europe. It is estimated that a mutant cystic fibrosis gene is carried by 1 in 29 persons of European descent, 1 in 65 persons of African American descent, and 1 in 150 persons of Asian descent. Individuals with cystic fibrosis are highly susceptible to bacterial infections in their lungs. Antibiotic treatment of patients who have these recurring infections eventually leads to the selection of antibiotic-resistant bacteria. The presence of bacteria, some alive and some lysed, contributes to the accumulation of a thick mucus in the lungs of these patients, making breathing very difficult and acting as a source for further infection. The thick mucus in the lungs is the result of the combination of the alginate that is secreted by the living bacteria, the DNA that is released from lysed bacterial cells, and degenerating leukocytes that accumulate in response to the infection, as well as filamentous actin derived from the cytoskeletons of damaged epithelial cells (Fig. 9.22). To address this problem, scientists isolated the cDNA for the human enzyme deoxyribonuclease I (DNase I) and subsequently expressed the cDNA in CHO cells in culture. DNase I can hydrolyze long polymeric DNA chains into much shorter oligonucleotides. The purified enzyme is delivered in an aerosol mist to the lungs of patients with cystic fibrosis. The DNase I decreases the viscosity and adhesivity of the mucus in the lungs and makes it easier for these patients to breathe. While this treatment is not a cure for cystic fibrosis, it nevertheless relieves the most severe symptom of the disease in most patients. In 1994, the FDA approved the enzyme for human use; it had sales of more than $250 million in 2008.

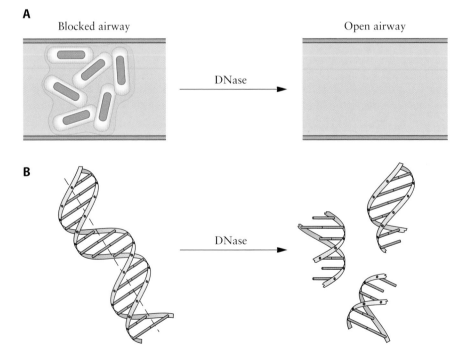

Figure 9.22 (A) Schematic representation of a portion of a human lung occluded by a combination of live alginate-secreting bacterial cells, lysed bacterial cells, and leukocytes and their released DNA being cleared by digestion by aerosol-delivered DNase I. (B) Digestion of DNA by DNase I.
doi:10.1128/9781555818890.ch9.f9.22

Alginate Lyase

Alginate is a polysaccharide polymer that is produced by a wide range of seaweeds and both soil and marine bacteria. Alginate is composed of chains of the sugars β-D-mannuronate and α-L-guluronate (Fig. 9.23). The properties of a particular alginate depend on the relative amounts and distribution of these two saccharides. For example, stretches of α-L-guluronate residues form both interchain and intrachain cross-links by binding calcium ions, and the β-D-mannuronate residues bind other metal ions. The cross-linked alginate polymer forms an elastic gel. In general, the structure and size of an alginate polymer determine its viscosity.

The excretion of alginate by mucoid strains of the bacterium *Pseudomonas aeruginosa* that infect the lungs of patients with cystic fibrosis significantly contributes to the viscosity of the mucus in the airways. Once mucoid strains of *P. aeruginosa* have become established in the lungs of cystic fibrosis patients, it is almost impossible to eliminate them by antibiotic treatment. This is because the bacteria form biofilms (Fig. 9.24A) in which the alginate prevents added antibiotics from coming into contact with the bacterial cells. In one experiment, it was shown that the addition of the enzyme alginate lyase, which can liquefy bacterial alginate, together with or prior to antibiotic treatment significantly decreased the number of bacteria found in biofilms (Fig. 9.24B). Thus, alginate lyase treatment not only decreases the viscosity of the mucus but also facilitates the ability of added antibiotics to kill the infecting bacterial cells. This result suggests that in addition to the DNase I treatment, depolymerization of the alginate might help clear blocked airways of individuals with cystic fibrosis.

An alginate lyase gene has been isolated from a *Flavobacterium* species, a gram-negative soil bacterium that is a strong producer of this enzyme. A *Flavobacterium* genomic DNA library was constructed in *E. coli* and screened for alginate lyase-producing clones by plating the entire library onto solid medium containing alginate. Following growth, colonies

Figure 9.23 A portion of an alginate molecule. The two residues on the left are α-(1→4)-linked L-guluronic acid (G), while the two residues on the right are β-(1→4)-linked D-mannuronic acid (M). Depending on its source, alginate may consist of blocks of similar alternating residues (e.g., MMMMMM, GGGGGG, and GMGMGMGM), all of which have different conformations and behaviors. doi:10.1128/9781555818890.ch9.f9.23

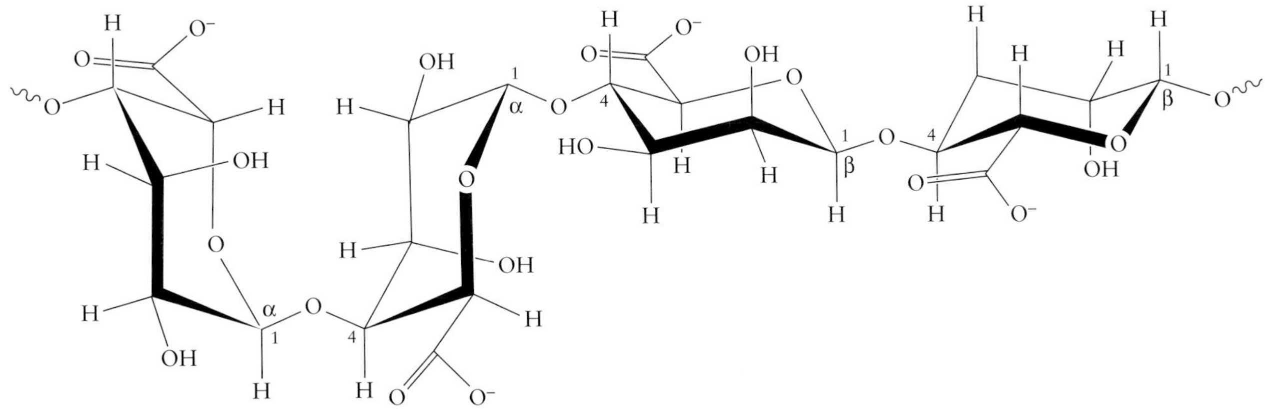

A

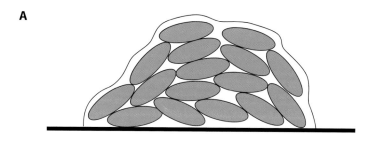

B

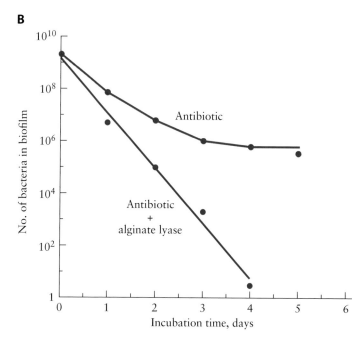

Figure 9.24 Schematic representation of the formation of a biofilm. (**A**) Shown is a bacterial microcolony encased in a thick layer of exopolysaccharide. (**B**) Time courses of the killing of bacteria in biofilms with and without treatment with alginate lyase. doi:10.1128/9781555818890.ch9.f9.24

that produced alginate lyase formed a halo around the colony when calcium was added to the plate (Fig. 9.25A). In the presence of calcium, all of the alginate in the medium, except in the immediate vicinity of an alginate lyase-positive clone, becomes cross-linked and opaque. Since hydrolyzed alginate chains do not form cross-links, the medium surrounding an alginate lyase-positive clone is transparent. Analysis of a cloned DNA fragment from one of the positive colonies revealed an open reading frame encoding a polypeptide with a molecular mass of approximately 69,000 Da. Detailed biochemical and genetic studies indicated that this polypeptide is a precursor of the three different alginate lyases produced by the *Flavobacterium* sp. After the 69,000-Da precursor is produced, a proteolytic enzyme cleaves off an N-terminal peptide of about 6,000 Da (Fig. 9.25B). The 63,000-Da protein can lyse both bacterial and seaweed alginates. Cleavage of the 63,000-Da protein yields a 23,000-Da enzyme that depolymerizes seaweed alginate and a 40,000-Da enzyme that is effective against bacterial alginate (Fig. 9.25B). To produce large amounts

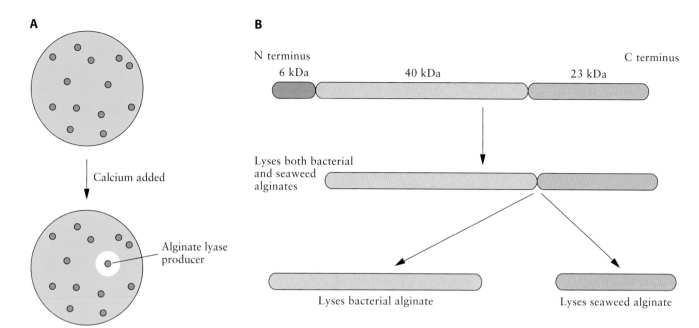

Figure 9.25 (**A**) Schematic representation of the detection of an alginate lyase-producing clone from a genomic DNA library of a *Flavobacterium* sp. in *E. coli*. The alginate that is present in the growth medium is digested by alginate lyase secreted by an *E. coli* clone. The alginate in the vicinity of such a colony is not cross-linked when calcium is added and instead produces a clear zone (halo) surrounding the colony. (**B**) Processing of the recombinant *Flavobacterium* alginate lyase protein precursor in *E. coli*. A 6-kDa leader peptide is removed from the N terminus of the 69-kDa precursor to yield a 63-kDa protein that can depolymerize alginate from both seaweed and bacteria. A second cleavage event converts the 63-kDa protein into a 23-kDa protein that is active against seaweed alginate and a 40-kDa protein that hydrolyzes bacterial alginate. doi:10.1128/9781555818890.ch9.f9.25

of the 40,000-Da enzyme, the DNA corresponding to the enzyme may be subcloned and expressed in bacterial cells. When bacteria that express and secrete the 40,000-Da enzyme are grown in liquid medium, the recombinant alginate lyase efficiently liquefies alginates produced by mucoid strains of *P. aeruginosa* isolated from the lungs of patients with cystic fibrosis.

Unfortunately, since most sources of alginate lyase are nonhuman, these proteins are not necessarily a good choice for use as a therapeutic agent, as they are likely to generate an immune response in the patient, thereby causing a range of complications. To significantly decrease the antigenicity of the alginate lyase, the enzyme may be chemically modified with polyethylene glycol (i.e., pegylation). However, nonspecific modification of alginate lyase with polyethylene glycol results in a large decrease in the activity of the enzyme. Nevertheless, it is possible to avoid the inactivation of the enzyme while still decreasing its antigenicity by specifically modifying alginate lyase with polyethylene glycol. To do this, several specific mutants of alginate lyase were created by directed mutagenesis in which one amino acid residue at a time was changed to cysteine (Fig. 9.26). Then, the modified alginate lyases were each reacted with a modified form of polyethylene glycol that derivatized only the newly created cysteine

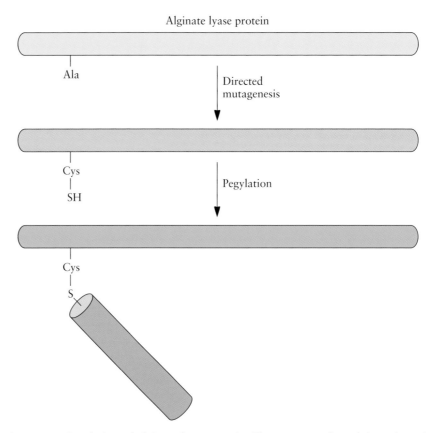

Figure 9.26 Pegylation of alginate lyase protein. The gene encoding alginate lyase is mutagenized so that the alanine residue in position 53 (from the N-terminal end) is changed to cysteine. The cysteine residue is modified by the addition of a polyethylene glycol derivative so that a covalent linkage is formed. The pegylated derivative is less immunogenic than the native form of the enzyme and more rapidly dissolves bacterial alginate. doi:10.1128/9781555818890.ch9.f9.26

residue. The modified (pegylated) enzymes were then tested for alginate lyase activity, antigenicity, and the ability to disrupt mucoid biofilms produced by the bacterium *P. aeruginosa*. One of the modified enzymes (in which an alanine residue at position 53 was changed to a cysteine residue) displayed behavior that was superior to those of both all of the other modified enzymes and the native enzyme. Specifically, the selected modified enzyme degraded bacterial alginate 80% more rapidly than the native enzyme, disrupted 94% of an established *P. aeruginosa* biofilm (compared to 75% disruption of the biofilm by the native enzyme), and had reduced immunogenicity. This result, although preliminary, is very encouraging. The next step will be to test this modified enzyme in animal models of cystic fibrosis.

Phenylalanine Ammonia Lyase

The human genetic disease phenylketonuria results from the impaired functioning of the enzyme phenylalanine hydroxylase. In the United States, about 1 of every 12,000 newborns has phenylketonuria. Approximately

500 different mutations in the gene encoding phenylalanine hydroxylase can lead to an enzyme with impaired function, preventing the disposal of excess phenylalanine. When phenylalanine hydroxylase, which oxidizes phenylalanine to tyrosine, is deficient, the normal cognitive development of an individual is impaired and mental retardation ensues due to a buildup of phenylalanine (Fig. 9.27A). Following diagnosis of phenylketonuria, either prenatally or shortly after birth, the treatment entails a controlled semisynthetic diet, first developed in the mid-1950s, with low levels of phenylalanine through infancy and sometimes for life. A possible alternative treatment would be the administration of phenylalanine hydroxylase. Unfortunately, phenylalanine hydroxylase is a multienzyme

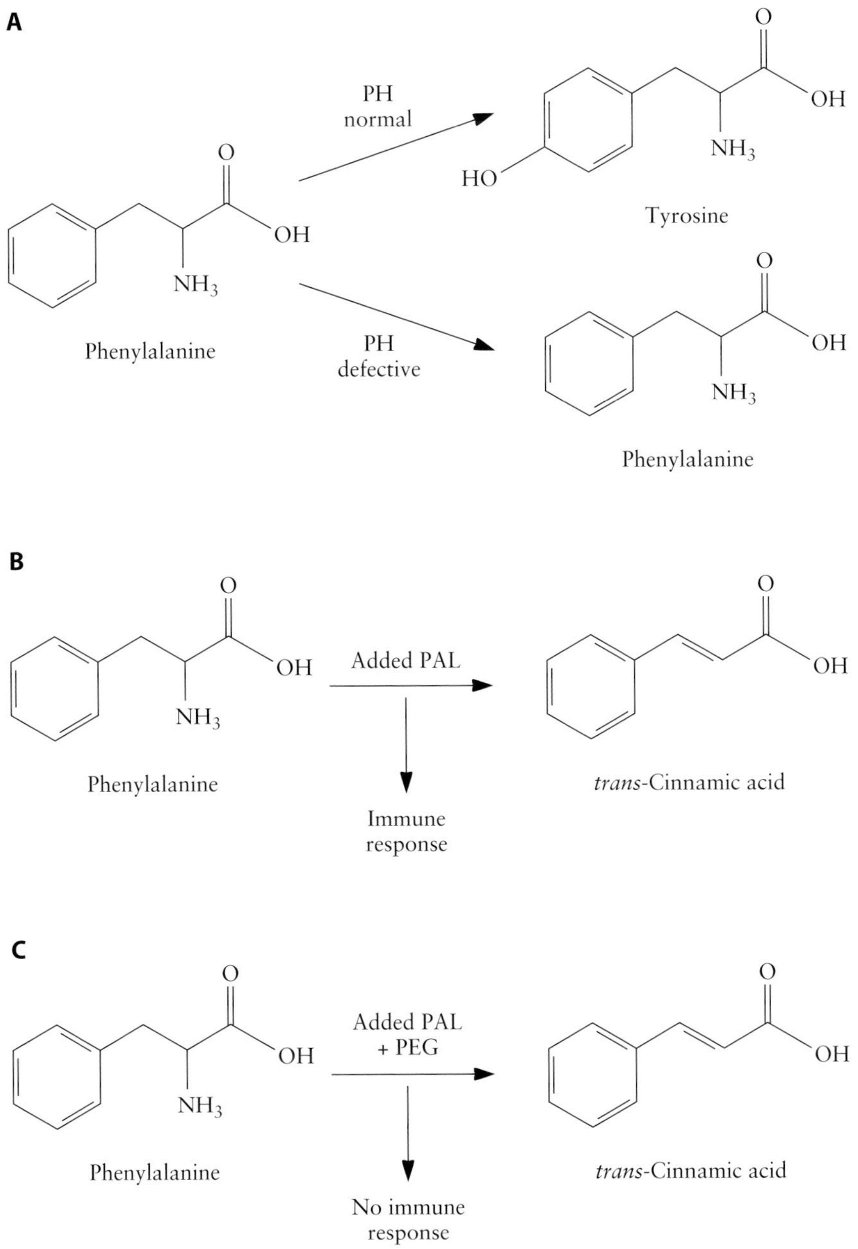

Figure 9.27 Fate of dietary phenylalanine in different individuals. Individuals with normal levels of phenylalanine hydroxylase (PH) activity convert phenylalanine to tyrosine, while those with a defective version of this enzyme activity do not (**A**). Individuals who are given phenylalanine ammonia lyase (PAL) can convert phenylalanine to ammonia and *trans*-cinnamic acid; however, they may mount an immune response directed against PAL (**B**). Immunogenicity is abolished when the PAL is pegylated (**C**). doi:10.1128/9781555818890.ch9.f9.27

complex that is not very stable and requires a cofactor for activity. This notwithstanding, some patients have successfully been treated with a pegylated version of phenylalanine hydroxylase. On the other hand, phenylalanine ammonia lyase, which converts phenylalanine to ammonia and *trans*-cinnamic acid (Fig. 9.27B), is a stable enzyme that does not require a cofactor and could potentially prevent the accumulation of phenylalanine in phenylketonuria patients. To test this concept, the gene for phenylalanine ammonia lyase from the yeast *Rhodosporidium toruloides* was cloned and overexpressed in *E. coli*. Preclinical studies were conducted with mice that were defective in producing phenylalanine ammonia lyase and therefore accumulated phenylalanine. With these mice, plasma phenylalanine levels were lowered when phenylalanine ammonia lyase was injected intravenously or encapsulated enzyme was administered orally. Thus, in short-term experiments in mice, phenylalanine ammonia lyase is an effective substitute for phenylalanine hydroxylase, and the orally delivered enzyme is sufficiently stable to survive the mouse gastrointestinal tract and still function.

Unfortunately, in humans, intravenous or subcutaneous injection of phenylalanine ammonia lyase results in an immune response, and oral delivery is less effective than is necessary to make this a useful ongoing therapeutic approach. To overcome the antigenicity of phenylalanine ammonia lyase, a series of formulations of linear and branched polyethylene glycols chemically conjugated to the enzyme was created. Following in vitro characterization of a series of pegylated enzyme derivatives, the most promising formulations were tested in vivo in mice that were defective in producing phenylalanine ammonia lyase. One linear 20-kDa pegylated phenylalanine ammonia lyase was found to no longer be immunogenic but to still retain full catabolic activity with phenylalanine, suggesting that it may have potential as a novel therapeutic agent for the treatment of phenylketonuria. This is an important step in the development of enzyme replacement therapy using phenylalanine ammonia lyase; however, a considerable amount of additional testing is still required.

α_1-Antitrypsin

The processing of a number of different pathogenic bacterial or viral precursor proteins by human proteases occurs when the protease recognizes the amino acid sequence Arg-X-Lys/Arg-Arg↓, with peptide bond cleavage on the C-terminal side of the C-terminal Arg (as indicated by the arrow) and where X is any of the 20 standard amino acids. Since a number of infectious agents contain this amino acid sequence and are therefore activated by this processing step, a therapeutic agent that targeted the processing enzyme and blocked its activity might act as a broad-spectrum antipathogenic (antibacterial and antiviral) agent (Fig. 9.28). When a variant of human α_1-antitrypsin was genetically engineered and tested in tissue culture experiments, the protein blocked the processing of HIV type 1 glycoprotein gp160, as well as measles virus protein F_0, and consequently, in both cases, the production of infectious viruses. When the α_1-antitrypsin variant was added to cell cultures, it blocked the production

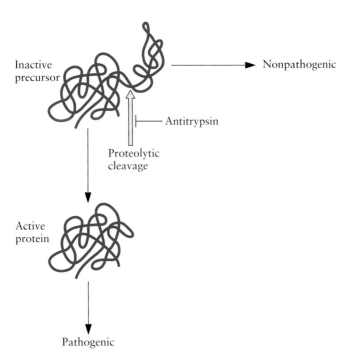

Figure 9.28 Schematic representation of α_1-antitrypsin inhibiting the proteolytic cleavage (activation) of pathogenic precursor proteins by human proteases. doi:10.1128/9781555818890.ch9.f9.28

of human cytomegalovirus, a major cause of illness and death in organ transplant recipients and AIDS patients. The α_1-antitrypsin variant is both potent and selective. Against human cytomegalovirus, it is at least 10-fold more effective than any currently used viral inhibitory agent. Its efficacy has been demonstrated in cell culture, but it remains to be determined if the strategy is effective with whole animals.

Glycosidases

The ABO blood group system is based upon the presence or absence of specific carbohydrate residues on the surfaces of erythrocytes, endothelial cells, and some epithelial cells. The monosaccharide that determines blood group A is a terminal α-1,3-linked *N*-acetylgalactosamine, while the corresponding monosaccharide of blood group B is α-1,3-linked galactose (Fig. 9.29). Group O cells lack both of these monosaccharides at the ends of their oligosaccharide chains and instead contain α-1,2-linked fucose, which is designated the H antigen. Plasma from blood group A individuals contains antibodies against the B antigen, blood group B individuals have antibodies against the A antigen, and blood group O individuals have antibodies against both the A and B antigens. In practice, this means that individuals with either anti-A or anti-B antibodies cannot safely receive a blood transfusion containing the incompatible antigen, since this is likely to cause a severe immune response (Table 9.3). As a consequence, blood group AB individuals are said to be universal recipients, while those from blood group O are universal donors. Thus, when a blood transfusion is

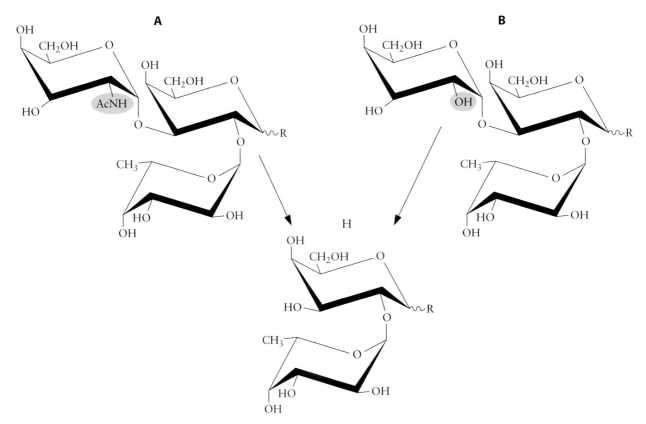

Figure 9.29 Digestion of the surface carbohydrates of red blood cells with specific glucosidases to remove the monosaccharides that determine blood groups A and B to obtain the H antigen (i.e., blood group O). AcNH, acetyl moiety covalently bound to a nitrogen atom. doi:10.1128/9781555818890.ch9.f9.29

required, it is advantageous to have a large supply of plasma that is from blood group O (e.g., in an emergency situation, there may not be sufficient time to check a patient's blood group). Fortunately, digestion of blood cells from either type A or B with specific glycosidases can cause types A, B, and AB to be converted into type O (Fig. 9.29). These enzymes were found following an extensive screening process of 2,500 fungal and bacterial isolates. Eventually, an active α-*N*-acetylgalactosamidase, which converts group A to group O, was found in the gram-negative bacterium *Elizabethkingia meningoseptica*, and one with α-galactosidase A, which

Table 9.3 Compatible and incompatible blood groups

Donor blood type	Recipient blood type			
	A	B	AB	O
A	Compatible	Incompatible	Compatible	Incompatible
B	Incompatible	Compatible	Compatible	Incompatible
AB	Incompatible	Incompatible	Compatible	Incompatible
O	Compatible	Compatible	Compatible	Compatible

Individuals from one blood group may safely receive a blood transfusion from individuals from a compatible blood group but not from someone from an incompatible blood group.

converts group B to group O, was found in *Bacteroides fragilis* (also a gram-negative bacterium). The genes were isolated, and the proteins were characterized. Both of the enzymes have high specificity for cleaving the appropriate monosaccharide under conditions that maintain the integrity and functioning of the treated red blood cells. Moreover, each enzyme could readily be removed from the treated red blood cells following treatment. While this is a very recent and still preliminary experiment, if this novel approach works effectively in a clinical setting, then it should become a boon for all types of blood transfusions.

Lactic Acid Bacteria

Lactic acid bacteria are widely used in the production and preservation of fermented foods, and many have been given the designation "generally regarded as safe" within the food industry. Many of these organisms are members of the indigenous microflora of the human gut and have been recognized for their health-promoting properties. Some strains of lactic acid bacteria, notably lactobacilli, are used in probiotic products that are often sold either in pharmacies or in health food stores. A probiotic is a live microorganism that is claimed to confer a health benefit by altering the indigenous microflora of the intestinal tract. Lactic acid bacteria have also been used to treat several gastrointestinal disorders, including lactose intolerance, traveler's diarrhea, antibiotic-associated diarrhea, infections caused by various bacterial and viral pathogens, and immunopathological disorders, such as Crohn disease and ulcerative colitis. Not only is oral vaccination easy to deliver, but also it has the potential to elicit both mucosal and systemic immune responses.

In the past few years, lactic acid bacteria have been used as a host system to express various foreign genes (Table 9.4), with the idea that these bacteria facilitate the delivery of the proteins encoded by the genes to the human gut. In particular, *Lactococcus lactis* has been developed as a host for this purpose. *L. lactis* is a nonpathogenic, noninvasive, noncolonizing gram-positive bacterium that is often used in the production of fermented foods and has been used for many years as a human probiotic. Moreover, unlike for *E. coli*, eukaryotic proteins produced in *L. lactis* generally do not form insoluble inclusion bodies. When genetically engineered *L. lactis* that can secrete the target protein is used to deliver protein antigens to humans or animals, it is not necessary to purify the target protein. Finally, it is worth noting that there is a distinction between the use of *L. lactis* to deliver therapeutic agents and its use as a live vaccine. In the former instance, the therapeutic agent typically has biological activity of its own, while in the latter case, the bacterium delivers a protein that elicits an immune response.

Interleukin-10

Ulcerative colitis and Crohn disease, both diseases of the intestinal tract, affect approximately 1 in every 500 to 1,000 people in the developed countries of the world. Ulcerative colitis is associated with excess type 2

Table 9.4 Some therapeutic proteins that have been expressed in *Lactococcus lactis*

Therapeutic protein	Target pathogen or disease
Tetanus toxin fragment C	Tetanus
β-Toxoid	*Clostridium perfringens*
GroEL heat shock protein	*Brucella abortus*
LcrV antigen	*Yersinia pseudotuberculosis*
MrpA structural fimbrial protein	*Proteus mirabilis*
Outer membrane protein Cag12	*Helicobacter pylori*
Urease subunit B	*Helicobacter pylori*
PspA, PsaA or PppA antigen	*Streptococcus pneumoniae*
Pili proteins	Group B *Streptococcus*
C-repeat region of M protein	*Streptococcus pyogenes*
Listeriolysin O	Listeriosis
E7 of human papillomavirus 16	Cervical cancer
L1 capsid protein of human papillomavirus 16	Cervical cancer
NSP4 or VP7 antigen	Rotavirus
VP2 and VP3 antigens	Infectious bursal disease virus
Envelope protein	HIV
Nucleocapsid protein	SARS virus
Envelope protein E domain III	Dengue virus
MSP-1	*Plasmodium yoelii* (malaria)
Cyst wall protein 2	*Giardia lamblia*
Murine IL-10	Inflammatory bowel disease, food allergy
Human IL-10	Crohn disease
Murine trefoil factors	Inflammatory bowel disease
Mig and IP-10 chemokines	Cancer immunotherapy
Human leptin	Weight control
Human IFN-β	Ulcerative colitis
Ovine IFN-ω	Enteric viral infections
Catalase	Gut cancer and inflammatory diseases
Rat heme oxygenase-1	Hemorrhagic shock or endotoxemia
Major birch pollen allergen	Allergy
Murine IL-12	Cancer immunotherapy, allergy, asthma
LcrV protein	Inflammatory bowel disease
Cyanovirin	HIV
Epidermal growth factor	Intestinal development in newly weaned infants

SARS, severe acute respiratory syndrome; IL-10, interleukin-10.

T helper cell cytokines, including interleukin-4 and interleukin-5, whereas in Crohn disease, type 1 T helper cell cytokines, including TNF-α, IFN-α, and interleukin-2, are overproduced. The treatment for Crohn disease often includes trying to lower the levels of cytokines, especially TNF-α. One approach has been the administration of antibodies against TNF-α. Other workers have targeted interleukin-10 as a means of controlling Crohn disease because it modulates the regulatory T cells that control inflammatory responses to intestinal antigens. However, interleukin-10 is not clinically acceptable because it needs to be administered by either frequent injections or rectal enemas. To overcome this problem, the bacterium *L. lactis* was engineered to synthesize and secrete interleukin-10.

Experiments were performed with mice to test whether interleukin-10-secreting *L. lactis* could be used to treat inflammatory bowel disease (Fig. 9.30A). First, interleukin-10-secreting *L. lactis* was fed to mice with ulcerative colitis that had been induced by 5% dextran sulfate in their drinking water. Second, strains of mice that are genetically incapable of synthesizing interleukin-10 and provide an animal model for ulcerative colitis were tested. In both of these cases, the engineered *L. lactis* significantly alleviated the symptoms of the disease, establishing that this approach works in principle. However, these mouse models for inflammatory bowel disease are not identical to the disease in humans, and a large number of questions remain before the treatment is used with humans.

One concern about the use of an interleukin-10-secreting *L. lactis* strain as a therapeutic approach is the possibility that the genetically modified bacterium will be released to the environment. If this were to happen, the plasmid carrying the interleukin-10 gene and any plasmid-borne antibiotic resistance marker genes could be spread to other bacteria in the environment. To prevent this from occurring, a synthetic human interleukin-10 gene that replaced the *L. lactis* thymidylate synthase gene, *thyA*, which is essential for the growth of the bacterium, was inserted into the bacterial chromosome of *L. lactis* by homologous recombination

Figure 9.30 (**A**) Schematic representation of the effects of intestinal interleukin-10 (IL-10)-secreting bacteria on inflammatory bowel disease in mice. (**B**) Genetic construct integrated into the chromosomal DNA of *L. lactis* in place of its thymidylate synthase gene. The promoter (p^{thyA}) is from the thymidylate synthase gene. The IL-10 gene was chemically synthesized so that its codon usage was optimized for *L. lactis*, thereby ensuring a high level of protein expression.
doi:10.1128/9781555818890.ch9.f9.30

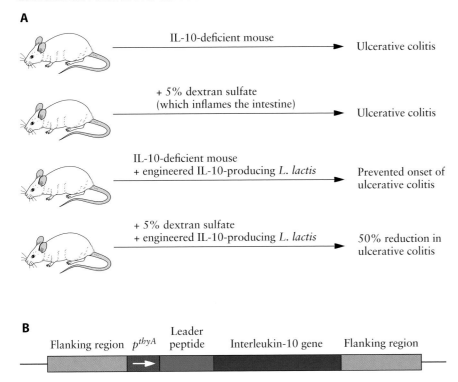

A

IL-10-deficient mouse → Ulcerative colitis

+ 5% dextran sulfate (which inflames the intestine) → Ulcerative colitis

IL-10-deficient mouse + engineered IL-10-producing *L. lactis* → Prevented onset of ulcerative colitis

+ 5% dextran sulfate + engineered IL-10-producing *L. lactis* → 50% reduction in ulcerative colitis

B

Flanking region p^{thyA} Leader peptide   Interleukin-10 gene   Flanking region

(Fig. 9.30B). This strain produced interleukin-10 and grew well in the laboratory when either thymidine or thymine was added to the medium. However, when the bacterium was deprived of thymidine and thymine, its viability declined by several orders of magnitude. When this modified bacterium was tested in pigs, whose digestive tract is similar to that of humans, it thrived and actively produced interleukin-10. In addition, laboratory experiments demonstrated that the modified *L. lactis* strain was extremely unlikely to acquire a thymidylate synthase gene from other bacteria in the environment, confirming both the safety and efficacy of this approach.

Recently, stage I clinical trials with this *L. lactis* strain were initiated. To date, 10 patients with Crohn disease have been treated. So far, a significant decrease in disease activity has been observed, with only minor adverse events. Moreover, engineered *L. lactis* bacteria isolated from the patients' feces were not able to grow without the addition of thymidine. In other words, the engineered *L. lactis* did not acquire a thymidylate synthase gene, indicating that the containment strategy was effective. Thus, initial indications are that this strategy appears to be working as well in humans as it did with small animals.

Leptin

It has been estimated that approximately 30% of the North American and 20% of the European populations are overweight. Moreover, North Americans annually spend tens of billions of dollars on various weight reduction schemes, most of which are unsuccessful. However, it is possible that real weight reduction may be obtained by administration of the protein leptin, which, in simple terms, communicates to the brain regarding the nutritional status of the body. Leptin, the product of the *obese (ob)* gene, is a 167-amino-acid protein with a molecular mass of approximately 16 kDa. Leptin is synthesized as a precursor with a 21-amino-acid-long signal peptide that is removed when leptin is secreted. Treatment with recombinant leptin can reduce food intake and correct metabolic perturbations in (homozygous) leptin-deficient mice. Leptin also helps to overcome human **congenital** leptin deficiency. However, when it is introduced subcutaneously, leptin is not particularly effective in obese patients unless their serum leptin concentrations reach levels 20- to 30-fold higher than normal. This poor response has been attributed to the inefficient transport of leptin across the blood–brain barrier. To overcome this problem, a scheme for the intranasal delivery of leptin has been devised. The nasal mucosa is highly vascularized, so delivery of a thin layer of medication across its broad surface area can result in rapid absorption of the medication into the bloodstream.

When leptin is produced in *E. coli*, it typically forms insoluble inclusion bodies that must be solubilized and renatured before the active protein is generated. This is a time-consuming, inefficient, and expensive process. In one study, the 462-base-pair (bp) cDNA for human leptin without its signal peptide was cloned and expressed under the control of the **nisin** promoter in *L. lactis* (Fig. 9.31). Nisin is a polycyclic peptide

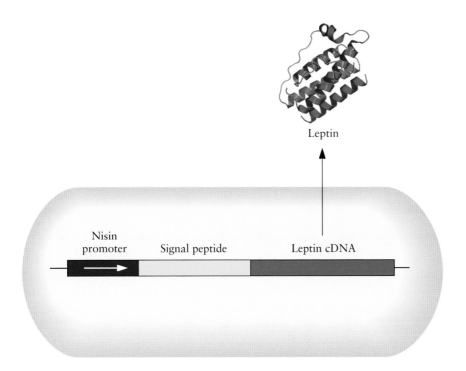

Leptin

Figure 9.31 Genetic construct used to secrete leptin from *L. lactis*.
doi:10.1128/9781555818890.ch9.f9.31

from *L. lactis* that has antibacterial activity and is used as a food preservative. In *L. lactis*, leptin was produced efficiently without the formation of an inclusion body and was secreted from the recombinant bacteria. When the leptin-producing *L. lactis* strain was administered intranasally in obese mice, the mice significantly reduced their food intake and body weight. This approach opens up the possibility that if delivered properly, leptin might act as an effective weight loss treatment in humans.

An HIV Inhibitor

Worldwide, the predominant mode of HIV transmission is by heterosexual contact. One possible way to protect women, who currently comprise about half of all new cases of HIV/AIDS, against HIV infection is a topical microbicide, delivered by a live vaginal *Lactobacillus* strain, that prevents HIV infection directly at mucosal surfaces. This strategy seems reasonable because naturally occurring vaginal *Lactobacillus* strains play a protective role in preventing urogenital infections.

The compound cyanovirin N, isolated from the cyanobacterium *Nostoc ellipsosporum*, blocks several steps of HIV infection, preventing virus entry into human cells. Consequently, cyanovirin N is a candidate for a topical microbicide to prevent HIV infections. To ensure that cyanovirin N would be expressed at a sufficiently high level in a vaginal strain of *Lactobacillus jensenii*, the gene was chemically synthesized to reflect the codon usage found in the bacterium rather than in *N. ellipsosporum*. Typically, the GC content of lactobacilli is about 36%, while the GC content of *N. ellipsosporum* is 44.5%. In addition, during the chemical synthesis of the gene, the codon for proline 51 was replaced by a codon for

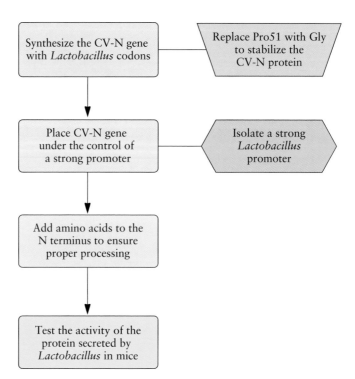

Figure 9.32 Flowchart of the scheme used to develop a *Lactobacillus* strain that produces and secretes cyanovirin N (CV-N). doi:10.1128/9781555818890.ch9.f9.32

a glycine residue to stabilize the cyanovirin N protein, and four amino acids were added to the N terminus to ensure proper cleavage of the signal sequence (Fig. 9.32). The modified cyanovirin N gene was fused to a strong and constitutive *Lactobacillus* promoter. The final construct was integrated into the chromosomal DNA of a strain of *L. jensenii*, with the result that about 4 μg of cyanovirin N per ml was released into the culture medium. When the *L. jensenii* strain that synthesized the modified form of cyanovirin N was tested for efficacy, it was found to be highly effective at preventing HIV infections in mice.

Insulin

Oral administration of insulin would greatly simplify the lives of most insulin-dependent diabetics. However, this has not been possible due to insulin's poor stability during its passage through the gastrointestinal tract. As an alternative to various unsuccessful strategies to produce encapsulated insulin that could be delivered orally, one group of researchers developed a single-chain insulin that can be delivered to diabetic patients using *L. lactis*.

The peptide hormone insulin is produced in animal pancreatic cells as a single polypeptide, preproinsulin, that is processed to proinsulin following its secretion and then to insulin following proteolytic cleavage (Fig. 9.33A). The insulin A chain includes 21 amino acid residues, while the B chain consists of 30 amino acid residues. Given the presence of two interchain and one intrachain disulfide bond (not shown in Fig. 9.33),

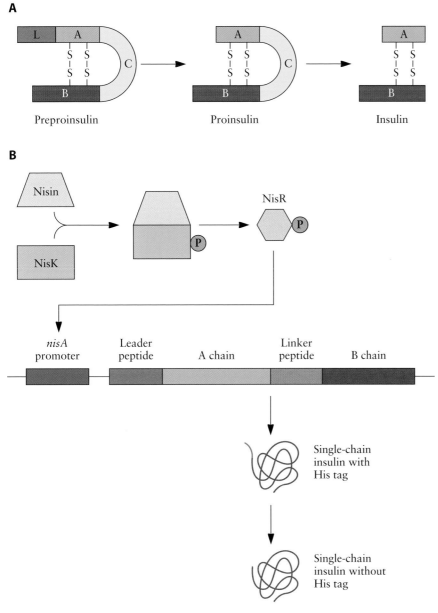

Figure 9.33 (**A**) Cleavage of inactive preproinsulin, first to proinsulin and then to yield active insulin. Proteases remove the leader peptide (L) and an internal peptide (C) to yield a peptide that consists of chains A and B. (**B**) Genetic regulation of the expression of a single-chain (sc) form of insulin in *L. lactis*. The binding of nisin, which is added to the medium, to NisK (a histidine kinase) causes NisK to autophosphorylate. NisK then transfers its phosphate to NisR, which activates the *nisA* promoter, inducing transcription of the single-chain insulin gene. The leader peptide ensures that the single-chain insulin will be secreted. Following secretion and removal of the leader peptide, a histidine (His) tag that is part of the protein facilitates its purification before the His tag is removed by proteolytic cleavage to yield the single-chain form of insulin. doi:10.1128/9781555818890.ch9.f9.33

it is technically quite difficult to produce insulin in *L. lactis*. Therefore, it was decided to synthesize a single-chain insulin in which the A and B chains would be joined covalently by a 6-amino-acid-long linker peptide. To ensure that the single-chain insulin was as active as native insulin, four additional amino acid changes were introduced into the protein. Insulin derivatives with two of these amino acid changes were already in clinical use when this work was undertaken, making the researchers confident that the proposed four amino acid changes would yield an active and stable form of insulin. In fact, the resulting single-chain insulin is more stable at high temperature and is less likely to aggregate than native insulin. The single-chain insulin gene was expressed in *L. lactis* (Fig. 9.33B) under the control of the *nisA* promoter that is activated by a phosphorylated version of the transcription factor NisR in the presence of nisin. The NisR protein is phosphorylated by a phosphate residue from a NisK protein bound to the protein nisin, a pore-forming toxin produced by *L. lactis* that is used commercially in the preparation of certain prepared foods. To date, it has been demonstrated that genetically engineered *L. lactis* can produce and secrete a single-chain insulin that has in vitro insulin-like biological activity. Whether this system will function effectively as a delivery system, replacing several-times-daily insulin injections for human diabetics, remains to be determined.

summary

A large number of proteins that have potential as therapeutic agents has been synthesized from cloned cDNA in bacteria. Because most of these proteins are from eukaryotic organisms, the strategy for their isolation involves synthesizing a cDNA library and subcloning the selected target cDNA into an appropriate expression vector (see chapter 1). In some instances, novel and useful variants of these proteins can be constructed either by shuffling functional domains of related genes or by directed replacement of functional domains of the cloned gene. In addition, long-acting and stable variants of some therapeutic proteins have been synthesized.

The development of recombinant DNA and monoclonal antibody technologies, combined with an understanding of the molecular structure and function of immunoglobulin molecules, has provided specific antibodies as therapeutic agents to treat various diseases. Antibody genes can be readily manipulated because the various functions of an antibody molecule are confined to discrete domains.

Drugs, prodrugs, or enzymes can be coupled to monoclonal antibodies or Fv fragments that are specific for proteins found only on the surfaces of certain cells, e.g., tumor cells. These antibody–drug or antibody–enzyme combinations act as therapeutic agents. However, if the therapy requires multiple treatments, the antibody component should be from a human source to prevent immunological cross-reactivity and sensitization of the patient. To achieve this, rodent monoclonal antibodies are "humanized" by substituting into human antibodies only the CDRs of the rodent monoclonal antibodies. In addition, it has become possible to produce and select human monoclonal antibodies in *E. coli* and in transgenic mice.

In some instances, genetically engineered enzymes may be used as therapeutic agents. For example, both recombinant DNase I and alginate lyase have been used in an aerosol form to decrease the viscosity of the mucus found in the lungs of patients with cystic fibrosis. In addition, phenylalanine ammonia lyase may help patients with phenylketonuria as a replacement for phenylalanine hydroxylase, α_1-antitrypsin may be used to limit some infections, and glycosidases may be utilized to convert blood groups A, B, and AB to type O.

Certain therapeutic agents may be delivered directly to their target cells by expressing the genes for these proteins in a bacterium that is normally associated with human tissues and has been shown to be safe, such as *Lactococcus lactis*. In one study, *L. lactis* was used to deliver interleukin-10 to the human intestinal tract as a means of treating individuals with Crohn disease.

review questions

1. What is the Fc portion of an antibody molecule? The Fab portion? The Fv portion? The CDR portion? The FR portion?

2. How are antibody L and H chains coordinately synthesized in *E. coli*?

3. How would you modify growth hormone to make it longer acting?

4. Why would DNase I and alginate lyase be useful for treating cystic fibrosis?

5. How is the production of alginate lyase from a cloned gene detected in *E. coli* transformants?

6. What is a combinatorial cDNA library?

7. How is bacteriophage M13 used to select Fv fragments that bind to specific target antigens?

8. How would you express foreign genes in mitochondria?

9. How are enzymes that are coupled to monoclonal antibodies or Fv fragments used as therapeutic agents?

10. How are mouse monoclonal antibodies humanized? Discuss the reasons for creating humanized monoclonal antibodies.

11. Describe a protocol for producing a therapeutic agent that targets and kills a specific cell type.

12. How would you engineer TNF-α to be a more specific and effective anticancer agent?

13. What would you do to make interleukin-10 more effective for treating inflammatory bowel disease?

14. Besides pegylation, how would you extend the in vivo half-life of therapeutic proteins?

15. How would you develop a strategy to protect at-risk women from HIV infection?

16. How might low levels of phenylalanine be attained, other than with a phenylalanine-free diet, in patients with the human genetic disease phenylketonuria?

17. What types of genetic manipulations can be used to generate a very large bacterial library of highly specific single-chain human monoclonal antibodies?

18. How would you engineer a mouse so that it produces only human antibodies?

19. How might bacteriophages be used to increase the sensitivity of some bacterial pathogens to antibiotics?

20. How would you design a short peptide so that it retains the antigen-binding specificity of an entire immunoglobulin molecule?

21. How would you develop antibodies that protect people against anthrax?

22. How would you develop antibodies that facilitate weight loss?

23. How can you use a chemotherapy agent to facilitate the targeting of tumor cells with monoclonal antibodies?

24. How would you develop therapeutic antibodies with an extended in vivo half-life?

25. How would you develop a single-chain version of insulin that can be delivered to diabetic patients using *L. lactis*?

references

Adams, G. P., and L. M. Weiner. 2005. Monoclonal antibody therapy of cancer. *Nat. Biotechnol.* **23**:1147–1157.

Alkawash, M. A., J. S. Soothill, and N. L. Schiller. 2006. Alginate lyase enhances antibiotic killing of mucoid *Pseudomonas aeruginosa* in biofilms. *APMIS* **114**:131–138.

Bahey-El-Din, M., C. G. M. Gahan, and B. T. Griffin. 2010. *Lactococcus lactis* as a cell factory for delivery of therapeutics. *Curr. Gene Ther.* **10**:34–45.

Barbas, C. F., III, and D. R. Burton. 1996. Selection and evolution of high-affinity human anti-viral antibodies. *Trends Biotechnol.* **14**:230–234.

Bermúdez-Humarán, L. G., S. Nouaille, V. Zilberfarb, G. Corthier, A. Gruss, P. Langella, and T. Issad. 2007. Effect of intranasal administration of a leptin-secreting *Lactococcus lactis* recombinant on food intake, body weight, and immune response of mice. *Appl. Environ. Microbiol.* **73**:5300–5307.

Braat, H., P. Rottiers, D. W. Hommes, N. Huyghebaert, E. Remaut, J.-P. Remon, S. J. H. van Deventer, S. Neirynck, M. P. Peppelenbosch, and L. Steidler. 2006. A phase I trial with transgenic bacteria expressing interleukin-10 in Crohn's disease. *Clin. Gastroenterol. Hepatol.* **4**:754–759.

Brideau-Andersen, A. D., X. Huang, S.-C. C. Sun, T. T. Chen, D. Stark, I. J. Sas, L. Zadik, G. N. Dawes, D. R. Guptill, R. McCord, S. Govindarajan, A. Roy, S. Yang, J. Gao, Y. H. Chen, N. J. Ø. Skartved, A. K. Pedersen, D. Lin, C. P. Locher, I. Rebbarpragada, A. D. Jensen, S. H. Bass, T. L. S. Nissen, S. Viswanathan, G. R. Foster, J. A. Symons, and P. A. Patten. 2007. Directed evolution of gene-shuffled IFN-α molecules with activity profiles tailored for treatment of chronic viral diseases. *Proc. Natl. Acad. Sci. USA* **104**:8269–8274.

Chen, Z., M. Moayeri, H. Zhao, D. Crown, S. H. Leppla, and R. H. Purcell. 2009. Potent neutralization of anthrax edema toxin by a humanized monoclonal antibody that competes with calmodulin for edema factor binding. *Proc. Natl. Acad. Sci. USA* **106:** 13487–13492.

Chester, K. A., and R. E. Hawkins. 1995. Clinical issues in antibody design. *Trends Biotechnol.* **13:**294–300.

Collet, T. A., P. Roben, R. O'Kennedy, C. F. Barbas III, D. R. Burton, and R. A. Lerner. 1992. A binary plasmid system for shuffling combinatorial antibody libraries. *Proc. Natl. Acad. Sci. USA* **89:** 10026–10030.

Curnis, F., A. Sacchi, L. Borgna, F. Magni, A. Gasparri, and A. Corti. 2000. Enhancement of tumor necrosis factor α antitumor immunotherapeutic properties by targeted delivery to aminopeptidase N (CD13). *Nat. Biotechnol.* **18:** 1185–1190.

de Vos, W. M., and J. Hugenholtz. 2004. Engineering metabolic highways in lactococci and other lactic acid bacteria. *Trends Biotechnol.* **22:**72–79.

Dübel, S. 2007. Recombinant therapeutic antibodies. *Appl. Microbiol. Biotechnol.* **74:**723–729.

Gámez, A., C. N. Sarkissian, L. Wang, W. Kim, M. Straub, M. G. Patch, L. Chen, S. Striepeke, P. Fitzpatrick, J. F. Lemontt, C. O'Neill, C. R. Scriver, and R. C. Stevens. 2005. Development of pegylated forms of recombinant *Rhodosporidium toruloides* phenylalanine ammonia-lyase for the treatment of classical phenylketonuria. *Mol. Ther.* **11:**986–989.

Gram, H., L. A. Marconi, C. F. Barbas III, T. A. Collet, R. A. Lerner, and A. S. Kang. 1992. In vitro selection and affinity maturation of antibodies from a naive combinatorial immunoglobulin library. *Proc. Natl. Acad. Sci. USA* **89:** 3576–3580.

Hanniffy, S., U. Wiedermann, A. Repa, A. Mercenier, C. Daniel, J. Fioamonti, H. Tlaskolova, H. Kozakova, H. Israelsen, S. Madsen, A. Vrang, P. Hols, J. Delcour, P. Bron, M. Kleerebezem, and J. Wells. 2004. Potential and opportunities for use of recombinant lactic acid bacteria in human health. *Adv. Appl. Microbiol.* **56:**1–64.

Holliger, P., and P. J. Hudson. 2005. Engineered antibody fragments and the rise of single domains. *Nat. Biotechnol.* **23:**1126–1136.

Huennekens, F. M. 1994. Tumor targeting: activation of prodrugs by enzyme-monoclonal antibody conjugates. *Trends Biotechnol.* **12:**234–239.

Huse, W. D., L. Sastry, S. A. Iverson, A. S. Kang, M. Alting-Mees, D. R. Burton, S. J. Benkovic, and R. A. Lerner. 1989. Generation of a large combinatorial library of the immunoglobulin repertoire in phage lambda. *Science* **246:**1275–1281.

Jakobovits, A., R. G. Amado, X. Yang, L. Roskos, and G. Schwab. 2007. From XenoMouse technology to panitumumab, the first fully human antibody product from transgenic mice. *Nat. Biotechnol.* **25:**1134–1143.

Jean, F., L. Thomas, S. S. Molloy, G. Liu, M. A. Jarvis, J. A. Nelson, and G. Thomas. 2000. A protein-based therapeutic for human cytomegalovirus infection. *Proc. Natl. Acad. Sci. USA* **97:**2864–2869.

Kreitman, R. J. 1999. Immunotoxins in cancer therapy. *Curr. Opin. Immunol.* **11:**570–578.

Kufer, P., R. Lutterbrüse, and P. A. Baeuerle. 2004. A revival of bispecific antibodies. *Trends Biotechnol.* **22:** 238–244.

Lamppa, J. W., M. E. Ackerman, J. I. Lai, T. C. Scanlon, and K. E. Griswold. 2011. Genetically engineered alginate lyase-PEG conjugates exhibit enhanced catalytic function and reduced immunoreactivity. *PLoS One* **6**(2):e17402.

Leader, B., Q. J. Baca, and D. E. Golan. 2008. Protein therapeutics: a summary and pharmacological classification. *Nat. Rev. Drug Discov.* **7:**21–39.

Little, M., F. Breitling, S. Dübel, P. Fuchs, and M. Braunagel. 1995. Human antibody libraries in *Escherichia coli. J. Biotechnol.* **41:**187–195.

Liu, Q. P., G. Sulzenbacher, H. Yuan, E. P. Bennett, G. Pietz, K. Saunders, J. Spence, E. Nudelman, S. B. Levery, T. White, J. M. Neveu, W. S. Lane, Y. Bourne, M. L. Olsson, B. Henrissat, and H. Clausen. 2007. Bacterial glycosidases for the production of universal red blood cells. *Nat. Biotechnol.* **25:** 454–464.

Liu, X., L. A. Lagenaur, D. A. Simpson, K. P. Essenmacher, C. L. Frazier-Parker, Y. Liu, S. S. Rao, D. H. Hamer, T. P. Parks, P. P. Lee, and Q. Xu. 2006. Engineered vaginal *Lactobacillus* strain for mucosal delivery of the human immunodeficiency virus inhibitor cyanovirin-N. *Antimicrob. Agents Chemother.* **50:**3250–3259.

Lonberg, N. 2005. Human antibodies from transgenic animals. *Nat. Biotechnol.* **23:**1117–1125.

Lorenzen, N., P. M. Cupit, K. Einer-Jensen, E. Lorenzen, P. Ahrens, C. J. Secombes, and C. Cunningham. 2000. Immunoprophylaxis in fish by injection of mouse antibody genes. *Nat. Biotechnol.* **18:**1177–1180.

Lu, T. K., and J. J. Collins. 2009. Engineered bacteriophage targeting gene networks as adjuncts for antibiotic therapy. *Proc. Natl. Acad. Sci. USA* **106:** 4629–4634.

Marks, J. D., A. D. Griffiths, M. Malmqvist, T. P. Clackson, J. M. Bye, and G. Winter. 1992. By-passing immunization: building high affinity antibodies by chain shuffling. *Bio/Technology* **10:** 779–783.

Mayorov, A. V., N. Amara, J. Y. Chang, J. A. Moss, M. S. Hixon, D. I. Ruiz, M. M. Meijler, E. P. Zorrilla, and K. D. Janda. 2008. Catalytic antibody degradation of ghrelin increases whole-body metabolic rate and reduces refeeding in fasting mice. *Proc. Natl. Acad. Sci. USA* **105:**17487–17492.

Mazor, Y., T. Van Blarcom, R. Mabry, B. L. Iverson, and G. Georgiou. 2007. Isolation of engineered full-length antibodies from libraries expressed in *Escherichia coli. Nat. Biotechnol.* **25:** 563–565.

Molina, A. 2008. A decade of rituximab: improving survival outcomes in non-Hodgkin's lymphoma. *Annu. Rev. Med.* **59:**237–250.

Murata, K., T. Inose, T. Hisano, S. Abe, Y. Yonemoto, T. Yamashita, M. Takagi, K. Sakaguchi, A. Kimura, and T. Imanaka. 1993. Bacterial alginate lyase: enzymology, genetics and application. *J. Ferment. Bioeng.* **76:**427–437.

Nagata, S., H. Taira, A. Hall, L. Johnsrud, M. Streuli, J. Ecsödi, W. Boll, K. Cantell, and C. Weissmann. 1980. Synthesis in *E. coli* of a polypeptide with human leukocyte interferon activity. *Nature* **284:**316–320.

Ng, D. T. W., and C. A. Sarkar. 2011. Nisin-inducible secretion of a biologically active single-chain insulin analog by *Lactococcus lactis* NZ9000. *Biotechnol. Bioeng.* **108**:1987–1996.

Osborn, B. L., L. Sekut, M. Corcoran, C. Poortman, B. Sturm, G. Chen, D. Mather, H. L. Lin, and T. J. Parry. 2002. Albutropin: a growth hormone-albumin fusion with improved pharmacokinetics and pharmacodynamics in rats and monkeys. *Eur. J. Pharmacol.* **456**: 149–158.

Papadopoulou, L. C., and A. S. Tsiftoglou. 2011. Transduction of human recombinant proteins into mitochondria as a protein therapeutic approach for mitochondrion disorders. *Pharm. Res.* **28**:2639–2656.

Qiu, X.-Q., H. Wang, B. Cai, L.-L. Wang, and S.-T. Yue. 2007. Small antibody mimetics comprising two complementarity-determining regions and a framework region for tumor targeting. *Nat. Biotechnol.* **25**:921–929.

Queen, C., W. P. Schneider, H. E. Selick, P. W. Payne, N. F. Landolf, J. F. Duncan, N. M. Avdalovic, M. Levitt, R. P. Junghans, and T. A. Waldmann. 1989. A humanized antibody that binds to the interleukin 2 receptor. *Proc. Natl. Acad. Sci. USA* **86**:10029–10033.

Rapoport, M., A. Saada, O. Elpeleg, and H. Lorberboum-Galski. 2008. TAT-mediated delivery of LAD restores pyruvate dehydrogenase complex activity in the mitochondria of patients with LAD deficiency. *Mol. Ther.* **16**:691–697.

Rapoport, M., L. Salman, O Sabag, M. S. Patel, and H. Lorberboum-Galski. 2011. Successful TAT-mediated enzyme replacement therapy in a mouse model of mitochondrial E3 deficiency. *J. Mol. Med.* **89**:161–170.

Reichert, J. M. 2006. Trends in US approvals: new biopharmaceuticals and vaccines. *Trends Biotechnol.* **24**: 293–298.

Reichert, J. M., C. J. Rosensweig, L. B. Faden, and M. C. Dewitz. 2005. Monoclonal antibody success in the clinic. *Nat. Biotechnol.* **23**:1073–1078.

Rubinfeld, B., A. Upadhyay, S. L. Clark, S. E. Fong, V. Smith, H. Koeppen, S. Ross, and P. Polakis. 2006. Identification and immunotherapeutic targeting of antigens induced by chemotherapy. *Nat. Biotechnol.* **24**:205–209.

Sampson, J. H., L. E. Crotty, S. Lee, G. E. Archer, D. M. Ashley, C. J. Wikstrand, L. P. Hale, C. Small, G. Dranoff, A. H. Friedman, H. S. Friedman, and D. D. Bigner. 2000. Unarmed, tumor-specific monoclonal antibody effectively treats brain tumors. *Proc. Natl. Acad. Sci. USA* **97**:7503–7508.

Sarkissian, C. H., and A. Gámez. 2005. Phenylalanine ammonia lyase, substitution therapy for phenylketonuria, where are we now? *Mol. Genet. Metab.* **86**: S22–S26.

Sarkissian, C. H., Z. Shao, F. Blain, R. Peevers, H. Su, R. Heft, T. M. S. Chang, and C. R. Scriver. 1999. A different approach to treatment of phenylketonuria: phenylalanine degradation with recombinant phenylalanine ammonia lyase. *Proc. Natl. Acad. Sci. USA* **96**: 2339–2344.

Schellenberger, V., C.-W. Wang, N. C. Geething, B. J. Spink, A. Campbell, W. Tu, M. D. Scholle, Y. Yin, Y. Yao, O. Bogin, J. I. Cleland, J. Silverman, and W. P. C. Stemmer. 2009. A recombinant polypeptide extends the in vivo half-life of peptides and proteins in a tunable manner. *Nat. Biotechnol.* **27**: 1186–1190.

Söderlind, E., L. Strandberg, P. Jirholt, N. Kobayashi, B. Alexeiva, A.-M. Åberg, A. Nilsson, B. Jansson, M. Ohlin, C. Wingren, L. Danielsson, R. Carlsson, and C. A. K. Borrebaeck. 2000. Recombining germline-derived CDR sequences for creating diverse single-framework antibody libraries. *Nat. Biotechnol.* **18**: 852–856.

Steidler, L., W. Hans, L. Schotte, S. Neirynck, F. Obermeier, W. Falk, W. Fiers, and E. Remaut. 2000. Treatment of murine colitis by *Lactococcus lactis* secreting interleukin-10. *Science* **289**: 1352–1355.

Steidler, L., S. Neirynck, N. Huyghebaert, V. Snoeck, A. Vermeire, B. Goddeeris, E. Cox, J. P. Remon, and E. Remaut. 2003. Biological containment of genetically modified *Lactococcus lactis* for intestinal delivery of human interleukin 10. *Nat. Biotechnol.* **21**: 785–789.

Subramanian, G. M., M. Fiscella, A. Lamousé-Smith, S. Zeuzem, and J. G. McHutchison. 2007. Albuferon α-2b: a generic fusion protein for the treatment of chronic hepatitis C. *Nat. Biotechnol.* **25**:1411–1419.

Ulmer, G. S., A. Herzka, K. J. Toy, D. L. Baker, A. H. Dodge, D. Sinicropi, S. Shak, and R. A. Lazarus. 1996. Engineering actin-resistant human DNase I for treatment of cystic fibrosis. *Proc. Natl. Acad. Sci. USA* **93**:8225–8229.

Vaughan, T. J., A. J. Williams, K. Pritchard, J. K. Osbourn, A. R. Pope, J. C. Earnshaw, J. McCafferty, R. A. Hodits, J. Wilton, and K. S. Johnson. 1996. Human antibodies with sub-nanomolar affinities isolated from a large non-immunized phage display library. *Nat. Biotechnol.* **14**:309–314.

Veronese, F. M., and G. Pasut. 2005. PEGylation, successful approach to drug delivery. *Drug Discov. Today* **10**: 1451–1458.

Weisser, N. E., and J. C. Hall. 2009. Applications of single-chain variable fragment antibodies in therapy and diagnostics. *Biotechnol. Adv.* **27**:502–520.

Wilkenson, I. R., E. Ferrandis, P. J. Artymiuk, M. Teillot, C. Soulard, C. Touvay, S. L. Pradananga, S. Justice, Z. Wu, K. C. Leung, C. J. Strasburger, J. R. Sayers, and R. J. Ross. 2007. A ligand-receptor fusion of growth hormone forms a dimer and is a potent long-acting agonist. *Nat. Med.* **13**: 1108–1113.

Winter, G., and W. J. Harris. 1993. Humanized antibodies. *Trends Pharmacol. Sci.* **14**:139–143.

Wu, C., H. Ying, C. Grinnell, S. Bryant, R. Miller, A. Clabbers, S. Bose, D. McCarthy, R.-R. Zhu, L. Santora, D. Davis-Taber, Y. Kunes, E. Fung, A. Schwartz, P. Sakorafas, J. Gu, E. Tarcsa, A. Murtaza, and T. Ghayur. 2007. Simultaneous targeting of multiple disease mediators by a dual-variable-domain immunoglobulin. *Nat. Biotechnol.* **25**:1290–1297.

Zahm, J.-M., C. Debordeaux, C. Maurer, D. Hubert, D. Dusser, N. Bonnet, R. A. Lazarus, and E. Puchelle. 2001. Improved activity of an actin-resistant DNase I variant on the cystic fibrosis airway secretions. *Am. J. Respir. Crit. Care Med.* **163**:1153–1157.

Zalevsky, J., A. K. Chamberlain, H. M. Horton, S. Karki, I. W. L. Leung, T. J. Sproule, G. A. Lazar, D. C. Roopenian, and J. R. Desjarlais. 2010. Enhanced antibody half-life improves in vivo activity. *Nat. Biotechnol.* **28**:157–160.

Zivin, J. A. 2000. Understanding clinical trials. *Sci. Am.* **282**(4):69–75.

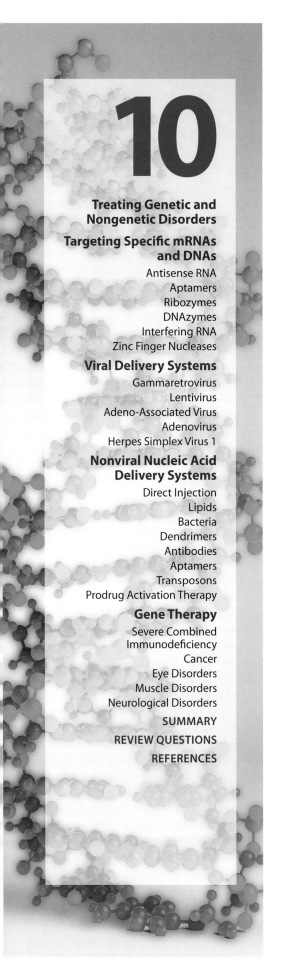

10

Nucleic Acid Therapeutic Agents and Human Gene Therapy

Treating Genetic and Nongenetic Disorders

It has been especially difficult to develop treatments for genetic diseases. In many instances prior to the 1990s, the gene that caused a particular disorder was unknown, and consequently, the root cause of the disorder was a mystery. Moreover, many genetic diseases have multiple effects that are difficult to treat in a straightforward manner and may require diverse therapies. Because of the physiological complexity of genetic diseases, curative measures have consisted mainly of treating the symptoms by administering drugs, performing surgery, restricting dietary intake, transplanting organs or bone marrow, or transfusing blood.

To many observers, the use of genes as therapeutic agents **(gene therapy)** seemed to be a logical extension of other therapies. In its simplest form, gene therapy was envisioned as a way to correct a defect at its biological source, with a single treatment that would alleviate all symptoms of the disorder. By 1990, after thorough reviews by many different regulatory panels in the United States, the first sanctioned human gene therapy trial was initiated with two young girls with adenosine deaminase (ADA)-deficient severe combined immunodeficiency (SCID).

Since 1990, there has been a proliferation of strategies designed to treat a variety of human genetic disorders. In this context, the Roman poet Ovid (43 BCE–17 CE) aptly asserted, "A thousand ills require a thousand cures." This adage is especially true for human gene therapies. Currently, gene therapy entails not only providing cells that have a genetic defect with a gene sequence or cDNA that overrides or dominates the mutant state but also additional strategies to correct gene **mutations** at the DNA level, regulate the extent of production of a protein, and destroy tumor cells.

From 1989 to 2010, more than 1,900 human gene therapy clinical trials were conducted at the phase I and phase II levels, with thousands of individuals (Fig. 10.1A). Treatments for various cancers, hemophilia,

doi:10.1128/9781555818890.ch10

519

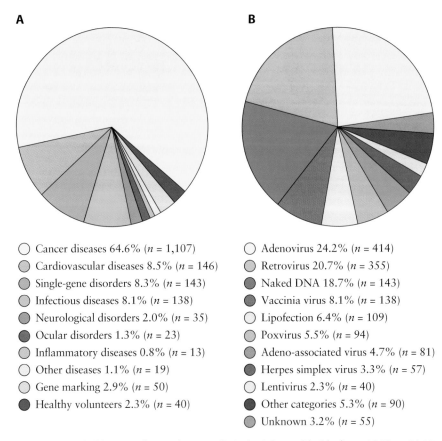

A

○ Cancer diseases 64.6% (*n* = 1,107)
◐ Cardiovascular diseases 8.5% (*n* = 146)
◑ Single-gene disorders 8.3% (*n* = 143)
◔ Infectious diseases 8.1% (*n* = 138)
● Neurological disorders 2.0% (*n* = 35)
● Ocular disorders 1.3% (*n* = 23)
◑ Inflammatory diseases 0.8% (*n* = 13)
○ Other diseases 1.1% (*n* = 19)
◐ Gene marking 2.9% (*n* = 50)
● Healthy volunteers 2.3% (*n* = 40)

B

○ Adenovirus 24.2% (*n* = 414)
◑ Retrovirus 20.7% (*n* = 355)
● Naked DNA 18.7% (*n* = 143)
● Vaccinia virus 8.1% (*n* = 138)
○ Lipofection 6.4% (*n* = 109)
◐ Poxvirus 5.5% (*n* = 94)
◑ Adeno-associated virus 4.7% (*n* = 81)
● Herpes simplex virus 3.3% (*n* = 57)
○ Lentivirus 2.3% (*n* = 40)
● Other categories 5.3% (*n* = 90)
◑ Unknown 3.2% (*n* = 55)

Figure 10.1 (**A**) Targets of gene therapy clinical trials worldwide from 1989 to 2010. Based on http://www.wiley.com/legacy/wileychi/genmed/clinical/. (**B**) Delivery systems used in clinical gene therapy trials from 1989 to 2010. Based on http://www.wiley.com/legacy/wileychi/genmed/clinical/. doi:10.1128/9781555818890.ch10.f10.1

ADA
adenosine deaminase

SCID
severe combined immunodeficiency

AIDS, cystic fibrosis, hypercholesterolemia, amyotrophic lateral sclerosis, and many other disorders have been tested for safety and efficacy (Fig. 10.1B). At present, all research on human gene therapy is directed toward correcting genetic defects of somatic cells, i.e., cells that do not contribute to the next generation. For ethical, safety, and technical reasons, human **germ line gene therapy** that entails the introduction of DNA into egg cells, sperm, or early embryonic cells **(blastomeres)** that can form part of the human germ line and be passed on to successive generations is not being examined experimentally at this time. In many countries, human germ line gene therapy experiments are prohibited by law. However, research on genetically modifying germ line cells in nonhuman organisms such as rodents, monkeys, and domesticated animals is being carried out.

Although the concept of human **somatic cell gene therapy** is straightforward, there are a number of critical biological considerations. How will the cells that are to be targeted for correction be accessed? How will the therapeutic (remedial) gene be delivered? What proportion of the target cells must acquire the input gene to counteract the disease? Does transcription of the input gene need to be precisely regulated to be effective? Will overexpression of the input gene cause alternative physiological problems? Will the cells with the input gene be maintained indefinitely, or will repeated treatments be required?

Somatic cell gene therapy is implemented in one of two ways: in vivo or ex vivo. For **in vivo gene therapy**, a therapeutic gene is introduced by either a viral or nonviral delivery system into a targeted tissue of the patient (Fig. 10.2A). On the other hand, **ex vivo gene therapy** entails collecting and culturing stem cells from an affected individual, introducing the therapeutic gene into these cultured cells, growing the cells with the therapeutic gene, and then either infusing or transplanting these cells back into the patient (Fig. 10.2B). The use of a patient's own cells (autologous cells) ensures that after their reintroduction no adverse immunological responses will occur. Moreover, the use of a particular type of **stem cell** ensures that a differentiated progeny cell that otherwise might be inaccessible to genetic modification carries the therapeutic gene. Under these conditions, the genetically modified stem cells provide a continual source of "corrected" cells that function properly.

Human disorders such as cancer, inflammatory conditions, and both viral and parasitic infections often result from the overproduction of a normal protein. Therapeutic systems using nucleotide sequences are being devised to treat these types of conditions. Theoretically, a small oligonucleotide could hybridize to a specific gene or mRNA and diminish

Figure 10.2 Schematic representation of in vivo gene therapy. (**A**) The therapeutic gene is introduced into a specific tissue by physical methods such as injection or systemically though the circulatory system. Various delivery systems are described in the text. (**B**) Schematic representation of ex vivo gene therapy. A cell sample containing stem cells is taken from an individual with a genetic disorder (1) and grown in culture (2). The isolated cells are transfected with a therapeutic gene (3). Cells carrying the therapeutic gene are grown (4) before transplantation or transfusion back into the patient (5). doi:10.1128/9781555818890.ch10.f10.2

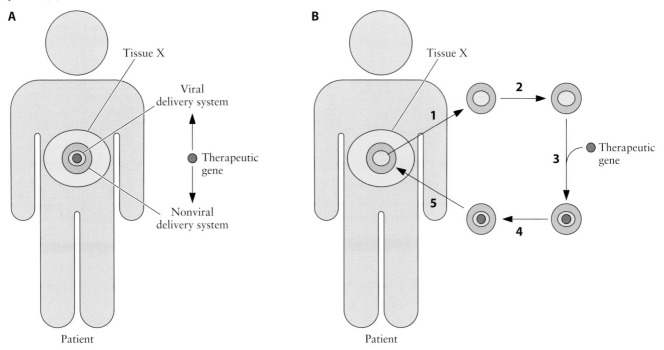

transcription or translation, respectively, thereby decreasing the amount of protein that is synthesized. An oligonucleotide that is designed to bind to a gene and block transcription is called an antigene oligonucleotide, and one that base-pairs with a specific mRNA is called an antisense oligonucleotide. Also, some synthetic RNA/DNA molecules called **aptamers** that bind to specific regions of proteins, and are not naturally nucleic acid-binding proteins, can prevent the targeted protein from functioning. These aptamers may be created and selected to bind with a high affinity to a wide range of proteins. **Ribozymes**, which are natural RNA sequences that bind and cleave specific RNA molecules, can be engineered to target an mRNA and subsequently decrease the amount of a particular protein that is synthesized. In addition, interfering RNAs, small double-stranded RNA molecules that direct the sequence-specific degradation of targeted mRNAs, may be used instead of either antisense RNAs (or oligonucleotides) or ribozymes.

Targeting Specific mRNAs and DNAs
Antisense RNA

An antisense RNA is an RNA sequence that is complementary to all or part of a functional RNA (usually an mRNA). To be an effective therapeutic agent, an antisense RNA or an RNA oligonucleotide must bind to a specified mRNA and prevent translation of the protein encoded by the target mRNA (Fig. 10.3).

The effectiveness of chemically synthesized antisense oligodeoxynucleotides (Fig. 10.3B) relies on hybridization to an accessible nucleotide sequence on the target mRNA, resistance to degradation by cellular nucleases, and ready delivery into cells. Oligonucleotides of about 15 to 24 nucleotides in length have sufficient specificity to hybridize to a unique mRNA. Unfortunately, there are no general rules for predicting the best target sites in various RNA transcripts. Antisense oligonucleotides that are directed to the 5′ and 3′ ends of mRNAs, intron–exon boundaries, and regions that are naturally double stranded have all been effective.

Since oligodeoxynucleotides are susceptible to degradation by intracellular nucleases, it was important to find ways to synthesize molecules that are resistant to attack by nucleases without affecting the ability of the antisense oligonucleotide to hybridize to a target sequence. With this in mind, the backbone, pyrimidines, and sugar moiety have been modified. Clinical trials with several phosphorothioate antisense oligonucleotides, which are considered to be "first-generation" therapeutic agents, have been initiated. Second-generation antisense oligonucleotides typically contain alkyl modifications at the 2′ position of the ribose and are generally less toxic and more specific than phosphorothioate-modified molecules. Third-generation antisense oligonucleotides contain a variety of modifications within the ribose ring and/or the phosphate backbone, as well as being less toxic than either first- or second-generation antisense oligonucleotides.

The use of an expression vector to produce an antisense RNA that suppresses a pathogenic condition has been tested. Episomally based

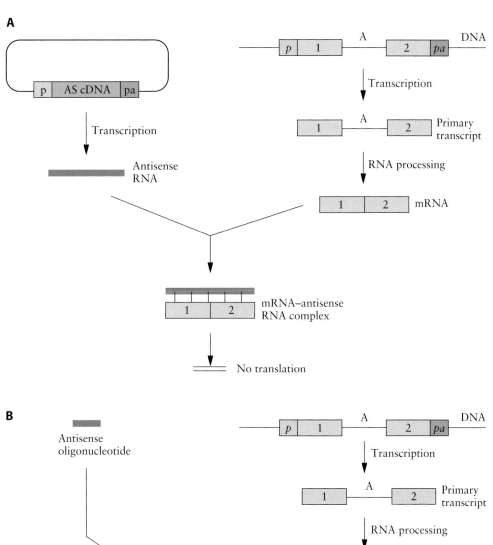

Figure 10.3 Inhibition of translation of specific mRNAs by antisense (AS) nucleic acid molecules. The promoter and polyadenylation regions are marked by *p* and *pa*, respectively; the intron is indicated by the letter A; and the exons are indicated by numbers (1 and 2). (**A**) A cDNA (AS gene) is cloned into an expression vector in reverse orientation, and the construct is transfected into a cell, where the AS RNA is synthesized. The AS RNA hybridizes to the target mRNA, and translation is blocked. (**B**) An AS oligonucleotide is introduced into a cell, and after it hybridizes with the target mRNA, translation is blocked. doi:10.1128/9781555818890.ch10.f10.3

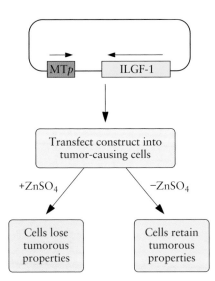

Figure 10.4 A cDNA for human insulin-like growth factor 1 (ILGF-1) cloned on a vector in the antisense orientation under the transcriptional control of a metallothionein promoter (MT*p*). Following transfection into tumor-causing cells, when low levels of ZnSO$_4$ are added, the cells have decreased tumorigenicity. The arrows above the gene and promoter indicate the normal direction of transcription. The origin of replication and other plasmid sequences have been omitted for clarity.
doi:10.1128/9781555818890.ch10.f10.4

FDA
U.S. Food and Drug Administration

expression vectors that carry the cDNA sequence for either insulin-like growth factor 1 or insulin-like growth factor 1 receptor were constructed with the cloned sequences oriented so that the transcripts were antisense rather than mRNA (sense) sequences. Insulin-like growth factor 1 is prevalent in malignant glioma, which is the most common form of human brain tumor. Excess production of insulin-like growth factor 1 receptor occurs in prostate carcinoma, which is a common type of cancer in males. In both vectors, the reverse-oriented cDNAs are under the control of the metallothionein promoter, which is induced by low levels of ZnSO$_4$.

Cultured glioma cells were transfected with the vector that produces the antisense version of the insulin-like growth factor 1 mRNA. In the absence of ZnSO$_4$, the tumor properties were retained; in contrast, when ZnSO$_4$ was added to the culture medium, these properties were lost (Fig. 10.4). In another experiment, nontransfected glioma cells caused tumors after they were injected into rats, whereas glioma cells that had been transfected with antisense insulin-like growth factor 1 cDNA did not develop tumors.

When mice were injected with rat prostate carcinoma cells that were transfected with the insulin-like growth factor 1 receptor cDNA in the antisense orientation, they developed either small or no tumors, whereas large tumors were formed when mice were treated with either nontransfected or control-transfected rat prostate carcinoma cells. In both cases, it was assumed that the antisense RNA hybridized with its complementary mRNA sequence and blocked translation of insulin-like growth factor 1 and insulin-like growth factor 1 receptor, thus preventing the proliferation of the cancer cells.

One phosphorothioate antisense oligonucleotide has been approved by the U.S. Food and Drug Administration (FDA) to treat cytomegalovirus infections of the **retina** in patients with AIDS. This particular antisense oligonucleotide, called fomivirsen and sold as Vitravene, is administered by injection directly into an affected eye after the application of a topical or local anesthetic. Fomivirsen treatment is typically once every 2 weeks for 4 weeks, followed by once every 4 weeks. Before treatment with fomivirsen is started, it is essential that the presence of cytomegalovirus be absolutely confirmed, since several other infective agents produce similar symptoms.

It has recently been shown that it is possible to protect mice against retroviruses by injecting them, intravenously or intraperitoneally, with phosphorothioate antisense oligonucleotides that prevent the conversion of the viral RNA **genome** into double-stranded DNA. In this system, the added antisense oligonucleotide effectively blocks replication of the retrovirus. When an added antisense oligonucleotide binds to the junction of the polypurine tract (which is present in many retroviruses) and the U3 element, a structure is formed that mimics the normal substrate for the virus-encoded enzyme RNase H (Fig. 10.5). This causes premature cleavage of the viral RNA, resulting in the virus being destroyed before reverse transcription occurs. This strategy is effective in protecting mice from retroviruses. In principle, it should also work in humans and on a range of different retroviruses. However, a number of technical obstacles must be overcome before this approach is ready for testing in humans.

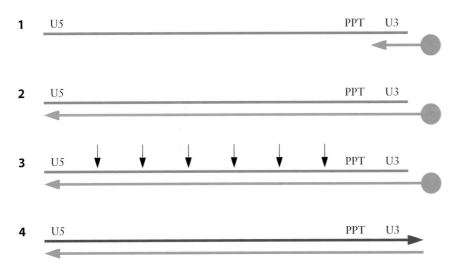

Figure 10.5 Schematic representation of the reverse transcription of retroviral RNA (red) to produce double-stranded DNA (the minus strand is light blue, and the plus strand is dark blue). In step 1, a tRNA (shown as a blue circle) primes synthesis of the minus strand. The RNA shows U3 and U5 elements and a polypurine tract (PPT). In step 2, a complete RNA–DNA duplex is formed. In step 3, RNase H digests the viral RNA in the RNA–DNA duplex into small pieces (the arrows indicate digestion sites). In step 4, a double-stranded DNA copy of the viral RNA is produced, which can exist as a provirus integrated into a cell's DNA. To block the formation of the minus strand and hence the synthesis of a double-stranded DNA version of the virus, an antisense oligonucleotide that hybridizes to the PPT region is added. doi:10.1128/9781555818890.ch10.f10.5

Aptamers

Aptamers are nucleic acid sequences, either RNA or DNA, that bind very tightly to portions of proteins, amino acids, drugs, or other molecules. They are typically 15 to 40 nucleotides long, have highly organized secondary and tertiary structures, and bind with very high affinity to their target molecules (i.e., $10^{-12} < K_d < 10^{-9}$, where K_d is the dissociation constant of the aptamer–target molecule complex). Aptamers are attractive as potential therapeutic agents because of their high specificity, relative ease of production, low level or absence of immunogenicity, and long-term stability. In fact, as a consequence of these properties, the use of aptamers may be preferred over the use of antibodies in applications where targeted protein inhibition is desired.

Aptamers that are directed against specific targets are typically selected by a procedure known as SELEX (systematic evolution of ligands by exponential enrichment), in which DNA or RNA ligands that bind to the target molecule are highly enriched during multiple rounds of selection (Fig. 10.6). In this procedure, a random DNA sequence is cloned between two particular DNA sequences. The 3′ region contains an attachment site for reverse transcriptase primers, and the 5′ region contains an attachment site for a polymerase chain reaction (PCR) primer. The double-stranded DNA is converted to RNA using T7 RNA polymerase. In the SELEX procedure, a large library of aptamers is added to a target molecule, with several different members of this collection of aptamers binding to the target molecule. The bound aptamer molecules are separated from the unbound aptamers, and the bound molecules are amplified

SELEX
systematic evolution of ligands by exponential enrichment

PCR
polymerase chain reaction

A

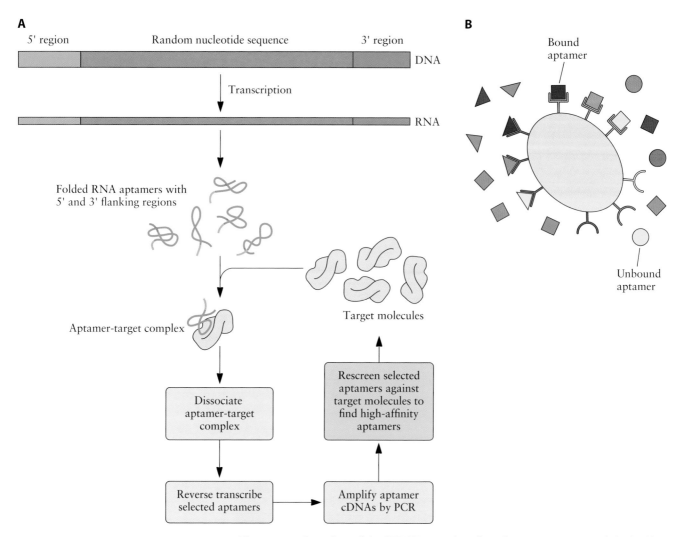

Figure 10.6 Overview of the SELEX procedure for selecting aptamers with high affinity to a target molecule (often a protein). (**A**) The selected aptamers are typically cycled through this procedure 5 to 15 times. (**B**) A view of the surface of the target molecule binding some aptamers while not binding others.
doi:10.1128/9781555818890.ch10.f10.6

and then retested for binding to the target molecule. The entire procedure includes several rounds of binding (to a target molecule), partitioning, and amplification of selected nucleotide sequences from an initial pool of up to 10^{16} nucleotide sequence variants. At each step, only the aptamers that bind most tightly to the target molecule are retained. The end result of this procedure is the selection of aptamers that bind to the target molecule with high affinity. Ultimately, the SELEX procedure yields one (or just a few) unique nucleic acid sequence(s) from the original mixture with a high affinity for the target molecule. To make aptamers less sensitive to nuclease digestion, OH residues at the 2′ positions of purines may be replaced with 2′-O-methyl residues. In addition, aptamers may be capped at their 3′ ends with a deoxythymidine residue. Table 10.1 lists some of the proteins against which aptamers have been generated, as well as the range of affinities of the aptamer for the target protein.

Table 10.1 Some proteins against which aptamers have been generated and the affinities of the aptamers for the proteins

Protein	K_d (nM)
Keratinocyte growth factor	0.0003
HIV-1 reverse transcriptase	0.02
Transforming growth factor β1	0.03
P-selectin	0.04
VEGF receptor	0.05
Platelet-derived growth factor	0.09
Immunoglobulin E	0.1
Extracellular signal-regulated kinase	0.2
CD4 antigen	0.5
HIV-1 RNase H	0.5
Factor IXa	0.58
Angiogenin	0.7
Complement factor 5	1.0
Transforming growth factor β2	1.0
Secretory phospholipase A2	2.0
Thrombin	2.0
Angiopoietin 2	2.2
γ-Interferon	2.7
L-selectin	3.0
Human neutrophil elastase	5.0
Tenascin C	5.0
Integrin	8.0
Hepatitis C virus NS3 protease	10.0
Factor VIIa	11.0
Yersinia pestis tyrosine phosphatase	18.0
Anti-insulin receptor antibody MA20	30.0
Trypanosoma cruzi cell adhesion receptor	172.0

The bioavailability of aptamers is a key issue that affects their application as in vivo diagnostic and therapeutic tools. The most common approaches to improve aptamer bioavailability include surrounding the aptamer with lipoproteins (i.e., liposomes) or the attachment of bulky groups, such as polyethylene glycol, cholesterol, or biotin–streptavidin, to the 5′ or 3′ ends, resulting in a decreased rate of renal clearance and therefore an increased plasma half-life.

An aptamer known as pegaptanib received approval from the FDA in December of 2004 for use as a human therapeutic agent. Pegaptanib is a 30-nucleotide-long aptamer that targets vascular endothelial growth factor (VEGF) and binds to the protein with extremely high affinity ($K_d = 0.05$ nM). This secreted protein promotes the growth of new blood vessels by stimulating the endothelial cells that not only form the walls of the blood vessels but also transport nutrients and oxygen to the tissues. When retinal pigment epithelial cells begin to senesce from lack of nutrition (ischemia), VEGF acts to stimulate the synthesis of new blood vessels (neovascularization). However, this process is imperfect, and often the blood vessels do not form properly and so leakage results, causing scarring in the **macular region** of the retina, with the eventual loss of central

VEGF
vascular endothelial growth factor

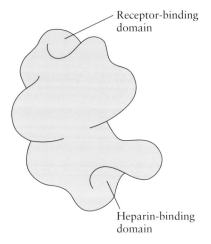

Figure 10.7 Schematic representation of protein VEGF, which contains both a receptor-binding domain and a heparin-binding domain.
doi:10.1128/9781555818890.ch10.f10.7

vision. These physiological changes contribute to the condition known as age-related macular degeneration, a leading cause of blindness in older adults. Pegaptanib was selected to bind to one of the four **isoforms** of the vascular endothelial growth factor protein (i.e., VEGF$_{165}$) that is responsible for age-related macular degeneration. The aptamer drug is injected directly into the eye every 6 weeks, or about nine times a year. All four forms of VEGF have a receptor-binding **domain** (Fig. 10.7), while only VEGF$_{165}$ has a heparin-binding domain, which is the specific target for pegaptanib. Pegaptanib is a new type of therapeutic agent with an unusual specificity that can effectively suppress age-related macular degeneration; to date, approximately 95% of the patients receiving pegaptanib have been at least 65 years old.

In some instances, the safety of aptamers used in clinical trials is a concern, especially when the optimal dose of a particular aptamer is not known. One way to overcome this problem is through the use of aptamer "antidotes." These molecules consist of short oligonucleotides whose sequences are complementary to the aptamers being tested. When antidotes are added to aptamers, they hybridize to the aptamers and inhibit their binding to the clinical target, thereby obviating concerns about too high a dose of the aptamer being used.

Ribozymes

Ribozymes are naturally occurring catalytic RNA molecules (RNA metalloenzymes) that are typically 40 to 50 nucleotides in length and have separate catalytic and substrate-binding domains. As is the case for most nucleic acid therapeutic agents, compared with protein therapeutics, an important advantage of ribozymes is that they are unlikely to evoke an immune response in a treated animal or human. The substrate-binding sequence combines by nucleotide complementarity and, possibly, nonhydrogen-bond interactions with its target sequence. The catalytic portion cleaves the target RNA at a specific site. By altering the substrate-binding domain, a ribozyme can be engineered to specifically cleave any mRNA sequence (Fig. 10.8). For therapeutic purposes, either hammerhead or hairpin ribozymes—named after the appearance of their secondary structure that results from intrastrand base-pairing—may be used (Fig. 10.8).

In practice, an indirect strategy is often used for creating a therapeutic ribozyme, since the large-scale production of synthetic RNA molecules is difficult and RNA molecules are susceptible to degradation after delivery to a target cell. One approach to overcome these drawbacks entails chemically synthesizing a double-stranded oligodeoxyribonucleotide with a ribozyme catalytic domain (~20 nucleotides) flanked by sequences that hybridize to the target mRNA after it is transcribed. The double-stranded form of the ribozyme oligodeoxyribonucleotide is cloned into a eukaryotic expression vector (usually a retrovirus). Cells are transfected with the construct, and the transcribed ribozyme cleaves the target mRNA, thereby suppressing the translation of the protein that is responsible for a disorder. To do this, target cells may be removed from a patient and then grown and transfected in culture before they are returned to the original tissue.

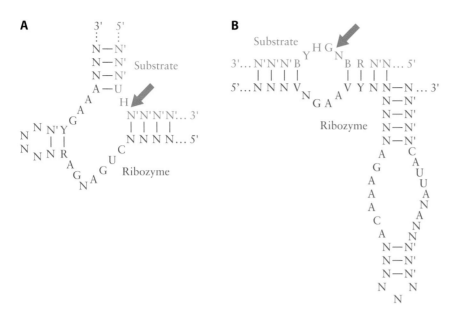

Figure 10.8 Two-dimensional representation of hammerhead (**A**) and hairpin (**B**) ribozyme–mRNA substrate complexes. The mRNA substrates and ribozymes are shown in red and blue, respectively. Y, a pyrimidine nucleotide (C or U); R, a purine nucleotide (A or G); H, any nucleotide except G; B, any nucleotide except A; V, any nucleotide except U; N and N′, any complementary nucleotides. The arrows indicate the points of mRNA cleavage. doi:10.1128/9781555818890.ch10.f10.8

As an alternative to intracellular ribozyme production, ribozymes may be delivered directly to cells by injection or with liposomes, spherical particles of lipid molecules in which the hydrophobic portions of the molecule are facing inward that can carry nucleic acids, drugs, or other therapeutic agents in its center (Fig. 10.9). Directly delivered ribozymes may be chemically modified to protect them from rapid breakdown by nucleases—the 2′ hydroxyl groups may be modified by alkylation or by substitution with either an amino group or a fluorine atom. These modifications increase the half-life of ribozymes in serum from minutes to days.

Under laboratory conditions, ribozymes can inhibit the expression of a variety of viral genes and significantly inhibit the proliferation of numerous organisms. For example, in cell culture, ribozymes inhibit the expression of (i) human cytomegalovirus transcriptional regulatory proteins, resulting in a 150-fold decrease in viral growth; (ii) human herpes simplex virus 1 (HSV-1) transcriptional activator, resulting in a reduction of around 1,000-fold in viral growth; and (iii) a reovirus mRNA encoding a protein required for viral proliferation. Moreover, a hammerhead ribozyme was designed to treat collagen-induced arthritis in mice. A plasmid-encoded ribozyme directed against the mRNA for tumor necrosis factor alpha, which is involved in rheumatoid arthritis, was intravenously injected into affected mice. Following this treatment, there was a significant reduction in the level of the tumor necrosis factor alpha mRNA, as well as a decrease in collagen-induced arthritis. This type of ribozyme, used so far only in mice, has the potential to become a therapeutic agent for humans.

HSV-1
herpes simplex virus 1

Figure 10.9 Schematic representation of a liposome carrying a nucleic acid. doi:10.1128/9781555818890.ch10.f10.9

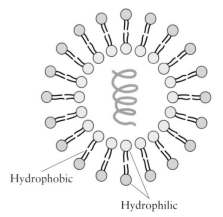

Hydrophobic

Hydrophilic

The development of resistance in humans to various chemical treatments is a persistent problem for the pharmaceutical industry. Generally, a single mutation that alters the target site is sufficient to impair the action of a drug. However, with appropriate ribozyme-based therapeutics, ribozymes for several different sites on a single viral gene could be used simultaneously, thereby cleaving an mRNA at different sites and making it less likely that any single viral mutation will confer resistance.

DNAzymes

Following the discovery of ribozymes, scientists speculated whether DNA might carry out some catalytic functions similar to what had been observed for RNA. Notwithstanding the fact that DNA does not contain a 2′ hydroxyl group on the sugar moiety, in the mid-1990s it was demonstrated that some short single-stranded DNAs could cleave RNA strands at specific sites. The kinetic parameters of the DNAzymes on RNA targets were superior to those of their RNA counterparts. Other advantages of DNAzymes versus ribozymes are their greater ease of chemical synthesis, broad target recognition properties, and high catalytic turnover. Since their initial development, numerous applications of these catalytic molecules have been tested, in particular as anticancer therapeutics. Despite some encouraging results, with DNAzymes, like many other nucleic acid molecules being developed as therapeutic agents, delivery to target sites has been problematic.

Recently, a DNAzyme that was complexed with a lipid carrier was injected directly into a mouse tumor that acts as a model of skin cancer. The DNAzyme targeted c-*jun* mRNA (where the c-jun protein is part of a transcription factor whose overproduction is associated with several human cancers) and profoundly inhibited skin cancer cell proliferation and metastases (Fig. 10.10). Moreover, the injection of the DNAzyme–lipid complex triggered additional antitumor immune responses. The DNAzyme significantly reduced concentrations of both c-*jun* mRNA and protein. And this occurred in the absence of any nonspecific toxicities or side effects, a major advantage compared to some other nucleic acid-based therapeutic agents. Whether DNAzymes will turn out to be as effective and as free of side effects in the treatment of human tumors remains to be determined.

Interfering RNA

The addition of double-stranded RNA to animal or plant cells reduces the expression of the gene from which the double-stranded RNA sequence is derived. This "gene silencing," which can specifically reduce the concentration of a target mRNA by up to 90%, is reversible, since there is no change in the target cells' DNA content or composition. This phenomenon has been termed RNA interference (RNAi) and occurs naturally in virtually all eukaryotic organisms. Although its various biological roles remain to be established, RNAi may act to protect both animals and plants from viruses and from the accumulation of transposons (a

RNAi
RNA interference

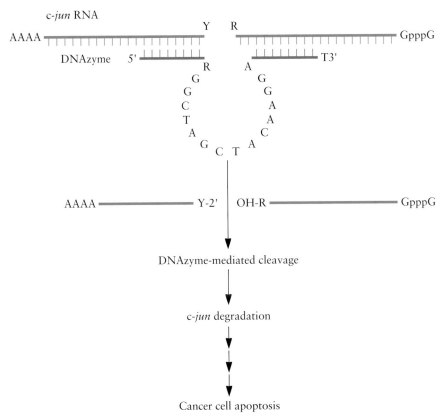

Figure 10.10 Schematic representation of a DNAzyme interacting with the c-*jun* mRNA. A purine (R = G or A) at the beginning of the DNAzyme catalytic core base-pairs with a pyrimidine (Y = C or U) in the target c-*jun* mRNA. Subsequent cleavage takes place between the pyrimidine and purine in the target sequence. Once cleaved, the mRNA is degraded by cellular ribonucleases. Inhibition of c-*jun* expression triggers a cascade of events eventually leading to apoptosis of the cancer cell. doi:10.1128/9781555818890.ch10.f10.10

transposon is a DNA sequence capable of independently replicating itself and inserting the copy into a new position within the same or a different chromosome). A working model for RNAi has been formulated based on experimental analyses (Fig. 10.11). Following the introduction of a double-stranded RNA molecule into a cell, the double-stranded RNA is cleaved by the RNase III-like enzyme Dicer into single-stranded pieces of RNA, approximately 21 to 23 nucleotides in length, that have been called small interfering RNAs (siRNAs). The antisense strand of an siRNA is incorporated into an RNA-induced silencing complex (RISC) that includes several other proteins and binds to and then cleaves the mRNA. The specific binding of the siRNA to the mRNA that occurs is based on the complementarity of the two RNA sequences. The site of cleavage of the targeted mRNA is between nucleotides 10 and 11 relative to the 5′ end of the siRNA (antisense) guide strand. Consistent with this model, the **transfection** of mammalian cells in culture with duplexes of 21-nucleotide RNA can also mediate RNAi activity.

Despite the fact that many aspects of RNAi are still not completely understood, it could form the basis for new therapeutic agents. The use

siRNA
small interfering RNA

RISC
RNA-induced silencing complex

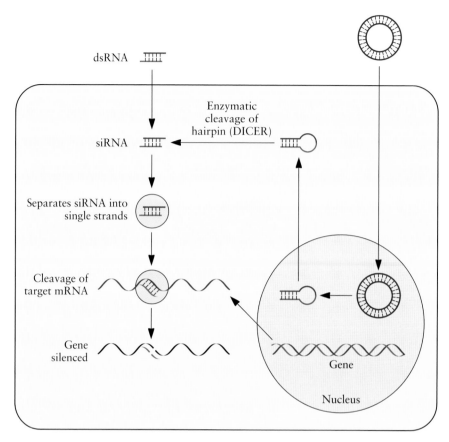

Figure 10.11 Schematic representation of the silencing of the expression of a specific gene by either siRNA or shRNA. dsRNA, double-stranded RNA. doi:10.1128/9781555818890.ch10.f10.11

of short RNA duplexes may eventually provide an alternative approach to gene silencing to the use of antisense oligonucleotides, aptamers, or ribozymes.

The phenomenon of RNAi is expected to facilitate the development of a wide range of antiviral compounds and therapies utilizing specific siRNAs delivered to the appropriate target cell. Similarly, the expression of endogenous eukaryotic genes may be inhibited by plasmid-driven expression of short hairpin RNAs (shRNAs), which are similar in structure to the microRNAs that often normally regulate gene expression in eukaryotic cells. In this case, the Dicer enzyme first removes the hairpin loop from the shRNA to convert it into an siRNA. In fact, there are a number of reports of the use of either siRNA or shRNA to suppress virus replication in tissue culture. Moreover, RNAi is effective in vivo (in mice), suggesting that all viruses may be inactivated by RNAi.

Independently of how an siRNA or shRNA is introduced into a cell, it may have nonspecific effects. For example, introduction of these molecules may inadvertently activate innate cellular immune responses, such as the interferon response. In addition, siRNA or shRNA may also be complementary to nontarget mRNAs. However, several experimental approaches may be utilized to avoid these problems. (i) Off-target effects are most often observed when the siRNA or shRNA concentration is ≥100

shRNA
short hairpin RNA

nM. By lowering the concentration as much as possible (often to 20 nM or less), off-target effects are often avoided. (ii) Since the interferon response can be induced by double-stranded RNAs with as few as 11 base pairs (bp) that are perfectly complementary, siRNA or shRNA is designed to contain at least a 1-nucleotide bulge (where the bases on opposing strands are noncomplementary) near the center of the molecule (typically 21 bp). (iii) Since siRNA or shRNA can exert a toxic effect when it contains the nucleotide sequence 5′-UGGC-3′, this sequence should be avoided. (iv) Blunt-ended 27-bp RNA duplexes or 29-bp shRNAs with 2 nucleotides overhanging at the 3′ end are much more potent inducers of RNAi than 21-mer siRNAs. The greater level of effectiveness of the slightly longer RNAs may reflect the fact that they are first bound and cleaved by Dicer, which facilitates their entry into the RISC. Using these slightly longer RNAs at low concentrations should avoid side reactions associated with 21-mer siRNAs.

Unlike siRNAs, shRNAs are expressed in vivo as part of a genetic construct that includes a promoter sequence. This means that shRNAs need to be introduced by using strategies different than those used with siRNAs. Thus, shRNAs are typically delivered to their target cells by using viral vectors. Viral vectors that integrate into the chromosomal DNA are generally used when persistent long-term knockdown of gene expression is desired; the most popular choice is lentiviruses.

Interfering RNAs have already found widespread use as tools in research that is directed toward understanding how gene expression is regulated in natural systems. One company that specializes in producing RNAi directed against human mRNAs advertises, "For each target (human) gene, we provide four plasmids each with a different short hairpin RNAi sequence (shRNA). Our experimentally verified design algorithm minimizes the risk of off target effects and ensures the maximum knock-down. At least one of the four shRNA plasmids will reduce the target mRNA levels in the transfected cells by >70%." With the ready availability of human shRNA libraries, new insights and understanding of many fundamental and disease processes should be rapidly forthcoming. It is hoped that this, in turn, will lead to a variety of new therapeutic agents and approaches.

In addition to the many reports of successful modification of the gene expression of cells in culture with RNAi, there is an increasing number of reports of the in vivo effectiveness of RNAi. Thus, siRNAs against a wide range of proteins, viruses, and diseases have been successfully expressed in mice, including siRNAs directed against HSV-2, hepatitis B virus, hepatitis C virus, Huntington disease, metastatic Ewing sarcoma (a form of cancer), respiratory syncytial virus, hepatic cancer, transforming growth factor receptor 2, severe acute respiratory syndrome, heme oxygenase 1, keratinocyte-derived chemokine, tumor necrosis factor alpha, and human epidermal growth factor receptor 2. In addition, by mid-2013, several different RNAi therapeutics were being tested in clinical trials, although no RNAi-based product has yet been approved. Three of these therapeutics target vascular endothelial growth factor protein or its receptor, a cause of age-related macular degeneration (see "Aptamers" above), and all have successfully completed either phase I or phase II clinical trials.

Zinc Finger Nucleases

Instead of treating a genetic disease with a therapeutic agent or by providing an additional "corrective" gene, it might be advantageous to directly repair a cell's defective genes. This may become possible with the use of zinc finger nucleases that bind to a specific gene and cleave its DNA. The cell then "heals" the broken DNA strands using copies of all or part of a replacement gene that researchers supply. In the case of gene therapy, the copies lack the disease-causing mutation that is found in the original.

The class of proteins that contain unique small protein structural domains that each bind a single molecule of Zn^{2+} are called zinc finger proteins. Zinc fingers can be classified into several different structural families and typically function as interaction modules that bind DNA, RNA, proteins, or small molecules. These proteins bind to DNA in a sequence-specific manner by inserting a protein α-helical region into the major groove of the DNA double helix, with each zinc finger interacting with a specific DNA triplet codon. Moreover, since the zinc fingers bind to the DNA independently of one another, they can be linked together in a peptide by genetic engineering, so that they will bind to a predetermined sequence on a DNA fragment. Thus, it is possible to engineer nucleases that can cut DNA at specific sites by fusing several zinc finger-encoding sequences with the portion of the gene for the nonspecific nuclease FokI from the bacterium *Flavobacterium okeanokoites* (Fig. 10.12). The DNA-binding zinc finger domain directs the nonspecific FokI cleavage domain to a specific DNA target site. The FokI cleavage domain is enzymatically active only as a dimer, so the creation of a specific nuclease that makes double-stranded breaks is dependent upon dimerization of zinc finger nucleases. Therefore, two zinc finger nuclease subunits are typically designed to recognize the target sequence in a tail-to-tail conformation, with each monomer binding to half-sites that are separated by a spacer sequence (Fig. 10.12).

Figure 10.12 (**A**) Schematic representation of the binding of a zinc finger nuclease to DNA targeted to be cleaved; (**B**) genetic construct used to produce a zinc finger nuclease in *E. coli* under the control of the T7 promoter. doi:10.1128/9781555818890.ch10.f10.12

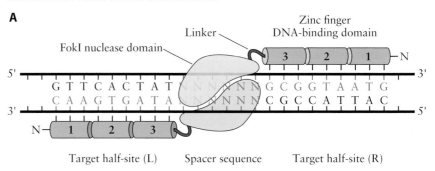

Bacteria that produce restriction enzymes protect their own DNA from being cleaved by these restriction enzymes by synthesizing other enzymes that bind to and methylate the restriction enzyme recognition sites on the DNA. However, a host genome would not be protected from digestion by the synthetic FokI hybrid restriction endonuclease. Consequently, during cell growth, the expression of the hybrid enzyme must be prevented; this is achieved by placing it under the control of the bacteriophage T7 expression system. Following production of zinc finger nucleases in bacteria, after purification of the expressed protein, the N-terminal histidine residues may be removed by treatment with thrombin.

Repair of double-strand breaks in the DNA of animal cells can occur by either nonhomologous end joining or homologous recombination (see chapter 3). With nonhomologous end joining, insertions or deletions may result, leading to disruption of the target gene. On the other hand, homologous recombination between the endogenous target locus and an exogenously introduced homologous DNA fragment (the donor DNA) allows for the possibility of gene targeting. This is because the presence of a double-strand break within the target locus stimulates recombination with the exogenous donor DNA by several orders of magnitude. Importantly, the sequence of the donor DNA determines the outcome of the gene targeting event. Thus, donor DNA can be designed to either create or correct a mutation in a specific gene locus. In addition, the donor DNA can contain either an entire expression cassette or a complementary DNA fragment of the gene to be corrected. To date, the possibility of using zinc fingers to correct a wide range of genetic mutations has been limited by the lack of simple procedures for producing and selecting specific zinc fingers. Nevertheless, scientists are optimistic that these procedures will soon be developed and the promise of this approach will be realized.

Viral Delivery Systems

Given the potential of introducing corrective human genes as well as using different types of nucleic acids to prevent the functioning of various pathogenic agents, the major limitation in converting this possibility into a reality is the development of simple, reliable, and safe delivery systems for these therapeutic molecules. **Systemic** introduction of a therapeutic agent often leads to the accumulation of very high levels in tissues where the agent is not required and may result in serious side effects. Consequently, viral vectors that deliver small nucleic acids to specific cellular targets have been developed. Although virus-based gene delivery has been successful, a number of safety concerns have arisen with regard to the use of these vectors. As a result, several nonviral methods have been developed as an alternative to virus-based systems; these are discussed later in this chapter.

Gammaretrovirus

The first viral delivery system to be developed was based on a mouse RNA virus (Moloney murine leukemia virus [MMLV]), which is a gammaretrovirus of the family *Retroviridae*. Some of the features that favored a

MMLV
Moloney murine leukemia virus

gammaretrovirus as a therapeutic gene delivery system included the small size of the genome (8,332 nucleotides), which could be handled readily with recombinant DNA techniques, and the relative ease of growing the virus in cell culture. In addition, as part of its life cycle, it integrates into **chromosomes**, which suggested that long-term gene expression of an introduced therapeutic gene was possible. The mature gammaretrovirus **virion** is a complex structure about 100 nm in diameter (Fig. 10.13A). The outermost lipid bilayer that envelops the virion is derived from the host cell. All other components are viral in origin. Embedded within the **cell membrane** are protrusions (spikes) composed of two proteins (surface glycoprotein and transmembrane protein). The inner surface of the lipid bilayer is coated with the matrix protein. Separated from the matrix layer, there is a protein shell made up of the capsid protein that encases the core. The core contains nucleocapsid molecules bound to the viral genome and the enzymes reverse transcriptase, integrase, and protease.

The genetic complement of a mature virion has two single-stranded RNA molecules with coding (*gag*, *pol*, and *env*) and noncoding sequences (Fig. 10.13B). Of the noncoding elements, there are an **R sequence** and a **U5 sequence** at the 5′ end and a **U3 sequence** and another R sequence at the 3′ end. Promoter and enhancer sequences are located in the U3 region. One of the other noncoding sequences is the packaging signal (Ψ) that is essential for incorporating RNA molecules into virus particles. The

Figure 10.13 (A) Schematic representation of a mature gammaretrovirus virion; (B) organization of a gammaretrovirus RNA genome. The noncoding elements shown here are the R and U5 sequences at the 5′ end and the U3 and R sequences at the 3′ prime end and the packaging signal site, which is designated with the Greek letter Ψ. The *gag* gene encodes a polyprotein that is cleaved to produce the matrix, capsid, nucleocapsid, and p12 proteins. The *pol* gene codes for a polyprotein that gives rise to the enzymes protease, reverse transcriptase, and integrase. The *env* gene encodes a polyprotein carrying the outer surface glycoprotein and transmembrane protein. The Env polyprotein is translated from a spliced RNA derived from an intact gammaretrovirus RNA genome. (C) Organization of the gammaretrovirus provirus. The noncoding 5′ and 3′ LTRs consist of U3, R, and U5 elements. The U3 region contains promoter and enhancer sequences. doi:10.1128/9781555818890.ch10.f10.13

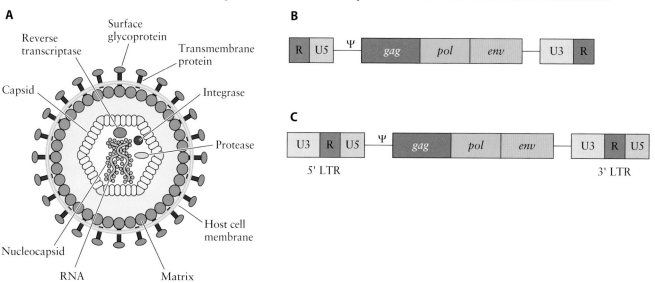

unique feature of the gammaretrovirus life cycle is that after infection, the RNA genome acts as a template for a reverse transcriptase complex that synthesizes a double-stranded DNA molecule that, in turn, integrates into a host cell chromosome. As a result of the replication and integration processes, an integrated viral DNA molecule **(provirus)** has a U3 element at the 5′ end and a U5 sequence at the 3′ end (Fig. 10.13C). Transcription of the provirus starts at the beginning of the the 5′ R sequence and terminates at the end of the 3′ R sequence. This transcript acts as the mRNA for the production of viral proteins and is the sequence that is incorporated into virus particles.

The *gag*, *pol*, and *env* sequences encode polyproteins that after proteolytic cleavage yield functional protein units. The Gag polyprotein carries the structural components of the virus, which include the capsid, matrix, and nucleocapsid proteins. A fourth Gag protein, p12, plays a part in the release of the virus from the host cell as well as aiding the integration process. The Pol polyprotein has sequences for three enzymes: reverse transcriptase, protease, and integrase. The Env polyprotein gives rise to two proteins (surface glycoprotein and transmembrane protein) that facilitate entry of the virus into a host cell.

In the life cycle of a gammaretrovirus, following the binding of the viral surface glycoprotein to the host cell receptor, the viral membrane fuses with the host cell membrane and the capsid is released into the cytoplasm of the host cell (Fig. 10.14). After dissociation of the capsid, a reverse transcriptase complex copies viral RNA to synthesize a double-stranded DNA molecule which forms part of a preintegration complex. Integration of viral DNA into a host cell chromosome only occurs when the target cell is undergoing mitosis. The provirus DNA is transcribed and the viral RNA is transported to the cytoplasm, where it is translated to produce the Gag and Pol polyproteins. The Env polyprotein is synthesized from a spliced RNA, and after proteolytic cleavage, the two Env components combine and embed in the host cell membrane. The Gag polyprotein wraps around both an RNA dimer and an intact Pol polyprotein before the assemblage is budded off with host cell membrane studded with surface glycoprotein and transmembrane protein units. Finally, both the Gag and Pol polyproteins are processed and a mature virion is fashioned.

Some of the requirements for constructing an effective viral delivery system include making space for a therapeutic gene in the viral genome, creating a system for packaging the therapeutic gene–virus constructs into a virus particle, and ensuring that the virus carrying the therapeutic gene is not harmful to either the targeted cell (unless it is a cancer cell) or the recipient organism. The provirus of MMLV, for example, was isolated and cloned into a plasmid, and all of the coding regions were removed and replaced by a therapeutic gene (Fig. 10.15). Care was taken to avoid leaving any partial viral open reading frames that might be translated in a recipient cell and, consequently, induce an immunogenic response in the host organism. Noncoding regions that were retained included the 5′ and 3′ long terminal repeats (LTRs), the packaging signal (Ψ), primer-binding sequence, and the polypurine tract (Fig. 10.15). The primer-binding sequence and the polypurine tract are necessary for reverse transcription

LTR
long terminal repeat

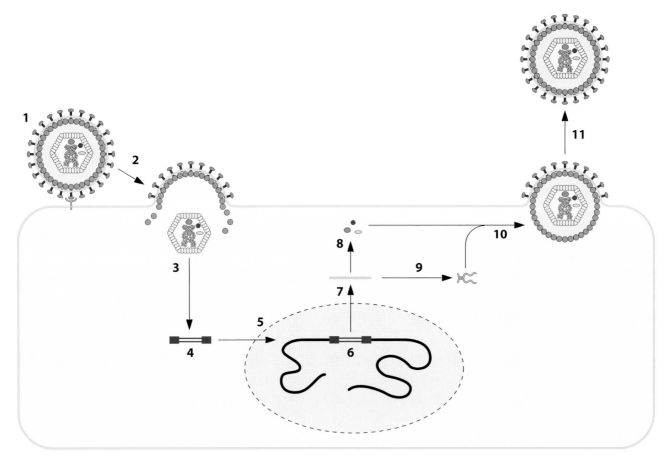

Figure 10.14 Life cycle of a gammaretrovirus. (**1**) A mature virion binds to a surface receptor on the host cell. (**2**) The envelope of the virus fuses with the cell membrane of the host cell, and the capsid compartment enters the cell. (**3**) The capsid disassociates. (**4**) RNA strands are copied into double-stranded DNA by reverse transcriptase. (**5**) The viral double-stranded DNA enters the nuclear area only during mitosis, when the nuclear membrane has broken down. (**6**) Integration of double-stranded viral DNA into a host cell chromosome. (**7**) The integrated double-stranded viral DNA (provirus) is transcribed. (**8**) Viral polyproteins are synthesized. (**9**) RNA genomes dimerize. (**10**) For each viral particle, a RNA dimer and Pol polyprotein are encased in Gag polyprotein molecules. This complex makes contact with regions of the cell membrane that have embedded Env proteins. (**11**) An immature virus is budded off from the host cell, and the protease cleaves the Gag and Pol proteins to form a mature virion. doi:10.1128/9781555818890.ch10.f10.14

and integration of the double-stranded DNA. The Ψ sequence is for packaging, and the U3 sequence in the 5′ LTR provides a promoter region to transcribe the therapeutic gene (Fig. 10.15).

For one type of packaging system (transient), the *gag/pol* gene unit and the *env* gene are cloned into separate plasmids, with each coding sequence under the control of the nonviral promoter (Fig. 10.16A). Both of these constructs lack a packaging signal. Three separate plasmids with the *gag/pol*, *env*, and therapeutic gene coding sequences are transfected into cells in culture. All of the viral components are produced within the cell. The only RNA that has a packaging signal is transcribed from the U3 promoter (LTR-driven promoter) of the therapeutic gene–virus construct.

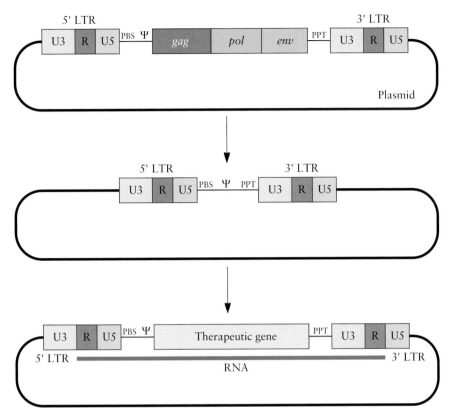

Figure 10.15 Constructing a therapeutic gene vector based on a gammaretrovirus provirus. After the gammaretrovirus is cloned into a plasmid, the coding genes *gag*, *pol*, and *env* are removed and replaced with a therapeutic gene. The noncoding regions that are represented include the LTRs, primer-binding site (PBS), Ψ (packaging signal), and polypurine tract (PPT). The RNA transcribed from the therapeutic gene construct is shown in blue. doi:10.1128/9781555818890.ch10.f10.15

This RNA is packaged and then released as part of a virion that has reverse transcriptase and integrase in its core (Fig. 10.16B). When these virions infect cells, the RNA is reverse transcribed, the resulting double-stranded DNA is integrated into a host chromosome, and the therapeutic gene is expressed (Fig. 10.16C). Generally, a gammaretrovirus-based delivery system can carry about 6 to 9 kilobase pairs (kb) of cloned DNA.

Stable packaging systems have been devised for some therapeutic gene–virus constructs. In these cases, the *gag/pol* and *env* viral gene constructs are integrated on separate chromosomes of a cell line (Fig. 10.16D). The reason either for using separate plasmids carrying different viral genes with a transient package system or for integrating these genes on distinct chromosomes of a permanent **packaging cell line** is to minimize the probability of exchanges among the constructs that might create molecules that carry the packaging signal with a full set of viral genes and therefore be infectious. On the other hand, the virus particles that are generated by various packaging protocols are replication defective and not able to form progeny viruses.

Initially, it was thought that LTR-driven transcription would be sufficient for the expression of a therapeutic gene in a gammaretrovirus.

Figure 10.16 (**A**) Production of gammaretrovirus particles with a therapeutic gene and synthesis of a therapeutic protein in a target cell. Three plasmids are used for transient gammaretrovirus packaging of a therapeutic gene. The *gag/pol* gene unit and the *env* gene that are under the control of nonviral promoters (*p*) are cloned into separate plasmids. The third plasmid carries the therapeutic gene with gammaretrovirus noncoding sequences including the packaging signal (Ψ). None of the plasmid genes are noted here or in other similar figures in this chapter. PBS, primer-binding site; PPT, polypurine tract. (**B**) Production of packaged therapeutic gene virions. The viral polyproteins (color-coded circles) are synthesized from the *gag/pol* gene unit and the *env* gene. The RNA from the therapeutic gene construct is synthesized from the U3 promoter (i.e., LTR driven), has a packaging signal (Ψ), and is packaged into virus particles. The arrow points to an enlarged representation of the RNA genomes of a virion that are produced by a packaging cell. (**C**) Production of a therapeutic protein in a target cell. (1) Entry of the capsid complex into the target cell. (2) Removal of the capsid protein. (3) Reverse transcription of the RNA version of the therapeutic gene–viral genome. (4) Entry into the nucleus and integration of the DNA version of the therapeutic gene–viral genome. (5) Transcription of the integrated therapeutic gene–viral sequence from the U3 promoter. Transcription occurs in cells that are not in the process of dividing. (6) Transport of the therapeutic gene–viral RNA to the cytoplasm. (7) Synthesis of the therapeutic protein (yellow circles). (**D**) Stable packaging cell line. The *gag/pol* gene unit and the *env* gene that are expressed by nonviral promoters are integrated into different packaging cell line chromosomes. Plasmids carrying the therapeutic gene with gammaretrovirus noncoding sequences including the packaging signal (Ψ) are introduced into a stable packaging cell line for the production of therapeutic gene-packaged virions (not shown here). doi:10.1128/9781555818890.ch10.f10.16

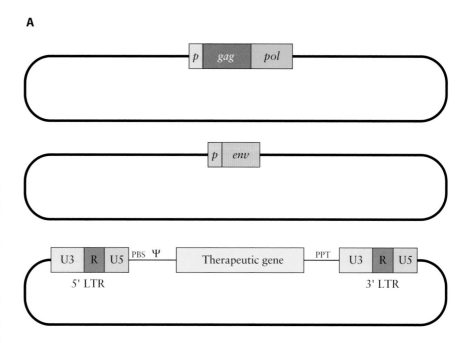

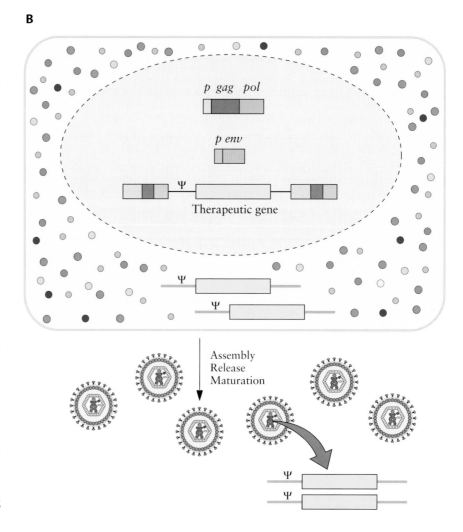

C

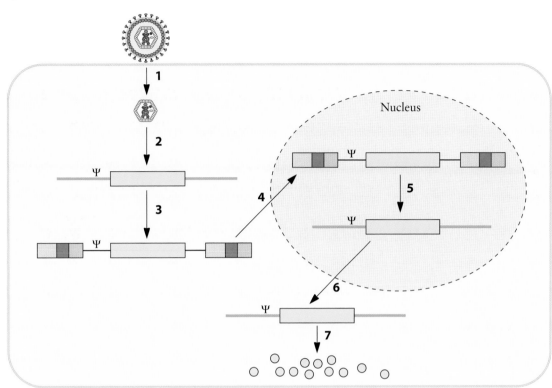

D

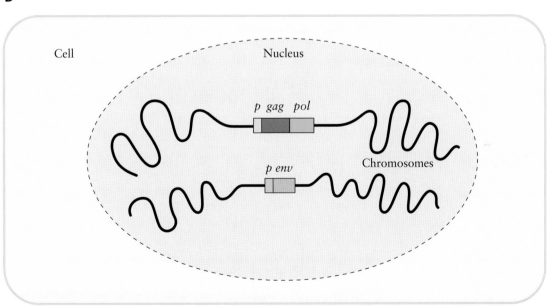

However, due to the strong U3 promoter, unwanted transcription (read-through) of an adjacent gene occurred frequently after integration, creating the possibility of tumor formation. Integration of DNA into a site that causes harmful consequences is called insertional mutagenesis. In addition, placing the therapeutic gene under the control of a cell-specific promoter may increase both safety and effectiveness where therapeutic gene activity occurs only in a targeted cell type. Another concern was that integration of the provirus into a compacted chromosome region might silence the therapeutic gene. Similarly, neighboring chromosome elements may block therapeutic gene transcription. Consequently, modifications were devised to address these issues.

Briefly, the U3 promoter and enhancer sequences were deleted from the 3′ LTR of the vector, and a cell-specific promoter was inserted to drive the expression of the therapeutic gene; the U3 promoter and enhancer sequences drive the transcription of the RNA that is to be packaged. After introduction into a host cell, the integrated construct does not have U3 promoter and enhancer sequences and transcription of the therapeutic gene is controlled by the cell-specific promoter. A vector with deleted U3 promoter and enhancer sequences in its 3′ LTR is called a self-inactivating (SIN) vector. To decrease transcription readthrough from the integrated construct, a strong transcriptional terminator was inserted downstream from the therapeutic gene. Also, the woodchuck hepatitis virus posttranscriptional regulatory element that enhances mRNA stability, export, and translation was added to the vector. Finally, to shield the therapeutic gene from transcription silencing, an insulator sequence was inserted into the 3′ LTR of the vector which flanks the therapeutic gene after integration into target cells (Fig. 10.17).

SIN
self inactivating

Lentivirus

Lentiviruses are members of the *Retroviridae* family. Unlike gammaretroviruses, they infect and integrate into nondividing cells. In addition, lentiviruses have long incubation times, suggesting that stable therapeutic gene expression over a long period is likely. As well, lentiviruses rarely induce immunological response in a host organism. Despite the overall similarities between gammaretroviruses and lentiviruses, there are differences. Lentiviruses are about 20% larger, their capsid is a rounded trapezoid, and lentiviruses have a number of additional genes that encode accessory and regulatory proteins that participate in various aspects of the life cycle.

HIV-1
human immunodeficiency virus type 1

The most thoroughly studied lentivirus at the molecular level is human immunodeficiency virus type 1 (HIV-1), which is responsible for AIDS. This disease has caused millions of deaths, and currently, millions of people worldwide are HIV-1 positive. Therefore, why would HIV-1, which is a dangerous human virus, be considered as a therapeutic gene viral delivery system? The primary reason is that nondividing cells such as **neurons** and differentiated cells can be targeted for gene therapy. Safety is not a major concern because all of the viral genes are removed from the therapeutic gene vector construct and the packaging systems do not produce any active HIV-1 virions.

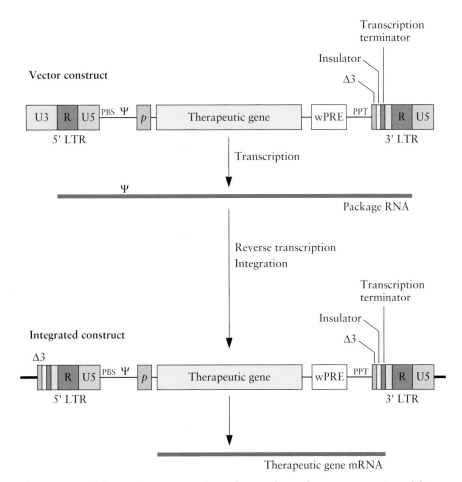

Figure 10.17 Schematic representation of an enhanced gammaretrovirus delivery system. A cell-specific promoter (*p*) that drives the therapeutic gene is added. The woodchuck posttranscription regulatory element (wPRE) and both insulator and transcription terminator sequences are inserted into the vector construct. The U3 promoter and enhancer sequences are deleted (ΔU3) from the 3′ LTR of the vector construct. Transcription of the therapeutic gene of the integrated construct in the host cell is under the control of a cell-specific promoter.
doi:10.1128/9781555818890.ch10.f10.17

Similar to what has been described for the gammaretroviruses, the HIV-1 genome encodes Gag, Pol, and Env polyproteins as well as other gene products (Fig. 10.18A). The accessory proteins (**Vif, Vpu, Vpr,** and **Nef**) are translated from various internal initiation sites and terminated at specific stop codons in the viral mRNA. The regulatory proteins, **Tat** and **Rev**, are encoded by sequences that are created by RNA splicing. The conversion of HIV-1 into a viral vector entails removing all the HIV-1 coding genes and replacing them with a therapeutic gene under the control of a cell-specific promoter (Fig. 10.18B). The **carrying capacity** of this vector is about 9 kb. The U3 promoter and enhancer sequences of the 5′ LTR are replaced with the promoter from the human cytomegalovirus in order that transcription of the HIV-1-based vector is independent of the Tat protein. Also, removal of the U3 sequences lowers the probability of reversion during the packaging stage that may lead to the formation of

A

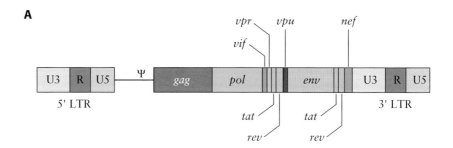

B

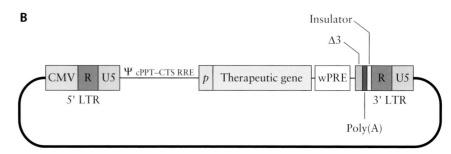

Figure 10.18 (**A**) Schematic representation of the genome of the lentivirus HIV-1. The *gag*, *pol*, and *env* genes are indicated. The only structural elements that are depicted are the LTRs and the packaging signal (Ψ). The accessory genes include *vif*, encoding viral infectivity factor, which promotes infectivity; *vpu*, encoding viral protein U, which enhances release of virions from the plasma membrane; *vpr*, encoding viral protein R, which forms part of the preintegration complex; and *nef*, encoding negative regulation factor, which enhances the spread of virus particles. After RNA splicing, the *tat* and *rev* genes encode regulatory proteins that modulate aspects of transcription and posttranscription. (**B**) Lentivirus vector. The U3 sequence of the 5′ LTR has been replaced by a human cytomegalovirus (CMV) promoter. The packaging signal (Ψ) is present. The central polypurine tract–central termination sequence (cPPT–CTS) facilitates transport of the vector DNA into the nucleus of the target cell. The Rev protein binds to the Rev-responsive element (RRE) and enhances the export of viral RNA from the nucleus to the cytoplasm. The therapeutic gene is under the control of a cell-specific promoter (*p*). The woodchuck hepatitis virus posttranscriptional regulatory element (wPRE) increases virus production, promotes RNA processing, and prevents transcriptional readthrough. The 3′ LTR, which forms the 5′ LTR of the provirus, has the U3 promoter and enhancer sequences deleted (ΔU3), a polyadenylation enhancer element [poly(A)], and a chromatin insulator sequence (insulator). doi:10.1128/9781555818890.ch10.f10.18

infectious viral particles. The packaging signal (Ψ) guarantees encapsidation of only the vector RNA. The 3′ LTR does not have a U3 region (ΔU3), creating a self-inactivating (SIN) vector. A strong polyadenylation signal is inserted into the 3′ LTR to prevent transcriptional readthrough into adjacent genes. Finally, the addition of an insulator sequence lowers the likelihood that after integration the therapeutic gene will be silenced. Once established, the therapeutic gene is transcribed by the cell-specific promoter.

Various lentiviral packaging systems have been developed. Here, a four-plasmid protocol is illustrated (Fig. 10.19). The plasmid with the *rev* gene supplies the Rev protein that binds to the *rev*-responsive element and increases the transport of the *gag/pol* mRNA into the cytoplasm, which, in turn, leads to higher yields of packaged viral particles carrying the

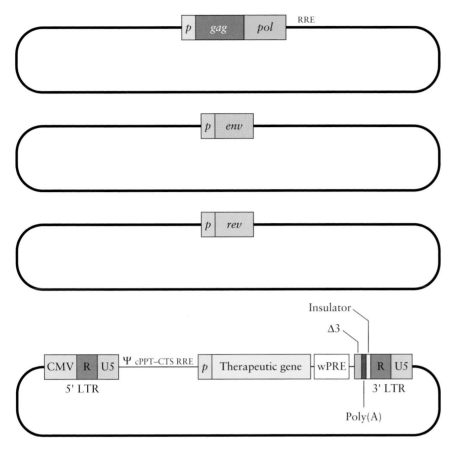

Figure 10.19 Lentivirus packaging plasmids. The gammaretrovirus and lentivirus packaging systems are similar, and both produce replication-incompetent virions. The Rev protein combines with the RRE sequence and facilitates nuclear export of an mRNA transcript. Abbreviations are as in Fig. 10.18.
doi:10.1128/9781555818890.ch10.f10.19

therapeutic gene construct. Lentiviral delivery systems are highly efficient in producing large numbers of packaged virions that readily transduce target cells. As a consequence, the HIV-1-based platform is being used to deliver a variety of genes for many preclinical and clinical studies. Nevertheless, integration of gammaretroviruses and lentiviruses into unwanted chromosome locations is an ongoing concern. The integration sites are not totally random. MMLV tends to incorporate into regions that precede transcriptionally active genes and HIV-1 into exons of transcribed genes. However, vectors that direct therapeutic genes to specific chromosome sites are currently being devised.

Adeno-Associated Virus

Adeno-associated virus (AAV) is a small (~20-nm-diameter), nonpathogenic, nonenveloped, single-stranded human DNA virus (Fig. 10.20). The production of AAV particles requires not only AAV gene products but also both host cell proteins and the gene products of another virus (helper virus), which may be adenovirus, herpesvirus, or another virus. In other

AAV
adeno-associated virus

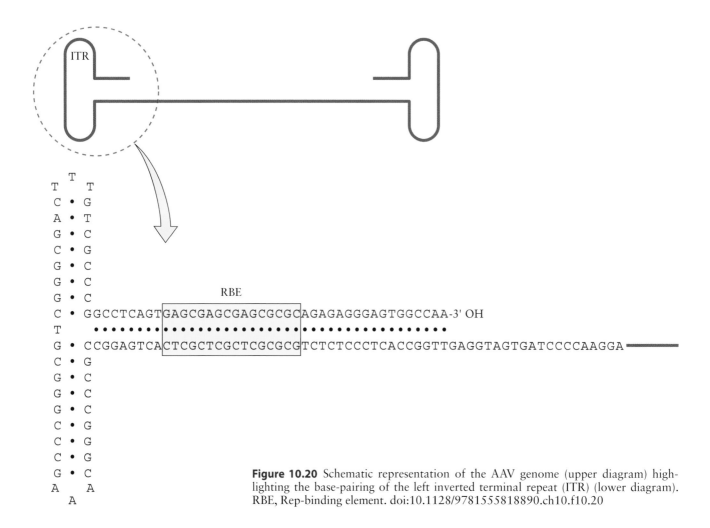

Figure 10.20 Schematic representation of the AAV genome (upper diagram) highlighting the base-pairing of the left inverted terminal repeat (ITR) (lower diagram). RBE, Rep-binding element. doi:10.1128/9781555818890.ch10.f10.20

words, the AAV genome does not contain all the genes that are necessary for the propagation of its own virions. Thus, AAV is assigned to the genus *Dependovirus*. The name adeno-associated virus was coined initially because AAV was found in an adenovirus sample.

The size of the AAV genome is about 4.6 kb, with two repeated regions called inverted terminal repeats at the ends. The inverted terminal repeats contain segments that base-pair with each other and form a T-shaped structure at each end of the viral genome. The coding component of the AAV genome consists of two open reading frames, *rep* and *cap*. The *rep* gene encodes four proteins (Rep78, Rep68, Rep52, and Rep40). Rep78 is encoded in a full-length RNA transcribed from the *p*5 promoter, and Rep68 is derived from a spliced version of this RNA (Fig. 10.21). A second promoter (*p*19) produces a transcript that encodes Rep52, and a spliced derivative of this RNA carries the sequence for Rep40. The *cap* gene is transcribed by a third promoter (*p*40), and internal translation start sites give rise to three proteins (VP1, VP2, and VP3). The Rep proteins are multifunctional. They bind to elements of the inverted terminal repeat and are critical for viral DNA replication, packaging, and integration and excision of viral genomes. The 3′ OH group of the left-hand inverted

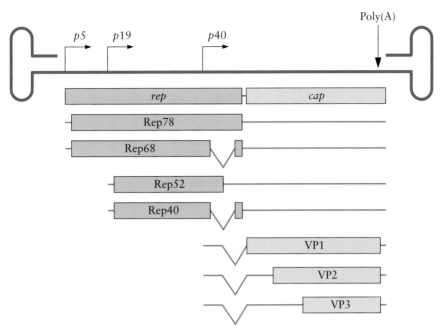

Figure 10.21 Schematic representation of the AAV genome with the *rep* and *cap* genes; transcripts (blue) derived from *p5*, *p19*, and *p40* (right-angled arrows); and the translated viral proteins (labeled colored boxes within the transcripts). The blue V's indicate spliced RNAs. doi:10.1128/9781555818890.ch10.f10.21

terminal repeat provides the primer site for the initiation of replication of the second viral DNA strand by a host cell DNA polymerase in conjunction with Rep proteins. The Cap proteins make up the units (facets) of the symmetrical capsid in a ratio of 1:1:8 for VP1, VP2, and VP3, respectively.

Currently, there are 12 known AAV strains (AAV1 through AAV12) due to genetic variants of the *cap* gene. From the perspective of human gene therapy, the various AAV **serotypes** provide viruses that are specific for certain cell types and many kinds of tumors. For example, AAV1 is specific for skeletal muscle and cardiac tissue, while AAV8 is specific for liver cells. In addition, it is possible to combine VP sequences from different strains and by genetic modification of *cap* genes to create capsids that bind to novel cell surface receptors.

The first step of the AAV infection cycle is the binding of the virus to surface receptors of either a dividing or nondividing cell (Fig. 10.22). Next, the virus particle is internalized by formation of a clathrin-coated vesicle. The clathrin layer is removed and the vesicle fuses with an early endosome which, in turn, is routed along microtubules to the cell **nucleus**, where the virus is released and where uncoating of the capsid occurs. The *rep* gene is expressed and second-strand DNA synthesis takes place with the aid of a host cell DNA polymerase. At this point, in the absence of a helper virus, integration of the viral genome into a specific site on the short arm of human chromosome 19 and, to a lesser extent, at a few other chromosome sites occurs. Alternatively, the presence of a helper virus provides proteins necessary for the production of AAV particles. Assembly of virions occurs in the nucleus, and the virus particles pass through the

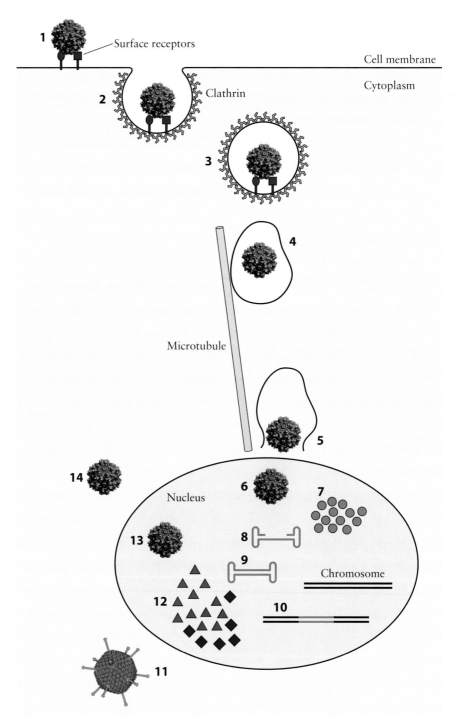

Figure 10.22 Schematic representation of the life cycle of AAV. The virus binds to specific surface receptors (**1**), and receptor-mediated endocytosis is initiated (**2**). The virus is internalized and wrapped in a clathrin-coated vesicle (**3**). Within an endosome, the virus is routed along microtubules to the nucleus (**4**), from which it escapes (**5**). More than likely, an intact virus enters the nucleus (**6**), where it is uncoated (**7**), releasing the viral genome (**8**), which is duplicated (**9**). In the absence of a helper virus, integration into a chromosome site is likely (**10**). In the presence of a helper virus (**11**), both AAV and helper virus proteins are synthesized (**12**) and lead to the assembly of virus particles (**13**) which leave the nucleus (**14**) and exit the cell (not shown). doi:10.1128/9781555818890.ch10.f10.22

cytoplasm and eventually escape from the cell. A provirus can be excised, and the AAV life cycle will be completed if the host cell is infected by an appropriate helper virus.

In the AAV vector packaging system, the AAV genome is cloned into a plasmid and all the viral DNA except for the inverted terminal repeats is removed and replaced with an expression unit (cassette). The cassette includes a therapeutic gene under the control of the cell-specific promoter, DNA elements that enhance transcription and translation, and a strong transcription stop signal (Fig. 10.23). The *rep* and *cap* genes are cloned into another plasmid under the control of a mammalian promoter. Three of the helper genes from adenovirus virus, *E2A*, *E4*, and *VA RNA*, are inserted into a third plasmid under the control of another promoter. The additional adenovirus helper genes, *E1A* and *E1B*, are integrated into the chromosomes of the packaging cell line (Fig. 10.23).

After cotransfection with the three plasmids, the components necessary for the formation of virus particles are synthesized (Fig. 10.24). Under these conditions, single-stranded DNA is replicated from the plasmid with the inverted terminal repeats, and only this DNA is taken into virions because it has a packaging sequence located within an inverted terminal repeat. The AAV vector particles are released and harvested. After **transduction** of target cells, the AAV vector DNA is delivered to the nucleus, where it forms, by recombination, duplex linear and circular molecules with two or more linked copies of the therapeutic gene cassette–inverted terminal repeat DNA (Fig. 10.24). Since these DNA molecules lack origins of replication, they are maintained for short periods in dividing cells and, potentially, for a long time in nondividing cells. Since no Rep protein is synthesized in the target cells, integration at the preferred AAV chromosome sites is unlikely. However, some integration of the AAV vector DNA does occur, but it is usually rare and random. To date, no instances of **insertional mutagenesis** have been observed with animal studies or in human trials. Many AAV-derived therapeutic gene vectors have been tested, with no significant adverse effects. Moreover, in some instances, expression of the therapeutic gene has been observed for a number of years after introduction into the host tissue.

Adenovirus

Adenoviruses are nonenveloped viruses (~100-nm-diameter particles) that infect the upper respiratory tract, with usually mild consequences. The capsid has 20-sided symmetry with protruding knobbed fibers anchored to a penton protein at the junction of the individual units of the capsid (Fig. 10.25). The protein that comprises the edges of the virion has been designated hexon and is the most abundant component of the capsid (Fig. 10.25). The core carries proteins that bind to the genome, which is a linear, double-stranded DNA molecule that is about 36 kb in length.

Adenovirus and AAV enter a cell and reach the **nuclear membrane** in similar ways. The knob of the fiber protein of adenovirus makes contact with a cell surface receptor, and then the penton protein binds to surface proteins. Importantly, the knob component can be genetically engineered to target an adenovirus vector to specific cell surface receptors. After

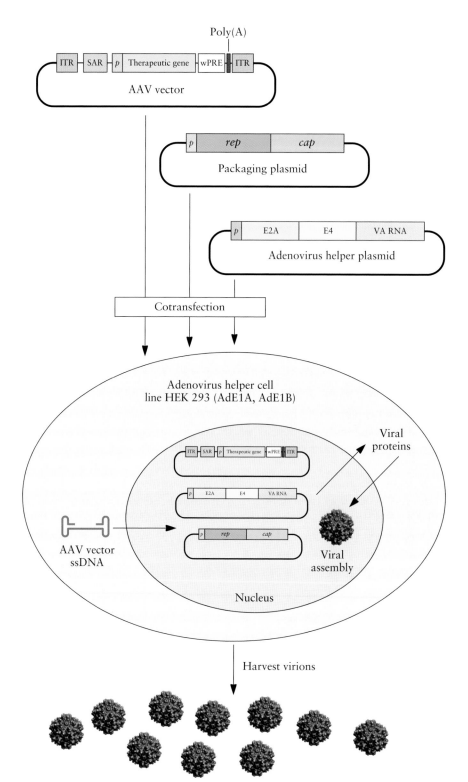

Figure 10.23 Schematic representation of an AAV packaging system. A three-plasmid system is shown here, viz., AAV vector, packaging plasmid with *rep* and *cap* genes, and a helper plasmid with adenovirus *E2A*, *E4*, and *VA RNA* genes. The plasmids are cotransfected into cell line HEK 293, which has integrated adenovirus genes *E1A* and *E1B*, which are expressed constitutively. Once in the nucleus of the cell, the plasmid-based adenovirus helper genes and the AAV genes are expressed. After translation, these proteins enter the nucleus and a double-stranded DNA version of the AAV vector is assembled into viral particles. The left inverted terminal repeat (ITR) of the AAV vector has sequences that enable replication and packaging of the AAV vector. The virus particles are harvested and concentrated. wPRE, woodchuck hepatitis virus posttranscriptional regulatory element; ssDNA, single-stranded DNA; TG, therapeutic gene.
doi:10.1128/9781555818890.ch10.f10.23

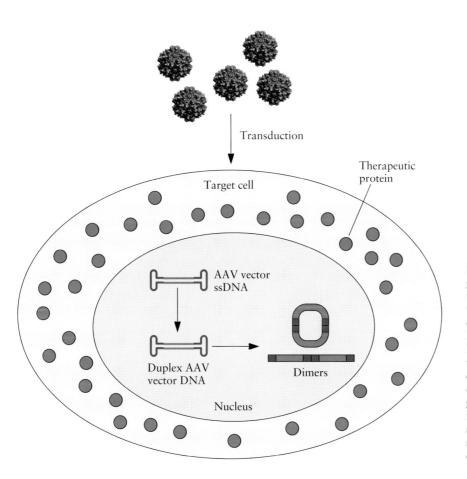

Figure 10.24 Delivery of a therapeutic gene as part of an AAV vector into a target cell. The target cell is transduced with viruses carrying a therapeutic gene expression cassette. The viral DNA is released and duplicated within the nucleus. DNA molecules with multiple units of the AAV vector DNA are formed by recombination. Here, linear and circular dimers are depicted with the therapeutic gene expression cassettes (yellow) and the inverted terminal repeats (gray). The therapeutic protein (yellow circles) is synthesized in the cytoplasm of the target cell. doi:10.1128/9781555818890.ch10.f10.24

Figure 10.25 Structure and proteins of adenovirus. (**A**) Three-dimensional representation of adenovirus. Image source: David S. Goodsell/RCSB Protein Data Bank [http://www.rcsb.org/pdb/101/motm.do?momID=132]). (**B**) Schematic representation of adenovirus. Labels on the right indicate the major capsid proteins, and those on the left mark some of the core proteins. doi:10.1128/9781555818890.ch10.f10.25

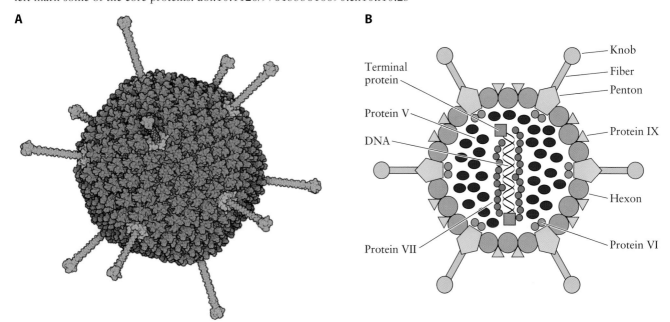

binding, the virion is enclosed in a clathrin-coated vesicle that fuses with an endosome, and the virus is routed to the nucleus. At a nuclear pore, the released virus is disassembled and adenoviral DNA enters the nucleus.

The adenoviral genome has inverted terminal repeats at the ends, with a packaging signal (Ψ) adjacent to the the 5′ inverted terminal repeat (Fig. 10.26). The adenoviral DNA encodes a large number of proteins, and transcription from both strands is carried out in two main phases: early and late (Fig. 10.26). Proteins required for viral DNA replication are synthesized early, while late synthesis entails the production of capsid and core components.

Immediately after entry into the nucleus, the *E1A* gene is expressed. The E1A RNA is spliced at different sites to produce three mRNAs. One of the encoded proteins from these RNAs activates the transcription of the early genes. Each of the adenoviral transcripts shown in Fig. 10.26 undergoes splicing to produce mRNAs for several viral proteins. The terminal protein binds to the 5′ end of the adenoviral DNA and primes the synthesis of the complementary DNA strand. After a number of rounds of DNA replication, a long primary RNA molecule, i.e., the major late transcriptional unit, is transcribed. This transcript is cleaved into five late gene sequences (L1 to L5), each of which is processed further; the resulting mRNAs encode capsid and core proteins. The formation of virions takes place in the nucleus. Once assembled, about 10,000 virions at a time exit the cell, destroying it in the process.

Adenovirus is well suited as a therapeutic gene vector. It infects a large number of different dividing and nondividing cells. Pathogenicity is mild. There is no concern about insertional mutagenesis because it does not insert into host chromosomes.

An effective adenovirus vector was created by removal of all the coding sequences, leaving only the inverted terminal repeats and the packaging signal (Fig. 10.27A). This vector has been dubbed as gutless, gutted, or helper dependent. In order to be packaged, an adenovirus vector must be about 36 kb in length (one or more genes can be delivered to target cells, or large genes can be used as therapeutic agents). If the therapeutic gene expression cassette is less than the standard length, extra DNA (stuffer DNA) must be added to the vector, as long as the additional DNA does not contain coding sequences that might lead to the synthesis of immunogenic peptides (Fig. 10.27B).

Figure 10.26 Schematic representation of the adenovirus genome with an abridged transcription map. The early (E) and late (L) genes are presented in green and blue, respectively. ITR, inverted terminal repeat; Ψ, packaging signal. doi:10.1128/9781555818890.ch10.f10.26

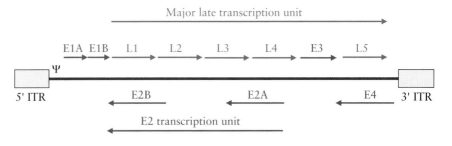

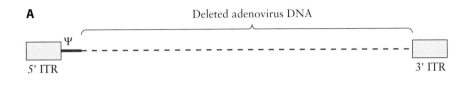

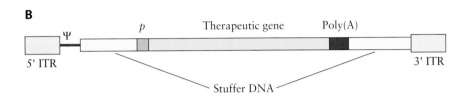

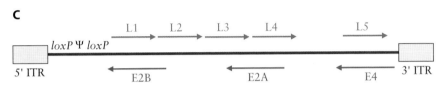

Figure 10.27 (**A**) DNA components of a therapeutic gene–adenovirus vector system. (**B**) Adenoviral elements that make up the gutless (gutted, helper-dependent) vector. (**C**) Adenovirus vector with a therapeutic gene expression cassette and stuffer DNA. Helper virus with *E1* and *E3* genes deleted and *lox*P sites flanking the packaging signal (Ψ). ITR, inverted terminal repeat; Ψ, packaging signal; *p*, promoter; poly(A), polyadenylation signal. doi:10.1128/9781555818890.ch10.f10.27

One packaging system for adenovirus vectors employs a cell line that has chromosomally integrated *E1A* and *E1B* genes and a helper virus that carries all the adenovirus genes except the *E1A*, *E1B*, and *E3* genes (Fig. 10.27C). Deletion of the *E1A* and *E1B* genes ensures that the helper virus will not replicate on its own. The *E3* gene proteins are not required for either viral DNA replication or capsid and core protein synthesis. However, to maintain a supply of the helper virus, it must have a packaging signal, which means that in the producer cell both the helper virus and vector DNAs will be packaged, lowering the yield of packaged therapeutic gene–adenovirus vector. To overcome this problem, the packaging signal of the helper virus is removed by the Cre–*loxP* system (see chapter 6) so that the adenovirus vector DNA is preferentially packaged. However, the system is not totally efficient, with about 0.1 to 10% of the progeny virions carrying the helper virus genome. Packaged helper viruses should not proliferate, to any extent, in a targeted tissue due to the deletion of the *E1* gene. To date, no adverse effects due to helper virus have been observed. Once packaged, the therapeutic gene is delivered to the nucleus of the target cell, where expression is transient in dividing cells and longlasting in nondividing cells.

Adenovirus vectors can be used to test the therapeutic effects of novel gene combinations and determine if subunits of a multimeric protein that are encoded on a single DNA molecule assemble more accurately than when they are delivered by separate viruses. The simplest multigene organization is separate expression cassettes for each gene (Fig. 10.28A). Another type of construct entails placing a protein cleavage site sequence

A

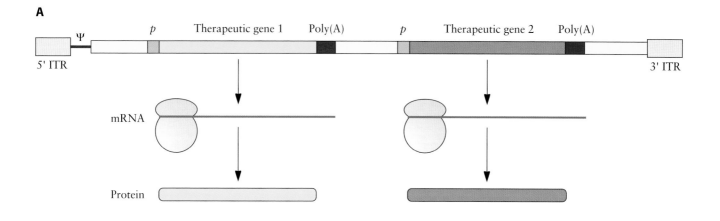

B

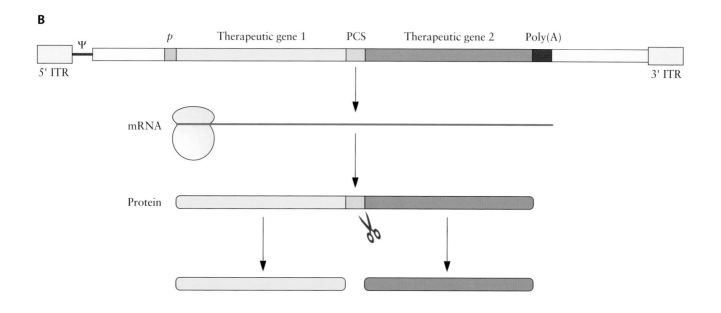

C

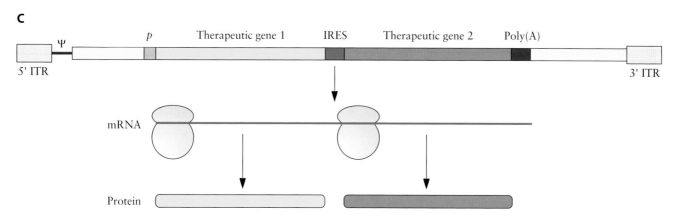

Figure 10.28 (**A**) Organization of two-gene, "gutless" adenovirus vectors. Two expression cassettes that are driven by separate cell-specific promoters (*p*) and terminated by individual poly(A) signaling sequences are incorporated into an adenovirus vector. (**B**) In the target tissue, the genes of each cassette are transcribed and translated. Two therapeutic genes that are separated by a protease cleavage site (PCS) are under the control of the cell-specific promoter (*p*) with a poly(A) signaling site. This arrangement is transcribed as a single mRNA. (**C**) After translation, the primary protein product is cleaved (scissors) to yield two separate proteins. Two therapeutic genes are separated by an internal ribosome entry site (IRES) sequence and under the control of the cell-specific promoter (*p*) with a poly(A) signaling site. This arrangement is transcribed as a single mRNA. Translation occurs at the 5′ end of the mRNA and stops at the IRES sequence and, for the second protein, at the IRES sequence. Two proteins are formed. Stuffer DNA is in yellow. doi:10.1128/9781555818890.ch10.f10.28

between two genes at the DNA level. After translation, the protein cleavage site is cut and the two proteins are released (Fig. 10.28B). With this approach, for example, an active monoclonal antibody was produced in a targeted tissue. Finally, an internal ribosome entry site sequence may be placed between two genes. In this case, ribosomes bind to the 5′ end of the mRNA, and translation proceeds to the internal ribosome entry site sequence. Ribosomes also bind at the internal ribosome entry site and translation goes to the end of the mRNA, so that two proteins are synthesized (Fig. 10.28C). In sum, high-capacity, "gutless" adenovirus vectors are an excellent resource for devising novel, multifunctional therapeutic strategies.

Herpes Simplex Virus 1

HSV-1 is a common human pathogen the causes periodic cold sores and, rarely, fatal encephalitis; it infects and persists within nondividing neurons. The virus usually remains quiescent in the neurons, although the viral production cycle can be initiated by stress or hormonal changes.

There are many different disorders that affect the central and peripheral nervous systems, including tumors, metabolic and immunological defects, and neurodegenerative syndromes such as Alzheimer disease and Parkinson disease. Because of its preference for neurons, HSV-1 is a suitable candidate as a gene therapy vector for treating these kinds of illnesses.

HSV-1 is an enveloped virus with a diameter of about 150 nm (Fig. 10.29). There are at least 13 different glycoproteins that protrude from the envelope, forming about 700 spikes of various lengths. Beneath

Figure 10.29 Organization and structure of HSV-1. (**A**) Schematic representation of HSV-1; (**B**) three-dimensional structure of HSV-1 capsid. Panel B is reproduced with permission from the Chimera Image Gallery at http://plato.cgl.ucsf.edu/chimera/ImageGallery/entries/herpes/herpes.html; image produced with UCSF Chimera. doi:10.1128/9781555818890.ch10.f10.29

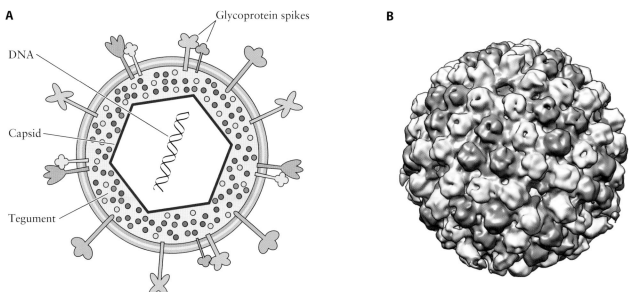

the envelope, there is an amorphous matrix of about 20 proteins (tegument) that surrounds a capsid (Fig. 10.29). The HSV-1 genome is a 152-kb, double-stranded DNA molecule that contains about 90 genes. Entry of the virus into a cell is facilitated by the glycoproteins that bind to surface receptors. These glycoproteins trigger fusion of the virus with either the host cell plasma membrane or directly to endocytic vesicle membranes, bypassing the formation of a clathrin-coated vesicle. The capsid–tegument complex is routed to the edge of the nucleus, where the protein components are removed and the viral DNA enters the nucleus. Once inside the nucleus, the viral DNA circularizes. At this point, the production of virus particles is initiated (lytic phase); if it is not, then the viral DNA is not replicated, nor are the mRNAs for the structural components transcribed (latent phase). Latency, which is due principally to an active HSV-1 promoter (latency-associated promoter 1 [LAP1]) that transcribes a latency-associated transcript, can be maintained in neurons as extrachromosomal DNA for extended periods. A second latency-associated transcript is transcribed by LAP2. HSV-1 DNA does not integrate into host cell chromosomes.

The HSV-1 lytic cycle entails DNA replication, successive waves of transcription, and the production of viral proteins. The capsid with its viral genome is assembled and surrounded by the tegument proteins in the nucleus. To pass from the nucleus to the outside the cell, the capsid–tegument complex is wrapped and unwrapped in various membranes. Briefly, the capsid–tegument complex is initially enveloped by the inner nuclear membrane, which is studded with some of the viral glycoproteins. This envelope allows the capsid–tegument complex to enter the cytoplasm through the outer nuclear membrane. Then, the virus becomes encased with a Golgi apparatus membrane that, in turn, enables it to exit through the cell membrane.

The HSV-1 genome is used as a platform for three types of vector systems: attenuated replication competent, replication incompetent (replication defective), and amplicon. Although it seems counterintuitive, proliferating HSV-1 particles that are replication competent with low pathogenicity and little immunogenicity can be used to selectively destroy neuronal tumor cells while not damaging healthy neurons. Thus, HSV-1 proliferation is confined to dividing cells by deleting HSV-1 genes that enable the virus to be produced in nondividing cells. As a result, dividing tumor cells are infected and destroyed by viral production, while nondividing neurons do not support the HSV-1 lytic cycle and therefore are spared (Fig. 10.30). Generally, replication-incompetent HSV-1 particles are injected directly into the brain tumor mass, and the released progeny viruses spread the zone of cell lysis by infecting neighboring tumor cells. The viruses and cell debris are cleared from the brain. The term "attenuated" is used to describe replication-incompetent HSV-1, which signifies that the normal ability to proliferate in nondividing neurons is constrained.

The primary purpose of replication-incompetent vectors is to deliver a therapeutic gene about 10 to 15 kb in length to the nucleus of a cell, usually a neuron, and establish its long-term expression. Several variants of replication-incompetent vectors and their packaging systems have been

LAP1
latency-associated promoter 1

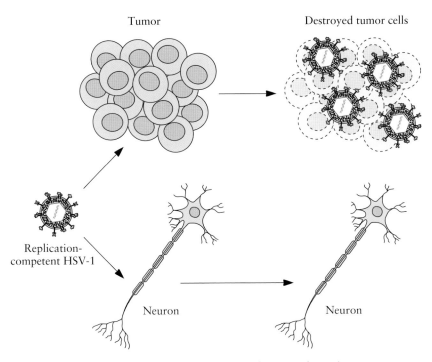

Tumor   Destroyed tumor cells

Replication-
competent HSV-1

Neuron   Neuron

Figure 10.30 Schematic illustration of the mode of action of a replication-competent HSV-1 vector. doi:10.1128/9781555818890.ch10.f10.30

devised. In all cases, genes that are essential for viral replication are deleted from the HSV-1 genome. For the system described here, removal of both copies of the *ICP4* (ICP stands for "infected-cell protein") gene and the *ICP27* gene prevents both viral DNA replication and production of viral structural components. In addition, the bacterial *lacZ* gene that encodes β-galactosidase under the control of an HSV-1 promoter and a mammalian polyadenylation signal is inserted into an HSV-1 gene that is not necessary for viral production in cell culture, such as *UL41* (Fig. 10.31). The *lacZ* gene cassette is flanked by PacI restriction endonuclease sites that are not present anywhere else in the HSV-1 genome. The *lacZ* gene provides a convenient marker for identifying cells that produce viruses with this HSV-1 genome. The *lacZ* gene gene is transcribed during viral production, and its presence in the cell debris after viral release can be visualized as a blue spot (plaque) following cleavage of the synthetic substrate 5-bromo-4-chloro-3-indolyl-β-D-galactopyranoside (X-Gal).

The therapeutic gene with a cell-specific promoter and a polyadenylation signal is inserted between UL41 sequences that were incorporated into a plasmid. In the example described here, the packaging cell line is engineered with *ICP4* and *ICP27* genes (Fig. 10.31). Briefly, the HSV-1 vector DNA is first cleaved with the restriction endonuclease PacI and then transfected along with the plasmid–therapeutic gene construct into packaging cells. The UL41 sequences of the vector and the plasmid–therapeutic gene cassette pair with each other, and homologous recombination results in the formation of a packageable DNA molecule carrying the therapeutic gene. As a result, both the recombinant vector with the therapeutic gene cassette and any uncut, full-length vector genomes are

X-Gal
5-bromo-4-chloro-3-indolyl-β-D-galactopyranoside

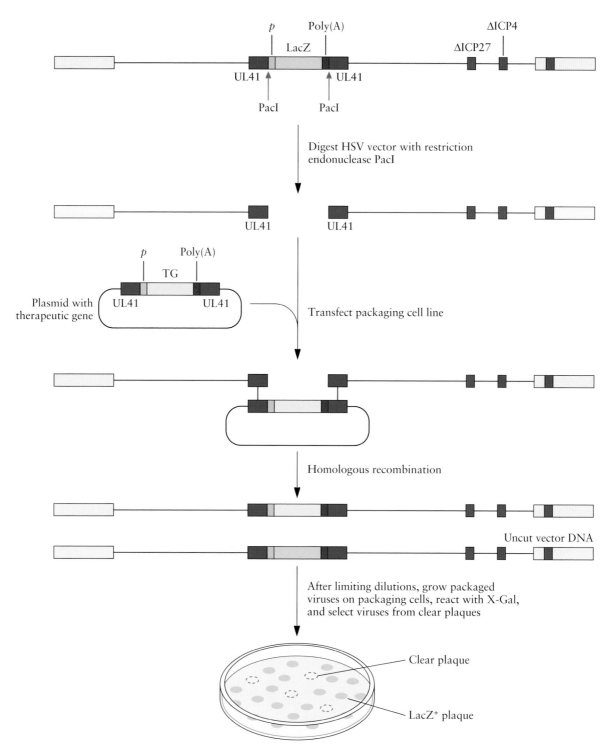

Figure 10.31 Schematic representation of the formation and isolation of an HSV-1 replication-incompetent vector with a therapeutic gene. *p*, promoter; poly(A), polyadenylation signal; PacI, restriction endonuclease site; UL41, nonessential HSV-1 gene; Δ, deleted gene; TG, therapeutic gene. doi:10.1128/9781555818890.ch10.f10.31

packaged. To distinguish between these two types of viruses, the initially packaged viruses are serially diluted to decrease the concentration and then added to colonies of packaging cells. After viral production, the lysed colonies are reacted with X-Gal. Blue plaques contain viruses that do not carry the therapeutic gene, whereas clear plaques have viruses with the therapeutic gene. Viruses from the clear plaques are selected and propagated to provide a stock of the vector carrying the therapeutic gene.

HSV-1 amplicon vectors are unusual therapeutic gene delivery systems with a number of distinct advantages. These vectors carry up to 150 kb of input DNA with either multiple copies of one or more therapeutic genes or very large multigene constructs. Moreover, HSV-1 amplicons are neither toxic nor pathogenic because they carry no viral coding genes. Packaged amplicons infect neurons and other cell types with the therapeutic gene(s) and become localized in the nucleus of the targeted cell. The key feature of the amplicon strategy is based on the mode of HSV-1 DNA replication, which is a rolling-circle mechanism. In this way, multiple lengths **(concatemers)** of the original DNA molecule are formed (Fig. 10.32). The HSV-1 packaging sequences are spaced about 150 kb apart. The mechanism for loading HSV-1 DNA into an empty capsid requires cleaving the concatemeric DNA at the packaging sites.

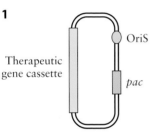

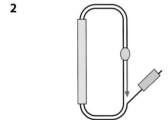

Figure 10.32 Rolling-circle DNA replication of an HSV-1 amplicon plasmid. (**1**) HSV-1 amplicon plasmid with HSV-1 sequences (OriS, blue; *pac*, green) and a therapeutic gene cassette (yellow) which consists of a therapeutic gene with a cell-specific promoter and an effective polyadenylation signal; (**2**) initial phase of discontinuous DNA synthesis (red) and early strand displacement; (**3**) later stage of rolling-circle DNA replication with multiple copies of the HSV-1 amplicon DNA. doi:10.1128/9781555818890.ch10.f10.32

For the production of packageable DNA with the HSV-1 amplicon system, a therapeutic gene(s) is cloned into a plasmid equipped with the HSV-1 origin (OriS) and HSV-1 packaging sequence (Fig. 10.33). If an amplicon plasmid is 5 kb, then about 30 copies of the therapeutic gene will be packaged into a capsid; if 25 kb, then 6 copies are packaged, and so on. For loading DNA into a capsid, only packaging signals that are 150 kb apart are cleaved. The intervening packaging sequences are ignored.

Initially, HSV-1 was used to supply the components for viral production, which led to packaging of both HSV-1 genomes and amplicon DNA molecules. This required the two viruses to be separated from one

Figure 10.33 Schematic representation of the packaging of a therapeutic gene–HSV-1 amplicon vector. In this case, the HSV genes that are required for producing virions that can be packaged with amplicon vector DNA are incorporated into a bacterial artificial chromosome (BAC). The packaging signals are deleted from the helper HSV-1 genome (Δ*pac*), and the *ICP27* gene is replaced with additional *ICP0* genes (+ICP0 [purple]). The *ICP27* gene (blue) is cloned into another plasmid. The three plasmids are transfected into a cell line that supports viral production. With a 25-kb amplicon plasmid, six copies of the therapeutic gene cassette are incorporated into an HSV-1 capsid. doi:10.1128/9781555818890.ch10.f10.33

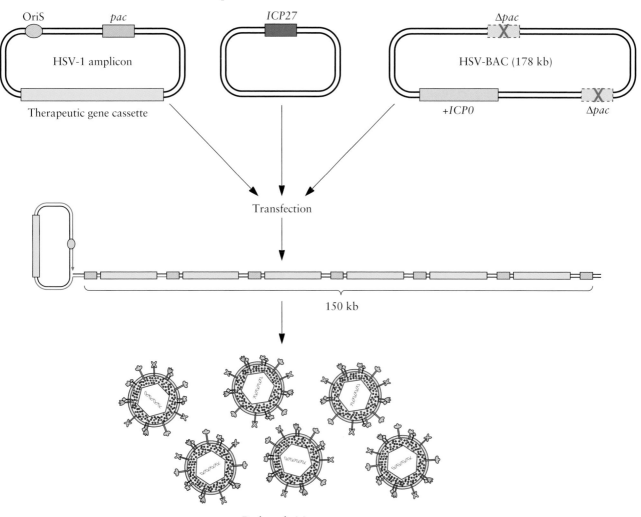

another. To avoid this problem, helper virus-independent strategies were devised. By cloning the HSV-1 genome without packaging sequences into a large-capacity cloning vehicle such as a bacterial artificial chromosome, a helper virus was no longer required for packaging amplicon DNA. Further modifications entailed deleting the *ICP27* gene and inserting additional copies of the *ICP0* gene into the HSV-1 DNA. These changes have two effects. First, the bacterial artificial chromosome-based HSV-1 genome is increased to 178 kb, which makes it less likely that it will be packaged. Second, extra copies of the early-acting *ICP0* gene help boost the production of viral components in the packaging cells. The *ICP27* gene, which is essential for the production of the viral components, is added to another plasmid. For this packaging system, three plasmids are tranfected into the packaging cells and 150 kb of amplicon DNA is packaged into virions (Fig. 10.33).

To date, HSV-1 amplicons have been tested in a number of **preclinical trials** with different therapeutic gene constructs. To expand the capabilities of HSV-1 amplicons, hybrid amplicons with sequences from other viruses have been created. For example, adding genes from Epstein–Barr virus to the HSV-1 amplicon enables it to replicate in the nuclei of dividing cells, with the consequence that a therapeutic gene can be maintained in growing tissue.

Nonviral Nucleic Acid Delivery Systems

Direct Injection

To alter the expression of either a defective endogenous gene or the gene of a pathogenic agent, naked DNA or RNA may be introduced directly into animal tissues. Nucleic acids may be injected on a macroscale into specific tissues, such as muscle, or on a microscale into specific cells or portions of those cells. Conceptually, microinjection of nucleic acids is the simplest possible nucleic acid delivery system; however, in practice this technique is fraught with problems. Naked nucleic acids are difficult to deliver due to their rapid clearance, the presence of nucleases that limit their serum half-life, a lack of organ-specific distribution, and the low efficiency of cellular uptake following systemic delivery. To be effective, the direct injection of nucleic acids requires exceptionally high levels of the input nucleic acid, with the possibility of unwanted side effects being very real. Thus, while the effectiveness of siRNAs for specifically inhibiting gene expression in animal cells in culture has been demonstrated on numerous occasions, it is difficult to efficiently deliver these RNAs to tissues in vivo.

Lipids

One approach used to overcome some of the problems associated with direct injection of therapeutic nucleic acids has been to chemically couple a therapeutic molecule such as an siRNA (at the terminal hydroxyl group of the sense strand RNA) to a lipid molecule such as cholesterol (Fig. 10.34). When this was done experimentally, the siRNA that was used

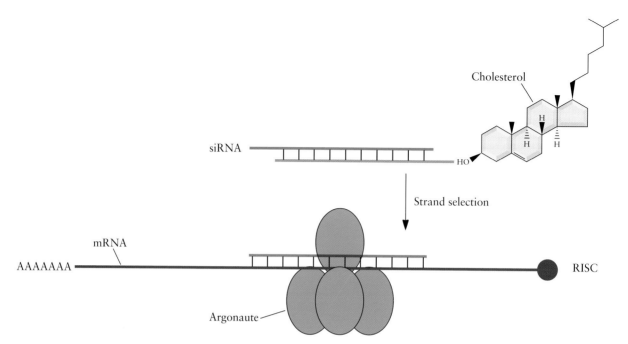

Figure 10.34 A conjugate of cholesterol and an siRNA in which the cholesterol is coupled through the 5′ OH of the sense strand of the siRNA. The cholesterol facilitates uptake of the siRNA into specific tissues. The antisense strand becomes part of the RISC and specifies where the mRNA is to be cleaved. doi:10.1128/9781555818890.ch10.f10.34

was complementary to an mRNA that encodes apolipoprotein B, a protein involved in the metabolism of cholesterol. Following intravenous introduction of the siRNA–cholesterol complex by injection into mice, the complex was specifically taken up by the liver, jejunum (part of the small intestine), heart, kidneys, lungs, and fat tissue cells. Once the complex was inside the tissue, the sense strand of the siRNA was destroyed and the antisense strand bound to the target mRNA.

With this approach, the level of apolipoprotein B was reduced by more than 50% in the liver and by 70% in the jejunum. This resulted in a significant decrease in the plasma apolipoprotein B level, as well as the total amount of cholesterol. This strategy is an important first step in the development of a method to therapeutically lower cholesterol levels in humans.

Several other molecules, including some long-chain fatty acids and bile acids, have been used in place of cholesterol to mediate the uptake of siRNAs into cells. A critical factor in mediating the interaction between fatty acid-conjugated siRNAs and lipoprotein particles is the length of the fatty acid alkyl chain. Thus, docosanyl (C_{22}) and stearoyl (C_{18}) conjugates bind more tightly to high-density lipoprotein and subsequently silence gene expression more effectively in vivo (in mice) than lauroyl (C_{12}) and myristoyl (C_{14}) conjugates. Studies are under way to improve the delivery of lipid-conjugated siRNAs to treat a wide range of diseases.

As mentioned briefly earlier in this chapter, scientists have had some success delivering nucleic acids using both synthetic and biological liposomes (Fig. 10.9). Biological liposomes are phospholipid-based spherical particles derived from human cells. If biological liposomes are derived from patients, they are likely to be recognized as self by one's immune system and thereby avoid an immune attack. However, biological liposomes tend to be difficult to harvest and maintain and are therefore not currently used to any significant extent. Notwithstanding some of the above-mentioned limitations of liposomal delivery, when apolipoprotein B siRNA-carrying liposomal particles were intravenously injected into monkeys, greatly reduced apolipoprotein B expression and serum cholesterol levels were observed.

Bacteria

Nonpathogenic bacteria that are normally found in association with various mammalian tissues and cells may be isolated, attenuated, and then genetically engineered to deliver therapeutic nucleic acid molecules. To date, bacterial nucleic acid delivery has been used mainly in cancer gene therapy and for DNA vaccination; it is also being tested for the treatment of genetic diseases, including cystic fibrosis. Strains of bacteria currently used include *Listeria monocytogenes*, certain *Salmonella* strains, *Bifidobacterium longum*, and modified *Escherichia coli*. The engineered bacteria are used as vectors to deliver these therapeutic agents directly to the affected tissues. For example, a nonpathogenic strain of *E. coli* was transformed with the plasmid vector TRIP, containing the gene that encodes the protein **invasin**, which permits *E. coli* to bind to and enter β1-**integrin**-positive mammalian cells, and the gene *hlyA*, which encodes listeriolysin O, a protein that enables genetic material to escape from entry vesicles (Fig. 10.35). In addition, the TRIP vector carries an shRNA

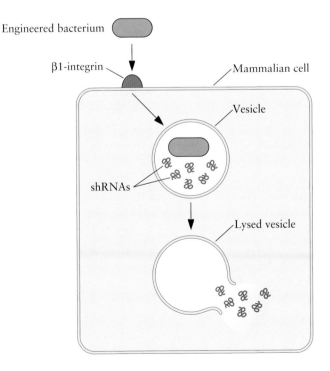

Figure 10.35 Use of a nonpathogenic strain of *E. coli* to deliver siRNAs to certain tissues. The bacterium was engineered to produce the protein invasin, which permits *E. coli* to enter β1-integrin-positive mammalian cells, as well as the gene *hlyA*, encoding listeriolysin O, which permits the shRNAs synthesized by the bacterium to be released inside the mammalian cell. doi:10.1128/9781555818890.ch10.f10.35

molecule under the control of a bacterial promoter directed against the mRNA produced by the cancer gene *CTNNB1*. As long as a bacterium is able to enter target mammalian cells and release shRNAs that are then processed to siRNAs (Fig. 10.11), the siRNAs may be directed against any specific mRNAs. The *E. coli* cells act as a vector to transport the shRNAs to where they are required, e.g., cancer cells. One important reason for the effectiveness of this approach is that bacterium-mediated RNAi expression may evade the host defense against exogenous DNA because the DNA is enclosed within the bacteria, potentially allowing for long-term expression of the transgene. This approach has been shown to work both for cancer cells in culture and with mice. Using live animals, the bacteria can be administered orally. The intrinsic toxicity of bacteria is the most significant risk in using bacteria as a delivery vehicle (i.e., bactofection). However, using attenuated strains of *Salmonella* in preliminary clinical gene transfer trials, no significant side effects were observed.

Another way in which bacteria may be used to target specific mammalian cells includes the use of bacterial minicells (Fig. 10.36A). In this case, a normal-size bacterial cell carries a mutation that upon cell division

Figure 10.36 Transformation of bacterial cells with plasmid DNA encoding either a mammalian cellular toxin or an shRNA. Plasmid DNA is present in multiple copies, while chromosomal DNA is present in a single copy. Upon cell division, both normal-size cells containing both chromosomal and plasmid DNA and minicells containing only plasmid DNA are formed (**A**). Normal-size cells and minicells are separated by centrifugation and cross-flow filtration. Then minicells may be linked to the targeted mammalian cells using antibodies directed against proteins on the minicell and mammalian cell surfaces. The Fc portions of the antibodies are linked by the addition of protein A/G (shown in red) (**B**). doi:10.1128/9781555818890.ch10.f10.36

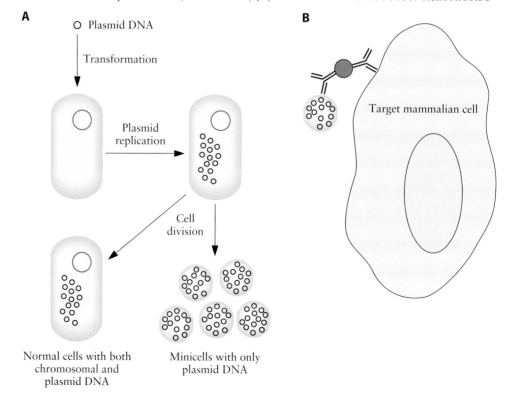

causes it to produce some minicells (only a fraction of the size of normal bacteria) that do not contain any chromosomal DNA. If minicell-forming bacteria are first transformed with plasmid DNA, the minicells, although they are devoid of chromosomal DNA, will contain plasmid DNA. Minicells may be physically separated from normal-size cells by centrifugation and/or cross-flow filtration. Minicells may be targeted to cancer cells using two different antibodies, one that targets a protein on the surface of the minicell and another targeting the surface of a cancer cell, with the two types of antibodies being linked to one another through their Fc regions by protein A/G (Fig. 10.36B). When the target cell is a cancer cell, the minicell, which can produce protein encoded by the plasmid, may be engineered to produce a compound that is toxic to the cancer cell. Alternatively, plasmids carried by minicells may be designed to produce specific shRNAs that are eventually converted into siRNAs. In practice, minicells have been shown to be able to successfully deliver a variety of chemotherapeutic agents to target cells, suggesting that this will also be an effective means of delivery of plasmid-encoded products to target cells.

Dendrimers

In the past 5 to 10 years, numerous polycations and polymer micelles have been used for formulating genes, shRNAs, and siRNAs into complexes that have been termed "polyplexes." The polycations that have been used include histones, polylysine, cationic oligopeptides, polyethyleneimine, polypropyleneimine, dendrimers, poly(2-(dimethylamino)ethyl methacrylate), and chitosan.

Dendrimers are biocompatible, nonimmunogenic, water-soluble nanoparticles with sizes ranging from 1 to 15 nm. They possess terminal modifiable amine functional groups as the "sensors" for binding various targeting or guest molecules. Unlike classical polymers, dendrimers have a high degree of molecular uniformity, a narrow molecular weight distribution, specific size and shape characteristics, and a large number of potentially interactive terminal amino moieties (Fig. 10.37). Dendrimers are produced in an iterative sequence of reaction steps, with each additional iteration leading to a higher-generation dendrimer. Each new layer creates a new "generation," with double the number of active sites (i.e., end groups) and approximately double the molecular weight of the previous generation. With dendrimers it is relatively easy to precisely control the size, composition, and chemical reactivity.

Dendrimers can form complexes with nucleic acid drugs including plasmid DNA, shRNA, and siRNA through electrostatic interactions, and they also bind to glycosaminoglycan molecules, such as heparan sulfate, hyaluronic acid, and chondroitin sulfate, that are typically found on cell surfaces. In addition, dendrimers have been shown to be more efficient and safer than either cationic liposomes or other cationic polymers for in vitro gene transfer. It has been suggested that the high transfection efficiency of dendrimers is a consequence of both their well-defined shape and the low pK$_a$s of the amines. One dendrimer-based formulation with activity against HSV (VivaGel) has successfully completed phase II clinical trials.

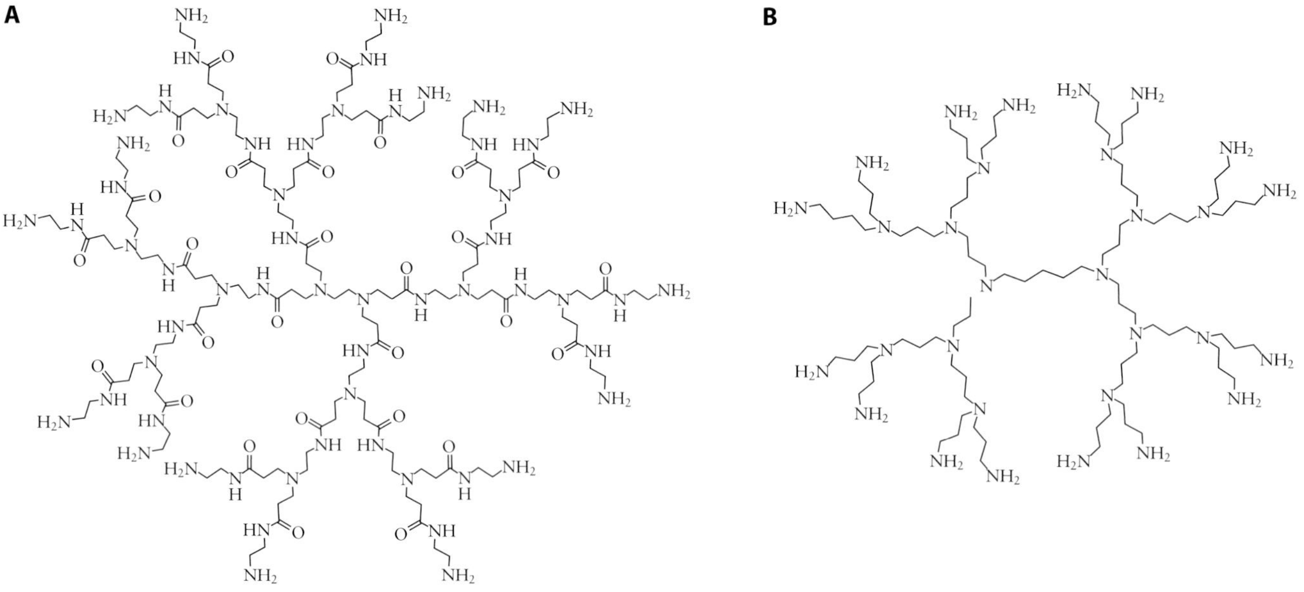

A

B

Figure 10.37 Structures of two commercially available dendrimers.
doi:10.1128/9781555818890.ch10.f10.37

Antibodies

Monoclonal antibodies may now be generated against nearly any target protein. These antibodies, or their variable regions, may be humanized and then produced in heterologous host cells. Moreover, at the DNA level, it is easy to fuse the antibody gene with the gene for another protein. With this in mind, the gene encoding a single-chain Fab fragment that binds specifically to an oncogenic protein called ErbB2, which is found on the surfaces of breast cancer cells, was fused to the gene for the positively charged nucleic acid-binding protein protamine. The Fab portion of the fusion protein (at the N terminus) binds to the surfaces of cells expressing ErbB2, while the C-terminal end of the fusion protein carries the 51-amino-acid-long protamine, which readily binds to added siRNAs (Fig. 10.38). In one test of this system, an anti-HIV envelope Fab and an siRNA that is designed to cleave the HIV *gag* mRNA were employed. Using cells in culture, it was possible to reduce the amount of secreted Gag protein (the protein of the nucleocapsid shell around the RNA of a retrovirus) by >70%. This system also works in vivo in mice when the construct is injected either intravenously or directly into tumors. The hope is that by using a combination of specific antibodies (or antibody fragments) that direct siRNAs only to certain cells, and siRNAs that selectively cleave specific target mRNAs, this system can be used to treat a wide range of diseases.

Aptamers

The binding specificity to a target antigen that is a central feature of the functioning of antibodies is also a property of aptamers. Thus, conjugating aptamers, which bind to specific cell surface proteins, to siRNAs that are designed to reduce the expression of certain mRNAs should provide

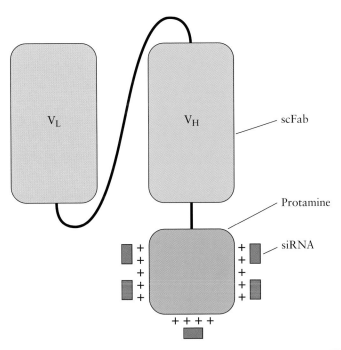

Figure 10.38 A single-chain Fab (scFab) fragment directed against a mammalian cell surface protein is fused to the positively charged polypeptide protamine, which binds noncovalently to negatively charged siRNAs. The Fab fragment acts to deliver the siRNA to specific cells. Note that a conventional (two-chain) Fab fragment has also been used to deliver siRNAs. doi:10.1128/9781555818890.ch10.f10.38

another method of targeting siRNAs to specific tissues or cells. Also, since both aptamers and siRNAs are chemically synthesized RNA oligonucle-otides, it should be simple and straightforward to synthesize chimeric RNA molecules that include both the binding specificity of an aptamer and an siRNA that targets a specific mRNA (Fig. 10.39). For example, an aptamer that binds selectively to a prostate-specific membrane anti-gen (found on cancerous prostate cells) was first selected. Then, a 21-bp siRNA directed against mRNAs encoded by either of two genes that are necessary for prostate cells to survive was added to the aptamer sequence. When this was done, both activities (i.e., aptamer binding and RNAi) were maintained in the chimeric molecule. The targeting aptamer did not impair the ability of the siRNA to silence the target gene, and the presence of the siRNA did not affect the ability of the aptamer to bind to its target. This simple but highly effective approach should be amenable to treating a wide range of human diseases provided that (i) silencing specific genes in a population produces therapeutic benefits, and (ii) there are surface receptors (usually proteins) that distinguish the target cell population and allow the siRNA to be internalized by the cell.

Transposons

Transposons are discrete DNA elements that have the ability to move from one chromosomal location to another. They are typically excised from one chromosomal locus and then integrated into another location, and this

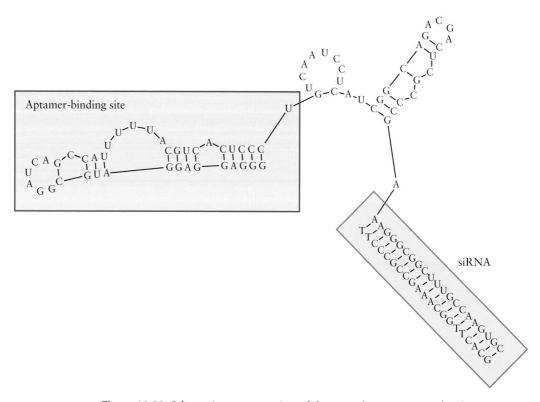

Figure 10.39 Schematic representation of the secondary structure of a chimeric RNA molecule consisting of an aptamer and an siRNA. The portion of the aptamer that binds to the target protein and the siRNA portion of the molecule are in colored boxes. doi:10.1128/9781555818890.ch10.f10.39

movement is catalyzed by an enzyme called a transposase. Being able to move DNA from one site to another makes transposons attractive as potential foreign gene delivery tools. The transposase can act in *trans* (i.e., it can reside on a separate DNA fragment) on virtually any target DNA sequence that is flanked by the terminal repeat DNA sequences, normally found at each end of the transposon. To use DNA transposons for gene delivery, a simple system has been developed (Fig. 10.40). This system consists of a plasmid that encodes and expresses the transposase as well as a plasmid containing the target DNA to be integrated, which is flanked by the transposon terminal repeat sequences that are required for transposition. Once it is synthesized, the transposase binds to the terminal repeat DNA sequences and catalyzes the excision of the gene of interest from the plasmid. The transposase then catalyzes the insertion of the target DNA into the genome of the host cell. Physically separating the transposase gene from the transposon-containing plasmid optimizes the stoichiometry of both components and hence the efficiency of the process.

At present, three different transposon systems are being developed as gene therapy delivery vectors: *Sleeping Beauty*, *Tol2*, and *PiggyBac*. *Sleeping Beauty*, which has been used most frequently, was constructed by combining fragments of silent and defective Tc1/mariner elements from salmonid fish and shows significant transposition efficiencies in vertebrate cells. When tested, the original *Sleeping Beauty* transposon was at least 10-fold more efficient than other Tc1/mariner transposons. Moreover, to

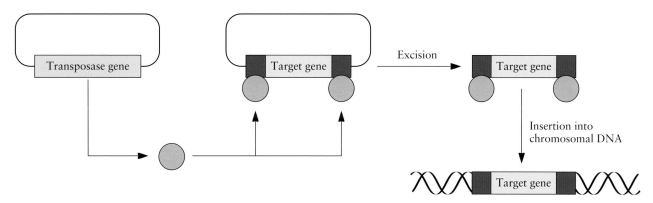

Figure 10.40 Schematic representation of transposase-catalyzed movement of a target gene into cellular chromosomal DNA. The transposase gene and the target gene flanked by transposon sequences (shown in blue) are encoded on separate plasmids. The transposase (depicted as an orange circle) binds to the transposon sequences and catalyzes the excision of the transposon-flanked target gene from the plasmid and its subsequent insertion into chromosomal DNA. doi:10.1128/9781555818890.ch10.f10.40

increase its activity, almost every amino acid in the transposase has been changed by in vitro molecular evolution by DNA shuffling and PCR. Following the screening of a large number of variants of the original transposon, one version, SB100X, was found to be more than 100-fold more active at catalyzing transposition in HeLa cells than the original *Sleeping Beauty*.

Since *Sleeping Beauty* may integrate into any portion of the chromosomal DNA, there is a risk of disrupting an important gene. However, in contrast to the case with many viral vectors, the terminal repeat sequences of *Sleeping Beauty* have very low intrinsic promoter/enhancer activity. Thus, *Sleeping Beauty* cannot readily activate chromosomal genes that flank the transposon integration sites. However, the internal promoter/enhancer sequences that are used to drive the expression of the target gene may potentially activate expression of neighboring genes in proximity of the integration site. To prevent this from happening, the expression cassette in a *Sleeping Beauty* transposon was flanked with insulator DNA sequences that significantly reduce the risk of activating the expression of neighboring genes.

Clinical trials using the SB100X version of the *Sleeping Beauty* transposon have recently been approved. In these trials, human T cells will be genetically altered with *Sleeping Beauty* transposons and then transferred for **adoptive immunotherapy** (a way of using the immune system to fight cancer) into patients. The phase I clinical trial will determine the feasibility, safety, and persistence of *Sleeping Beauty* transposon-modified T cells in vivo.

Prodrug Activation Therapy

Prodrug activation therapy (**suicide gene** therapy) is designed for localized eradication of tumor cells with minimal debilitating biological effects that often accompany chemo- or radiotherapy. In practice, prodrug activation therapy entails introducing a gene for an enzyme that is not

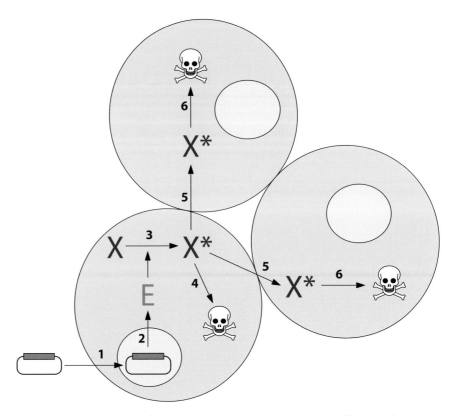

Figure 10.41 Overview of prodrug activation therapy. Tumor cells are either transfected or transduced with the gene (red) that encodes an enzyme (E) that activates a prodrug (X) (**1**). The introduced prodrug-activating gene is expressed in a recipient cell (**2**), and the enzyme converts the prodrug to its cytotoxic form (X*), which, in turn, destroys the cell (skull and crossbones) (**3** and **4**). The activated prodrug (X*) enters adjacent cells (**5**) and destroys them (**6**), creating a bystander effect. doi:10.1128/9781555818890.ch10.f10.41

part of the human repertoire and is under the control of a tumor-specific promoter into tumor cells. A prodrug is then added to the tumor mass. It is expected that the enzyme will be synthesized in recipient cells and will convert the prodrug to its active form, thereby leading to tumor cell death. When a gene causes the death of its own cell, it is called a suicide gene. Moreover, the activated molecule can enter into neighboring cells, which very likely did not receive the suicide gene, and kill them (Fig. 10.41). The latter process has been designated the bystander effect.

Many prodrug-modifying enzyme combinations have been devised. For example, the ganciclovir [2-amino-9-(1,3-dihydroxypropan-2-yloxymethyl)-3H-purin-6-one]–HSV thymidine kinase (HSV TK) duo has been thoroughly examined. Briefly, HSV TK phosphorylates ganciclovir to form ganciclovir monophosphate. Although host cell kinases do not efficiently phosphorylate ganciclovir, they readily add phosphate groups to ganciclovir monophosphate to form ganciclovir triphosphate. Ganciclovir triphosphate inhibits DNA polymerase activity and terminates DNA synthesis, causing the death of a proliferating cell. Moreover, phosphorylated ganciclovir can pass into unmodified cells through gap junctions and kill these cells as well. Generally, one HSV TK-expressing tumor cell destroys up to 10 unmodified neighbor cells.

HSV TK
herpes simplex virus thymidine kinase

Various strategies have been devised to improve both the efficiency of gene transfer into tumor cells and the extent of penetration into a tumor mass as well as augmenting the bystander effect of prodrug activation systems. For example, by creating a double suicide gene combination of 5-fluorouracil–cytosine deaminase and ganciclovir–HSV TK with both of the enzyme-coding genes under the control of a cancer-specific promoter, tumor destruction was increased with little effect on noncancerous cells. Clinical trials have shown that the ganciclovir–HSV TK treatment and other prodrug activation combinations can be effective in purging tumor cells but not to the extent of entirely eradicating a cancer. That is, prodrug activation therapy must be used in combination with other cancer therapies.

Gene Therapy
Severe Combined Immunodeficiency

SCID, which is caused by different gene mutations, is characterized by the loss of function of B and T cells. The B cells are responsible for the production of antibodies, while the T cells mediate the activation of macrophages and other cell types that protect against pathogens. As a result, individuals with SCID, which occurs in about one in 500,000 live births, are susceptible to severe infections of the bloodstream, lungs, and nervous tissue that are life threatening. One type of SCID, which makes up about 15% of the cases, is due to mutations in the ADA gene and is called ADA-deficient SCID. ADA is part of the purine salvage pathway, and when it is impaired, the levels of adenosine and adenine deoxyribonucleotides accumulate intracellularly. Since B and T cells actively undergo cell division, they are extremely susceptible to excess amounts of adenosine and adenine deoxyribonucleotides. Consequently, the populations of B and T cells are inadequate for the immune system to respond fully to pathogens and other foreign entities.

One of the strategies for treating SCID entails hematopoietic stem cell transplantation using bone marrow cells. This procedure is effective if there is a complete human leukocyte antigen (HLA; see chapter 4) match between the transplanted cells and those of the recipient. Perfect and near-perfect matches are hard to find. In many instances, matches are partial and the transplant is less successful and can be rejected or, more seriously, result in graft-versus-host disease, in which the acquired immune cells attack the cells of the patient. If transplantation is not feasible, patients with ADA-deficient SCID may be administered purified bovine ADA that is coated with polyethylene glycol (pegylated ADA) to prolong its half-life. While pegylated ADA is an effective treatment for a number of patients with ADA-deficient SCID, some recipients develop an immune reaction to the enzyme. Moreover, this therapy is lifelong. In light of the difficulties of traditional treatments for SCID, gene therapy offers a promising alternative.

The first human gene therapy clinical trial was conducted in 1990 on two young girls with ADA-deficient SCID. Peripheral blood lymphocytes were collected from each patient, transduced with a gammaretrovirus vector carrying a human ADA cDNA sequence, and cultured before

HLA
human leukocyte antigen

reintroduction into the patients. A 12-year follow-up study showed that for one patient about 20% of her lymphocytes expressed ADA. This result shows that T-cell expression of an introduced gene can be maintained for a very long period. On the other hand, the enzyme was not evident in the second patient. The procedure had no apparent side effects. However, since this trial was designed, for the most part, to ascertain whether the treatment was safe, both patients continued to be treated with pegylated ADA. As a result, it was not possible to judge definitively whether the therapeutic gene had any beneficial effect. In other clinical trials with patients with ADA-deficient SCID in the 1990s, a short-term, therapeutic benefit was observed.

By 2000, isolation and transduction of hematopoietic stem cells (HSCs), which give rise to all blood cells, became feasible. Transduced HSCs produce progenitor lymphoid cells that, in turn, provide a supply of transduced lymphocytes, in contrast to transduced peripheral blood lymphocytes that were used in the early SCID gene therapy trials. Consequently, ex vivo gene therapy clinical trials for ADA-deficient SCID with transduced HSCs were initiated. In this case, 12 of 16 patients no longer required enzyme replacement therapy after 9 years. However, four of these patients developed T-cell acute lymphocytic leukemia. The leukemia was likely due to insertional mutagenesis that upregulated the *LM02* proto-oncogene. Probably, transduced cells with these particular insertions had a selective advantage after infusion and the excessive proliferation gave rise to T-cell acute lymphocytic leukemia. Future SCID gene therapy trials will use SIN vectors or other gene delivery systems to avoid LM02-induced leukemia.

Cancer

About two-thirds of all gene therapy trials have been directed at treating a large number of different cancers using a variety of gene delivery strategies with dozens of therapeutic genes. Generally, these trials have been found to be safe. However, eradication of any particular cancer by these agents alone has proven to be elusive. The crucial objective for managing cancers in this way is to determine how to deliver an effective gene-based anticancer system to a large number of cells of a particular cancer that would lead to its remission.

One category of anticancer gene therapy involves the generation of nonreplicating viruses that deliver one or more genes encoding products that specifically destroy cancer cells. Generally, these vectors are used in combination with other cancer treatments such as chemotherapy or radiation to maximize the overall effectiveness. Many of these vectors, such as Gendicine (officially approved by the State Food and Drug Administration of China as part of a treatment repertoire for head and neck squamous cell carcinoma), are based on the adenovirus platform. In these cases, the adenovirus *E1* gene is deleted, which makes the vector incapable of replicating. An expression cassette that encodes an anticancer protein and has no demonstrable effect on normal cells is inserted into the vector. With Gendicine and other similar vectors, the anticancer agent is often either the *TP53* gene or its cDNA (Fig. 10.42).

HSC
hematopoietic stem cell

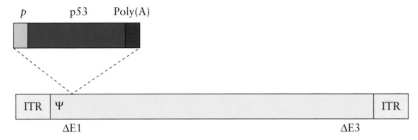

Figure 10.42 Anticancer replication-defective adenovirus vector. The deletion of the adenovirus *E1* and *E3* genes (ΔE1 and ΔE3) abolishes the ability of a packaged vector to produce virions in targeted cells. After the introduction of the p53 coding sequence into a cancer cell, the production of p53 induces cell arrest and apoptosis. Additional p53 in normal cells has no adverse effect. *p*, promoter ITR, inverted terminal repeat; ψ, packaging signal. doi:10.1128/9781555818890.ch10.f10.42

The p53 protein is referred to as the "guardian of the genome." It is a ubiquitous transcription factor that interacts with more than 100 different proteins and is mutated in over 50% of all cancers. In normal cells, activation of p53 occurs in response to DNA damage, oxidative stress, or other adverse cellular conditions. Through complex sets of interacting proteins, p53 induces DNA repair proteins and arrest of the cell cycle and initiates programmed cell death **(apoptosis)** to remove cells that are damaged beyond repair (Fig. 10.43). In the absence of p53, these pathways are not initiated, and consequently, cell cycling proceeds normally

Figure 10.43 Simplified overview of the p53 pathways. doi:10.1128/9781555818890.ch10.f10.43

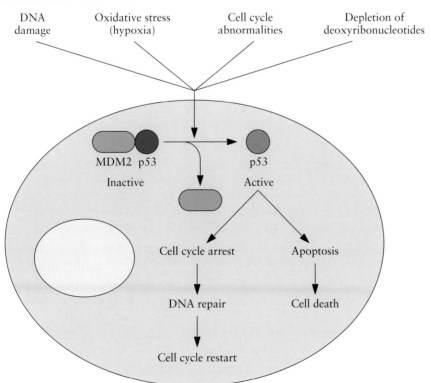

and apoptosis is not triggered. These features are the hallmarks of cancer cells. The production of p53 in cancer cells by an introduced p53 DNA sequence generally leads to apoptosis, whereas in normal cells, the extra p53 production has no effect (Fig. 10.44). In addition, destroyed cancer cells release tumor-specific antigens which may initiate an immune response against the cancer.

A second class of anticancer vectors relies on the selective replication of a virus in cancer cells and not in normal cells. These vectors lyse cancer cells and release virions that attack adjacent cancer cells (Fig. 10.45). One strategy, using an adenovirus vector, is to place the *E1* gene under the

Figure 10.44 Overview of the mode of action of a replication-defective p53 gene vector. The transduced p53 gene produces p53 in p53-deficient (p53⁻) cancer cells, which leads to cell lysis. In turn, tumor-specific antigens induce an immune response to the cancer cells. The transduced p53 gene in normal cells produces p53 that has no adverse effect. doi:10.1128/9781555818890.ch10.f10.44

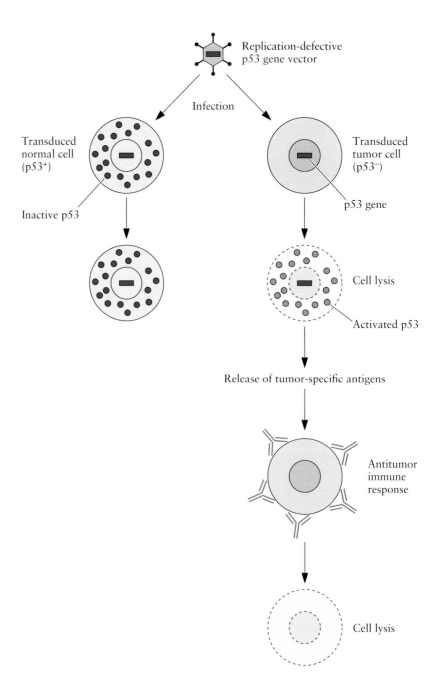

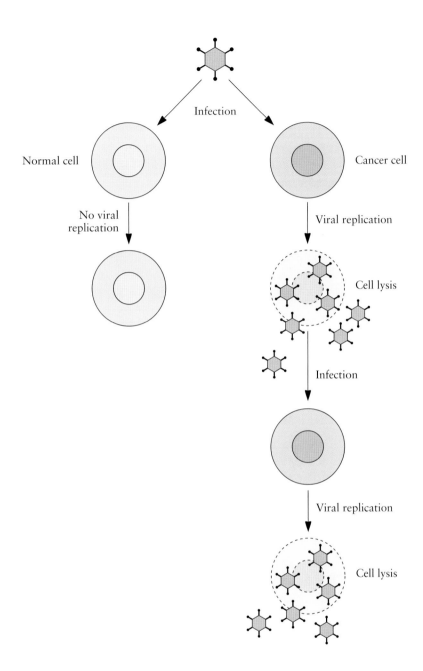

Figure 10.45 Overview of the mode of action of a replication-competent virus that produces virions only in cancer cells, with no viral replication in normal cells. doi:10.1128/9781555818890.ch10.f10.45

control of a cancer-specific promoter such as the human telomerase reverse transcriptase promoter (hTERTp). Telomerase reverse transcriptase is a ribonucleoprotein that adds repeat units (TTAGGG) to the ends of chromosomes, i.e., **telomeres**. As normal cells undergo successive cell divisions, the telomeres become shorter due to the absence of telomerase activity. Eventually, after about 60 cell cycles in vitro, mitosis ceases due, in part, to the loss of telomeric DNA. By contrast, part of the cancer process is the activation of the telomerase reverse transcriptase gene, which keeps elongating telomeres and ensures tumor progression. Most cancers have elevated levels of telomerase activity. A number of hTERTp–*E1* adenovirus vectors have been developed. Some of these rely solely on viral production to kill cancer cells (Fig. 10.46A). Others have an added gene

hTERTp
human telomerase reverse transcriptase promoter

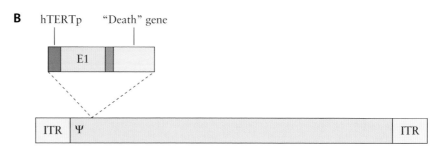

Figure 10.46 Anticancer adenovirus vectors that replicate only in cancer cells. The human telomerase reverse transcriptase promoter (hTERTp) drives the expression of the *E1* gene (**A**) or both the *E1* gene and a "death" gene, such as that encoding apoptin, that specifically destroys cancer cells (**B**). ITR, inverted terminal repeat. doi:10.1128/9781555818890.ch10.f10.46

also under the control of the hTERT promoter that triggers apoptosis in cancer cells and not in normal cells (Fig. 10.46B). An example of the latter type of "death" gene is the gene for apoptin, which is found in a chicken anemia virus and induces apoptosis only in tumor cells. Oncolytic virus is the term used for a virus that exclusively infects and lyses cancer cells. In addition to adenovirus, many oncolytic viruses have been created using HSV, lentivirus, reovirus, vaccinia virus, and others.

Eye Disorders

The photoreceptors of the retina of the eye comprise about 70% of the total complement of human sensory receptors, and about 30% of all nerve fibers going to the **central nervous system** are contained in the optic nerves. The structural components of the eye (Fig. 10.47A), such as the cornea, iris, and lens, have distinctive properties that direct light onto the retina. The photoreceptors (**rods** and **cones**) of the retina (Fig. 10.47B) transduce this radiant energy into nerve signals by a series of activation steps called the phototransduction cascade. These impulses are transmitted from the ganglia of the retina through the fibers of the optic nerves to the visual cortex of the brain, where the signals are processed and a conscious interpretation of the retinal image is created.

In addition to injury, infections, and nonhereditary diseases, many inherited conditions also disrupt ocular function. Over 300 different gene loci are associated with eye defects. Some genetic disorders affect different tissues and organ systems, including the eye, at the same time. Alternatively, several genetically determined, eye-specific (**nonsystemic** and **nonsyndromic**) diseases modify a single ocular structure and/or adjacent components within the eye. Because a single gene is often the basis of many nonsyndromic eye disorders, they are excellent candidates for gene

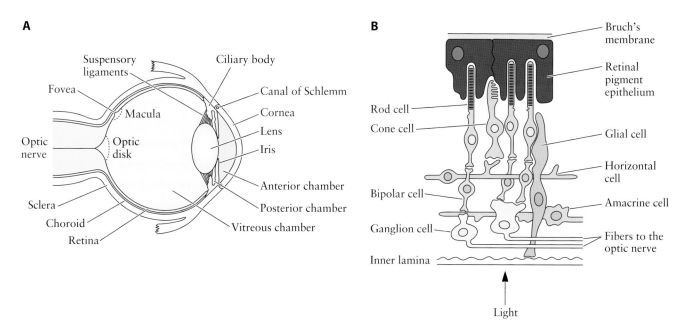

Figure 10.47 (**A**) Structure of the human eye. (**B**) Schematic representation of the retina. The structural relationships among the retinal pigment epithelial cells, rods, cones, and neuronal cells of the retina are depicted. The arrow marks the direction of incoming light. doi:10.1128/9781555818890.ch10.f10.47

therapy. Moreover, the human eye is not subject to immune attack; i.e., it is immune privileged. The eye has discrete and accessible components, and noninvasive techniques can be used to assess the effects of treatment. These features have facilitated a large number of studies designed to alleviate eye diseases using gene therapy strategies with animal models and in human clinical trials.

The commonly used vectors for ocular gene therapy are derived from AAV and adenovirus. These agents are injected either into the vitreous (intravitreal) or into the base of the retina (subretinal). In animal studies, a mouse strain that lacks the *Bbs4* gene, which simulates Bardet-Biedl syndrome in humans, has been created. In humans with Bardet-Biedl syndrome, extreme retinal degeneration occurs by the mid-teens. The human *BBS4* gene controls transport of the visual protein pigments (**opsins**) from the inner segment to the outer segment of rods and cones. Failure to establish opsins in the disks of the photoreceptors leads to apoptosis of these cells. The addition of the murine *Bbs4* gene by subretinal injection in mice lacking this gene was examined using an AAV vector. Under these conditions, cell death did not occur in regions where the vector was administered. The photoreceptor cells functioned normally, indicating that gene therapy may be eventually be used to treat some types of blindness in humans.

Thus, after a successful series of animal model gene therapy experiments, phase I clinical trials were initiated with individuals with Leber **congenital** amaurosis (LCA). This disorder starts with severe visual impairment in infants, and total blindness occurs during the third decade. At least 11 different genes are associated with LCA. Of these, LCA2 is due to

LCA
Leber congenital amaurosis

loss of the retinal pigment epithelium-specific 65-kilodalton (kDa)-protein gene (*RPE65*). This gene was administered as part of an AAV vector to nine patients in three different phase I trials. No significant adverse effects were noted despite different protocols. About 6 to 12 months after gene therapy, many treated patients showed various degrees of improvement in various visual parameters. However, more trials, especially with younger patients, are required to determine whether this form of gene therapy is effective over a long period.

Muscle Disorders

Many activities that require movement (smiling, writing, running, drawing, walking, gripping, chewing, kissing, swallowing, breathing, etc.) depend on one or more of the approximately 650 different muscles in the human body. Muscles are responsible for locomotion, upright posture, balancing on two legs, support of internal organs, controlling valves and body openings, production of heat, and movement of materials, including blood, along internal tubes. There are three types of muscle tissue: skeletal, cardiac, and smooth. Skeletal muscles comprise about 40% of our body mass, are attached to the skeleton, and are responsible for locomotion; both contraction and relaxation are controlled voluntarily (i.e., consciously). Cardiac (heart) muscle pumps blood and functions involuntarily. Smooth muscle, which lines the walls of internal systems, is involuntary and propels material through internal passageways. Both voluntary and involuntary muscle contractions are stimulated by nerve impulses.

The hallmarks of skeletal muscle disorders are chronic or acute pain and muscle weakening, often with loss of muscle mass that can be fatal. Gene mutations of muscle structural components are among the causes of muscle disorders and include a group of about 30 inherited disorders that have been designated as muscular dystrophies and are characterized by progressive muscle wasting and weakness.

Duchenne **muscular dystrophy**, an X-linked trait (i.e., the gene is located on the X chromosome), is the most common of the muscular dystrophies and occurs in about 1 in 3,500 males. Onset of Duchenne muscular dystrophy muscle weakness begins in childhood and the condition progressively worsens, with a wheelchair being required at about 12 years of age, and often death occurs in the early twenties. Duchenne muscular dystrophy mainly affects specific muscle groups. As the disease progresses, cardiac muscles can also be affected. Other muscular dystrophies have different times of onset and patterns of progression as well as affecting particular muscle groups. There are no effective treatments for the different muscular dystrophies.

The genetic defect in Duchenne muscular dystrophy is due to mutations of the dystrophin gene. This gene encodes a structural protein that is part of a dystrophin–glycoprotein complex, which maintains the structural integrity of the plasma membrane of muscle cells, among other functions (Fig. 10.48). The dystrophin–glycoprotein complex regulates muscle cell regeneration and maintains calcium levels in muscle cells.

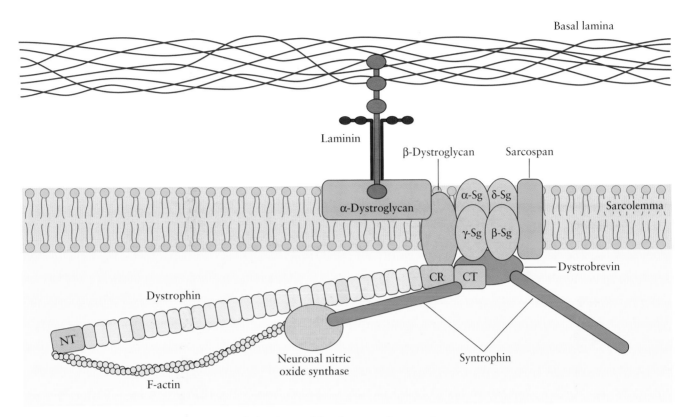

Figure 10.48 Schematic representation of the dystrophin–glycoprotein complex. Dystrophin consists of an N-terminal domain (NT), four hinge domains (gray), 24 spectrin-like repeats (light blue), a cysteine-rich domain (CR), and a C-terminal domain (CT). The sarcoglycan proteins are labeled as α-Sg, β-Sg, δ-Sg, and γ-Sg. doi:10.1128/9781555818890.ch10.f10.48

With Duchenne muscular dystrophy, the muscle cells are damaged after every contraction and the normal repair process is ineffective due to the absence of functional dystrophin molecules. In a mild form of Duchenne muscular dystrophy, called Becker muscular dystrophy, a partially defective dystrophin is synthesized and often carries a single-site mutation. In contrast, dystrophin that has a large deletion or is severely truncated usually occurs with Duchenne muscular dystrophy.

An initial problem that researchers faced with preclinical gene therapy trials using animal models for Duchenne muscular dystrophy was the size of the dystrophin gene. In humans, it is the largest gene in the genome, covering about 2.6×10^3 kb, with the coding sequence comprising about 11 kb. There are a number of repeated **spectrin**-like units that form part of the dystrophin molecule. Detailed molecular analyses of individuals with mild Duchenne muscular dystrophy symptoms revealed that a large portion of the dystrophin gene may be deleted without a total loss of function. Based on these observations, truncated versions of the dystrophin gene without introns were constructed. One class of these smaller coding sequences was designated as minidystrophins, which range from about 4 to 7 kb in length (Fig. 10.49). The minidystrophins are suitable as potential therapeutic agents to be used with viral delivery systems.

A

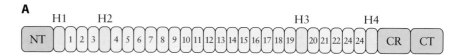

B

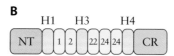

Figure 10.49 Structures of dystrophin (**A**) and a minidystrophin (**B**). Hinge domains are gray, and spectrin-like repeats are numbered and light blue. NT, N-terminal domain; CR, cysteine-rich domain; CT, C-terminal domain.
doi:10.1128/9781555818890.ch10.f10.49

A phase I clinical trial was conducted with a minidystrophin gene delivered by a novel AAV vector. Previously, this minidystrophin gene was shown to be effective with a Duchenne muscular dystrophy mouse model system. The AAV vector was generated by DNA shuffling among the *cap* genes for a number of AAV serotypes. After screening, a novel hybrid *cap* gene was identified that enabled the vector to transduce skeletal muscle cells exclusively. Also, the vector was not susceptible to preexisting antibodies.

The minidystrophin gene trial was judged to be safe. Unfortunately, the minidystrophin protein was not detected by conventional assays. However, the minidystrophin protein elicited a cellular immune response in four of the six patients tested without causing any adverse effects, indicating that some expression had occurred. These results suggest that in future clinical trials, expression of the minidystrophin gene needs to be enhanced, possibly with a muscle cell-specific promoter.

Neurological Disorders

The nervous system is a complex, extensive cellular communication network that is divided into two parts: the central and peripheral nervous systems. The central nervous system consists of the brain and spinal cord, while all the nerves extending from and going to the central nervous system comprise the peripheral nervous system. Information is detected and transmitted by impulses along nerves to the central nervous system, where it is processed and interpreted. Response signals then are conducted back via nerves to nerve, muscle, or gland cells. The complete nervous system has more than 1×10^{11} individual neurons and about 5 to 10 times as many nonneuronal cells that are closely associated with neurons.

Neurons are highly specialized biochemically differentiated cells that transmit electrical impulses and have a variety of shapes, sizes, and roles. Neurons may synthesize acetylcholine, dopamine, serotonin, γ-aminobutyric acid, somatostatin, or other compounds important for the transmission of nerve impulses. A neuron has three basic structural elements: a cell body, **dendrites**, and an **axon** (Fig. 10.50).

Masses of axons in the brain form the white matter, and clusters of neuronal cell bodies comprise the gray matter. Over 50 different white matter disorders have been identified, with many of them due to defects in **myelin sheath** formation.

Regions within the brain are demarcated by concentrations of cell bodies (nuclei and ganglia) and their specific functions. Several cell body clusters that are found within the middle region of the brain are designated the **basal ganglia** (Fig. 10.51A). Of these, the substantia nigra inhibits forced involuntary movements. Neurodegeneration of a portion of the substantia nigra, called the pars compacta, occurs in patients with Parkinson disease. The **hippocampus**, which lies deep within the brain, contributes to emotional states, such as fear, anger, rage, pleasure, and sorrow, and is also associated with learning and memory capabilities. This cluster of neurons undergoes neuronal degeneration in individuals with Alzheimer disease (Fig. 10.51B).

Parkinson Disease

Parkinson disease affects about 1 million people in the United States, with about 60,000 new cases occurring annually. This **neurodegenerative disorder** is chronically progressive, with an average age at onset of about 57 years. The characteristic symptoms involve motor functions that include an extended time to initiate normal movements (bradykinesia), involuntary shaking (resting tremor), stiff gait, stooped posture, and muscle rigidity. The nonmotor component of Parkinson disease entails sleep disorder, rapid eye movements, tingling of the skin, and loss of the sense of smell. Psychiatric disturbances often accompany the disease.

The primary effect of Parkinson disease is a progressive loss of the dopamine-releasing neurons, which affects the transmission of signals between other brain components and results in motor deficiencies. While various genes have been implicated in Parkinson disease, about 85% of the patients do not have a mutation in any of the identified Parkinson disease genes. Consequently, there is no specific gene that might be used therapeutically. Levodopa therapy, which is an attempt to make up for the loss of dopamine, is effective for short periods, but eventually motor fluctuations develop, voluntary movements diminish, and the frequency of tremors increases. Nevertheless, levodopa therapy and other treatments provide some relief from the symptoms of the disease.

Another option for controlling the motor problems associated with Parkinson disease is deep brain stimulation. Briefly, thin-wire electrodes are implanted on both sides of the brain. A battery-operated, electronic-pulse neurostimulator (brain pacemaker) that regulates the delivery of electric pulses to the target regions is implanted in the chest with a wire connection to the electrodes. It is thought that the electronic pulses normalize the Parkinson disease-induced signal pathway. Deep brain stimulation does not cause significant brain tissue damage. Rather, this stimulation reduces motor-related symptoms such as walking problems, tremors, rigidity, and slow movements, although it has no appreciable impact on the nonmotor aspects of the disease.

Based on the deep brain stimulation procedure, it was reasoned that delivery of a gene that would dampen the overactive subthalamic nucleus in Parkinson disease patients would, in turn, reduce the signal

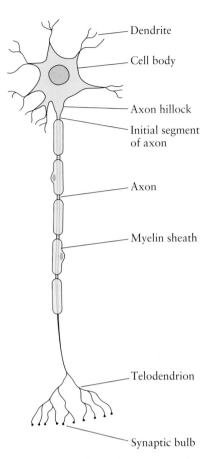

Figure 10.50 Schematic representation of a neuron. The cell body (soma) contains the nucleus, the components for protein synthesis, and the enzymes for metabolic pathways. Most neurons have large numbers of multibranched arrays of dendrites that receive stimuli and transmit nerve impulses toward the cell body. Each neuron has one axon (nerve fiber) that carries a nerve impulse away from the cell body. Axons are encased in a multilayed myelin sheath that increases the speed of electrical impulses down an axon and prevents dissipation of the nerve signal.
doi:10.1128/9781555818890.ch10.f10.50

A

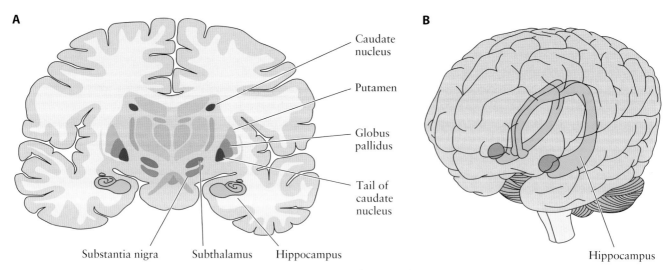

Caudate nucleus

Putamen

Globus pallidus

Tail of caudate nucleus

Substantia nigra   Subthalamus   Hippocampus

B

Hippocampus

Figure 10.51 (**A**) Coronal section through the anterior portion of the cerebrum of the human brain showing some of the elements of the basal ganglia and limbic system; (**B**) schematic representation of the hippocampus within the human brain. doi:10.1128/9781555818890.ch10.f10.51

reaching the ventral anterior and ventral lateral nuclei of the thalamus and, as a result, increase the signals to the motor cortex. The reduction of γ-aminobutyric acid input to the subthalamic nucleus is part of the consequences of Parkinson disease. Therefore, inserting the gene for the production of γ-aminobutyric acid directly into the subthalamic nucleus was considered a likely treatment strategy.

γ-Aminobutyric acid is synthesized by glutamic acid decarboxylase (GAD) (Fig. 10.52). In humans, there are two GAD genes, *GAD1* and *GAD2*, which encode the GAD67 and GAD65 enzymes, respectively. The GAD65 enzyme is found mostly in axons and synthesizes the **neurotransmitter** γ-aminobutyric acid. The GAD67 isoform is localized in nerve cell bodies and is involved in metabolic processes. The two human *GAD* genes have been cloned into separate AAV2 vectors. After promising

GAD
glutamic acid decarboxylase

Figure 10.52 Synthesis of γ-aminobutyric acid. doi:10.1128/9781555818890.ch10.f10.52

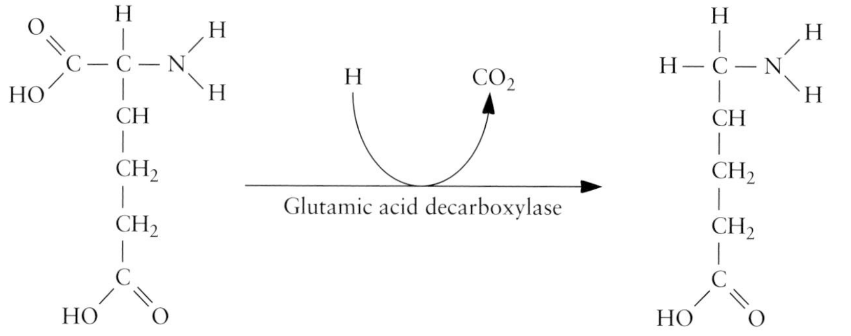

Glutamic acid

Glutamic acid decarboxylase

γ-Aminobutyric acid

preclinical trials, a phase I trial was initiated with both GAD65– and GAD67–AAV2 vectors being infused into one subthalamic nucleus of Parkinson disease patients. This study established the safety of the therapy with alleviation of symptoms for up to 1 year in some patients. Subsequently, a phase II trial with a larger number of subjects was conducted in which the GAD65– and GAD67–AAV2 vectors were delivered to both subthalamic nuclei. The patients infused with the vectors, in comparison to those who had a sham operation, showed statistically significant improvement in their condition. A phase III trial is required to definitively determine if this gene therapy will become a common procedure for treating Parkinson disease.

Alzheimer Disease

Alzheimer disease, named for the German neurologist Alois Alzheimer, who described its clinical and neuropathological features in 1907, currently accounts for two-thirds or more of all diagnosed cases of **dementia**. There are currently more than 4 million people with Alzheimer disease in the United States, where ~100,000 people die from this disease every year. As the population ages, this number will increase, as Alzheimer disease predominantly affects older people.

Many different clinical features are associated with Alzheimer disease, including the inability to create new memories, the loss of short-term memory, and the inability to concentrate. Alzheimer disease patients often have no accurate sense of time, they often become disoriented, and both sentence formation and coherent verbal communication diminish and are eventually lost. Some Alzheimer disease patients become extremely passive, others become very hostile, some are abnormally suspicious, and ~50% become delusional. In the end, death is often the result of respiratory failure. For Alzheimer disease patients diagnosed at 65 years or older, the disease lasts from about 8 to 20 years. In cases of onset before 65 years of age, the course of the disease is rapid and death occurs within 5 to 10 years from the time of diagnosis.

Mutations in three genes—those for **amyloid** precursor protein (*APP*), presenilin-1 (*PSEN1*), and presenilin-2 (*PSEN2*)—account for less than 5% of all cases of Alzheimer disease and often cause early onset of the disease in individuals who are 60 years old or younger. Thus, despite considerable research effort, the root cause of Alzheimer disease in the vast majority of cases is unknown.

Histological analysis of autopsied brains of Alzheimer disease patients shows losses of **synapses** and neurons in the hippocampus, a region of the **cerebral cortex** (neocortex) beneath the hippocampus, and the nucleus basalis of Meynert (Fig. 10.53). Many cerebral cortex neurons that connect with other cortical neurons also degenerate. By the final stage of the disease, the overall width of most of the neocortex is dramatically reduced and brain activity is severely affected. Furthermore, dense spherical structures (20 to 200 mm in diameter), called **senile plaques**, are prevalent outside the neurons of the hippocampus and other regions of the brain. When senile plaques are surrounded by

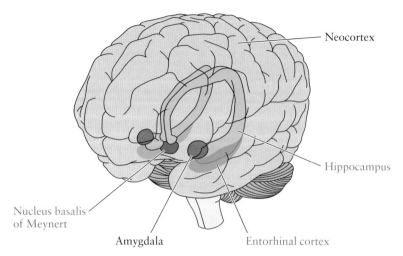

Figure 10.53 Schematic representation of some of the regions of the brain that undergo neuronal loss as Alzheimer disease progresses. The names with blue letters denote regions of the brain that undergo a loss of neurons during the early stages of Alzheimer disease. doi:10.1128/9781555818890.ch10.f10.53

cellular debris from disintegrated axons and dendrites, they are called neuritic plaques. Finally, aggregations of fibrils (neurofibrillary tangles) accumulate within the cell bodies and dendritic processes of the neurons of the hippocampus, the neocortex, the amygdala, and other parts of the brain.

The core of a senile plaque is a densely packed fibrous structure known as an amyloid body. The principal protein of Alzheimer disease amyloid bodies is a 4-kDa peptide (Aβ protein; also called β-protein, A4, and β/A4). Of these, the two main isoforms contain 40 and 42 amino acids and are designated $A\beta_{40}$ (Aβ40) and $A\beta_{42}$ (Aβ42), respectively. The Aβ proteins are cleavage products of a precursor protein called amyloid precursor protein. Under normal conditions, an enzyme called α-secretase cleaves the amyloid precursor protein after amino acid residue 687. Neither of the fragments produced by this cleavage gives rise to the formation of fibrils or amyloid. Cleavage by β-secretase after amino acid residue 671 also produces two nonamyloidogenic fragments. When some amyloid precursor protein molecules are doubly cleaved by β-secretase and γ-secretase, then $A\beta_{40}$ and $A\beta_{42}$ isoforms are released.

An additional hallmark of Alzheimer disease is reduced synthesis of the neurotransmitter acetylcholine. The major source of acetylcholine in the brain is the cluster of neurons (cholinergic neurons) comprising the nucleus basalis of Meynert. These neurons project into the cortex and other centers (e.g., the hippocampus). Destruction of these cholinergic neurons occurs during the initial stage of Alzheimer disease, likely causing memory dysfunction and the deterioration of the ability of Alzheimer disease patients to process concepts. Accompanying the **atrophy** of cholinergic neurons is the absence of nerve growth factor activity. The nerve growth factor protein is essential for the maintenance and survival of neurons, axonal growth, enhancing neuronal metabolism, repair of injured neurons, and prevention of nerve cell death. Based on these features, the nerve growth factor gene has been considered as a therapeutic agent for Alzheimer disease and other neurodegenerative disorders.

Animal studies have established that nerve growth factor can prevent the degeneration of cholinergic neurons due to amyloid accumulation and other injurious events. Also, nerve growth factor boosts cholinergic neuron activity and can reverse some memory loss. A phase I trial was therefore initiated for the ex vivo gene delivery of the nerve growth factor gene into the brains of eight Alzheimer disease patients. Fibroblasts were removed from individuals with mild Alzheimer disease, transduced with a retrovirus carrying the nerve growth factor gene, and introduced into each recipient. No adverse effects were noted after 2 years, and the rate of cognitive decline was slowed. As well, positron emission tomography scans showed increased brain activity. However, expression of the nerve growth factor gene declined after about 18 months. Nevertheless, these results were encouraging enough to warrant an additional phase I trial. At this point, the researchers switched the protocol from an ex vivo gene therapy strategy to one that used a nerve growth factor cDNA–AAV2 vector system. Not only is this viral system easier to administer to patients, but also gene expression may be maintained for longer periods. In a phase I trial, this vector was infused directly into the nucleus basalis of Meynert on both sides of the brain. After 12 months, no adverse occurrences were recorded. Importantly, the rate of cognitive decline was curtailed and brain activity was enhanced. A future phase II trial will determine whether this strategy effectively limits the progress of Alzheimer disease.

X-Linked Adrenoleukodystrophy

In X-linked adrenoleukodystrophy, as a result of a defective *ABCD1* gene, which encodes the adrenoleukodystrophy protein (located in the **peroxisome** plasma membrane), there is an excess accumulation of hexacosanoic acid ($C_{26:0}$) (Fig. 10.54A) and tetracosanoic acid ($C_{24:0}$) in the plasma of affected individuals. The buildup of these very-long-chain fatty acids causes a range of clinical effects with various degrees of severity. The adrenoleukodystrophy protein is an ATP cassette transporter that delivers very-long-chain fatty acids into peroxisomes, where they are degraded (Fig. 10.54B and C). Mutations of the *ABCD1* gene that interfere with the transport of very-long-chain fatty acids into peroxisomes lead to the accumulation of these molecules in neurons, adrenal glands, and the blood system (Fig. 10.54D).

In about 35% of the cases of this disease, onset is between 3 and 10 years of age, with progressive behavioral, learning, and neurological impairment within 3 years of onset, followed by coma, and then death in the early teens (childhood cerebral adrenoleukodystrophy). The most common type of X-linked adrenoleukodystrophy is designated adrenomyeloneuropathy, which occurs in about 45% of the cases and begins in individuals who are about 30 years old. The course of this version of X-linked adrenoleukodystrophy is usually slow, with loss of mobility due

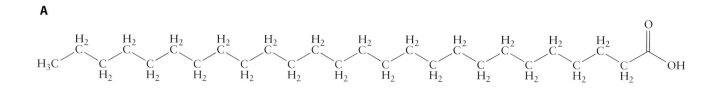

B

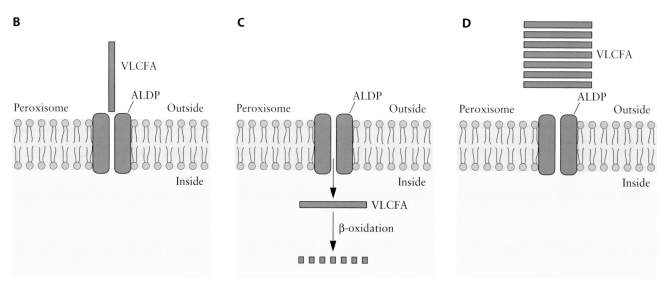

Figure 10.54 (**A**) Hexacosanoic acid. (**B**) Entry of a very-long-chain fatty acid (VLCFA) by means of the adrenoleukodystrophy protein (ALDP) into a peroxisome. (**C**) Breakdown of a VLCFA by β-oxidation after entry into a peroxisome. (**D**) Failure of entry of VLCFAs into a peroxisome due to a defect in ALDP leads to the accumulation of VLCFAs in neurons, adrenal cells, and plasma. doi:10.1128/9781555818890.ch10.f10.54

to spinal cord and peripheral nerve demyelination. In about 20% of these individuals, cerebral demyelination leads to early death.

There is no general treatment that significantly affects the neurological symptoms of X-linked adrenoleukodystrophy. Successful bone marrow or hematopoietic stem cell transplants can slow the progress of this disease due to cells entering the central nervous system and forming microglial cells. These brain cells have functioning adrenoleukodystrophy protein molecules that enable very-long-chain fatty acids to be degraded. To overcome the lack of a matched donor, ex vivo gene therapy with autologous hematopoietic stem cells that were transduced with a self-inactivating lentivirus carrying an *ABCD1* cDNA sequence was conducted on two young boys with symptomatic X-linked adrenoleukodystrophy. After 30 months, expression of adrenoleukodystrophy protein was observed in different types of blood and immune cells. In addition, from about 12 months onward, demyelination within the brains of both patients had demonstrably declined. Further testing will determine if ex vivo gene therapy routinely alleviates the effects of X-linked adrenoleukodystrophy.

summary

Several human disorders that result from the overproduction of a normal protein may be treated by using (i) nucleotide sequences that bind to a specific mRNA and prevent its translation, i.e., an antisense oligonucleotide; (ii) small RNA molecules, i.e., aptamers, that assume highly organized secondary and tertiary structures and bind tightly to a wide range of molecules, including proteins and amino acids; (iii) RNA or DNA sequences that bind and cleave specific RNA molecules, i.e., ribozymes or DNAzymes; or (iv) small double-stranded RNA molecules that direct the sequence-specific degradation of mRNA, i.e., interfering RNAs. In addition, zinc finger nucleases may be used to directly repair a cell's defective genes. These techniques may also be used to lessen or prevent diseases caused by pathogenic viruses and other disease-causing organisms.

The development of effective treatments for genetic diseases has been elusive because, in many instances, the appropriate gene product cannot be provided to a patient. However, when a normal version of a gene is identified and cloned, either the gene or a cDNA derivative may be used to correct the defect in affected individuals. Numerous viral systems have been developed for the delivery of therapeutic genes. Viral vectors take advantage of the ability of a virus to penetrate a specific cell, protect the DNA from degradation, and direct it to the cell nucleus. Several viruses have been engineered for gene therapy applications. Packaging cell lines for some viral systems ensure that virtually no infectious viruses are present in a sample of vector viruses.

The key requirements for gene therapy are the efficient and safe delivery of a therapeutic gene to a specific tissue in sufficient numbers such that the introduced gene can alleviate a particular disorder. Somatic cell gene therapy is implemented in one of two ways: in vivo or ex vivo. For in vivo gene therapy, a therapeutic gene is introduced into a targeted tissue of the patient. Ex vivo gene therapy entails collecting and culturing stem cells from an affected individual, introducing the therapeutic gene into these cultured cells, growing the cells with the therapeutic gene, and then either infusing or transplanting these cells back into the patient.

Construction of an effective viral delivery system includes making space for a therapeutic gene within the viral genome, creating a system for packaging the therapeutic gene–virus constructs into a virus particle, and ensuring that the virus carrying the therapeutic gene is not harmful to either the targeted cell, unless it is a cancer cell, or the recipient organism. For the gammaretrovirus and lentivirus vectors, SIN systems were devised to avoid unwanted transcription of adjacent chromosome regions after integration. Other viruses, such as AAV, adenovirus, and HSV, have been altered to create vectors that deliver therapeutic genes to targeted cells.

The greatest impediment to the development of nucleic acid-based therapeutic agents is the difficulty in delivering these agents to their target tissue(s). Nonviral approaches for the delivery of nucleic acid-based therapeutic agents include intravenous injection, local injection at the site of the pathology, packaging the nucleic acid into cationic liposomes, physical methods (e.g., electroporation), and conjugating the nucleic acid to another molecule (e.g., lipid, cholesterol, antibody fragment, or aptamer).

Although they are still at an early stage of development, gene therapy clinical trials have been initiated for a wide range of diseases. These include SCID and many different cancers, as well as eye, muscle, and neurological disorders. In many instances, the results of phase I and phase II clinical studies have been encouraging.

review questions

1. How can antisense oligonucleotides be used as a therapeutic agent?

2. What is an aptamer, how is it selected, and how is it used as a therapeutic agent?

3. What are ribozymes, and how can they be used as human therapeutic agents?

4. What are DNAzymes, and how do they compare to ribozymes?

5. What are interfering RNAs, and how might they be used as human therapeutic agents?

6. How can the interferon response, which is usually induced by double-stranded RNA, be avoided when utilizing siRNAs as therapeutic agents?

7. How can the progression of age-related macular degeneration be limited using RNA therapeutics?

8. How can zinc finger nucleases be used to repair a cell's defective genes?

(continued)

review questions *(continued)*

9. What is ex vivo gene therapy?

10. What is in vivo gene therapy?

11. What are the features of an idealized viral vector delivery system?

12. What are the potential problems that might prevent the success of a gene therapy strategy?

13. What is a packaging cell line?

14. Describe how an AAV gene therapy vector is packaged.

15. What are the advantages and disadvantages of gammaretrovirus, adenovirus, and herpesvirus vector systems?

16. Describe the basic features of an HSV-based vector system.

17. What is a self-complementary AAV vector?

18. Describe how the AIDS virus can be used as a vector system. What are the advantages of a lentivirus-based vector for gene therapy?

19. A variety of novel delivery strategies have been developed to introduce nucleic acids to their cellular targets. These include the use of lipids, dendrimers, antibodies, and aptamers. Briefly explain how these strategies may be used in practice.

20. Describe how the prodrug strategy works.

21. How are bacteria and bacterial minicells used to specifically deliver therapeutic nucleic acids to their cellular targets?

22. Discuss how transposons may be used as gene delivery agents.

23. Describe a gene therapy strategy that has been tested for alleviating the motor effects of Parkinson disease.

references

Aiuti, A., I. Brigida, F. Ferrua, B. Cappelli, R. Chiesa, S. Marktel, and M. G. Roncarolo. 2009. Hematopoietic stem cell gene therapy for adenosine deaminase deficient-SCID. *Immunol. Res.* **44:** 150–159.

Aiuti, A., F. Cattaneo, S. Galimberti, U. Benninghoff, B. Cassani, L. Callegaro, S. Scaramuzza, G. Andolfi, M. Mirolo, I. Brigida, A. Tabucchi, F. Carlucci, M. Eibl, M. Aker, S. Slavin, H. Al-Mousa, A. Al Ghonaium, A. Ferster, A. Duppenthaler, L. Notarangelo, U. Wintergerst, R. H. Buckley, M. Bregni, S. Marktel, M. G. Valsecchi, P. Rossi, F. Ciceri, R. Miniero, C. Bordignon, and M. G. Roncarolo. 2009. Gene therapy for immunodeficiency due to adenosine deaminase deficiency. *N. Engl. J. Med.* **360:**447–458.

Alba, R., A. Bosch, and M. Chillon. 2005. Gutless adenovirus: last-generation adenovirus for gene therapy. *Gene Ther.* **12**(Suppl. 1):S18–S27.

Argnani, R., M. Lufino, M. Manservigi, and R. Manservigi. 2005. Replication-competent herpes simplex vectors: design and applications. *Gene Ther.* **12**(Suppl. 1):S170–S177.

Aronovich, E., R. S. McIvor, and P. B. Hackett. 2011. The Sleeping Beauty transposon system: a non-viral vector for gene therapy. *Hum. Mol. Genet.* **20:** R14–R20.

Baum, C., A. Schambach, J. Bohne, and M. Galla. 2006. Retrovirus vectors: toward the plentivirus? *Mol Ther.* **13:** 1050–1063.

Biffi, A., P. Aubourg, and N. Cartier. 2011. Gene therapy for leukodystrophies. *Hum. Mol. Genet.* **20:**R42–R53.

Bishop, K. M., E. K. Hofer, A. Mehta, A. Ramirez, L. Sun, M. Tuszynski, and R. T. Bartus. 2008. Therapeutic potential of CERE-110 (AAV2-NGF): targeted, stable, and sustained NGF delivery and trophic activity on rodent basal forebrain cholinergic neurons. *Exp. Neurol.* **211:**574–584.

Blankinship, M., P. Gregorevic, and J. Chamberlain. 2006. Gene therapy strategies for Duchenne muscular dystrophy utilizing recombinant adeno-associated virus vectors. *Mol. Ther.* **13:**241–248.

Boztug, K., M. Schmidt, A. Schwarzer, P. P. Banerjee, I. A. Diez, R. A. Dewey, M. Bohm, A. Nowrouzi, C. R. Ball, H.

Glimm, S. Naundorf, K. Kuhlcke, R. Blasczyk, I. Kondratenko, L. Marodi, J. S. Orange, C. von Kalle, and C. Klein. 2010. Stem-cell gene therapy for the Wiskott-Aldrich syndrome. *N. Engl. J. Med.* 363:1918–1927.

Bramlage, B., E. Luzi, and F. Eckstein. 1998. Designing ribozymes for the inhibition of gene expression. *Trends Biotechnol.* **16:**434–438.

Bunka, D. H. J., and P. G. Stockley. 2006. Aptamers come of age—at last. *Nat. Rev. Microbiol.* **4:**588–596.

Cai, H., F. S. Santiago, L. Prado-Lourenco, B. Wang, M. Patrikakis, M. P. Davenport, G. J. Maghzal, R. Stocker, C. R. Parish, B. H. Chong, G. J. Lieschke, T.-W. Wong, C. N. Chesterman, D. J. Francis, F. J. Moloney, R. St. C. Barnetson, G. M. Halliday, and L. M. Khachigian. 2012. DNAzyme targeting c-jun suppresses skin cancer growth. *Sci. Transl. Med.* **4:**1–12.

Cairns, M. J., T. M. Hopkins, C. Witherington, L. Wang, and L.-Q. Sun. 1999. Target site selection for an RNA-cleaving catalytic DNA. *Nat. Biotechnol.* **17:**480–486.

Campochiaro, P. A. 2007. Gene therapy for ocular neovascularization. *Curr. Gene Ther.* 7:25–33.

Carrillo-Tripp, M., C. Shepherd, I. Borelli, S. Venkataraman, G. Lander, P. Natarajan, J. E. Johnson, C. L. Brooks III, and V. S. Reddy. 2009. VIPERdb: an enhanced and web API enabled relational database for structural virology. *Nucleic Acids Res.* 37:D436–D442.

Cartier, N., S. Hacein-Bey-Abina, C. C. Bartholomae, G. Veres, M. Schmidt, I. Kutschera, M. Vidaud, U. Abel, L. Dal-Cortivo, L. Caccavelli, N. Mahlaoui, V. Kiermer, D. Mittelstaedt, C. Bellesme, N. Lahlou, F. Lefrere, S. Blanche, M. Audit, E. Payen, P. Leboulch, B. l'Homme, P. Bougneres, C. Von Kalle, A. Fischer, M. Cavazzana-Calvo, and P. Aubourg. 2009. Hematopoietic stem cell gene therapy with a lentiviral vector in X-linked adrenoleukodystrophy. *Science* 326:818–823.

Cathomen, T., and J. K. Joung. 2008. Zinc-finger nucleases: the next generation emerges. *Mol. Ther.* 16:1200–1207.

Chu, T. C., K. Y. Twu, A. D. Ellington, and M. Levy. 2006. Aptamer mediated siRNA delivery. *Nucleic Acids Res.* 34: e73.

Connolly, S. A., J. O. Jackson, T. S. Jardetzky, and R. Longnecker. 2011. Fusing structure and function: a structural view of the herpesvirus entry machinery. *Nat. Rev. Microbiol.* 9:369–381.

Cullen, B. R. 2006. Enhancing and confirming the specificity of RNAi experiments. *Nat. Methods* 3:677–681.

de Fougerolles, A., H.-P. Vornlocher, J. Maraganore, and J. Lieberman. 2007. Interfering with disease: a progress report on siRNA-based therapeutics. *Nat. Rev. Drug Discov.* 6:443–453.

de Silva, S., and W. J. Bowers. 2009. Herpes virus amplicon vectors. *Viruses* 1:594–629.

Dillon, C. P., P. Sandy, A. Nencioni, S. Kissler, D. A. Rubinson, and L. Van Parijs. 2005. RNAi as an experimental and therapeutic tool to study and regulate physiological and disease processes. *Annu. Rev. Physiol.* 67:147–173.

Durand, S., and A. Cimarelli. 2011. The inside out of lentiviral vectors. *Viruses* 3:132–159.

Eager, R. M., and J. Nemunaitis. 2011. Clinical development directions in oncolytic viral therapy. *Cancer Gene Ther.* 18:305–317.

Elbashir, S. M., J. Harborth, W. Lendeckel, A. Yalcin, K. Weber, and T. Tuschi. 2001. Duplexes of 21-nucleotide RNAs mediate RNA interference in cultured mammalian cells. *Nature* 411:494–498.

Elbashir, S. M., W. Lendeckel, and T. Tuschi. 2001. RNA interference is mediated by 21- and 22-nucleotide RNAs. *Genes Dev.* 15:188–200.

Emery, A. E. 1998. The muscular dystrophies. *BMJ* 317:991–995.

Faria, M., D. G. Spiller, C. Dubertret, J. S. Nelson, M. R. H. White, D. Scherman, C. Hélène, and C. Giovannangeli. 2001. Phosphor-amidate oligonucleotides as potent antisense molecules in cells and in vivo. *Nat. Biotechnol.* 19: 40–44.

Feigin, A., M. G. Kaplitt, C. Tang, T. Lin, P. Mattis, V. Dhawan, M. J. During, and D. Eidelberg. 2007. Modulation of metabolic brain networks after subthalamic gene therapy for Parkinson's disease. *Proc. Natl. Acad. Sci. USA* 104: 19559–19564.

Fichou, Y., and C. Férec. 2006. The potential of oligonucleotides for therapeutic applications. *Trends Biotechnol.* 24:563–570.

Field, A. K. 1998. Viral targets for antisense oligonucleotides: a mini review. *Antivir. Res.* 37:67–81.

Garcia, L., G. D'Alessandro, B. Bioulac, and C. Hammond. 2005. High-frequency stimulation in Parkinson's disease: more or less? *Trends Neurosci.* 28:209–216.

Gaspar, H. B., S. Cooray, K. C. Gilmour, K. L. Parsley, S. Adams, S. J. Howe, A. Al Ghonaium, J. Bayford, L. Brown, E. G. Davies, C. Kinnon, and A. J. Thrasher. 2011. Long-term persistence of a polyclonal T cell repertoire after gene therapy for X-linked severe combined immunodeficiency. *Sci. Transl. Med.* 3:97ra79.

Gaspar, H. B., S. Cooray, K. C. Gilmour, K. L. Parsley, F. Zhang, S. Adams, E. Bjorkegren, J. Bayford, L. Brown, E. G. Davies, P. Veys, L. Fairbanks, V. Bordon, T. Petropolou, C. Kinnon, and A. J. Thrasher. 2011. Hematopoietic stem cell gene therapy for adenosine deaminase-deficient severe combined immunodeficiency leads to long-term immunological recovery and metabolic correction. *Sci. Transl. Med.* 3:97ra80.

Ghosh, A., and D. Duan. 2007. Expanding adeno-associated viral vector capacity: a tale of two vectors. *Biotechnol. Genet. Eng. Rev.* 24:165–177.

Goins, W. F., D. Wolfe, D. M. Krisky, Q. Bai, E. A. Burton, D. J. Fink, and J. C. Glorioso. 2004. Delivery using herpes simplex virus: an overview. *Methods Mol. Biol.* 246:257–299.

Gonçalves, M. 2005. Adeno-associated virus: from defective virus to effective vector. *Virol. J.* 2:43.

Goyenvalle, A., J. Seto, K. Davies, and J. Chamberlain. 2011. Therapeutic approaches to muscular dystrophy. *Hum. Mol. Genet.* 20:R69–R78.

Guo, P., O. Coban, N. M. Snead, J. Trebley, S. Hoeprich, S. Guo, and Y. Shu. 2010. Engineering RNA for targeted siRNA delivery and medical application. *Adv. Drug Deliv. Rev.* 62: 650–666.

Haasnoot, E., M. Westerhout, and B. Berkhout. 2007. RNA interference against viruses: strike and counterstrike. *Nat. Biotechnol.* 25:1435–1443.

Hacein-Bey-Abina, S., J. Hauer, A. Lim, C. Picard, G. P. Wang, C. C. Berry, C. Martinache, F. Rieux-Laucat, S. Latour, B. H. Belohradsky, L. Leiva, R. Sorensen, M. Debre, J. L. Casanova, S. Blanche, A. Durandy, F. D. Bushman, A. Fischer, and M. Cavazzana-Calvo. 2010. Efficacy of gene therapy for X-linked severe combined immunodeficiency. *N. Engl. J. Med.* 363:355–364.

Heilbronn, R., and S. Weger. 2010. *Viral Vectors for Gene Transfer: Current Status of Gene Therapeutics*, p. 143–170. *In* M. Schäfer-Korting (ed.), *Drug Delivery*. Springer, Heidelberg, Germany.

Isaacs, F. J., D. J. Dwyer, and J. J. Collins. 2006. RNA synthetic biology. *Nat. Biotechnol.* 24:545–554.

Izsvák, Z., P. B. Hackett, L. J. N. Cooper, and Z. Ivics. 2010. Translating *Sleeping Beauty* transposition into cellular therapies: victories and challenges. *Bioessays* **32**:756–767.

Johnson, D. C., and J. D. Baines. 2011. Herpesviruses remodel host membranes for virus egress. *Nat. Rev. Microbiol.* **9**: 382–394.

Kang, E. M., U. Choi, N. Theobald, G. Linton, D. A. Long Priel, D. Kuhns, and H. L. Malech. 2010. Retrovirus gene therapy for X-linked chronic granulomatous disease can achieve stable long-term correction of oxidase activity in peripheral blood neutrophils. *Blood* **115**:783–791.

Kaplitt, M. G., A. Feigin, C. Tang, H. L. Fitzsimons, P. Mattis, P. A. Lawlor, R. J. Bland, D. Young, K. Strybing, D. Eidelberg, and M. J. During. 2007. Safety and tolerability of gene therapy with an adeno-associated virus (AAV) borne GAD gene for Parkinson's disease: an open label, phase I trial. *Lancet* **369**: 2097–2105.

Kemp, S., and R. Wanders. 2010. Biochemical aspects of X-linked adrenoleukodystrophy. *Brain Pathol.* **20**:831–837.

Kim, D.-H., M. A. Behlke, S. D. Rose, M.-S. Chang, S. Choi, and J. J. Rossi. 2005. Synthetic dsRNA Dicer substrates enhance RNAi potency and efficacy. *Nat. Biotechnol.* **23**:222–226.

Kim, Y.-G., J. Cha, and S. Chandrasegaran. 1996. Hybrid restriction enzymes: zinc finger fusions to *Fok*I cleavage domain. *Proc. Natl. Acad. Sci. USA* **93**: 1156–1160.

Knipe, D. M., and A. Cliffe. 2008. Chromatin control of herpes simplex virus lytic and latent infection. *Nat. Rev. Microbiol.* **6**:211–221.

Kono, K., Y. Torikoshi, M. Mitsutomi, T. Itoh, N. Emi, H. Yanagie, and T. Takagishi. 2001. Novel gene delivery systems: complexes of fusigenic polymer-modified liposomes and lipoplexes. *Gene Ther.* **8**:5–12.

Kyritsis, A. P., C. Sioka, and J. S. Rao. 2009. Viruses, gene therapy and stem cells for the treatment of human glioma. *Cancer Gene Ther.* **16**:741–752.

Lares, M. R., J. R. Rossi, and D. L. Ouellet. 2010. RNAi and small interfering RNAs in human disease therapeutic applications. *Trends Biotechnol.* **28**: 570–579.

Lee, J.-H., M. D. Canny, A. De Erkenez, D. Krilleke, Y.-S. Ng, D. T. Shima, A. Pardi, and F. Jucker. 2005. A therapeutic aptamer inhibits angiogenesis by specifically targeting the heparin binding domain of VEGF165. *Proc. Natl. Acad. Sci. USA* **102**:18902–18907.

LeWitt, P. A. 2008. Levodopa for the treatment of Parkinson's disease. *N. Engl. J. Med.* **359**:2468–2476.

LeWitt, P. A., A. R. Rezai, M. A. Leehey, S. G. Ojemann, A. W. Flaherty, E. N. Eskandar, S. K. Kostyk, K. Thomas, A. Sarkar, M. S. Siddiqui, S. B. Tatter, J. M. Schwalb, K. L. Poston, J. M. Henderson, R. M. Kurlan, I. H. Richard, L. Van Meter, C. V. Sapan, M. J. During, M. G. Kaplitt, and A. Feigin. 2011. AAV2-GAD gene therapy for advanced Parkinson's disease: a double-blind, sham-surgery controlled, randomised trial. *Lancet Neurol.* **10**:309–319.

Li, W., A. Asokan, Z. Wu, T. Van Dyke, N. DiPrimio, J. Johnson, L. Govindaswamy, M. Agbandje-McKenna, S. Leichtle, D. Redmond, Jr., T. McCown, K. Petermann, N. Sharpless, and R. Samulski. 2008. Engineering and selection of shuffled AAV genomes: a new strategy for producing targeted biological nanoparticles. *Mol. Ther.* **16**: 1252–1260.

Liu, B. L., M. Robinson, Z. Q. Han, R. H. Branston, C. English, P. Reay, Y. McGrath, S. K. Thomas, M. Thornton, P. Bullock, C. A. Love, and R. S. Coffin. 2003. ICP34.5 deleted herpes simplex virus with enhanced oncolytic, immune stimulating, and anti-tumour properties. *Gene Ther.* **10**:292–303.

Liu, M. M., J. Tuo, and C. C. Chan. 2011. Gene therapy for ocular diseases. *Br. J. Ophthalmol.* **95**:604–612.

Luo, X. R., J. S. Li, Y. Niu, and L. Miao. 2011. Targeted killing effects of double CD and TK suicide genes controlled by survivin promoter on gastric cancer cells. *Mol. Biol. Rep.* **38**:1201–1207.

MacDiarmid, J. A., N. B. Mugridge, J. C. Weiss, L. Phillips, A. L. Burn, R. P. Paulin, J. E. Haasdyk, K.-A. Dickson, V. M. Brahmbhatt, S. T. Pattison, A. C. James, G. Al Bakri, R. C. Straw, B. Stillman, R. M. Grahman, and A. Brahmbhatt. 2007. Bacterially derived 400 nm particles for encapsulation and cancer cell targeting of chemotherapeutics. *Cancer Cell* **11**:431–445.

Maetzig, T., M. Galla, C. Baum, and A. Schambach. 2011. Gammaretroviral vectors: biology, technology and application. *Viruses* **3**:677–713.

Maguire, A. M., F. Simonelli, E. A. Pierce, E. N. Pugh, Jr., F. Mingozzi, J. Bennicelli, S. Banfi, K. A. Marshall, F. Testa, E. M. Surace, S. Rossi, A. Lyubarsky, V. R. Arruda, B. Konkle, E. Stone, J. Sun, J. Jacobs, L. Dell'Osso, R. Hertle, J. X. Ma, T. M. Redmond, X. Zhu, B. Hauck, O. Zelenaia, K. S. Shindler, M. G. Maguire, J. F. Wright, N. J. Volpe, J. W. McDonnell, A. Auricchio, K. A. High, and J. Bennett. 2008. Safety and efficacy of gene transfer for Leber's congenital amaurosis. *N. Engl. J. Med.* **358**:2240–2248.

Manservigi, R., R. Argnani, and P. Marconi. 2010. HSV recombinant vectors for gene therapy. *Open Virol. J.* **4**:123–156.

Marconi, P., R. Manservigi, and A. L. Epstein. 2010. HSV-1-derived helper-independent defective vectors, replicating vectors and amplicon vectors, for the treatment of brain diseases. *Curr. Opin. Drug Discov. Devel.* **13**:169–183.

Matzen, K., L. Elzaouk, A. A. Matskevich, A. Nitzsche, J. Heinrich, and K. Moelling. 2007. RNase H-mediated retrovirus destruction in vivo triggered by oligodeoxynucleotides. *Nat. Biotechnol.* **25**:669–674.

McCarty, D. M. 2008. Self-complementary AAV vectors; advances and applications. *Mol. Ther.* **16**: 1648–1656.

McNamara, J. O., II, E. R. Andrechek, Y. Wang, K. D. Viles, R. E. Rempel, E. Gilboa, B. A. Sullenger, and P. H. Giangrande. 2006. Cell type-specific delivery of siRNAs with aptamer-siRNA chimeras. *Nat. Biotechnol.* **24**:1005–1015.

Mendell, J., L. Rodino-Klapac, X. Rosales-Quintero, J. Kota, B. Coley, G. Galloway, J. Craenen, S. Lewis, V. Malik, C. Shilling, B. Byrne, T. Conlon, K. Campbell, W. Bremer, L. Viollet, C. Walker, Z. Sahenk, and K. Clark. 2009. Limb-girdle muscular dystrophy type 2D gene therapy restores α-sarcoglycan and associated proteins. *Ann. Neurol.* **66**:290–297.

Mendell, J. R., K. Campbell, L. Rodino-Klapac, Z. Sahenk, C. Shilling, S. Lewis, D. Bowles, S. Gray, C. Li, G. Galloway, V. Malik, B. Coley, K. R. Clark, J. Li, X. Xiao, J. Samulski, S. W. McPhee, R. J. Samulski, and C. M. Walker. 2010. Dystrophin immunity in Duchenne's muscular dystrophy. *N. Engl. J. Med.* **363**:1429–1437.

Mendell, J. R., L. R. Rodino-Klapac, X. Q. Rosales, B. D. Coley, G. Galloway, S. Lewis, V. Malik, C. Shilling, B. J. Byrne, T. Conlon, K. J. Campbell, W. G. Bremer, L. E. Taylor, K. M. Flanigan, J. M. Gastier-Foster, C. Astbury, J. Kota, Z. Sahenk, C. M. Walker, and K. R. Clark. 2010. Sustained alpha-sarcoglycan gene expression after gene transfer in limb-girdle muscular dystrophy, type 2D. *Ann. Neurol.* **68:** 629–638.

Minati, L., T. Edginton, M. G. Bruzzone, and G. Giaccone. 2009. Current concepts in Alzheimer's disease: a multidisciplinary review. *Am. J. Alzheimers Dis. Other Demen.* **24:**95–121.

Mingozzi, F., and K. A. High. 2011. Therapeutic in vivo gene transfer for genetic disease using AAV: progress and challenges. *Nat. Rev. Genet.* **12:** 341–355.

Muul, L. M., L. M. Tuschong, S. L. Soenen, G. J. Jagadeesh, W. J. Ramsey, Z. Long, C. S. Carter, E. K. Garabedian, M. Alleyne, M. Brown, W. Bernstein, S. H. Schurman, T. A. Fleisher, S. F. Leitman, C. E. Dunbar, R. M. Blaese, and F. Candotti. 2003. Persistence and expression of the adenosine deaminase gene for 12 years and immune reaction to gene transfer components: long-term results of the first clinical gene therapy trial. *Blood* **101:**2563–2569.

Naldini, L. 2011. Ex vivo gene transfer and correction for cell-based therapies. *Nat. Rev. Genet.* **12:**301–315.

Nathwani, A. C., E. Tuddenham, S. Rangarajan, C. Rosales, J. McIntosh, D. Linch, P. Chowdary, A. Riddell, A. Pie, C. Harrington, J. O'Beirne, K. Smith, J. Pasi, B. Glader, P. Rustagi, C. Ng, M. Kay, J. Zhou, Y. Spence, C. Morton, J. Allay, J. Coleman, S. Sleep, J. Cunningham, D. Srivastava, E. Basner-Tschakarjan, F. Mingozzi, K. High, J. Gray, U. Reiss, A. Nienhuis, and A. Davidoff. 2011. Adenovirus-associated virus vector–mediated gene transfer in hemophilia B. *N. Engl. J. Med.* **365:**2357–2365.

Nimjee, S. M., C. P. Rusconi, and B. A. Sullenger. 2005. Aptamers: an emerging class of therapeutics. *Annu. Rev. Med.* **56:**555–583.

Odom, G. L., P. Gregorevic, J. M. Allen, and J. S. Chamberlain. 2011. Gene therapy of *mdx* mice with large truncated dystrophins generated by recombination using rAAV6. *Mol. Ther.* **19:**36–45.

Oliveira, S., G. Storm, and R. M. Schiffelers. 2006. Targeted delivery of siRNA. *J. Biomed. Biotechnol.* **2006:** 1–9.

O'Rorke, S., M. Keeney and A. Pandit. 2010. Non-viral polyplexes: scaffold mediated delivery for gene therapy. *Prog. Polym. Sci.* **35:**441–458.

Palmer, P. Ng, P. R. Lowenstein, and M. G. Castro. 2010. A novel bicistronic high-capacity gutless adenovirus vector that drives constitutive expression of herpes simplex virus type 1 thymidine kinase and Tet-inducible expression of Flt3L for glioma therapeutics. *J. Virol.* **84:**6007–6017.

Pendergast, P. S., H. N. Marsh, D. Grate, J. M. Healy, and M. Stanton. 2005. Nucleic acid aptamers for target validation and therapeutic applications. *J. Biomol. Technol.* **16:**224–234.

Rossi, J. J. 2004. A cholesterol connection in RNAi. *Nature* **432:**155–156.

Rossi, J. J. 2005. Receptor-targeted siRNAs. *Nat. Biotechnol.* **23:**682–683.

Roth, J. A. (ed.). 2010. *Gene-Based Therapies for Cancer.* Springer, New York, NY.

Schaffer, D. V., J. Koerber, and K. Lim. 2008. Molecular engineering of viral gene delivery vehicles. *Annu. Rev. Biomed. Eng.* **10:**169–194.

Schiffmann, R., and M. S. van der Knaap. 2009. Invited article: an MRI-based approach to the diagnosis of white matter disorders. *Neurology* **72:** 750–759.

Schultz, B. R., and J. S. Chamberlain. 2008. Recombinant adeno-associated virus transduction and integration. *Mol. Ther.* **16:**1189–1199.

Seow, Y., and M. J. Wood. 2009. Biological gene delivery vehicles: beyond viral vectors. *Mol. Ther.* **17:**767–777.

Shahi, S., G. K. Shanmugasundaram, and A. C. Banerjea. 2001. Ribozymes that cleave reovirus genome segment Si also protect cells from pathogenesis caused by reovirus infection. *Proc. Natl. Acad. Sci. USA* **98:**4101–4106.

Shim, M. S., and Y. J. Kwon. 2010. Efficient and targeted delivery of siRNA *in vivo.* *FEBS J.* **277:**4814–4827.

Simons, D. L., S. L. Boye, W. W. Hauswirth, and S. M. Wu. 2011. Gene therapy prevents photoreceptor death and preserves retinal function

in a Bardet-Biedl syndrome mouse model. *Proc. Natl. Acad. Sci. USA* **108:** 6276–6281.

Siolas, D., C. Lerner, J. Burchard, W. Ge, P. S. Linsley, P. J. Paddison, G. J. Hannon, and M. A. Cleary. 2005. Synthetic shRNAs as potent RNAi triggers. *Nat. Biotechnol.* **23:**227–231.

Snøve, O., and J. J. Rossi. 2006. Expressing short hairpin RNAs *in vivo.* *Nat. Methods* **3:**689–695.

Song, E., P. Zhu, S.-K. Lee, D. Chowdhury, S. Kussman, D. M. Dykxhoorn, Y. Feng, D. Palliser, D. B. Weiner, P. Shankar, W. A. Marasco, and J. Lieberman. 2005. Antibody mediated *in vivo* delivery of small interfering RNAs via cell-surface receptors. *Nat. Biotechnol.* **23:**709–717.

Soutschek, J., A. Akinc, B. Bramlage, K. Charisse, R. Constien, M. Donoghue, S. Elbashir, A. Geick, P. Hadwiger, J. Harborth, M. John, V. Kesavan, G. Lavine, R. K. Pandey, T. Racie, K. G. Rajeev, I. Röhl, I. Toudjarska, G. Wang, S. Wuschko, D. Bumcrot, V. Koteliansky, S. Limmer, M. Manoharan, and H.-P. Vornlocher. 2004. Therapeutic silencing of an endogenous gene by systematic administration of modified siRNAs. *Nature* **432:**173–178.

Springer, C. J. 2004. Introduction to vectors for suicide gene therapy. *Methods Mol. Med.* **90:**29–45.

Thathiah, A., and B. De Strooper. 2009. G protein-coupled receptors, cholinergic dysfunction, and Abeta toxicity in Alzheimer's disease. *Sci. Signal.* **2:**re8.

Toscano, M. G., Z. Romero, P. Munoz, M. Cobo, K. Benabdellah, and F. Martin. 2011. Physiological and tissue-specific vectors for treatment of inherited diseases. *Gene Ther.* **18:**117–127.

Trepel, M., C. A. Stoneham, H. Eleftherohorinou, N. D. Mazarakis, R. Pasqualini, W. Arap, and A. Hajitou. 2009. A heterotypic bystander effect for tumor cell killing after adeno-associated virus/phage-mediated, vascular-targeted suicide gene transfer. *Mol. Cancer Ther.* **8:**2383–2391.

Tuszynski, M. H., L. Thal, M. Pay, D. P. Salmon, H. S. U, R. Bakay, P. Patel, A. Blesch, H. L. Vahlsing, G. Ho, G. Tong, S. G. Potkin, J. Fallon, L. Hansen, E. J. Mufson, J. H. Kordower, C. Gall, and J. Conner. 2005. A phase 1 clinical trial of nerve growth factor gene therapy for Alzheimer disease. *Nat. Med.* **11:** 551–555.

Vanden Driessche, T., Z. Ivics, Z. Izsvák, and M. K. L. Chuah. 2009. Emerging potential of transposons for gene therapy and generation of induced pluripotent stem cells. *Blood* **114:** 1461–1468.

Van Rij, R. P., and R. Andino. 2006. The silent treatment: RNAi as a defense against virus infection in mammals. *Trends Biotechnol.* **24:**186–193.

Volpers, C., and S. Kochanek. 2004. Adenoviral vectors for gene transfer and therapy. *J. Gene Med.* **6**(Suppl. 1): S164–S171.

Wang, B., J. Li, and X. Xiao. 2000. Adeno-associated virus vector carrying human minidystrophin genes effectively ameliorates muscular dystrophy in mdx mouse model. *Proc. Natl. Acad. Sci. USA* **97:**13714–13719.

Wang, S., P. Liu, L. Song, L. Lu, W. Zhang, and Y. Wu. 2011. Adeno-associated virus (AAV) based gene therapy for eye diseases. *Cell Tissue Bank.* **12:**105–110.

Wang, Z., J. S. Chamberlain, S. J. Tapscott, and R. Storb. 2009. Gene therapy in large animal models of muscular dystrophy. *ILAR J.* **50:**187–198.

Wolfrum, C., S. Shi, N. K. Jayaprakash, M. Jayaraman, G. Wang, R. J. Pandey, K. G. Rajeev, T. Nakayama, K. Charrise, E. M. Ndungo, T. Zimmermann, V. Koteliansky, M. Manoharan, and M. Stoffel. 2007. Mechanisms and optimization of in vivo delivery of lipophilic siRNAs. *Nat. Biotechnol.* **25:** 1149–1157.

Wu, Z., A. Asokan, and R. J. Samulski. 2006. Adeno-associated virus serotypes: vector toolkit for human gene therapy. *Mol. Ther.* **14:**316–327.

Xiang, S., J. Fruehauf, and C. J. Li. 2006. Short hairpin RNA-expressing bacteria elicit RNA interference in mammals. *Nat. Biotechnol.* **24:** 697–702.

Yamamoto, M., and D. T. Curiel. 2009. Current issues and future directions of oncolytic adenoviruses. *Mol. Ther.* **18:** 243–250.

Yin, Y., F. Lin, Q. Zhuang, L. Liu, and C. Qian. 2009. Generation of full-length functional antibody against preS2 of hepatitis B virus in hepatic cells in vitro from bicistrons mediated by gutless adenovirus. *BioDrugs* **23:** 391–397.

Zhou, S., D. Mody, S. S. DeRavin, J. Hauer, T. Lu, Z. Ma, S. Hacein-Bey Abina, J. T. Gray, M. R. Greene, M. Cavazzana-Calvo, H. L. Malech, and B. P. Sorrentino. 2010. A self-inactivating lentiviral vector for SCID-X1 gene therapy that does not activate LMO2 expression in human T cells. *Blood* **116:**900–908.

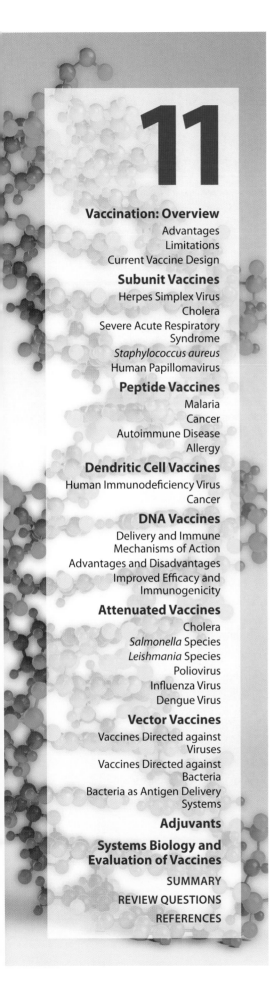

11

Vaccines

Vaccination: Overview

It has become increasingly evident that our understanding of how to stimulate an optimum immune response to combat a specific pathogen that induces an infectious disease, whether the parameters of such a response are shared by all pathogens, and if they can be induced by a vaccine at the optimum time and location will require novel rational design and approaches. This chapter presents the advantages, limitations, and future challenges of the use of various types of vaccines to protect against a host of many different types of existing and emerging bacterial, viral, and parasitic infectious diseases, autoimmune diseases, and cancers in the 21st century.

Advantages

The attainment and maintenance of health, a major goal of our society, provide a high quality of life in high-income countries and an improved quality of life in poor countries. Much recent progress in raising the standard of health care has resulted from the significant contribution of vaccination to the prevention of infectious diseases. Vaccines have revolutionized health care during the 20th century. They have provided one of the most effective interventions in modern medicine and have eliminated most of the childhood diseases that previously caused millions of deaths. Two years ago, Bill Gates said that "vaccines are one of the best investments we can make in the future because healthy people can drive thriving economies."

Beginning with Edward Jenner's first use of a vaccine against smallpox in 1796, vaccines have become the sine qua non choice for the eradication of disease (see chapter 5). While smallpox claimed about 375 million lives in the 20th century, it is noteworthy that no deaths from smallpox have been registered since a successful eradication campaign in 1978. Today, a large number of people are alive and well due to the more than 70 vaccines

doi:10.1128/9781555818890.ch11

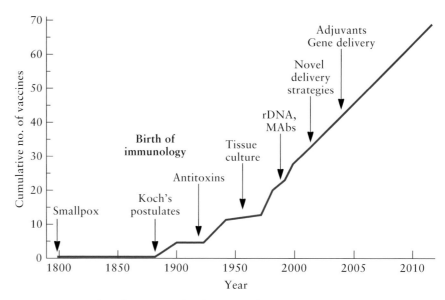

Figure 11.1 Timeline for vaccine development. Shown are major milestones and advances in vaccine development since Edward Jenner's first use of vaccination against smallpox in 1796, including the birth of immunology in the 1880s. rDNA, recombinant DNA. Adapted from Nabel, *N. Engl. J. Med.* **368**:551–560, 2013, with permission from the Massachusetts Medical Society (original figure copyright © 2013 Massachusetts Medical Society). doi:10.1128/9781555818890.ch11f.11.1

now licensed for use against approximately 30 microbes (Fig. 11.1 and 11.2). Diseases including diphtheria, measles, mumps, poliomyelitis, rubella, smallpox, and others gave rise to more than 39 million infections in the 20th century in Canada and the United States. Despite this alarming number of infections, vaccines have since reduced considerably the frequencies of these diseases (Table 11.1). This public health intervention owes its success not only to the identification of effective vaccines but also to an efficient infrastructure for vaccine production, regulatory and safety procedures, and vaccine delivery. Vaccines are the most economical and beneficial way to protect against serious epidemics. Society benefits

Table 11.1 Comparison of estimated number of cases with infectious diseases in the United States in the 20th century pre- and post-vaccine usage

Disease	No. of prevaccine cases in the 20th century (millions)	No. of deaths in 2002
Diphtheria	17.6	2
Haemophilus influenzae type b	2.00	22
Measles	5.03	36
Mumps	1.52	236
Pertussis	1.47	6,632
Poliomyelitis	1.63	0
Rubella	4.77	20
Smallpox	4.81	0
Tetanus	0.13	13

Data are from the Centers for Disease Control and Prevention, *MMWR Morb. Mortal. Wkly. Rep.* **48**: 243–248, 1999, and Roush and Murphy, *JAMA* **298**:2155–2163, 2007.

The vaccines used were against the causative agents of the diseases.

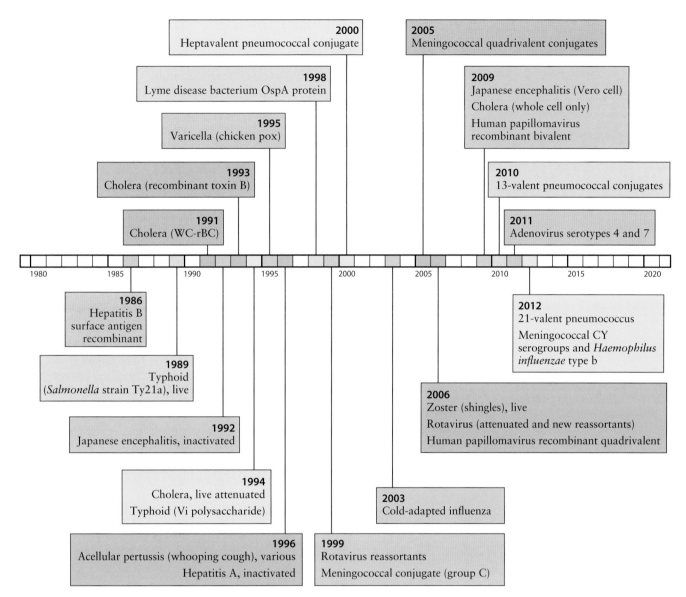

Figure 11.2 Timeline for commercial vaccines licensed from 1980 to 2012 against the indicated pathogens. OspA, outer surface protein A; WC, whole-cell *Vibrio cholerae* O1; rBC, recombinant B subunit of cholera toxin. Adapted from Nabel, *N. Engl. J. Med.* **368**:551–560, 2013, with permission from the Massachusetts Medical Society (original figure copyright © 2013 Massachusetts Medical Society). doi:10.1128/9781555818890.ch11f.11.2

economically by reduced hospitalization, avoidance of long-term disability, and diminished absence from work. Thus, vaccines provide the most cost-effective means to save lives, preserve good health, and maintain a high quality of life.

Limitations

Infectious diseases still have a negative impact on too many human lives. This is partially due to the fact that vaccines have not yet realized their full potential. Effective vaccines are often not available in developing

Table 11.2 Some infections for which vaccines are not yet available

Disease	Estimated no. of annual deaths worldwide
Malaria	1,272,000
Schistosomiasis	15,000
Intestinal worm manifestation	12,000
Tuberculosis	1,566,000
Diarrheal disease	1,798,000
Respiratory infections	3,963,000
HIV/AIDS	2,777,000
Measles	611,000

The estimated mortality data shown for 2002 were obtained from Murray et al., p. 1–38, in Ezzati et al. (ed.), *Comparative Quantification of Health Risks: Global and Regional Burden of Disease Attributable to Selected Major Risk Factors* (World Health Organization, Geneva, Switzerland, 2004).

Note that current measles vaccines are effective but are heat sensitive, which makes their use difficult in tropical countries.

HIV
human immunodeficiency virus

countries. The Global Alliance for Vaccines and Immunization estimates that more than 1.5 million children die every year from vaccine-preventable diseases. Second, effective vaccines are also not yet available for many leading diseases, including human immunodeficiency virus (HIV)/AIDS, tuberculosis, and malaria, which incur more than 5 million deaths globally each year (Table 11.2). Natural immunity to infection and protective immune responses have been reported for most vaccines licensed since 1990 (Fig. 11.2). However, it has proven difficult thus far to show preventative immunity for vaccines used to treat HIV infection, tuberculosis, and malaria.

Table 11.3 Licensure of commercial vaccines between 1920 and 1990

Vaccine(s)	Year
Diphtheria toxoid	1923
Pertussis	1926
Tetanus toxoid	1923
Tuberculosis (BCG)	1927
Yellow fever	1935
Influenza	1936
Typhus	1938
Polio (injected inactivated)	1955
Polio (oral live), measles (live)	1963
Mumps (live)	1967
Rubella (live)	1969
Anthrax (secreted proteins)	1970
Meningococcal polysaccharides	1974
Pneumococcus polysaccharides	1977
Adenovirus (live), rabies (cell culture)	1980
Hepatitis B (plasma derived), tick-borne encephalitis	1981
H. influenzae type b polysaccharide	1985
Hepatitis B surface antigen recombinant	1986
H. influenzae conjugate	1987
Typhoid (salmonella strain Ty21a, live)	1989

Data were obtained from Nabel, *N. Engl. J. Med.* **368**:551–560, 2013. Adapted with permission from the Massachusetts Medical Society (original figure copyright © 2013 Massachusetts Medical Society).

The year of licensure of several vaccines for various pathogens during the period from 1970 to 1990 is shown.

Additional drawbacks in vaccine development arise from many vaccine technologies being outdated and not well suited to rapidly address emerging outbreaks. For example, the production of influenza vaccines relies on technology developed 50 years ago (growth in chicken embryo eggs) and the use of *Haemophilus influenzae* strains licensed more than 20 years ago (Table 11.3). Current seasonal influenza vaccines may not be well matched or effective against circulating viral strains. Moreover, the unanticipated emergence of new strains of influenza virus, e.g., from an animal reservoir in the 2009 influenza A (H1N1) pandemic, prohibited vaccine developers from rapidly developing a new vaccine strain. Thus, while much success has resulted from the synthesis of many vaccines in the 20th century, many challenges remain for the production of efficacious vaccines in the 21st century. Today, more than 2 billion humans suffer from diseases that theoretically could be reduced significantly by vaccination. In addition, new diseases for which vaccines might be useful continue to emerge.

Current vaccines typically consist of either an **inactivated** (killed) (Fig. 11.3A) or an **attenuated** (live, nonvirulent) pathogen (Fig. 11.3B). Traditionally, the pathogen is grown in culture, purified, and either

Figure 11.3 Inactivated and attenuated forms of a vaccine. The methods of production of inactivated and attenuated forms of a vaccine are shown. The main difference between these forms is that while an inactivated vaccine stimulates immunity and cannot multiply, a live attenuated vaccine can both stimulate immunity and also multiply. doi:10.1128/9781555818890.ch11f.11.3

A   Heat-killed or formalin-inactivated vaccine

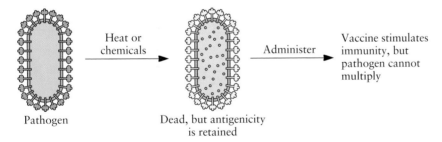

B   Live attenuated vaccine

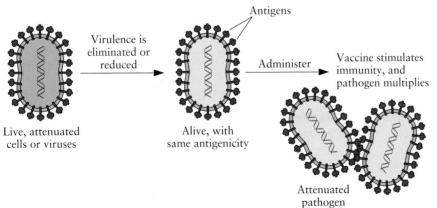

box 11.1
Limitations to the Current Mode of Vaccine Production

- As not all infectious agents can be grown in culture, vaccines have not been developed for several diseases.
- Production of animal and human viruses requires animal cell culture, which is expensive.
- Both the yield and rate of production of animal and human viruses in culture are often quite low, making vaccine production costly.

- Extensive safety precautions are necessary to ensure that laboratory and production personnel are not exposed to a pathogenic agent.
- Batches of vaccine may not be killed or may be insufficiently attenuated during the production process, thereby introducing virulent organisms into the vaccine and inadvertent spread of the disease.

- Attenuated strains may revert, a possibility that requires continual testing to ensure that the reacquisition of virulence has not occurred.
- Not all diseases (e.g., AIDS) are preventable by the use of traditional vaccines.
- Most current vaccines have a limited shelf life and often require refrigeration to maintain potency. This requirement creates storage problems in countries with rural areas not equipped with electrical wires and facilities.

inactivated or attenuated without losing its ability to evoke an immune response that is effective against the virulent form of the pathogen. Although considerable success has been achieved in creating effective vaccines against many infectious diseases (Table 11.1), there are several limitations to the current mode of vaccine production. Adaptation to growth in culture, requirement for large-scale growth, production and yield, relevant safety issues, insufficient attenuation and introduction of virulent organisms, attenuation reversion, limited shelf life, and need for refrigeration pose some of the limitations associated with viral growth and vaccine production (Box 11.1).

Current Vaccine Design

During the last two decades, recombinant DNA technology has provided a means of creating a new generation of vaccines that overcome the drawbacks of traditional vaccines. The availability of gene cloning has enabled researchers to contemplate various novel strategies for vaccine development, as summarized in Box 11.2.

Because of less stringent regulatory requirements, the first vaccines that were produced by recombinant DNA techniques were for animal diseases, such as foot-and-mouth disease, rabies, and scours, a diarrheal disease of pigs and cattle. In addition, many more animal vaccines are currently being developed. For human diseases, a large number of recombinant vaccines are currently in various stages of development, including clinical trials (Table 11.4).

In addition to the use of recombinant DNA technology, the following observations may facilitate our goal to construct new and improved vaccines for the 21st century. By 6 years of age, children in Canada and the United States receive 10 different vaccines that protect against 14 different viral and bacterial diseases. Nine of these vaccines are delivered intramuscularly into an arm or leg. Several contain attenuated (live, nonvirulent)

box 11.2
Application of Gene Cloning Technology
to the Development of Novel Vaccines

Virulence genes may be deleted from a pathogen, with the proviso that the modified pathogen can still stimulate a vigorous immune response. Thus, the genetically modified pathogen may be used as a live vaccine without concern about reversion to virulence, as it is not possible for a whole gene to be reacquired spontaneously during growth in culture.

Live nonpathogenic carrier systems that carry discrete antigenic determinants of an unrelated pathogen may be created. In this form, the carrier system facilitates the induction of a strong immune response directed against the pathogen.

For pathogens that cannot be maintained in culture, the genes for the proteins that have critical antigenic determinants can be isolated, cloned, and expressed in an alternative host system, such as *Escherichia coli* or a mammalian cell line. These cloned gene proteins can be formulated into a vaccine.

Some pathogens do not damage host cells directly. Rather, disease results when the host immune system attacks its own (infected) cells. For such a disease, it may be possible to induce the activity of pathogen-specific targeted CTLs. Although not a true vaccine, these CTLs attack only infected cells, thereby removing the source of the adverse immune response. In these cases, the gene for a fusion protein is typically constructed. After one part of this fusion protein binds to an infected cell, the other part kills the infected cell.

or inactivated (killed) viruses, whereas others are composed of bacterial sugars, virus-like particles, or purified proteins. Only some vaccines may be formulated with aluminum-based **adjuvants** (agents that stimulate immunity). Despite these differences, upon injection they all induce protective immunity against an array of pathogens that target different tissues and use different strategies of immune evasion.

Taken together, the above-mentioned findings suggest that these various vaccines may use common mechanisms to induce protection and that these mechanisms may be identified and applied to the development of future vaccines. However, while significant advances have been made to rationally develop vaccines by optimization of a few factors, including delivery systems, size of particulate vaccines, targeting of antigen-presenting dendritic cells, and addition of components (T-cell epitopes) to improve vaccine efficacy, the following problems must still be solved to successfully develop these vaccines. Systems biology approaches have identified molecular signatures of the immune responses to influenza and yellow fever vaccines, but these signatures do not correlate with protection conferred by the vaccines. Preclinical studies of new adjuvants have revealed novel insight into their stimulatory effects on innate immune cells, although for many of these adjuvants, the key pathways by which they induce protective immunity remain unknown. One can now predict T-cell receptor (TCR) and B-cell receptor epitopes and their immunogenicity, but the accuracy of these methods still requires improvement for routine application. Moreover, the recent experiences with the malaria and dengue vaccines show that predicting immunogenicity is not equivalent to predicting protection. Different pathogens have different degrees of antigenicity, mutation rates, and mechanisms to bypass protective immunity. A given pathogenic infection might be blocked if the right magnitude, breadth, and potency of immune response are rapidly recruited upon

TCR
T-cell receptor

Table 11.4 Human disease agents for which recombinant vaccines are currently being developed

Pathogenic agent	Disease
Viruses	
Varicella-zoster viruses	Chicken pox
Cytomegalovirus	Infection in infants and immunocompromised patients
Dengue virus	Hemorrhagic fever
Hepatitis A virus	High fever, liver damage
Hepatitis B virus	Long-term liver damage
HSV-2	Genital ulcers
Influenza A and B viruses	Acute respiratory disease
Japanese encephalitis	Encephalitis
Parainfluenza virus	Inflammation of the upper respiratory tract
Rabies virus	Encephalitis
Respiratory syncytial virus	Upper and lower respiratory tract lesions
Rotavirus	Acute infantile gastroenteritis
Yellow fever virus	Lesions of heart, kidney, and liver
HIV	AIDS
Bacteria	
Vibrio cholerae	Cholera
E. coli enterotoxin strains	Diarrheal disease
Neisseria gonorrhoeae	Gonorrhea
Haemophilus influenzae	Meningitis, septicemic conditions
Mycobacterium leprae	Leprosy
Neisseria meningitidis	Meningitis
Bordetella pertussis	Whooping cough
Shigella strains	Dysentery
Streptococcus group A	Scarlet fever, rheumatic fever, throat infection
Streptococcus group B	Sepsis, urogenital tract infection
Streptococcus pneumoniae	Pneumonia, meningitis
Clostridium tetani	Tetanus
Mycobacterium tuberculosis	Tuberculosis
Salmonella enterica serovar Typhi	Typhoid fever
Parasites	
Onchocerca volvulus	River blindness
Leishmania spp.	Internal and external lesions
Plasmodium spp.	Malaria
Schistosoma mansoni	Schistosomiasis
Trypanosoma spp.	Sleeping sickness
Wuchereria bancrofti	Filariasis

initial infection. The objective is to achieve active immunity, which refers to antibody and/or cell-mediated immune responses elicited by administration of the vaccine (see chapters 2 and 4). Both active immunity and immune memory (see chapters 2 and 4) induced by a vaccine resemble the type of immunity observed upon natural infection but without risk of disease. The various strategies currently used for the development of different types of vaccines and immunity (Table 11.5) address how these problems may be overcome. It remains to be determined whether some or all of the requirements for an effective vaccine can be met (Box 11.3).

Table 11.5 Vaccine strategies

Type of vaccine	Example(s)	Form of protection
Subunit (antigen) vaccines	Tetanus toxoid, diphtheria toxoid	Antibody response
Conjugate (peptide) vaccines	*Haemophilus influenzae* infection	Th cell-dependent antibody response
DNA vaccines	Clinical trials ongoing for several infections	Antibody and cell-mediated immune responses
Live attenuated or killed bacteria	BCG, cholera	Antibody response
Live attenuated viruses	Polio, rabies	Antibody and cell-mediated immune responses
Vector vaccines (viruses, bacteria)	Clinical trials of HIV and vaccinia virus and tuberculosis bacteria	Antibody and cell-mediated immune responses

Examples of different types of vaccines are provided, and the nature of the protective immune responses induced by these vaccines is summarized. Toxoid is a modified bacterial toxin that is altered so that it is nontoxic but still retains the ability to stimulate the formation of protective antibodies.

Subunit Vaccines

Traditional vaccines generally consist of either inactivated or attenuated forms of a pathogen. The antibodies elicited by these vaccines stimulate an immune response to inactivate (neutralize) pathogens by binding to proteins on the outer surface of the pathogen. For disease-causing viruses, purified outer surface viral proteins, either capsid or envelope proteins (Fig. 11.4), are often sufficient for eliciting neutralizing antibodies in the host organism. Vaccines that use components of a pathogen rather than the whole pathogen are called **subunit vaccines**.

Subunit vaccines offer the potential to develop safe and highly characterized vaccines that target immune responses toward specific epitopes. Since they are composed of individual components (e.g., antigens, adjuvants, delivery systems, and targeting moieties), they enable the production of custom vaccines beyond that provided by traditional whole-attenuated-pathogen approaches. Using recombinant DNA technology, the generation of libraries of different vaccine components permits the selection of vaccine constituents to stimulate immune responses

---

box 11.3
Requirements for an Effective Vaccine

Safe and protective
- The vaccine must not elicit illness or death, and it must protect against illness resulting from exposure to live pathogen. Protection against illness must be sustained for several years (T- and B-cell-mediated immune memory).

Nature of infectious pathogen
- Extracellular: must induce protective neutralizing antibodies
- Intracellular: must induce a protective CD8$^+$ CTL response

Host defense at point of entry of infectious agent
- Mucosal immunity: an important goal of vaccination against many organisms that enter through mucosal surfaces (e.g., oral and nasal)

Preexisting antibodies at the time of exposure to the infection
- Antibodies against extracellular pathogens (diphtheria and tetanus exotoxins) may not protect against infection by these exotoxins and may require vaccination for protection.

- Antibodies to an intracellular pathogen (poliovirus) may not be protective and may require vaccination to activate T cells for protection.

Antibodies and T cells directed against the correct epitopes
- Antibodies against many epitopes may be elicited, but only some of these epitopes may confer protection.

- T-cell epitopes recognized can also affect the nature of the response.

Practical considerations
- Cost-effective (low cost per dose); biologically stable; ease of administration, few (or no) adverse side effects

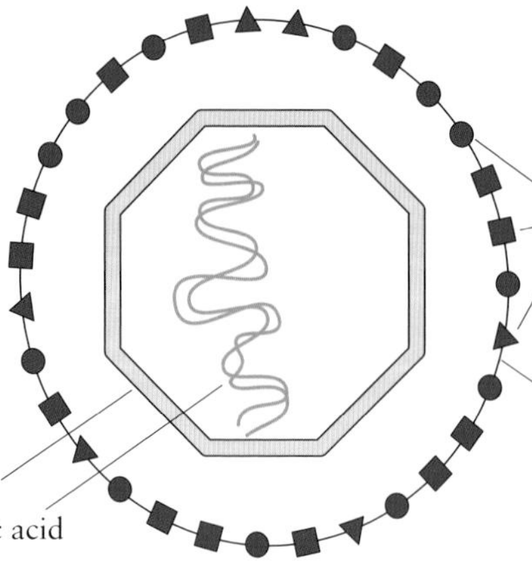

Figure 11.4 Schematic representation of an animal virus. Viruses consist of a small nucleic acid genome (3 to 200 kb of either double- or single-stranded DNA or RNA) present in a viral protein capsid. Depending on the virus, the capsid may be surrounded by a protein-containing viral envelope (membrane). doi:10.1128/9781555818890.ch11f.11.4

CTL
cytotoxic T lymphocyte

that yield protective immunity against a specific pathogen. In addition to the identification of antigens and their components that can elicit protective or therapeutic immunity, the use of appropriate adjuvants that stimulate potent, antigen-specific immune responses is equally important. Although very few adjuvants are currently used in licensed vaccine formulations, several have been assessed in preclinical studies or are currently undergoing human vaccine trials. Adjuvants that enable the generation of potent cytotoxic T lymphocyte (CTL) (see chapter 4) responses are of particular interest and are essential for the development of therapeutic vaccines. The development of new subunit vaccine components is a rather important area of investigation upon which future vaccine development will be based.

Nonetheless, there are advantages and disadvantages to the use of subunit vaccines (Box 11.3). The advantages include the use of a purified protein(s) as an immunogen that ensures that the preparation is stable and safe, is precisely defined chemically (improving lot-to-lot consistency), is free of extraneous proteins and nucleic acids that may initiate undesirable side effects in the host organism, enables incorporation of unnatural components, and can be freeze-dried (permitting nonrefrigerated transport and storage). The disadvantages include the high cost of protein purification and the fact that an isolated protein or carbohydrate antigen (e.g., capsid or envelope derived) may differ in its conformation from that achieved in situ and thereby reduce the immunity directed towards this antigen. Thus, the decision to produce a subunit vaccine depends on an assessment of several biological and economic factors. The current emphasis and future direction of subunit vaccine development are discussed below, with a focus on the described components and their potential to stimulate a desirable and vigorous immune response.

Herpes Simplex Virus

HSV
herpes simplex virus

Herpes simplex virus (HSV) has been implicated as a cancer-causing (oncogenic) agent, in addition to its more common roles in causing sexually transmitted disease, severe eye infections, and encephalitis. Therefore, prevention of HSV infection by vaccination with either killed or attenuated virus may put the recipient at risk for cancer. Current evidence suggests that protection from HSV infection may be best achieved by treatment with a subunit vaccine that is not oncogenic.

The primary requirement for creating any subunit vaccine is identification of the component(s) of the infectious agent that elicits antibodies that react against the intact form of the infectious agent. HSV-1 envelope glycoprotein D (gD) is such a component, because after injection into mice, it elicits antibodies that neutralize intact HSV. The HSV-1 gD gene was isolated and then cloned into a mammalian expression vector and expressed in **Chinese hamster ovary (CHO)** cells (Fig. 11.5), which, unlike the *Escherichia coli* system, properly glycosylate foreign eukaryotic proteins. The complete sequence of the gD gene encodes a protein that becomes bound to the mammalian host cell membrane (Fig. 11.6A). However, since a membrane-bound protein is much more difficult to purify than a soluble one, the gD gene was modified by removing the nucleotides encoding the C-terminal transmembrane-binding domain (Fig. 11.6B). The modified gene was then transformed into CHO cells, where the product was glycosylated and secreted into the external medium (Fig. 11.5). In laboratory trials, the modified form of gD was effective against both HSV-1 and HSV-2 infections.

However, the successful prevention of HSV infection observed in animal models has not yet been achieved in many clinical trials of HSV-1 or HSV-2 whole or subunit vaccines. Before designing more powerful treatments to treat HSV infection, it is important to identify (i) mechanisms of nonprotective immunity associated with natural infection, (ii) major effectors of immunity that control the acute and latent phases of herpesvirus infection, (iii) immune evasion strategies employed by HSV-1 and

gD
glycoprotein D

CHO
Chinese hamster ovary

Figure 11.5 Schematic representation of the development of a subunit vaccine against HSV. The isolated HSV gD gene is used to transfect CHO cells, after which the cells are grown in culture to produce HSV gD. Mice injected with the purified HSV gD are protected against infection by HSV. doi:10.1128/9781555818890.ch11f.11.5

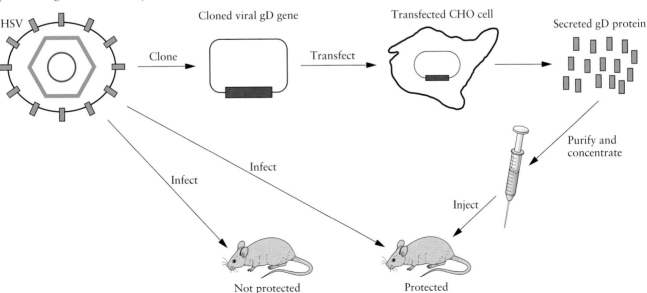

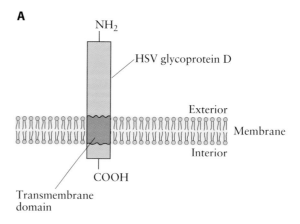

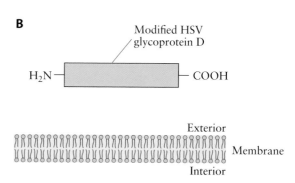

Figure 11.6 (**A**) Location of HDV-1 gD with the transmembrane domain in the envelope. (**B**) Extracellular location of soluble gD protein without the transmembrane domain. doi:10.1128/9781555818890.ch11f.11.6

HSV-2 to dampen the immune response, (iv) protective versus pathogenic protein (such as glycoprotein K) antigens among the more than 80 immunogenic HSV proteins, and (v) a safe antigen delivery system.

Cholera

Cholera is a major global public health problem and remains an important threat in developing countries, particularly in areas of population overcrowding and poor sanitation. Cholera can lead to death within 24 h if left untreated. Without treatment, severe infection has a mortality rate of 30 to 50%. In 2007, the **World Health Organization (WHO)** recorded 177,963 cholera cases and 4,031 deaths worldwide. However, the estimated actual burden of cholera is about 3 to 5 million cases and 100,000 to 130,000 deaths per year. The disease is endemic to parts of Africa, Asia, the Middle East, and South America.

    The causative agent of cholera, the bacterium *Vibrio cholerae*, colonizes the small intestine and secretes large amounts of the pathogenic hexameric enterotoxin. This toxin consists of one subunit, the A subunit, that has ADP-ribosylation activity and stimulates adenylate cyclase and five identical B subunits that bind specifically to an intestinal mucosal cell receptor (see Fig. 7.13A in chapter 7). The A subunit has two functional

WHO
World Health Organization

domains: the A1 peptide, which contains the toxic activity, and the A2 peptide, which joins the A and B subunits. Until recently, a traditional cholera vaccine consisting of phenol-killed *V. cholerae* was commonly used. This vaccine generated only moderate protection, typically lasting only 3 to 6 months. More recently, a combination vaccine (Dukoral) consisting of the heat-inactivated *V. cholerae* Inaba classic strain, heat-inactivated Ogawa classic strain, formalin-inactivated Inaba El Tor strain, formalin-inactivated Ogawa classic strain, and a recombinant cholera toxin B subunit has been used. The vaccine is taken orally (two doses 1 to 2 weeks apart), and recent evidence indicates that this vaccine is immunogenic, as it elicits high-titer immunoglobulin G (IgG) and IgA antibodies in children. The IgA antibodies found in the gut after oral immunization inhibit bacterial colonization and the binding of the toxins to intestinal epithelial cells. An additional booster immunization is not required until about 3 years later.

IgG
immunoglobulin G

An important feature of the combination cholera vaccine is its ability to stimulate **herd immunity** (protection against spread of infection in a population based on a critical mass of successfully vaccinated individuals). This is a hallmark of orally or nasally administered mucosal vaccines, as pathogen-specific IgA antibodies prevent infection and therefore reduce the virulence and spread of infection. Furthermore, even if the efficacy of a mucosal vaccine is lower than that of most vaccines administered systemically, it might effectively prevent the spread of infection in a given population. It follows that a better understanding of how IgA-dependent herd immunity can be exploited may result in the enhanced prevention of mucosal infections.

Severe Acute Respiratory Syndrome

The first new infectious disease of the 21st century was identified as **severe acute respiratory syndrome (SARS)**, and the first case of SARS was reported in the Guangdong province of China in November 2002. In 2003, there were simultaneous outbreaks of SARS in several major cities, including Hong Kong, Singapore, and Toronto, Canada. Given the high frequency of air travel, the disease rapidly spread to 29 countries on five continents. With the assistance of the WHO, authorities in affected regions immediately implemented strict infection control procedures, so that by 23 September 2003, the outbreak was effectively contained. However, this was not before a total of 8,096 SARS cases and 774 associated deaths were reported. Within a very short time, scientists had identified a novel coronavirus, SARS coronavirus (SARS-CoV), as the causative agent of the disease.

SARS
severe acute respiratory syndrome

SARS-CoV
SARS coronavirus

The overall fatality rates of SARS are currently about 10% in the general population and more than 50% in patients 65 years of age and older. The most recent epidemic of SARS occurred in Beijing and Anhui in China in April 2004 and originated from laboratory contamination. Since then, not one new case of SARS has been reported, possibly because of continued global vigilance and surveillance and laboratory biosafety practices, as well as the euthanizing or quarantining of animals that may have been exposed to SARS-CoV. Although the outbreaks of SARS seem to have terminated, SARS remains a safety concern because of the possible reintroduction of a SARS-like CoV into humans and the risk of an escape of SARS-CoV from laboratories.

SARS-CoV is enveloped and contains a single-stranded plus-sense RNA genome of about 30 kilobases (kb). The viral spike (S) protein, which is inserted into the viral membrane, binds to a receptor protein that is present on the surfaces of mammalian host cells (Fig. 11.7). After binding of the virus to the receptor, the viral and cell membranes can fuse and facilitate entry of the virus into the cell.

The spikes of SARS-CoV are composed of trimers of S protein, which belongs to a group of class I viral fusion glycoproteins that also includes HIV glycoprotein 160 (Env), influenza virus hemagglutinin (HA), paramyxovirus F, and Ebola virus glycoprotein. The SARS-CoV S protein encodes a surface glycoprotein precursor predicted to be 1,255 amino acids in length, and the amino terminus and most of the protein are predicted to be on the outside of the cell surface or the virus particles. The predicted S protein consists of a signal peptide (amino acids 1 to 12) located at the N terminus, an extracellular domain (amino acids 13 to 1195), a transmembrane domain (amino acids 1196 to 1215), and an intracellular domain (amino acids 1216 to 1255). The S protein is composed of two subunits; the S1 subunit contains a receptor-binding domain that engages with the host cell receptor angiotensin-converting enzyme 2, and the S2 subunit mediates fusion between the viral and host cell membranes. The S protein is essential to induce neutralizing antibody and T-cell responses, as well as protective immunity, during infection with SARS-CoV. Because the S protein mediates receptor recognition as well as virus attachment and entry, it is an important target for the development of SARS vaccines and therapeutics.

Thus, the extracellular domain of the S protein is an attractive candidate for the development of a subunit vaccine. In particular, amino acids 318 to 510 bind efficiently to the host cell receptor protein. Determination of the whole nucleotide sequence of the SARS virus in 2003 was soon followed by the expression of a codon-optimized version of this 192-amino-acid peptide in CHO cells. In addition to encoding the

S protein
spike protein

HA
hemagglutinin

Figure 11.7 Schematic representation of the binding of the SARS virus glycosylated S protein to its cellular outer surface protein receptor angiotensin-converting enzyme 2 (ACE2). doi:10.1128/9781555818890.ch11f.11.7

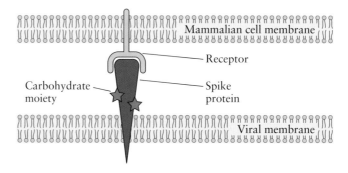

Figure 11.8 Characteristics of part of the recombinant plasmid construct used to transfect CHO cells and to produce the fragment of the SARS virus S protein that interacts with the host cell receptor. doi:10.1128/9781555818890.ch11f.11.8

192-amino-acid S peptide, the DNA construct introduced into the CHO cells also included a mammalian secretion signal, an N-terminal (*Staphylococcus aureus*) protein A purification tag, and a tobacco etch virus protease cleavage site (Fig. 11.8). The recombinant protein synthesized in CHO cells was secreted into the growth medium, purified by immunoaffinity chromatography on an IgG column, and then digested with tobacco etch virus protease to remove the protein A purification tag. Using this construct, the S protein fragment was synthesized and purified. To date, the fully glycosylated form of this subunit vaccine candidate has been shown to elicit a strong immune response in mice and rabbits. Moreover, it was recently shown that this vaccine candidate can protect immunized animals against infection with the SARS virus.

Despite the success with current vaccine candidates in the effective neutralization of SARS-CoV in young-animal replication models without clinical symptoms, it is important to note that they may not protect an elderly human population against SARS-CoV infection. Thus, it is essential to test the vaccine candidates in robust lethal-challenge models using aged animals. Future vaccines should effectively protect both the young and the elderly populations from infection by either human or animal SARS-CoV strains that may cause future SARS epidemics. However, clinical trials of such candidate vaccines are difficult to initiate due to a lack of SARS-CoV-infected subjects and insufficient financial support by the large pharmaceutical companies.

Staphylococcus aureus

The gram-positive bacterium *Staphylococcus aureus* colonizes the skin or mucous membranes of about 30% of the human population. It is a major cause of localized and invasive infections that occur in skin and soft tissue, muscle and liver abscesses, septic arthritis, osteomyelitis, pneumonia, pleural pus, bloodstream infections (about 88% of all *S. aureus* infections), endocarditis, and toxin-mediated syndromes such as toxic shock syndrome and food poisoning. *S. aureus* also elicits many health care-associated surgical site infections. **Methicillin-resistant *S. aureus* (MRSA)** strains have emerged and spread rapidly in the community and are now the most common cause of community-associated skin and soft tissue infections. MRSA-induced infections have fewer effective treatment options. The morbidity and mortality rates of patients with health care-associated MRSA bloodstream infections are about twice those of patients with infections caused by methicillin-susceptible strains. The frequently used antistaphylococcal agents (e.g., vancomycin, linezolid, and daptomycin) are

MRSA
methicillin-resistant
Staphylococcus aureus (MRSA)

not very effective in treating MRSA infections, and 12 strains of *S. aureus* are fully resistant to vancomycin. Thus, in view of the many limitations in the availability of effective therapy for serious *S. aureus* infections, the discovery of a safe and effective *S. aureus* vaccine for current prevention strategies has the potential for great benefit in public health care.

To address the challenge of treating *S. aureus* infections, whole-cell attenuated or killed vaccines have been developed. However, these vaccines have not been particularly effective. Similarly, subunit vaccines composed of individual bacterial surface proteins generate immune responses that afford only partial protection when tested in experimental animals.

However, a more effective subunit vaccine has recently been developed to protect individuals against *S. aureus* by combining several of the bacterium's antigens (Fig. 11.9). Starting with one disease-causing strain of *S. aureus*, 23 bacterial outer surface proteins were identified from genomic DNA sequence data. Then, the coding regions of these proteins, minus the signal sequences, were amplified by polymerase chain reaction (PCR) and cloned into plasmid vectors that enabled the proteins to be expressed in *E. coli* with a poly-His tag at the N terminus of the protein (to facilitate the purification of the overexpressed protein). The proteins were expressed and purified, and mice were separately immunized with each of the 23 purified proteins. The immunized mice were subsequently challenged by injections of live disease-causing *S. aureus*. Many of the recombinant surface proteins generated an immune response that afforded partial protection against staphylococcal disease, with some proteins affording more

PCR
polymerase chain reaction

Figure 11.9 Development of a vaccine against *S. aureus*. The coding regions of several bacterial outer surface proteins from a disease-causing strain of *S. aureus* were cloned into plasmid vectors. The proteins were expressed and purified, and mice were separately immunized with each of the purified proteins. The mice were then injected with live disease-causing *S. aureus*. Many recombinant surface proteins generated an immune response that afforded partial protection against staphylococcal disease, with some proteins affording more protection than others. Mice that were not injected with any of the surface proteins but were infected with *S. aureus* were not protected from disease. doi:10.1128/9781555818890.ch11f.11.9

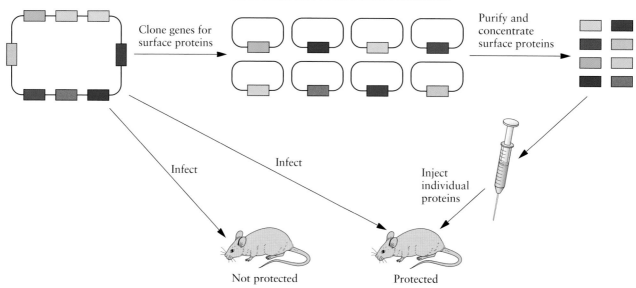

protection than others. However, over the long term, immunization with individual surface proteins afforded only modest protection. A mixture of the four proteins that individually generated the most effective antibodies was used to immunize mice and was found to completely protect against the pathogen. The experimental design ensured that only common, and not strain-specific, *S. aureus* surface proteins were used to immunize mice. Thus, it is not surprising that the tetravalent subunit vaccine that was developed was effective against five different clinical isolates (strains) of *S. aureus*. This work represents an important first step in the development of an *S. aureus* vaccine.

Human Papillomavirus

Human papillomavirus (HPV) infections cause many common sexually transmitted diseases. While most of these infections are benign and often asymptomatic, persistent infection with some strains of HPV is associated with the development of cervical and related cancers, as well as genital warts. Over 100 different types of HPV exist, and 15 types have carcinogenic potential. HPV type 16 (HPV-16) and HPV-18 account for about 70% of invasive cervical cancers. HPV-16 is also strongly associated with cancers in both the anogenital and oropharyngeal (oral part of the throat in the digestive and respiratory systems) regions.

HPV
human papillomavirus

HPV-16
HPV type 16

Importantly, the recent high rate of increase in the diagnosis of oropharyngeal cancer predicts that this form of HPV-induced cancer will surpass cervical cancers caused by HPV by 2020 in the United States. The fact that HPV-16 and HPV-18 alone account for most of the aggressive infections suggests that intervention directed specifically at these types may reduce the burden of disease appreciably. In particular, as HPV-16 is associated with approximately 50% of cervical cancers, a vaccine that prevents HPV-16 infection may significantly decrease the incidence of cervical cancer. Moreover, a vaccine directed against several different types of HPVs could effectively prevent nearly all HPV-induced cervical cancers. Such a vaccine may provide a significant public health care benefit, as HPV-associated cervical cancer is the third most commonly diagnosed cancer and second leading cause of cancer-related death worldwide among women, accounting for more than 250,000 deaths per year.

In June 2006, the U.S. Food and Drug Administration approved a vaccine that protects women against infection by HPV-6, -11, -16, and -18, the types most frequently associated with genital warts (HPV-6 and -11) and cervical cancer (HPV-16 and -18). At the end of 2006, this vaccine was approved for use in more than 50 countries worldwide. The vaccine, called Gardasil, is quadrivalent; i.e., it contains virus-like particles assembled from the major capsid (L1) proteins of the above-mentioned four types of HPV (Fig. 11.10 and Box 11.4). The L1 protein can self-assemble into virus-like particles that resemble papillomavirus virions, and these particles are highly immunogenic, inducing neutralizing antibodies directed against the whole live virus. The gene for the L1 protein from each of the four virus types was cloned and expressed in a recombinant *Saccharomyces cerevisiae* (yeast) strain. Following separate fermentations of

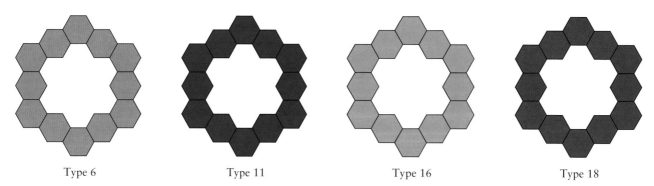

Type 6   Type 11   Type 16   Type 18

Figure 11.10 Schematic representation of the virus-like particles assembled from cloned and overproduced L1 proteins from the capsids of four different strains of HPV. These virus-like particles are the constituents of a commercial subunit vaccine against the virus. doi:10.1128/9781555818890.ch11f.11.10

the four yeast strains, the viral capsid proteins assembled into virus-like particles (consisting of the viral capsid without any other viral proteins or viral nucleic acid). These virus-like particles were then purified and combined to form the quadrivalent vaccine.

Unfortunately, recent data from the Centers for Disease Control and Prevention reveal that only 32% of teenagers who qualify for

box 11.4
A Vaccine To Prevent Cervical Cancer

On 15 September 2007, the headline on the front page of *The Globe and Mail*, a Toronto, Canada, newspaper, read, "Should your daughter get the needle?" The article that followed related how the Canadian federal government, in conjunction with the Ontario provincial government, was funding a program that would offer free vaccinations against HPV to girls in grade 8 (typically 12- and 13-year-olds). The vaccine, which had received approval in the United States a year earlier, provides inoculated women with immunity against the viruses that are responsible for approximately 70% of all cervical cancers and 90% of genital warts. Grade 8 was chosen because, according to officials, it is before most girls become sexually active. The vaccine, sold under the brand name Gardasil, is given by needle in three doses over 6 months and is approved for females

between the ages of 9 and 26 years. The rationale for giving the vaccine at such a young age is related to the fact that once a woman has been exposed to the four strains of the virus for which the vaccine provides protection, the vaccine will no longer be effective. The three doses of vaccine cost about $300 to $400, although those inoculated through this program received the vaccine free of charge. Boys can also get HPV infections, but testing is still under way to determine whether the vaccine works as well for them.

According to *The Globe and Mail*, "For many parents it's a no-brainer: anything that will protect their daughters from cancer . . . is worth the risks." However, at the same time, a small but vocal minority has expressed serious reservations about this program. On one hand, there are individuals who do not trust the medical establishment, the pharmaceutical companies, and/or

the government. Others have expressed concerns that some girls will naïvely believe that this vaccine will protect them against any and all sexually transmitted diseases and use this as a rationale or excuse for becoming sexually active at an early age. Still others have questioned the potential side effects from the vaccine, despite the fact that extensive clinical trials have shown that they are quite rare. Notwithstanding the concerns of some individuals, the vaccine was initially offered through school inoculation programs to young females in the Canadian provinces of Newfoundland and Labrador, Prince Edward Island, Nova Scotia, and Ontario. However, by September 2008, all of the other provinces in Canada had decided to implement this program. The real benefits of the program (hopefully an enormous reduction in cervical cancer) may not be known for several decades; in the meantime, the debate will continue. Also, since the vaccine does not protect against all strains of HPV, it is essential that women continue to get an annual Pap test.

immunization are getting all three of the recommended doses of the quadrivalent vaccines in the United States, suggesting that the majority of teens remain susceptible to infection with the high-risk HPV-16 and -18 and thus susceptible to high-grade disease. In addition, patients already infected with either of these types have not received any therapeutic effect with these vaccines, indicating the absence of any benefit of these vaccines for preexisting cervical infection or the presence of cervical lesions.

Therefore, a gap in immune-based coverage of established cervical disease currently exists in patients who were infected prior to approval of the prophylactic vaccines, as well as those who were approved for vaccination but declined to initiate or finish the three-dose regimen prior to exposure and infection. It is evident that the development of an immune-based interventional therapy may significantly benefit this group of patients, by providing them the possibility of a noninvasive, nonsurgical option for treatment of HPV infection. Indeed, several recent clinical reports of the control and regression of HPV-related disease highlight the associations and correlations of HPV-specific T-cell responses in peripheral blood and tissues into which lymphocytes were recruited. These findings and others further support the notion that a strong HPV-specific Th1 cell-biased (see chapters 2 and 4) immune response mediated by γ-interferon (IFN-γ) may be important not only for the control or elimination of infection in precancerous states of HPV-16 and -18 infection but also for the control or elimination of HPV-induced cervical cancer. Recent data show that the IFN-γ-induced activation of $CD8^+$ cytolytic T cells that are recruited to tumor tissue and sites of infection is essential for these protective T-cell responses and partial clearance of HPV-positive lesions. Ongoing clinical trials in which strong Th1- and $CD8^+$ T-cell-dependent responses to HPV-16 and -18 are induced will reveal whether this approach is beneficial.

IFN-γ
γ-interferon

Despite the above-mentioned issues that remain to be overcome with global HPV vaccination programs in the United States, there is cause to celebrate the extraordinary success of Australia's HPV vaccination program. Remarkable data were recently obtained for about 86,000 new patients vaccinated with the quadrivalent HPV vaccine beginning in 2007 from six sexual health clinics in Australia. Vaccine coverage rates were excellent and averaged about 80% for all three doses of the quadrivalent vaccine. The proportion of women under 21 years of age who had genital warts was reduced from 11.5% in 2007 to 0.85% in 2011. Only 13 cases of genital warts were observed in women younger than 21 years in these six health clinics in 2011. The proportion of women aged 21 to 30 years who presented with genital warts was reduced from 11.3% in 2007 to 3.1% in 2011. Unfortunately, the rate of diagnoses of genital warts in women over 30 was unaffected. The proportion of men under 21 years old with genital warts also decreased appreciably, from 12.1% in 2007 to 2.2% in 2011. However, for both heterosexual and homosexual men over 21 years old, the percentage of men with genital warts was not reduced. Overall, these findings are to be regarded as a major public health achievement that may have a very significant impact on the costs of health care for sexually transmitted diseases.

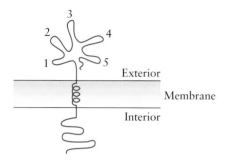

Figure 11.11 Generalized envelope-bound protein with extracellular epitopes (1 to 5) that may elicit an immune response.
doi:10.1128/9781555818890.ch11f.11.11

Peptide Vaccines

Does a small peptide fragment of a protein act as an effective subunit vaccine and induce the production of neutralizing antibodies? Intuitively, one would expect that only those peptides of a protein that are accessible to antibody binding, e.g., those on the exterior surface of the virus, are immunologically recognized by an antibody and that those located in inaccessible regions inside the virus particle will be ignored if they do not contribute to the conformation of the immunogenic peptide (Fig. 11.11). If this is the case, it may be anticipated that short peptides that mimic epitopes (antigenic determinants) are immunogenic and may be used as peptide vaccines.

However, there are certain limitations to the use of short peptides as vaccines: (i) to be effective, an epitope must consist of a short stretch of contiguous amino acids, which does not always occur naturally; (ii) the peptide must be able to assume the same conformation as the epitope in the intact viral particle; and (iii) a single epitope may not be sufficiently immunogenic.

Malaria

Malaria is a parasitic disease transmitted to humans and other vertebrates by mosquitoes and is potentially fatal. Historically, the malaria parasite is thought to have killed more humans than any other single cause. Today, along with HIV and tuberculosis, malaria remains one of the "big three" infectious diseases, every year exacting a heavy toll on human life and health in parts of Central and South America, Asia, and sub-Saharan Africa, where up to 90% of malaria deaths occur. Malaria parasites are bigger, more complicated, and wilier than the viruses and bacteria that have been controlled with vaccines. Despite decades of research toward a vaccine for malaria, this goal has remained elusive. Nevertheless, recent advances provide optimism that a licensed malaria vaccine may be available in the near future.

The genus *Plasmodium* consists of approximately 125 known species of parasitic protozoa, 5 of which (*Plasmodium falciparum*, *P. vivax*, *P. ovale*, *P. malariae*, and *P. knowlesi*) infect humans and cause malaria. Because *P. falciparum* elicits the most severe malaria disease and deaths, it is the target of most vaccine development efforts. The *Plasmodium* life cycle is very complex. Sporozoites from the saliva of a biting female mosquito are transmitted to either the blood or the lymphatic system and then migrate to the liver and invade liver cells (hepatocytes) (Fig. 11.12). The parasite buds off the hepatocytes in merosomes containing hundreds or thousands of merozoites. These merosomes lodge in pulmonary capillaries, and while they slowly disintegrate there during a period of 2 to 3 days, they release merozoites. The merozoites invade red blood cells (erythrocytes), in which the parasite divides several times to produce new merozoites, which then leave the erythrocytes and travel in the bloodstream to invade new erythrocytes. The parasite eventually forms gametocytes, which may be ingested by feeding mosquitoes. Fusion of the gametes that

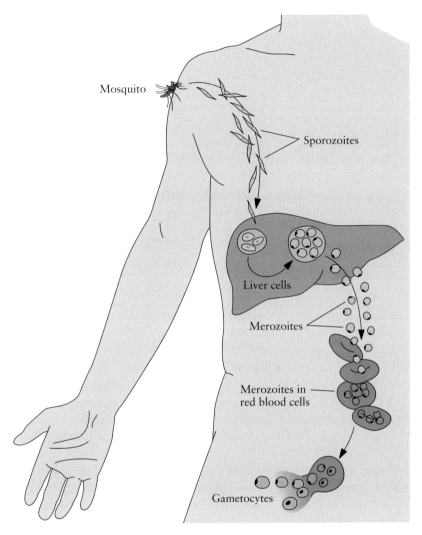

Figure 11.12 Infection of an individual with *Plasmodium falciparum* (a malaria-causing parasite) introduced by a mosquito. doi:10.1128/9781555818890.ch11f.11.12

develop from gametocytes leads to the formation of new sporozoites in the mosquito that can infect new individuals, spreading the disease. A highly efficacious pre-erythrocyte-stage vaccine would be expected to completely block infection and block transmission, thereby preventing parasites from reaching the blood and causing disease.

In the life cycle of the malaria parasite, it is the asexual blood-stage multiplication that is responsible for most of the acute symptoms of the disease. Thus, malaria vaccines that target the blood stage of the parasite may prevent or reduce clinical illness without preventing infection. In areas where malaria is endemic (maintained in the population), some individuals show considerable resistance to the disease despite the fact that when their blood is examined, they are found to carry the parasite. This resistance to the worst symptoms of malaria results from an antibody-dependent cellular cytotoxicity mechanism that inhibits parasite development. Thus, some individuals who are infected with the malaria parasite make antibodies against merozoite surface protein 3 that prevent parasite growth. While the N-terminal part of this protein varies considerably among different *Plasmodium* strains, its C-terminal end is highly conserved in these various strains.

Therefore, peptides that correspond to regions of the C terminus of merozoite surface protein 3 were synthesized. Human serum antibodies from individuals who were resistant to the parasite were affinity purified based upon their interaction with one or more of these peptides and were then tested in an antibody-dependent cellular-cytotoxicity assay. Antibodies directed against peptides B, C, and D (Fig. 11.13) had a major inhibitory effect on parasite growth. Based on the ability of peptides B, C, and D to bind to and select protective antibodies, peptide 181–276 of merozoite surface protein 3 was synthesized and is currently being tested in clinical trials as a novel malaria vaccine. While more research needs to be done, current evidence suggests that synthetic peptide vaccines may become highly specific, relatively inexpensive, safe, and effective alternatives to traditional vaccines.

Despite the promise that peptide vaccines hold for the prevention of malaria infection, it is important to note that both humoral and cellular factors contribute to acquired immune protection against malaria. While cellular immune responses are more effective in controlling the pre-erythrocyte stages of malaria infection, antibodies are better at blocking erythrocyte invasion to suppress blood-stage infection. For these reasons, cellular immune responses are emphasized more in the development of preerythrocyte vaccines and antibody responses more in the development of blood-stage vaccines.

However, the basis of protective immunity against malaria is poorly understood and requires further experimentation, as clinical protection does not seem to correlate closely with a specific immune response as yet. Moreover, since the host immune response to malaria infection contributes to the pathogenesis of malaria, a vaccine may increase the risk of harmful inflammatory responses to subsequent infection, especially for a

Figure 11.13 Schematic representation of *Plasmodium falciparum* merozoite surface protein 3 and peptides corresponding to portions of the C terminus. The peptides, labeled A to F, are drawn to scale, with the numbers above the whole protein indicating the amino acid number (counting from the N terminus). The final peptide is currently being tested in clinical trials for efficacy as a malaria vaccine. doi:10.1128/9781555818890.ch11f.11.13

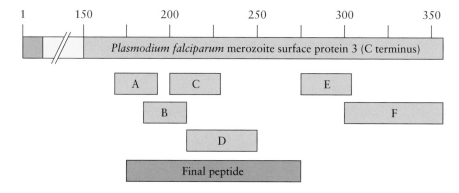

vaccine directed against the blood stages that mediate pathology. Thus, the monitoring and prevention of such adverse inflammatory responses to malaria vaccines or to postvaccination malaria infection are important aspects of clinical malaria vaccine development, especially for blood-stage vaccines.

In November 2012, a large phase IIIb trial of the most advanced malaria vaccine in development, RTS,S/AS01, was reported. RTS,S contains the repeat and T-cell epitope regions of the immunodominant circumsporozoite, which covers the exterior of the parasite when it first enters the body after the bite of an infected mosquito (Fig. 11.12). The repeat and T-cell epitope regions are fused to the hepatitis B virus surface antigen to form protein particles in the presence of an additional S antigen and are administered with the liposomal-based AS01 adjuvant (adjuvant system 01, a proprietary adjuvant system from GSK Biologicals) to boost the immune response (Fig. 11.14). AS01 contains 3-deacylated **monophosphorylated lipid A (MPL)**, a detoxified product of the Re595 strain of *Salmonella enterica* serovar Minnesota. MPL binds to and stimulates Toll-like receptor 4 (TLR4), which activates both humoral and cellular immune responses. AS01 also consists of the QS21 compound, a saponin derived from the bark of the South American *Quillaja saponaria* tree.

In African infants, this trial showed only 31% protection against clinical malaria compared with a control nonmalaria vaccine. The result was in contrast to the promising finding of 55.8% protection in an earlier and smaller phase IIIa trial with RTS,S in children aged 5 to 17 months at trial enrollment in the first 12 months after vaccination. In addition,

MPL
monophosphorylated lipid A

TLR4
Toll-like receptor 4

Figure 11.14 Schematic representation of the RTS,S hybrid vaccine particle used in the treatment of malaria. The central repeat (R) region is derived from the *Plasmodium falciparum* circumsporozoite protein (CSP) and is genetically fused to the T-cell epitope (T) and surface antigen (S) regions derived from the hepatitis B virus. The hybrid protein, RTS, expressed in transformed *Saccharomyces cerevisiae* yeast cells, self-assembles into virus-like particles similar to native hepatitis B surface antigen (HBsAg), with exposure of the circumsporozoite epitope on the exterior surface of these particles. The glycosylphosphatidylinositol (GPI) anchor attaches the CSP to the lipid bilayer of a cell membrane. Adapted from Regules et al., *Expert Rev. Vaccines* **10:** 589–599, 2011, with permission. doi:10.1128/9781555818890.ch11f.11.14

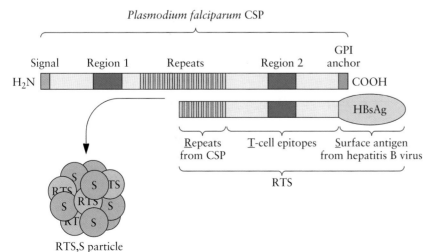

in the initial trial of RTS,S in adults, the three formulations used induced similar antibody and T-cell responses, but only one formulation protected against malaria. The finding that the kind and quality of immune response required to prevent infection are variable compromises somewhat our attempts to rationally approach vaccine design.

Cancer

Peptides corresponding to immunogenic tumor antigens or antigen epitopes have been administered as cancer vaccines. Peptide vaccines have the advantage that they do not require manipulation of patient tissues, whose availability may be limited. A potential mechanism of tumor elimination by peptide vaccine therapy is shown in Figure 11.15. Importantly, many peptide vaccines are designed to enhance CD8$^+$ CTL responses, since these responses elicit tumor destruction and elimination.

Several peptide vaccines have been successfully used to produce antigen-specific responses in pancreatic cancer patients. In a phase I study, vaccination with a peptide of 100 amino acids in length (100-mer) from the extracellular domain of the **mucin 1 (MUC-1)** cell surface-associated

MUC-1
mucin 1

Figure 11.15 Possible mechanism of tumor elimination by peptide vaccine therapy. The steps of this potential antitumor immune response are numbered in order as follows: (**1**) introduction of vaccine to the bloodstream; (**2**) processing and presentation of the peptide by a dendritic cell (DC) in a lymph node, resulting in activation of CD4$^+$ Th cells and CD8$^+$ CTLs; (**3**) interaction between an MHC-I molecule on a dendritic cell and TCR during antigen presentation facilitated by CD8; (**4**) generation of tumor-specific CD8$^+$ CTLs that lyse tumor cells; (**5**) degranulation of CTLs following recognition of tumor antigen; (**6**) Fas-FasL interaction-mediated transduction of a death signal to the tumor; and (**7**) tumor destruction and elimination. Adapted from Bartnik et al., *Vaccines* 1:1–16, 2013. doi:10.1128/9781555818890.ch11f.11.15

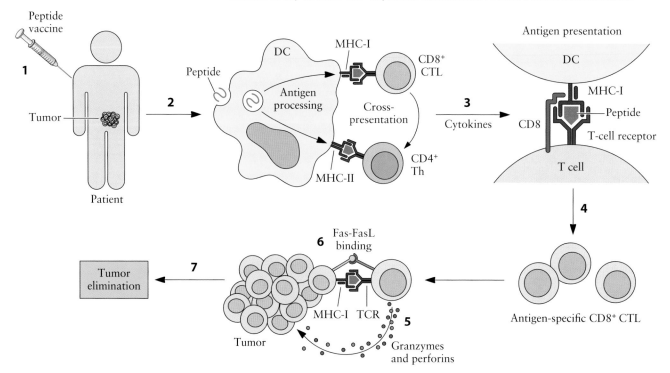

glycoprotein generated a MUC-1-specific T-cell response in some patients with resected or advanced pancreatic cancer. Two of the 15 patients were alive at 61 months postinjection. In another phase I clinical trial using the same 100-mer peptide vaccine, the production of anti-MUC-1 circulating antibodies was detected in the sera of patients with inoperable pancreatic or biliary cancer, although no significant impact on survival was discovered. A telomerase-based vaccine, consisting of a human telomerase reverse transcriptase peptide, induced a telomerase-specific T-cell-mediated immune response in 63% of the patients monitored, as measured by delayed-type hypersensitivity (DTH; see chapter 4) in nonresectable pancreatic cancer. Those with a positive DTH response lived longer than those that did not have a positive DTH response. In other vaccine studies, a peptide containing amino acids 2 to 169 of the **vascular endothelial growth factor receptor (VEGFR)** was administered with **gemcitabine** (a nucleoside analogue used in chemotherapy) to patients with advanced pancreatic cancer. A total of 83% of patients receiving the vaccine had an antigen-specific DTH response, and VEGFR 2–169-specific CD8$^+$ T cells were detected in 61% of those vaccinated with a median survival time of 8.7 months. A randomized, placebo-controlled, multicenter, phase II/III clinical trial of this VEGFR 2–169 peptide vaccine, combined with gemcitabine, is being conducted with patients with advanced nonresectable or recurrent pancreatic cancer.

Very encouraging results were provided by studies of **K-Ras** (a GTPase member of the Ras family that mediates many signal transduction pathways)-targeted peptide vaccines. A mutation of a *K-RAS* gene is essential for the development of many cancers. In a pilot vaccine study, pancreatic and colorectal cancer patients were vaccinated with K-Ras peptides containing patient-specific mutations. Three of five pancreatic cancer patients displayed an antigen-specific immune response to K-Ras, and these responders had no evidence of disease. Disease progression was observed in two pancreatic cancer patients that did not respond to the vaccine. Of the pancreatic cancer patients, a mean disease-free survival of ≥35.2 months and a mean overall survival of ≥44.4 months were observed. In a longer-term study, patients were monitored for up 10 years after surgical resection of pancreatic adenocarcinoma and vaccination with a K-Ras peptide vaccine administered concomitantly with the granulocyte-macrophage colony-stimulating factor (GM-CSF) cytokine. Remarkably, 20% of patients who received the vaccine were still alive after 10 years, and a memory T-cell response was still present in 75% of the survivors.

To increase the immunogenicity of peptide vaccines, key anchor residues in the peptides were mutated to increase binding to major histocompatibility complex class I (MHC-I) molecules and presentation to CD8$^+$ T cells (see chapters 2 and 4). This is very important when vaccinating against tumor (self) epitopes, as they are usually weak or intermediate binders to HLA molecules. In a murine model of pancreatic cancer, this strategy increased survival when applied to a peptide vaccine derived from the murine **mesothelin** protein (present on normal mesothelial cells and overexpressed in several human tumors, including mesothelioma and ovarian and pancreatic adenocarcinoma). A similarly modified MUC-1

DTH
delayed-type hypersensitivity

VEGFR
vascular endothelial growth factor receptor

GM-CSF
granulocyte-macrophage colony-stimulating factor

MHC-I
major histocompatibility complex class I

peptide vaccine increased IFN-γ production by patient and normal donor T cells. Interestingly, vaccination with a modified HLA-A2-restricted peptide of the protein **survivin** resulted in remission of liver metastasis in one individual with pancreatic cancer. Survivin, an inhibitor of apoptosis, is highly expressed in most cancers associated with chemotherapy resistance, increased tumor recurrence, and shorter patient survival. Thus, antisurvivin therapy may be an attractive cancer treatment strategy.

Autoimmune Disease

The immune system of a healthy person represents a delicate balance between pathogenic T cells that can elicit tissue damage and cause autoimmune disease and T cells that regulate immune responses and prevent or limit tissue damage (see chapter 4) (Fig. 11.16). The characterization of T regulatory (Treg) cells has provided the stimulus to discover safe and effective ways to induce these cells. Induction of Treg cells is a major goal of immunotherapy of autoimmune disease, since antigen-specific Treg cells can confer specificity to this therapy, reduce or eliminate any use of nonspecific drugs, and, most importantly, restore self-tolerance.

Studies of several chronic inflammatory autoimmune diseases in different experimental animal models indicate that the administration of

Treg
T regulatory

Figure 11.16 Benefit of antigen-specific immunotherapy in the immunopathogenesis of an autoimmune disease. During the pathogenesis of an autoimmune disease such as T1D, pancreatic islet β cell destruction is mediated by interactions between proinflammatory Th1 cells and CTLs that are primed against an islet autoantigen (e.g., insulin) by inflammatory dendritic cells (DCs). The main beneficial effects of antigen-specific immunotherapy are (i) deletion of proinflammatory Th1 cells and CTLs, (ii) induction of regulation mediated by Treg cells, and (iii) regulation mediated by deviation from a Th1 to a Th2 cell phenotype. Immune regulation and immune deviation are anti-inflammatory mechanisms designed to prevent or downregulate inflammatory responses. The major benefit of Treg induction is linked suppression during which Treg cells induced to suppress a response to one autoantigen can also suppress a response to another autoantigen(s) presented by the same dendritic cell. Adapted from Peakman and von Herrath, *Diabetes* **59**:2087–2093, 2010, with permission. doi:10.1128/9781555818890.ch11f.11.16

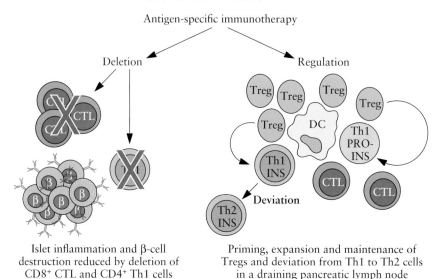

Islet inflammation and β-cell
destruction reduced by deletion of
CD8+ CTL and CD4+ Th1 cells

Priming, expansion and maintenance of
Tregs and deviation from Th1 to Th2 cells
in a draining pancreatic lymph node

autoantigens or autoantigen peptides (epitopes), at different stages of a disease by various routes, can both sustain health and protect from inflammatory autoimmune disease (see chapter 4). Antigen-specific immunotherapy can effectively control the autoimmune response via the induction (pre-disease onset) or restoration (post-disease onset) of immune tolerance to a tissue-specific target cell, e.g., the pancreatic islet β cell in type 1 diabetes (T1D). Furthermore, this immunotherapy may alleviate the concerns about safety and antigen-nonspecific immunosuppression in the host.

Despite its many advantages, the successful application of peptide immunotherapy in humans has proven difficult. Predicting the outcome of immunization with autoantigens, e.g., islet β cell autoantigens, is complex. The resulting immune response depends not only on the dose, frequency, and route of administration of an autoantigen (or peptide) but also on the use of suitable adjuvants, as well as several host factors (age, gender, and severity of inflammation). In patients with T1D, some islet-reactive T cells are likely preactivated at the time of immunization, and their avidities for autoantigens may vary depending on previous cell interactions in central (thymus) or peripheral (lymph node or spleen) lymphoid sites. These conditions influence the magnitude and cytokine production of the resulting antigen-specific response.

It is encouraging that studies with human T cells in vivo recently showed that desirable Treg cell responses are elevated and deleterious islet-specific effector T cells are deleted by antigen-specific immunotherapy (Table 11.6). Both outcomes may prevent T1D, which is manifested by both a Treg cell deficiency and the resistance of effector T cells to immunoregulation (Fig. 11.16). Treg cells have been generated in humans, as evidenced by the production of the interleukin-10 (IL-10) cytokine induced by a proinsulin peptide after low-dose intradermal administration of the peptide in T1D patients. IL-10 is an immunosuppressive cytokine that has direct effects on lymphocytes and antigen-presenting cells, such as dendritic cells. Repeated stimulation of lymphocytes in vitro in the presence of large amounts of IL-10 leads to the differentiation of Treg cells. Moreover, the Treg cells induced by autoantigen therapy seem to be active only in tissues in which the autoantigen is expressed. Thus, this type of therapy may offer a site-specific approach to antigen-regulated

T1D
type 1 diabetes

IL-10
interleukin-10

Table 11.6 Mechanism of action of antigen-specific therapy and predicted outcomes

Mechanism	Predicted outcome	Induction of tolerance	Durability of responses
Immune regulation against target antigen (mediated by Treg cells, IL-10, etc.)	T-cell responses detected (cell proliferation, cytokine production, other functions) Should facilitate linked suppression of responses to other target cell antigens	Yes	Several months to 1 yr
Immune deviation of dominant antigen-specific T-cell phenotype (e.g., from Th1 to Th2)	T-cell responses detected (cell proliferation, cytokine production, other functions) May facilitate linked suppression of responses to other target cell antigens	Yes	Several months
Immune deletion of antigen-specific T cells	May not detect deletion No role for linked suppression	Maybe	Short (weeks/months)

For the proposed mechanism of these responses, the example of an autoantigen, such as insulin in T1D, is used.

Table 11.7 Novel approaches to enhance the delivery and potency of peptide vaccines for antigen-specific immunotherapy

Direct route of therapy	Indirect route of therapy
Intranasal (antigen or peptide)	Gene modified probiotic delivery (antigen or peptide)
Oral (antigen)	Injection of tolerogenic dendritic cells, blood cells, or inert particles as carriers (antigen or peptide)
Intradermal, subcutaneous, or intramuscular (antigen, peptide or DNA in presence or absence of adjuvants)	Injection of antigen-specific Treg cells (primed ex vivo; expanded ex vivo, gene-modified Treg cells)

Adapted from Peakman and von Herrath, *Diabetes* **59**:2087–2093, 2010, with permission.

Shown are several direct and indirect approaches currently under investigation for the optimal delivery of antigen-specific immunotherapy. Antigen refers to a given autoantigen(s) under study for a specific autoimmune disease. Dendritic cells, may be tested in dendritic cell-targeted peptide vaccines.

immunomodulation that can provide clinical benefit. It is recommended that optimization of dosing and delivery regimens for antigen-specific therapy in parallel with the development of suitable adjunct therapies receive priority.

A clinical disadvantage may be that antigen-specific therapies may have less potency than generalized immunosuppression, as they tend to work only prior to the onset of disease. Additional concerns with the use of antigen-specific therapy include the acceleration of disease and possible induction of life-threatening DTH responses (see chapter 4). However, clinical studies have shown that the use of native (but not altered) peptide sequences of autoantigens in T1D (insulin and proinsulin), multiple sclerosis (myelin basic protein), and rheumatoid arthritis (DnaJ) does not exacerbate disease. Similarly, the repeated injection of appropriate doses of native peptide sequences into patients with autoimmune disease has not elicited any DTH, allergy, or anaphylaxis responses to date.

Finally, the introduction of new strategies to enhance the delivery and potency of peptide vaccines may prove beneficial for antigen-specific immunotherapy (Table 11.7). These strategies may include the delivery of multiple epitopes from one or more autoantigens to duplicate the success realized in clinical allergy; the use of steroid hormone adjuvants (glucocorticoids and vitamin D) to modulate dendritic cells that present autoantigen peptides and enhance tolerance induction in the skin, the delivery of antigens to the intestine using the gene from the gram-positive bacterium *Lactococcus lactis* modified to deliver islet autoantigens and cytokines, the use of soluble TCRs specific for islet peptides and antigens coupled to inert cells, and the development of relevant combinations of native autoantigen peptides with immunomodulators selected to maximize dendritic cell-activated Treg cell function and expansion while reducing the effector T-cell load.

Allergy

Allergy is a hyperimmune response based on IgE production against harmless environmental antigens, i.e., allergens (see chapter 4). Interestingly, vaccination with allergens, termed allergen-specific immunotherapy, is the only disease-modifying therapy of allergy with long-lasting effects. New forms of allergy diagnosis and allergy vaccines based on peptides, recombinant allergen derivatives, and allergen genes have emerged from the characterization of allergens. Such vaccines enable targeting of the immune system with the benefit of reducing and eliminating side effects. Successful clinical trials performed with the new vaccines indicate that broad allergy vaccination promises to enhance the control of the allergy pandemic.

Allergen-specific immunotherapy with T-cell epitope-containing allergen peptides can regulate allergen-specific T-cell responses. The administration of peptides from the major cat allergen Fel d 1 decreases T-cell-mediated DTH responses, possibly by the induction of IL-10-producing Treg cells and linked immunosuppression. In a TCR-transgenic model, Th1 cells and Treg cells suppress the effector functions of Th2 cells, suggesting a role for immune deviation. However, the effects of T-cell peptide therapy on allergen-induced mast cell degranulation and acute allergic inflammation observed in clinical trials have been modest thus far. After mapping of T-cell epitopes of allergens became feasible, the concept of inducing T-cell tolerance with non-IgE-reactive but T-cell-epitope-containing peptides emerged as the first targeted allergen-specific immunotherapy approach. Currently, several immunotherapy studies with allergen-derived T-cell peptides are ongoing; a recent study has shown relief from the late-phase allergic symptoms to cat allergens. Although there is evidence for linked immunosuppression, the T-cell peptide approach will likely require that several different allergen-derived peptides be included in the vaccine to cover the diverse spectrum of T-cell epitopes recognized by allergic patients. Note that these concepts of T-cell-dependent **immune regulation** via linked suppression and the recognition of multiple epitopes are similar to those approaches illustrated above for the therapy of autoimmune disease (Table 11.6 and Fig. 11.16).

To reduce IgE-mediated side effects during allergen-specific immunotherapy, recombinant hypoallergenic allergen derivatives were developed. These derivatives reduce IgE reactivity while preserving T-cell epitopes in the natural allergens, using recombinant DNA technologies including mutation, reassembly, or fragmentation. Successful phase III clinical trials conducted with recombinant hypoallergenic derivatives of the major birch pollen allergen Bet v 1 revealed that these derivatives work by inducing allergen-specific IgG antibodies that block the action of the allergen. Similar clinical success was realized for recombinant grass pollen allergens that also generated allergen-specific IgG blocking antibodies. Recombinant hypoallergens for the treatment of allergy to house dust mites and cat dander have been successful in experimental animal models and await further analysis in clinical trials.

During clinical studies performed with non-IgE-reactive T-cell peptides and recombinant hypoallergenic allergen derivatives, it was recognized that allergen-derived T-cell epitopes induce late-phase, non-IgE-mediated side effects. To eliminate these side effects, a new type of allergy vaccine was developed based on the identification of allergen peptides that form part of the major IgE binding sites on allergens but lack allergenic activity.

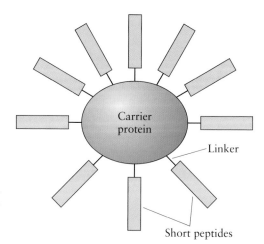

Figure 11.17 Structure of a peptide vaccine composed of identical short peptides of a protein antigen bound via a linker peptide to a large carrier protein.
doi:10.1128/9781555818890.ch11f.11.17

These peptides, which are typically 20 to 35 amino acids in length, can either be coupled chemically via a linker to a carrier protein or produced as recombinant fusion proteins with a carrier protein (Fig. 11.17). Immunization with carrier-bound allergen peptides induces and focuses IgG antibodies towards the IgE epitopes on allergens with the T-cell help of the carrier protein and thus do not activate allergen-specific T cells. Since the presence of allergen-specific T-cell epitopes is reduced in these vaccines, they should bypass T-cell-mediated side effects. Viral carrier proteins have been suggested to induce a beneficial antiviral immunity in addition to protection against allergens. Allergen-specific immunotherapy approaches using T-cell-reactive allergen peptides, carrier-bound allergen peptides, recombinant allergen, and hypoallergens are currently in clinical evaluation and hold promise that allergen-based vaccines will be soon available for broad, safe, and durable therapeutic vaccination.

Dendritic Cell Vaccines

Human Immunodeficiency Virus

Critical to the modulation of the immune response is the presentation of specific antigens to the immune system by dendritic cells. Three subgroups of dendritic cells, including two forms of myeloid dendritic cells and one plasmacytoid dendritic cell, each with distinct sets of TLRs (see chapters 2 and 4), modulate the response to specific antigens and adjuvants. This modulation is achieved via a feedback loop between dendritic cells and Treg cells that maintains the physiological numbers of these two cell types (see Fig. 4.5). Traditionally, vaccines have relied on live attenuated or inactivated organisms, attenuated bacteria or capsules, or inactivated toxins and have generally worked by inducing protective antibodies. However, in many infections like HIV, malaria, and tuberculosis as well as cancers, it is necessary to maintain a durable and protective T-cell immunity. Recent progress has been made in enhancing T-cell immunity through an improved understanding of the biology of dendritic cells and their response to adjuvants.

Thus, many efforts have been made to develop a safe T-cell-based protein vaccine that exploits the pivotal role of dendritic cells in initiating adaptive immunity. One approach has involved a focus on the HIV Gag p24 protein antigen, an HIV group-specific antigen that makes up the viral capsid. HIV Gag p24 was introduced into a monoclonal antibody (MAb) that efficiently and specifically targets the DEC-205 antigen uptake receptor on dendritic cells. Attempts were made to coadminister the Gag p24 antigen and anti-DEC-205 MAb with a synthetic double-stranded RNA, polyriboinosinic acid·polyribocytidylic acid [poly(I·C)] or its analogue poly-ICLC [poly(I·C) stabilized with carboxymethyl cellulose and poly-L-lysine]. It was reasoned that the double-stranded RNA, a viral mimic, would bind to TLR3 and thereby function as an adjuvant (see chapter 4). Indeed, HIV Gag p24 with anti-DEC-205 MAb was found to be highly immunogenic in mice, rhesus macaques, and healthy human volunteers. In mice primed with both anti-DEC-Gag p24 protein and poly-ICLC, there was a very large and rapid boost in Gag-specific CD8$^+$ T cells upon a single injection of the vaccine, which previously had been very difficult to achieve. This finding yielded a very desirable result; i.e., protein vaccination enables CD8$^+$ T cells in the vaccinated host to respond quickly and well, even to a low-dose antigenic challenge. More recently, human subjects injected with a single low dose of HIV Gag p24 and anti-DEC-205 MAb plus poly-ICLC were observed to form both T- and B-cell responses to dendritic cell-targeted protein. These data indicate that the response to adjuvant was very rapid and suggest that adaptive immunity may be achieved in the host within a few weeks to months. A randomized dose escalation of the vaccine study is under way, as well as efforts to design the vaccine to include HIV envelope protein and to evaluate other adjuvants (see "Adjuvants" below).

MAb
monoclonal antibody

poly(I·C)
polyriboinosinic acid · polyribocytidylic acid

poly-ICLC
poly(I·C) stabilized with carboxymethyl cellulose and poly-L-lysine

Cancer

Approaches similar to that use for HIV protein vaccines can also be applied to other important clinical targets such as cancer. It has been reported that tumor immune T cells stimulated in response to a protein vaccine targeted by DEC-205 (antigen uptake receptor on dendritic cells) are able to recognize endogenous processed antigen in tumor cells, and upon seeing specific antigen, the T cells can proliferate and make many cytokines (Fig. 11.18). Although tumors and HIV represent distinct challenges, new protein vaccine approaches for both targets require that strong, specific, and durable T-cell immunity be generated. Thus, the concept of a dendritic cell-targeted vaccine seems relevant for immunization against highly expressed human tumor antigens. Candidate tumor antigens include mesothelin (overexpressed in mesothelioma and ovarian and pancreatic adenocarcinoma tumors), survivin (a protein inhibitor of apoptosis, highly expressed in most cancers and associated with chemotherapy resistance, increased tumor recurrence, and shorter patient survival), **carcinoembryonic antigen (CEA)** (a glycoprotein involved in cell adhesion; expressed on metastatic colon carcinoma cells), and human epidermal growth factor receptor 2 (mediates the pathogenesis and progression of breast cancer).

CEA
carcinoembryonic antigen

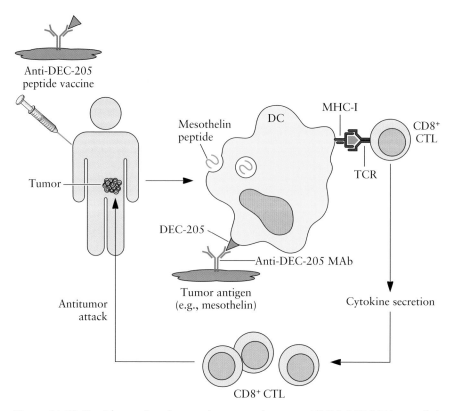

Figure 11.18 Peptide vaccine therapy for cancer by an anti-DEC-205 MAb-coupled tumor antigen targeted to dendritic cells. In this CD8[+] CTL-mediated antitumor response, a tumor protein antigen (or peptide) such as mesothelin is chemically coupled to the anti-DEC-205 MAb. This MAb binds to the DEC-205 antigen uptake receptor on the surface of a dendritic cell (DC) and in this way efficiently and specifically directs the mesothelin antigen to the dendritic cell for antigen processing. A processed mesothelin peptide(s) binds to MHC-I and is presented to a TCR on a CD8[+] antigen-specific CTL, which is activated by a dendritic cell to secrete many cytokines following antigen activation. These cytokines stimulate the proliferation and expansion of mesothelin peptide-specific CD8[+] T cells to mount an attack against the mesothelial tumor cells and destroy them. doi:10.1128/9781555818890.ch11f.11.18

Using a dendritic cell-targeted approach to enhance the presentation of tumor antigen peptides, an anti-murine DEC-205 antibody fused to either human mesothelin or survivin enhanced immune responses by directing delivery of tumor antigen to activated murine dendritic cells (Fig. 11.18). Similarly, the HIV Tat protein, which facilitates entry into cells, was fused to antigens to enhance uptake by dendritic cells. To date, this technique has been used mainly to introduce antigens into dendritic cells and not as a peptide vaccine alone. Another protein-based vaccine approach utilized an antibody vaccine that functionally mimics the tumor antigen CEA. This 3H1 murine MAb stimulated anti-CEA antibody and T-cell responses that rejected CEA-expressing tumor in a murine model of colon cancer. It is anticipated that the results of two recently completed phase II trials involving the 3H1 MAb in lung and colorectal cancer will be published in the near future.

Recent efforts have also been made to generate personalized peptide vaccines based on dendritic cell-targeted tumor antigen epitopes that are most immunogenic for a particular patient, a move towards "personalized medicine." In combination with gemcitabine therapy, this approach was used to treat pancreatic cancer in a phase I clinical trial. Prior to vaccination, peripheral blood T cells from patients were screened against a panel of tumor antigen-derived peptides, and patients were then vaccinated only with the anti-DEC-205-bound peptides to which they responded. An increase in tumor antigen-specific T-cell responses was observed from the 13 patients analyzed, but there was no correlation between these T-cell responses to clinical responses or antibody responses postvaccination. However, 11 patients did experience a significant reduction in tumor size, and the median survival time was 7.6 months. In the absence of treatment, the median survival time was about 2 to 3 months. Thus, despite some setbacks, dendritic cell-targeted protein (peptide) vaccines are a potential new vaccine platform, either alone or in combination with highly attenuated viral vectors, to induce integrated immune responses against microbial or cancer antigens, with improved ease of clinical use. As discussed above, the efficacy of dendritic cell-targeted autoantigen peptide vaccines is also being tested in clinical trials of autoimmune disease (e.g., T1D and multiple sclerosis) (Table 11.7).

DNA Vaccines

When tumor cells or antigens are injected into the body as a vaccine, they may elicit a vigorous immune response initially but subsequently may become less effective over time. This is because the immune system may recognize them as foreign and seek to rapidly destroy them. Without further antigen stimulation, the immune system often returns to its normal state of activity. Thus, to increase the durability of an antitumor response after tumor vaccination, a steady supply of an antigenic stimulus is required. To facilitate this requirement, plasmid vectors containing a gene(s) that encodes tumor antigens have been used to provide a constant source of tumor antigen to stimulate an immune response. Such vaccines are termed DNA vaccines.

Delivery and Immune Mechanisms of Action

Early studies showed that the method of DNA delivery affected the cell types that were transfected. Gene gun or **biolistic** (high-pressure bombardment of the skin with plasmid DNA used to coat gold microbeads) delivery tended to directly transfect epidermal keratinocytes and also skin dendritic cells, which migrated rapidly to regional lymph nodes. In this case, dendritic cells were transfected directly and behaved as the source of antigen presentation. Alternatively, intramuscular injection of plasmid predominantly led to transfection of muscle cells (myocytes). Myocytes lack expression of MHC-II and costimulatory molecules (see chapter 4) and thus would not be expected to prime T cells directly. Instead, immune priming likely occurs by dendritic cells that migrate to the site of DNA inoculation in response to inflammatory or chemotactic signals following vaccination. Gene gun immunization generally induces a greater CD8$^+$ T-cell response and requires less vaccine to achieve tumor immunity.

Thus, the site and procedure used for injection of plasmid DNA influence the extent of immunity induced, with transfected muscle and skin cells able to act as a reservoir of antigen but unable to prime the immune response. It is likely that cross-presentation of antigens on the muscle and skin cells by dendritic cells is the major route to priming (Fig. 11.15). However, gene gun delivery to skin sites may also lead to the direct transfection of dendritic cells; in this case, the host-synthesized antigen is processed and presented by dendritic cells in the context of both MHC-I and MHC-II molecules (Fig. 11.15). Antigen-loaded dendritic cells travel to the draining lymph nodes, where they present peptide antigens to naïve T cells, thereby eliciting both humoral and cellular immune responses (see chapter 4). Although plasmid DNA vaccines can induce $CD4^+$ T helper cell (Th cell) and antibody responses, they elicit even stronger $CD8^+$ CTL responses because they express antigens intracellularly. This introduces the antigens into the MHC-I antigen processing and presentation pathway (Fig. 11.15). Importantly, these features of DNA vaccines, which produce only low levels of antigen, induce both antibody and T-cell immune responses.

Th cell
T helper cell

Advantages and Disadvantages

The use of DNA vectors represents an important platform for clinical applications, in which large-scale vaccine production is not easily manageable with other forms of vaccine including recombinant protein, whole tumor cells, or viral vectors. Although virus-mediated gene transfer by genetically modified lentiviruses, adenoviruses, adeno-associated viruses, and retroviruses is advantageous because of its high transfection efficiency and stability, the largest hurdles using viral vectors are to overcome the immunogenicity of the viral packaging proteins. In addition, viral methods are disadvantageous because of their high expense, toxic side effects, limits on transgene size, and potential for insertional mutagenesis (mutagenesis of DNA by the insertion of one or more bases).

In comparison to viral vectors, nonviral vectors can encode several immunological components and are less cytotoxic, relatively more stable, and more cost-effective for production and storage (Table 11.8). Injected

Table 11.8 Advantages and disadvantages of DNA vaccines

Process or characteristic	Advantage	Disadvantage
Construction	May consist of several immunological components; permits synthetic- and PCR-based modifications; one plasmid may encode several antigens, or several plasmids could be admixed and administered simultaneously	
Production	Viral or bacterial strains are not used, excluding the possibility that attenuated strains will revert to virulence and cause disease; rapid and inexpensive production and formulation, since synthesis and purification of a protein(s) are not required; easily engineered and reproducible; facilitates large-scale production and isolation, since growth of dangerous organisms is not required	
Safety	Absence of pathogenic infection in vivo; no significant adverse side effects in clinical trials; neutralizing antibody-mediated responses are rare; permits an immune boost strategy	
Stability	Stability of DNA permits long-term storage; relatively insensitive to temperature changes	
Immunogenicity		Weakly immunogenic

Adapted from Fioretti et al., *J. Biomed. Biotechnol.* **2010**:174378, 2010, and from Glick et al., *Molecular Biotechnology: Principles and Applications of Recombinant DNA*, 4th ed. (ASM Press, Washington, DC, 2010).

nonviral vectors have proven safe and have not resulted in any significant adverse side effects in animal models and human clinical trials. Monitoring of the safety and efficacy in a trial of a DNA vaccine against HIV type 1 (HIV-1) infection demonstrated that DNA plasmid vaccines are safe and can induce detectable cellular and antibody immune responses. Gene manipulation and a simple plasmid backbone allow the incorporation of genes, which may then be expressed by cells transfected in vivo. Although this transfection is inefficient and varies with the target tissue and means of delivery, sufficient DNA is generally taken up to prime the immune response. Plasmid DNA vaccines are free of the problems associated with producing recombinant protein vaccines, and they are also safer than live attenuated viruses that can cause pathogenic infection in vivo. Interestingly, even after multiple immunizations with DNA vaccines, anti-DNA antibodies are not produced. The ability to introduce antigen to the host immune system and enable it to elicit strong $CD4^+$ Th1 cell and $CD8^+$ CTL responses is a unique feature of DNA vaccines that distinguishes them from conventional protein or peptide vaccines. As a result of this feature, DNA vaccines can readily induce both humoral and cellular immune responses. In the case of DNA vaccine treatment for cancer, the following hurdles must be overcome. Tumor antigens are often weakly immunogenic, and the immune system of cancer patients may be severely compromised. Thus, DNA vaccine-guided immune responses must be sufficiently strong to suppress tumor growth and/or cause tumor regression.

| **HIV-1** |
| human immunodeficiency virus type 1 |

Improved Efficacy and Immunogenicity

To overcome the generally low immunogenicity of DNA vaccines in large animal models and humans, several approaches have been tested, including plasmid design, immunomodulatory molecules, delivery techniques, and **prime–boost** strategy (Table 11.9).

Table 11.9 Strategies to improve the efficacy and immunogenicity of DNA vaccines

Strategy	Approach	Type of cancer
Route of administration	Intramuscular/electroporation	Prostate, B-cell lymphoma
	Intradermal/electroporation	Prostate, colon
	Biolistic	Cervical
	Intratumor	Melanoma, renal carcinoma
	Liquids under high pressure	Colon, B-cell lymphoma
Immune modulators as adjuvants	Cytokine	Liver, prostate, melanoma, B-cell lymphoma
	Chemokine	Prostate, B-cell lymphoma, follicular lymphoma
	Th cell epitopes	Colon, B-cell lymphoma
	TLR ligands	Lung
	Heat shock proteins	Cervical
Prime–boost regimen	Plasmid DNA or plasmid DNA + electroporation	Prostate, colon
	Plasmid DNA or recombinant protein antigen	Prostate, breast
	Plasmid DNA or viral vector	Liver, prostate, melanoma
	Viral vector or plasmid DNA	Prostate

Adapted from Fioretti et al., *J. Biomed. Biotechnol.* **2010**:174378, 2010.

Plasmid Design

The induction of robust immune responses by DNA vaccines requires that the expression of the encoded antigen be maximal. Strong viral promoters, such as the cytomegalovirus promoter intron A region (CMV-intA), are favored overregulated or endogenous eukaryotic promoters. Nuclear targeting sequences are introduced to increase the efficiency of nuclear plasmid uptake from the cytoplasm after intramuscular injection. The use of codon-optimized (not wild-type) sequences significantly improves vaccination. Optimal coding sequences are determined from the amino acid sequence of the antigen using algorithms that account for relative tRNA abundance in the cytosol of human cells and the predicted mRNA structure. The selected gene sequence, constructed in vitro with synthetic oligonucleotides, is optimal for expression and induction of an immune response.

CpG
cytosine–phosphate–guanine

Cytosine–phosphate–guanine (CpG) unmethylated regions in bacterial DNA can function as adjuvants in innate immunity and stimulate an immune response to DNA vaccines. The inclusion of CpG motifs in plasmid DNA vaccines induces the expression of proinflammatory cytokines, e.g., IL-12 or IFN type I. CpGs are recognized by the TLR9 expressed on dendritic cells, which stimulate the function of $CD8^+$ CTLs in immune responses. The coadministration of genes that encode TLR ligands with a gene that encodes a tumor antigen improves the immunogenicity of DNA vaccines.

Engineering DNA vaccine design for maximizing T-cell epitope-specific immunity has allowed epitope enhancement by sequence modification. Epitope sequences can be modified to increase the affinity of a peptide epitope for an MHC molecule. The knowledge of sequence motifs for peptide binding is the key to improve the primary and/or secondary anchor residues that mediate the specificity of binding to MHC-I or MHC-II. This strategy can greatly increase the potency of a vaccine and can convert a weakly immunogenic epitope into a dominant one by making it more competitive to bind to available MHC molecules.

Immunomodulatory Molecules as Adjuvants

Although the intramuscular or intradermal injection of a high dose (5 to 10 mg) of naked DNA vaccines can induce specific antibody and CTL responses in clinical trials, these responses are rather weak. Modifying the microenvironment of the vaccinated site by coadministration of DNA plasmids coding for immunostimulatory molecules, protein, or chemical adjuvants improves the low immunogenicity of DNA vaccines (Table 11.9). The encapsulation of plasmid DNA with liposomes, polymers, and microparticles has been attempted, but this approach has not enhanced the immunogenicity of the DNA vaccine relative to that observed with nonencapsulated plasmid DNA vaccines and has behaved poorly in clinical trials.

As many biological adjuvants are now available, several clinical trials are under way in which a given adjuvant(s) is tailored and encoded in the same DNA vector as encodes the antigen under study (Table 11.10). These adjuvants include chemokines to attract dendritic cells, cytokines,

Table 11.10 Adjuvants being tested in ongoing cancer trials to enhance immunity

Form of delivery	Adjuvant	Phase
DNA	Cytokines (GM-CSF, IL-2, IL-6, IL-12, IL-15, IL-21)	I/II
	Bacterial toxins (pDOM/tetanus toxin FrC)	I/II
	Immune modulators (HSP65)	I/II
Protein	Cytokines (GM-CSF, IL-2)	I/II

Adapted from Fioretti et al., *J. Biomed. Biotechnol.* **2010**:174378, 2010.

pDOMFrC is a plasmid that contains a domain (DOM) of fragment C (FrC) of tetanus toxin. FrC is a non-self-antigen that activates CD4$^+$ T-cell help to reverse tolerance and induce high levels of immunity.

costimulatory molecules, dendritic cell-targeting antibodies, and molecules to manipulate antigen presentation and/or processing. A commonly used cytokine is GM-CSF, which induces the proliferation, maturation, and migration of dendritic cells as well as the expansion and differentiation of T and B cells.

Route of Administration

The method of delivery of a DNA vaccine significantly influences its immunogenicity. In a mouse melanoma model, DNA vaccination was administered together with intratumoral delivery of antiangiogenic plasmids encoding angiostatin and endostatin. Combined melanoma vaccination resulted in 57% tumor-free survival over 90 days after challenge. In a modest proportion of patients with malignant disease, intratumoral injection of DNA led to regression of tumor at distant sites.

As mentioned above, physical methods of delivery are superior to other delivery methods that administer DNA in solution. Biolistic gene gun immunization of mice with a plasmid DNA vaccine induces a greater CD8$^+$ T-cell response and requires less vaccine to achieve tumor immunity. A promising strategy to increase the increase the transfection of DNA encoding target antigens is **electroporation**, in which DNA is injected intramuscularly and the skeletal muscle is immediately electrically stimulated with a pulse generator. Although this procedure causes some patient discomfort and may lead to local tissue injury and inflammation, patients tolerate this procedure without the need for any anesthesia and without any long-term negative side effects. Delivery by electroporation increases both the level and duration of immune responses in primates. Electroporation overcomes the difficulty in translating the effectiveness of DNA vaccination from rodents to large animals to human subjects. This method of DNA delivery enhances the cellular uptake of DNA vaccines appreciably, perhaps by causing some tissue damage. The tissue damage caused by electroporation results in inflammation and recruits dendritic cells, macrophages, and lymphocytes to the injection site, inducing significant antibody- and T-cell-mediated immune responses. Moreover, it is tolerable without anesthetic and does not induce unwanted immune responses against the delivery mechanism; thus, it can be used for repeat administrations.

Until now, the intramuscular or intradermal route of injection has been used for most DNA vaccines. Although these vaccines can induce a potent immune response, they do not induce mucosal immunity. Mucosal immunity can prevent pathogens from entering the body, while systemic immunity deals with pathogens only once they are inside the body. This is an important consideration because mucosal surfaces, including the respiratory, intestinal, and urogenital tracts, are the major sites of transmission of many infectious diseases. However, because of the protective barriers of the mucosal surfaces, traditional antigen-based vaccines are largely ineffective unless they are administered with specific agents that penetrate or bind to the mucosa, e.g., mucosal adjuvants.

Mucosal immunity induces a separate and distinct response from systemic immunity. The antibodies produced as part of the mucosal immune response restrict not only mucosal pathogens but also microorganisms that initially colonize mucosal surfaces and then cause systemic disease. Many mucosal vaccines are live attenuated organisms that infect mucosal surfaces and are effective at inducing mucosal responses. Of these, oral polio vaccines and both attenuated *Salmonella enterica* serovar Typhi Ty21a and *Vibrio cholerae* vaccines are licensed for use in humans.

Another important route of delivery of a vaccine to mucosal surfaces is the oral route. Oral vaccines are safe and easy to administer and convenient for all ages. In recent animal model studies, formulation of oral vaccines in a nanoparticle-releasing microparticle delivery system selectively induced large intestinal protective immunity against infections at mucosal sites in the rectal and genital regions. These large intestine-targeted oral vaccines area may potentially substitute for intracolorectal immunization, which is effective for rectal and genital infections but not for mass vaccination. Interestingly, this novel delivery system may be modified to selectively target either the small or large intestine for immunization and thereby achieve region-specific immunity at mucosal surfaces in the gut.

DNA vaccines designed for delivery to mucosal surfaces are similar to those used for intramuscular or intradermal delivery. To increase plasmid uptake and decrease its subsequent degradation, various methods of formulating DNA have been attempted. Cationic (positively charged) liposomes have been used to deliver DNA (has a negatively charged phosphate backbone) to the respiratory tract, and DNA entrapment in biodegradable microparticles has been used for the oral delivery of foreign DNA. To improve the potency of DNA vaccines for humans, several strategies have been designed. These strategies include the use of plasmids, which in addition to encoding a target gene (e.g., tumor antigen) also express a cytokine(s) such as IL-2, IL-10, or IL-12, which functions as an intercellular immunomodulator of immune responses. Additional systems, including liposomes, live vectors (bacteria and viruses), and several adjuvants that increase the immune response (bacterial toxins, carboxymethyl cellulose, lipid derivatives, aluminum salts, and saponins), have been examined for delivery of DNA to different cell types. Such approaches are generally optimized in mice before analysis in larger animals and humans, with the caveat that an approach that is optimal in mice may not prove successful in humans.

Prime–Boost Regimens

Vaccination schedules based on combined prime–boost regimens using different vector systems to deliver the desired antigen have successfully improved the outcome of DNA vaccination. The DNA prime–viral vector boost approach, in which the same antigen is used to prime and boost the response, focuses on the induction of T-cell immune responses. Viral vectors that have been tested as booster vaccines include adenovirus, vaccinia virus, fowlpoxvirus, and recombinant vesicular stomatitis virus.

When the priming and boosting vectors used are different, this strategy allows for greater expansion of disease antigen-specific T-cell populations and elicits the most potent cellular immune responses. These antigen-specific T-cell responses are 4-fold to 10-fold higher than those obtained when the same priming and boosting vectors are used. Due to the strong Th1-biased and MHC-I-restricted immune responses that DNA vaccines can induce, they are attractive priming agents in prime–boost regimens for clinical trials of infectious diseases in which this type of immunity correlates with protection. A summary of results obtained in recent clinical trials of cancer and tuberculosis is presented below.

Clinical Trials

Injection of plasmid DNA vaccines into human patients is well tolerated and safe. DNA vaccines currently being tested do not show relevant levels of integration into host cellular DNA. Preclinical studies with nonhuman primates and humans did not detect increases in antinuclear or anti-DNA antibodies. No evidence of induction of autoimmunity against the plasmid DNA has yet been reported for a DNA vaccine.

The earliest phase I clinical trial was conducted with a DNA vaccine for HIV-1 used in individuals infected by HIV-1 and in volunteers who were not infected by HIV-1. Other prophylactic and therapeutic DNA vaccine trials followed, including trials that tested DNA vaccines against cancer, tuberculosis, influenza, malaria, hepatitis B, and HIV-1. These trials demonstrated that the DNA vaccine platform is well tolerated and safe, as no adverse events were reported and all trials were carried to completion. The evidence of the safety of DNA vaccines led to a relaxation of the requirements for approval by both the U.S. Food and Drug Administration and the national regulatory authorities in Europe. Thus, many phase I and phase II clinical trials are currently being conducted.

Cancer

Since tumor antigens are weakly immunogenic, they often induce a low level of spontaneous immunity and in some cases result in tolerance. To increase the immunogenicity of a DNA vaccine, the use of adjuvants (e.g., cytokines) and immunomodulatory molecules has been extensively employed in clinical trials. Clinical trials conducted over the last few years have provided promising results, particularly when DNA vaccines were used in combination with other forms of vaccines, as demonstrated in prostate and liver cancer clinical trials. Delivery of gene-based vaccines by physical methods (electroporation and gene gun) has amplified the

Table 11.11 Use of DNA vaccines in phase I or II clinical trials of cancer

Tumor	Objectives	Status	Results	Side effects
Lymphoma	Safety, dose, immunogenicity	Completed	Safe, nontoxic, antibody and cellular immune responses observed	
Prostate	Safety, efficacy	Open, recruiting		
	Safety, feasibility, immunogenicity	Open	Safe, nontoxic, PAP-specific CD8$^+$ T-cell secretion of IFN-γ	Transient pain at injection site
	Safety	Completed	PSAP-specific CD4$^+$ and CD8$^+$ T-cell proliferation	
	Antibody response to PSMA	Completed	Anti-PSMA antibodies detected	
Melanoma	Immune and clinical responses	Completed	No regression of established disease	Grade I or II toxicity
	Safety, tolerability Antitumor response	Open, not recruiting	PAP-specific T-cell proliferative responses observed	Grade I toxicity
Cervical	Safety, feasibility, detection of changes in lesion size and HPV load, immune responses and clinical responses	Open, not recruiting	Safe, nontoxic, no significant differences between vaccinated and nonvaccinated cohorts	Transient pain at injection site
Liver	Dose-limiting toxicity and maximum tolerated dose	Completed	Nontoxic	
Breast	Memory T-cell response to HER2 ICD	Open, recruiting	Nontoxic	

Adapted from Fioretti et al., *J. Biomed. Biotechnol.* **2010**:174378, 2010.

PSMA, prostate-specific membrane antigen, also known as glutamate carboxypeptidase II; PAP, prostatic acid phosphatase; PSAP, prostate-specific acid phosphatase (an enzyme produced by the prostate in males); HER2 ICD, human epidermal growth factor receptor 2 intracellular domain.

immune responses induced by therapeutic vaccines against cancer. Several completed and ongoing clinical trials of different types of cancer in which DNA vaccines have been used are summarized in Table 11.11.

Tuberculosis

Novel vaccines are required to replace *Mycobacterium bovis* BCG vaccines to achieve enhanced protection against *Mycobacterium tuberculosis* infection worldwide. However, to date, a novel vaccine of this type for clinical use has not yet been developed. Recently, two novel tuberculosis vaccines were developed based on a plasmid DNA platform. A DNA vaccine was produced using mycobacterial heat shock protein 65 (HSP65) and IL-12 by using the hemagglutinating virus of Japan (HVJ) liposome (HSP65 + IL-12/HVJ) (Fig. 11.19). A mouse IL-12 expression vector (mIL-12 DNA) encoding single-chain IL-12 proteins comprised of the p40 and p35 subunits was constructed. In a mouse model, a single gene gun vaccination with the combination of HSP65 DNA and mIL-12 DNA provided significant protection against challenge with virulent *M. tuberculosis*; bacterial numbers were 100-fold lower in the lungs of these mice than in those of control mice vaccinated with only BCG.

Clinically, a comparison of the HVJ liposome-encapsulated HSP65 DNA and mIL-12 DNA (HSP65 + mIL-12/HVJ) DNA vaccine to BCG treatment revealed that the HVJ liposome method improved the protective efficacy of the HSP65 DNA vaccine in mice and guinea pigs much more than the gene gun method of vaccination. This increased protection was associated with the ability of the HSP65 + IL-12/HVJ vaccine to induce CD8$^+$ CTL activity and IFN-γ secretion in response to HSP65. In a cynomolgus monkey model, currently the best animal model of human

HSP65
heat shock protein 65

HVJ
hemagglutinating virus of Japan

mIL-12
mouse interleukin-12

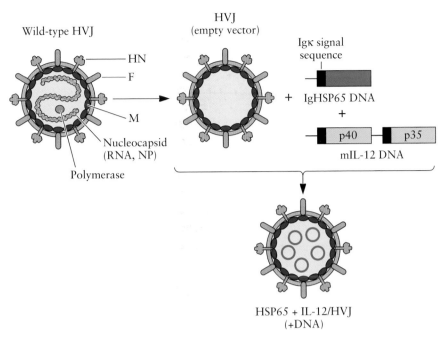

Figure 11.19 Schematic diagram of an *M. tuberculosis* DNA vaccine. The HSP65 gene was amplified from *M. tuberculosis* H37Rv genomic DNA and cloned into pcDNA3.1 to generate pcDNA-hsp65 (designated HSP65 DNA). The *hsp65* gene was fused with the mouse Ig κ secretion signal sequence to generate pcDNA-Ighsp65 (designated IgHSP65 DNA). For construction of the mIL-12 *p40* and *p35* single-chain genes, *mIL12p35* and *mIL12p40* genes were cloned from pcDNAp40p35, fused, and cloned into pcDNA3.1 to generate pcDNA-mIL12p40p35-F (designated mIL-12 DNA). IgHSP65 DNA and mIL-12 DNA were incorporated into an empty vector (nonviral vector) containing the lipid envelope of HVJ. DNA vaccination was performed with the HSP65 + mIL-12/HVJ liposomal vector [HVJ (+ DNA)]. HN, hemagglutinin-neuraminidase protein; F, fusion protein; M, matrix protein; NP, nucleoprotein. Adapted from Okada et al., *Clin. Dev. Immunol.* **2011:**549281, 2011. doi:10.1128/9781555818890.ch11f.11.19

tuberculosis, vaccination with HSP65 + IL-12/HVJ provided significantly more protection than BCG. Most importantly, HSP65 + IL-12/HVJ resulted in an increased survival of the host monkeys for more than 1 year. This is the first report of successful DNA vaccination against *M. tuberculosis* in the monkey model. These results suggest that the HSP65 + IL-12/HVJ DNA vaccine may be a promising candidate for a new tuberculosis DNA vaccine, which is superior to the currently available BCG vaccine. It remains to optimize the HSP65 + IL-12/HVJ DNA vaccine in nonhuman primates and then proceed to test the vaccine in human clinical trials.

The rationale behind the booster vaccines administered to individuals receiving only BCG is that BCG-induced immunity wanes with time and lasts only 10 to 15 years. These boosting protocols increase MHC-II-restricted CD4$^+$ T-cell responses but not MHC-I-restricted CD8$^+$ T cells (poorly induced by the BCG vaccine). A high level of CD8$^+$ T-cell activity is required to control latent TB and the prevention of its reactivation. Use of attenuated, live *M. tuberculosis* mutants may generate CD8$^+$ more efficiently and requires further experimentation. Plasmid DNA vaccines are the most powerful subunit vaccines capable of triggering

CD8$^+$ T-cell responses. In contrast to viral vectors, they do not induce vector immunity and can be used for repeated boosting when needed. Combination of plasmid DNA encoding early secreted and particularly latency-associated antigens with the existing BCG vaccine may provide a novel and much improved vaccination approach that may overcome the weak potential of BCG to trigger MHC-I-restricted immune responses and achieve significantly enhanced protection against *M. tuberculosis* infection.

Attenuated Vaccines

Genetic manipulation may be used to construct modified organisms (bacteria or viruses) that are used as live recombinant vaccines. These vaccines are either nonpathogenic organisms engineered to carry and express antigenic determinants from a target pathogen or engineered strains of pathogenic organisms in which the virulence genes have been modified or deleted. In these instances, as part of a bacterium or a virus, the important antigenic determinants are presented to the immune system with a conformation that is very similar to the form of the antigen in the disease-causing organism. Although successful in some cases, purified antigen alone often lacks the native conformation and elicits a weak immunological response.

Cholera

It is usually advantageous to develop a live vaccine, because they are generally much more effective than killed or subunit vaccines. The major requirement for a live vaccine is that no virulent forms be present in the inoculation material. With this objective in mind, a live cholera vaccine has been developed. **Cholera**, caused by the gram-negative bacterium *V. cholerae*, is a fast-acting intestinal disease characterized by fever, dehydration, abdominal pain, and diarrhea. It is transmitted by drinking water contaminated with fecal matter. In developing countries, the threat of cholera is a real and significant health concern whenever water purification and sewage disposal systems are inadequate.

The burden of cholera is greatest in sub-Saharan Africa and South Asia. Incidence rates for children less than 5 years of age are significantly higher than for older children or adults in areas where cholera is endemic. Among the 1.4 billion people at risk of cholera, 2.8 million cholera cases are estimated to occur in countries where it is endemic (maintained) each year and 87,000 cases in countries where it is not endemic. These estimates exclude the recent Haitian outbreak during which an estimated 500,000 cases and 7,000 deaths occurred less than 18 months after the outbreak started in October 2010. These findings validate the high public health need to vaccinate against cholera to help eradicate the global cholera burden.

Since *V. cholerae* colonizes the surface of the intestinal mucosa, it was reasoned that an effective cholera vaccine should be administered orally and directed to this structure. With this in mind, a strain of *V. cholerae* was created with part of the coding sequence for the A1 peptide deleted.

This strain cannot produce active enterotoxin; therefore, it is nonpathogenic and is a good candidate for a live vaccine. Specifically, a tetracycline resistance gene was incorporated into the A1 peptide DNA sequence on the *V. cholerae* chromosome. This insertion inactivated the A1 peptide activity and also made the strain resistant to tetracycline. Although the A1 peptide sequence has been disrupted, the strain is not acceptable as a vaccine because the inserted tetracycline resistance gene can be excised spontaneously, thereby restoring enterotoxin activity.

Consequently, it was necessary to engineer a strain carrying a defective A1 peptide sequence that could not revert (Fig. 11.20). In this strategy, a plasmid containing the cloned DNA segment for the A1 peptide

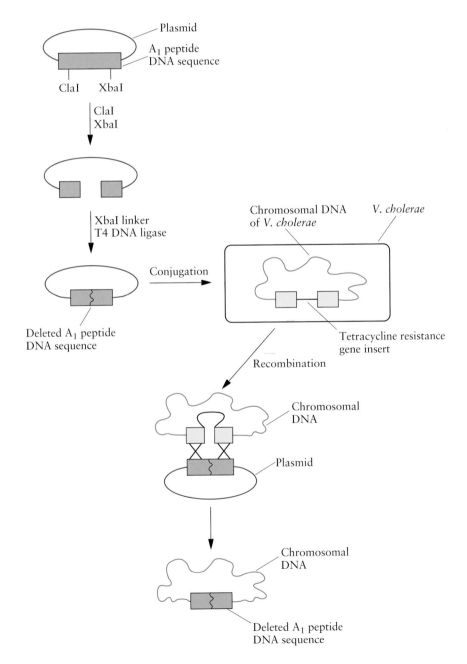

Figure 11.20 Strategy for deleting part of the cholera toxin A1 peptide DNA sequence from a strain of *V. cholerae*. Note that the tetracycline resistance gene is introduced into the gene for the A1 peptide as part of a transposon. This construct no longer makes A1 peptide; however, it cannot be used as a vaccine because the transposon may be excised and thereby restore A1 synthesis and pathogenicity.
doi:10.1128/9781555818890.ch11f.11.20

was digested with the restriction enzymes ClaI and XbaI, each of which cut only within the A1 peptide-coding sequence of the insert. To recircularize the plasmid, an XbaI linker was added to the ClaI site and then cut with XbaI. T4 DNA ligase was used to join the plasmid at the XbaI sites, thereby deleting a 550-base-pair (bp) segment from the middle of the A1 peptide-coding region. This deletion removed 183 of the 194 amino acids of the A1 peptide. By conjugation, the plasmid containing the deleted A1 peptide-coding sequence was then transferred into the *V. cholerae* strain carrying the tetracycline resistance gene in its A1 peptide DNA sequence. Recombination (a double crossover) between the remaining A1 coding sequence on the plasmid and the tetracycline resistance gene-disrupted A1 peptide gene on the chromosome replaced the chromosomal A1 peptide-coding sequence with the homologous segment on the plasmid carrying the deletion. After growth for a number of generations, the extrachromosomal plasmid, which is unstable in *V. cholerae*, was spontaneously lost. Cells with an integrated defective A1 peptide were selected on the basis of their tetracycline sensitivity. The desired cells no longer had the tetracycline resistance gene but carried the A1 peptide sequence with the deletion.

A stable strain with an A1 peptide sequence containing a deletion was selected in this way. This strain did not produce active enterotoxin but nevertheless retained all the other biochemical features of the pathogenic form of *V. cholerae*; i.e., *V. cholerae* with an A1 peptide containing a deletion is a good vaccine candidate because the bacterium that synthesizes only the A2 and B peptides is as immunogenic as the native bacterium. When this strain was evaluated in clinical trials to test its effectiveness as a cholera vaccine, the results were equivocal. While the vaccine conferred nearly 90% protection against diarrheal disease in volunteers, it induced side effects in some of those who were tested. This strain may require modification at another chromosomal locus before it can be used as a vaccine.

The whole-cell cholera vaccine which is administered parenterally is not recommended by the WHO due to its low protective efficacy and diarrhea-inducing tendency. However, currently available oral cholera vaccines are safe and offer good protection (more than 70%) for an acceptable period (at least 1 year). The use of oral cholera vaccines is considered a public health tool additional to usually recommended cholera control measures, such as provision of safe water and adequate sanitation. Oral cholera vaccine use is recommended for populations to limit the risk of occurrence of cholera outbreaks in displaced populations in areas where cholera is endemic and of the spread and incidence of cholera during an outbreak.

Several experimental live oral candidate vaccines are currently being developed. Live attenuated vaccines confer greater and longer-term protection due to rapid intestinal colonization and eliminate the need for repeated dosing. They have the potential to be administered as a single dose and can be easily integrated into a suitable vaccine schedule. CVD 103 HgR (Orochol or Mutachol) was the first licensed live attenuated vaccine to reach the market. It was given orally together with a buffer in a single-dose schedule. In placebo-controlled trials, the vaccine proved to be safe, immunogenic, and effective. However, in a subsequent large-scale trial conducted in an area where cholera is endemic, the vaccine failed to demonstrate protection. As a result, the production of this vaccine was stopped in 2004.

Three live attenuated vaccines are now in the active stage of development. The first one is trehalose-reformulated Peru-15 (Choleragarde), which is derived from the vaccine strain Peru-15, which was created from a *V. cholerae* O1 El Tor Inaba strain isolated in Peru in 1991. The strain is genetically engineered to be nontoxigenic (Fig. 11.20) and is now undergoing phase II clinical trials. The second and third vaccines are known as Cuban 638 and VA 1.4, respectively. These three vaccines are stored in lyophilized form, solubilized in buffer before use, and then administered orally.

Salmonella Species

Other attempts to engineer nonpathogenic strains of pathogenic bacteria that could be used as live vaccines have involved deletions in chromosomal regions that code for independent and essential functions. At least two deletions are preferred, because the probability that both sets of functions can be simultaneously reacquired is very small. It is assumed that a "doubly deleted" strain would have a limited ability to proliferate when it is used as a vaccine, thereby restricting its pathogenicity while allowing it to stimulate an immunological response.

Strains of the genus **Salmonella** cause enteric fever, infant death, typhoid fever, and food poisoning. Therefore, an effective vaccine against these organisms is needed. Deletions in a number of different genes have been used to attenuate various *Salmonella* strains (Table 11.12). These mutations can be grouped into three basic categories: mutations in

Table 11.12 Deleted genes and their functions in the development of attenuated strains of *Salmonella* spp.

Deleted gene(s)	Gene function(s)
galE	Synthesis of LPS; decrease toxicity from galactose
aroA, *aroC*, or *aroD*	Synthesis of chorismate, an aromatic amino acid precursor and a PABA precursor; PABA is involved in the synthesis of iron chelators
purA or *purE*	Synthesis of purines
asd	Peptidoglycan and lysine biosynthesis
phoP and *phoQ*	Regulation of acid phosphatases and genes necessary for survival in the microphage
cya	Encodes adenylate cyclase, which is involved in cAMP synthesis
crp	Enclosed camp receptor; regulates expression of proteins involved in transport and breakdown of carbohydrates and amino acids
cdt	Involved in tissue colonization by the bacterium
dam	Encodes DNA methylase; appears to be a master switch for 20–40 different virulence genes
htrA	Encodes a stress-induced polypeptide; results in significantly reduced persistence in human tissues

cAMP, cyclic AMP; PABA, *p*-aminobenzoic acid.

(i) biosynthetic genes, (ii) regulatory genes, and (iii) genes involved in virulence. In addition, strains with more than one deletion have been constructed. For example, one double-deletion strain has deletions in the *aro* genes, which encode enzymes involved in the biosynthesis of aromatic compounds, and in the *pur* genes, which encode enzymes involved in purine metabolism. These double-deletion strains, which can be grown on a complete and enriched medium that supplies the missing nutrients, generally establish only low-level infections, since their host cells contain only a very low level of the metabolites that they require for growth. Typically, their virulence is reduced 100-fold or greater. The attenuated *Salmonella* strains are effective oral vaccines for mice, sheep, cattle, chickens, and humans.

Deletion of the *dam* gene, which encodes DNA methylase, may be a highly effective approach to produce avirulent *Salmonella* strains. The *dam* gene is a master switch that regulates the expression of 20 to 40 different *Salmonella* regulatory proteins. Thus, when mice were immunized with Dam-negative strains of *Salmonella*, they tolerated up to 10,000 times the normally lethal dose. Generally, pathogenic bacteria turn on many of their genes as briefly as possible to avoid detection and attack by the host's immune system. However, with Dam-negative strains, these genes are expressed for much longer periods, making it easier for the host immune system to detect and destroy the invading bacteria. Because many other gut-colonizing bacteria have *dam* genes, if this approach with *Salmonella* turns out to be as effective as is expected, it may be possible to utilize a similar protocol with a range of pathogenic bacteria.

As discussed above, live attenuated *Salmonella* strains are excellent carriers or vectors for prokaryotic or eukaryotic antigens, as they can stimulate strong systemic and local immune responses against the expressed antigens. With this aim in mind, three *Salmonella enterica* serovar Typhi strains were engineered to express a gene encoding the α-helical domain of the *Streptococcus pneumoniae* surface protein, PspA. The latter strains served as live biological vaccine vectors in a recent phase I clinical trial to evaluate maximum safe and tolerable single-dose levels after their oral administration to subjects. The objectives of the study were (i) to evaluate maximum safe tolerable single-dose levels of the three recombinant attenuated *S. enterica* serovar Typhi vaccine vectors using dose escalation, dose-sequential studies in healthy adult subjects and (ii) to evaluate immunogenicity of the three recombinant attenuated *S. enterica* serovar Typhi vaccine vectors with regard to their abilities to induce mucosal and systemic antibody responses to the *S. pneumoniae* PspA and *S. enterica* serovar Typhi antigens. The vaccines are not anticipated to prevent disease. Although the immune responses generated by the vaccine vectors may confer some degree of protection against future infection with *S. pneumoniae* and *S. enterica* serovar Typhi, such protection is incidental. The primary goals of this study were (i) to select the *S. typhi* vector that provides optimal delivery of the PspA antigen in a safe and immunogenic manner and (ii) to demonstrate that the regulated attenuation strategy results in highly immunogenic antigen delivery vectors for oral vaccination. The trial has been completed, and the results are forthcoming in the near future.

Leishmania **Species**

Although the human immune system can respond to infections by protozoan parasites of the genus ***Leishmania***, it has been difficult to develop an effective vaccine against these organisms. Attenuated strains of *Leishmania* are sometimes effective as vaccines; however, they often revert to virulence. Also, the attenuated parasite can persist for long periods in an infected but apparently asymptomatic individual. Such individuals can act as reservoirs for the parasite, which can be transferred to other people by an intermediate host. To overcome these problems, an attenuated strain of *Leishmania* that is unable to revert to virulence was created by targeted deletion of an essential metabolic gene, such as dihydrofolate reductase–thymidylate synthase. In one of these attenuated strains, *Leishmania major* E10-5A3, the two dihydrofolate reductase–thymidylate synthase genes present in wild-type strains were replaced with the genes encoding resistance to the antibiotics G-418 and hygromycin. For growth in culture, it is necessary to add thymidine to the medium that is used to propagate the attenuated (but not the wild-type) strain. In addition, unlike the wild type, the attenuated strain is unable to replicate in macrophages in tissue culture unless thymidine is added to the growth medium (Fig. 11.21). Importantly, the attenuated strain survives for only a few days when inoculated into mice; in that time, it does not cause any disease. Moreover, this period is sufficient to induce substantial immunity against *Leishmania* in BALB/c mice after administration of the wild-type parasite. Since the attenuated parasite did not establish a persistent infection or cause disease, even in the most susceptible mouse strains tested, it is considered to be a strong candidate vaccine. Following additional experiments with animals, it will be possible to test whether this attenuated parasite is effective as a vaccine in humans.

Poliovirus

Two effective vaccines, the inactivated polio vaccine and oral polio vaccine, which consists of the live attenuated Sabin strains, have been used for the control of paralytic **poliomyelitis** since the 1950s. Both vaccines

Figure 11.21 Schematic representation of the *Leishmania major* DHFR-TS E10-5A3 auxotrophic cell line. The DHFR-TS E10-5A3 line was produced by targeted deletion of two copies of an essential metabolic gene, *dhfr-ts* (which encodes the bifunctional dihydrofolate reductase–thymidylate synthase), in *L. major*. This line does not cause disease when injected into highly susceptible mouse strains, and vaccination confers significant and specific protective immunity. In line E10-5A3, the two *dhfr-ts* genes were replaced by the *hph* (hygromycin phosphotransferase [HYG]) gene, which confers resistance to hygromycin B, and the *aph* (neomycin phosphotransferase [NEO]) gene, which confers resistance to aminoglycosides such as G-418. doi:10.1128/9781555818890.ch11f.11.21

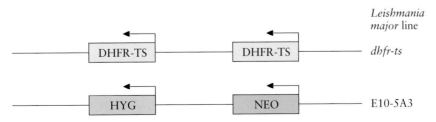

are effective in providing individual protection against poliomyelitis, but the ease of oral administration and lower costs of the attenuated oral polio vaccine favored its use in most countries. The Global Poliomyelitis Eradication Initiative has been very successful since its launch in 1988, with annual reported cases caused by wild-type polioviruses decreasing by about 96% in 2010 (from 35,251 cases appearing in 125 countries in 1988 to 1,352 cases appearing in four countries—Afghanistan, Pakistan, India, and Nigeria—in 2010). The availability of monovalent and bivalent oral vaccines against poliovirus types 1 and 3 could help to accelerate eradication in countries where polio is endemic and act as possible tools for future outbreak control.

Use of an attenuated oral vaccine is complicated by the rare outcomes of vaccine-associated paralytic poliomyelitis, chronic shedding of revertant poliovirus by B-cell-immunodeficient individuals, and the emergence of circulating vaccine-derived poliovirus strains, which are biologically equivalent to wild-type polioviruses. The accumulation of mutations in the determinants of the thermosensitive phenotype of attenuated Sabin oral vaccine strains, in addition to recombination events during their replication in the human gut, has been associated with the reversion of attenuated oral vaccine strains to neurovirulent ones. Thus, it is important that the use of the attenuated oral vaccine strains be halted as soon as possible to eradicate polioviruses. Inactivated vaccine used in routine vaccination schedules is an optimal solution to prevent the oral-vaccine-related risks of vaccine-associated paralytic poliomyelitis and vaccine-derived poliovirus strains.

To avoid recurrent paralytic poliomyelitis, a sequential inactivated virus or an inactivated and attenuated oral virus vaccination schedule was implemented in most countries. The problems associated with the worldwide use of inactivated virus include the high cost, the injectable route of administration, the induction of diminished mucosal intestinal immunity, and the need for biocontainment during its production. However, current research is being directed to overcome these disadvantages to maximize the chance for rapid progress. It is projected that the generation of specific drugs against polioviruses along with the new inactivated virus will help to address future polio epidemics and achieve global eradication of polioviruses. These drugs, now in the early stages of development, include replication inhibitors acting on viral replication protein 2C or 3A, capsid inhibitors acting on virus uncoating and the release of viral RNA from the capsid into the cell, protease inhibitors acting on 3C viral protease, and several nucleoside inhibitors of viral polymerases.

Influenza Virus

Influenza vaccines are formulated annually to protect against strains expected to circulate globally during the upcoming influenza seasons. Recommendations regarding strain composition are issued by the WHO and national regulatory authorities. All licensed influenza vaccines are currently trivalent, containing two type A strains (A/H1N1 and A/H3N2) and one type B strain; these strains have circulated worldwide since 1977.

Vaccine strains in licensed **live attenuated influenza vaccines (LAIVs)** are generally comparable in immunogenicity and efficacy, whether administered in monovalent or trivalent formulations. For a LAIV, reduced immunogenicity of one or more vaccine strains is not an accurate measure of clinically significant viral interference. This is because the immune mechanisms by which the vaccine exerts its effects are not well known, and an anti-HA antibody response is not an absolute correlate of protection for LAIVs. Furthermore, some influenza HAs are intrinsically less immunogenic than others owing to variations in sequence, independent of whether they are presented to the immune system as a wild-type virus, a live attenuated vaccine, or an inactivated vaccine.

Currently used LAIVs are reassorted, cold-adapted and temperature-sensitive master strain viruses with recommended seasonal strains (donate relevant HA and neuraminidase [N] protein genes) or are reverse genetic engineered (analysis of the phenotype of specific gene sequences obtained by DNA sequencing) viruses that express those cold-adapted and temperature-sensitive mutations as well as other attenuating mutations. **Reassortment** occurs when two similar viruses, such as influenza viruses, that infect the same cell exchange DNA or RNA. If a single host (human or animal) is infected by two different influenza virus strains, new assembled viral particles may be created from segments whose origins are mixed, some derived from one strain and some from the other strain. The new strain will share properties of both of its parental lineages.

The reassorted or reverse genetically engineered viruses are limited in their ability to replicate at the warmer temperatures of the lower respiratory tract after intranasal administration. Live attenuated vaccines have superior efficacy to inactivated virus without adjuvant in young children, but they are not licensed for use in children less than 2 years of age and have restricted use in children under 5 years of age who commonly wheeze. The efficacy of LAIVs falls by young adulthood, probably because acquired, cross-reactive immunity limits replication of the attenuated virus after intranasal administration. The risk of wheezing in children and reduced efficacy in older individuals are also likely to pose challenges for the new live attenuated vaccines in development.

In view of such challenges, novel approaches to generating a live attenuated influenza vaccine by deleting or altering the NS1 gene are being explored. The resultant virus, ΔNS1-H1N1, contains the surface glycoproteins from A/NC/20/99, whereas the remaining gene segments are from the influenza virus strain IVR-116. In addition, ΔNS1-H1N1 lacks the complete NS1 open reading frame (Fig. 11.22).

The NS1 protein inhibits host resistance to infection by several mechanisms including the blocking of IFN-α production. If NS1 function is lost or reduced, a greater host native immune response to vaccination leads to a more robust cellular and humoral immune response and inhibits replication of the ΔNS1-mutated virus. Limited replication attenuates the infection and, importantly from a clinical safety perspective, limits viral shedding and avoids transmission of the vaccine virus with the potential for reassortment. Human volunteer recipients of a monovalent ΔNS1 vaccine candidate produced both nasal wash and serum antibodies,

LAIV
live attenuated influenza vaccine

N
neuraminidase

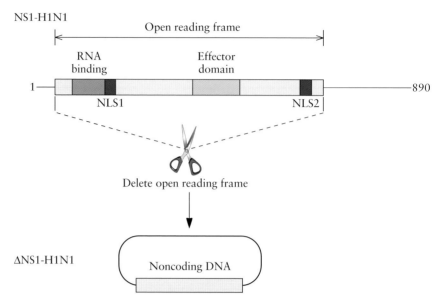

Figure 11.22 Diagram of the DNA sequence present in the NS1-HIN1 and ΔNS1-H1N1 influenza viruses. The variant vaccine virus, ΔNS1-H1N1, was generated by deletion of the open reading frame from the wild-type NS1-H1N1 influenza virus. ΔNS1-H1N1 virus expresses surface glycoproteins and various gene segments inherited from other influenza virus strains. The open reading frame of the nonstructural protein NS1 contains the RNA-binding domain, effector domain, and the two nuclear localization signals (NLS1 and NLS2). Deletion of this open reading frame totally inactivates the NS1 viral protein, which is desirable since this deletion eliminates the ability of NS1 to counteract the IFN-mediated immune response of the host. Instead, upon infection of host cells, the ΔNS1/H1N1 virus elicits high levels of IFN, which promotes strong B- and T-cell-mediated immune responses in the host. The latter responses protect the host against NS1-containing wild-type influenza virus challenge. ΔNS1 viruses do not replicate or form viral progeny, which prevents vaccinated hosts from shedding vaccine virus. Adapted from Egorov et al., *J. Virol.* **72:**6437–6441, 1998. doi:10.1128/9781555818890.ch11f.11.22

with almost no viral shedding, and experienced local symptoms similar to those experienced by placebo recipients. Serum antibodies that were induced despite the intranasal route of administration also neutralized antigenically different viruses. A live attenuated trivalent ΔNS1 formulation is under clinical evaluation. Viruses attenuated with more limited deletions or mutations, coupled with mutations in other gene segments or in NS2, are in preclinical development.

Four strains may be recommended for inclusion in influenza vaccine formulations in the future, based on epidemiology and the inability of trivalent formulations containing a single influenza B virus strain to protect against the two distinct B lineages that frequently cocirculate. The lineage responsible for annual epidemics caused by influenza B virus has been difficult to predict. In the United States, in 5 of the past 10 influenza seasons, the predominant circulating influenza B virus lineage differed from that present in the vaccine. Thus, the inclusion of another B strain in the annual influenza vaccine formulations would increase the chance of the vaccine matching the circulating epidemic strains of influenza.

Human clinical trials in adults and children of quadrivalent LAIVs (multivalent, Ann Arbor strain) and inactivated vaccines containing influenza B virus strains representing both lineages are currently under way, and safety and immunogenicity data in comparison with trivalent vaccines will be available in the near future. For LAIVs, because vaccine strains for each influenza B virus lineage (B/Victoria and B/Yamagata) are efficacious in the trivalent formulation, the main question is whether the addition of a second B lineage LAIV strain interferes with the replication of the other vaccine strains. This question was evaluated in nonclinical studies in seronegative ferrets, with no evidence of significant interference. In humans, demonstration of quadrivalent vaccine efficacy in randomized controlled field studies with all four vaccine components is not feasible because it would require multiple years and because in many countries, placebo-controlled studies are no longer ethical owing to recommendations for the annual vaccination of children against influenza. As a result, the clinical development of a quadrivalent LAIV will involve assessment of safety and immunogenicity in children and adults. The goal is to demonstrate comparable safety and immunogenicity between a quadrivalent LAIV and two trivalent LAIV formulations, each including a B strain of a different influenza B virus lineage. A quadrivalent vaccine would be expected to have safety and effectiveness comparable to those of the currently licensed trivalent vaccines, with additional protection against the alternate influenza B lineage. Quadrivalent seasonal influenza vaccines are expected to enter the commercial market within 2 to 5 years.

In a recent study of 88,468 participants vaccinated with different influenza vaccines, live attenuated vaccines performed better than inactivated vaccines in children (80% versus 48%), whereas inactivated vaccines performed better than live attenuated vaccines in adults (59% versus 39%). There was a large difference (20%) in efficacy against influenza A (69%) and influenza B (49%) virus types for unmatched strains. The conclusions reached are that influenza vaccines are efficacious, but efficacy estimates depend on many variables, including type of vaccine and age of persons vaccinated, degree of matching of the circulating strains to the vaccine, influenza type (relevant to unmatched strains), and methods used to determine endpoints of analysis.

Dengue Virus

Dengue is one of the most important and widespread mosquito-borne viral diseases of humans, with about half the world's population now at risk. The WHO estimates that 50 million to 100 million dengue virus infections occur each year in more than 100 countries and that half a million people develop severe dengue necessitating hospital admission. A specific treatment is not available, and in absence of a vaccine, prevention relies on individual protection against mosquitoes and vector control strategies.

The major challenges facing the development of a dengue vaccine include the existence of four pathogenic dengue virus serotypes (serotypes 1 to 4) that compete and interact immunologically. Additional problems are associated with the lack of suitable animal models or a correlate of protection. During the last 50 years, various vaccine approaches have been attempted, and several candidate vaccines are in early clinical or preclinical development. One candidate vaccine, CYD-TDV, is a recombinant, live, attenuated, tetravalent dengue vaccine based on the yellow fever 17D vaccine strain and produced in Vero cells (monkey kidney epithelial cells are used as host cells for growing virus) (Fig. 11.23). Previously, phase I and II trials conducted in southeast Asia and Latin America with cohorts of adults and children, who did not have immunity against dengue virus, showed that a three-dose regimen given over 12 months is well tolerated and elicits balanced neutralizing antibody responses against the four serotypes. More recently, a randomized, controlled, single-center, phase IIb, proof-of-concept trial was conducted with healthy 4- to 11-year-old Thai schoolchildren, who received three injections of the CYD-TDV live

Figure 11.23 Schematic diagram of the CYD-TDV dengue virus vaccine. In 2010, the first phase III clinical trial to investigate the CYD tetravalent dengue vaccine (TDV) was initiated. The CYD-TDV candidate is composed of four recombinant, live, attenuated vaccines (CYD 1 to 4) based on a yellow fever virus vaccine 17D (YFV 17D) backbone, each expressing the premembrane (prM) and envelope (E) genes of one of the four wild-type dengue virus serotypes. These wild-type viruses were the PUO-359/TVP-1140 Thai strain for serotype 1, PUO-218 Thai strain for serotype 2, PaH881/88 Thai strain for serotype 3, and 1228 (TVP-980) Indonesian strain for serotype 4. The CYD-TDV vaccine is produced by combination of the four CYD viruses into a single preparation. The vaccine is freeze-dried, contains no adjuvant or preservative, and is presented in a single-dose vial or in a five-dose multidose vial. Preclinical studies have demonstrated that CYD-TDV induces controlled stimulation of human dendritic cells and significant immune responses in monkeys. Adapted from Guy et al., *Vaccine* **29**: 7229–7241, 2011. doi:10.1128/9781555818890.ch11f.11.23

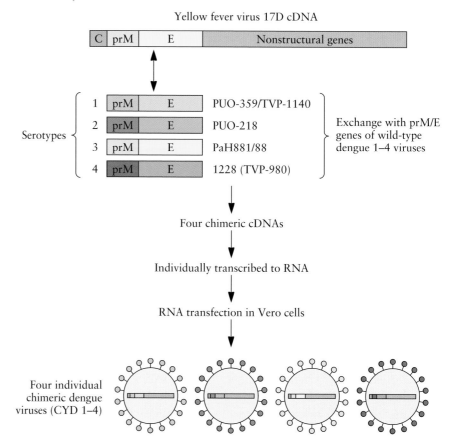

attenuated tetravalent dengue vaccine (2,452 children) or a control (rabies vaccine or placebo; 1,221 children) at 0, 6, and 12 months. The primary objective was to assess protective efficacy after three injections against virus-positive confirmed symptomatic dengue, irrespective of severity or serotype. Efficacy was found to be 30.2% and differed by serotype. Protection was limited to three of the four serotypes of dengue virus, but the vaccine did not confer protection against the most prevalent serotype (serotype 2). Nonetheless, dengue vaccine was well tolerated, with no adverse side effects after 2 years of follow-up after the first dose. These data are the first to show that the generation of a safe vaccine against dengue may be possible. Ongoing large-scale phase III studies in various epidemiological settings may further demonstrate the efficacy of a new live attenuated dengue vaccine. Further, this trial may provide proof of concept for a very significant public health care advance given that about half the world's population is currently at risk for dengue virus infection.

Vector Vaccines
Vaccines Directed against Viruses

The use of a live vaccinia virus vaccine helped to eradicate smallpox globally. The vaccinia poxvirus consists of a double-stranded DNA genome (187 kb) that encodes about 200 proteins. Cytoplasmic, rather than nuclear, replication and transcription of vaccinia virus genes are possible because vaccinia virus DNA contains genes for DNA polymerase, RNA polymerase, and the enzymes to cap, methylate, and polyadenylate mRNA. Thus, if a foreign gene is inserted into the vaccinia virus genome under the control of a vaccinia virus promoter, it will be expressed independently of host regulatory and enzymatic functions. Vaccinia virus infects humans and many other vertebrates and invertebrates.

Since vaccinia virus has a broad host range, is well characterized molecularly, is stable for years in lyophilized (freeze-dried) form, and is usually benign, this virus is an excellent candidate for a vector vaccine. A vector vaccine delivers and expresses cloned genes encoding antigens that elicit neutralizing antibodies against pathogens. The large size of the vaccinia virus genome and its lack of unique restriction sites prevent the insertion of additional DNA into this genome.

The genes for specific antigens must be introduced into the viral genome by in vivo homologous recombination. The DNA sequence that encodes a specific antigen, such as **hepatitis B core antigen (HBcAg)**, is inserted into a plasmid vector immediately downstream of a cloned vaccinia virus promoter and in the middle of a nonessential vaccinia virus gene, such as thymidine kinase (Fig. 11.24A). This plasmid is used to transfect thymidine kinase-negative animal cells in culture, usually chicken embryo fibroblasts previously infected with wild-type vaccinia virus to produce a functional thymidine kinase. Recombination between promoter-flanking DNA sequences and the neutralizing antigen gene on the plasmid with the homologous sequences on the viral genome incorporates the cloned gene into the viral DNA (Fig. 11.24B). The use of thymidine kinase-negative host cells and disruption of the thymidine kinase gene in the recombined

HBcAg
hepatitis B core antigen

A

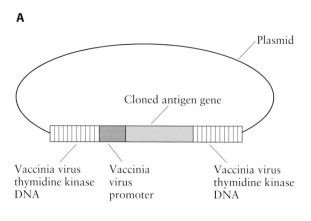

Plasmid

Cloned antigen gene

Vaccinia virus
thymidine kinase
DNA

Vaccinia
virus
promoter

Vaccinia virus
thymidine kinase
DNA

B

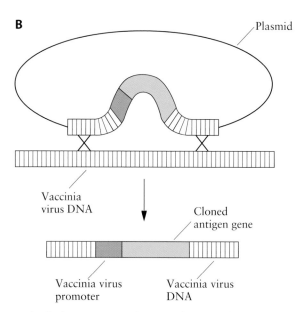

Plasmid

Vaccinia
virus DNA

Cloned
antigen gene

Vaccinia virus
promoter

Vaccinia virus
DNA

Figure 11.24 Method of integration of a gene that encodes a viral antigen into vaccinia virus. (**A**) Plasmid carrying a cloned expressible antigen gene. (**B**) A double-crossover event results in the integration of the antigen gene into vaccinia virus DNA. doi:10.1128/9781555818890.ch11f.11.24

virus render the host cells resistant to the toxic effects of bromodeoxyuridine. This selection enriches for cell lines that carry a recombinant vaccinia virus, which may be confirmed by DNA hybridization with a probe for the antigen gene. Since thymidine kinase-negative mutants of vaccinia virus arise spontaneously at a relatively high frequency, a selectable marker is often cotransferred with the target gene. This facilitates the detection of a spontaneous thymidine kinase mutant from a mutant deliberately generated by homologous recombination. A virus with a spontaneous mutation does not carry the selectable marker, while a virus that underwent homologous recombination does. The *neo* gene, which encodes the enzyme neomycin phosphotransferase II and confers resistance to the kanamycin analogue G-418, is often used as the selectable marker. The *neo* gene is relatively stable after insertion into the vaccinia virus genome.

To avoid disruption of any vaccinia virus genes or screening for selectable markers, a novel system was devised in which every recombinant
virus that forms a plaque contains and expresses the target gene. Wild-type
vaccinia virus contains the *vp37* gene, responsible for plaque formation
when the virus is grown on an animal cell monolayer (Fig. 11.25A). Deleting the *vp37* gene and replacing it with an *Escherichia coli* marker gene
(Fig. 11.25B) creates a vaccinia virus mutant that does not form plaques
after 2 to 3 days of growth in vitro. Target genes are introduced into
the mutant vaccinia virus by homologous recombination with a transfer
vector that carries *vp37* and the target gene (Fig. 11.25C). If homologous
recombination between the non-plaque-forming mutant and the transfer vector occurs, the viruses that can form plaques have acquired the
vp37 gene. Also, the target gene is inserted into the vaccinia virus genome,
and the selectable marker gene is lost. Since the *vp37* gene is deleted in
the mutant vaccinia virus, this mutation cannot revert to the wild type.
Therefore, every virus that forms a plaque carries the desired construct.
This simple procedure is applicable to the cloning and expression of any

Figure 11.25 (A) Portion of a wild-type vaccinia virus genome that contains the *vp37*
gene that is responsible for plaque formation in host cells. (B) Portion of a mutant
vaccinia virus genome in which the *vp37* gene has been replaced by a marker gene. (C)
Portion of a vaccinia virus transfer vector. "Left flank" and "right flank" refer to the
DNA sequences that immediately precede and follow the *vp37* gene in the wild-type
vaccinia virus genome. The native *vp37* promoter is part of the *vp37* gene sequence
(not shown). MCS is a multiple-cloning site with seven unique restriction enzyme
sites. *p7.5* is a strong early/late vaccinia virus promoter. The target gene is inserted
into the multiple-cloning site. Subsequently, homologous recombination between the
transfer vector (**C**) and the genomic DNA of the mutant virus (**B**) results in the replacement of the *E. coli* marker gene with the *vp37* gene, together with a target gene.
doi:10.1128/9781555818890.ch11f.11.25

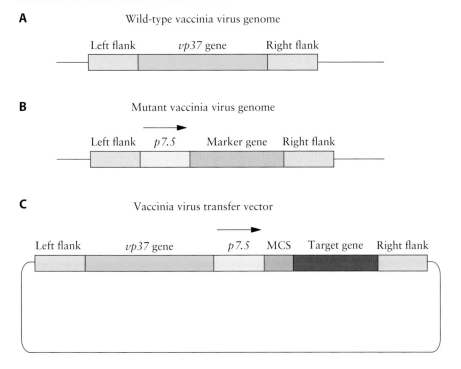

target gene, does not require extra marker genes, and does not disrupt any vaccinia virus genes.

Several antigen genes have been inserted into the vaccinia virus genome and subsequently expressed in animal cells in culture. These antigens include rabies virus G protein, hepatitis B virus surface antigen, Sindbis virus surface proteins, influenza virus nucleoprotein and HA, vesicular stomatitis virus N and G proteins, and HSV glycoproteins. Several recombinant vaccinia virus vehicles have been shown to be effective vaccines. For example, a recombinant vaccinia virus that expresses the HSV-1 gD gene prevents herpesvirus infections in mice. Another recombinant vaccinia virus that expresses the rabies virus surface antigen gene can elicit neutralizing antibodies in foxes, the major carriers of rabies in Europe. The vaccinia virus–rabies virus glycoprotein recombinant virus vaccine presently on the market (Raboral) is a live viral vaccine containing 108 plaque-forming units (PFU), or live viral particles, per dose. It is constructed by insertion of a rabies virus glycoprotein G gene into a vaccinia virus thymidine kinase gene. Upon ingestion, the vaccinia virus replicates and expresses rabies virus glycoprotein G, which elicits rabies virus glycoprotein-specific neutralizing antibodies. This immunity typically lasts about 12 months in cubs and 18 months in adult foxes.

The use of **vaccinia virus vector vaccines** carrying cloned genes encoding several different antigens enables the simultaneous vaccination of individuals against many different diseases. The timing of production of a foreign protein gene carried in a vaccinia virus-derived vector depends on whether a vaccinia virus promoter functions during the early or late phase of the infection cycle, and the strength of the promoter determines the amount of an antigen that is produced. Usually, late promoters for an 11-kilodalton (kDa) protein (p11) and the cowpox virus A-type inclusion protein (pCAE) are used to achieve high levels of foreign-gene expression. When genes encoding several different foreign proteins are inserted into one vaccinia virus vector, each is placed under the control of a different vaccinia virus promoter to avoid homologous recombination between different regions of the virus genome and deletion of the cloned genes.

A live recombinant viral vaccine has several advantages over killed virus or subunit vaccines. First, the virus can express the authentic antigen(s) in a manner that closely resembles a natural infection. Second, the virus can replicate within the host, thereby amplifying the amount of antigen that activates humoral and cellular immune responses. A disadvantage of using a live recombinant viral vaccine is that vaccination of an immunosuppressed host, such as an individual with AIDS, can lead to a serious viral infection. One way to avoid this problem may be to insert the gene encoding human IL-2 into the viral vector. IL-2 enhances the response of activated T cells and permits the recipient to limit the proliferation of the viral vector, which decreases the possibility of an unwanted infection. If the proliferation of vaccinia virus has deleterious effects in certain patients, it would be helpful to kill or inhibit it after vaccination. This may be achieved by creating an IFN-sensitive vaccinia virus (wild-type vaccinia virus is relatively resistant to IFN) whose proliferation is blocked. Such a virus vector would be susceptible to drug intervention if complications from vaccination with vaccinia virus vectors arose.

PFU
plaque-forming unit(s)

The basis of resistance of vaccinia virus to IFN was not known until a vaccinia virus open reading frame (K3L) was found to encode a 10.5-kDa protein that has an amino acid sequence very similar to that of a portion of the 36.1-kDa host cell eukaryotic initiation factor 2a (eIF-2a). The N-terminal regions of both of these proteins contain 87 amino acids that are nearly identical. Moreover, this shared sequence contains a serine at residue 51, which in eIF-2a is normally phosphorylated by IFN-activated P1 kinase. Phosphorylation of Ser51 in eIF-2a blocks protein synthesis and viral replication in IFN-treated cells. Vaccinia virus may therefore avoid inhibition by IFN because the K3L protein is a competitive inhibitor of eIF-2a phosphorylation (Fig. 11.26). Deletion of all or a portion of the K3L gene from vaccinia virus should make the virus sensitive to IFN. Indeed, when wild-type and a K3L-negative mutant of vaccinia virus were tested for sensitivity to IFN, the mutant was found to be 10 to 15 times more sensitive to IFN than the wild-type virus. Reinsertion of the wild-type K3L sequence into the mutant virus restored the level of IFN sensitivity found in the wild type. This indicates that K3L controls the IFN resistance phenotype of vaccinia virus. This finding is important for the development of safer vaccinia virus vectors, as other IFN-resistant viruses may contain sequences comparable to K3L and be amenable to the construction of IFN-sensitive deletion mutants.

Currently, several veterinary vaccinia virus-based vaccines are licensed, and clinical studies to test their efficacies in preventing a number of human infectious diseases are under way. This technology is based in part on the development of an attenuated vaccinia virus strain previously used to eradicate smallpox. To avoid any risk of the vector inducing a disease, some genetic information was removed from the virus genome so that the viral vector was highly attenuated. This attenuated virus has

eIF-2a
eukaryotic initiation factor 2a

Figure 11.26 Competitive inhibition of the IFN-stimulated phosphorylation (inhibition) of eIF-2a by protein K3L, which is encoded by vaccinia virus and is nearly identical to a portion of eIF-2a. (**A**) In the presence of IFN, a kinase is activated that phosphorylates eIF-2a molecules and thereby prevents them from functioning. (**B**) When vaccinia virus protein K3L is also present, it is phosphorylated instead of eIF-2a, which remains active. The thickness of the arrows represents the relative flux through each pathway. doi:10.1128/9781555818890.ch11f.11.26

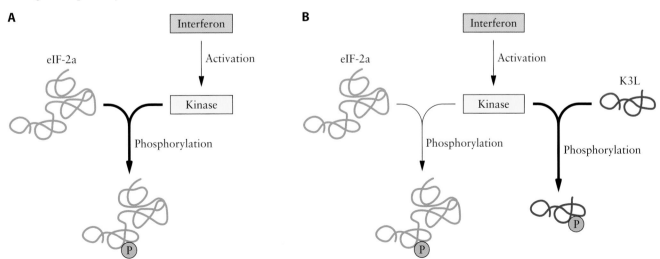

been used to express many viral antigens with the expectation that the recombinant virus would be an effective live vaccine. Protection has been achieved by cloning glycoproteins from porcine pseudorabies virus, HA glycoproteins from equine influenza virus, a spike protein from the SARS virus, and a polyprotein of Japanese swine encephalitis virus. People were then vaccinated to prevent transmission of these viruses. Based on the success of these attenuated vaccinia virus vaccines, it was proposed that this virus be considered as a general delivery system for a wide range of proteins.

For mass-vaccination campaigns in developing countries, it would be advantageous to be able to deliver live vaccines in a simple, expeditious, and cost-effective manner. In addition, with mucosally transmitted pathogens, such as HIV, traditional parenteral vaccination routes may not induce mucosal immune responses sufficient to provide protective immunity. An alternative to traditional vaccination may be aerosol immunization, which is potentially safer, easier, and less expensive to administer. Interestingly, two attenuated vaccinia virus-based vectors delivered by aerosol immunization were shown to be safe and effective and yielded long-lasting systemic and mucosal immune responses in rhesus macaques. It remains to test this approach in humans, with the hope that it may offer an effective means of inoculating large numbers of people in the future.

MVA
modified vaccinia Ankara virus

There is also considerable interest in modified vaccinia Ankara virus (MVA) as a vector system, which consists of a highly attenuated strain of vaccinia virus produced by frequent passage of vaccinia virus in chicken cells. MVA has lost about 10% of the vaccinia virus genome and cannot replicate efficiently in primate cells. MVA is considered the vaccinia virus strain of choice for clinical investigation because of its high safety profile (in monkeys, mice, swine, sheep, cattle, elephants, and humans). Studies with mice and nonhuman primates have further demonstrated the safety of MVA under conditions of immunosuppression. A recent clinical trial of a recombinant MVA–pandemic influenza A/H5N1 vaccine was reported to be successful. Currently, the use of MVA as a recombinant HIV vaccine (MVA-B) is being tested in approximately 300 volunteers in several phase I studies conducted by the International AIDS Vaccine Initiative.

Although much work on the development of live attenuated viral vaccines has been done with vaccinia virus, other viruses, such as HSV, poliovirus, and influenza virus (see previous section) as well as adenovirus and varicella–zoster virus (see chapter 10), are also being tested as potential vaccine vectors. As discussed above, live attenuated poliovirus can be delivered orally, and such a mucosal vaccine directed to receptors in the lungs or gastrointestinal tract might also be useful against a range of diseases, including cholera, typhoid fever, influenza, pneumonia, mononucleosis, and rabies. In addition, adenoviruses naturally target mucosal receptors, and mucosal administration of the vector can overcome the antivector immunity seen with parenteral administration. Defective adenovirus 5 particles expressing HA delivered intranasally can induce innate and adaptive heterosubtypic (cross-protection to infection with an adenovirus serotype other than the one used for primary infection) responses. In mice, defective adenoviruses may also induce a transient influenza-resistant state, simulating a protective drug effect, without interfering with the development of persistent immunity. More conserved viral genes and HAs of various adenovirus subtypes also have been incorporated to produce more broadly protective candidate vaccines. Orally delivered adenovirus 4 and 5 vectored vaccines are currently in development.

Vaccines Directed against Bacteria

Since the discovery and subsequent widespread dissemination of antibiotics, only a modest amount of research has been directed toward the development of vaccines for bacterial diseases. However, there is considerable need for the development of **bacterial vaccines** for several reasons. First, not all bacterial diseases are readily treated with antibiotics. Second, the use of antibiotics over the last 40 years has resulted in the proliferation of bacterial strains that are resistant to several antibiotics. Third, reliable refrigeration facilities for the storage of antibiotics are not commonly available in many tropical countries. Fourth, it is often difficult to ensure that individuals receiving antibiotic therapy undergo the full course of treatment.

Given the need to produce vaccines that will be effective against bacterial diseases, an important question is, which strategies are likely to be most effective? In instances where the disease-causing bacterium does not grow well in culture, the development of an attenuated strain is not feasible. For these types of bacteria, alternative approaches must be used. For example, *Rickettsia rickettsii*, a gram-negative obligately intracellular bacterium that causes Rocky Mountain spotted fever, does not grow in culture. In this case, a cloned 155-kDa protein that is a major surface antigen of *R. rickettsii* was used as a subunit vaccine and was found to protect immunized mice against infection by this disease-causing bacterium.

Tuberculosis

Today, along with HIV and malaria, tuberculosis remains one of the "big three" infectious diseases globally. This disease is caused by the bacterium *M. tuberculosis*, which can form lesions that lead to cell death in any tissue or organ. The lungs are most commonly affected. Patients suffer fever and loss of body weight, and without treatment, tuberculosis is often fatal. It is estimated that approximately 2 billion people are currently infected with the organism and that about 2 million to 3 million deaths per year result from these infections. During the past 50 years, antibiotics have been used to treat patients infected with *M. tuberculosis*. However, numerous multidrug-resistant strains of *M. tuberculosis* are now prevalent. In the United States, among HIV patients infected with an antibiotic-resistant strain of *M. tuberculosis*, there is a 50% mortality rate within 60 days. Consequently, a bacterial disease that was thought to be under control has again become a serious global public health problem.

Currently, in some countries, BCG, an attenuated strain of *Mycobacterium bovis* that was developed between 1906 and 1919, is still used as a vaccine against tuberculosis. However, the overall efficacy of BCG is controversial, and the use of this vaccine has some serious limitations.

While the BCG vaccine can prevent disseminated disease and mortality in newborns and children, it cannot prevent chronic infection or protect against pulmonary tuberculosis in adults. Consequently, *M. tuberculosis* establishes a latent chronic infection that reactivates when there are diminished immune responses, e.g., in aged people, in individuals with genetic immune defects, and in those whose medication reduces their immune responses, such as patients treated with antibodies against tumor necrosis factor alpha. Immunosuppression caused by HIV is now an extremely important factor in the reactivation of tuberculosis, and in the 15 million people coinfected by HIV and *M. tuberculosis*, it is the major cause of mortality in this population. About 2 billion people carry a latent *M. tuberculosis* infection, and approximately 10% progress to active disease at some time. Additional limitations to the use of BCG as a vaccine are that individuals treated with BCG respond positively to a common tuberculosis diagnostic test, which makes it impossible to distinguish between individuals infected with *M. tuberculosis* and those inoculated with BCG cells. For these reasons, the BCG strain is not approved for use in several countries, including the United States.

To determine whether a safer and more effective vaccine against tuberculosis might be developed, the extent of the immunoprotection elicited by purified *M. tuberculosis* extracellular proteins was examined. Following growth of the bacterium in liquid culture, 6 of the most abundant of the approximately 100 secreted proteins (Fig. 11.27) were purified. Each of these proteins was used separately and then in combination to immunize guinea pigs. The immunized animals were then challenged with an aerosol containing approximately 200 cells of live *M. tuberculosis*—a large dose for these animals. The animals were observed for 9 to 10 weeks before their lungs and spleens were examined for the presence of disease-causing organisms. In these experiments, some of the purified protein combinations provided a lower level of protection against weight loss, death, and infection of lungs and spleen than the live BCG

Figure 11.27 Schematic representation of the development of a multiprotein subunit vaccine for tuberculosis. The six most abundant secreted proteins from *M. tuberculosis* are purified from the growth medium and then tested for the ability to induce antibodies in guinea pigs. The immunized animals are subsequently challenged with *M. tuberculosis*. doi:10.1128/9781555818890.ch11f.11.27

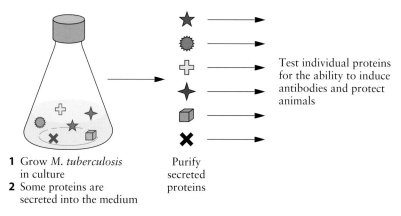

vaccine. A major protein that provided protection was the *M. tuberculosis* secretory protein, a 30-kDa mycolyl transferase also known as α-antigen, or antigen 85B (Ag85B). However, a DNA vaccine encoding this protein was even less effective than the purified secreted protein.

While this and possibly other *M. tuberculosis*-secreted proteins might eventually be part of a safe and efficacious vaccine for the prevention of tuberculosis in humans, it is necessary to develop a suitable delivery system for them. In theory, the optimal delivery system for an antigen that provides protection against tuberculosis should be (i) able to multiply in the mammalian host, (ii) nonpathogenic, and (iii) able to express and secrete the protective antigen. All of these requirements are satisfied by the available BCG strain. Therefore, an *E. coli*–mycobacterium shuttle vector that contained the gene for the 30-kDa protein (α-antigen) under the control of its own promoter was introduced into two different BCG strains (Fig. 11.28). Transformed cells produced 2.0- to 5.4-fold more 30-kDa protein than did nontransformed cells. Despite the fact that the introduced genes were plasmid carried and therefore potentially unstable, transformed cells continued to express a high level of 30-kDa protein after the vaccination of a test animal. In agreement with the hypothesis that the extracellular proteins of intracellular organisms are key immunoprotective molecules, guinea pigs immunized with transformed BCG strains had significantly fewer bacilli in their lungs and spleens. In addition, there were smaller and fewer lesions in their lungs, spleens, and livers, and the survival of the animals was significantly increased, compared with findings in animals vaccinated with a nontransformed BCG strain. This was the first report of a vaccine against tuberculosis that is more potent than the currently available commercial vaccine. This vaccine is currently in clinical trials; if it is successful, it could save tens of thousands of lives. Moreover, it is possible to prepare dried preparations of BCG that may serve as the basis for a live bacterial vaccine that is delivered as an aerosol, thereby facilitating the inoculation of newborn infants.

It is encouraging that there are 12 different vaccines against tuberculosis currently in clinical trials. Several of them are subunit vaccines consisting of recombinant antigens such as the Mtb72F fusion protein or the Ag85B–ESAT-6 fusion protein delivered with the adjuvant AS02, the Ag85–TB10.4 fusion protein delivered with the adjuvant IC31, the fusion of Ag85B–ESAT-6–Rv2660c, and a variety of antigens delivered via DNA or viral vectors. Other subunit vaccines have boosted BCG immunity in preclinical studies. These subunit vaccines could be used to boost BCG vaccination in infants in the hope of preventing chronic infection. These vaccines could also be used in adolescents and adults to boost immunity induced by BCG or natural infection to delay or avoid reactivation.

Another approach to improving tuberculosis vaccines is to reengineer BCG to achieve better priming. For example, the rBCG30 strain was engineered to overexpress Ag85B to make it more immunogenic. Clinical trials of rBCG30 were found to induce better CD4$^+$ T-cell responses against Ag85B than wild-type BCG. Another engineered BCG strain was designed to engage the MHC-I antigen presentation pathway based on the assumption that CD8$^+$ T cells are important for protection by killing

Ag85B
antigen 85B

Figure 11.28 Plasmid construct used to transform BCG to make it a more effective vaccine. The plasmid is isolated from *E. coli* cells and then introduced into BCG by electroporation. *ori* E, *E. coli* origin of replication; *ori* M, *Mycobacterium* origin of replication; Hyg^r gene, hygromycin resistance gene (and its promoter); P and α-antigen, the promoter and the coding region, respectively, of the 30-kDa secreted protein. doi:10.1128/9781555818890.ch11f.11.28

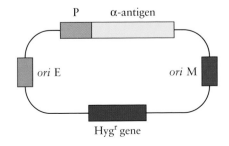

M. tuberculosis-infected cells. This strain was therefore engineered to express the cytolysin of *Listeria monocytogenes*, a protein that enables the mycobacterium to escape from the vacuole to the cytosol, where it can be presented via the MHC-I antigen presentation pathway. The vaccine strain rBCGDUreC:Hly also has an inactivated urease gene that allows better acidification of the vacuole and improves the release of the bacterium. Preclinical studies demonstrated that this vaccine was more attenuated and more protective than BCG, and it is now being tested in phase I clinical studies.

It is interesting that after a century of tuberculosis vaccine development, and after immunizing more than 3 billion people with BCG, we still know relatively little about immunity to *M. tuberculosis*. For example, we do not know why BCG induces protection, why immunity does not prevent persistent infection, or what immune responses are needed to achieve sterile immunity or to prevent reactivation of latent infection. Information about immunity to tuberculosis can, however, be obtained by studying infected individuals. Two recent studies used systems approaches to compare the transcripts in the blood of individuals with active infection to those of individuals who were latently infected. Subsets of genes were identified that correlated with the extent of the disease. Although these genome signatures are not related to tuberculosis vaccine efficacy or immunogenicity, identification of the essential cellular pathways associated with tuberculosis disease progression may help to define molecular components of these pathways that can be targeted in novel vaccines.

Autism

Clostridium bolteae is a bacterium that controls the development of gastrointestinal (gut) disorders. The number of these bacteria is frequently elevated in the gastrointestinal tracts of autistic children compared to those in healthy children. Among children with **autism spectrum disorders** (a group of developmental brain disorders), more than 90% suffer from chronic, severe gastrointestinal symptoms, and of those, about 75% have constipation and diarrheal disease.

The number of cases of autism has increased about 6-fold over the past 20 years, and we have little knowledge about the factors that predispose autistic children to *C. bolteae*. In 2012, the Centers for Disease Control and Prevention in the United States reported that the prevalence rate of autism spectrum disorders was 1 in 88 based on combined data from 14 monitoring sites obtained from 2000 to 2008. Given that autism spectrum disorders are the most common form of any neurological disorder or severe developmental disability of childhood, identification of the root cause(s) of autism has become a central focus of neurobiology research today. Although environmental factors are implicated in the development of autism, current evidence suggests that toxins and/or metabolites produced by gut bacteria, including *C. bolteae*, may also be associated with the symptoms and severity of autism, particularly **regressive autism**. Regressive autism occurs when a child appears to develop normally but then, between 15 and 30 months of age, begins to lose speech and social skills and is subsequently diagnosed with autism.

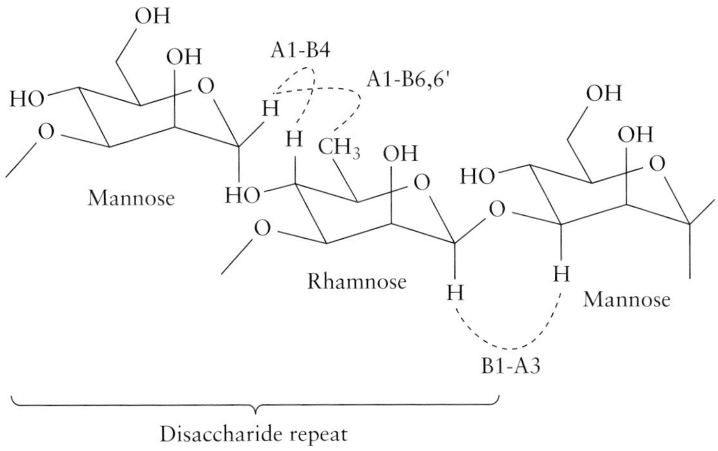

Figure 11.29 Structure of *C. bolteae* capsular polysaccharide. An analysis of the monosaccharide composition of the *C. bolteae* capsular polysaccharide surface antigen revealed the presence of rhamnose (Rha) and mannose (Man), with Man present as a 3-monosubstituted pyranose unit [→3)-Man-(→1] and Rha as a 4-monosubstituted pyranose unit [→4)-Rha-(→1]. This structure is the first to be reported for a *C. bolteae* surface antigen and presents the possibility that this polysaccharide may be used as a vaccine to reduce or prevent *C. bolteae* infection of the gastrointestinal tract in autistic patients. Adapted from Pequegnat et al., *Vaccine* **31**:2787–2790, 2013, with permission from Elsevier. doi:10.1128/9781555818890.ch11f.11.29

Although most *C. bolteae* infections are treated with antibiotics, the expectation is that a vaccine will improve current treatment. In this regard, it is of considerable interest that scientists have very recently developed a carbohydrate-based vaccine against *C. bolteae*, which has the potential to be the first vaccine designed to control constipation and diarrhea caused by this bacterium. It is possible that this vaccine may also control autism-related symptoms associated with this bacterium. The anti-*C. bolteae* vaccine targets the specific complex polysaccharides, or carbohydrates, present on the surface of the bacterium. *C. bolteae* produces a conserved specific capsular polysaccharide comprised of rhamnose and mannose units: [→3)-α-D-Manp-(1→4)-β-D-Rhap-(1→] (Fig. 11.29). When rabbits were immunized with the *C. bolteae* vaccine, the rabbits produced *C. bolteae*-specific antibodies directed against the *C. bolteae* polysaccharide. Additional experiments in animal models and clinical studies are being pursued to determine whether injection of the *C. bolteae* vaccine induces such antipolysaccharide antibodies that both detect *C. bolteae* and protect against *C. bolteae* infection, i.e., reduce or prevent *C. bolteae* colonization of the intestinal tract in autistic patients. If this vaccine proves effective and protective, this will be the first vaccine for several gut bacteria common in autistic children and may also be of benefit in the control of some autism symptoms.

Bacteria as Antigen Delivery Systems

Antigens that are located on the outer surface of a bacterial cell are generally more immunogenic than are those in the cytoplasm. Thus, localization of a neutralizing antigen from a pathogenic bacterium on the surface of a live nonpathogenic bacterium is expected to increase the immunogenicity of the antigen. Flagella consist of filaments of a single protein called **flagellin**, which microscopically appear as threadlike structures on the outer surfaces of some bacteria. If the flagella of a nonpathogenic organism could be configured to carry a specific epitope from a pathogenic bacterium, protective immunogenicity might be easily achieved.

This strategy was used to engineer a cholera vaccine (Fig. 11.30). A synthetic oligonucleotide specifying an epitope of the cholera toxin B subunit was inserted into a portion of the *Salmonella* flagellin gene that varies considerably from one strain to another (hypervariable segment). The construct was then introduced into a flagellin-negative strain of *Salmonella*. The epitope (residues 50 to 64) of the cholera toxin B subunit elicits antibodies directed against intact cholera toxin. The chimeric flagellin was found to function normally, and the epitope was expressed on the surface of the flagellum. Immunization of mice by intraperitoneal injections of approximately 5×10^6 live or formalin-killed "flagellum-engineered" bacteria elicited high levels of antibodies directed against both the peptide (residues 50 to 64) and intact cholera toxin. Two or three different epitopes can be inserted into a single *Salmonella* flagellin gene, thereby creating a multivalent bacterial vaccine.

Attenuated *Salmonella* strains can be administered orally, which would enable them to deliver a range of bacterial, viral, and parasite antigens to the mucosal immune system. For this purpose, the choice of the promoter that drives the transcription of the foreign antigen is important. If too strong a promoter is used, the metabolic load might constrain bacterial proliferation. Moreover, unlike a closed system, such as a fermentation vessel, shifting the temperature or adding specific metabolites to induce foreign-gene expression is not possible when the bacterial vector is added to a host animal. On the other hand, promoters that respond to environmental signals may provide effective means of controlling the expression of the foreign antigen gene. For example, the *E. coli nirB* promoter, which is regulated by both nitrite and the oxygen tension of the environment, is activated under anaerobic conditions. The *nirB* promoter has been used to direct the expression of the nontoxic immunogenic

Figure 11.30 Using *Salmonella* as an antigen delivery system and a flagellin–antigen fusion protein for presenting the antigen to the host immune system. A flagellin negative strain of *Salmonella* was transformed with a plasmid containing a synthetic oligonucleotide specifying an epitope of the cholera toxin B subunit inserted into a hypervariable region of a *Salmonella* flagellin gene. doi:10.1128/9781555818890.ch11f.11.30

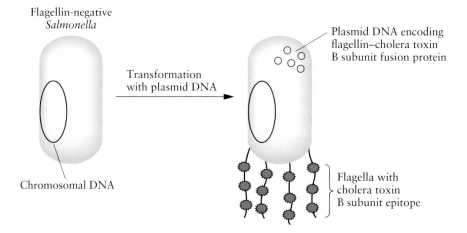

fragment C of *Clostridium tetani* toxin (tetanus toxin) in an attenuated strain of *Salmonella*. More than 1 million deaths per year in the developing world result from *C. tetani* infections. When the engineered *Salmonella* strain was grown aerobically in culture, tetanus toxin fragment C was not synthesized. However, following oral administration of the bacterium to test mice, fragment C was produced, and the animals generated antibodies against the peptide. Thus, the engineered *Salmonella* strain has potential as a live oral tetanus vaccine.

Helicobacter pylori is a gastrointestinal, microaerophilic (requires a less-than-atmospheric level of oxygen to survive) gram-negative bacterium that is widely distributed among human populations. It is believed to be the causative agent for several gastrointestinal diseases, including chronic gastritis, peptic ulcers, gastric lymphoma, and gastric cancer. Among infected individuals, which include more than half of the world's population, about 10% are at risk of developing peptic ulcers. In recent years, the medical treatment for peptic ulcers has changed from antacids to antibiotics and proton pump inhibitors. The antibiotics eradicate the *H. pylori* infection, while the proton pump inhibitors block the enzyme hydrogen–potassium ATPase, preventing the production of acid from the parietal cells at the gastric mucosa, which facilitates healing of the mucosa.

Unfortunately, *H. pylori* is resistant to many commonly used antibiotics, including metronidazole, amoxicillin, erythromycin, and clarithromycin. Treatment of *H. pylori* requires multidrug regimens because the organism resides in a layer of mucus that acts as a barrier to antibiotic penetration. In addition, the necessary course of antibiotic treatment is too expensive for populations of less developed countries.

Colonization of the gastrointestinal tract by *H. pylori* is facilitated by the action of an *H. pylori*-encoded urease. This enzyme hydrolyzes urea to carbon dioxide and ammonia, thereby neutralizing stomach acid, and enables the bacterium to survive, bind, and function in the host. Urease is a cytosolic and surface-exposed nickel metalloenzyme and is one of the most abundantly expressed proteins in *H. pylori*. The enzyme comprises two subunits, A and B, that assemble into a complex [(αβ)3]4 supramolecular structure. Subunit B is more antigenic, making it a possible vaccine candidate. To develop a vaccine that protects individuals against *H. pylori* infections, the genes encoding *H. pylori* urease subunits A and B were constitutively expressed under the control of a *Salmonella* promoter in a genetically deleted (attenuated) strain of *S. enterica* serovar Typhi (Fig. 11.31). Neither immunization with urease-expressing *S. enterica* serovar Typhi alone nor immunization with the purified urease enzyme plus an adjuvant protected against a challenge with a mouse-adapted strain of *H. pylori*. In contrast, a combined vaccination with urease-expressing *S. enterica* serovar Typhi and urease plus an adjuvant was protective. While the success of this approach remains to be established in humans, these initial results are encouraging and provide the impetus to develop a human vaccine against *H. pylori* in the near future.

Another approach to the regulation of antigen delivery by bacteria is the use of bacterial components that elicit vigorous immune responses. One such component is **lipopolysaccharide (LPS)**, which consists of a lipid

LPS
lipopolysaccharide

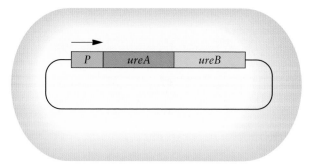

Figure 11.31 Schematic representation of an attenuated strain of *S. enterica* serovar Typhi transformed with a plasmid encoding *H. pylori* urease subunits A and B under the transcriptional control of a *Salmonella* promoter (*P*). The arrow indicates the direction of transcription. doi:10.1128/9781555818890.ch11f.11.31

covalently linked to a polysaccharide. LPS is an endotoxin that possesses high therapeutic potential by means of its adjuvant properties. On the surface of gram-negative bacteria, the hydrophobic anchor of LPS is the endotoxin lipid A that is recognized by the innate immune system. Recently, the Food and Drug Administration in the United States approved an adjuvant comprised of a less toxic combination of MPL species. Whereas wild-type *E. coli* LPS activates strong inflammatory MyD88 (myeloid differentiation primary response gene 88)-mediated TLR4 signaling, MPL elicits much lower inflammatory responses. To further determine whether structural modification of LPS molecules influences the nature and magnitude of innate immune responses, a recent study used 61 diverse *E. coli* strains generated to express unique lipid A regions (Fig. 11.32). The objective was to analyze whether this collection of variants stimulates distinct TLR4 agonist activities and cytokine induction. The reasoning was that if targeting a specific immune response through the administration of engineered LPS is possible, improved vaccine adjuvants could be developed.

Presently, the available adjuvants cannot trigger certain types of immune responses, e.g., innate immune responses. On the other hand, LPS can induce significant innate immune responses; however, it requires modification to avoid stimulation of undesired inflammatory responses too robust for safe use. The experimental results obtained showed that mice immunized with engineered lipid A–antigen emulsions exhibited robust IgG titers, indicating the efficacy of these molecules as adjuvants. Thus, this approach demonstrates how combinatorial engineering of lipid A can be exploited to generate a spectrum of immunostimulatory molecules for vaccine and therapeutic development.

The engineered library of *E. coli* strains with LPS variants described above offers a wide range of innate immune responses that may allow selection of appropriate adjuvants for many different types of vaccines. First, MPL is a combination of lipid A species from *Salmonella enterica* serovar Minnesota R595 that must be detoxified chemically by successive acid and base hydrolysis. Alternatively, one may produce and purify sufficient MPL by using an MPL-producing strain of *E. coli*. Second, these engineered *E. coli* strains provide access to whole bacteria, intact LPS (lipid and polysaccharide), and lipid A. This type of access may provide an improved vaccine delivery system. Third, many bacteria that produce

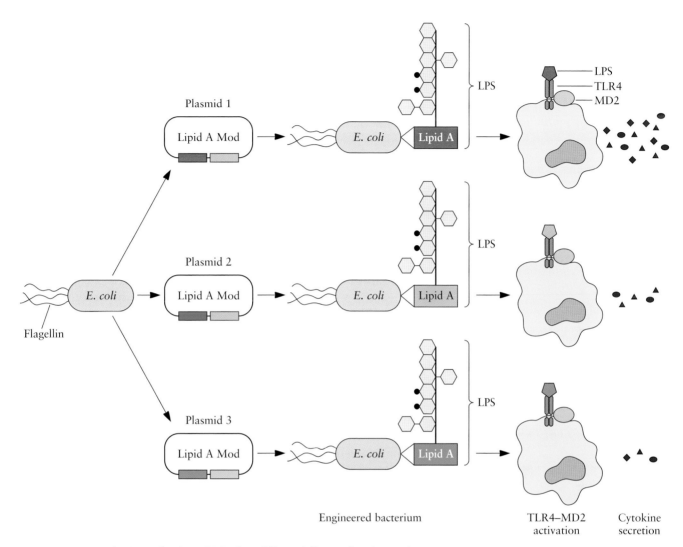

Figure 11.32 Production of unique LPSs that differentially regulate innate immune responses and vaccination. The variability in LPS structures (indicated by different colors) in the outer surface of *E. coli* strains is shown. Plasmids are engineered to contain combinations of up to five lipid A-modifying (Lipid A Mod) enzymes. Lipid A (endotoxin) is recognized by the innate immune system through the conserved PRR, the TLR4–myeloid differentiation factor 2 (TLR4–MD2) complex, which initiates a signaling pathway that elicits the production of cytokines that mediate clearance of infection. The altered LPS molecules differ in their binding and activation of the TLR4–MD2 complex. This changes the nature of downstream cytokine production, which is illustrated by shapes that indicate the different types and quantity of cytokines released. Adapted from Needham et al., *Proc. Natl. Acad. Sci. USA* **110**:1464–1469, 2013, with permission. doi:10.1128/9781555818890.ch11f.11.32

LPS are difficult to work with in the laboratory. In contrast, *E. coli* is well characterized, is easily grown in culture, and can generate numerous distinct LPS surfaces. These advantages of the use of *E. coli* strains to elicit diverse immune responses further support the idea that bacterial components may be utilized as custom-made adjuvants of the immune system. Such adjuvants, in combination with antigens from a bank of disease-causing organisms, should enhance and expedite future vaccine and therapeutic development for many existing and emerging infectious diseases.

Adjuvants

Adjuvants were discovered serendipitously more than 90 years ago in the early years of vaccine production. Those batches of vaccine that were accidentally contaminated during production frequently stimulated stronger immune responses than those obtained in the absence of the contaminant(s). Among immunologists, these contaminants were known as "the dirty little secret of immunology." The main ingredient of the vaccine, a killed or inactivated version of the bacterium or virus, protected against the pathogen and stimulated the production of specific antibacterial or antiviral antibodies. Identification of the structure of the contaminating bacterium- or virus-derived immunostimulants in the vaccine demonstrated that adjuvants are substances that function to enhance immunity to vaccines and antigens with which they are coadministered.

For about 70 years, the adjuvant of choice in nearly every vaccine worldwide was **alum**, an aluminum salt precipitate (Table 11.13). In addition to alum, vaccines containing AS04 adjuvant (a combination of alum and MPL) have been used to treat patients with hepatitis B virus or HPV infections. Although alum salts have been the most common adjuvant used, they are generally weak adjuvants and may not increase the poor immunogenicity of certain viral or bacterial antigens. Immunization with alum-adsorbed antigens is associated with $CD4^+$ Th2 cell- and antibody-mediated immunity and is generally unsuitable for eliciting the $CD8^+$ CTL-mediated immunity critical for vaccines that target cancers and intracellular pathogens. The latter deficit in CTL-mediated immunity arises frequently in the rapidly growing aging population and makes this demographic more vulnerable to many infections against which they were previously immune. Susceptibility to infections such as those caused by influenza virus, meningococci, group B streptococci, pneumococci, respiratory syncytial virus, and varicella–zoster virus is elevated in this age group. Predictably, this group of people need more frequent booster vaccinations, in many cases with vaccines enhanced by adjuvants specifically designed to stimulate the aging immune system to respond more strongly to vaccination. An example of a licensed adjuvant that has effectively boosted immune responses in the elderly is the squalene-containing oil-in-water emulsion MF59, which is licensed for use as an adjuvanted

Table 11.13 Adjuvants licensed for use in human vaccines

Adjuvant (company, year licensed)	Class	Component(s)	Vaccines (disease)
Alum (1924)	Mineral salts	Aluminum phosphate or aluminum hydroxide	Several
MF59 (Novartis, 1997)	Oil-in-water emulsion	Squalene, polysorbate 80, sorbitan trioleate	Fluad (seasonal influenza), Focetria (pandemic influenza), Aflunov (prepandemic influenza)
AS03 (GlaxoSmithKline, 2009)	Oil-in-water emulsion	Squalene, polysorbate 80, α-tocopherol	Pandremix (pandemic influenza), Prepandrix (prepandemic influenza)
Virosomes (Berna Biotech, 2000)	Liposomes	Lipids, HA	Inflexal (seasonal influenza), Epaxal (hepatitis A)
AS04 (GlaxoSmithKline, 2005)	Alum-adsorbed TLR4 agonist	Aluminum hydroxide, MPL	Fendrix (hepatitis B), Cervarix (HPV)

Adapted by permission from Macmillan Publishers Ltd. (Rappuoli et al., *Nat. Rev. Immunol.* **11**:865–872, 2011).
Alum and AS04 are adjuvants licensed in the United States.

seasonal influenza vaccine in the European Union (Table 11.13). In several other countries, MF59 treatment has reduced hospitalization in the elderly.

The treatment of emerging infections has also benefitted from the use of the water-in-oil emulsion adjuvants, MF59 and AS03. During the 2009 pandemic influenza outbreak caused by the H1N1 virus, the availability of MF59 and AS03 facilitated the production of more doses of vaccine by reducing the amount of antigen needed per dose of vaccine administered to all age groups. This experience demonstrated that it may be possible to prevent the risk of avian influenza in the future by priming the population with a vaccine containing an H1N1 strain plus an adjuvant. Nevertheless, it may still prove to be a daunting task to produce sufficient vaccine to prime and boost all or most of the seven billion people in the world in a timely manner to prevent or reduce infection by a new and rapidly spreading pandemic disease agent. Thus, it is essential that new technologies continue to be developed to produce novel, fast-acting types of vaccines and adjuvants.

The two general classes of adjuvants are immunopotentiators and delivery systems. Many immunopotentiators are pathogen-associated molecular patterns (PAMPs), which are invariant molecular structures (e.g., cell wall components and nucleic acids) found exclusively in pathogens but not humans (see chapter 2). Receptors for these PAMPs, called pattern recognition receptors (PRRs), enable the cells of the innate immune system to distinguish pathogen-derived versus self-derived molecules. By comparison, adjuvant delivery systems are typically particulate systems designed to improve antigen uptake and presentation. They also function to stabilize antigens against degradation, sustain antigen release, target specific cells of the immune system (e.g., dendritic cells), and codeliver antigen with adjuvant. Combinations of both classes of adjuvants, i.e., immunopotentiators and delivery systems, are expected to generate the most efficient adjuvants.

Delivery systems allow the development of "single-shot" vaccines in which a prime–boost regimen may be achieved with a single injection, which may increase patient compliance with vaccination. Moreover, these systems exert greater control over the size and shape of administered vaccine components. For example, particles of a size similar to that of viruses or bacteria can mimic pathogens, potentially improving their uptake by dendritic cells. While 20- to 200-nm particles efficiently enter the lymphatic system and are internalized by dendritic cells, this is not the case for particles less than 10 nm in size. Larger particles (up to 20 mm) do not efficiently enter lymph capillaries, and they require cellular transport to the lymph. Examples of delivery systems include lipid-based systems (emulsions, immune-stimulating complexes, liposomes, and virosomes), polymer-based systems (nanoparticles and microparticles), and virus-like particles.

The conjugation (covalent linkage) of antigens to PAMPs can significantly enhance antigen-specific immune responses by targeting dendritic cells and providing costimulation to the same cell. Relevant PAMPs and/or a combination of adjuvants may be selected to direct a specific immune response toward a given pathogen. For example, different adjuvants may

PAMP
pathogen-associated molecular pattern

PRR
pattern recognition receptor

influence the differentiation of T cells toward different lineages (e.g., Th1, Th2, and Th17). Alternatively, a particular combination of adjuvants may block the activity of one or more T-cell subsets. Adjuvant mixtures should therefore be assessed both individually and in combination. Furthermore, because vaccines are generally administered to healthy individuals, the safety of novel adjuvants also needs careful assessment. This is best reflected by the reports of Bell's palsy (a form of facial paralysis) noted after the intranasal administration of vaccines containing *E. coli* heat-labile enterotoxin (LPS) as the adjuvant and the association of the AS03 adjuvant with cases of narcolepsy in Finland.

Several classes of PRRs have been described, including the 10 TLRs in humans expressed by B cells, dendritic cells, T cells, monocytes, and macrophages (see chapters 2 and 4). The distinguishing characteristics of these receptors are their subcellular localization, intracellular signaling pathways, and ligands. Cell surface TLRs (TLR1, -2, and -4 through -6) mainly sense pathogen-associated cell wall components, whereas intracellular TLRs (TLR3, -7, -8, and -9) sense nucleotide-based components. Recognition of PAMPs by TLRs leads to transcription factor activation followed by secretion of proinflammatory cytokines, type I IFN, and expression of IFN-inducible gene products. In addition, TLRs are divided according to their ability to sense lipopeptides (TLR1, TLR2, and TLR6), nucleotide-derived materials (TLR3, TLR7, TLR8, and TLR9), LPS (TLR4), and flagellin (TLR5). A list of vaccine adjuvants that target different TLRs, and which have been tested in humans but are not yet licensed for use, is shown in Table 11.14. Several novel adjuvants that target TLR2, TLR7, and TLR9 are in the advanced developmental stage, either alone or in combination with alum, emulsions, saponins, and liposomes. Such novel adjuvants may lead to the commercial availability of a number of novel vaccines for both elderly patients and various diseases in the near future.

In conclusion, most of the unlicensed adjuvants in development stimulate several different components of the innate immune system. Adjuvants currently used in humans enhance humoral immunity, but many new adjuvants in clinical or preclinical development are focused on enhancing specific types of T-cell responses and generating the multifaceted immune responses required for many infectious diseases, including malaria, HIV, and tuberculosis.

One caveat is that the rate of adoption of new adjuvants as licensed vaccines has been slow, for two main reasons. The first is a safety concern: a potentially increased risk of autoimmune disease may be triggered or exacerbated by infection. Secretion of type I IFNs induced by TLR4 agonists can initiate the pathogenesis of systemic lupus erythematosus, an autoimmune disease, and disease flares are often triggered by viral infections. Second, in animal models, TLR agonists can break tolerance by overcoming immunosuppression by Treg cells. Repeated injection with IFN-inducing TLR agonists can also enhance the growth and pathogenicity of *M. tuberculosis* in mouse models. Despite these concerns, it is encouraging that engineered adjuvants can enhance the immune response to foreign microbial antigens. Few, if any, adjuvants render a self-antigen

Table 11.14 Adjuvants tested in humans but not yet licensed for use

Adjuvant(s)	Class	Components
CpG7909, CpG1018	TLR9 agonist	CpG oligonucleotides alone or combined with alum or emulsions
Imidazoquinolines	TLR7 + TLR8 agonists	Small molecules
Poly(I·C)	TLR3 agonist	Double-stranded RNA analogues
Pam3Cys	TLR2 agonist	Lipopeptide
Flagellin	TLR5 agonist	Bacterial protein linked to antigen
Iscomatrix	Combination	Saponin, cholesterol, dipalmitoylphosphatidylcholine
AS01	Combination	Liposome, MPL, saponin (QS21)
AS02	Combination	Oil-in-water emulsion, MPL, saponin (QS21)
AF03	Oil-in-water emulsion	Squalene, Montane 80, Eumulgin B1 PH
CAF01	Combination	Liposome, DDA, TDB
IC31	Combination	Oligonucleotide, cationic peptides

Adapted by permission from Macmillan Publishers Ltd. (Rappuoli et al., *Nat. Rev. Immunol.* 11:865–872, 2011).

AF03, adjuvant formulation 03; CAF01, cationic adjuvant formulation 01; DDA, dimethyldioctadecylammonium; Pam3Cys, tripalmitoyl-*S*-glyceryl cysteine; TDB, trehalose dibehenate.

sufficiently immunogenic to trigger an autoimmune disease, even if autoreactive T cells are present. Furthermore, although many of the widely used and safest vaccines (live, attenuated viral and bacterial vaccines) rely on activation through multiple TLRs, these vaccines have not been linked to an increased risk of any autoimmune disease in humans. Similarly, the large human safety databases obtained with the MF59 or AS04 vaccines, both licensed for human use in several countries, as well as more limited experience with several advanced experimental vaccines have not yielded an increase in autoimmune or infectious diseases. It is hoped that similar preclinical and clinical results will be obtained during the future testing of combinations of adjuvants. Such outcomes will predictably accelerate the licensing of additional novel adjuvants and provide new therapeutics to fight preexisting and chronic and emerging infectious diseases.

Systems Biology and Evaluation of Vaccines

It is commonly held that the sine qua non to accelerate vaccine development is an increase of our understanding of the molecular and cellular components of the human immune system. Conventionally, the approach to evaluate vaccines during the last 70 years has been to measure the concentration of antibodies in blood that neutralize a pathogen (neutralizing antibodies). Although this has been a reliable indicator of the effectiveness of a vaccine, current evidence suggests that the availability of novel adjuvants and analysis of additional markers of immune response (i.e., innate immune response and T-cell immune response) may be more relevant to vaccine efficacy. This interest in analyzing other arms of the immune response, in addition to neutralizing antibody, has been expedited by the development of several new technologies that allow many parameters of the immune system to be measured at one time in a single blood sample. These technologies include gene expression microarrays (see chapter 3), multiplex cytokine assays (see chapters 2 and 4), synthesis of peptide–MHC tetramers (see chapter 4), and fluorescence-activated cell sorting (FACS) analysis (see chapters 2 and 4).

For gene expression microarrays, manufacturing techniques at the nanoscale have enabled the synthesis of DNA probes for all expressed genes in the human genome (more than 25,000) and the arrangement of these probes on a single silicon chip. This chip can then be used to analyze the expression of any of these genes in RNA from peripheral blood lymphocytes (Table 11.15). Importantly, this technology was used to analyze the human response to yellow fever vaccine, one of the most successful vaccines known. In these studies, numerous significant gene expression patterns correlated closely with the response to this vaccine across many types of immune cells. These studies provided much insight into what makes a successful immune response and have charted a road map for future studies.

Collectively, more than 100 cytokines that mediate cell–cell interactions in the various arms of the immune system have been identified. The expression and quantity of these many cytokines may change during innate and adaptive immune responses, particularly in response to a vaccine, indicating the significance of being able to monitor the expression of very many cytokines during such responses. To assay many cytokines at one time (multiplex cytokine assay), antibodies specific to these molecules are attached to beads and then analyzed for their binding to more than 50 of the different cytokines found in the blood, and their relative concentrations are then measured. The increase and decrease in expression of these cytokines can signal the onset or decline in an immune response, an important parameter of vaccine design.

Cells of the immune system can express about 350 known cell surface molecules, called CD antigens, or secrete one or more of the approximately 100 known cytokines. FACS analyses that use a fluorescence activation-based flow cytometer equipped with multiple high-intensity lasers to detect fluorescent dyes of many colors (32 colors are now possible) can catalogue many of these molecules, and the new mass spectrometry-based machine, which uses lanthanide metal labels, can provide significantly more information about cell types in the blood, their relative activation state, and their frequency and functional properties (e.g., what cytokines they are secreting) (see chapter 4).

FACS
fluorescence-activated cell sorting

Table 11.15 Systems biology approach to vaccine development

Method of analysis	No. and parameter detected
Gene expression microarray	>25,000 genes
Multiplex cytokine assay	>50 secreted cytokines
FACS analysis	350 cell surface-associated CD antigens
FACS isolation and quantitation using pMHC tetramers	Antigen-specific T cells present in low frequency and with low TCR avidity for antigen

The listed methods of analysis may be performed at a single time on one sample of blood from an individual that received a single dose of a vaccine and/or adjuvant under study. CD, cluster determinant.

T cells mediate and regulate a wide spectrum of immune responses. Detection and isolation of the low-frequency T cells that contribute to a specific response, which need to be monitored during a response to a vaccine, have proven to be difficult. This is due to the very low affinity of a TCR for its ligand, an antigenic peptide bound to MHC (pMHC). This problem was solved in recent years owing to the use of tetramers of a particular pMHC containing a biotinylation site on the MHC and the tetrameric nature of streptavidin, in which each of the four subunits has its own biotin binding site (see chapter 2). Such pMHC tetramers bind with higher affinity and stability to T cells and have been used clinically to determine the magnitude of T-cell responses to systemic viral infections in humans (e.g., HIV-1, influenza virus, hepatitis B virus, cytomegalovirus, Epstein–Barr virus, and hantavirus). In human vaccine clinical trials, tetramer analyses have had the most impact in epitope-targeted vaccines, such as those designed to elicit responses to well-defined tumor antigens.

Figure 11.33 Progress in technologies for vaccine development. Conventionally, for about last 70 years, vaccines have been developed by empirical approaches consisting primarily of killed or live attenuated microorganisms, partially purified components of pathogens (subunit vaccines), detoxified toxins, or polysaccharides. These vaccines have been very successful in eliminating many devastating diseases. However, since empirical approaches were limited in the types of vaccines they could generate, newer technologies developed during the past 30 years have surpassed the previously used empirical ones. The latter technologies include recombinant DNA technology, glycoconjugation, reverse vaccinology, and many emerging next-generation technologies, such as novel adjuvants, systems biology, and structure-based vaccine design (structural vaccinology), that promise a successful future for vaccines. MMRV, measles, mumps, rubella, varicella. Adapted by permission from Macmillan Publishers Ltd. (Rappuoli et al., *Nat. Rev. Immunol.* **11**:865–872, 2011). doi:10.1128/9781555818890.ch11f.11.33

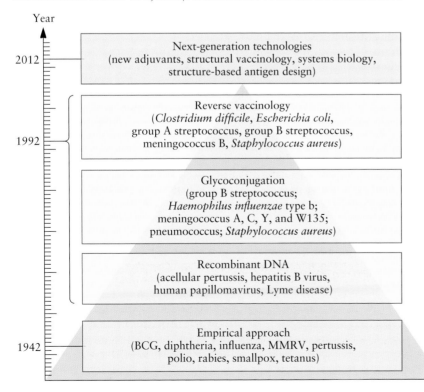

Thus, application of these many new systems biology methods of analysis to vaccine research is rapidly changing the modes of evaluation of existing and novel vaccines in the 21st century. A much more comprehensive view of the innate and adaptive immune responses to a given vaccine is now attainable, as revealed initially by the report of the successful yellow fever vaccine. This arsenal of new methodologies (Fig. 11.33) provides a more reliable and more rapid way to assess and improve efficacy in smaller numbers of people, which is expected to reduce the time and expense, and at the same time increase the success rate, of vaccine development.

summary

Traditional vaccines have demonstrated considerable success in preventing human infectious diseases and preserving public health by reducing or eradicating human death and suffering. The success of such therapies has ushered in a new era for vaccines in the 21st century. This new era is attributable largely to an increased understanding of the mechanisms of human immunity and microbial pathogenesis. Importantly, this new knowledge has catalyzed remarkable advances that can be translated into an improved state of public health. Such scientific, medical, and biotechnological advances promise to improve the use of existing vaccines and expand the list of vaccines developed in this century.

A primary driver of vaccine research in the 21st century is the development of vaccines to prevent cancer and several chronic infectious diseases, including HIV, tuberculosis, and malaria. Although vaccines against these various diseases are still at an early stage of development, it is encouraging that due to the revolutionary technologies of the past 20 years, vaccines have become much safer and may now be developed against infectious agents or diseases that could not be effectively targeted using early vaccination methods. An example of a new innovative technology is the method of manufacturing of a vaccine that allowed a shift from egg-based methods to cell-based or recombinant methods, including production from insect or plant cells. Recombinant DNA technology made possible the development and large-scale manufacture of the hepatitis B vaccine. Glycoconjugation technology (covalent linkage of carbohydrates to proteins, peptides, lipids, or saccharides) made possible the development of vaccines against *H. influenzae* type b, pneumococci, and meningococci. Genome-wide sequencing of viruses and bacteria has permitted the selection of targets for vaccine development. The expression and evaluation of these microbial gene products have allowed the discovery of new antigens by reverse vaccinology and made possible the development of a vaccine against meningococcus B. Following this example, the genome approach provided promising antigen targets for vaccines against group B streptococci, group A streptococci, and pneumococci, as well as for antibiotic-resistant bacteria, such as *S. aureus*. Novel technology has enabled the use of virus-like particles in the development of a vaccine against HPV. DNA vaccines and vector vaccines directed against viruses or bacteria may expedite the development of new vaccines for cancer, HIV, tuberculosis, and malaria.

Considerable progress has been made in understanding the human innate immune system, which should facilitate the production and licensing of novel TLR agonist-based adjuvants in the next decade. Thus, the rapid emergence of new next-generation technologies (Fig. 11.33) should promote the development of effective vaccines against many new pathogens, improve the safety and efficacy of the existing vaccines, and begin to address novel targets for the treatment of chronic infectious diseases, autoimmune diseases, and cancer. It is recommended that new vaccines be systematically analyzed for their potential to confer protection against a wide spectrum of diseases in people of all ages, particularly in young children and the aging population.

Finally, to accelerate vaccine development and increase our understanding of the human immune system, it is important to (i) implement innovative clinical trials in which several vaccines or vaccination regimes are tested in parallel and (ii) obtain information early (1 to 2 weeks) after vaccination using sensitive high-throughput technologies that allow for systems biology analyses of gene expression patterns and cytokine production in both T and B cells as well as in microbes and in particular systems biology methodologies (Table 11.15). A total of more than 25,000 human genes and more than 50 human cytokines can be analyzed at one time in a single blood sample. Such information not only identifies susceptible microbial targets but also has the potential to define new biomarkers of protective immune responses. This approach is termed systems vaccinology. Such information enables an accurate and comprehensive analysis of immune activation, minimizes undesirable adverse side effects, and maximizes clinical efficacy. Depending on the vaccine, dose, regimen, recipient, and many other associated factors, successful protection may require neutralizing antibodies, effective T-cell responses, or a combination of antibodies and T cells.

review questions

1. Outline the advantages and limitations of traditional approaches to vaccine production. Briefly describe how these approaches have resulted in adapting new technologies to current vaccine design.

2. Define a subunit vaccine, and discuss the advantages and disadvantages of the use of such vaccines. Compare the efficiency and efficacy of live attenuated vaccines versus subunit vaccines.

3. To date, clinical trials of HSV-1 or HSV-2 whole or subunit vaccines have not been successful. Describe what information is required before designing more powerful treatments to treat HSV infection that might enable such vaccines to prevent HSV infection in humans.

4. What is the hallmark of orally or nasally administered mucosal vaccines that gives rise to herd immunity and prevents infection?

5. As part of your work with an international animal health organization, you are given the task of developing a vaccine against a bovine virus that is the cause of tens of thousands of cattle deaths around the world annually. The viral genome consists of a 10-kb linear piece of single-stranded RNA with a poly(A) tail that encodes eight different proteins. The virus does not have a viral envelope, and the major antigenic determinant is the capsid protein viral protein 2. Outline an experimental strategy to develop a vaccine against this virus.

6. Briefly describe a protocol for developing a vaccine against an enterotoxin-producing bacterium, such as *V. cholerae*.

7. Discuss some of the different strategies that have been used to produce vaccines against cholera.

8. Describe how neutralizing antibody and T-cell responses were induced to provide protective immunity against infection with SARS-CoV. How did this result lead to the rapid production of a SARS vaccine?

9. How would you develop a vaccine against *S. aureus* and particularly against the infectious antibiotic-resistant strains of *S. aureus* encountered in hospitals?

10. How would you develop a vaccine against HPV? Describe why the HPV vaccine produced in 2006 was a milestone in vaccine development.

11. Discuss the development of peptide vaccines that are directed against viruses. Use foot-and-mouth disease virus as an example.

12. Describe how malaria, a parasitic disease, is transmitted to humans. Discuss what recent advances provide optimism that a licensed malaria vaccine may be available in the near future.

13. What is known about a potential mechanism of tumor elimination by peptide vaccine therapy? How can peptide vaccines be designed to enhance CD8$^+$ CTL responses and thereby elicit tumor destruction and elimination?

14. Describe how injection of autoantigen peptides (epitopes) can both sustain health and protect from an inflammatory autoimmune disease in an antigen-specific manner when administered at different stages of a disease and by various routes. Outline the important role that dendritic cells play in controlling the outcomes of antigen-specific therapy for autoimmune disease.

15. How does allergen-specific immunotherapy with T-cell epitope-containing allergen peptides regulate allergen-specific T-cell responses and susceptibility to allergy?

16. How has our improved understanding of the biology of dendritic cells and their response to adjuvants aided us in achieving sufficient immunity to reduce and/or prevent HIV infection?

17. Are personalized peptide vaccines designed based on dendritic cell targeted tumor antigen epitopes immunogenic and effective in the treatment of cancer?

18. How has biolistic delivery contributed to tumor immunity?

19. How may the immunogenicity and efficacy of DNA vaccines be improved? Discuss relevant outcomes in the treatment of cancer and tuberculosis.

20. Are prime–boost regimens effective for the successful outcome of DNA vaccination?

21. Describe the three live attenuated vaccines now in the active stage of development for protection against *V. cholerae* infection.

22. Outline how and why attenuated influenza vaccines are formulated annually to protect against strains expected to circulate globally during upcoming influenza seasons.

23. What is vaccinia virus, and how can it be used to produce unique live recombinant vaccines?

24. As an employee of the WHO, you have to decide on the best strategy for eradicating rabies in wild animal populations. Assuming that you must choose between a peptide- and a vaccinia virus-based vaccine, select one type of vaccine and justify your choice.

(continued)

review questions (continued)

25. How can bacteria be used as part of a DNA vaccine delivery system?

26. How can vaccinia virus be made more sensitive to IFN? Explain.

27. Suggest several methods that you could use to deliver DNA for genetic immunization to animal cells.

28. How would you improve the traditional vaccine against tuberculosis?

29. Describe attempts to develop a carbohydrate-based vaccine against *C. bolteae* that may control autism-related symptoms.

30. Describe what adjuvants are, why they have taken on increased importance in the generation of immune responses, and which adjuvants are either licensed or now being tested for approval for licensure.

31. Define the term "systems biology," and explain why this approach is expected to rapidly improve and advance the development of vaccines in the 21st century.

references

Ali, A., B. Donovan, H. Wand, T. R. H. Read, D. G. Regan, A. E. Grulich, C. K. Fairley, and R. J. Guy. 2013. Genital warts in young Australians five years into national human papillomavirus vaccination program: national surveillance data. *BMJ* **346**:12032.

Audran, R., M. Cachat, F. Lurati, S. Soe, O. Leray, G. Corradin, P. Druilhe, and F. Spertini. 2005. Phase I malaria vaccine trial with a long synthetic peptide derived from the merozoite surface protein 3 antigen. *Infect. Immun.* **73**: 8017–8026.

Bandell, A., J. Woo, and K. Coelingh. 2011. Protective efficacy of live attenuated influenza vaccine (multivalent, Ann Arbor strain): a literature review addressing interference. *Expert Rev. Vaccines* **10**:1131–1141.

Bartnik, A., A. J. Nirmal, and S. Y. Yang. 2013. Peptide vaccine therapy in colorectal cancer. *Vaccines* **1**:1–16.

Blasco, R., and B. Moss. 1995. Selection of recombinant vaccinia viruses on the basis of plaque formation. *Gene* **149**:157–162.

Cao, Y., Z. Liu, P. Li, P. Sun, Y. Fu, X. Bai, H. Bao, Y. Chen, D. Li, and Z. Liu. 2012. Improved neutralising antibody response against foot-and-mouth-disease virus in mice inoculated with a multi-epitope peptide vaccine using polyinosinic and poly-cytidylic acid as an adjuvant. *J. Virol. Methods* **185**:124–128.

Centers for Disease Control and Prevention. 1999. Impact of vaccines universally recommended for children—United States, 1900–1998. *MMWR Morb. Mortal. Wkly. Rep.* **48**:243–248.

Charles, I., and G. Dougan. 1990. Gene expression and the development of live enteric vaccines. *Trends Biotechnol.* **8**: 117–121.

Chentoufi, A. A., E. Kritzer, D. M. Yu, A. B. Nesburn, and L. BenMohamed. 2012. Towards a rational design of an asymptomatic clinical herpes vaccine: the old, the new, and the unknown. *Clin. Dev. Immunol.* **2012**:187585. doi:10.1155/2012/187585.

Chentoufi, A. A., X. Dervillez, P. A. Rubbo, T. Kuo, X. Zhang, N. Nagot, E. Tuaillon, P. Van De Perre, A. B. Nesburn, and L. Benmohamed. 2012. Current trends in negative immuno-synergy between two sexually transmitted infectious viruses: HIV-1 and HSV-1/2. *Curr. Trends Immunol.* **13**:51–68.

Cichutek, K. 2000. DNA vaccines: development, standardization and regulation. *Intervirology* **43**:331–338.

Clarke, B. E., S. E. Newton, A. R. Carroll, M. J. Francis, G. Appleyard, A. D. Syred, P. E. Highfield, D. J. Rowlands, and F. Brown. 1987. Improved immunogenicity of a peptide epitope after fusion to hepatitis B core protein. *Nature* **330**:381–384.

Davis, M. M., and J. D. Altman. 2012. New methods for analyzing vaccine responses. *Jordan Rep.* **2012**:46–52.

DiazGranados, C. A., M. Denis, and S. Plotkin. 2012. Seasonal influenza vaccine efficacy and its determinants in children and non-elderly adults: a systematic review with meta-analyses of controlled trials. *Vaccine* **31**:49–57.

Dodson, L. F., W. G. Hawkins, and P. Goedegebuure. 2011. Potential targets for pancreatic cancer immunotherapeutics. *Immunotherapy* **3**:517–537.

Du, L., Y. He, Y. Zhou, S. Liu, B. J. Zheng, and S. Jiang. 2009. The spike protein of SARS-CoV—a target for vaccine and therapeutic development. *Nat. Rev. Microbiol.* **7**:226–236.

Egorov, A., S. Brandt, S. Sereinig, J. Romanova, B. Ferko, D. Katinger, A. Grassauer, G. Alexandrova, H. Katinger, and T. Muster. 1998. Transfectant influenza A viruses with long deletions in the NS1 protein grow efficiently in Vero cells. *J. Virol.* **72**:6437–6441.

Fioretti, D., S. Iurescia, V. M. Fazio, and M. Rinaldi. 2010. DNA vaccines: developing new strategies against cancer. *J. Biomed. Biotechnol.* **2010**:174378. doi:10.1155/2010/174378.

Flexner, C., A. Hugin, and B. Moss. 1987. Prevention of vaccinia virus infection in immunodeficient mice by vector-directed IL-2 expression. *Nature* **330**:259–262.

Gaucher, D., R. Therrien, N. Kattaf, B. R. Angermann, G. Boucher, A. Filali-Mouhim, J. M. Moser, R. S. Mehta, D. R. Drake III, E. Castro, R. Akondy, A. Rinfret, B. Yassine-Diab, E. A. Said, Y. Chouikh, M. J. Cameron, R. Clum, D. Kelvin, R. Somogyi, L. D. Greller, R. S. Balderas, P. Wilkinson, G. Pantaleo, J. Tartaglia, E. K. Haddad, and R. P. Sékaly. 2008. Yellow fever vaccine induces integrated multilineage and polyfunctional immune responses. *J. Exp. Med.* **205**:3119–3131.

Guy, B., B. Barrere, C. Malinowski, M. Saville, R. Teyssou, and J. Lang. 2011. From research to phase III: preclinical, industrial and clinical development of the Sanofi Pasteur tetravalent dengue vaccine. *Vaccine* **29**:7229–7241.

Heinsbroek, E., and E. J. Ruitenberg. 2010. The global introduction of inactivated polio vaccine can circumvent the oral polio vaccine paradox. *Vaccine* **28**:3778–3783.

Horwitz, M. A., G. Harth, B. J. Dillon, and S. Maslesa-Galic. 2000. Recombinant bacillus Calmette-Guérin (BCG) vaccines expressing the *Mycobacterium tuberculosis* 30-kDa major secretory protein induce greater protective immunity against tuberculosis than conventional BCG vaccines in a highly susceptible animal model. *Proc. Natl. Acad. Sci. USA* **97**:13853–13858.

Horwitz, M. A., and G. Harth. 2003. A new vaccine against tuberculosis affords greater survival after challenge than current vaccine in the guinea pig model of pulmonary tuberculosis. *Infect. Immun.* **71**:1672–1679.

Kalinski, P., R. Muthuswamy, and J. Urban. 2013. Dendritic cells in cancer immunotherapy: vaccines and combination immunotherapies. *Expert Rev. Vaccines* **12**:285–295.

Kaper, J. B., J. G. Morris, Jr., and M. M. Levine. 1995. Cholera. *Clin. Microbiol. Rev.* **8**:48–86.

Kirnbauer, R., F. Booy, N. Cheng, D. R. Lowy, and J. T. Schiller. 1992. Papillomavirus L1 major capsid protein self-assembles into virus-like particles that are highly immunogenic. *Proc. Natl. Acad. Sci. USA* **89**:12180–12184.

Kita, Y., T. Tanaka, S. Yoshida, N. Ohara, Y. Kaneda, S. Kuwayama, Y. Muraki, N. Kanamaru, S. Hashimoto, H. Takei, C. Okada, Y. Fukunaga, Y. Sakaguchi, I. Furukara, K. Yamada, Y. Inoue, Y. Takemoto, M. Naito, T. Yamada, M. Matsumoto, D. N. McMurray, E. C. Cruz, E. V. Tan, R. M. Abalos, J. A. Burgos, R. Gelber, Y. Sheiky, S. Reed, M. Sakatani, and M. Okada. 2005. Novel vaccines against *M. tuberculosis*. *Vaccine* **23**:2132–2135.

Koutsky, L. A., K. A. Ault, C. M. Wheeler, D. R. Brown, E. Barr, F. B. Alvarez, L. M. Chiacchierini, and K. U. Jansen. 2002. A controlled trial of a human papillomavirus type 16 vaccine. *N. Engl. J. Med.* **347**:1645–1650.

Landry, S., and C. Heilman. 2005. Future directions in vaccines: the payoffs of basic research. *Health Aff.* (Millwood) **24**:758–769.

Lauring, A. S., J. O. Jones, and R. Andino. 2010. Rationalizing the development of live attenuated virus vaccines. *Nat. Biotechnol.* **28**:573–579.

Leitner, W. W., H. Ying, and N. P. Restifo. 2000. DNA and RNA-based vaccines: principles, progress and prospects. *Vaccine* **18**:765–777.

Li, Z., Y. Yi, X. Yin, Y. Zhang, M. Liu, H. Liu, X. Li , Y. Li, Z. Zhang, and J. Liu. 2012. Development of a foot-and-mouth disease virus serotype A empty capsid subunit vaccine using silkworm (*Bombyx mori*) pupae. *PLoS One* **7**:e43849.

Linhart, B., and R. Valenta. 2012. Vaccines for allergy. *Curr. Opin. Immunol.* **24**:354–360.

Manoj, S., L. A. Babiuk, and S. van Drunen Littel-van den Hurk. 2004. Approaches to enhance the efficacy of DNA vaccines. *Crit. Rev. Clin. Lab. Sci.* **41**:1–39.

Miner, J. N., and D. E. Hruby. 1990. Vaccinia virus: a versatile tool for molecular biologists. *Trends Biotechnol.* **8**:20–25.

Morrow, M. P., J. Yan, and N. Y. Sardesai. 2013. Human papillomavirus therapeutic vaccines: targeting viral antigens as immunotherapy for precancerous disease and cancer. *Expert Rev. Vaccines* **12**:271–283.

Moss, B. 1991. Vaccinia virus: a tool for research and vaccine development. *Science* **252**:1662–1667.

Moyle, P. M., and I. Toth. 2013. Modern subunit vaccines: development, components, and research opportunities. *Chem. Med. Chem.* **8**:360–376.

Murray, C. J. L., M. Ezzati, A. D. Lopez, A. Rodgers, and S. Vander Hoorn. 2004. Comparative quantification of health risks: conceptual framework and methodological issues, p. 1–38. *In* M. Ezzati, A. D. Lopez, A. Rodgers, and C. J. L. Murray (ed.), *Comparative Quantification of Health Risks: Global and Regional Burden of Disease Attributable to Selected Major Risk Factors.* World Health Organization, Geneva, Switzerland.

Nabel, G. J. 2013. Designing tomorrow's vaccines. *N. Engl. J. Med.* **368**:551–560.

Needham, B. D., S. M. Carroll, D. K. Giles, G. Georgiou, M. Whiteley, and M. S. Trent. 2013. Modulating the innate immune response by combinatorial engineering of endotoxin. *Proc. Natl. Acad. Sci. USA* **110**:1464–1469.

Okada, M., Y. Kita, T. Nakajima, N. Kanamaru, S. Hashimoto, T. Nagasawa, Y. Kaneda, S. Yoshida, Y. Nishida, H. Nakatani, K. Takao, C. Kishigami, S. Nishimatsu, Y. Sekine, Y. Inoue, D. N. McMurray, and M. Sakatani. 2011. Novel prophylactic vaccine using a prime-boost method and hemagglutinating virus of Japan-envelope against tuberculosis. *Clin. Dev. Immunol.* **2011**:549281. doi:10.1155/2011/549281.

Olotu, A., G. Fegan, J. Wambua, G. Nyangweso, K.O. Awuondo, A. Leach, M. Lievens, D. Leboulleux, P. Njuguna, N. Peshu, K. Marsh, and P. Bejon. 2013. Four-year efficacy of RTS,S/AS01E and its interaction with malaria exposure. *N. Engl. J. Med.* **368**:1111–1120.

Peakman, M., and M. von Herrath. 2010. Antigen-specific immunotherapy for type 1 diabetes: maximizing the potential. *Diabetes* **59**:2087–2093.

Pequegnat, B., M. Sagermann, M. Valliani, M. Toh, H. Chow, E. Allen-Vercoe, and M. A. Monteiro. 2013. A vaccine and diagnostic target for *Clostridium bolteae*, an autism-associated bacterium. *Vaccine* **31**:2787–2790.

Pliaka, V., Z. Kyriakopoulou, and P. Markoulatos. 2012. Risks associated with the use of live-attenuated vaccine poliovirus strains and the strategies for control and eradication of paralytic poliomyelitis. *Expert Rev. Vaccines* **11**: 609–628.

Pulendran, B. 2009. Learning immunology from the yellow fever vaccine: innate immunity to systems vaccinology. *Nat. Rev. Immunol.* **9**:741–747.

Querec, T. D., R. S. Akondy, E. K. Lee, W. Cao, H. I. Nakaya, D. Teuwen, A. Pirani, K. Gernert, J. Deng, B. Marzolf, K. Kennedy, H. Wu, S. Bennouna, H. Oluoch, J. Miller, R. Z. Vencio, M. Mulligan, A. Aderem, R. Ahmed, and B. Pulendran. 2009. Systems biology approach predicts immunogenicity of the yellow fever vaccine in humans. *Nat. Immunol.* **10**:116–125.

Rappuoli, R., C. W. Mandl, S. Black, and E. De Gregorio. 2011. Vaccines for the twenty-first century. *Nat. Rev. Immunol.* **11**:865–872.

Rappuoli, R., and A. Aderem. 2011. A 2020 vision for vaccines against HIV, tuberculosis and malaria. *Nature* **473**: 463–469.

Regules, J. A., J. F. Cummings, and C. F. Ockenhouse. 2011. The RTS,S vaccine candidate for malaria. *Expert Rev. Vaccines* **10**:589–599.

Rice, J., C. H. Ottensmeier, and F. K. Stevenson. 2008. DNA vaccines: precision tools for activating effective immunity against cancer. *Nat. Rev. Cancer* **8**: 108–120.

Rimmelzwaan, G. F., and G. Sutter. 2009. Candidate influenza vaccines based on recombinant modified vaccinia virus Ankara. *Expert Rev. Vaccines* **8**:447–454.

Romano, M., and K. Huygen. 2012. An update on vaccines for

tuberculosis—there is more to it than just waning of BCG efficacy with time. *Expert Opin. Biol. Ther.* **12**:1601–1610.

Roush, S. W., and T. V. Murphy. 2007. Historical comparisons of morbidity and mortality for vaccine-preventable diseases in the USA. *JAMA* **298**: 2155–2163.

Sabchareo, A., D. Wallace, C. Sirivichayakul, K. Limkittikul, P. Chanthavanich, S. Suvannadabba, V. Jiwariyavej, W. Dulyachai, K. Pengsaa, T. A. Wartel, A. Moureau, M. Saville, A. Bouckenooghe, S. Viviani, N. G. Tornieporth, and J. Lang. 2012. Protective efficacy of the recombinant, live-attenuated, CYD tetravalent dengue vaccine in Thai schoolchildren: a randomized, controlled phase 2b trial. *Lancet* **380**: 1559–1567.

Saha, A., M. I. Chowdhury, M. Nazim, M. M. Alam, T. Ahmed, M. B. Hossain, S. K. Hore, G. N. Sultana, A. M. Svennerholm, and F. Qadri. 2013. Vaccine specific immune response to an inactivated oral cholera vaccine and EPI vaccines in a high and low arsenic area in Bangladeshi children. *Vaccine* **31**: 647–652.

Trumpfheller, C., M. P. Longhi, M. Caskey, J. Idoyaga, L. Bozzacco, T. Keler, S. J. Schlesinger, and R. M. Steinman. 2012. Dendritic cell-targeted protein vaccines: a novel approach to induce T-cell immunity. *J. Intern. Med.* **271**: 183–192.

Ulivieri, C., and C. T. Baldari. 2013. T-cell-based immunotherapy of autoimmune diseases. *Expert Rev. Vaccines* **12**: 297–310.

Van Kampen, K. R., Z. Shi, P. Gao, J. Zhang, K. W. Foster, D. T. Chen, D. Marks, C. A. Elmets, and D. C. Chang. 2005. Safety and immunogenicity of adenovirus-vectored nasal

and epicutaneous influenza vaccines in humans. *Vaccine* **23**:1029–1036.

Verma, R., P. Khanna, and S. Chawla. 2012. Cholera vaccine: new preventive tool for endemic countries. *Hum. Vaccines Immunother.* **8**:682–684.

Wacheck, V., A. Egorov, F. Groiss, A. Pfeiffer, T. Fuereder, D. Hoeflmayer, M. Kundl, T. Popow-Kraupp, M. Redlberger-Fritz, C. A. Mueller, J. Cinatl, M. Michaelis, J. Geiler, M. Bergmann, J. Romanova, E. Roethl, A. Morokotti, M. Wolschek, B. Ferko, J. Seipelt, R. Dick-Gudenus, and T. Muster. 2010. A novel type of influenza vaccine: safety and immunogenicity of replication-deficient influenza virus created by deletion of the interferon antagonist NS1. *J. Infect. Dis.* **201**: 354–362.

Wraith, D. C., M. Goldman, and P. H. Lambert. 2003. Vaccination and autoimmune disease: what is the evidence? *Lancet* **362**:1659–1666.

Yamada, A., T. Sasada, M. Noguchi, and K. Itoh. 2013. Next-generation peptide vaccines for advanced cancer. *Cancer Sci.* **104**:15–21.

Zakhartchouk, A. N., C. Sharon, M. Satkunarajah, T. Auperin, S. Viswanathan, G. Mutwiri, M. Petric, R. H. See, R. C. Brunham, B. B. Finlay, C. Cameron, D. J. Kelvin, A. Cochrane, J. M. Rini, and L. A. Babiuk. 2007. Immunogenicity of a receptor-binding domain of SARS coronavirus spike protein in mice: implications for a subunit vaccine. *Vaccine* **25**:136–143.

Zhu, Q., and J. A. Berzofsky. 2013. Oral vaccines: directed safe passage to the front line of defense. *Gut Microbes* **4**:246–252.

Societal Issues

MEDICAL BIOTECHNOLOGY IS currently driving advances in medicine. Many of the products of the molecular biotechnologies described in the preceding chapters represent a substantial proportion of treatments recently made commercially available for human diseases. In addition to new drugs, innovative methods for efficient delivery of drugs to target tissues are being developed. Our increasing understanding of the genetic and molecular basis of disease is leading to methods for early disease detection and to individualized treatments, both of which have the potential to improve disease prognosis.

All new treatments and diagnostic tests must be proven effective and safe before they are used in human medicine. There is some level of risk of unintended, harmful effects associated with all medical interventions. For example, all medications may cause side effects, regardless of how they are produced. The risk is reduced as much as possible by the requirement for testing on small groups of individuals in clinical trials. Most governments have regulations in place to ensure that these requirements are met before new human therapeutic agents are made available to the public.

Pharmaceutical companies take financial risks in developing new medicines. Only a small fraction of new drugs are successful in clinical trials and are made available to patients. To protect investments and to encourage research and development of new medicines, most nations have implemented a patenting system that gives the inventor exclusive rights to use and sell the drug for a specified period. However, patenting is not without controversy, as many argue that the system actually discourages innovation, restricts options for patients, and gives patent holders rights to profit from substances that are not inventions, for example, human genes. The significant investment of capital, time, and human and technical resources to bring a small number of drugs to market contributes to the high costs of drugs that are unaffordable for many patients. These issues associated with medical biotechnology are the focus of this chapter.

doi:10.1128/9781555818890.ch12

Safety and Ethical Issues

Regulation of New Drugs

There is consensus among most countries that regulations for the approval of new pharmaceuticals for commercial use are sufficient to ensure the efficacy and safety of a drug regardless of how it is produced. In the United States, the Food and Drug Administration (FDA) is responsible for regulating the introduction of foods, drugs, and medical devices into the marketplace. Similar regulations regarding the commercialization of new therapeutic products are developed and enforced in Canada by Health Canada under the Food and Drugs Act. In the European Union (EU), new drugs can be authorized by individual national regulatory agencies or by the European Medicines Agency, which has a centralized procedure for authorization of drugs marketed in all EU member states and in European Economic Area countries. The centralized procedure is mandatory for treatments developed for diseases such as cancers, diabetes, immune dysfunction, and viral infections, including rare diseases, and those produced by biotechnologies.

The FDA's Center for Drug Evaluation and Research evaluates new drugs under the Federal Food Drug and Cosmetics Act before they are made available to the public. Along with a diverse range of chemically synthesized drugs, this includes biological molecules, whether recombinant or not, such as hormones (e.g., insulin and human growth hormone), therapeutic proteins (e.g., interferon and hyaluronidase), and monoclonal antibodies. Although the FDA assesses applications for drug approval, the producer must test the drug and provide evidence that it is safe and effective. The drugs are first tested in small (mice and rats) and then large (e.g., monkeys) laboratory animals, and those drugs that show promise are then tested in humans in clinical trials for safety, including toxicity and the severity of side effects (phase I), treatment efficacy in those affected with the disease (phase II), and optimal dosage, interaction with other drugs, and efficacy in a larger number of patients and in different populations (phase III) (see chapter 9, Box 9.1).

Approval is usually limited to specific applications for which the drug was shown to be effective in clinical trials. For example, the FDA recently (2013) approved Kadcyla (ado-trastuzumab emtansine), a new treatment for breast cancer. The drug is comprised of the monoclonal antibody trastuzumab, which specifically targets HER2, a protein that is elevated in 20% of breast cancers and contributes to cancer cell proliferation, chemically linked to the chemotherapeutic agent emtansine (DM1), which inhibits cell growth. The new therapy is approved only for HER2-positive patients with metastatic breast cancer who have been previously treated with trastuzumab (not conjugated to DM1) and a taxane, a member of a group of chemotherapeutic drugs. Because it is intended for use in patients for whom there is no satisfactory alternative treatment available, the drug was reviewed under an expedited FDA program.

Although all new drugs undergo rigorous evaluation, there may be consequences that cannot be predicted from testing in clinical trials.

FDA
U.S. Food and Drug Administration

EU
European Union

Adverse reactions may occur due to an inappropriate dose of a drug, an unforeseeable reaction to a drug such as an allergic reaction, reactions with other drugs, or the state of health or ethnic background of a patient. Thus, health care professionals and consumers should continue to monitor new therapies after they are released and report adverse effects. Drug producers may be required to conduct postmarket studies to further evaluate drug safety and efficacy.

Premarket regulation of therapeutic agents derived from living organisms has been the mandate of the FDA's Center for Biologics Evaluation and Research since 1972 under the Public Health Service Act. These products are known as **biologics** and include vaccines, blood and blood products (including recombinant plasma proteins, such as clotting factors), tissue- and cell-based treatments, and gene therapies (Table 12.1). To date, no human gene therapies have been approved. On the other hand, many vaccines have been licensed for immunization against bacterial and viral pathogens in the United States. All of these have undergone rigorous testing, including preclinical and clinical testing, to ensure that they are safe and free of contaminants and effectively protect people against the targeted infection.

The FDA Center for Biologics Evaluation and Research also regulates stem cell therapies. There are three types of stem cells: **multipotent adult stem cells** found throughout the body (e.g., hematopoietic, mesenchymal, and neural stem cells), **pluripotent human embryonic stem cells** derived from the inner cell mass of a blastocyst, and **induced pluripotent stem cells** that are generated in the laboratory by reprogramming adult somatic cells such as skin cells. All are undifferentiated cells that respond to specific environmental cues to differentiate into cells with specialized functions. The only stem cell-based therapies that have been approved by the FDA to date are those that use hematopoietic stem cells, a type of adult stem cell found in bone marrow or umbilical cord blood (collected by mothers' consent), to treat some blood

Table 12.1 Examples of biologics approved by the FDA in 2012

Proper name	Trade name	Application
Varicella–zoster immunoglobulin	Varizig	To reduce the severity of chicken pox in high-risk individuals
Human immunoglobulin	Bivigam	To treat primary humoral immunodeficiency (antibody deficiency)
Fibrin sealant patch (human fibrinogen and thrombin used to coat on a backing layer)	Evarrest	To stop soft tissue bleeding during certain types of surgery
Influenza virus vaccine (produced in cultured animal cells)	Flucelvax	To prevent seasonal infections with influenza viruses A and B
Hematopoietic progenitor cells (cord blood)	Ducord	For transplantation in patients with disorders affecting the hematopoietic system
Meningococcal and *Haemophilus* combination vaccine	Menhibrix	To prevent *Neisseria meningitidis* (serotypes C and Y) and *Haemophilus influenzae* (type b) infections in young children
Human T-lymphotropic virus enzyme immunoassay	Avioq HTLV-I/II	To screen blood for the presence of anti-human T-lymphotropic virus (type I and II) antibodies
Cellular sheet containing human keratinocytes, dermal fibroblasts and extracellular matrix proteins, and bovine collagen	GINTUIT	To treat oral mucogingival conditions during surgery

From http://www.fda.gov/BiologicsBloodVaccines/DevelopmentApprovalProcess/BiologicalApprovalsbyYear/default.htm.

cancers and inherited blood disorders that prevent normal blood formation (Table 12.1). After transplantation into a patient, the precursor cells mature into different types of blood cells in the bone marrow and are released into the bloodstream to restore missing functions. There are, however, other stem cell therapies that are currently under review. Several clinical trials have been initiated or completed to study adult stem cell therapies for treatment of liver cirrhosis, Crohn disease, and heart disease. In most cases, the source of the stem cells is the patient's own body. Currently, there are two active phase I clinical trials to evaluate the safety of human embryonic stem cell-derived retinal cells to treat specific eye diseases (Stargardt macular dystrophy and age-related macular degeneration); a third trial using human embryonic stem cells to restore spinal cord function was halted. There are several safety concerns about the transplantation of stem cells that must be addressed. These include the ability of the cells to differentiate into many different, unintended cell types and their ability to proliferate inappropriately, which can lead to tumor formation.

In the United States, drugs derived from genetically engineered organisms must be approved by the FDA Center for Biologics Evaluation and Research before they are made commercially available. Genetically engineered animals must also receive approval under the new animal drug provisions of the Federal Food Drug and Cosmetic Act. Approval must be obtained each time a gene is introduced into an animal, even when the same gene is introduced into a new animal. It is generally recognized that the site of insertion into the animal's genome is difficult to control and may impact the level of expression of the transgene or the animal's native genes and that it could therefore have an impact on the health of the animal. All animals containing a recombinant DNA construct are regulated, including subsequent generations of animals that contain the construct that were derived from the original manipulated parent animals by breeding with a nontransgenic animal. Approval may not be required for low-risk transgenic animals such as laboratory animals that are used for research.

Recently, the FDA approved the use of the human protein antithrombin produced in goat's milk in individuals with a hereditary deficiency in the production of this protein and who are undergoing surgery or giving birth. Antithrombin is a protease inhibitor that acts as an anticlotting factor by inhibiting the activity of thrombin and other coagulation proteases and thereby prevents the excessive formation of blood clots and promotes the clearing of clotting factors. It also has anti-inflammatory activity. Approximately 1 in 5,000 people are unable to produce this protein naturally, which puts them at risk for heart attacks and strokes. While antithrombin can be extracted from the plasma of donated blood, the supply is not sufficient to meet the needs of patients. Human antithrombin expressed from a mammary gland-specific promoter is secreted into the milk of transgenic goats. Milk extraction is more efficient and less costly and has a lower risk of contamination with human pathogens than blood extraction. The milk of transgenic goats is a significant source of human antithrombin, with yields of 2 to 10 g per liter of milk. It has been estimated that 75 transgenic goats are required to meet the annual worldwide demand for antithrombin.

Much of the research with transgenic animals has been devoted to developing the mammary glands of these animals as bioreactors for the production of pharmaceutical proteins. Recombinant antithrombin is the first commercial product derived from a transgenic animal to receive FDA approval, and it is likely that several other human therapeutic proteins produced in transgenic goats will be available soon, including other blood proteinase inhibitors such as antitrypsin and human clotting factors such as factor IX for the treatment of hemophilia. Because these recombinant proteins are intended for injection into humans, there may be some concern by the public that viruses, prions (the causative agent of scrapie), or other infectious agents may be transmitted from goats to humans. However, the goats used as bioreactors must be tested, monitored, and certified as free of specific pathogens by the FDA. Moreover, the recombinant proteins are highly purified by several filtration and chromatographic steps and are exposed to dry heat that inactivates viruses.

Generally, there is greater economic incentive, public acceptability, and ethical justification associated with the use of transgenic plants and animals for the production of pharmaceuticals than for their production for food. However, transgenic animals and plants carrying human genes must be contained. Transgenic animals that are used for non-food research purposes (e.g., laboratory mice) or to produce human therapeutic proteins are maintained within specific-pathogen-free controlled environments. The transgenic goats that produce recombinant proteins in milk must be maintained as a closed herd. Therefore, they pose a lower risk for unintended release (of animals or transgenes) into the environment than do those that are raised in open environments for food production. No transgenic animals intended as food for human consumption have received regulatory approval.

Transgenic plants that produce pharmaceutical proteins are frequently agricultural crop plants such as maize, tobacco, potato, rice, and safflower. To date, they have mainly been developed for production in contained facilities, for example, in cell cultures. Pharmaceutical plants grown in greenhouses, cell culture systems, and other contained facilities are regulated as drugs. The plants are genetically manipulated, often with several different genes, to maximize yields of proteins that are intended to be biologically active in humans or other animals. Thus, the plants potentially pose a greater risk to human health and the environment when grown in the field, and they are therefore given special consideration by regulators. For example, there are concerns that material derived from the plants could inadvertently end up in the food chain, perhaps through seed dispersal. Prevention of such problems may require special confinement measures such as dedicated farm machinery or containment netting to separate pharmaceutical crops from food crops. Employing nonfood crops such as tobacco to produce pharmaceutical proteins would circumvent some of these problems and also provide tobacco farmers, who are facing income losses due to a decline in tobacco consumption, with an alternate market.

Regulation of Genetic and Genomic Testing

Genetic tests analyze the sequence or expression of genes for the purposes of diagnosing human disease and predicting disease susceptibility, prognosis, or response to therapy (see chapter 8). They are currently regulated as medical devices in the United States, EU, and Canada. In the United States, these are regulated under the Food, Drug and Cosmetic Act and in some cases as biological products under the Public Health Services Act. Quality standards for clinical laboratories that perform these tests have also been established. Medical devices are classified in one of three groups depending on the potential risk of harm to patients from their intended use and are subject to different levels of control depending on their classification. In this context, the definition of harm includes the potential consequences to a patient's health from the interpretation of the test results. Most molecular diagnostic assays for human diseases are considered to be of either moderate risk (class II), for example, a test that is substantially equivalent to an existing test for influenza virus infection, or high risk (class III), for example, tests that utilize new technologies or screen for tumor markers, or for highly virulent pathogens. Both require premarket evaluation and either premarket notification (class II) or FDA approval (class III) before they are made available to the general public. Scientific evidence, usually including clinical studies, must be presented to show that the assay is safe and effective for its intended use. The analytical performance of a molecular diagnostic assay that uses antibodies, receptor proteins, nucleic acids, and other biological molecules is based on its specificity and sensitivity. Assessment of risk takes into consideration the consequences of an assay's performance, i.e., the consequences of false-positive or -negative results to a patient's health. Several nucleic acid-based diagnostic tests that analyze a genetic sequence or measure levels of gene expression have been approved by the FDA for a variety of human diseases and conditions (Table 12.2).

A decade ago, when the first human genome sequence was published, considerable discussion focused on the next revolution in medicine: personalized genomic medicine. It was predicted that a patient's genome sequence could be used to determine susceptibility to disease, to tailor interventions to reduce the risk of developing a disease, and to prescribe effective drugs at the correct dosage that would yield fewer side effects and be more cost-effective. While there is little doubt that genome sequencing has contributed to our understanding of human diseases (see chapter 3), there is greater uncertainty of whether genome sequence data are useful in clinical medicine. Genome-wide association studies have identified more than 2,000 loci that are associated with susceptibility to many common diseases, but for the moment, the predictive power is too low to be useful in the clinical setting. The genetic variants that have been identified explain at best 20%, and more typically less than 10%, of the risk of developing a common heritable disease. Thus, the absence of a disease-associated allele does not necessarily indicate reduced risk. On the other hand, interpretation of genome sequences may lead to false-positive conclusions, i.e., the diagnosis of a disease that is not present. For some

Table 12.2 Some FDA-approved nucleic acid-based tests for human genetic and infectious diseases

Test name	Disease or condition detected	Type of test	Intended use
Vysis EGR1 FISH probe kit	Acute myeloid leukemia	Fluorescence in situ hybridization of probes to detect deletion of EGR1 on chromosome 5q	To determine disease prognosis
MammaPrint	Breast cancer	Microarray-based gene expression profiling of tumors	To assess risk of metastasis
xTAG cystic fibrosis kit	Cystic fibrosis	Multiplex PCR/allele-specific primer extension to detect mutations in the cystic fibrosis transmembrane regulator gene	To identify carriers of gene variants associated with the disease and for screening newborns
Infiniti CYP2C19 assay	Drug metabolism	Microarray-based assay to detect variants of the cytochrome P450 2C19 gene associated with drug metabolism	To determine therapeutic strategies and treatment dose for drugs that are metabolized by the CYP2C19 gene product
AlloMap molecular expression testing	Heart transplant	Quantitative real-time PCR assay to determine expression profiles (RNA levels) of several genes	To identify heart transplant recipients with a low probability of moderate to severe acute rejection
NADiA ProsVue	Prostate cancer	Immunoassay to measure prostate-specific antigen	To determine the prognosis and assess risk of recurrence of prostate cancer
Verigene *Clostridium difficile* nucleic acid test	*Clostridium difficile* infection	Multiplex PCR to detect *C. difficile* toxin genes in stool samples	To diagnose *C. difficile* infection
Cobas CT/NG test	*Chlamydia trachomatis* and *Neiserria gonorrhoeae* infection	TaqMan assay to detect specific *C. trachomatis* and *N. gonorrhoeae* DNA sequences	To diagnose chlamydial and gonococcal disease
MultiCode-RTx HSV-1 and -2 kit	HSV-1 and -2 infection	Real-time PCR to detect specific HSV-1 and -2 DNA sequences	To diagnose HSV-1 and -2 infections
JBAIDs influenza virus subtyping kit	Influenza A virus infection	Reverse transcriptase real-time PCR assay targeting the viral hemagglutinin and nucleocapsid protein genes	To detect and differentiate among seasonal influenza A/H1 and A/H3 and 2009 H1N1 influenza virus infections
BD Max MRSA assay	MRSA	TaqMan assay to detect MRSA-specific DNA sequences	To prevent and control MRSA in health care settings

From http://www.fda.gov/MedicalDevices/ProductsandMedicalProcedures/InVitroDiagnostics.

EGR1, early growth response 1 gene; PCR, polymerase chain reaction; HSV-1 and -2, herpes simplex viruses 1 and 2, respectively; MRSA, methicillin-resistant *Staphylococcus aureus*.

diseases, the presence of an allele associated with high risk of the disease in the genome sequence of a healthy individual may have predictive value only when other characteristics are present that contribute to the development of that disease. For example, the C282Y mutation in the *HFE* gene is associated with high risk for hemochromatosis, an iron metabolism disorder, only in individuals who exhibit elevated blood iron levels. Moreover, often nothing is known about the biological function of a disease-associated locus in the disease process, and therefore, the information cannot be used beyond determining associations. For example, the locus may not be an effective target for treatment.

Although the cost of sequencing a human genome is now low enough to be accessible to many individuals, few clinical personnel have the skills to interpret the sequences. Genome sequences among individuals are highly polymorphic, containing approximately 150,000 single nucleotide variants in hundreds of genes associated with many different diseases (see chapter 3). This is a large amount of information to process, interpret, and relay to patients, both immediately following sequence acquisition and later as our ability to interpret genome sequences improves. And how should this information be disclosed to patients? There is the potential to reveal risks for disease for which there are no treatments. For example, while identification of *APC* (adenomatous polyposis coli) gene alleles associated with colon cancer may lead to regular monitoring by colonoscopy, there are no interventions available for individuals who possess alleles associated with Huntington disease. Patients must be counseled about the medical, emotional, and social risks and benefits of genomic testing, as well as understand the scope and limitations of the predictive ability of genome sequence data. Many are also concerned about the potential for discrimination against individuals who carry deleterious alleles. In 2008, the U.S. Genetic Information Nondiscrimination Act was passed to prevent health insurance providers and employers from requesting genetic information and discriminating against individuals based on that information.

Patenting Biotechnology

Biotechnology products are produced by companies with the expectation that they will profit from their sale. Many of these products require substantial funds and many years to develop and receive regulatory approval, and such high-risk investments are not made without assurance of legal protection from competition. Society also has a large stake in technological advancements that improve the quality of life, which are facilitated by disclosure of the details of new technologies. Intellectual property rights are granted by governments to satisfy the interests of both groups. That is, they provide inventors with exclusive rights to the novel products or processes that they develop while encouraging innovation. For medical biotechnology, the most important form of intellectual property is a **patent**. A patent is a legal document that gives the patent holder exclusive rights to make, use, or sell the described invention for a defined period. The period of exclusivity is 20 years from the date that the application is filed. A patent is also a public document that contains a detailed description of the invention. However, while the patent holder may develop other products from the original invention, competitors would have to license the right to use the invention in order to develop a product based on it.

Patenting

Product and process patents are the two major categories of patents (Table 12.3). Products include homogeneous substances, complex mixtures, and various devices, while processes include preparative procedures, methodologies, or actual uses. Generally, for either a product or a process to be patentable, it must satisfy four fundamental requirements:

1. The invention must be novel. The product or process must not already exist or be described in another patent or publication, even in another country, prior to the submission of the patent application.

Table 12.3 Common types of patent categories with examples of biotechnology inventions

Category	Examples
Product patents	
Substance	Cloned genes, recombinant proteins, monoclonal antibodies, plasmids, promoters, vectors, cDNA sequences, antigens, peptides, RNA constructs, antisense oligonucleotides, peptide nucleic acids, ribozymes, and fusion proteins
Composition of matter	Multivalent vaccines, biofertilizers, bioinsecticides, host cells, microorganisms, transformed cell lines, and transgenic organisms
Devices	Pulsed-field gel electrophoresis apparatus, DNA sequencing units, and microprojectile gene transfer machine, magnetic resonance imaging
Process patents	
Process of preparation	DNA isolation, synthesizing double-stranded DNA, vector–insert construction, PCR applications, and purification of recombinant protein
Method of working	Nucleic acid hybridization assays, diagnostic procedures, and mutation detection systems using PCR
Use	Applying biofertilizers and bioinsecticides, fermentation of genetically modified microorganisms, and nontherapeutic animal treatment systems

2. The invention must be nonobvious. A patent cannot be granted for something that was merely previously unknown, i.e., a discovery. Rather, the invention must contain an inventive step and be sufficiently different from products or processes already in use or previously described.
3. The invention must be useful in some way, whether it is a process, an instrument, a compound, a microorganism, or a multicellular organism.
4. The invention must be adequately described in the patent application such that a person knowledgeable in the same field can implement it.

A patent cannot be granted for anything that is "a product of nature." The notion here is that it is not appropriate for society to give a monopoly to someone for something that occurs naturally, that has merely been discovered, and therefore belongs to the public. Often companies and individuals skirt this constraint by applying for a patent that covers the process of purification of a product, thereby avoiding the direct question of ownership of either a natural substance or an organism that produces the product. According to the U.S. Supreme Court, virtually "anything under the sun that is man-made" is patentable; however, in other countries, including member states of the EU, therapeutic and diagnostic procedures are not patentable (although therapeutic substances are patentable).

In many countries, such as Canada, Australia, and member states of the EU, a patent application must be filed before the invention is disclosed to the public. In the United States, an inventor has 1 year following public disclosure in which to apply for a patent. A patent application must include the background of the invention that describes the current "state of the art" in the field of the invention; an explanation of the nature of the invention and a description of how the invention works, including figures and schematic representations where necessary; and a list of claims about the invention and how the invention may be used. The application is sent to the U.S. Patent and Trademark Office (PTO), where it is reviewed by an examiner for novelty, nonobviousness, utility, feasibility, and general acceptability as a patentable invention.

PTO
U.S. Patent and Trademark Office

If an examiner agrees that the invention meets all the criteria for patentability, then a patent is awarded. Usually it takes 2 to 5 years following the filing of the initial application before a patent is granted. However, the receipt of a patent is not a license to produce and sell the invention. All statutory regulations must be met before any product can be marketed. For example, if a patent is granted for a genetically engineered microorganism, the manufacturer must satisfy the regulations for its production, distribution, and release. Protection of patent rights is the responsibility of the patent holder, and generally that means bringing a lawsuit(s) against those who are presumed to be infringing on the patent. These disputes are decided by the courts and not the patent office. Similarly, if a person or company feels that an awarded patent is inappropriate, the legitimacy of the patent can be challenged by a lawsuit.

Patenting in Different Countries

The rights given by a patent extend only throughout the country in which the application was filed. Therefore, to protect an invention, a patent application must be filed separately in each country. Although the World Intellectual Property Organization is attempting to develop international standards, patent offices in different countries often reach quite different conclusions about the same patent application. For example, in 1989, the biotechnology company Genentech applied for a patent in the United Kingdom for, among other things, the production of human tissue plasminogen activator (tPA) by recombinant DNA processes. This protein exists in small amounts in the human body and converts plasminogen to plasmin. Plasmin is an active enzyme that degrades the fibrin of a blood clot. Consequently, human tPA may be used as a therapeutic agent for the prevention and treatment of coronary thrombosis. Genentech assembled a complete version of a human tPA cDNA and cloned this cDNA into *Escherichia coli* for the production of large amounts of pure tPA. In its patent application, Genentech claimed rights to human tPA as a product based on certain procedures of recombinant DNA technology that they developed, the cloning vector system, and the transformed microorganism. As a part of the "process" category, protection for the use of human tPA as a pharmaceutical agent was also sought by Genentech. A total of 20 claims were presented in the original patent application. Some of these were broad and others narrow in scope. The patent was rejected by the United Kingdom's patent office. Genentech then appealed to the United Kingdom's Court of Appeals, which, after considerable deliberation, invalidated all of the claims for a variety of reasons. The judgment

tPA
tissue plasminogen activator

concluded that the patent was novel, but some of the judges argued that the submission was obvious; therefore, it could not be patented.

In contrast, Genentech was readily awarded a patent for human tPA in the United States. The U.S. patent not only protected the form of human tPA that was to be marketed by Genentech but also gave Genentech exclusive rights to all similar, but not identical, active forms of human tPA. Genentech won a lawsuit against two other biotechnology companies that were found to be infringing on its tPA patent, although they were selling nonidentical forms of tPA.

The Japanese version of Genentech's tPA patent was limited to the amino acid sequence of the human tPA that was cloned and patented by Genentech. In Japan, other companies could sell variant forms of human tPA. Thus, basically the same patent application was rejected, approved and given a broad interpretation, and approved and given a narrow interpretation by three different patent offices. This illustrates the divergent views about what is or is not a patentable invention.

Patenting DNA Sequences

Since 1980, thousands of patent applications for DNA, RNA, or cDNA sequences have been approved by patent offices throughout the world. In the United States, more than 40,000 DNA-related patents have been issued, at an annual rate of almost 4,000 over the last decade, and about 20% of human genes have been patented. Some of these human genes are used to make therapeutic proteins, such as recombinant erythropoietin. Many of the patented gene sequences are used as diagnostic probes to detect disease-related genes. One example results from the discovery that particular mutations in the human genes *BRCA1* and *BRCA2* are linked to breast and ovarian cancer. A patent, issued to Myriad Genetics Inc., claims methods to detect mutations in *BRCA1* and *BRCA2* to diagnose a predisposition to these cancers.

The issue of patenting DNA sequences was first broached in 1991 when scientists from the U.S. National Institutes of Health filed for the patent rights for 315 partially sequenced human cDNAs. Two additional filings brought the total number of partial sequences to 6,869. In 1994, in a preliminary ruling, the U.S. PTO notified the National Institutes of Health that it would reject the patent application on the grounds that the functions of the sequences were not known. In other words, partial sequences by themselves did not fulfill the requirement of utility and were not patentable. However, by 1997, over 350 patent applications for more than 500,000 partial DNA sequences had been filed, mostly by private companies, which purportedly met the standard for usefulness. One of these patent proposals sought protection for about 18,500 expressed sequence tags. Serious concerns were raised about granting patents for large numbers of sequenced genes and partially sequenced DNA fragments with broadly based applications. The consortium that was sequencing the human genome around this time considered that while some of the human genome sequences might eventually be useful, it was premature and speculative to award patents without additional information about the functions of the sequences. The sequence information generated from the Human Genome Project was quickly deposited in databases that were readily accessible to the public.

The central argument of the debate is whether isolated DNA sequences are discoveries or products of nature, and therefore not patentable, or inventions, and therefore patentable. In part as a consequence of the large-scale sequencing projects, the U.S. PTO reexamined the issues and concluded that genes and partial DNA sequences were patentable. In 2001, a set of guidelines for patenting DNA sequences in the United States was released. A key requirement is that each DNA sequence must have "specific, substantial and credible utility." The written specifications and claims for each sequence must be thorough and demonstrate the actual use of each sequence and not merely a potential function. Similarly, in the EU and Canada, DNA sequences are patentable as long as there is an indication of function.

The patenting of DNA sequences found in humans and other organisms remains controversial. For example, the validity of the *BRCA1* and *BRCA2* diagnostic patents was challenged by the American Civil Liberties Union and the Public Patent Foundation on the basis that Myriad Genetics' exclusive rights to the genes made it impossible for women to confirm the test results using a similar diagnostic test and that the high cost of the test precluded many women from determining their risk for breast and ovarian cancers. In 2010, a U.S. district court invalidated the patent on the grounds that isolated genes are not patentable; however, a federal appeals court overturned the decision in 2011, finding that the chemical nature of isolated DNA is different from DNA in cells. For example, cDNA generated from mRNA in vitro lacks intron sequences found in natural DNA, and isolated DNA is cleaved from the original molecule and thus is modified. In other words, the act of isolating the sequences requires human intervention and therefore the molecules are "engineered." Moreover, the isolated DNA sequences are used in a manner that is fundamentally different from their natural use, in much the same way that the use of purified antibiotics to treat bacterial infections is different from their natural use by soil microorganisms. However, in 2013, the U.S. Supreme Court determined that "Myriad did not create anything . . . separating that gene from its surrounding genetic environment is not an act of invention" and therefore the DNA sequences are not patentable. On the other hand, the ruling stated that cDNA is patentable because it is made in the laboratory.

Also controversial is the increased scope of the DNA patent claims in recent years. Often included in the claims to DNA sequences and the proteins they encode are antibodies against the protein, even when the antibody has not actually been produced. Antibodies against human proteins are important from a commercial perspective because they are used for diagnostic purposes and as therapeutic agents; therefore, there is considerable incentive to include these in patent applications. However, some argue that they do not meet the criteria for patentability because in many cases the antibodies have not been produced, and therefore, their characteristics are not specifically described and working examples are not provided.

Patenting Living Organisms

Cell lines and genetically engineered unicellular organisms are considered to be patentable. For example, in 1873, Louis Pasteur received two patents (U.S. patents 135,245 and 141,072) for a process for fermenting beer that included the living organism (yeast) used in the process. However, the first time a scientist attempted to patent a genetically modified microorganism that was engineered by introducing different plasmids, the case was highly controversial. Each of the plasmids carried the genes for a separate hydrocarbon degradative pathway enabling the genetically modified bacterium to break down many of the components of crude oil and therefore potentially to be used to clean up oil spills. The bacterium was developed by A. Chakrabarty, an employee of the General Electric Corporation at the time. The patent application for this bacterium was rejected by the U.S. PTO on the grounds that microorganisms are products of nature and, as living things, are not patentable. In 1980, in a landmark decision, the U.S. Supreme Court decided that this organism was patentable according to the U.S. Patent Statute, arguing that "a live, human-made microorganism is patentable subject matter . . . as a manufacture or composition of matter."

The argument against patenting this genetically engineered microorganism tended to center on how the organism was developed. In the past, induced mutation followed by selection for novel properties was an acceptable way to create a patentable living organism. However, genetic engineering was considered by some to be "tampering with nature." Consequently, it was argued that no inventor should benefit from manipulating products of nature. This position was not upheld. In the United States from 1980 onward and later in other countries, organisms, regardless of the means that were used to develop them, were mandated to be judged by the standard criteria of novelty, nonobviousness, and utility to determine if they are patentable.

Living multicellular organisms that meet these criteria are also eligible for patenting in the United States and most other developed countries, although this continues to raise ethical and social concerns. The exclusive ownership of multicellular organisms is not a new concept. Since 1930, the U.S. Plant Patent Act has given plant breeders the right to own various plant varieties. In 1988, the first genetically engineered animal, a transgenic mouse, was patented. The transgene consisted of a cancer-causing gene (oncogene) driven by a viral promoter that increased the probability that the mouse would develop tumors. The **"oncomouse"** can be used to test whether a compound either causes or prevents cancer. The granting of the oncomouse patent (U.S. patent 4,736,866) was contentious, especially on moral grounds. However, from a historical perspective, it is unlikely that a position based on ethical considerations will be completely successful in preventing the patenting of all transgenic animals. For example, if an invention purports to facilitate a new treatment for human disease, the currently prevalent view in most countries is that human rights and needs supersede those of animals. Since the decision on the patentability of the oncomouse, hundreds of patents have been granted in the United States for various transgenic animals, including those that act as models for human diseases or produce human therapeutic proteins.

Patenting and Fundamental Research

One of the major goals of the patent system is to encourage innovation by providing incentive to engage in research and disclose inventions that others can improve on. However, some opponents believe that awarding a monopoly restricts competition, leads to higher prices, curtails new inventions, and favors large corporations at the expense of individual inventors or small companies. For example, one of the concerns over gene patenting is that it will impede the development of new medical biotechnologies because researchers must obtain permission to use the genes, which can be expensive and cumbersome if multiple patents are involved. Researchers may avoid research that requires use of patented materials or processes. Others counter that there is no evidence to support the notion that patents are serious impediments to innovation. For example, U.S. patent 4,237,224, which was granted to Stanley Cohen and Herbert Boyer in 1980 for recombinant DNA technology, for both the use of viral and plasmid vectors and the cloning of foreign genes, has obviously not seriously constrained the development of recombinant DNA technology.

Some members of the academic scientific community are concerned that patents and the consequences of patenting may be detrimental to established scientific values. Traditionally, university-based scientific research has been an open system with a free exchange of ideas and materials. The belief is that the growth of scientific knowledge and the development of technologies benefit from public availability of research articles and databases. However, some scientists feel that the integrity of traditional scientific inquiry has become secondary to commercial interest in and financial gain from innovations. Many scientists are now advised by patent lawyers not to disclose the results of their research until a patent is filed. Licensing fees and royalties from patents can be significant sources of income for financially constrained, nonprofit institutions, such as universities. For example, during its lifetime from 1980 to 1997, the Cohen-Boyer patent for recombinant DNA technology earned about $45 million for Stanford University and the University of California. Also, the Massachusetts Institute of Technology files more than 100 patents annually in all research fields and generates about $5.5 million per year from licensing patent rights. Most universities have established patent policies and offices that facilitate patenting and the transfer of technology, at a price, to industry. Faculty members usually receive a portion of the income from their inventions. Clearly, entrepreneurial activity is a fact of life at many universities.

Economic Issues

According to the World Health Organization, $50 billion is spent globally on pharmaceuticals each year, with the United States as the largest market. Many biologic drugs are costly. For example, in 2009, the cost to treat a patient for rheumatoid arthritis with Enbrel, a fusion protein that inhibits the cytokine tumor necrosis factor and is produced by cultured mammalian cells, was $26,000/year. Herceptin, a monoclonal antibody used to treat some forms of breast cancer, costs patients $37,000/year; Rebif, recombinant beta-interferon used to treat multiple sclerosis, $40,000/year;

Humira (monoclonal antibody adalimumab) for Crohn disease, $51,000/year; and Cerezyme (recombinant enzyme imiglucerase) for Gaucher disease, $200,000/year, one of the most expensive drugs.

Why are the costs so high? Contributing factors are the human and technological resources, as well as the strict and lengthy regulatory process required to bring new drugs to market. It is a risky business. The pharmaceutical industry estimates that the average cost to develop a new drug is between $4 and $11 billion. This is calculated by dividing the total annual research and development budget of a pharmaceutical company by the number of drugs approved. Thus, included in the cost is the investment in drugs that did not reach the market. A major factor is the high price of clinical trials, which typically involve thousands of patients (most in phase III) and can cost tens of millions of dollars each. This takes into account the costs to produce the drug, to recruit and manage a large number of patients and medical personnel, for medical procedures and tests, and for data analysis. On average, 65% of new drugs are successful following phase I clinical trials. Phase II and III trials have success rates of 40% and 64%, respectively, and 93% of these are finally approved by the regulatory agency. This means that only about 16% of new drugs are approved for commercial use. When only biologics such as therapeutic recombinant proteins and monoclonal antibodies are considered, the success rate is higher (32%). On average, it takes about 8 years to bring a new drug to market, although this varies depending on the type of drug. For example, new drugs developed to treat diseases of the central nervous system take a longer time to reach approval (about 10 years) than do antivirals for treatment of AIDS (about 5 years).

Some are concerned that the high cost and low success rate will slow the development of new drugs, especially the so-called **orphan drugs** that can be used to treat rare diseases or diseases found mainly in low-income populations. In the United States, an orphan drug is defined as one used to treat, diagnose, or prevent a disease that affects fewer than 200,000 people or that will not generate enough sales to recover the costs of development and marketing (Table 12.4). More than 7,000 diseases are considered to be rare, including multiple sclerosis, cystic fibrosis, and Duchenne muscular dystrophy. Most rare diseases are genetic disorders. To encourage the development of orphan drugs, many countries facilitate the approval process and provide financial incentives. The average time for clinical testing and regulatory approval for orphan drugs is about 6.6 years, compared to 8.0 years for nonorphan drugs. Financial incentives may include tax credits or special grant programs for orphan drug producers, a reduction in the number of patients required for clinical trials, or an extended period of patent protection. These incentives have resulted in a greater number of drugs to treat rare diseases. Since the enactment of the U.S. Orphan Drug Act in 1983, more than 400 orphan drugs have been developed, in contrast to 10 in the previous decade. In 2011, one-third of FDA new drug approvals were for orphan drugs. Among these are treatments for rare cancers such as metastatic melanoma, including the drugs vemurafenib, an inhibitor of the kinase B-Raf, and the monoclonal antibody ipilimumab, which targets a cytotoxic T-lymphocyte antigen (CTLA-4).

Table 12.4 Some orphan drugs recently approved by the FDA

Orphan drug	Drug description	Orphan disease
Canakinumab	Monoclonal antibody targets interleukin-1β	Pediatric juvenile rheumatoid arthritis
Regorafenib	Small molecule inhibits multiple kinases	Gastrointestinal stromal tumors
Mipomersen	Antisense oligonucleotide targets apolipoprotein B mRNA	Homozygous familial hypercholesterolemia
Bedaquiline	Diarylquinoline targets ATP synthase in *Mycobacterium tuberculosis*	Active tuberculosis
Teduglutide	Recombinant glucagon-like peptide 2 analogue promotes mucosal growth	Short bowel syndrome
Raxibacumab	Monoclonal antibody targets *Bacillus anthracis* toxins	Anthrax
Pasireotide	Decreases cortisol overproduced by the adrenal glands	Cushing disease
Cabozantinib	Small molecule inhibits the activity of multiple tyrosine kinases	Metastatic medullary thyroid cancer
Bosutinib	Blocks signaling by tyrosine kinase	Chronic myelogenous leukemia
Taliglucerase alpha	Recombinant enzyme produced in plant (carrot) cells that replaces missing glucocerebrosidase	Gaucher disease
Ivacaftor	Small molecule increases chloride transport through the cystic fibrosis transmembrane regulator protein	Cystic fibrosis
Glucarpidase	Recombinant enzyme inactivates methotrexate	Toxic plasma methotrexate levels in patients with impaired renal function
Gabapentin enacarbil	γ-Aminobutyric acid analogue (prodrug; activated in vivo)	Postherpetic neuralgia (nerve pain associated with shingles)

From http://www.accessdata.fda.gov/scripts/opdlisting/oopd/index.cfm.

Has the high cost of drugs forced some patients to seek cheaper, **illegitimate drugs** and companies to manufacture them? In this context, illegitimate drugs are those made to deliberately deceive the consumer; they include **substandard drugs** and **falsified drugs**, for which the quality, source, or identity of the drug is falsely represented. Many different types of drugs have been counterfeited, ranging from treatments for life-threatening diseases to analgesics (Table 12.5). The contents may contain incorrect, low concentrations of, or no active ingredients. Thus, the medicines may be ineffective, impure, or toxic. Furthermore, illegitimate antimicrobial agents often contain insufficient concentrations of antibiotics that select for the growth of drug-resistant bacteria. Low-dose antibiotics are a major contributing factor to the emergence of drug-resistant strains of *Mycobacterium tuberculosis* and *Plasmodium* spp. that cause tuberculosis and malaria, respectively.

Table 12.5 Some examples of recently seized illegitimate drugs

Drug(s)	Intended use	Illegitimate drug	Year and place of seizure
Avastin	Cancer	Lacked active ingredient bevacizumab	2012, United States
Viagra and Cialis	Erectile dysfunction	Undeclared active ingredients	2012, United Kingdom
Truvada and Viread	HIV/AIDS	Authentic product in falsified packaging	2011, United Kingdom
Zidolam-N	HIV/AIDS	Falsified drug; discolored, molding	2011, Kenya
Alli	Weight loss	Undeclared active ingredient	2010, United States
Antidiabetic traditional medicine	Reduce blood sugar levels	Contained 6 times the normal dose of glibenclamide	2009, China
Metakelfin	Antimalarial	Low levels of active ingredient	2009, Tanzania

From http://www.who.int/mediacentre/factsheets/fs275/en/index.html.

Illegitimate drugs can be found throughout the world but are particularly problematic in developing countries where marketing of pharmaceuticals is poorly regulated or the regulations are not enforced, and adherence to international quality standards is financially difficult for many small manufacturers. In developed countries such as the United States, most EU countries, Canada, Australia, New Zealand, and Japan, illegitimate drugs represent less than 1% of the market. But in many African, Asian, and Latin American countries, the problem is much greater. However, many unlicensed and unregulated vendors sell illegitimate drugs over the Internet, mainly to individuals in middle- to high-income countries. In the United States, online pharmacies must be accredited by the National Association of Boards of Pharmacies and comply with standards of quality. In addition, globalization of the pharmaceutical production and distribution system contributes to the problem. Ingredients for medicines are often produced, combined, and packaged in different countries and therefore pass through many hands before reaching the patient. This provides many opportunities for substandard and falsified products to enter into the manufacturing and distribution chain. International cooperation is necessary to establish and enforce regulations for manufacturing quality and marketing to ensure a safe supply of drugs for patients.

summary

Most countries have regulations in place to ensure that new drugs and diagnostic tests are safe and effective before they are made commercially available. A key regulatory requirement is the series of clinical trials, in which treatments that show promise after testing in animals are tested further on different populations of humans. Clinical trials are performed to determine drug toxicity (phase I), efficacy (phase II), optimal dosage, and potential for interactions with other drugs (phase III). A treatment is usually approved only for the specific, tested application. The regulatory requirements apply to all new drugs whether they are produced through chemical synthesis or molecular biotechnologies. This includes biologics (vaccines, blood products, cell-based treatments, and nucleic acid therapies) that are derived from living organisms, including genetically engineered organisms. The FDA recently approved a human therapeutic protein (antithrombin) produced in the milk of a transgenic goat, and it is likely that more pharmaceutical proteins will be produced in this way in the near future. Genetically engineered animals that produce recombinant proteins must be approved each time a gene is introduced and must be confined to controlled environments. Containment measures must also be taken to prevent transgenic plants that produce pharmaceutical proteins from ending up in the food supply.

In many countries, including the United States, commercial genetic and molecular tests to diagnose or predict susceptibility to a disease, and the clinical laboratories that perform them, are regulated. Consideration is given to the potential for harm to a patient's health from the interpretation of the test, for example, false-positive (diagnosis of a disease that is not present) or false-negative (erroneous indication of low disease risk) results. Currently, there are limitations to the clinical application of human genome sequences for predicting disease susceptibility and tailoring therapies to reduce disease risk or increase treatment efficacy. One issue is the low predictive power of genetic sequences. Although many disease-associated alleles have been identified, at best these explain only 20% of the risk of developing a common heritable disease. A better understanding of the genetic basis of disease is required. Other concerns are the lack of clinical personnel skilled in the interpretation of highly polymorphic genomic sequence data and counseling patients about the implications of the results, and the potential for discrimination against individuals who carry deleterious alleles by employers or insurance providers.

Companies that produce biotechnology products often invest a great deal of time and financial resources to develop the

(continued)

summary *(continued)*

products to the commercial-use stage. Patents are a means to protect their investment by giving the patent holder exclusive rights to make, use, or sell the product for a specific period. At the same time, public disclosure of the invention is expected to encourage innovation; however, opponents believe that awarding a monopoly curbs new inventions, as other researchers may be concerned about infringing on a patent or must obtain permissions to use patented materials or processes. For a patent to be granted, an invention must be novel, not obvious, and useful. The patenting of nucleic acid sequences is controversial. Recently, the U.S. Supreme Court ruled that DNA sequences are not inventions and therefore not patentable, although cDNA, which is produced in the laboratory from RNA, is patentable. Cell lines, and unicellular and multicellular organisms, including genetically engineered plants and animals, may be patented as long as they meet the standard criteria of novelty, nonobviousness, and utility.

Development of a new drug requires a substantial financial investment, in large part due to the long and costly regulatory process, especially clinical trials. To encourage pharmaceutical companies to develop orphan drugs to treat rare diseases, many governments provide financial incentives and facilitate the approval process. Some human therapeutic monoclonal antibodies and recombinant proteins cost patients more than $25,000/year. The high drug costs have forced some patients to seek illegitimate drugs that may contain no or low concentrations of active ingredients, or toxic contaminants, and are therefore ineffective and unsafe. This is a global problem that will require international cooperation to establish and enforce drug manufacturing and marketing regulations.

review questions

1. Who is responsible for providing evidence of the safety and efficacy of a new drug before it is made commercially available?

2. Briefly describe the stages of testing required for a new drug to be approved by the FDA.

3. What are biologics? How are they regulated?

4. Visit the U.S. National Institutes of Health registry and database of human clinical trials at http://www.clinicaltrials.gov. Search for studies using key words such as "stem cells" and "gene therapy." Briefly summarize the purpose of the clinical trial, including the treatment, procedure, or test under investigation and its potential application.

5. List some issues regarding the production of human therapeutic proteins in genetically engineered animals.

6. What are some of the issues regarding genetic testing to determine disease susceptibility?

7. What is the benefit of a patent to the patent holder? What is the benefit of a patent to researchers who do not hold the patent?

8. What are the essential criteria for patenting an invention?

9. What are some arguments for and against the patenting of human genes?

10. Contact the U.S. PTO website (http://patents.uspto.gov/) and conduct a search for medical biotechnology patents. Use various combinations of search terms, such as "human gene" AND "therapeutic," "antibody" AND "diagnostic," and so on. Summarize the inventions in some recent patents related to human medicine.

11. Summarize some of the factors that may contribute to the high cost of a new biologic.

12. What is an orphan drug? What are some of the incentives to produce these drugs?

13. What are some of the factors that contribute to the use and availability of illegitimate drugs?

references

Brunham, L. R., and M. R. Hayden. 2012. Whole-genome sequencing: the new standard of care? *Science* **336**: 1112–1113.

Committee for Orphan Medicinal Products and the European Medicines Agency Scientific Secretariat. 2011. European regulation on orphan medicinal products: 10 years of experience and future perspectives. *Nat. Rev. Drug Discov.* **10**:341–349.

Cook-Deegan, R. 2012. Law and science collide over human gene patents. *Science* **338**:745–747.

Dickson, M., and J. P. Gagnon. 2004. Key factors in the rising cost of new drug discovery and development. *Nat. Rev. Drug Discov.* **3**:417–429.

DiMasi, J. A., L. Felman, A. Seckler, and A. Wilson. 2010. Trends in risks associated with new drug development: success rates for investigational drugs. *J. Clin. Pharm. Ther.* **87**:272–277.

Drmanac, R. 2011. The ultimate genetic test. *Science* **336**:1110–1112.

Gibbs, J. N. 2011. Regulating molecular diagnostic assays: developing a new regulatory structure for a new technology. *Expert Rev. Mol. Diagn.* **11**:367–381.

Holman, C. M. 2007. Patent border wars: defining the boundary between scientific discoveries and patentable inventions. *Trends Biotechnol.* **25**: 539–543.

Holman, C. M. 2012. Debunking the myth that whole-genome sequencing infringes thousands of gene patents. *Nat. Biotechnol.* **30**:240–244.

Jensen, K., and F. Murray. 2005. Intellectual property landscape of the human genome. *Science* **310**:239–240.

Kaitin, K. I., and J. A. DiMasi. 2011. Pharmaceutical innovation in the 21st century: new drug approvals in the first decade, 2000–2009. *Clin. Pharmacol. Ther.* **89**:183–188.

Melnikova, I. 2012. Rare diseases and orphan drugs. *Nat. Rev. Drug Discov.* **11**:267–268.

National Institutes of Health. 2013. *Guidelines for Research Involving Recombinant or Synthetic Nucleic Acid Molecules (NIH Guidelines).* Department of Health and Human Services, National Institutes of Health, Washington, DC. http://oba.od.nih.gov/rdna/ nih_guidelines_oba.html.

Spök, A., R. M. Twyman, R. Fischer, J. K. C. Ma, and P. A. C. Sparrow. 2008. Evolution of a regulatory framework for pharmaceuticals derived from genetically modified plants. *Trends Biotechnol.* **26**:506–517.

Tambuyzer, E. 2010. Rare diseases, orphan drugs and their regulation: questions and misconceptions. *Nat. Rev. Drug Discov.* **9**:921–929.

U.S. Food and Drug Administration. 15 January 2009. Guidance for industry no. 187—regulation of genetically engineered animals containing heritable recombinant DNA constructs. http:// www.fda.gov/cvm. Accessed 17 June 2013.

Glossary

A–B exotoxin A protein toxin secreted by some bacteria that contains an A domain with toxin activity and a B domain that binds to a host cell receptor.

Acquired immunodeficiency syndrome A disease caused by a CD4$^+$ T-cell-mediated infection with HIV. Abbreviated AIDS.

Active immunity An immune reaction(s) that may be induced in an individual by infection or vaccination.

Acute infection A short-term infection characterized by rapid reproduction of the pathogen and rapid onset of symptoms.

ADA *See* Adenosine deaminase.

Adaptive immune system The immune system in which T cells and B cells confer immunity during a later, more antigen-specific and more efficient immune response.

Adaptive immunity A form of immunity that develops more slowly, confers specificity against a foreign antigen, and mediates the later and more vigorous defense against infections.

Adaptor (**1**) A synthetic double-stranded oligonucleotide that is blunt ended at one end and at the other end has a nucleotide extension that can base-pair with a cohesive end created by cleavage of a DNA molecule with a specific type II restriction endonuclease. After blunt-end ligation of the adaptor to the ends of a target DNA molecule, the construct can be cloned into a vector by using the cohesive ends of the adaptor. (**2**) A synthetic single-stranded oligonucleotide that, after self-hybridization, produces a molecule with cohesive ends and an internal restriction endonuclease site. When the adaptor is inserted into a cloning vector by means of the cohesive ends, the internal sequence provides a new restriction endonuclease site.

Adenosine deaminase An enzyme that catalyzes the breakdown of purines. A deficiency in ADA activity can lead to the accumulation of toxic purine metabolites in cells that are dividing and actively synthesizing DNA. Abbreviated ADA.

Adenosine diphosphate (ADP)-ribosyltransferase A bacterial toxin that catalyzes the transfer of an ADP-ribose group from nicotinamide adenine dinucleotide (NAD$^+$) to a host protein, thereby disrupting the activity of that protein.

Adenosine triphosphate A nucleotide that is an energy source for many cellular processes. During cell catabolism, the breakdown of large molecules releases energy in the form of ATP, which is required for many anabolic processes. This switch from catabolism to anabolism leads to the activation, proliferation, and differentiation of a T cell. Abbreviated ATP.

Adhesin A molecule on the surface of a bacterial cell that mediates adherence to other cells.

Adjuvant An agent that stimulates immunity.

Adoptive immunotherapy Isolation of tumor-specific immune cells, enriching them outside the body and transfusing them back into the patient. The transfused cells are highly cytotoxic to the cancer cells.

ADP-ribosylation The addition of one or more ADP-ribose moieties to a protein. These reactions are involved in cell signaling.

Affinity The relative strength of a noncovalent interaction between an antigen-binding site of an antibody and its specific determinant or epitope on an antigen. This strength is represented by a dissociation constant (K_d), where a low K_d indicates a strong or high-affinity interaction.

Agglutination Clumping of cells, especially when bound by an antibody. Also, a type of immunological test to detect the presence of a specific antigen.

AIDS *See* Acquired immunodeficiency syndrome.

AIRE The autoimmune regulator protein that controls the expression of some peripheral tissue-restricted protein antigens (e.g., insulin) in the thymus. Many self-proteins normally expressed in peripheral tissues can also be expressed in some epithelial cells of the thymus.

Allele Either of two different DNA sequences at a given site (or gene).

Allergens Antigens that stimulate immediate hypersensitivity (allergic) reactions.

Allergy An immediate hypersensitivity reaction to an allergen. Also called atopy.

ALS *See* Autoimmune lymphoproliferative syndrome.

Alum An aluminum-based salt; sometimes used as an adjuvant in a vaccine.

Amyloid An abnormal fibrillar aggregation of proteins, which, after staining with Congo red, produces a green color under polarized light. Also called amyloid body.

Amyloplast A nonpigmented organelle found in some plant cells that is responsible for the synthesis and storage of starch.

Anaphylaxis A severe, often life-threatening, whole-body allergic reaction to the presence of an antigen. The most severe form of immediate hypersensitivity.

Anergy A state of long-term hyporesponsiveness in T cells that is characterized by an inhibition of TCR signaling and interleukin-2 expression.

Aneuploidy A situation in which the chromosome number is not an exact multiple of the haploid number that results from the failure of paired chromosomes (at first meiosis) or sister chromatids (at second meiosis) to separate at anaphase. Thus, two cells are produced, one with a missing copy of a chromosome and the other with an extra copy of that chromosome.

Angiosperm A flowering, seed-producing plant.

Annotation Assignment of functions to sequence features in genomic sequences; this may include identification of protein-coding and regulatory sequences and prediction of gene function.

Antibiotic A substance that is produced naturally by some fungi and bacteria, or is chemically synthesized, and that can inhibit the growth of, or kill, bacteria.

Antibodies A family of blood-derived glycoproteins known as immunoglobulins that are synthesized by B cells in response to a specific antigen.

Antibody microarray A type of protein microarray consisting of antibodies immobilized on a solid support that is used to detect and quantify proteins in a complex sample.

Antigenic drift Variation in viruses that arises from mutations in genes encoding surface proteins that are normally recognized by antibodies. The antibodies present no longer recognize the variant proteins and therefore do not effectively protect against infections by the new viral strains.

Antigenic shift Variation in viruses that arises when two or more different viral strains, for example, two different strains of influenza A viruses, infect a cell and package new combinations of genomic molecules into virions. Major genetic changes lead to new viral strains that can spread rapidly in a population that lacks immunity.

Antigenic variation Alteration in the structure of proteins on the surface of an infectious microorganism such that they are no longer recognized by the host immune system.

Antigenomic RNA An RNA molecule that is complementary to the single-stranded genomic RNA molecule of a virus. It is produced by replication of the genomic RNA using an RNA-dependent RNA polymerase and serves as a template for synthesis of more genomic RNA.

Antigen-primed lymphocyte A lymphocyte previously exposed to and activated by an antigen.

Antiserum Serum containing antibodies against specific antigens.

Apoptosis A controlled process leading to the death of the cell that occurs normally during the development of a multicellular organism or in response to cell damage or infection. Also known as programmed cell death.

APS *See* Autoimmune polyendocrine syndrome.

Aptamer A synthetic nucleic acid, typically 15 to 40 nucleotides long, that has highly organized secondary and tertiary structures and binds with high affinity to a protein that normally does not bind to a nucleic acid.

Artificial chromosome An artificially constructed yeast or human chromosome that includes a telomere, centromere, and sequences for autonomous replication and maintenance in a yeast or human cell, respectively; usually used to clone large fragments of DNA. Bacterial artificial chromosomes are derived from plasmids and are often used to clone large fragments of genomic DNA for sequencing.

Atopy An immediate hypersensitivity reaction that leads to allergy. Individuals with a strong propensity to develop such reactions are "atopic." These reactions may include hay fever, food allergies, bronchial asthma, and anaphylaxis.

ATP *See* Adenosine triphosphate.

Atrophy Deterioration of a cell, tissue, or organ.

Attenuated Inactivated; used to describe an inactivated form of a pathogen.

Auristatin E A synthetic, highly toxic antineoplastic agent that because of its toxicity cannot be used as a drug itself; rather, it is linked to a monoclonal antibody.

Autism spectrum disorders A group of developmental brain disorders that are associated with an elevated level of the gut-derived bacterium *Clostridium bolteae*.

Autoimmune disease A disease caused by the inability to block an attack by the immune system on an individual's own cells and tissues.

Autoimmune lymphoproliferative syndrome A rare disease rare that is the only known example of a defect in apoptosis that causes a complex autoimmune phenotype in humans. Abbreviated ALS.

Autoimmune polyendocrine syndrome A recessive genetic disorder in which the function of many endocrine glands is altered. This rare autoimmune disorder is caused by mutations in the *AIRE* gene. Abbreviated APS.

Autoimmunity An attack by the immune system on an individual's own cells and tissues.

Autosomal dominant A type of genetic inheritance. All humans have two chromosomal copies of each gene or allele, one allele per each member of a chromosome pair. Dominant monogenic diseases involve a mutation in only one allele of a disease related gene. In autosomal dominant inheritance, an affected person has at least one affected parent, and affected individuals have a 50% chance of passing the disorder onto their children.

Autosomal recessive A type of genetic inheritance. Recessive monogenic diseases occur due to a mutation in both alleles of a disease-related gene. In autosomal recessive inheritance, affected children are usually born to unaffected parents. Parents of affected children usually do not have disease symptoms, but each carries a single copy of the mutated gene. There is an increased incidence of autosomal recessive disorders in families in which parents are related. Children of parents who are both heterozygous for the mutated gene have a 25% chance of inheriting the disorder, and the disorder affects either sex.

Autosomal SCID A type of SCID that is not X linked. About 50% of SCID cases are autosomal. *See* Severe combined immunodeficiency.

Autosomes Chromosomes that are not sex chromosomes. Typically, humans have 22 pairs of autosomes.

Auxin A plant hormone which stimulates both rapid responses, such as increases in cell elongation, and long-term effects, such as increases in cell division and differentiation. Also called indole-3-acetic acid.

Avidity The overall strength of binding between two proteins, such as an antigen and an antibody. The avidity of binding between an antigen and an antibody depends on the affinity and valency of interactions between these two proteins.

Axon A single cellular extension emanating from the cell body of a neuron that transmits a nerve impulse to the synaptic bulb. Also called nerve fiber.

Bacmid A shuttle vector that can be propagated in both *Escherichia coli* and insect cells.

Bacteriocin An antibacterial compound, such as colicin, produced by a strain of certain bacteria and harmful to other strains within the same family.

Bacteriophage A virus that infects bacterial cells. Also called phage.

Bacteriophage λ An *Escherichia coli* virus with a 49-kilobase-pair double-stranded DNA genome that has been modified for use as a cloning vector.

Baculovirus A member of a family of DNA viruses that infects only invertebrate animals. Some have a very specific insect host and may be used in biological pest control.

Basal ganglia (singular, basal ganglion) Clusters of nerve cell bodies within the cerebral hemispheres.

B-cell antigen receptor A membrane-bound form of an antibody, expressed specifically by B cells, that functions as a receptor that binds to soluble antigens and antigens on the surface of microbes and other cells. Upon engagement of antigens by relevant B-cell antigen receptors, B cells are triggered to become activated for antibody synthesis and secretion, and they elicit humoral immune responses against specific antigens.

B cells Bone marrow-derived cells that produce antibody molecules and mediate humoral immunity. Mature B cells localize in lymphoid follicles in peripheral lymphoid tissues (spleen and lymph nodes), in bone marrow and in the blood circulation.

Binary vector system A two-plasmid system in *Agrobacterium* spp. for transferring a T-DNA region that carries cloned genes into plant cells. The virulence genes are on one plasmid, and the engineered T-DNA region is on the other plasmid.

Biofilm A community of microorganisms growing on a surface and encased in an extracellular matrix composed of polysaccharides, DNA, and proteins.

Bioinformatics Research, development, and application of computational tools to acquire, store, organize, analyze, and visualize data for biological, medical, behavioral, and health sciences.

Biolistics Delivery of DNA to plant and animal cells and organelles by means of DNA-coated pellets that are fired under pressure at high speed. Also called microprojectile bombardment.

Biologics Therapeutic agents derived from living organisms (including genetically engineered organisms). Includes proteins, vaccines, blood products, cells, and tissues.

Biomarker A biological feature that is used to measure either the progress of a disease or the effect of a treatment.

Biopsy Surgical removal of a tissue sample from a patient for diagnostic purposes.

Bioreactor A vessel in which cells, cell extracts, or enzymes carry out a biological reaction. Often refers to a growth chamber (fermenter or fermentation vessel) for cells or microorganisms.

Blastomere A cell derived from the cleavage of a fertilized egg during early embryonic development.

Bone marrow The central cavity of bone where B-cell development and maturation occur. This is also the site of production of all circulating blood cells and immature lymphocytes in adults.

Broad-host-range plasmid A plasmid that can replicate in a number of different bacterial species.

Budded virus A virus that acquires its envelope by budding through the the cell membrane.

Budding A process during which a virus is enveloped in a membrane as it exits a host cell.

Burkitt lymphoma A rare form of aggressive non-Hodgkin lymphoma that primarily affects children.

CAG trinucleotide repeats Motifs which appear many times at the 5′ end of the Huntington disease gene; this motif encodes glutamine. A highly variable number of these repeats accounts for the Huntington disease gene mutation, which leads to the expression of an abnormally long polyglutamine tract at the N terminus of the huntingtin protein. Such polyglutamine tracts increase protein aggregation, which may alter cell function.

Callus Undifferentiated tissue that develops on or around an injured or cut plant surface or in tissue culture.

Calmodulin A protein that binds calcium and is involved in regulating a variety of activities in cells.

Canola A plant whose seed (also called rapeseed) is used to produce high-quality cooking oil. Canola is Canada's most economically important crop.

Capsid A structure that is composed of the coat protein(s) of a virus and is external to the viral nucleic acids. The capsid often determines the shape of the virus.

Cap snatching A process by which some viruses acquire a 5′ 7-methyl-guanosine capped RNA primer to initiate transcription by cleaving the first 10 to 13 nucleotides with an attached cap from host mRNA.

Capsule An external slime layer, often composed of polysaccharides, that surrounds the cell wall of some bacteria.

Carborundum The trademark name of an abrasive compound composed of silicon carbide.

Carcinoembryonic antigen A glycoprotein involved in cell adhesion and expressed on metastatic colon carcinoma cells. Abbreviated CEA.

Carrying capacity The maximum amount of DNA that can be inserted into a vector.

cDNA Complementary DNA, i.e., a double-stranded DNA complement of an mRNA sequence; synthesized in vitro by reverse transcriptase and DNA polymerase.

CDRs *See* Complementarity-determining regions.

CEA *See* Carcinoembryonic antigen.

Cell-mediated immunity The form of adaptive immunity that is provided by T cells and defends the body against such intracellular microbes. Some $CD4^+$ T cells activate phagocytes to destroy microbes that have been ingested by phagocytes into intracellular vesicles. Other $CD8^+$ T cells kill various host cells that harbor infectious microbes in the cytoplasm.

Cell membrane The trilamellar envelope (membrane) surrounding a cell. Also called plasma membrane, plasmalemma.

Central nervous system All the neurons of the brain and spinal cord. Abbreviated CNS.

Central tolerance A form of immunological tolerance to self antigens that is induced when developing lymphocytes encounter these antigens in central (primary) lymphoid organs, such as the thymus and bone marrow.

Centromere The most condensed and constricted region of a chromosome, to which the spindle fiber is attached during mitosis. The region that joins the two sister chromatids.

Cerebral cortex The neuronal layer covering the cerebrum.

Cerebrum The portion of the brain in the upper brain cavity that is divided into two hemispheres. The outer cortical layer of the cerebrum is densely packed with neurons and massive numbers of nerve connections. The basal ganglia are found in the interior of the cerebral hemispheres.

CGH *See* Comparative genomic hybridization.

Chaperone A protein complex that aids in the correct folding of nascent or misfolded proteins.

Chaperone–usher system The secretion apparatus for assembly of fimbriae in gram-negative bacteria; consists of chaperone proteins that facilitate transport of fimbrial subunits (pilin) across the cytoplasmic membrane and an usher complex for transport across and anchorage in the outer membrane.

Chaperonin A protein that uses energy from ATP hydrolysis to maintain cellular proteins in the correct folded configuration for proper function.

Chemoheterotroph An organism that utilizes organic molecules as energy and carbon sources for growth.

Chinese hamster ovary A type of cell maintained as a cell line in culture and used as host cells for transfection. Abbreviated CHO.

CHO *See* Chinese hamster ovary.

Cholera A fast-acting intestinal disease characterized by fever, dehydration, abdominal pain, and diarrhea; it is caused by the gram-negative bacterium *Vibrio cholerae*. This disease is transmitted by drinking water contaminated with fecal matter and is a significant health concern whenever water purification and sewage disposal systems are inadequate.

Chromatin Consists of DNA with its associated packaging proteins. The tightly packed regions of chromosomes are called heterochromatin, while other regions are less condensed and are called euchromatin. Less condensed packing of chromatin generally increases the transcription of genes in the region.

Chromatin conformation capture on chip An alternative technique to array painting in which many fragments across the breakpoints are captured by cross-linking physically close parts of the genome, followed by restriction enzyme digestion, locus-specific PCR, and hybridization to tailored microarrays. Abbreviated 4C.

Chloroplast A plastid that contains chlorophyll, where photosynthesis takes place.

Chromoplast A plastid that contains pigments other than chlorophyll, usually carotenoids.

Chromosome A microscopic thread-like structure in the nucleus of a eukaryotic cell, consisting of a single, intact DNA molecule and associated proteins in a compact structure. Chromosomes carry the genetic information of an organism. In humans, each somatic cell has 46 chromosomes.

Chromosome disorders Disorders caused by abnormalities in the number (increase or decrease of genes) or the structure of chromosomes.

Chromosome painting The specific visualization of an entire chromosome in metaphase spreads and in interphase nuclei by in situ hybridization with a mixture of sequences generated from that particular chromosome.

Chronic infection A type of persistent infection that is characterized by the continuous presence of low numbers of the infectious microorganism and is eventually cleared by the host immune system.

Chronic myeloid leukemia A specific form of blood cancer caused by a chromosomal translocation, in which portions of two chromosomes (chromosomes 9 and 22) are exchanged.

Clonal expansion The process by which, after antigen-induced activation, lymphocytes undergo proliferation and give rise to many thousands of clonal progeny cells, all with the same antigen specificity. This process ensures that adaptive immunity can balance and keep the rate of proliferation of microbes in check; otherwise, infection will occur.

Clonal selection hypothesis The theory proposed by Sir Macfarlane Burnet in 1957 which postulated that lymphocytes proliferate in response to antigens only if the antigen can be recognized by their specific receptors. This theory explains that each clonally derived lymphocyte arises from a single precursor and is capable of recognizing and responding to a distinct antigenic determinant. Upon encounter with a non-self-antigen, a specific preexisting lymphocyte clone is selected and activated. During ontogeny, potentially self-reactive lymphocytes are eliminated by a process of clonal deletion which involves programmed cell death, or apoptosis.

Clones Populations of cells that are derived from a single precursor cell. Cells in a clone express identical receptors and specificities.

CNS *See* Central nervous system.

Codominant expression Equal expression of the alleles of a given gene(s) inherited from both parents.

Codon A set of 3 nucleotides in mRNA that specifies a tRNA carrying a specific amino acid that is incorporated into a polypeptide chain during protein synthesis.

Codon optimization An experimental strategy in which codons within a cloned gene that are not the ones generally used by the host cell translation system are changed to the preferred codons without changing the amino acids of the synthesized protein.

Cointegrate vector system A two-plasmid system for transferring cloned genes to plant cells. The cloning vector has a T-DNA segment that contains cloned genes. After introduction into *Agrobacterium tumefaciens*, the cloning vector DNA undergoes homologous recombination with a resident disarmed Ti plasmid to form a single plasmid carrying the genetic information for transferring the genetically engineered T-DNA region.

Coleoptile A protective sheath enclosing the shoot tip and embryonic leaves of grasses.

Colostrum The first lacteal secretion produced by the mammary gland of a mother prior to the production of milk.

Combinatorial diversity A mechanism by which the immune system can generate antibodies of different antigen specificities by generating different combinations of H- and L-chain regions. Different combinations of variable, diversity, and joining (V, D, and J) segments may result from somatic recombination of DNA in the T-cell receptor and B-cell receptor gene loci during T-cell and B-cell development.

Comparative genomic hybridization A method of detection of copy number differences between two genomes. Abbreviated CGH.

Competence The ability of bacterial cells to take up DNA molecules and undergo genetic transformation.

Complementarity-determining regions Three regions which are the sites of an antibody molecule that recognize and bind antigens; these sites lie within the variable (V_H and V_L) domains at the amino-terminal ends of the two H and L chains. These regions display the highest variability in amino acid sequence of an antibody molecule. Abbreviated CDRs.

Complementary DNA A double-stranded DNA complement of an mRNA sequence; synthesized in vitro by reverse transcriptase. Abbreviated cDNA.

Complement system A system of serum- and cell surface-derived proteins that interact with each other and other molecules of the immune system. These interactions occur in three different pathways, each composed of a cascade of proteolytic enzymes produced to generate a host of effector molecules and cells that mediate innate and adaptive immune responses.

Concatemer A tandem array of repeating unit-length DNA elements.

Cone A photoreceptor of the retina with a cone-shaped outer segment. Also called cone cell, cone photoreceptor.

Congenital Present at birth, regardless of cause.

Conifer Any of various needle-leaved or scale-leaved, chiefly evergreen, cone-bearing gymnospermous trees or shrubs.

Conjugation The transfer of plasmid DNA from a donor bacterium to a recipient bacterium following cell-to-cell contact and formation of an intercellular pore.

Constant region The region of an antibody molecule that performs its effector functions. This region does not vary in its sequence and may be expressed in five different forms, each of which is specialized for activating different effector mechanisms. Abbreviated C region.

Contig A set of overlapping DNA segments that cover a region of a genome; a contiguous sequence of DNA produced by assembly of overlapping DNA fragments.

Copy number variants Arrays that identify recurrent chromosomal rearrangements more easily and may genotype copy number variants present in about 41% of the general population. The genome-wide coverage of these arrays now permits the discovery of copy number variants without prior knowledge of the DNA sequence.

Costimulator A molecule other than the T-cell receptor that stimulates the activation of T-cell signaling pathways and proliferation.

CpG *See* Cytosine–phosphate–guanine.

C region *See* Constant region.

Crohn disease A chronic inflammatory disease of the gastrointestinal tract that typically involves the distal portion of the ileum and is characterized by cramping and diarrhea.

Cross-presentation The ability of one cell type, the dendritic cell, to present the antigens of other cells, the infected cells, and activate naïve T cells specific for these antigens. The dendritic cells that engulf infected cells may also present microbial antigens to CD4$^+$ T helper cells. Thus, both CD4$^+$ and CD8$^+$ T cells specific for the same microbe may be activated.

Crown gall A bulbous growth that occurs at the bases of certain plants and that is due to infection of the plant by a member of the bacterial genus *Agrobacterium*. Also called crown gall tumor.

Cultivar A variety of plant that is (i) below the level of a subspecies taxonomically and (ii) found only under cultivation.

Cutaneous and mucosal lymphoid systems Lymphoid systems located under the skin epithelia and the gastrointestinal and respiratory tracts, respectively. Pharyngeal tonsils and Peyer's patches of the intestine are mucosal lymphoid tissues. Cutaneous and mucosal lymphoid tissues are sites of immune responses to antigens that breach epithelia.

Cytogenetics A field of genetics in which chromosomes present in cells at the metaphase stage of the cell cycle are differentially stained and then analyzed microscopically.

Cytokine Any of several regulatory proteins, such as the interleukins and lymphokines, that are released by cells of the immune system and act as intercellular regulators of the immune response.

Cytokinin A plant hormone that stimulates cell division.

Cytoplasm All of the contents outside the nucleus and enclosed within the cell membrane of a cell. In prokaryotes, all of the contents of the cell within the cell membrane.

Cytosine–phosphate–guanine A term describing unmethylated regions in bacterial DNA that function as adjuvants in innate immunity by binding to Toll-like receptor 9 and stimulating an immune response to DNA vaccines. Abbreviated CpG.

Cytotoxic or cytolytic T cells T cells that destroy a cell with a particular antigen on its surface. CD8$^+$ T cells are cytolytic T cells that are able to lyse cells harboring intracellular microbes.

Damage-associated molecular pattern A group of molecules that are released by stressed or necrotic cells and are recognized by the innate immune system. These cells are eliminated by the subsequent innate immune response. Germ line-encoded pattern recognition receptors for damage-associated molecular patterns have evolved as a protective mechanism against potentially harmful microbes. Abbreviated DAMP.

DAMP *See* Damage-associated molecular pattern.

Deletion The removal or absence of a portion of a chromosome.

Dementia Impairment, to various degrees, of short- and long-term memory, abstract thinking, judgment, sentence organization, and/or motor functions.

Denaturation (**1**) Separation of duplex nucleic acid molecules into single strands. (**2**) Disruption of the conformation of a macromolecule without breaking covalent bonds.

Dendrite One of the many slender, multibranched cellular extensions that emanate from the cell body of a neuron and carry nerve impulses toward the cell body.

Dendritic cells A type of phagocytic cell of the immune system that possesses long protruding finger-like processes similar to the dendrites of nerve cells. Immature dendritic cells and travel via the blood from the bone marrow into tissues. They take up particulate matter by phagocytosis and ingest large amounts of extracellular fluid by macropinocytosis.

Dengue One of the most important and widespread mosquito-borne viral diseases of humans, with about half the world's population now at risk. It is estimated that 50 million to 100 million dengue virus infections occur each year in more than 100 countries and that half a million people develop severe dengue necessitating hospital admission. No specific treatment is yet available.

Desensitization Repeated treatment of an allergic individual with small doses of an allergen(s). Many patients may benefit from this treatment, which may work by reducing Th2 responses and/or by inducing tolerance (anergy) in allergen-specific T cells.

Dicot A flowering plant that has two cotyledons or seed leaves; a dicotyledonous plant.

Dideoxynucleotide A nucleoside triphosphate that lacks a hydroxyl group on the 3′ carbon of the pentose sugar (as well as the 2′ carbon).

DiGeorge syndrome The most frequent defect in T-cell maturation; results from incomplete development of the thymus and a block in T-cell maturation.

Dimorphic Existing either in yeast form or as mold (mycelial form); this term is used to describe fungi.

Diploid Having two sets of chromosomes. In the human body, diploid cells contain 46 chromosomes ($n = 2$): 23 chromosomes are derived from the mother's egg cell, and the other 23 are from the father's sperm. These chromosomes appear as 22 homologous pairs of autosomes (nonsex chromosomes) and 1 pair of sex chromosomes, XX in females and XY in males.

DNA-dependent RNA polymerase An enzyme that catalyzes the synthesis of RNA using a DNA template; mediates transcription in all cellular organisms.

DNA ligase An enzyme that joins two DNA molecules by formation of phosphodiester bonds.

DNA methylation Addition of methyl groups to cytosine residues of DNA.

DNA microarray An array of thousands of gene sequences or oligonucleotide probes bound to a solid support. Also called DNA chip, gene array.

DNase I An enzyme that degrades DNA. It is used to remove DNA from RNA preparations and from cell-free extracts. Also called deoxyribonuclease I.

DNA shuffling In vitro random fragmentation and reassembly of a DNA sequence in an effort to create variants or mutants which when expressed as protein have different activities.

Domain A functional region of a protein.

Down syndrome An abnormality that displays 3 copies of chromosome 21; caused by trisomy 21.

Drug efflux pump A protein complex in the cytoplasmic membrane of prokaryotic cells that exports many different antibiotics from the cell; confers resistance to some antibiotics. Also called multidrug efflux pump.

Duplications Extra pieces of genetic material resulting from the copying of a portion of a chromosome.

Dystrophy A noninflammatory progressive breakdown of a tissue or organ.

Early-onset familial Alzheimer disease A form of Alzheimer disease that arises at a younger age (<65 years) in families. About 1 to 6% of all Alzheimer disease is early onset, and about 60% of early-onset Alzheimer disease runs in families.

Edema A systemic reaction of immune hypersensitivity characterized by the abnormal accumulation of fluid beneath the skin or in a cavity(ies) of the body.

Edward syndrome An abnormality that displays 3 copies of chromosome 18; caused by trisomy 18.

Effector cells Cells that carry out effector functions during an immune response, including cytokine secretion (e.g., T helper cells), microbe destruction (e.g., macrophages, neutrophils, and eosinophils), destruction of microbe-infected host cells (e.g., cytotoxic T cells), and antibody secretion (e.g., activated B cells).

Effector T cells T cells that are able to be activated and produce molecules capable of eliminating antigens. When naïve T cells recognize microbial antigens and also receive additional signals induced by microbes, the antigen-specific T cells proliferate and differentiate into effector T cells.

Elaioplast A plastid that is specialized for the storage of lipid.

Electrophoresis A technique that separates molecules (often DNA, RNA, or protein) on the basis of relative migration in a strong electric field.

Electroporation Treatment of cells with a short pulse of electrical current that induces transient pores, through which DNA is taken into the cell. Also used for intramuscular transfection of cells in live animals.

Embryonic stem cell A cell from an embryo in the blastula stage; these cells can develop into any cell in the human body. Also known as a human pluripotent stem cell.

Emulsion PCR A technique that uses PCR to produce tens of thousands of copies of a DNA template that is bound to a bead within a water-in-oil emulsion.

Endocytosis Entrance of foreign material into the cell without passing through the cell membrane. The membrane folds around the material, resulting in the formation of a vesicle containing the material.

Endomembrane A membrane that surrounds the organelles in the cytoplasm of eukaryotic cells. The endomembrane system includes the nuclear envelope, endoplasmic reticulum, Golgi apparatus, endosomes, and lysosome.

Endosome A vesicle formed by the invagination and pinching off of the cell membrane during endocytosis.

Endotoxic shock A massive inflammatory response induced by release of large amounts of endotoxin when gram-negative bacterial cells are lysed during a severe infection; a life-threatening condition characterized by low blood pressure and poor organ function.

Endotoxin A component (lipid A) of the cell wall of gram-negative bacteria that elicits an inflammatory response and fever in humans after the bacterial cell has lysed.

Enterotoxin A toxin secreted by some bacteria that, after its release into the small intestine, causes cramps, diarrhea, and nausea.

Enveloped virus A virus that possesses a membrane enclosing the capsid; the membrane is derived from the membrane of the host cell in which the virion was formed.

Enzyme replacement therapy A medical treatment replacing an enzyme in patients in whom that particular enzyme is deficient or absent.

Epidermal growth factor A polypeptide that promotes growth and differentiation, is essential in embryogenesis, and is important in wound healing; a mitogenic polypeptide produced by many different cell types and made in large amounts by some tumors.

Epigenetic Pertaining to chemical modifications to DNA or histone proteins associated with DNA. Modifications include methylation of DNA and addition of acetyl, ubiquitin, methyl, phosphate, or sumoyl groups to specific amino acids in histone proteins.

Episome A DNA element that can replicate independently of the chromosome(s), such as a plasmid, or can be inserted into and replicate with a chromosome.

Epitope The region of an antigen that is recognized by a specific antibody.

Epitope spreading An initial immune response against one or a few epitopes on a self-protein antigen that may expand to include responses against many more epitopes on this self-antigen.

Epstein–Barr virus A herpesvirus that is the causative agent of infectious mononucleosis.

Error-prone PCR Use of PCR under conditions that promote the insertion of an incorrect nucleotide at every few hundred or so nucleotides of the template. Used as a method of random mutagenesis.

Etiological agent The cause or origin of a disease. It may be a mutation that is responsible for a genetic disease, a microorganism that causes infectious disease, or an environmental factor, such as a toxin, that causes a disease.

Eukaryotic Relating to organisms, including animals, plants, fungi, and some algae, that have (i) chromosomes enclosed within a membrane-bounded nucleus and (ii) functional organelles, such as mitochondria and chloroplasts, in the cytoplasm of their cells.

Excisionase An enzyme encoded in genome of some bacteriophage that, together with the bacteriophage enzyme integrase and integration host factor, catalyzes sequence-specific recombination which results in removal of the bacteriophage genome from the host genome.

Exome sequencing Genome-wide exon sequencing (sequencing of coding DNA regions only); it is now used to map many disease loci.

Exotoxin A protein toxin secreted by a bacterial cell that damages host cells.

Expressed sequence tag A short sequence of cDNA that can be used to identify expressed genes. Abbreviated EST.

Extracellular matrix The organized polysaccharide and protein structure that is secreted by animal cells and makes up the connective tissue.

Ex vivo gene therapy Therapy consisting of removal of cells, often stem cells, from a patient, growth of the cells in culture, transduction of the cells with a therapeutic gene, and infusion of the transduced cells back into the patient.

Fab fragment A fragment in IgG. In each IgG molecule, the two H chains are identical and the two L chains are identical, and they give rise to two identical antigen-binding sites in the antigen-binding fragment (Fab) that can bind simultaneously to and cross-link two identical antigenic structures. The Fab fragment is produced upon cleavage of an antibody molecule with the protease papain. It does not bind to IgG Fc receptors on cells or with complement.

Falsified drug A medicine that has been deliberately mislabeled to deceive the consumer as to the contents and/or the manufacturer.

Fas (CD95) A receptor protein that is expressed on many cell types and mediates self-tolerance by a process of cell apoptosis.

FasL *See* Fas ligand.

Fas ligand A protein expressed mainly on activated T cells. Binding of FasL to its Fas death receptor induces the apoptosis of T and B cells exposed to self-antigens. Abbreviated FasL.

Fc fragment A crystallizable fragment of an antibody molecule that arises after cleavage by papain and that contains only the disulfide-linked carboxy-terminal regions of the two H chains. The Fc fragment of an antibody interacts with effector cells (phagocytes and NK cells) and molecules, and the functional differences between the various classes of H chains lie mainly in the Fc fragment.

Fiber-FISH A technique of fluorescence in situ hybridization to chromatin fibers. During mitosis, chromatin DNA fibers become coiled into chromosomes, with each chromosome having two chromatids joined at a centromere.

Fimbria *See* Pilus.

Fine mapping A method of analysis of translocation breakpoints using next-generation sequencing, whole-genome sequencing, and paired-end technology. These technologies are linked with large-insert paired-end libraries of about 3 kilobase pairs to increase physical coverage and to maximize the detection of read pairs that span a breakpoint.

FISH *See* Fluorescence in situ hybridization.

Flagellin The single protein which makes up flagella. Localization of a neutralizing antigen from a pathogenic bacterium on the surface of a live nonpathogenic bacterium is expected to increase the immunogenicity of the antigen. Flagella consist of filaments of flagellin, which microscopically appear as thread-like structures on the outer surfaces of some bacteria. If the flagella of a nonpathogenic organism could be configured to carry a specific epitope from a pathogenic bacterium, protective immunogenicity might be achieved more readily.

Fluorescence The emission of light from a molecule that has been excited by absorption of light of a shorter wavelength.

Fluorescence in situ hybridization A technique in which fluorescence microscopy reveals the presence and localization of defined labeled DNA probes binding to complementary DNA sequences on metaphase chromosome spreads. Abbreviated FISH.

Fluorophore The portion of a molecule that can fluoresce.

Follicles In the case of lymph nodes, discrete arrangements of B cells around the periphery, or cortex, of each node.

4C *See* Chromatin conformation capture on chip.

Fucose A hexose deoxy sugar found on N-linked glycans on mammalian, insect, and plant cell surfaces. Fucose lacks a hydroxyl group on the carbon at the 6-position; it is equivalent to 6-deoxy-L-galactose.

Fusion protein The product of two or more coding sequences from different genes that have been cloned together and that, after translation, form a single polypeptide sequence. Also called hybrid protein, chimeric protein.

Fv fragment The N-terminal half of an Fab fragment that contains all of the antigen-binding activity of the intact antibody molecule.

Gain-of-function variants Variants which result from a mutation that confers new or enhanced activity on a protein. Most mutations of this type are not heritable (germ line) but rather occur during development of an individual (somatic mutations).

Ganglia (singular, ganglion) A group of nerve cell bodies.

Gas chromatography A technique to separate gaseous molecules in a mixture based on their different affinities for a stationary-phase substance as they move in a mobile phase.

GAVI *See* Global Alliance for Vaccines and Immunisation.

G banding Staining of chromosomes with the Giemsa dye produces alternating light and dark bands that reveal differential chromosomal structures characteristic of each chromosomal pair. These light and dark bands of chromosomes result from the specificity of binding of the Giemsa stain for the phosphate groups of DNA, as the stain attaches to regions of DNA where there are large amounts of A-T bonding. Giemsa staining can identify different types of changes in chromosomal structure, such as gene rearrangements.

Gemcitabine A nucleoside analogue used in chemotherapy for cancer patients.

Gene conversion Recombination between homologous genes that results in the nonreciprocal transfer of a sequence from a donor site to a recipient site; expression of the recipient gene results in production of a variant protein.

Generative (or primary) lymphoid organs The sites of development of lymphocytes from immature precursors, such as the bone marrow and thymus for B cells and T cells, respectively.

Gene therapy Use of a gene or cDNA to treat a disease.

Genetic code The complete set of 64 codons that code for all 20 amino acids and 3 termination codons.

Genetic predisposition (genetic susceptibility) The condition of having inherited a risk for the development of a particular disease.

Genome (1) The entire complement of genetic material of an organism, virus, or organelle. (2) The haploid set of chromosomes (DNA) of a eukaryotic organism.

Genome-wide association study A genome-wide association study is an examination of genetic variation across the whole human genome. Humans have limited genetic variation. About 90% of heterozygous DNA sites (e.g., SNPs) in each individual are common variants. Common polymorphisms (minor allele frequency of >1%) have been found to contribute to susceptibility to common diseases, and genome-wide association study of common variants facilitates the mapping of loci that contribute to common diseases in humans. Abbreviated GWAS.

Genomic library A population of cells (usually cells of the bacterium *Escherichia coli*) in which each individual cell carries a DNA molecule that was inserted into a cloning vector. Ideally, all of the cloned DNA molecules represent the entire genome of another organism. Also called a gene library or clone bank. This term is also used to refer to all of the vector molecules, each carrying a piece of the genomic DNA of an organism.

Genomics The study of the genome of an organism.

Genomic variants Different types of chromosomal rearrangements that result from structural variation in the human genome. These genomic alterations involve segments of DNA that are larger than 50 base pairs.

Germinal center A central region that can occur in a follicle if the B cells in the follicle have recently responded to an antigen. Germinal centers contribute significantly to the production of antibodies. T cells localize to areas outside but adjacent to the follicles in the paracortex.

Germ line gene therapy Delivery of a therapeutic gene to a fertilized egg or an early embryonic cell so that all the cells of the mature individual, including the reproductive cells, acquire the gene, with the goal of curing a disorder.

Global Alliance for Vaccines and Immunisation An international body that gathers global information on vaccine production and the outcomes of immunization regimens. Abbreviated GAVI.

Glycosylation The enzymatic covalent addition of sugar or sugar-related molecules to proteins or polynucleotides.

Granulocytes White blood cells characterized by the densely staining granules in their cytoplasm. Also called polymorphonuclear leukocytes because they have irregularly shaped nuclei. The three types of granulocytes—neutrophils, eosinophils, and basophils—are distinguished by their different patterns of granule staining.

GWAS *See* Genome-wide association study.

Haploid Having one-half of the normal diploid chromosome content. Egg cells, spermatozoa, and mature erythrocytes each contain 23 chromosomes ($n = 1$).

HapMap Project An international organization formed to develop (2002 to 2009) a haplotype map (HapMap) of the human genome that describes the common patterns of human genetic variation. This project identified haplotypes, i.e., blocks of genes on haploid chromosomes that control the inheritance of this variation. The haplotypes on each chromosome formed the basis of the HapMap, a key resource for researchers to find genetic variants affecting health, disease, and responses to drugs and environmental factors. The information produced by the project is freely available to researchers around the world.

HBcAg *See* Hepatitis B core antigen.

H chain *See* Heavy chain.

Heavy chain One type of polypeptide chain of an antibody molecule that pairs by disulfide bonds and noncovalent interactions (hydrogen bonds, electrostatic forces, Van der Waals forces, and hydrophobic forces) with an L chain. Five classes of heavy chains (IgA, IgD, IgE, IgG, and IgM) exist, and each heavy chain consists of a variable region and three or four constant regions. Abbreviated H chain.

Hemagglutinin A protein in the envelope of influenza viruses that binds to a specific receptor on a host cell.

Hematopoietic stem cells Bone marrow-derived stem cells that can develop into different types of blood cells, including red blood cells, platelets, and leukocytes.

Hepatitis B core antigen An antigen present in vaccinia vector vaccines; it is obtained by insertion of the corresponding gene into a plasmid vector immediately downstream of a cloned vaccinia virus promoter and in the middle of a nonessential vaccinia virus gene. Abbreviated HBcAg.

Hepatitis C virus A virus that infects the liver and may cause cirrhosis of the liver and ultimately liver cancer or liver failure.

Her-2 Human epidermal growth factor receptor 2, which mediates the pathogenesis and progression of breast cancer.

Herd immunity A form of protection against the spread of infection in a population based on a critical mass of successfully vaccinated individuals.

Herpes simplex virus A virus that has been implicated as a cancer-causing agent, in addition to its roles in causing sexually transmitted disease, severe eye infections, and encephalitis. Prevention of herpes simplex virus infection by vaccination with either killed or attenuated virus may put the recipient at risk for cancer. Abbreviated HSV.

Heterochromatin Tightly compacted regions of chromatin that are usually transcriptionally silent.

High endothelial venules Specialized postcapillary venules through which naïve T cells that mature in the thymus enter the lymph nodes after migrating through the circulation.

Hippocampus A portion of the limbic system of the human brain, controlling memory, learning, emotions, and other functions.

HIV *See* Human immunodeficiency virus.

HLA *See* Human leukocyte antigen.

Holoenzyme The active enzyme that is formed by the combination of a cofactor (coenzyme) and an apoenzyme (inactive enzyme precursor).

Homeostasis The resting state attained after immune responses have eradicated an infection. Immune responses are designed to defend against different classes of microbes. All immune responses decline as an infection is eliminated, allowing the system to return to homeostasis, prepared to respond to another infection.

Homologous chromosome A chromosome paired with its matched chromosome copy.

Host tropism A process that determines the host cells that can become infected with a specific pathogen; often determined by the presence of specific host cell surface receptors that interact with adhesins on the pathogen.

Housekeeping gene A gene that is required for an essential cellular process and is therefore expressed at a constant level in all cells of an organism under all conditions.

HPV *See* Human papillomavirus.

HSV *See* Herpes simplex virus.

HSV-1 amplicon A helper-dependent HSV type 1 vector that carries multiple copies of a therapeutic gene or multiple gene complexes.

H-2 complex The set of genes that map to the mouse MHC region on chromosome 17.

Human Genome Project An international scientific research project developed (1988 to 2001) to define the DNA sequence of the human genome and to map and characterize the approximately 20,000 to 25,000 genes of the human genome with respect to structure and function.

Human immunodeficiency virus A human retrovirus that infects cells of the immune system, mainly $CD4^+$ T cells, and causes the progressive destruction of these cells. Abbreviated HIV.

Human leukocyte antigen One of a highly polymorphic group of human MHC proteins encoded by the human leukocyte antigen complex on human chromosome 6. These proteins were discovered as antigens of leukocytes that were identified by reactivity with specific antibodies. Abbreviated HLA.

Human papillomavirus A virus that is the cause of many common sexually transmitted diseases. While most of these infections are benign and often asymptomatic, persistent infection with some strains of human papillomavirus is associated with the development of cervical and related cancers, as well as genital warts. Over 100 different types of this virus exist, and 15 types have carcinogenic potential. Abbreviated HPV.

Human serum albumin The most abundant protein in human blood plasma. It is produced in the liver and is the transport protein for numerous small molecules.

Humoral immunity A form of adaptive immunity mediated by antibodies produced by B cells; the antibodies bind to antigens. This form of immunity provides the main type of defense against extracellular microbes and their toxins.

Huntingtin A protein encoded by the Huntington disease gene, which is located on the long arm of human chromosome 4. Huntingtin is quite variable in its structure.

Hybridomas Cell lines derived by cell fusion or somatic cell hybridization between a normal lymphocyte and an immortalized lymphocyte tumor line. B-cell hybridomas, created by fusion of normal B cells of defined antigen specificity with a myeloma cell line, are used to produce monoclonal antibodies. T-cell hybridomas, created by fusion of normal T cells of defined specificity with a T-cell tumor line, are frequently used to analyze T-cell function.

Icosahedron A geometric shape with 20 triangular faces; the shape of the capsid of some viruses.

IGF *See* Insulin growth-like factor.

Igs *See* Immunoglobulins.

Illegitimate drug A medicine that has not been approved by national regulatory agencies for marketing to the public. Falsified and substandard drugs (note that the definitions of these vary among countries) are considered illegitimate drugs.

Immediate hypersensitivity A rapid, IgE antibody- and mast cell-mediated vascular and smooth muscle reaction. This reaction is followed by an inflammatory response (e.g., hay fever, food allergies, bronchial asthma, and anaphylaxis) that occurs in individuals who encounter a foreign antigen(s) to which they have been exposed previously.

Immune effector functions The functions of the immune system that involve the complement system of blood proteins, enzymes, and antibodies together with certain T cells and B cells. These functions are required to contain and eliminate an infection.

Immune homeostasis A stable number of lymphocytes that occurs in the absence of antigen; naïve lymphocytes die by a process known as apoptosis and are replaced by new cells that have developed in the generative lymphoid organs. This balanced cycle of cell loss and replacement maintains the stable number of lymphocytes.

Immune hypersensitivity reactions Pathogenic immune reactions to an antigen that may elicit a sensitivity to a secondary challenge with that antigen and, as a result, induce tissue injury and disease.

Immune regulation Regulatory lymphocytes, such as regulatory T cells, control various immune responses against self-components. A failure in immune regulation may result in allergy and autoimmune disease.

Immune response An adaptive response to a foreign antigen(s) mediated by the cells and molecules of the immune system.

Immune surveillance The recognition of tumors by the immune system, which recognizes and destroys invading pathogens and host cells that give rise to cancer.

Immune system The molecules, cells, tissues, and organs that together collaborate to mount an immune response or protect against infectious agents.

Immunodeficiency diseases Disorders that arise from defective immunity.

Immunogenic A classification of an antigen that can stimulate lymphocytes to proliferate and differentiate into effector cells, resulting in a productive immune response.

Immunoglobulins One of the most polymorphic group of proteins in mammals, presumably because they recognize a rather large array of different antigens. All antibodies or immunoglobulins are constructed similarly from paired H and L chains. In each IgG molecule, the two H chains are identical and the two L chains are identical, and give rise to two identical antigen-binding sites in the antigen-binding fragment (Fab) that can bind simultaneously to and cross-link two identical antigenic structures. Abbreviated Ig.

Immunological ignorance A mechanism of induction of immunological tolerance which is mediated by the ability of lymphocytes to ignore (i.e., not recognize) the presence of an antigen.

Immunological memory The capacity of an adaptive immune system to elicit more rapid, larger, and effective responses upon repeat exposure to a microbe.

Immunological recognition A function of the immune system that is performed by leukocytes (e.g., neutrophils, macrophages, and NK cells) of the innate immune system during an early rapid response and by the lymphocytes (T cells and B cells) of the later, more antigen-specific, and more efficient adaptive immune system.

Immunological tolerance The inability of the immune system to mount an immune response to a self-antigen after exposure to that antigen.

Immunotherapies Therapies that use cells and/or antibodies that eradicate infectious pathogens by stimulation of relevant cellular pathways and molecules that generate long-lasting immunity to pathogens.

Inactivated Killed (in the case of a pathogen). Many current vaccines consist of an inactivated pathogen.

Inclusion body A protein that is overproduced in a recombinant bacterium and forms a crystalline array of mostly inactive protein inside the bacterial cell.

Indels Chromosomal insertions and deletions. Indels, as well as structural variants (450 base pairs), are detected by DNA sequence reads aligned to a reference genome.

Indirect ELISA *See* Indirect enzyme-linked immunosorbent assay.

Indirect enzyme-linked immunosorbent assay An immunological method to quantify antibodies produced against a specific antigen. The primary antibodies are captured using a standardized antigen that has been immobilized on a solid support and then detected and quantified using a labeled secondary antibody that binds with high affinity to a primary antibody. Abbreviated indirect ELISA.

Induced pluripotent stem cells Adult somatic cells, such as skin cells, that have been genetically reprogrammed in the laboratory to behave like embryonic stem cells.

Inflammasome A multiprotein complex which transmits signals that activate an enzyme that cleaves a precursor of the proinflammatory cytokine interleukin-1 to generate its biologically active form. Some cytoplasmic receptors that participate in innate immune reactions recognize microbes and components of dead cells. Such receptors can associate with an inflammasome.

Innate immune system A system in which immunological recognition is conducted by leukocytes (e.g., neutrophils, macrophages, and NK cells) that mediate an early rapid response.

Innate immunity A mechanism of host defense that mediates an initial and rapid protection against microbial infections.

Insertional mutagenesis Disruption of a gene by the insertion of DNA.

Insertion sequence element A small transposable element consisting of a gene encoding transposase flanked by inverted repeat sequences that are recognized by transposase.

Insulin growth-like factor A protein (IGF-1 or IGF-2) that integrates the signal input from upstream components in the mTOR signaling pathway. Abbreviated IGF.

Integrase An enzyme encoded in the genome of some viruses that catalyzes recombination between specific sequences in the viral and host genomes and results in insertion of the viral genome into the host genome.

Integration host factor A DNA bending protein that, among other structural and regulatory functions, acts with the bacteriophage enzymes integrase and excisionase to excise the bacteriophage genome from the host genome.

Integrin A mammalian protein that acts as a transmembrane receptor that mediates the attachment between a cell and the tissues around it.

Interferon A family of glycoproteins known as cytokine, the production of which can be stimulated by viral infection, that has the ability to inhibit virus replication.

Interleukin-2 A lymphokine secreted by certain T lymphocytes that stimulates T-cell proliferation.

Intron A segment of a gene that is transcribed but is excised from the primary transcript during processing into an mRNA molecule. Also called an intervening sequence.

Invasin A bacterial protein that allows enteric bacteria to penetrate mammalian cells.

Inversion An event that results when a portion of a chromosome is broken off, turned upside down, and reattached.

In vivo expression technology A method to identify bacterial genes that are expressed only during growth in a host by testing for promoters that are activated in vivo.

In vivo gene therapy Treating a disorder by delivering a therapeutic gene(s) to a tissue or organ.

In vivo-induced-antigen technology A method to identify proteins that are produced only during infection of a host using antibodies in serum from an infected host to screen a genomic expression library.

Irinotecan A DNA topoisomerase inhibitor used in the treatment of various types of cancer.

Isocaudomers Restriction endonucleases that produce the same nucleotide extensions but have different recognition sites.

Isoelectric point The pH at which the net charge on a protein or other molecule is zero.

Isoform (1) One of a group of variant gene products derived from the same gene either by exon skipping or by different transcription initiation sites. (2) Two or more proteins from different genes that have identical activities.

Isoschizomers Restriction enzymes that recognize and bind to the same nucleotide sequence in DNA and cut at the same site.

Isotype Five different classes of Igs—IgM, IgD, IgG, IgA, and IgE—that are distinguishable by their C region. The class and effector function of an antibody are defined by the structure of its H chain class or isotype, and the H chains of the five main isotypes are denoted μ, δ, γ, α, and ε, respectively.

Karyogram A photographic representation of stained chromosomes arranged in order of size, i.e., decreasing length.

Karyotype The number and form of chromosomes in a species. Genetic testing of a full set of chromosomes from an individual enables a comparison with a "normal" karyotype for the species. A chromosome disorder may be detected or confirmed in this manner. Chromosome disorders usually occur when there is an error in cell division following meiosis or mitosis.

Kinase An enzyme that catalyzes the transfer of a phosphate group from a donor, such as ADP or ATP, to an acceptor molecule.

Klenow polymerase A product of proteolytic digestion of the DNA polymerase I of *Escherichia coli* that retains both polymerase and 3′ exonuclease activities but not 5′ exonuclease activity.

Klinefelter syndrome A chromosomal disorder that occurs in males with the genotype 47,XXY.

K-Ras A GTPase member of the Ras family that mediates many signal transduction pathways. Inclusion of the *K-RAS* gene product in targeted peptide vaccines has proven successful for the treatment of pancreatic and colorectal cancers in pilot clinical trials.

LAIV *See* Live attenuated influenza vaccine.

Laser capture microdissection A technique that employs a laser to extract specific cells from tissues under a light microscope.

Latent infection A type of persistent infection during which a virus is present but remains inactive in the host cell and does not produce infectious virions.

Late-onset familial Alzheimer disease A complex polygenic disorder controlled by multiple susceptibility genes. The strongest association is with the *APOE* e4 allele at locus *AD2*.

Late-phase reaction Tissue injury and inflammation that is mediated by cytokines released from mast cells, which occurs several hours after repeated episodes of immediate hypersensitivity.

Leishmania A genus of protozoan parasites to which the human immune system can respond. Despite this response to *Leishmania* infections, the development of an effective vaccine against these organisms has proven difficult. Ongoing efforts are being carried out with attenuated (live) strains of *Leishmania*.

Leukocytes A class of cells which comprises neutrophils, macrophages, and NK cells. Inflammation depends on the recruitment and activation of leukocytes. Defense against intracellular viruses is mediated primarily by NK cells and select cytokines (interferons).

Ligation The covalent joining of two double-stranded pieces of DNA, usually by the enzyme DNA ligase.

Light chain One type of polypeptide chain of an antibody molecule that pairs by disulfide bonds and noncovalent interactions (hydrogen bonds, electrostatic forces, Van der Waals forces, and hydrophobic forces) with a heavy chain. Two types of light chains, κ and λ, exist, and each light chain consists of a V region and a C region. Abbreviated L chain.

Lipopolysaccharide An inflammatory protein that possesses high therapeutic potential by means of its adjuvant properties. On the surface of gram-negative bacteria, the hydrophobic anchor of lipopolysaccharide is the endotoxin lipid A recognized by the innate immune system. Abbreviated LPS.

Liquid chromatography A technique used to separate molecules that are dissolved in a solution. Separation occurs as the molecules in the solution (the mobile phase) pass through a column containing a second solution (the stationary phase) because the molecules interact with the second solution to different extents and therefore pass through the column at different rates.

Live attenuated influenza vaccine An influenza vaccine prepared from reassorted, cold-adapted, and temperature-sensitive viruses with recommended seasonal strains or from reverse-genetic-engineered viruses that express those cold-adapted and temperature-sensitive mutations and other attenuating mutations. Abbreviated LAIV.

Loss-of-function variants Variants that result from a point mutation that confers reduced or abolished activity on a protein. Most loss-of-function mutations are recessive, indicating that clinical signs are observed only when both chromosomal copies of a gene carry such a mutation.

LPS *See* Lipopolysaccharide.

Lymph nodes Nodular aggregates of lymphoid tissues located along lymphatic channels throughout the body. A fluid called lymph is drained by lymphatics from all epithelia and connective tissues and most parenchymal organs. These organs transport this fluid from tissues to lymph nodes, where antigen-presenting cells in the nodes sample the antigens of microbes that may enter through epithelia into tissues.

Lymphatic system The system of organs and vessels that function in immunity, including production and transport of white blood cells (leukocytes) and trapping of foreign particles (in lymph nodes).

Lymphocyte repertoire The total collection of lymphocyte specificities. The adaptive immune system can distinguish between millions of different antigens or portions of antigens. Specificity for many different antigens indicates that the lymphocyte repertoire is large and extremely diverse.

Lymphocytes A collection of white blood cells localized in the blood, lymphoid tissues, and organs that express receptors for antigens and mediate cell-dependent immune responses during innate immunity (NK cells) and adaptive immunity (T cells and B cells).

Lymphoma A cancer of the lymphatic system.

Lysogenic phase A stage in the infection cycle of some bacteriophage in which the viral genome (prophage) remains in the chromosome of the host bacterium and lytic functions are repressed.

Lysosomal storage diseases A group of approximately 50 inherited metabolic disorders that result from defects in lysosomal function.

Lysosome A membrane-enclosed organelle in the cytoplasm of a eukaryotic cell that contains degradative enzymes.

Lytic phase The stage of a viral infection cycle in which the virus reproduces and destroys (lyses) the host cell, releasing infectious virions.

MAbs *See* Monoclonal antibodies.

Macrophages Cells produced by the differentiation of moncytes that circulate in the blood and continually migrate into almost all tissues where they differentiate and reside. Macrophages function as scavenger cells that clear dead cells and cell debris from the body. They are a type of phagocyte.

Macular region A small, oval area of the central part of the retina. Also called macula, macula lutea.

Magnetic resonance imaging A medical diagnostic procedure that uses nuclear magnetic resonance to visualize the internal structures in the body.

Major histocompatibility complex A large genetic locus (on human chromosome 6 and mouse chromosome 17) that consists of many highly polymorphic genes, including the major histocompatibility complex class I and class II genes that encode the peptide-binding molecules recognized by T cells. This locus also contains genes that code for cytokines, molecules involved in antigen processing, and complement proteins. Abbreviated MHC.

Malaria A mosquito-borne infectious disease of humans and other animals, which is caused by a microorganism of the genus *Plasmodium*.

Massive parallelization Performing a large number of reactions, such as sequencing reactions, simultaneously.

Mass spectrometry A technique for measurement of the mass-to-charge ratio of ions.

M cell A type of phagocytic cell in the epithelium of the small intestines that transports organisms from the intestinal lumen to underlying immune cells. Also called microfold cell.

MCS *See* Multiple-cloning site.

Megabase A length of double-stranded DNA containing 2 million nucleotides, 1 million on each strand. Abbreviated Mb.

Meiosis A cell division event that results in cells with half the number of usual chromosomes, 23 instead of the normal 46. These are the eggs and sperm.

Melt curve A graph showing the temperature at which a double-stranded nucleic acid molecule denatures. The melting, or dissociation, temperature is dependent on the nucleotide sequence of the molecule. In real-time PCR, a melt curve is often performed after amplification as a quality control measure to determine if all of the products contain the same sequence or to detect single nucleotide polymorphisms.

Memory cells Long-lived cells that are induced during the primary immune response. Subsequent stimulation by the same antigen leads to a secondary immune response that eliminates an antigen more efficiently than a primary response. Secondary responses result from the activation of memory cells.

Meningitis Inflammation of the meninges, the thin, membranous covering of the brain and the spinal cord, caused by a bacterial or viral infection.

Meristematic tissue Plant tissue consisting of undifferentiated cells where active cell divison occurs.

Mesophilic bacteria Bacteria that grow best between 20°C and 55°C.

Mesothelin A protein that is present on normal mesothelial cells and overexpressed in several human tumors, including mesothelioma and ovarian and pancreatic adenocarcinoma.

Messenger RNA An RNA molecule carrying the information that, during translation, specifies the amino acid sequence of a protein molecule. Abbreviated mRNA.

Metabolic load The physiological and biochemical changes in a bacterium that occur as a consequence of overexpressing a foreign protein.

Metabolite (1) A low-molecular-weight biological compound that is usually synthesized by an enzyme. (2) A compound that is essential for a metabolic process.

Metabolomics The study of the complete repertoire of metabolites of a cell, tissue, or organism.

Metagenomic library A collection of cloned DNA fragments that represents the genomes of all organisms (usually microorganisms) in an environmental sample.

Metastatic disease Cancer that has spread from the initial tumor site to other parts of the body. Also called metastasis.

Methicillin-resistant *Staphylococcus aureus* A strain of the bacterium *Staphylococcus aureus* that is resistant to methicillin and most other β-lactam antibiotics. These strains have emerged and spread rapidly in the community and are now the most common cause of community-associated skin and soft tissue infections. Methicillin-resistant *S. aureus*-induced infections have few effective treatment options. Abbreviated MRSA.

M-FISH *See* Multiplex FISH.

MHC *See* Major histocompatibility complex.

MHC haplotype The set of MHC alleles present on each chromosome. In humans, each HLA allele is given a numerical designation; e.g., an HLA haplotype of an individual could be HLA-A2, HLA-B5, or HLA-DR3.

MHC restriction The ability of different T-cell clones in an individual to recognize peptides only when bound and displayed by that individual's MHC molecules on the surface of antigen-presenting cells.

Microarray DNA chips used to establish copy number changes and SNPs. The microarrays are also used to analyze DNA methylation, alternative splicing, microRNAs, and protein–DNA interactions. Each array consists of thousands of immobilized oligonucleotide probes or cloned sequences. Labeled DNA or RNA fragments are applied to the array surface, allowing the hybridization of complementary sequences between probes and targets. The advantages of microarray technology are its sensitivity, specificity, and scale, as it enables a rapid assay of thousands of genomic regions in a single experiment.

Microfold cell *See* M cell.

MicroRNAs Nucleotide sequences of about 22 nucleotides in length that function as posttranscriptional regulators and result in translational repression or target degradation and gene silencing.

Mitochondrial homeostasis The functional state of mitochondria achieved to maintain a metabolically active state of neurons that requires considerable energy.

Mitosis A cell division event that produces two cells, each of which ia a duplicate (46 chromosomes each) of the original cell. Mitosis occurs throughout the body, except in the reproductive organs. In both processes, the correct number of chromosomes appears in the daughter cells.

Molecular mimicry The process of cross-reaction between T cells reactive with microbial antigens or self-antigens.

Monoclonal antibodies Antibody molecules of homogeneous structure that are derived from a single clone of B cells and which possess known specificity for a specific target antigen. Such antibodies are produced in cultures of hybrid cells (hybridomas) formed between antibody-forming B cells and immortalized myeloma tumor plasma cells; all of the antibodies produced by the hybridoma cell line bind to the same unique epitope. Abbreviated MAbs.

Monocot A flowering plant that has only one cotyledon or seed leaf, i.e., a monocotyledonous plant.

Monophosphorylated lipid A A less toxic combination of lipid A, compared to wild-type *E. coli* LPS, which elicits much lower inflammatory responses. Abbreviated MPL.

MPL *See* Monophosphorylated lipid A.

M protein A protein that forms hair-like filaments on the surface of the bacterium *Streptococcus pyogenes* and binds to fibronectin on host cells.

mRNA *See* Messenger RNA.

MRSA *See* Methicillin-resistant *Staphylococcus aureus*.

M13 + strand The single-stranded DNA molecule that is present in the infective bacteriophage M13.

mTOR A serine/threonine protein kinase enzyme (whose name is derived from "*m*ammalian *t*arget *o*f *r*apamycin") that regulates cell growth, cell proliferation, cell motility, cell survival, protein synthesis, and gene transcription. Thus, mTOR can control the state of anergy in a cell, and the mTOR signaling pathway may be dysregulated in some human diseases, e.g., cancers.

Mucin 1 A heavily glycosylated cell surface-associated protein secreted by specialized epithelial cells that is recognized as a T-cell antigen in advanced pancreatic cancer. Abbreviated MUC-1.

MUC-1 *See* Mucin 1.

Mucosa The protective surface of tissues exposed to the external environment that consists of an epithelial layer covered with a mucous layer (containing mucins) and underlaid with a layer of proteinaceous connective tissue.

Multifactorial In the case of genetic disorders, those which are influenced by several different lifestyles and environmental factors.

Multiple-cloning site A synthetic DNA sequence that contains a number of different restriction endonuclease sites. Abbreviated MCS.

Multiple myeloma A malignant tumor of antibody-producing B cells that secretes an immunoglobulin or part of an immunoglobulin molecule.

Multiplex FISH A technique in which specific translocations and complex rearrangements are characterized by techniques adapted from chromosome painting. Abbreviated M-FISH.

Multipotent adult stem cells Adult cells found naturally in multicellular organisms that can divide and differentiate into a limited number of different types of specialized cells.

Muscular dystrophy Progressive weakening and wasting of muscle tissue.

Mutagenesis Chemical or physical treatment that changes the nucleotides of the DNA of an organism.

Mutant An organism that differs from the wild type because it carries one or more genetic changes in its DNA. Also called a variant.

Mutation A change of one or more nucleotide pairs of a DNA molecule.

Myelin sheath Multiple layers of cell membrane from an accessory nerve cell wrapped around an axon.

Myeloid The common myeloid progenitor is the precursor of the macrophages, granulocytes, mast cells, and dendritic cells of the innate immune system and also of megakaryocytes and red blood cells.

Myristoylation An irreversible, cotranslational protein modification found in eukaryotes in which a molecule of myristic acid is covalently attached to the alpha-amino group of an N-terminal amino acid of a nascent protein.

Naïve lymphocytes Lymphocytes that express antigen receptors but do not perform the functions required to eliminate antigens. These cells reside in and circulate between peripheral lymphoid organs and survive for several weeks or months, waiting to encounter and respond to antigen. The state of a lymphocyte not previously exposed to an antigen.

Narrow-host-range plasmid A plasmid that can replicate in one bacterial species or, at most, a few different bacterial species.

Natural killer cells Cells that kill microbe-infected host cells but do not express the kinds of clonally distributed antigen receptors that B cells and T cells do and are components of innate immunity, capable of rapidly attacking infected cells. Abbreviated NK cells.

Nef Negative regulatory factor; an HIV protein virulence factor that manipulates the host's cellular machinery, allowing infection, survival, and replication of the virus.

Negative selection A process in which the receptors on an immature T cell in the thymus bind with high affinity to a self-antigen displayed as a peptide–MHC ligand on an antigen-presenting cell, and this T cell receives a signal(s) that triggers apoptosis and dies before it is fully mature. This process is a major mechanism of establishing T-cell-mediated immunological tolerance.

Negative-sense RNA Viral RNA that has the opposite polarity to (is complementary to) mRNA and must be transcribed by an RNA-dependent RNA polymerase to produce mRNA for translation.

Neonatal sepsis The presence in a newborn baby (less than 90 days old) of a bacterial bloodstream infection.

Neoschizomers Restriction enzymes that recognize and bind to the same nucleotide sequence in DNA and cut at different sites.

Neuraminidase A protein in the envelope of influenza viruses that mediates release of virions from the host cell.

Neurodegenerative disorder One of many conditions (diseases) which affect neurons in the human brain.

Neuron A cell that specializes in initiating and transmitting electrical impulses. Also called nerve cell.

Neurotransmitter A chemical compound or peptide that is released by a presynaptic neuron and produces an excitatory or inhibitory response in a postsynaptic neuron.

Next-generation sequencing High-throughput DNA sequencing that enables determination of hundreds of millions of DNA sequences simultaneously and at low cost; does not require cloning of sequencing templates. Abbreviated NGS.

NGS *See* Next-generation sequencing.

Nisin A polycyclic pore-forming antibacterial peptide with 34 amino acid residues, used as a food preservative.

NK cells *See* Natural killer cells.

NMR *See* Nuclear magnetic resonance.

Nonsystemic (1) A condition that is confined to a specific part of the body or a single organ. (2) A disorder that affects one specific organ or tissue. Also called nonsyndromic.

Nuclear localization signal A short sequence of basic (positively charged) amino acids on a protein that target the protein to the nuclear pore complex for translocation into the nucleus.

Nuclear magnetic resonance A technique for identification and quantification of molecules (usually proteins or metabolites) based on their structure; based on the principle that in an applied magnetic field, molecules absorb and emit electromagnetic energy at a characteristic resonance frequency that is determined by their structure, and therefore differences in resonance frequency enable differentiation among molecules with different structures. Abbreviated NMR.

Nuclear membrane The double-layered envelope enclosing the chromosomes of eukaryotic cells. Also called nuclear envelope.

Nuclear pore complex A large complex of proteins in the nuclear membrane that form a pore through which molecules are transported into and out of the nucleus.

Nucleocapsid The nucleic acid genome and surrounding protein coat of a virus.

Nucleoside analogue A synthetic antiviral molecule that has a structure similar, but not identical, to that of a natural nucleoside and therefore can bind to and inhibit enzymes, such as DNA polymerase and reverse transcriptase, that are required for viral nucleic acid replication.

Nucleus (1) The organelle of a eukaryotic organism enclosing the chromosomes. (2) A cluster of cell bodies of neurons within the central nervous system. (3) The central portion of an atom containing protons and neutrons.

Occluded virus A virus enclosed in an inclusion body; produced in the latter stages of infection.

Oligonucleotide A short molecule (usually 6 to 100 nucleotides) of single-stranded DNA. Oligonucleotides are sometimes called oligodeoxyribonucleotides or oligomers and are usually synthesized chemically.

Oncogene A gene that plays a role in the cell division cycle. Often, mutated forms of oncogenes cause a cell to divide in an uncontrolled manner.

Oncomouse A transgenic mouse that carries an activatable gene that makes it susceptible to tumor formation.

Oncovirus A virus that can cause cancer. Also called oncogenic virus.

1000 Genomes Project An international research effort that established (2008 to 2012) the most detailed catalogue of human genetic variation.

Open reading frame A sequence of nucleotides in a DNA molecule that encodes a peptide or protein. This term is often used when, after the sequence of a DNA fragment has been determined, the function of the encoded protein is not known. Abbreviated ORF.

Operator The region of DNA that is upstream from a prokaryotic gene(s) and to which a repressor or activator binds.

Opine The condensation product of an amino acid with either a keto acid or a sugar.

Opportunistic pathogen A microorganism that causes disease only when a host's defenses are compromised.

Opsin The protein component of the visual pigments located in the disk membranes of the outer segments of the rods and cones of the retina.

ORF *See* Open reading frame.

Origin of DNA replication The nucleotide sequence at which DNA synthesis is initiated. Abbreviated *ori*.

Origin of transfer A specific sequence of nucleotides on a plasmid that is recognized by proteins that initiate transfer of a plasmid from a donor bacterium to a recipient bacterium via conjugation. Also called *oriT*.

Orphan drug A medicine that is used to treat rare diseases.

Packaging cell line A cell line designed for the production of replication-defective viral particles.

PAI *See* Pathogenicity island.

Paired-end read mapping A technique developed to study chromosome rearrangements. Sequence read pairs are short sequences from both ends of each of the millions of DNA fragments generated during preparation of a DNA library. Clustering of at least two pairs of reads that differ either in size or orientation suggests a chromosome rearrangement. When aligned to the reference genome, read pairs map at a distance corresponding to an average library insert size of 200 to 500 bp and up to 5 kilobase pairs for large-insert libraries.

Paired end reads Sequences obtained from both ends of a DNA fragment; the distance between the end sequences may be determined from the size of the fragment. This information can be used to assemble (order and orient) contiguous sequences (contigs) into larger scaffolds. Also called mate pairs.

Palindrome A DNA sequence that is the same when each strand is read in the same direction (5′ to 3′). These types of sequences serve as recognition sites for type II restriction endonucleases.

Palmitoylation Covalent attachment of fatty acids, such as palmitic acid, to cysteine and, less frequently, to serine and threonine residues of proteins.

PAMP *See* Pathogen-associated molecular pattern.

Paracortex The areas to which T cells localize in germinal center regions in lymph nodes. These areas are outside but adjacent to the follicles in the germinal centers.

Passive immunity Immunity conferred on an individual by the passive transfer of antibodies or lymphocytes from an actively immunized individual.

Patau syndrome A syndrome caused by trisomy 13, in which individuals carry 3 copies of chromosome 13.

Patent A legal document that gives exclusive rights to the patent holder to sell and use an invention for a specified period; awarded for inventions that are novel, nonobvious, and useful.

Pathogen A microorganism that is capable of causing disease.

Pathogen-associated molecular pattern Molecules associated with pathogenic bacteria that elicit a host innate immune response. Abbreviated PAMP.

Pathogenicity The ability to cause disease.

Pathogenicity island A region of a pathogen's genome containing a cluster of genes encoding virulence factors; sequence features present in the region often suggest that the genes were acquired from another organism (via horizontal gene transfer). Abbreviated PAI.

Pattern recognition receptors The receptors present on cells that mediate innate immunity and recognize the structures shared by the target molecules of innate immunity and microbes. Abbreviated PRRs.

PCR *See* Polymerase chain reaction.

Pegylation The process of covalent attachment of polyethylene glycol polymer chains to another molecule.

Penetrance The proportion of individuals in a population that carry a particular variant of a gene (allele or genotype) who also express an associated trait or phenotype. Full penetrance occurs when all individuals carrying a gene express the phenotype. Incomplete penetrance occurs when some individuals fail to express the phenotype, even though they carry the variant allele.

Peptide mass fingerprinting Identifying an unknown protein by comparing the masses of peptides obtained from protease digestion of the protein with the masses of peptides from known proteins; the peptide masses of the unknown protein are usually measured by mass spectrometry.

Peptidoglycan A polymer of sugars (*N*-acetylglucosamine and *N*-acetylmuramic acid) and cross-linked peptides that makes up the bacterial cell wall. Also called murein.

Peripheral (or secondary) lymphoid organs Organs in which adaptive immune responses to microbes are initiated. The spleen, lymph nodes, and the mucosal and cutaneous immune systems are examples of peripheral lymphoid organs, also called secondary lymphoid organs.

Peripheral tolerance A form of immunological tolerance to self-antigens that is induced when mature lymphocytes encounter these antigens in peripheral (secondary) lymphoid organs, such as the spleen and lymph nodes.

Periplasm The space (periplasmic space) between the cell (cytoplasmic) membrane of a bacterium or fungus and the outer membrane or cell wall.

Permeabilize To make a cell porous, thereby allowing substances to pass in or out.

Peroxisome A small organelle that is present in the cytoplasm of many cells (especially liver and kidney) of vertebrate animals and that contains enzymes such as catalase and oxidase.

Phagocytes Cells (e.g., macrophages) that are phagocytic in their action and can ingest and kill microbes by producing an array of toxic chemicals and degradative enzymes during an early innate immune response.

Phagosome A membrane-enclosed vesicle that forms from phagocytosis and may contain engulfed particles such as bacteria; may fuse with lysosomes for degradation of engulfed microorganisms.

Phenotype The combination of expressed traits of an individual, including morphology, development, behavior, and biochemical and physiological properties. Phenotypes result from the expression of genes and environmental factors that interact with these genes.

Phosphatase An enzyme that cleaves a phorphoryl group from a protein.

Phosphatidylinositol 3-kinase A kinase protein family related to mTOR that integrates the signal input from upstream pathways generated by insulin, insulin-like growth factors (IGF-1 and IGF-2), and amino acids. Abbreviated PI3K.

Phospholipase Any of a group of enzymes that catalyze the breakdown of phospholipids.

Pilin The proteins that polymerize to form a pilus.

Pilus A proteinaceous filament that protrudes from the bacteria cell wall and mediates attachment to host cells.

Plaque A clear area that is visible in a bacterial lawn on an agar plate and is due to lysis of the bacterial cells by bacteriophage.

Plasma cells Fully differentiated B cells that secrete immunoglobulins.

Plasmid A self-replicating extrachromosomal DNA molecule.

Plasmid incompatibility group A group of related plasmids that use the same mechanism of replication and therefore cannot coexist in the same host cell.

Plastid Any of several pigmented cytoplasmic organelles found in plant cells.

Pluripotent human embryonic stem cells Unspecialized cells that are derived from the inner cell mass of the blastocyst of an embryo and have the potential to differentiate into nearly all different types of specialized cells.

Poliomyelitis An infectious disease caused by poliovirus that can lead to paralysis. Two effective vaccines, the inactivated polio vaccine and oral polio vaccine, which consists of the live-attenuated Sabin strains, have been used for the control of paralytic poliomyelitis since the 1950s. Both vaccines are effective in providing individual protection against poliomyelitis virus, but the ease of oral administration and lower costs of the attenuated oral polio vaccine favored its use in most countries. The availability of monovalent and bivalent oral vaccines against poliovirus types 1 and 3 could help to accelerate eradication in countries where the virus is endemic.

Polyadenylation Addition of a poly(A) tail to an RNA molecule.

Polycistronic An mRNA found only in prokaryotes that encodes more than one protein.

Polyclonal antibodies Antibodies that are produced by a collection of several different B-cell clones and which react with different epitopes of a given antigen; even antibodies that bind the same epitope of an antigen can be heterogeneous.

Polygenic disease A genetic disorder that is causally associated with the effects of several genes.

Polyglutamine disorder A disorder that is due to CAG trinucleotide repeat expansion, a type of genetic mutation. Huntington disease is an example of a polyglutamine disorder. The Huntington disease gene contains many repeats of the CAG trinucleotide (which encodes glutamine). A highly variable number of CAG trinucleotide repeats accounts for the Huntington disease gene mutation, which leads to the expression of an abnormally long polyglutamine tract (sequence of glutamine residues) at the N terminus of the huntingtin protein beginning at residue 18. Such polyglutamine tracts increase protein aggregation, which may alter cell function. Thus, Huntington disease is one of 14 trinucleotide repeat disorders that cause neurological dysfunction in humans.

Polymerase chain reaction A technique for amplifying a specific segment of DNA by using a thermostable DNA polymerase, deoxyribonucleotides, and primer sequences in multiple cycles of denaturation, renaturation, and DNA synthesis. Abbreviated PCR.

Polymorphism Variation in phenotypic or genetic characteristics among individuals in a population.

Polymorphonuclear leukocytes Phagocytic cells, also called neutrophils, that possess a segmented multilobed nucleus and cytoplasmic granules filled with degradative enzymes. These cells are the most abundant circulating lymphocytes and the major cell type that mediates acute inflammatory responses to bacterial infections.

Polyploidy An abnormality in which the chromosome number is an exact multiple of the haploid number ($n = 23$) and is larger than the diploid number ($n = 46$). Polyploidy arises from fertilization of an egg by two sperm (total number of chromosomes increases to 69) or the failure in one of the divisions of either the egg or the sperm so that a diploid gamete is produced.

Polyprotein A single large polypeptide encoded in some viral genome that undergoes cleavage by proteases to produce individual proteins.

Pore-forming toxin A cytotoxin secreted by some bacterial pathogens that consists of proteins that insert into the host cell membrane and form a pore that disrupts osmoregulation.

Positive-sense RNA Viral RNA that has the same polarity as mRNA.

Prader–Willi syndrome A rare genetic disorder that results from the absence or nonexpression of a group of genes on chromosome 15.

Preclinical trials Tests of a drug or treatment in cell culture and both small and large animals to establish proof of principle and the absence of adverse effects. These tests are required before a phase I human clinical trial can be conducted.

Primary antibody In an immunological technique, an antibody that binds directly to an antigen.

Primary immune response An immune response elicited upon the first exposure to an antigen. This response is mediated by naïve lymphocytes that neither are experienced immunologically nor have previously responded to antigens.

Primary immunodeficiencies Diseases that result from genetic abnormalities in one or more components of the immune system.

Primary lymphoid organ An organ in which T and B cells are generated, mature, and become competent to respond to antigens. The thymus and bone marrow are primary or generative lymphoid organs for T and B cells, respectively.

Prime–boost Vaccination schedules based on combined prime–boost regimens using different vector systems to deliver the desired antigen have successfully improved the outcome of DNA vaccination. The DNA prime–viral vector boost approach, in which the same antigen is used to prime and boost the response, focuses on the induction of T-cell immune responses.

Prodrug An inactive compound converted into a pharmacological agent by an in vivo metabolic process.

Professional antigen-presenting cells Antigen-presenting cells for T cells that can display peptides bound to MHC proteins and express costimulator molecules on their surface. The most important professional antigen-presenting cells for initiating primary T-cell responses are dendritic cells.

Prognosis A prediction of the outcome of a disease.

Proinsulin A single-chain protein precursor of insulin.

Prokaryotes Organisms, usually bacteria, that have neither a membrane-bound nucleus enclosing their chromosomes nor functional organelles, such as mitochondria and chloroplasts.

Promoter A segment of DNA to which RNA polymerase attaches. It usually lies upstream of (5′ to) a gene. A promoter sequence aligns the RNA polymerase so that transcription initiates at a specific nucleotide.

Proofreading An activity of some DNA polymerases that may correct mismatched bases.

Propeptide A protein precursor that is generally inactive; also called a proprotein.

Prophage A repressed or inactive state of a bacteriophage genome that is maintained in a bacterial host cell as part of the chromosomal DNA.

Protease An enzyme that cleaves peptide bonds and therefore digests proteins into smaller peptides.

Protease inhibitor A protein that can form a tight complex with a protease and block its activity.

Protein microarray An array of a large number of different proteins for massively parallel analyses.

Proteome The complete repertoire of proteins of a cell, tissue, or organism.

Proteomics The study of the structure, function, and interactions of the members of a proteome.

Protocorm The tuberous mass of cells that is produced when a seed germinates.

Protoplasts Plant or bacterial cells, including the protoplasm and plasma membrane, after the cell wall has been removed.

Provirus A stage in the life cycle of a retrovirus in which the single-stranded RNA genome is converted into double-stranded DNA, which may then be integrated into the genome of a host cell.

PRRs *See* Pattern recognition receptors.

Psychrophilic bacterium A bacterium that grows best between 0 and 20°C.

Pyrophosphate Two covalently linked phosphate groups that are released during hydrolysis of a nucleoside triphosphate to a nucleoside monophosphate. Pyrophosphate is released following formation of a phosphodiester bond between nucleotides during DNA synthesis.

Pyrosequencing A sequencing method that detects the release of pyrophosphate when a known nucleotide is added to a growing DNA strand in a template dependent manner, that is, when DNA polymerase catalyzes the addition of the complementary nucleotide.

Quantitative PCR A technique based on PCR that is used to measure the levels of transcripts produced from a target gene; the absolute or relative number of transcripts can be determined. Abbreviated qPCR.

Quorum sensing system A bacterial signaling system that regulates population density-dependent gene expression; secreted chemical signals are known as autoinducers.

Reading frame A series of codons that code for amino acids in a nucleotide sequence.

Read length The number of nucleotides that can be determined in a single sequencing reaction.

Reassortment A phenomenon that occurs when two similar viruses infect the same cell and exchange DNA or RNA. New assembled viral particles may be created from segments whose origin is mixed, some derived from one strain and some from the other strain.

Recombinant DNA technology Insertion of a gene into a DNA vector to form a new DNA molecule that can be perpetuated in a host cell. Also called gene cloning, genetic engineering, molecular cloning.

Reference genome A genome sequence assembled from the genome sequences of a number of individuals that represents the genes of an organism; can be used to align sequencing reads from new genomes.

Regressive autism A type of autism that occurs when a child appears to develop normally but then, between 15 and 30 months of age, begins to lose speech and social skills. Although environmental factors are implicated in the development of autism, current evidence suggests that toxins and/or metabolites produced by gut bacteria, including *Clostridium bolteae*, may also be associated with the symptoms and severity of regressive autism.

Regulatory T cells T cells that suppress the potential exacerbation of antiself immune responses. The immune system reacts against a vast number and variety of microbes and foreign antigens, without reacting against the host's own antigens. If self-reactivity does occur, regulatory T cells suppress the possible exacerbation of antiself responses and resultant disease. Also called Treg cells.

Repressor A protein that binds to the operator or promoter region of a gene and prevents transcription by blocking the binding of RNA polymerase.

Restriction endonuclease An enzyme that recognizes a specific DNA sequence and cleaves the phosphodiester bonds on both strands between particular nucleotides.

Retina The inner layer of the eye, containing photoreceptors that transduce radiant energy into electric impulses and nerve cells that transmit these impulses to the brain.

Retrovirus A class of eukaryotic RNA viruses that can form double-stranded DNA copies of their genomes; the double-stranded forms can integrate into the genome of an infected cell.

Rev An HIV protein that allows fragments of viral mRNA that contain a Rev response element to be exported from the nucleus to the cytoplasm.

Reverse-phase microarray An array of multiprotein complexes of cell lysates or tissue specimens.

Reverse transcriptase An RNA-dependent DNA polymerase that uses an RNA molecule as a template for the synthesis of a complementary DNA strand.

Reversible chain terminator A nucleotide that has a blocking group at the 3′ carbon of the deoxyribose sugar to prevent subsequent addition of nucleotides to a growing DNA strand. In DNA sequencing by single nucleotide addition, the modified nucleotides are used to ensure that the DNA strand is extended by only a single nucleotide during each cycle. After identifying the incorporated nucleotide by detection of a unique fluorophore, the blocking group is removed to restore the 3′ hydroxyl group for the next cycle of nucleotide addition.

Ribonuclease An enzyme that cleaves RNA. Abbreviated RNase.

Ribosome-binding site A sequence of nucleotides near the 5′ phosphate end of a bacterial mRNA that facilitates the binding of the mRNA to the small ribosomal subunit. Also called a Shine–Dalgarno sequence.

Ribozyme An RNA molecule that has catalytic activity.

RNA-dependent RNA polymerase An enzyme that catalyzes the synthesis of RNA using an RNA template.

RNA interference A method to inhibit expression of a target gene. A small RNA binds to a complementary region of the mRNA of the target gene and prevents its translation into protein.

RNA sequencing A method to quantify expression of all of the genes in a cell, tissue, or organism by high-throughput sequencing of cDNA. Abbreviated RNA-Seq.

RNA splicing variants Alternate forms of mRNA that are produced from the same gene by differential splicing of introns from pre-mRNA.

RNA polymerase An enzyme that is responsible for making RNA from a DNA template.

RNase *See* Ribonuclease.

Rod A photoreceptor of the retina with a rod-shaped outer segment. Also called rod cell, rod photoreceptor.

Rolling-circle replication A mode of DNA replication that produces concatemeric duplex DNA.

R sequence A short repeated sequence at each end of a retroviral genome that acts during reverse transcription to ensure correct end-to-end transfer in the growing chain.

Salmonella A bacterium that may be used as an antigen delivery system in which a flagellin-derived antigen fusion protein presents the antigen to the host immune system. To construct this delivery system, a flagellin-negative strain of *Salmonella* is transformed with a plasmid containing a synthetic oligonucleotide specifying an epitope of the cholera toxin B subunit inserted into a hypervariable region of a *Salmonella* flagellin gene.

Sandwich enzyme-linked immunosorbent assay An immunological method to detect and quantify a specific antigen. The antigens are captured using antibodies that have been immobilized on a solid support and then detected and quantified using a labeled primary antibody that binds to the antigen.

SARS *See* Severe acute respiratory syndrome.

Scaffold Sequence contigs that are assembled in the correct order and orientation to reconstruct a portion of the sequence of a genome.

SCID *See* Severe combined immunodeficiency.

Secondary antibody In an immunological technique, an antibody that binds to a primary antibody.

Secondary immune response An adaptive immune response that occurs on second exposure to an antigen. When memory T cells encounter the antigen that induced their development, they rapidly respond to give rise to secondary immune responses, which occur more rapidly and with greater magnitude than primary immune responses.

Secondary immunodeficiencies Defects in the immune system that may result from infections, nutritional abnormalities, or medical treatments that cause loss or inadequate function of various components of the immune system.

Secretory immunoglobulin A An antibody that is secreted across mucous membranes by nearby B lymphocytes to prevent bacteria and viruses from attaching to epithelial surfaces; found in the mucus produced by the intestinal, respiratory, and genitourinary tracts. Abbreviated sIgA.

Selective toxicity A property of an antibiotic that kills or inhibits a bacterium but does not harm host cells.

Self-reactivity The binding of T and B cells to self-antigens. T and B cells recognize foreign antigens, respond to them, and when required eliminate them. Clonal expansion of these cells is generally highly efficient, but mutations can arise infrequently and generate T cells and B cells with receptors that bind to self-antigens and therefore display self-reactivity. Self-reactive cells may be either clonally deleted or functionally suppressed by other regulatory cells (Treg cells) of the immune system.

Senile plaque A dense spheroid amyloid body found outside of neurons and often associated with Alzheimer disease.

Sepsis The presence of bacteria or their toxins in the bloodstream.

Sequence coverage The average number of times that a given nucleotide is sequenced in a sequencing project.

Sequence read depth The number of reads mapping at each chromosomal position when the copy number of chromosomal gains or losses is determined.

Serotype A strain of microorganism that has a distinctive set of surface antigens or a unique surface antigen.

Serum sickness An immune response induced by the systemic administration of a protein antigen that triggers an antibody response followed by the formation of circulating immune complexes.

Severe acute respiratory syndrome The first new infectious disease of the 21st century. A novel coronavirus is the causative agent of the disease. Abbreviated SARS.

Severe combined immunodeficiency A disorder arising from defects in both B-cell and T-cell development. Abbreviated SCID.

Sex chromosomes Chromosomes which determine the sex of an individual and the inheritance of sex-linked traits. One pair of sex chromosomes, XX in females and XY in males, is present in diploid cells.

Shotgun cloning Construction of a library of small, overlapping fragments of genomic DNA to sequence the fragments. The overlapping sequences are then assembled to obtain the sequence of the entire genome.

Shuttle cloning vector A cloning vector, usually a plasmid, that can replicate in two different organisms because it carries two different origins or replication.

Sialic acid An N- or O-substituted derivative of neuraminic acid, which is a monosaccharide containing a nine-carbon backbone.

Siderophore A low-molecular-weight molecule that binds very tightly to iron. Siderophores are synthesized by a variety of soil microorganisms and plants to ensure that the organism can obtain sufficient amounts of iron from the environment.

sIgA *See* Secretory immunoglobulin A.

Signal peptide A segment of about 15 to 30 amino acids at the N terminus of a protein that enables the protein to be secreted (pass through a cell membrane). The signal sequence is removed as the protein is secreted. Also called signal peptide, leader peptide.

Signature-tagged mutagenesis A method based on transposon mutagenesis to identify bacterial genes that are essential for bacterial growth in a host.

Single-nucleotide polymorphism A chromosomal site at which one individual may have the adenosine (A) nucleotide and the other individual may have the guanosine (G) nucleotide. Abbreviated SNP.

Site-specific mutagenesis A technique to change one or more specific nucleotides in a cloned gene in order to create an altered form of a protein with a specific amino acid change(s). Also called oligonucleotide-directed mutagenesis.

SKY *See* Spectral karyotyping.

SNP *See* Single-nucleotide polymorphism.

Somatic cell gene therapy Introduction of a gene or cDNA by a viral or nonviral delivery system to non-germ line cells.

Somatic mutation A change in the structure of a gene that may arise during DNA replication and is not inherited from a parent and also not passed to offspring.

Somatic point mutation A somatic mutation that results in a single base substitution in DNA.

Sortase An enzyme in the cell membrane of gram-positive bacteria that catalyzes pilus assembly in the cell wall.

Spectral karyotyping A technique adapted from chromosome painting, in which specific chromosomal translocations and complex rearrangements are characterized by the differential coloring of all chromosomes in a single experiment. Abbreviated SKY.

Spectrin A cytoskeletal protein that lines the intracellular side of the plasma membrane in eukaryotic cells.

Spermidine A polyamine compound found in ribosomes and living tissues and having various metabolic functions.

Spike protein A glycoprotein found in the envelope of some viruses that may mediate attachment to a host cell.

Spleen An organ located in the abdomen that serves the same role in immune responses to blood-borne antigens as that of lymph nodes in responses to lymph-borne antigens. Blood entering the spleen flows through a network of channels (sinusoids), and blood-borne antigens are trapped and concentrated by splenic dendritic cells and macrophages. The spleen contains abundant phagocytes, which ingest and destroy microbes in the blood.

Split-read method A technique used to map chromosomal DNA breakpoints for small deletions (1 base pair to 10 kilobase pairs) in unique regions of the genome and in read lengths as low as 36 base pairs. All reads are first mapped to the reference genome. For each read pair, the location and orientation of the mapped read are used, and an algorithm is applied to search for the unmapped pair read (split read).

Stem cell A self-renewing progenitor cell that gives rise to specialized (differentiated) cells. Undifferentiated stem cells that can develop into different types of blood cells are known as pluripotent hematopoietic stem cells which can give rise to stem cells of more limited developmental potential, such as progenitors of red blood cells, platelets, and the myeloid and lymphoid lineages of leukocytes.

Streptavidin A tetrameric biotin-binding protein (subunit molecular weight, 14,500), produced by *Streptomyces avidinii*, that binds up to four molecules of biotin.

Streptokinase A bacterial enzyme that catalyzes the conversion of plasminogen to plasmin, thereby helping to dissolve blood clots.

Structural variants Two forms of structural gene variants in cell function that arise are gain-of-function variants and loss-of-function variants. Abbreviated SVs.

Subcutaneous Located or placed under the skin.

Substandard drug A medicine that contains no or low levels of active ingredients, wrong ingredients, and/or contaminants.

Subunit vaccines Vaccines incorporating only the microbial components responsible for protective immunity.

Suicide gene A gene that under certain conditions causes the death of its own cell.

Sumoyl A small protein (small ubiquitin-like modifier protein) that is covalently attached to other proteins.

Superantigen A peptide toxin secreted by some pathogenic bacteria or viruses that activates a large number of T cells by binding to both a T-cell receptor and MHC class II on a macrophage in the absence of an antigen.

Survivin A protein inhibitor of apoptosis which is highly expressed in most cancers. This protein is associated with chemotherapy resistance, increased tumor recurrence, and shorter patient survival, making antisurvivin therapy an attractive cancer treatment strategy.

SVs *See* Structural variants.

Synapse A specialized electrochemical junction between a neuron and another neuron, muscle cell, or gland cell. Chemical or electrical signals (nerve impulses) are transmitted across a synapse.

Synaptic Pertaining to a synapse.

Systemic Usually pertains to a disease condition, occurring throughout the body or affecting more than one tissue or organ. Also called syndromic.

Systemic infection An infection in which the pathogen has spread to several regions of the body.

Systems biology A methodological approach that evaluates information early (1 to 2 weeks) after vaccination using sensitive high-throughput technologies that allow for analysis of gene expression patterns and cytokine production in both T and B cells as well as in microbes and in particular methodologies. Such information not only identifies susceptible microbial targets but also has the potential to define new biomarkers of protective immune responses.

Tandem affinity purification A technique to capture, purify, and identify proteins that interact in multiprotein complexes.

Tat An HIV regulatory protein that enhances the efficiency of viral transcription.

T cell A type of lymphocyte that mediates cell-mediated immune responses in the adaptive immune system. Bone marrow-derived T cells mature in the thymus, circulate in the blood, localize to secondary lymphoid tissues, and are recruited to peripheral sites of antigen exposure. Functional subsets of CD4$^+$ T helper cells and CD8$^+$ cytotoxic T cells express TCRs that recognize foreign antigen peptides bound to self-MHC class II and self-MHC class I molecules, respectively.

T-cell antigen receptor A clonally expressed antigen receptor on CD4$^+$ and CD8$^+$ T cells. A T-cell antigen receptor recognizes complexes of foreign peptides bound to self-MHC class I and self-MHC class II molecules on the surface of antigen-presenting cells. It is a heterodimer of two disulfide-linked transmembrane polypeptide chains, termed TCRα and TCRβ. These chains each contain one N-terminal Ig-like V domain, one Ig-like C domain, a hydrophobic transmembrane region, and a short cytoplasmic region. Abbreviated TCR.

TCR *See* T-cell antigen receptor.

T-DNA The segment of a Ti plasmid that is transferred and integrated into chromosomal sites in the nuclei of plant cells.

Telomere The repetitive DNA sequences at the end of a eukaryotic chromosome.

Terminator A sequence of DNA at the 3′ end of a gene that stops transcription. Also called transcription terminator.

T helper cells Cells that help B cells to produce antibodies and help phagocytes to destroy ingested microbes.

Ti plasmid A large extrachromosomal element that is found in strains of *Agrobacterium* and is responsible for crown gall formation.

Tissue plasminogen activator A protein involved in dissolving blood clots. Abbreviated tPA.

TLRs *See* Toll-like receptors.

Tolerogenic A classification of an antigen that after recognition by lymphocytes in the immune system leads to a state of immunological tolerance.

Toll-like receptors Receptors belonging to a family of proteins (nine such receptors exist) that play a key role in the innate immune system and digestive system. They are single, membrane-spanning, noncatalytic receptors that mediate infection by recognition of different structurally conserved molecules derived from microbes. Abbreviated TLRs.

Toxoid A toxin that has been treated to destroy its toxicity but is able to stimulate antibody production.

tPA *See* Tissue plasminogen activator.

Transcription The process of RNA synthesis that is catalyzed by RNA polymerase; it uses a DNA strand as a template.

Transcriptomics The study of the complete repertoire of RNA molecules of a cell, tissue, or organism.

Transduction Transfer of DNA, often a coding sequence, to a recipient cell by a virus.

Transfection Introduction of foreign nucleic acids into animal cells.

Transformation (1) The uptake and establishment of DNA in a bacterium, yeast cell, or plant cell in which the introduced DNA often changes the phenotype of the recipient organism. (2) Conversion, by various means, of animal cells in tissue culture from controlled to uncontrolled cell growth.

Translation Protein synthesis; the process whereby proteins are made on ribosomes.

Translocation Exchange of DNA segments from two different chromosomes.

Transposase An enzyme that is encoded by a transposon gene and mediates the insertion of the transposon into a new chromosomal site and excision from a site.

Transposition The movement of a transposon from a site in a DNA molecule to a new site catalyzed by transposase.

Transposon A DNA sequence (mobile genetic element) that can insert randomly into a chromosome, exit the site, and relocate at another chromosomal site. For example, Tn5 is a bacterial transposon that carries the genes for resistance to the antibiotics neomycin and kanamycin and the genetic information for its insertion and excision. Also called transposable element.

Treg cells *See* Regulatory T cells.

Trinucleotide repeat disorders A type of genetic disorder. Among these disorders, 14 affect humans and elicit neurological dysfunction, i.e., impaired function of the brain that gives rise to neurodegenerative disorders. Trinucleotide CGG repeat expansions (200 to greater than 1,000 repeats) at a particular locus are very common mutations that may inactivate a gene at this locus.

Trisomy 18 *See* Edward syndrome.

Trisomy 21 The first human chromosomal disorder that was discovered (in 1959). It is an abnormality that displays an extra copy (total of 3 copies) of chromosome 21. The genes on all three copies of chromosome 21 are normal.

Tuberculosis A human infectious disease caused by *Mycobacterium tuberculosis*. The current vaccine is based on bacillus Calmette–Guérin (BCG). Novel vaccines are required to replace BCG vaccines to achieve enhanced protection against *M. tuberculosis* infection.

Tumorigenesis The generation of a tumor.

Tumor necrosis factor A protein produced by macrophages capable of inducing necrosis (death) of tumor cells.

Tumor suppressor gene A gene encoding a protein that regulates the cell cycle (cell proliferation) and thereby prevents the growth of tumors.

Turner syndrome An example of monosomy, a syndrome in which a girl is born with only one X chromosome.

Two-hybrid system An assay for identifying pairwise protein–protein interactions.

Type III secretion system A bacterial protein secretion system that forms a needle-like structure for transport of bacterial proteins, including toxins, directly into the host cytoplasm.

Type IV secretion system A bacterial protein secretion system that transports proteins or DNA across the bacterial membranes into the external environment or directly into a recipient bacterial or eukaryotic cell.

U5 sequence A unique 18-base sequence found in retroviruses that is between R and PBS (primer binding site) and is complementary to the 3′ end of the tRNA primer.

Ulcerative colitis A form of inflammatory bowel disease that causes swelling, ulcerations, and loss of function of the large intestine.

Upstream The stretch of DNA base pairs that lie in the 5′ direction from the site of initiation of transcription. Usually, the first transcribed base is designated +1 and the upstream nucleotides are indicated with minus signs, e.g., −1 and −10. Also, to the 5′ side of a particular gene or sequence of nucleotides.

Uropathogenic *Escherichia coli* A pathogenic strain of *E. coli* that causes urinary tract infections.

U3 sequence A sequence found in retroviruses between PPT and R, which has a signal that the provirus uses in transcription.

Vaccinia virus vector vaccines Vaccines made with vaccinia virus, a large enveloped virus belonging to the poxvirus family that was used as a vaccine in the eradication of smallpox. If a foreign gene is inserted into the vaccinia virus genome under the control of a vaccinia virus promoter, it will be expressed independently of host regulatory and enzymatic functions. Vaccinia virus infects humans and many other vertebrates and invertebrates and make an excellent host to form vector vaccines.

Valence The number of antigen-binding sites on an antibody that are available for binding to the epitopes on an antigen.

Variable region The antigen-binding region of an antibody molecule that varies extensively in its amino acid sequence between antibody molecules. This region endows an antibody with exquisite specificity of binding to a particular antigenic epitope. Abbreviated V region.

Vascular endothelial growth factor A protein currently being used (peptide from residues 2 to 169) in combination with gemcitabine in a randomized phase II/III clinical trial in patients with unresectable advanced or recurrent pancreatic cancer. Abbreviated VEGFR.

VEGFR *See* Vascular endothelial growth factor.

Very-long-chain fatty acid A carboxylic acid with an unbranched aliphatic tail with more than 22 carbons.

Vif Viral infectivity factor, an HIV protein that disrupts the antiviral activity of a human cytidine deaminase enzyme that mutates viral nucleic acids.

Viral matrix Structural proteins produced by some viruses that connect the viral membrane to the nucleocapsid.

***vir* genes** A set of genes on a Ti plasmid that prepare the T-DNA segment for transfer into a plant cell.

Virion (1) The infective form of a virus. (2) A complete virus particle.

Virulence The degree of pathogenicity (ability to cause disease) of an organism.

Virulence factor A molecule produced by a pathogen that contributes to its ability to cause disease.

Vpr Viral protein R; a 96-amino-acid HIV protein that is involved in regulating nuclear import of the HIV-1 preintegration complex.

Vpu Viral protein unique; an HIV protein involved in viral budding, thereby enhancing virion release from the cell.

V region *See* Variable region.

WHO *See* World Health Organization.

World Health Organization The directing and coordinating authority for health within the United Nations system. It is responsible for providing leadership on global health matters, shaping the health research agenda, setting norms and standards, articulating evidence-based policy options, providing technical support to countries, and monitoring and assessing health trends. Abbreviated WHO.

XenoMouse A transgenic mouse strain that has been engineered to produce human rather than murine antibodies.

X linked Linked to mutations in genes on the X chromosome. The X-linked alleles can also be dominant or recessive. These alleles are expressed in both men and women but more so in men, as they carry only one copy of the X chromosome (XY), whereas women carry two (XX). About half of the genetic changes that cause SCID disorders affect only male children.

X-linked agammaglobulinemia The most common clinical disorder caused by a block in B-cell maturation. Precursor B cells in the bone marrow do not mature beyond this stage, causing a relative absence of mature B cells and serum immunoglobulins.

Yeast artificial chromosome A vector used to clone DNA fragments larger than 100 kilobase pairs and up to 3,000 kilobase pairs.

Y linked In terms of inheritance, affecting only males. Disease-affected males always have an affected father, and all sons of an affected man have the disease.

Index